Methods in Enzymology

Volume 354
ENZYME KINETICS AND MECHANISM
Part F
Detection and Characterization of
Enzyme Reaction Intermediates

METHODS IN ENZYMOLOGY

EDITORS-IN-CHIEF

John N. Abelson   Melvin I. Simon

DIVISION OF BIOLOGY
CALIFORNIA INSTITUTE OF TECHNOLOGY
PASADENA, CALIFORNIA

FOUNDING EDITORS

Sidney P. Colowick and Nathan O. Kaplan

Methods in Enzymology

Volume 354

Enzyme Kinetics and Mechanism

*Part F
Detection and Characterization of
Enzyme Reaction Intermediates*

EDITED BY

Daniel L. Purich

DEPARTMENT OF BIOCHEMISTRY AND MOLECULAR BIOLOGY
UNIVERSITY OF FLORIDA COLLEGE OF MEDICINE
GAINESVILLE, FLORIDA

ACADEMIC PRESS

An imprint of Elsevier Science

Amsterdam Boston London New York Oxford Paris
San Diego San Francisco Singapore Sydney Tokyo

This book is printed on acid-free paper. ∞

Copyright © 2002, Elsevier Science (USA).

All Rights Reserved.
No part of this publication may be reproduced or transmitted in any form or by any means, electronic or mechanical, including photocopy, recording, or any information storage and retrieval system, without permission in writing from the Publisher.

The appearance of the code at the bottom of the first page of a chapter in this book indicates the Publisher's consent that copies of the chapter may be made for personal or internal use of specific clients. This consent is given on the condition, however, that the copier pay the stated per copy fee through the Copyright Clearance Center, Inc. (222 Rosewood Drive, Danvers, Massachusetts 01923), for copying beyond that permitted by Sections 107 or 108 of the U.S. Copyright Law. This consent does not extend to other kinds of copying, such as copying for general distribution, for advertising or promotional purposes, for creating new collective works, or for resale. Copy fees for pre-2002 chapters are as shown on the title pages. If no fee code appears on the title page, the copy fee is the same as for current chapters.
0076-6879/2002 $35.00

Explicit permission from Academic Press is not required to reproduce a maximum of two figures or tables from an Academic Press chapter in another scientific or research publication provided that the material has not been credited to another source and that full credit to the Academic Press chapter is given.

Academic Press
An imprint of Elsevier Science.
525 B Street, Suite 1900, San Diego, California 92101-4495, USA
http://www.academicpress.com

Academic Press
84 Theobalds Road, London WC1X 8RR, UK
http://www.academicpress.com

International Standard Book Number: 0-12-182257-5

PRINTED IN THE UNITED STATES OF AMERICA
02 03 04 05 06 07 SB 9 8 7 6 5 4 3 2 1

Table of Contents

CONTRIBUTORS TO VOLUME 354 ix

PREFACE xiii

VOLUMES IN SERIES xv

1. Covalent Enzyme–Substrate Compounds: Detection and Catalytic Competence — DANIEL L. PURICH — 1

2. Rapid Mix-Quench MALDI-TOF Mass Spectrometry for Analysis of Enzymatic Systems — JEFFREY W. GROSS AND PERRY A. FREY — 27

3. Pre-Steady-State Kinetics of Enzymatic Reactions Studied by Electrospray Mass Spectrometry with On-Line Rapid-Mixing Techniques — LARS KONERMANN AND DONALD J. DOUGLAS — 50

4. Trapping of α-Glycosidase Intermediates — RENÉE M. MOSI AND STEPHEN G. WITHERS — 64

5. Trapping Covalent Intermediates on β-Glycosidases — JACQUELINE WICKI, DAVID R. ROSE, AND STEPHEN G. WITHERS — 84

6. 2-Hydroxy-6-keto-nona-2,4-diene 1,9-Dioic Acid 5,6-Hydrolase: Evidence from ^{18}O Isotope Exchange for gem-Diol Intermediate — TIMOTHY D. H. BUGG, SARAH M. FLEMING, THOMAS A. ROBERTSON, AND G. JOHN LANGLEY — 106

7. Nucleoside-Diphosphate Kinase: Structural and Kinetic Analysis of Reaction Pathway and Phosphohistidine Intermediate — JOËL JANIN AND DOMINIQUE DEVILLE-BONNE — 118

8. Galactose-1-Phosphate Uridylyltransferase: Kinetics of Formation and Reaction of Uridylyl-Enzyme Intermediate in Wild-Type and Specifically Mutated Uridylyltransferases — SANDARUWAN GEEGANAGE AND PERRY A. FREY — 134

9. Kinetic Evidence for Covalent Phosphoryl-Enzyme Intermediate in Phosphotransferase Activity of Human Red Cell Pyrimidine Nucleotidases — ADOLFO AMICI, MONICA EMANUELLI, SILVERIO RUGGIERI, NADIA RAFFAELLI, AND GIULIO MAGNI — 149

10. Characterization of α(2→6)-Sialyltransferase Reaction Intermediates: Use of Alternative Substrates to Unmask Kinetic Isotope Effects — BENJAMIN A. HORENSTEIN AND MICHAEL BRUNER — 159

11. Use of Sodium Borohydride to Detect Acyl-Phosphate Linkages in Enzyme Reactions — DANIEL L. PURICH — 168

12. Evidence for Phosphotransferases Phosphorylated on Aspartate Residue in N-Terminal DXDX(T/V) Motif — JEAN-FRANÇOIS COLLET, VINCENT STROOBANT, AND EMILE VAN SCHAFTINGEN — 177

13. MurC and MurD Synthetases of Peptidoglycan Biosynthesis: Borohydride Trapping of Acyl-Phosphate Intermediates — AHMED BOUHSS, SÉBASTIEN DEMENTIN, JEAN VAN HEIJENOORT, CLAUDINE PARQUET, AND DIDIER BLANOT — 189

14. Transaldolase B: Trapping of Schiff Base Intermediate between Dihydroxyacetone and ε-Amino Group of Active-Site Lysine Residue by Borohydride Reduction — GUNTER SCHNEIDER AND GEORG A. SPRENGER — 197

15. T4 Endonuclease V: Use of NMR and Borohydride Trapping to Provide Evidence for Covalent Enzyme–Substrate Imine Intermediate — M. L. DODSON, ANDREW J. KURTZ, AND R. STEPHEN LLOYD — 202

16. Detection of Covalent Tetrahedral Adducts by Differential Isotope Shift ^{13}C NMR: Acetyl-Enzyme Reaction Intermediate Formed by 3-Hydroxy-3-methylglutaryl-CoA Synthase — HENRY M. MIZIORKO AND DMITRIY A. VINAROV — 208

17. Detection of Intermediates in Reactions Catalyzed by PLP-Dependent Enzymes: O-Acetylserine Sulfhydrylase and Serine–Glyoxalate Aminotransferase — WILLIAM E. KARSTEN AND PAUL F. COOK — 223

18. Protein Tyrosine Phosphatases: X-Ray Crystallographic Observation of Cysteinyl-Phosphate Reaction Intermediate — DAVID BARFORD — 237

19. GTP:GTP Guanylyltransferase: Trapping Procedures for Detecting and Characterizing Chemical Nature of Enzyme–Nucleotide Phosphoramidate Reaction Intermediate — JARED L. CARTWRIGHT AND ALEXANDER G. MCLENNAN — 251

20. γ-Glutamyl Thioester Intermediate in Glutaminase Reaction Catalyzed by *Escherichia coli* Asparagine Synthetase B — HOLLY G. SCHNIZER, SUSAN K. BOEHLEIN, JON D. STEWART, NIGEL G. J. RICHARDS, AND SHELDON M. SCHUSTER — 260

21. γ-Glutamyltranspeptidase and γ-Glutamyl Peptide Ligases: Fluorophosphonate and Phosphonodifluoromethyl Ketone Analogs as Probes of Tetrahedral Transition State and γ-Glutamyl-Phosphate Intermediate — JUN HIRATAKE, MAKOTO INOUE, AND KANZO SAKATA — 272

22. Stoichiometric Redox Titrations of Complex Metalloenzymes	PAUL A. LINDAHL	296
23. Urate Oxidase: Single-Turnover Stopped-Flow Techniques for Detecting Two Discrete Enzyme-Bound Intermediates	PETER A. TIPTON	310
24. Nitric Oxide Synthase: Use of Stopped-Flow Spectroscopy and Rapid-Quench Methods in Single-Turnover Conditions to Examine Formation and Reactions of Heme–O_2 Intermediate in Early Catalysis	CHIN-CHUAN WEI, ZHI-QIANG WANG, AND DENNIS J. STUEHR	320
25. Myeloperoxidase: Kinetic Evidence for Formation of Enzyme-Bound Chlorinating Intermediate	H. BRIAN DUNFORD AND LEAH A. MARQUEZ-CURTIS	338
26. Time-Resolved Resonance Raman Spectroscopy of Intermediates in Cytochrome Oxidase	DENIS L. ROUSSEAU AND SANGHWA HAN	351
27. Porphobilinogen Deaminase: Accumulation and Detection of Tetrapyrrole Intermediates Using Enzyme Immobilization	ALCIRA BATLLE	368
28. Adenosylcobalamin-Dependent Glutamate Mutase: Pre-Steady-State Kinetic Methods for Investigating Reaction Mechanism	HUNG-WEI CHIH, IPSITA ROYMOULIK, MARJA S. HUHTA, PRASHANTI MADHAVAPEDDI, AND E. NEIL G. MARSH	380
29. Ribonucleotide Reductase: Kinetic Methods for Demonstrating Radical Transfer Pathway in Protein R2 of Mouse Enzyme in Generation of Tyrosyl Free Radical	ASTRID GRÄSLUND	399
30. Galactose Oxidase: Probing Radical Mechanism with Ultrafast Radical Probe	BRUCE P. BRANCHAUD AND B. ELIZABETH TURNER	415
31. Kinetic Characterization of Transient Free Radical Intermediates in Reaction of Lysine 2,3-Aminomutase by EPR Lineshape Analysis	PERRY A. FREY, CHRISTOPHER H. CHANG, MARCUS D. BALLINGER, AND GEORGE H. REED	426
32. Demonstration of Peroxodiferric Intermediate in M-Ferritin Ferroxidase Reaction Using Rapid Freeze-Quench Mössbauer, Resonance Raman, and XAS Spectroscopies	CARSTEN KREBS, DALE E. EDMONDSON, AND BOI HANH HUYNH	436
33. A Survey of Covalent, Ionic, and Radical Intermediates in Enzyme-Catalyzed Reactions	R. DONALD ALLISON AND DANIEL L. PURICH	455

AUTHOR INDEX 471

SUBJECT INDEX 503

Contributors to Volume 354

*Article numbers are in parentheses following the names of contributors.
Affiliations listed are current.*

R. DONALD ALLISON (33), *Department of Biochemistry and Molecular Biology, University of Florida College of Medicine, Gainesville, Florida 32610*

ADOLFO AMICI (9), *Institute of Biochemistry, University of Ancona, 60100 Ancona, Italy*

MARCUS D. BALLINGER (31), *Sunesis Pharmaceuticals, Inc., Redwood City, California 94013*

DAVID BARFORD (18), *Section of Structural Biology, Institute of Cancer Research, Chester Beatty Laboratories, London SW3 6JB, United Kingdom*

ALCIRA BATLLE (27), *Centro de Investigaciones sobre Porfirinas y Porfirias, Argentine National Research Council and FCEN, University of Buenos Aires, 1056 Buenos Aires, Argentina*

DIDIER BLANOT (13), *Enveloppes Bactériennes et Antibiotiques, UMR 8619 CNRS, Université de Paris-Sud, 91405 Orsay, France*

SUSAN K. BOEHLEIN (20), *Department of Chemistry, University of Florida, Gainesville, Florida 32611*

AHMED BOUHSS (13), *Enveloppes Bactériennes et Antibiotiques, UMR 8619 CNRS, Université de Paris-Sud, 91405 Orsay, France*

BRUCE P. BRANCHAUD (30), *Department of Chemistry, University of Oregon, Eugene, Oregon 97403*

MICHAEL BRUNER (10), *Department of Chemistry, University of Florida, Gainesville, Florida 32611*

TIMOTHY D. H. BUGG (6), *Department of Chemistry, University of Warwick, Coventry CV4 7AL, United Kingdom*

JARED L. CARTWRIGHT (19), *School of Biological Sciences, University of Liverpool, Liverpool L69 7ZB, United Kingdom*

CHRISTOPHER H. CHANG (31), *Department of Chemistry, University of Florida, Gainesville, Florida 32611*

HUNG-WEI CHIH (28), *Department of Chemistry, University of Michigan, Ann Arbor, Michigan 48109*

JEAN-FRANÇOIS COLLET (12), *Laboratory of Physiological Chemistry, Catholic University of Louvain, Christian de Duve Institute of Cellular Pathology, B-1200 Brussels, Belgium*

PAUL F. COOK (17), *Department of Chemistry and Biochemistry, University of Oklahoma, Norman, Oklahoma 73019*

SÉBASTIEN DEMENTIN (13), *Enveloppes Bactériennes et Antibiotiques, UMR 8619 CNRS, Université de Paris-Sud, 91405 Orsay, France*

DOMINIQUE DEVILLE-BONNE (7), *Unité de Régulation Enzymatique des Activités Cellulaires, CNRS URA 1773, Institut Pasteur, 75724 Paris 15, France*

M. L. DODSON (15), *Sealy Center for Molecular Science, Department of Human Biological Chemistry and Genetics, University of Texas Medical Branch, Galveston, Texas 77555*

DONALD J. DOUGLAS (3), *Department of Chemistry, University of British Columbia, Vancouver, British Columbia, Canada V6T 1Z1*

H. BRIAN DUNFORD (25), *Department of Chemistry, University of Alberta, Edmonton, Alberta, Canada T6G 2G2*

DALE E. EDMONDSON (32), *Departments of Chemistry and Biochemistry, Emory University, Atlanta, Georgia 30322*

MONICA EMANUELLI (9), *Institute of Biochemistry, University of Ancona, 60100 Ancona, Italy*

SARAH M. FLEMING (6), *Department of Chemistry, University of Warwick, Coventry CV4 7AL, United Kingdom*

PERRY A. FREY (2, 8, 31), *Department of Biochemistry, University of Wisconsin, Madison, Wisconsin 53726*

SANDARUWAN GEEGANAGE (8), *Cancer Research Division, Eli Lilly and Company, Indianapolis, Indiana 46285*

ASTRID GRÄSLUND (29), *Department of Biochemistry and Biophysics, Stockholm University, S-106 91 Stockholm, Sweden*

JEFFREY W. GROSS (2), *GlaxoSmith Kline, King of Prussia, Pennsylvania 19406*

SANGHWA HAN (26), *Department of Biochemistry, Kangwon National University, Chunchon, Kangwon-Do 200-701, Korea*

JUN HIRATAKE (21), *Institute for Chemical Research, Kyoto University, Uji, Kyoto 611-0011, Japan*

BENJAMIN A. HORENSTEIN (10), *Department of Chemistry, University of Florida, Gainesville, Florida 32611*

MARJA S. HUHTA (28), *Department of Chemistry, University of Michigan, Ann Arbor, Michigan 48109*

BOI HANH HUYNH (32), *Department of Physics, Emory University, Atlanta, Georgia 30322*

MAKOTO INOUE (21), *Institute for Chemical Research, Kyoto University, Uji, Kyoto 611-0011, Japan*

JOËL JANIN (7), *Laboratory of Enzymology and Structural Biochemistry, CNRS UPR 9063, 91198 Gif-sur-Yvette Cedex, France*

WILLIAM E. KARSTEN (17), *Department of Chemistry and Biochemistry, University of Oklahoma, Norman, Oklahoma 73019*

LARS KONERMANN (3), *Department of Chemistry, University of Western Ontario, London, Ontario, Canada N6A 5B7*

CARSTEN KREBS (32), *Department of Biochemistry and Molecular Biology, Pennsylvania State University, University Park, Pennsylvania 16802*

ANDREW J. KURTZ (15), *Sealy Center for Molecular Science, Department of Human Biological Chemistry and Genetics, University of Texas Medical Branch, Galveston, Texas 77555*

G. JOHN LANGLEY (6), *Department of Chemistry, University of Southampton, Highfield, Southampton SO17 1BJ, United Kingdom*

PAUL A. LINDAHL (22), *Departments of Chemistry and Biochemistry and Biophysics, Texas A&M University, College Station, Texas 77843*

R. STEPHEN LLOYD (15), *Sealy Center for Molecular Science, Department of Human Biological Chemistry and Genetics, University of Texas Medical Branch, Galveston, Texas 77555*

PRASHANTI MADHAVAPEDDI (28), *Department of Chemistry, University of Michigan, Ann Arbor, Michigan 48109*

GIULIO MAGNI (9), *Institute of Biochemistry, University of Ancona, 60100 Ancona, Italy*

LEAH A. MARQUEZ-CURTIS (25), *Research and Development, Canadian Blood Services, Edmonton, Alberta, Canada T6G 2R8*

E. NEIL G. MARSH (28), *Department of Chemistry, University of Michigan, Ann Arbor, Michigan 48109*

ALEXANDER G. MCLENNAN (19), *School of Biological Sciences, University of Liverpool, Liverpool L69 7ZB, United Kingdom*

HENRY M. MIZIORKO (16), *Department of Biochemistry, Medical College of Wisconsin, Milwaukee, Wisconsin 53226*

RENÉE M. MOSI (4), *AnorMED, Inc., Langley, British Columbia, Canada V2Y 1N5*

CLAUDINE PARQUET (13), *Enveloppes Bactériennes et Antibiotiques, UMR 8619 CNRS, Université de Paris-Sud, 91405 Orsay, France*

DANIEL L. PURICH (1, 33), *Department of Biochemistry and Molecular Biology, University of Florida College of Medicine, Gainesville, Florida 32610*

NADIA RAFFAELLI (9), *Department of Agricultural Biotechnology, University of Ancona, 60100 Ancona, Italy*

GEORGE H. REED (31), *Department of Biochemistry, University of Wisconsin, Madison, Wisconsin 53726*

NIGEL G. J. RICHARDS (20), *Department of Chemistry, University of Florida, Gainesville, Florida 32611*

THOMAS A. ROBERTSON (6), *Department of Chemistry, University of Warwick, Coventry CV4 7AL, United Kingdom*

DAVID R. ROSE (5), *Ontario Cancer Institute/Princess Margaret Hospital, University of Toronto, Toronto, Ontario, Canada M5G 2M9*

DENIS L. ROUSSEAU (26), *Department of Physiology and Biophysics, Albert Einstein College of Medicine, Bronx, New York 10461*

IPSITA ROYMOULIK (28), *Department of Chemistry, University of Michigan, Ann Arbor, Michigan 48109*

SILVERIO RUGGIERI (9), *Department of Agricultural Biotechnology, University of Ancona, 60100 Ancona, Italy*

KANZO SAKATA (21), *Institute for Chemical Research, Kyoto University, Uji, Kyoto 611-0011, Japan*

GUNTER SCHNEIDER (14), *Department of Medical Biochemistry and Biophysics, Karolinska Institutet, S-171 77 Stockholm, Sweden*

HOLLY G. SCHNIZER (20), *Department of Biochemistry, University of Florida College of Medicine, Gainesville, Florida 32610*

SHELDON M. SCHUSTER (20), *Department of Biochemistry and Molecular Biology, University of Florida College of Medicine, Gainesville, Florida 32610*

GEORG A. SPRENGER (14), *Institute for Biotechnology 1, Forschungszentrum Jülich GmbH, D-52425 Jülich, Germany*

JON D. STEWART (20), *Department of Chemistry, University of Florida, Gainesville, Florida 32611*

VINCENT STROOBANT (12), *Ludwig Institute of Cancer Research, B-1200 Brussels, Belgium*

DENNIS J. STUEHR (24), *Department of Immunology, Lerner Research Institute, Cleveland Clinic, Cleveland, Ohio 44195*

PETER A. TIPTON (23), *Department of Biochemistry, University of Missouri, Columbia, Missouri 65211*

B. ELIZABETH TURNER (30), *Department of Biochemistry, Angiogenix, Inc., Burlingame, California 94010*

JEAN VAN HEIJENOORT (13), *Enveloppes Bactériennes et Antibiotiques, UMR 8619 CNRS, Université de Paris-Sud, 91405 Orsay, France*

EMILE VAN SCHAFTINGEN (12), *Laboratory of Physiological Chemistry, Catholic University of Louvain, Christian de Duve Institute of Cellular Pathology, B-1200 Brussels, Belgium*

DMITRIY A. VINAROV (16), *Department of Biochemistry, Medical College of Wisconsin, Milwaukee, Wisconsin 53226*

ZHI-QIANG WANG (24), *Department of Immunology, Lerner Research Institute, Cleveland Clinic, Cleveland, Ohio 44195*

CHIN-CHUAN WEI (24), *Department of Immunology, Lerner Research Institute, Cleveland Clinic, Cleveland, Ohio 44195*

JACQUELINE WICKI (5), *Department of Chemistry, University of British Columbia, Vancouver, British Columbia, Canada V6T 1Z1*

STEPHEN G. WITHERS (4, 5), *Department of Chemistry, University of British Columbia, Vancouver, British Columbia, Canada V6T 1Z1*

Preface

Enzyme catalysis is best described as a multistep chemical reaction cycle consisting of at least two and frequently more enzyme-bound reaction intermediates. During catalysis, the relative abundance of each intermediate species depends on both kinetic and thermodynamic constraints. Brief reaction cycle times on the submillisecond time-scale, as indicated by high enzymatic turnover numbers, require many intermediates to be formed for only fleetingly short periods. Unstable intermediates frequently form in almost vanishingly small amounts. Even when otherwise unstable intermediates are found to accumulate, their spectral and chemical properties, especially their susceptibility to undergo side-reactions, often challenge the sensitivity of even the most modern detection methods. This situation is particularly problematic when enzymes are used in catalytic rather than reagent quantities.

Despite such obstacles, enzyme chemists have created a thriving enterprise in detecting and characterizing enzyme reaction intermediates. This achievement attests to their ingenuity in adopting highly effective experimental approaches, augmented by the sensitivity of certain spectroscopic and/or isotopic methods, by the development of rapid mixing-quenching devices, and by recombinant DNA and bacterial expression techniques that provide wild-type and site-specific mutant enzymes in large quantities. While other chapters in Parts A–E of "Enzyme Kinetics and Mechanism" provide related information about reaction intermediates, every chapter in Part F is devoted to the task of detecting and characterizing enzyme reaction intermediates. This volume contains detailed descriptions of a broad range of experimental approaches that are applied to a diversified group of enzymes and reaction types. Some methods merely provide indirect kinetic evidence attesting to the likelihood that certain intermediates occur. Other trapping procedures directly identify the chemical nature of a particular intermediate. More robust techniques offer the opportunity to characterize the rates and extents of intermediate formation/breakdown or to reveal more subtle features of catalysis. Only through the design of such penetrating experiments can investigators draw useful inferences and insights concerning the structure and reactivity of enzymes.

I am especially pleased that so many highly experienced enzyme chemists were willing to share their expertise to produce a volume that promises to be of great practical importance. Such a compendium should be a valuable resource for anyone seeking to apply a particular technique to characterize their favorite enzyme as well as for those considering the properties of mechanistically related enzymatic reactions. Others may find that the chapters presented in Part F can form the nucleus for graduate-level special topics courses or for research seminars focusing on rate-enhancing strategies at play in biological catalysis.

DANIEL L. PURICH

METHODS IN ENZYMOLOGY

VOLUME I. Preparation and Assay of Enzymes
Edited by SIDNEY P. COLOWICK AND NATHAN O. KAPLAN

VOLUME II. Preparation and Assay of Enzymes
Edited by SIDNEY P. COLOWICK AND NATHAN O. KAPLAN

VOLUME III. Preparation and Assay of Substrates
Edited by SIDNEY P. COLOWICK AND NATHAN O. KAPLAN

VOLUME IV. Special Techniques for the Enzymologist
Edited by SIDNEY P. COLOWICK AND NATHAN O. KAPLAN

VOLUME V. Preparation and Assay of Enzymes
Edited by SIDNEY P. COLOWICK AND NATHAN O. KAPLAN

VOLUME VI. Preparation and Assay of Enzymes (*Continued*)
Preparation and Assay of Substrates
Special Techniques
Edited by SIDNEY P. COLOWICK AND NATHAN O. KAPLAN

VOLUME VII. Cumulative Subject Index
Edited by SIDNEY P. COLOWICK AND NATHAN O. KAPLAN

VOLUME VIII. Complex Carbohydrates
Edited by ELIZABETH F. NEUFELD AND VICTOR GINSBURG

VOLUME IX. Carbohydrate Metabolism
Edited by WILLIS A. WOOD

VOLUME X. Oxidation and Phosphorylation
Edited by RONALD W. ESTABROOK AND MAYNARD E. PULLMAN

VOLUME XI. Enzyme Structure
Edited by C. H. W. HIRS

VOLUME XII. Nucleic Acids (Parts A and B)
Edited by LAWRENCE GROSSMAN AND KIVIE MOLDAVE

VOLUME XIII. Citric Acid Cycle
Edited by J. M. LOWENSTEIN

VOLUME XIV. Lipids
Edited by J. M. LOWENSTEIN

VOLUME XV. Steroids and Terpenoids
Edited by RAYMOND B. CLAYTON

VOLUME XVI. Fast Reactions
Edited by KENNETH KUSTIN

VOLUME XVII. Metabolism of Amino Acids and Amines (Parts A and B)
Edited by HERBERT TABOR AND CELIA WHITE TABOR

VOLUME XVIII. Vitamins and Coenzymes (Parts A, B, and C)
Edited by DONALD B. MCCORMICK AND LEMUEL D. WRIGHT

VOLUME XIX. Proteolytic Enzymes
Edited by GERTRUDE E. PERLMANN AND LASZLO LORAND

VOLUME XX. Nucleic Acids and Protein Synthesis (Part C)
Edited by KIVIE MOLDAVE AND LAWRENCE GROSSMAN

VOLUME XXI. Nucleic Acids (Part D)
Edited by LAWRENCE GROSSMAN AND KIVIE MOLDAVE

VOLUME XXII. Enzyme Purification and Related Techniques
Edited by WILLIAM B. JAKOBY

VOLUME XXIII. Photosynthesis (Part A)
Edited by ANTHONY SAN PIETRO

VOLUME XXIV. Photosynthesis and Nitrogen Fixation (Part B)
Edited by ANTHONY SAN PIETRO

VOLUME XXV. Enzyme Structure (Part B)
Edited by C. H. W. HIRS AND SERGE N. TIMASHEFF

VOLUME XXVI. Enzyme Structure (Part C)
Edited by C. H. W. HIRS AND SERGE N. TIMASHEFF

VOLUME XXVII. Enzyme Structure (Part D)
Edited by C. H. W. HIRS AND SERGE N. TIMASHEFF

VOLUME XXVIII. Complex Carbohydrates (Part B)
Edited by VICTOR GINSBURG

VOLUME XXIX. Nucleic Acids and Protein Synthesis (Part E)
Edited by LAWRENCE GROSSMAN AND KIVIE MOLDAVE

VOLUME XXX. Nucleic Acids and Protein Synthesis (Part F)
Edited by KIVIE MOLDAVE AND LAWRENCE GROSSMAN

VOLUME XXXI. Biomembranes (Part A)
Edited by SIDNEY FLEISCHER AND LESTER PACKER

VOLUME XXXII. Biomembranes (Part B)
Edited by SIDNEY FLEISCHER AND LESTER PACKER

VOLUME XXXIII. Cumulative Subject Index Volumes I-XXX
Edited by MARTHA G. DENNIS AND EDWARD A. DENNIS

VOLUME XXXIV. Affinity Techniques (Enzyme Purification: Part B)
Edited by WILLIAM B. JAKOBY AND MEIR WILCHEK

VOLUME XXXV. Lipids (Part B)
Edited by JOHN M. LOWENSTEIN

VOLUME XXXVI. Hormone Action (Part A: Steroid Hormones)
Edited by BERT W. O'MALLEY AND JOEL G. HARDMAN

VOLUME XXXVII. Hormone Action (Part B: Peptide Hormones)
Edited by BERT W. O'MALLEY AND JOEL G. HARDMAN

VOLUME XXXVIII. Hormone Action (Part C: Cyclic Nucleotides)
Edited by JOEL G. HARDMAN AND BERT W. O'MALLEY

VOLUME XXXIX. Hormone Action (Part D: Isolated Cells, Tissues, and Organ Systems)
Edited by JOEL G. HARDMAN AND BERT W. O'MALLEY

VOLUME XL. Hormone Action (Part E: Nuclear Structure and Function)
Edited by BERT W. O'MALLEY AND JOEL G. HARDMAN

VOLUME XLI. Carbohydrate Metabolism (Part B)
Edited by W. A. WOOD

VOLUME XLII. Carbohydrate Metabolism (Part C)
Edited by W. A. WOOD

VOLUME XLIII. Antibiotics
Edited by JOHN H. HASH

VOLUME XLIV. Immobilized Enzymes
Edited by KLAUS MOSBACH

VOLUME XLV. Proteolytic Enzymes (Part B)
Edited by LASZLO LORAND

VOLUME XLVI. Affinity Labeling
Edited by WILLIAM B. JAKOBY AND MEIR WILCHEK

VOLUME XLVII. Enzyme Structure (Part E)
Edited by C. H. W. HIRS AND SERGE N. TIMASHEFF

VOLUME XLVIII. Enzyme Structure (Part F)
Edited by C. H. W. HIRS AND SERGE N. TIMASHEFF

VOLUME XLIX. Enzyme Structure (Part G)
Edited by C. H. W. HIRS AND SERGE N. TIMASHEFF

VOLUME L. Complex Carbohydrates (Part C)
Edited by VICTOR GINSBURG

VOLUME LI. Purine and Pyrimidine Nucleotide Metabolism
Edited by PATRICIA A. HOFFEE AND MARY ELLEN JONES

VOLUME LII. Biomembranes (Part C: Biological Oxidations)
Edited by SIDNEY FLEISCHER AND LESTER PACKER

VOLUME LIII. Biomembranes (Part D: Biological Oxidations)
Edited by SIDNEY FLEISCHER AND LESTER PACKER

VOLUME LIV. Biomembranes (Part E: Biological Oxidations)
Edited by SIDNEY FLEISCHER AND LESTER PACKER

VOLUME LV. Biomembranes (Part F: Bioenergetics)
Edited by SIDNEY FLEISCHER AND LESTER PACKER

VOLUME LVI. Biomembranes (Part G: Bioenergetics)
Edited by SIDNEY FLEISCHER AND LESTER PACKER

VOLUME LVII. Bioluminescence and Chemiluminescence
Edited by MARLENE A. DELUCA

VOLUME LVIII. Cell Culture
Edited by WILLIAM B. JAKOBY AND IRA PASTAN

VOLUME LIX. Nucleic Acids and Protein Synthesis (Part G)
Edited by KIVIE MOLDAVE AND LAWRENCE GROSSMAN

VOLUME LX. Nucleic Acids and Protein Synthesis (Part H)
Edited by KIVIE MOLDAVE AND LAWRENCE GROSSMAN

VOLUME 61. Enzyme Structure (Part H)
Edited by C. H. W. HIRS AND SERGE N. TIMASHEFF

VOLUME 62. Vitamins and Coenzymes (Part D)
Edited by DONALD B. MCCORMICK AND LEMUEL D. WRIGHT

VOLUME 63. Enzyme Kinetics and Mechanism (Part A: Initial Rate and Inhibitor Methods)
Edited by DANIEL L. PURICH

VOLUME 64. Enzyme Kinetics and Mechanism (Part B: Isotopic Probes and Complex Enzyme Systems)
Edited by DANIEL L. PURICH

VOLUME 65. Nucleic Acids (Part I)
Edited by LAWRENCE GROSSMAN AND KIVIE MOLDAVE

VOLUME 66. Vitamins and Coenzymes (Part E)
Edited by DONALD B. MCCORMICK AND LEMUEL D. WRIGHT

VOLUME 67. Vitamins and Coenzymes (Part F)
Edited by DONALD B. MCCORMICK AND LEMUEL D. WRIGHT

VOLUME 68. Recombinant DNA
Edited by RAY WU

VOLUME 69. Photosynthesis and Nitrogen Fixation (Part C)
Edited by ANTHONY SAN PIETRO

VOLUME 70. Immunochemical Techniques (Part A)
Edited by HELEN VAN VUNAKIS AND JOHN J. LANGONE

VOLUME 71. Lipids (Part C)
Edited by JOHN M. LOWENSTEIN

VOLUME 72. Lipids (Part D)
Edited by JOHN M. LOWENSTEIN

VOLUME 73. Immunochemical Techniques (Part B)
Edited by JOHN J. LANGONE AND HELEN VAN VUNAKIS

VOLUME 74. Immunochemical Techniques (Part C)
Edited by JOHN J. LANGONE AND HELEN VAN VUNAKIS

VOLUME 75. Cumulative Subject Index Volumes XXXI, XXXII, XXXIV–LX
Edited by EDWARD A. DENNIS AND MARTHA G. DENNIS

VOLUME 76. Hemoglobins
Edited by ERALDO ANTONINI, LUIGI ROSSI-BERNARDI, AND EMILIA CHIANCONE

VOLUME 77. Detoxication and Drug Metabolism
Edited by WILLIAM B. JAKOBY

VOLUME 78. Interferons (Part A)
Edited by SIDNEY PESTKA

VOLUME 79. Interferons (Part B)
Edited by SIDNEY PESTKA

VOLUME 80. Proteolytic Enzymes (Part C)
Edited by LASZLO LORAND

VOLUME 81. Biomembranes (Part H: Visual Pigments and Purple Membranes, I)
Edited by LESTER PACKER

VOLUME 82. Structural and Contractile Proteins (Part A: Extracellular Matrix)
Edited by LEON W. CUNNINGHAM AND DIXIE W. FREDERIKSEN

VOLUME 83. Complex Carbohydrates (Part D)
Edited by VICTOR GINSBURG

VOLUME 84. Immunochemical Techniques (Part D: Selected Immunoassays)
Edited by JOHN J. LANGONE AND HELEN VAN VUNAKIS

VOLUME 85. Structural and Contractile Proteins (Part B: The Contractile Apparatus and the Cytoskeleton)
Edited by DIXIE W. FREDERIKSEN AND LEON W. CUNNINGHAM

VOLUME 86. Prostaglandins and Arachidonate Metabolites
Edited by WILLIAM E. M. LANDS AND WILLIAM L. SMITH

VOLUME 87. Enzyme Kinetics and Mechanism (Part C: Intermediates, Stereochemistry, and Rate Studies)
Edited by DANIEL L. PURICH

VOLUME 88. Biomembranes (Part I: Visual Pigments and Purple Membranes, II)
Edited by LESTER PACKER

VOLUME 89. Carbohydrate Metabolism (Part D)
Edited by WILLIS A. WOOD

VOLUME 90. Carbohydrate Metabolism (Part E)
Edited by WILLIS A. WOOD

VOLUME 91. Enzyme Structure (Part I)
Edited by C. H. W. HIRS AND SERGE N. TIMASHEFF

VOLUME 92. Immunochemical Techniques (Part E: Monoclonal Antibodies and General Immunoassay Methods)
Edited by JOHN J. LANGONE AND HELEN VAN VUNAKIS

VOLUME 93. Immunochemical Techniques (Part F: Conventional Antibodies, Fc Receptors, and Cytotoxicity)
Edited by JOHN J. LANGONE AND HELEN VAN VUNAKIS

VOLUME 94. Polyamines
Edited by HERBERT TABOR AND CELIA WHITE TABOR

VOLUME 95. Cumulative Subject Index Volumes 61–74, 76–80
Edited by EDWARD A. DENNIS AND MARTHA G. DENNIS

VOLUME 96. Biomembranes [Part J: Membrane Biogenesis: Assembly and Targeting (General Methods; Eukaryotes)]
Edited by SIDNEY FLEISCHER AND BECCA FLEISCHER

VOLUME 97. Biomembranes [Part K: Membrane Biogenesis: Assembly and Targeting (Prokaryotes, Mitochondria, and Chloroplasts)]
Edited by SIDNEY FLEISCHER AND BECCA FLEISCHER

VOLUME 98. Biomembranes (Part L: Membrane Biogenesis: Processing and Recycling)
Edited by SIDNEY FLEISCHER AND BECCA FLEISCHER

VOLUME 99. Hormone Action (Part F: Protein Kinases)
Edited by JACKIE D. CORBIN AND JOEL G. HARDMAN

VOLUME 100. Recombinant DNA (Part B)
Edited by RAY WU, LAWRENCE GROSSMAN, AND KIVIE MOLDAVE

VOLUME 101. Recombinant DNA (Part C)
Edited by RAY WU, LAWRENCE GROSSMAN, AND KIVIE MOLDAVE

VOLUME 102. Hormone Action (Part G: Calmodulin and Calcium-Binding Proteins)
Edited by ANTHONY R. MEANS AND BERT W. O'MALLEY

VOLUME 103. Hormone Action (Part H: Neuroendocrine Peptides)
Edited by P. MICHAEL CONN

VOLUME 104. Enzyme Purification and Related Techniques (Part C)
Edited by WILLIAM B. JAKOBY

VOLUME 105. Oxygen Radicals in Biological Systems
Edited by LESTER PACKER

VOLUME 106. Posttranslational Modifications (Part A)
Edited by FINN WOLD AND KIVIE MOLDAVE

VOLUME 107. Posttranslational Modifications (Part B)
Edited by FINN WOLD AND KIVIE MOLDAVE

VOLUME 108. Immunochemical Techniques (Part G: Separation and Characterization of Lymphoid Cells)
Edited by GIOVANNI DI SABATO, JOHN J. LANGONE, AND HELEN VAN VUNAKIS

VOLUME 109. Hormone Action (Part I: Peptide Hormones)
Edited by LUTZ BIRNBAUMER AND BERT W. O'MALLEY

VOLUME 110. Steroids and Isoprenoids (Part A)
Edited by JOHN H. LAW AND HANS C. RILLING

VOLUME 111. Steroids and Isoprenoids (Part B)
Edited by JOHN H. LAW AND HANS C. RILLING

VOLUME 112. Drug and Enzyme Targeting (Part A)
Edited by KENNETH J. WIDDER AND RALPH GREEN

VOLUME 113. Glutamate, Glutamine, Glutathione, and Related Compounds
Edited by ALTON MEISTER

VOLUME 114. Diffraction Methods for Biological Macromolecules (Part A)
Edited by HAROLD W. WYCKOFF, C. H. W. HIRS, AND SERGE N. TIMASHEFF

VOLUME 115. Diffraction Methods for Biological Macromolecules (Part B)
Edited by HAROLD W. WYCKOFF, C. H. W. HIRS, AND SERGE N. TIMASHEFF

VOLUME 116. Immunochemical Techniques (Part H: Effectors and Mediators of Lymphoid Cell Functions)
Edited by GIOVANNI DI SABATO, JOHN J. LANGONE, AND HELEN VAN VUNAKIS

VOLUME 117. Enzyme Structure (Part J)
Edited by C. H. W. HIRS AND SERGE N. TIMASHEFF

VOLUME 118. Plant Molecular Biology
Edited by ARTHUR WEISSBACH AND HERBERT WEISSBACH

VOLUME 119. Interferons (Part C)
Edited by SIDNEY PESTKA

VOLUME 120. Cumulative Subject Index Volumes 81–94, 96–101

VOLUME 121. Immunochemical Techniques (Part I: Hybridoma Technology and Monoclonal Antibodies)
Edited by JOHN J. LANGONE AND HELEN VAN VUNAKIS

VOLUME 122. Vitamins and Coenzymes (Part G)
Edited by FRANK CHYTIL AND DONALD B. MCCORMICK

VOLUME 123. Vitamins and Coenzymes (Part H)
Edited by FRANK CHYTIL AND DONALD B. MCCORMICK

VOLUME 124. Hormone Action (Part J: Neuroendocrine Peptides)
Edited by P. MICHAEL CONN

VOLUME 125. Biomembranes (Part M: Transport in Bacteria, Mitochondria, and Chloroplasts: General Approaches and Transport Systems)
Edited by SIDNEY FLEISCHER AND BECCA FLEISCHER

VOLUME 126. Biomembranes (Part N: Transport in Bacteria, Mitochondria, and Chloroplasts: Protonmotive Force)
Edited by SIDNEY FLEISCHER AND BECCA FLEISCHER

VOLUME 127. Biomembranes (Part O: Protons and Water: Structure and Translocation)
Edited by LESTER PACKER

VOLUME 128. Plasma Lipoproteins (Part A: Preparation, Structure, and Molecular Biology)
Edited by JERE P. SEGREST AND JOHN J. ALBERS

VOLUME 129. Plasma Lipoproteins (Part B: Characterization, Cell Biology, and Metabolism)
Edited by JOHN J. ALBERS AND JERE P. SEGREST

VOLUME 130. Enzyme Structure (Part K)
Edited by C. H. W. HIRS AND SERGE N. TIMASHEFF

VOLUME 131. Enzyme Structure (Part L)
Edited by C. H. W. HIRS AND SERGE N. TIMASHEFF

VOLUME 132. Immunochemical Techniques (Part J: Phagocytosis and Cell-Mediated Cytotoxicity)
Edited by GIOVANNI DI SABATO AND JOHANNES EVERSE

VOLUME 133. Bioluminescence and Chemiluminescence (Part B)
Edited by MARLENE DELUCA AND WILLIAM D. MCELROY

VOLUME 134. Structural and Contractile Proteins (Part C: The Contractile Apparatus and the Cytoskeleton)
Edited by RICHARD B. VALLEE

VOLUME 135. Immobilized Enzymes and Cells (Part B)
Edited by KLAUS MOSBACH

VOLUME 136. Immobilized Enzymes and Cells (Part C)
Edited by KLAUS MOSBACH

VOLUME 137. Immobilized Enzymes and Cells (Part D)
Edited by KLAUS MOSBACH

VOLUME 138. Complex Carbohydrates (Part E)
Edited by VICTOR GINSBURG

VOLUME 139. Cellular Regulators (Part A: Calcium- and Calmodulin-Binding Proteins)
Edited by ANTHONY R. MEANS AND P. MICHAEL CONN

VOLUME 140. Cumulative Subject Index Volumes 102–119, 121–134

VOLUME 141. Cellular Regulators (Part B: Calcium and Lipids)
Edited by P. MICHAEL CONN AND ANTHONY R. MEANS

VOLUME 142. Metabolism of Aromatic Amino Acids and Amines
Edited by SEYMOUR KAUFMAN

VOLUME 143. Sulfur and Sulfur Amino Acids
Edited by WILLIAM B. JAKOBY AND OWEN GRIFFITH

VOLUME 144. Structural and Contractile Proteins (Part D: Extracellular Matrix)
Edited by LEON W. CUNNINGHAM

VOLUME 145. Structural and Contractile Proteins (Part E: Extracellular Matrix)
Edited by LEON W. CUNNINGHAM

VOLUME 146. Peptide Growth Factors (Part A)
Edited by DAVID BARNES AND DAVID A. SIRBASKU

VOLUME 147. Peptide Growth Factors (Part B)
Edited by DAVID BARNES AND DAVID A. SIRBASKU

VOLUME 148. Plant Cell Membranes
Edited by LESTER PACKER AND ROLAND DOUCE

VOLUME 149. Drug and Enzyme Targeting (Part B)
Edited by RALPH GREEN AND KENNETH J. WIDDER

VOLUME 150. Immunochemical Techniques (Part K: *In Vitro* Models of B and T Cell Functions and Lymphoid Cell Receptors)
Edited by GIOVANNI DI SABATO

VOLUME 151. Molecular Genetics of Mammalian Cells
Edited by MICHAEL M. GOTTESMAN

VOLUME 152. Guide to Molecular Cloning Techniques
Edited by SHELBY L. BERGER AND ALAN R. KIMMEL

VOLUME 153. Recombinant DNA (Part D)
Edited by RAY WU AND LAWRENCE GROSSMAN

VOLUME 154. Recombinant DNA (Part E)
Edited by RAY WU AND LAWRENCE GROSSMAN

VOLUME 155. Recombinant DNA (Part F)
Edited by RAY WU

VOLUME 156. Biomembranes (Part P: ATP-Driven Pumps and Related Transport: The Na, K-Pump)
Edited by SIDNEY FLEISCHER AND BECCA FLEISCHER

VOLUME 157. Biomembranes (Part Q: ATP-Driven Pumps and Related Transport: Calcium, Proton, and Potassium Pumps)
Edited by SIDNEY FLEISCHER AND BECCA FLEISCHER

VOLUME 158. Metalloproteins (Part A)
Edited by JAMES F. RIORDAN AND BERT L. VALLEE

VOLUME 159. Initiation and Termination of Cyclic Nucleotide Action
Edited by JACKIE D. CORBIN AND ROGER A. JOHNSON

VOLUME 160. Biomass (Part A: Cellulose and Hemicellulose)
Edited by WILLIS A. WOOD AND SCOTT T. KELLOGG

VOLUME 161. Biomass (Part B: Lignin, Pectin, and Chitin)
Edited by WILLIS A. WOOD AND SCOTT T. KELLOGG

VOLUME 162. Immunochemical Techniques (Part L: Chemotaxis and Inflammation)
Edited by GIOVANNI DI SABATO

VOLUME 163. Immunochemical Techniques (Part M: Chemotaxis and Inflammation)
Edited by GIOVANNI DI SABATO

VOLUME 164. Ribosomes
Edited by HARRY F. NOLLER, JR., AND KIVIE MOLDAVE

VOLUME 165. Microbial Toxins: Tools for Enzymology
Edited by SIDNEY HARSHMAN

VOLUME 166. Branched-Chain Amino Acids
Edited by ROBERT HARRIS AND JOHN R. SOKATCH

VOLUME 167. Cyanobacteria
Edited by LESTER PACKER AND ALEXANDER N. GLAZER

VOLUME 168. Hormone Action (Part K: Neuroendocrine Peptides)
Edited by P. MICHAEL CONN

VOLUME 169. Platelets: Receptors, Adhesion, Secretion (Part A)
Edited by JACEK HAWIGER

VOLUME 170. Nucleosomes
Edited by PAUL M. WASSARMAN AND ROGER D. KORNBERG

VOLUME 171. Biomembranes (Part R: Transport Theory: Cells and Model Membranes)
Edited by SIDNEY FLEISCHER AND BECCA FLEISCHER

VOLUME 172. Biomembranes (Part S: Transport: Membrane Isolation and Characterization)
Edited by SIDNEY FLEISCHER AND BECCA FLEISCHER

VOLUME 173. Biomembranes [Part T: Cellular and Subcellular Transport: Eukaryotic (Nonepithelial) Cells]
Edited by SIDNEY FLEISCHER AND BECCA FLEISCHER

VOLUME 174. Biomembranes [Part U: Cellular and Subcellular Transport: Eukaryotic (Nonepithelial) Cells]
Edited by SIDNEY FLEISCHER AND BECCA FLEISCHER

VOLUME 175. Cumulative Subject Index Volumes 135–139, 141–167

VOLUME 176. Nuclear Magnetic Resonance (Part A: Spectral Techniques and Dynamics)
Edited by NORMAN J. OPPENHEIMER AND THOMAS L. JAMES

VOLUME 177. Nuclear Magnetic Resonance (Part B: Structure and Mechanism)
Edited by NORMAN J. OPPENHEIMER AND THOMAS L. JAMES

VOLUME 178. Antibodies, Antigens, and Molecular Mimicry
Edited by JOHN J. LANGONE

VOLUME 179. Complex Carbohydrates (Part F)
Edited by VICTOR GINSBURG

VOLUME 180. RNA Processing (Part A: General Methods)
Edited by JAMES E. DAHLBERG AND JOHN N. ABELSON

VOLUME 181. RNA Processing (Part B: Specific Methods)
Edited by JAMES E. DAHLBERG AND JOHN N. ABELSON

VOLUME 182. Guide to Protein Purification
Edited by MURRAY P. DEUTSCHER

VOLUME 183. Molecular Evolution: Computer Analysis of Protein and Nucleic Acid Sequences
Edited by RUSSELL F. DOOLITTLE

VOLUME 184. Avidin-Biotin Technology
Edited by MEIR WILCHEK AND EDWARD A. BAYER

VOLUME 185. Gene Expression Technology
Edited by DAVID V. GOEDDEL

VOLUME 186. Oxygen Radicals in Biological Systems (Part B: Oxygen Radicals and Antioxidants)
Edited by LESTER PACKER AND ALEXANDER N. GLAZER

VOLUME 187. Arachidonate Related Lipid Mediators
Edited by ROBERT C. MURPHY AND FRANK A. FITZPATRICK

VOLUME 188. Hydrocarbons and Methylotrophy
Edited by MARY E. LIDSTROM

VOLUME 189. Retinoids (Part A: Molecular and Metabolic Aspects)
Edited by LESTER PACKER

VOLUME 190. Retinoids (Part B: Cell Differentiation and Clinical Applications)
Edited by LESTER PACKER

VOLUME 191. Biomembranes (Part V: Cellular and Subcellular Transport: Epithelial Cells)
Edited by SIDNEY FLEISCHER AND BECCA FLEISCHER

VOLUME 192. Biomembranes (Part W: Cellular and Subcellular Transport: Epithelial Cells)
Edited by SIDNEY FLEISCHER AND BECCA FLEISCHER

VOLUME 193. Mass Spectrometry
Edited by JAMES A. MCCLOSKEY

VOLUME 194. Guide to Yeast Genetics and Molecular Biology
Edited by CHRISTINE GUTHRIE AND GERALD R. FINK

VOLUME 195. Adenylyl Cyclase, G Proteins, and Guanylyl Cyclase
Edited by ROGER A. JOHNSON AND JACKIE D. CORBIN

VOLUME 196. Molecular Motors and the Cytoskeleton
Edited by RICHARD B. VALLEE

VOLUME 197. Phospholipases
Edited by EDWARD A. DENNIS

VOLUME 198. Peptide Growth Factors (Part C)
Edited by DAVID BARNES, J. P. MATHER, AND GORDON H. SATO

VOLUME 199. Cumulative Subject Index Volumes 168–174, 176–194

VOLUME 200. Protein Phosphorylation (Part A: Protein Kinases: Assays, Purification, Antibodies, Functional Analysis, Cloning, and Expression)
Edited by TONY HUNTER AND BARTHOLOMEW M. SEFTON

VOLUME 201. Protein Phosphorylation (Part B: Analysis of Protein Phosphorylation, Protein Kinase Inhibitors, and Protein Phosphatases)
Edited by TONY HUNTER AND BARTHOLOMEW M. SEFTON

VOLUME 202. Molecular Design and Modeling: Concepts and Applications (Part A: Proteins, Peptides, and Enzymes)
Edited by JOHN J. LANGONE

VOLUME 203. Molecular Design and Modeling: Concepts and Applications (Part B: Antibodies and Antigens, Nucleic Acids, Polysaccharides, and Drugs)
Edited by JOHN J. LANGONE

VOLUME 204. Bacterial Genetic Systems
Edited by JEFFREY H. MILLER

VOLUME 205. Metallobiochemistry (Part B: Metallothionein and Related Molecules)
Edited by JAMES F. RIORDAN AND BERT L. VALLEE

VOLUME 206. Cytochrome P450
Edited by MICHAEL R. WATERMAN AND ERIC F. JOHNSON

VOLUME 207. Ion Channels
Edited by BERNARDO RUDY AND LINDA E. IVERSON

VOLUME 208. Protein–DNA Interactions
Edited by ROBERT T. SAUER

VOLUME 209. Phospholipid Biosynthesis
Edited by EDWARD A. DENNIS AND DENNIS E. VANCE

VOLUME 210. Numerical Computer Methods
Edited by LUDWIG BRAND AND MICHAEL L. JOHNSON

VOLUME 211. DNA Structures (Part A: Synthesis and Physical Analysis of DNA)
Edited by DAVID M. J. LILLEY AND JAMES E. DAHLBERG

VOLUME 212. DNA Structures (Part B: Chemical and Electrophoretic Analysis of DNA)
Edited by DAVID M. J. LILLEY AND JAMES E. DAHLBERG

VOLUME 213. Carotenoids (Part A: Chemistry, Separation, Quantitation, and Antioxidation)
Edited by LESTER PACKER

VOLUME 214. Carotenoids (Part B: Metabolism, Genetics, and Biosynthesis)
Edited by LESTER PACKER

VOLUME 215. Platelets: Receptors, Adhesion, Secretion (Part B)
Edited by JACEK J. HAWIGER

VOLUME 216. Recombinant DNA (Part G)
Edited by RAY WU

VOLUME 217. Recombinant DNA (Part H)
Edited by RAY WU

VOLUME 218. Recombinant DNA (Part I)
Edited by RAY WU

VOLUME 219. Reconstitution of Intracellular Transport
Edited by JAMES E. ROTHMAN

VOLUME 220. Membrane Fusion Techniques (Part A)
Edited by NEJAT DÜZGUÜNES

VOLUME 221. Membrane Fusion Techniques (Part B)
Edited by NEJAT DÜZGÜNES

VOLUME 222. Proteolytic Enzymes in Coagulation, Fibrinolysis, and Complement Activation (Part A: Mammalian Blood Coagulation Factors and Inhibitors)
Edited by LASZLO LORAND AND KENNETH G. MANN

VOLUME 223. Proteolytic Enzymes in Coagulation, Fibrinolysis, and Complement Activation (Part B: Complement Activation, Fibrinolysis, and Nonmammalian Blood Coagulation Factors)
Edited by LASZLO LORAND AND KENNETH G. MANN

VOLUME 224. Molecular Evolution: Producing the Biochemical Data
Edited by ELIZABETH ANNE ZIMMER, THOMAS J. WHITE, REBECCA L. CANN, AND ALLAN C. WILSON

VOLUME 225. Guide to Techniques in Mouse Development
Edited by PAUL M. WASSARMAN AND MELVIN L. DEPAMPHILIS

VOLUME 226. Metallobiochemistry (Part C: Spectroscopic and Physical Methods for Probing Metal Ion Environments in Metalloenzymes and Metalloproteins)
Edited by JAMES F. RIORDAN AND BERT L. VALLEE

VOLUME 227. Metallobiochemistry (Part D: Physical and Spectroscopic Methods for Probing Metal Ion Environments in Metalloproteins)
Edited by JAMES F. RIORDAN AND BERT L. VALLEE

VOLUME 228. Aqueous Two-Phase Systems
Edited by HARRY WALTER AND GÖTE JOHANSSON

VOLUME 229. Cumulative Subject Index Volumes 195–198, 200–227

VOLUME 230. Guide to Techniques in Glycobiology
Edited by WILLIAM J. LENNARZ AND GERALD W. HART

VOLUME 231. Hemoglobins (Part B: Biochemical and Analytical Methods)
Edited by JOHANNES EVERSE, KIM D. VANDEGRIFF, AND ROBERT M. WINSLOW

VOLUME 232. Hemoglobins (Part C: Biophysical Methods)
Edited by JOHANNES EVERSE, KIM D. VANDEGRIFF, AND ROBERT M. WINSLOW

VOLUME 233. Oxygen Radicals in Biological Systems (Part C)
Edited by LESTER PACKER

VOLUME 234. Oxygen Radicals in Biological Systems (Part D)
Edited by LESTER PACKER

VOLUME 235. Bacterial Pathogenesis (Part A: Identification and Regulation of Virulence Factors)
Edited by VIRGINIA L. CLARK AND PATRIK M. BAVOIL

VOLUME 236. Bacterial Pathogenesis (Part B: Integration of Pathogenic Bacteria with Host Cells)
Edited by VIRGINIA L. CLARK AND PATRIK M. BAVOIL

VOLUME 237. Heterotrimeric G Proteins
Edited by RAVI IYENGAR

VOLUME 238. Heterotrimeric G-Protein Effectors
Edited by RAVI IYENGAR

VOLUME 239. Nuclear Magnetic Resonance (Part C)
Edited by THOMAS L. JAMES AND NORMAN J. OPPENHEIMER

VOLUME 240. Numerical Computer Methods (Part B)
Edited by MICHAEL L. JOHNSON AND LUDWIG BRAND

VOLUME 241. Retroviral Proteases
Edited by LAWRENCE C. KUO AND JULES A. SHAFER

VOLUME 242. Neoglycoconjugates (Part A)
Edited by Y. C. LEE AND REIKO T. LEE

VOLUME 243. Inorganic Microbial Sulfur Metabolism
Edited by HARRY D. PECK, JR., AND JEAN LEGALL

VOLUME 244. Proteolytic Enzymes: Serine and Cysteine Peptidases
Edited by ALAN J. BARRETT

VOLUME 245. Extracellular Matrix Components
Edited by E. RUOSLAHTI AND E. ENGVALL

VOLUME 246. Biochemical Spectroscopy
Edited by KENNETH SAUER

VOLUME 247. Neoglycoconjugates (Part B: Biomedical Applications)
Edited by Y. C. LEE AND REIKO T. LEE

VOLUME 248. Proteolytic Enzymes: Aspartic and Metallo Peptidases
Edited by ALAN J. BARRETT

VOLUME 249. Enzyme Kinetics and Mechanism (Part D: Developments in Enzyme Dynamics)
Edited by DANIEL L. PURICH

VOLUME 250. Lipid Modifications of Proteins
Edited by PATRICK J. CASEY AND JANICE E. BUSS

VOLUME 251. Biothiols (Part A: Monothiols and Dithiols, Protein Thiols, and Thiyl Radicals)
Edited by LESTER PACKER

VOLUME 252. Biothiols (Part B: Glutathione and Thioredoxin; Thiols in Signal Transduction and Gene Regulation)
Edited by LESTER PACKER

VOLUME 253. Adhesion of Microbial Pathogens
Edited by RON J. DOYLE AND ITZHAK OFEK

VOLUME 254. Oncogene Techniques
Edited by PETER K. VOGT AND INDER M. VERMA

VOLUME 255. Small GTPases and Their Regulators (Part A: Ras Family)
Edited by W. E. BALCH, CHANNING J. DER, AND ALAN HALL

VOLUME 256. Small GTPases and Their Regulators (Part B: Rho Family)
Edited by W. E. BALCH, CHANNING J. DER, AND ALAN HALL

VOLUME 257. Small GTPases and Their Regulators (Part C: Proteins Involved in Transport)
Edited by W. E. BALCH, CHANNING J. DER, AND ALAN HALL

VOLUME 258. Redox-Active Amino Acids in Biology
Edited by JUDITH P. KLINMAN

VOLUME 259. Energetics of Biological Macromolecules
Edited by MICHAEL L. JOHNSON AND GARY K. ACKERS

VOLUME 260. Mitochondrial Biogenesis and Genetics (Part A)
Edited by GIUSEPPE M. ATTARDI AND ANNE CHOMYN

VOLUME 261. Nuclear Magnetic Resonance and Nucleic Acids
Edited by THOMAS L. JAMES

VOLUME 262. DNA Replication
Edited by JUDITH L. CAMPBELL

VOLUME 263. Plasma Lipoproteins (Part C: Quantitation)
Edited by WILLIAM A. BRADLEY, SANDRA H. GIANTURCO, AND JERE P. SEGREST

VOLUME 264. Mitochondrial Biogenesis and Genetics (Part B)
Edited by GIUSEPPE M. ATTARDI AND ANNE CHOMYN

VOLUME 265. Cumulative Subject Index Volumes 228, 230–262

VOLUME 266. Computer Methods for Macromolecular Sequence Analysis
Edited by RUSSELL F. DOOLITTLE

VOLUME 267. Combinatorial Chemistry
Edited by JOHN N. ABELSON

VOLUME 268. Nitric Oxide (Part A: Sources and Detection of NO; NO Synthase)
Edited by LESTER PACKER

VOLUME 269. Nitric Oxide (Part B: Physiological and Pathological Processes)
Edited by LESTER PACKER

VOLUME 270. High Resolution Separation and Analysis of Biological Macromolecules (Part A: Fundamentals)
Edited by BARRY L. KARGER AND WILLIAM S. HANCOCK

VOLUME 271. High Resolution Separation and Analysis of Biological Macromolecules (Part B: Applications)
Edited by BARRY L. KARGER AND WILLIAM S. HANCOCK

VOLUME 272. Cytochrome P450 (Part B)
Edited by ERIC F. JOHNSON AND MICHAEL R. WATERMAN

VOLUME 273. RNA Polymerase and Associated Factors (Part A)
Edited by SANKAR ADHYA

VOLUME 274. RNA Polymerase and Associated Factors (Part B)
Edited by SANKAR ADHYA

VOLUME 275. Viral Polymerases and Related Proteins
Edited by LAWRENCE C. KUO, DAVID B. OLSEN, AND STEVEN S. CARROLL

VOLUME 276. Macromolecular Crystallography (Part A)
Edited by CHARLES W. CARTER, JR., AND ROBERT M. SWEET

VOLUME 277. Macromolecular Crystallography (Part B)
Edited by CHARLES W. CARTER, JR., AND ROBERT M. SWEET

VOLUME 278. Fluorescence Spectroscopy
Edited by LUDWIG BRAND AND MICHAEL L. JOHNSON

VOLUME 279. Vitamins and Coenzymes (Part I)
Edited by DONALD B. MCCORMICK, JOHN W. SUTTIE, AND CONRAD WAGNER

VOLUME 280. Vitamins and Coenzymes (Part J)
Edited by DONALD B. MCCORMICK, JOHN W. SUTTIE, AND CONRAD WAGNER

VOLUME 281. Vitamins and Coenzymes (Part K)
Edited by DONALD B. MCCORMICK, JOHN W. SUTTIE, AND CONRAD WAGNER

VOLUME 282. Vitamins and Coenzymes (Part L)
Edited by DONALD B. MCCORMICK, JOHN W. SUTTIE, AND CONRAD WAGNER

VOLUME 283. Cell Cycle Control
Edited by WILLIAM G. DUNPHY

VOLUME 284. Lipases (Part A: Biotechnology)
Edited by BYRON RUBIN AND EDWARD A. DENNIS

VOLUME 285. Cumulative Subject Index Volumes 263, 264, 266–284, 286–289

VOLUME 286. Lipases (Part B: Enzyme Characterization and Utilization)
Edited by BYRON RUBIN AND EDWARD A. DENNIS

VOLUME 287. Chemokines
Edited by RICHARD HORUK

VOLUME 288. Chemokine Receptors
Edited by RICHARD HORUK

VOLUME 289. Solid Phase Peptide Synthesis
Edited by GREGG B. FIELDS

VOLUME 290. Molecular Chaperones
Edited by GEORGE H. LORIMER AND THOMAS BALDWIN

VOLUME 291. Caged Compounds
Edited by GERARD MARRIOTT

VOLUME 292. ABC Transporters: Biochemical, Cellular, and Molecular Aspects
Edited by SURESH V. AMBUDKAR AND MICHAEL M. GOTTESMAN

VOLUME 293. Ion Channels (Part B)
Edited by P. MICHAEL CONN

VOLUME 294. Ion Channels (Part C)
Edited by P. MICHAEL CONN

VOLUME 295. Energetics of Biological Macromolecules (Part B)
Edited by GARY K. ACKERS AND MICHAEL L. JOHNSON

VOLUME 296. Neurotransmitter Transporters
Edited by SUSAN G. AMARA

VOLUME 297. Photosynthesis: Molecular Biology of Energy Capture
Edited by LEE MCINTOSH

VOLUME 298. Molecular Motors and the Cytoskeleton (Part B)
Edited by RICHARD B. VALLEE

VOLUME 299. Oxidants and Antioxidants (Part A)
Edited by LESTER PACKER

VOLUME 300. Oxidants and Antioxidants (Part B)
Edited by LESTER PACKER

VOLUME 301. Nitric Oxide: Biological and Antioxidant Activities (Part C)
Edited by LESTER PACKER

VOLUME 302. Green Fluorescent Protein
Edited by P. MICHAEL CONN

VOLUME 303. cDNA Preparation and Display
Edited by SHERMAN M. WEISSMAN

VOLUME 304. Chromatin
Edited by PAUL M. WASSARMAN AND ALAN P. WOLFFE

VOLUME 305. Bioluminescence and Chemiluminescence (Part C)
Edited by THOMAS O. BALDWIN AND MIRIAM M. ZIEGLER

VOLUME 306. Expression of Recombinant Genes in Eukaryotic Systems
Edited by JOSEPH C. GLORIOSO AND MARTIN C. SCHMIDT

VOLUME 307. Confocal Microscopy
Edited by P. MICHAEL CONN

VOLUME 308. Enzyme Kinetics and Mechanism (Part E: Energetics of Enzyme Catalysis)
Edited by DANIEL L. PURICH AND VERN L. SCHRAMM

VOLUME 309. Amyloid, Prions, and Other Protein Aggregates
Edited by RONALD WETZEL

VOLUME 310. Biofilms
Edited by RON J. DOYLE

VOLUME 311. Sphingolipid Metabolism and Cell Signaling (Part A)
Edited by ALFRED H. MERRILL, JR., AND YUSUF A. HANNUN

VOLUME 312. Sphingolipid Metabolism and Cell Signaling (Part B)
Edited by ALFRED H. MERRILL, JR., AND YUSUF A. HANNUN

VOLUME 313. Antisense Technology (Part A: General Methods, Methods of Delivery, and RNA Studies)
Edited by M. IAN PHILLIPS

VOLUME 314. Antisense Technology (Part B: Applications)
Edited by M. IAN PHILLIPS

VOLUME 315. Vertebrate Phototransduction and the Visual Cycle (Part A)
Edited by KRZYSZTOF PALCZEWSKI

VOLUME 316. Vertebrate Phototransduction and the Visual Cycle (Part B)
Edited by KRZYSZTOF PALCZEWSKI

VOLUME 317. RNA–Ligand Interactions (Part A: Structural Biology Methods)
Edited by DANIEL W. CELANDER AND JOHN N. ABELSON

VOLUME 318. RNA–Ligand Interactions (Part B: Molecular Biology Methods)
Edited by DANIEL W. CELANDER AND JOHN N. ABELSON

VOLUME 319. Singlet Oxygen, UV-A, and Ozone
Edited by LESTER PACKER AND HELMUT SIES

VOLUME 320. Cumulative Subject Index Volumes 290–319

VOLUME 321. Numerical Computer Methods (Part C)
Edited by MICHAEL L. JOHNSON AND LUDWIG BRAND

VOLUME 322. Apoptosis
Edited by JOHN C. REED

VOLUME 323. Energetics of Biological Macromolecules (Part C)
Edited by MICHAEL L. JOHNSON AND GARY K. ACKERS

VOLUME 324. Branched-Chain Amino Acids (Part B)
Edited by ROBERT A. HARRIS AND JOHN R. SOKATCH

VOLUME 325. Regulators and Effectors of Small GTPases (Part D: Rho Family)
Edited by W. E. BALCH, CHANNING J. DER, AND ALAN HALL

VOLUME 326. Applications of Chimeric Genes and Hybrid Proteins (Part A: Gene Expression and Protein Purification)
Edited by JEREMY THORNER, SCOTT D. EMR, AND JOHN N. ABELSON

VOLUME 327. Applications of Chimeric Genes and Hybrid Proteins (Part B: Cell Biology and Physiology)
Edited by JEREMY THORNER, SCOTT D. EMR, AND JOHN N. ABELSON

VOLUME 328. Applications of Chimeric Genes and Hybrid Proteins (Part C: Protein-Protein Interactions and Genomics)
Edited by JEREMY THORNER, SCOTT D. EMR, AND JOHN N. ABELSON

VOLUME 329. Regulators and Effectors of Small GTPases (Part E: GTPases Involved in Vesicular Traffic)
Edited by W. E. BALCH, CHANNING J. DER, AND ALAN HALL

VOLUME 330. Hyperthermophilic Enzymes (Part A)
Edited by MICHAEL W. W. ADAMS AND ROBERT M. KELLY

VOLUME 331. Hyperthermophilic Enzymes (Part B)
Edited by MICHAEL W. W. ADAMS AND ROBERT M. KELLY

VOLUME 332. Regulators and Effectors of Small GTPases (Part F: Ras Family I)
Edited by W. E. BALCH, CHANNING J. DER, AND ALAN HALL

VOLUME 333. Regulators and Effectors of Small GTPases (Part G: Ras Family II)
Edited by W. E. BALCH, CHANNING J. DER, AND ALAN HALL

VOLUME 334. Hyperthermophilic Enzymes (Part C)
Edited by MICHAEL W. W. ADAMS AND ROBERT M. KELLY

VOLUME 335. Flavonoids and Other Polyphenols
Edited by LESTER PACKER

VOLUME 336. Microbial Growth in Biofilms (Part A: Developmental and Molecular Biological Aspects)
Edited by RON J. DOYLE

VOLUME 337. Microbial Growth in Biofilms (Part B: Special Environments and Physicochemical Aspects)
Edited by RON J. DOYLE

VOLUME 338. Nuclear Magnetic Resonance of Biological Macromolecules (Part A)
Edited by THOMAS L. JAMES, VOLKER DÖTSCH, AND ULI SCHMITZ

VOLUME 339. Nuclear Magnetic Resonance of Biological Macromolecules (Part B)
Edited by THOMAS L. JAMES, VOLKER DÖTSCH, AND ULI SCHMITZ

VOLUME 340. Drug–Nucleic Acid Interactions
Edited by JONATHAN B. CHAIRES AND MICHAEL J. WARING

VOLUME 341. Ribonucleases (Part A)
Edited by ALLEN W. NICHOLSON

VOLUME 342. Ribonucleases (Part B)
Edited by ALLEN W. NICHOLSON

VOLUME 343. G Protein Pathways (Part A: Receptors)
Edited by RAVI IYENGAR AND JOHN D. HILDEBRANDT

VOLUME 344. G Protein Pathways (Part B: G Proteins and Their Regulators)
Edited by RAVI IYENGAR AND JOHN D. HILDEBRANDT

VOLUME 345. G Protein Pathways (Part C: Effector Mechanisms)
Edited by RAVI IYENGAR AND JOHN D. HILDEBRANDT

VOLUME 346. Gene Therapy Methods
Edited by M. IAN PHILLIPS

VOLUME 347. Protein Sensors and Reactive Oxygen Species (Part A: Selenoproteins and Thioredoxin)
Edited by HELMUT SIES AND LESTER PACKER

VOLUME 348. Protein Sensors and Reactive Oxygen Species (Part B: Thiol Enzymes and Proteins)
Edited by HELMUT SIES AND LESTER PACKER

VOLUME 349. Superoxide Dismutase
Edited by LESTER PACKER

VOLUME 350. Guide to Yeast Genetics and Molecular and Cell Biology (Part B)
Edited by CHRISTINE GUTHRIE AND GERALD R. FINK

VOLUME 351. Guide to Yeast Genetics and Molecular and Cell Biology (Part C)
Edited by CHRISTINE GUTHRIE AND GERALD R. FINK

VOLUME 352. Redox Cell Biology and Genetics (Part A)
Edited by CHANDAN K. SEN AND LESTER PACKER

VOLUME 353. Redox Cell Biology and Genetics (Part B)
Edited by CHANDAN K. SEN AND LESTER PACKER

VOLUME 354. Enzyme Kinetics and Mechanisms (Part F: Detection and Characterization of Enzyme Reaction Intermediates)
Edited by DANIEL L. PURICH

VOLUME 355. Cumulative Subject Index Volumes 321–354 (in preparation)

VOLUME 356. Laser Capture Microscopy and Microdissection (in preparation)
Edited by P. MICHAEL CONN

VOLUME 357. Cytochrome P450 (Part C) (in preparation)
Edited by ERIC F. JOHNSON AND MICHAEL R. WATERMAN

VOLUME 358. Bacterial Pathogenesis (Part C: Identification, Regulation, and Function of Virulence Factors) (in preparation)
Edited by VIRGINIA L. CLARK AND PATRIK M. BAVOIL

VOLUME 359. Nitric Oxide (Part D: Nitric Oxide Detection, Mitochondria and Cell Functions, and Peroxynitrite Reactions) (in preparation)
Edited by ENRIGUE CADENAS AND LESTER PACKER

VOLUME 360. Biophotonics (Part A) (in preparation)
Edited by GERARD MARRIOTT AND IAN PARKER

VOLUME 361. Biophotonics (Part B) (in preparation)
Edited by GERARD MARRIOTT AND IAN PARKER

[1] Covalent Enzyme–Substrate Compounds: Detection and Catalytic Competence

By DANIEL L. PURICH

Introduction

Although all enzyme reactions proceed through intermediate stages involving reversible enzyme–substrate and enzyme–substrate species, a smaller subset of enzymes forms one or more discrete covalent intermediates during their reaction cycles. Reactive intermediates facilitate catalysis: (a) by tethering substrates within an active site and reducing their degrees of freedom, (b) by preserving the group transfer potential of a reactant, (c) by making a better leaving group for nucleophilic catalysis, and (d) by transferring reactants between topologically remote active sites in multifunctional enzyme complexes. Although identification of covalent enzyme–substrate compounds remains a major focus in modern enzymology, an equally significant enterprise is the demonstration of their productive participation in catalysis. This chapter summarizes the experimental approaches now at the disposal of the enzyme chemist.

Initial Rate Kinetics

With few exceptions, initial rate steady-state kinetics is an unreliable method for discerning the participation of covalent intermediates in enzyme-catalyzed reactions.[1–3] This limitation arises from the insensitivity of initial rate kinetics to the occurrence of internal isomerization reactions. In fact, initial rate kinetics alone cannot distinguish among the following three cases:

Case I: Single, reversible central complex

$$E + S = EX \rightarrow E + P$$

Case II: Several reversible central complexes

$$E + S = ES \cdots [EX_i] \cdots EP \rightarrow E + P$$

Case III: Several central complexes, of which one (E~X) is covalent

$$E + S = ES = E{\sim}X = EP \rightarrow E + P$$

---

[1] H. J. Fromm, "Initial Rate Enzyme Kinetics." Springer-Verlag, Berlin and New York, 1975.
[2] W. W. Cleland, *Biochim. Biophys. Acta* **67,** 104 (1963).
[3] D. L. Purich and R. D. Allison, "Handbook of Biochemical Kinetics." Academic Press, New York, 2000.

In all three, the initial rate equation [$v = V_m[S]/(K_m + [S])$] is identical. This fact is clearly demonstrated by the kinetic properties of serine proteases and esterases operating by means of acyl-enzyme intermediates. The same is true for enediol intermediate formation during the triose-phosphate isomerase reaction or Schiff base formation in the aldolase reaction. A notable exception are phosphatase reactions, such as that catalyzed by glucose-6-phosphatase; the activated phosphorylhistidine (or P~E) intermediate can be detected in the presence of a second sugar that intercepts the P~E intermediate and is thereby phosphorylated by the transferase-like activity of the enzyme.

For multisubstrate reactions, steady-state kinetics only provides relatively limited information about covalent compounds or their kinetic competency. For example, initial rate kinetics provided no clue of a γ-glutamy phosphate intermediate in the glutamine synthetase reaction, a thiol ester intermediate in the CoA-linked aldehyde oxidoreductase reaction, or a 6-phospho-IMP intermediate in the adenylosuccinate synthetase reaction. Notable exceptions are double-displacement-type reactions, such as the well-known ping-pong Bi Bi reaction mechanism, where the enzyme cycles between an unmodified enzyme and a covalently modified enzyme species. Two-substrate ping-pong enzymes frequently possess a single substrate binding site and an active-site group (or cofactor) that undergoes modification. In the yeast nucleoside diphosphate kinase reaction, active-site histidyl residue is phosphorylated; in aminotransferase reactions, the tightly bound pyridoxal-5-phosphate coenzyme is converted to its pyridoxamine form, and a keto acid is released. Because the product of the first substrate departs before the second substrate arrives, ping-pong reactions result in parallel-line kinetics such as those observed with most aminotransferases as well as nucleoside-5'-diphosphate kinase.[4] In the absence of either product, the ping-pong Bi Bi mechanism[4] has a initial rate kinetic expression (i.e., $E_0/v = \Phi_0 + \Phi_1/[A] + \Phi_2/[B]$), where E_0, v, [A], [B], and the Φ terms refer to total enzyme concentration, initial velocity, substrate A level, substrate B level, and the combinations of rate constants defined by Dalziel.[5] This rate equation lacks the $\Phi_{12}/[A][B]$ term present in the initial rate equations of sequential bisubstrate mechanisms (i.e., $E_0/v = \Phi_0 + \Phi_1/[A] + \Phi_2/[B] + \Phi_{12}/[A][B]$). It is this term that imparts the convergence to v^{-1} versus [substrate]$^{-1}$ plots of initial rate data. As shown in Fig. 1, the parallel-line plots of ping-pong enzymes may be distinguished from the convergent-line plots of enzymes operating by sequential kinetic mechanisms. This kinetic behavior is exemplified by yeast nucleoside-diphosphate kinase.[4,6] It should also be noted that competitive inhibition patterns are unique for segregating ping-pong and sequential bisubstrate mechanisms.[1,3] One of the earliest documented problems with using initial rate data to demonstrate a ping-pong reaction scheme was

[4] E. Garces and W. W. Cleland, *Biochemistry* **8**, 633 (1969).
[5] K. Dalziel, *Acta Chem. Scand.* **11**, 1706 (1957).
[6] W. W. Cleland, "The Enzymes," 3rd Ed., Vol. 2, p. 1. Academic Press, 1970.

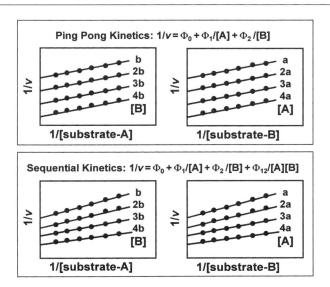

FIG. 1. Initial-rate kinetic behavior of two-substrate enzyme reactions obeying the rate law for ping-pong type or double-displacement mechanisms (*top*) or the rate law for sequential-type mechanisms (*bottom*), the latter including ternary complex ordered pathways, the Theorell–Chance pathway, or a rapid-equilibrium random pathway.

provided by the mammalian hexokinases. In their initial rate study of the mitochondrial form of bovine brain hexokinase, Fromm and Zewe[7,8] observed that the resulting Lineweaver–Burk plots (i.e., v^{-1} vs [D-glucose]$^{-1}$ at several constant ATP concentrations, and v^{-1} versus [ATP]$^{-1}$ at several constant D-glucose concentrations] consistently yielded parallel-line patterns, providing tentative evidence for the formation of a covalent enzyme–substrate intermediate. This rate behavior was also observed with one rat skeletal muscle hexokinase isozyme as well as with the Triton X-100 solubilized brain mitochondrial enzyme, whereas a second muscle hexokinase isozyme gave clear evidence of a sequential reaction scheme.[9] This discrepancy was later resolved when the alternative substrate D-fructose was substituted for glucose in the initial rate studies.[9–11] Control experiments demonstrated that fructose and glucose competed for the same site on the enzyme, but use of the alternative sugar substrate was also found to alter the relative magnitude of the $\Phi_{12}/[A][B]$ term, such that the slope effect was evident.

[7] H. J. Fromm and V. Zewe, *J. Biol. Chem.* **237,** 1661 (1962).
[8] T. L. Hanson and H. J. Fromm, *J. Biol. Chem.* **240,** 4133 (1965).
[9] H. J. Fromm and J. Ning, *Biochem. Biophys. Res. Commun.* **32,** 672 (1968).
[10] J. Ning, D. L. Purich, and H. J. Fromm, *J. Biol. Chem.* **244,** 3840 (1969).
[11] D. L. Purich and H. J. Fromm, *J. Biol. Chem.* **246,** 3456 (1971).

Cleland[6] proposed that a distinction might be possible by holding substrate A and B in a constant ratio (i.e., $[A] = \alpha[B]$), such that the sequential rate equations take on a quadratic form (i.e., $E_0/v = \Phi_0 + \Phi_1/\alpha[B] + \Phi_2/[B] + \Phi_{12}/[B]^2$). This experimental manipulation creates the condition that plots of v^{-1} versus $1/[B]$ become nonlinear as a result of the $1/\alpha[B]^2$ term. The major limitation is that curvature may still remain obscured by the dominance of the other terms in the expression. The slope term ($\Phi_1/\alpha + \Phi_2 + 2\Phi_{12}/\alpha[B]$) is the derivative, and the magnitude of the ϕ_{12} term relative to the other terms is not greatly amplified. Those adherents who stubbornly use Dalziel's Φ-method might wrongly conclude that the initial rate studies of the mammalian hexokinases should have been carried out at lower substrate concentrations, perhaps by taking advantage of the fluorescent properties of NADPH or NADH produced in the glucose-6-phosphate dehydrogenase or pyruvate kinase/lactate dehydrogenase reactions. The fact is that auxiliary enzyme systems have their own implicit limitations in terms of substrate sensitivity, and even the use of radioisotopes did not succeed in detecting convergence in double-reciprocal plots with glucose as the phosphoryl acceptor substrate in the reaction catalyzed by mammalian hexokinase.

A better way for detecting parallel-line data in a Lineweaver–Burk plot would be to use an entirely different plot. For a ping-pong reaction, a plot of $[A]/v$ versus $[A]$ at various constant concentrations of substrate B will convergence on the $[A]/v$-axis, and any deviation in this common point of intersection becomes quite apparent. Another practical application[12] of alternative substrate approaches for distinguishing sequential and ping-pong mechanisms recognizes that the kinetics of the second half-reaction in a ping-pong scheme requires only the formation of modified enzyme F. The term containing the second substrate contains rate constants that are only associated with the second half-reaction. If another substrate for A is used, only the rate constants in the first half-reaction will be changed (i.e., k_1, k_2, k_3, and k_4 become k'_1, k'_2, k'_3, and k'_4). Therefore, at identical concentrations of total enzyme and at a nonsaturating level of substrate A or A', plots of $1/v$ versus $1/[B]$ and of $1/v'$ versus $1/[B]$ must have *identical* slopes but possibly different intercepts. If such parallel-line data are not observed, the reaction mechanism must be of the sequential type. When this method, first suggested to the author by Professor Herbert Fromm, was used to probe acetate kinase,[12] the results were incompatible with a ping-pong mechanism.

Partial Exchange Reactions

Under favorable circumstances, the exchange of isotopically labeled substrate(s) and/or product(s) can also provide evidence for a ping-pong mechanism, which allows partial exchange reactions to occur. The ping-pong Bi Bi mechanism is the sum of two half-reactions: $E + A \rightleftharpoons E \cdot A \rightleftharpoons F \cdot P \rightleftharpoons F + P$ and

---

[12] J. A. Todhunter, K. B. Reichel, and D. L. Purich, *Arch. Biochem. Biophys.* **174**, 120 (1976).

F + B ⇌ F · B ⇌ E · Q ⇌ E + Q, where a covalent enzyme intermediate, designated F, is formed. Even in the absence of B and Q, the combination of E, A, and P will rapidly establish an equilibrium for the first half-reaction. Thus, addition of labeled substrate (A*) or product (P*) to a solution containing no B or Q will still permit the A ⇌ P exchange, as dictated for the first half-reaction. By symmetry, a B ⇌ Q exchange may occur in the presence of Enz, B, and Q, even though no A or P is added. This property of ping-pong systems is not shared by sequential reaction mechanisms,[1–3] which require the presence of *all* substrates and *all* products for any exchange reaction to be detectable. Failure to observe a partial exchange suggests (a) that there is no kinetically discernible covalent intermediate, or (b) that the intermediate does not form at all. Thus, the failure to observe any partial exchange reactions does not discount the catalytic participation of covalent intermediates; it only means that no product is released from the Enz' · P complex before the next substrate binds.

Although one may consider the ability of an enzyme to catalyze partial exchange reactions as an operational test for the formation of a covalent enzyme–substrate intermediate, observation of an A ⇌ P exchange in the absence of B and Q and a B ⇌ Q exchange in the absence of A and P may be the result of a contaminated reaction solution. If the enzyme or a substrate contains the second substrate, the observed half-reaction exchange may in reality represent an overall equilibrium exchange reaction. The greatest limitation using exchanges to infer mechanism is the potential failure to exclude other substrates and products from the reaction medium. This is a serious experimental obstacle, and many falsely positive indications of partial exchange reactions have been reported (see Table I). A good example is the phosphoglycerate kinase reaction, for which there was seemingly strong exchange evidence for a catalytically competent acyl-P intermediate. This evidence for a phosphoryl-enzyme intermediate was bolstered by: (a) the apparent demonstration of the competence of the intermediate to transfer a phosphoryl group to ADP; (b) the characteristic U-shaped pH stability profile of an acyl-P compound, including a signature hump attributable to the monoanion; (c) reaction of the intermediate with hydroxylamine; and (d) chemical trapping of the enzyme hydroxamate, followed by isolation of 2,4-diaminobutyric acid.[13,14] Later work by Johnson *et al.*[15] revealed that the enzyme was contaminated with tightly bound substrate. They demonstrated that isolation of the putative phosphoryl enzyme from ATP was quite variable, that the phosphoryl enzyme reacted only slowly with substrates, and that Enz · 1,3-bisphospho-D-glycerate had

---

[13] C. Roustan, A. Brevet, L.-A. Pradel, and N. van Thoai, *Eur. J. Biochem.* **37**, 248 (1973).
[14] A. Brevet, C. Roustan, G. Desvages, L.-A. Pradel, and N. van Thoai, *Eur. J. Biochem.* **39**, 141 (1974).
[15] P. E. Johnson, S. J. Abbott, G. A. Orr, M. Sémériva, and J. R. Knowles, *Biochemistry* **15**, 2893 (1976).

TABLE I
ERRONEOUS EXCHANGES IN ENZYME-CATALYZED REACTIONS

Enzyme	Observed exchange	Comment	Ref.
ATP phosphoribosyltransferase	PRPP ⇌ PP_i PR-ATP ⇌ ATP	Enzyme contains tightly bound PR-ATP formed from ATP impurity in commercial PRPP	a
Acetate kinase	Acetyl-P ⇌ Acetate	Some investigators report a slight exchange rate	b, c
Yeast hexokinase	ADP ⇌ ATP	Enzyme impurity or intrinsic ATPase activity of hexokinase	d
Liver pyruvate carboxylase	ADP ⇌ ATP	Exchange is due to an enzyme impurity	e
Phosphoglycerate kinase	ADP ⇌ ATP	Enzyme contains tightly bound cosubstrate	f
Phosphoribosyl-PP synthetase	ATP ⇌ AMP	Substrate contamination	g

[a] W. T. Brashear and S. M. Parsons, *J. Biol. Chem.* **250,** 6885 (1975).
[b] R. S. Anthony and L. B. Spector, *J. Biol. Chem.* **247,** 2120 (1972).
[c] M. T. Skarstedt and E. Silverstein, *J. Biol. Chem.* **251,** 6775 (1976).
[d] C. T. Walsh, Jr. and L. B. Spector, *Arch. Biochem. Biophys.* **145,** 1 (1971).
[e] M. F. Utter, R. E. Barden, and B. L. Taylor, *Adv. Enzymol.* **42,** 1 (1975).
[f] P. E. Johnson, S. J. Abbott, G. A. Orr, M. Semeriva, and J. R. Knowles, *Biochemistry* **15,** 2893 (1976).
[g] R. L. Switzer and P. D. Simcox, *J. Biol. Chem.* **249,** 5304 (1974).

a dissociation constant that was sufficiently low so as to remain intact even after gel filtration. To further demonstrate that contamination was a problem, Johnson et al.[15] boiled the presumptive phosphoryl enzyme, thereby hydrolyzing all of the bound bisphosphoglycerate into 3-phosphoglycerate, and the latter was then assayed quantitatively. They found that the amount of "E-P" compound paralleled the extent of enzyme contamination by bisphosphoglycerate. In view of the stereochemical investigations of Blätter and Knowles[16] on *Escherichia coli* acetate kinase, similarly observed partial exchange data falsely indicated participation of a phosphoryl-enzyme compound in that reaction.

While contamination of enzyme preparations with substrates is clearly a major source for erroneous exchange experiments, one must also consider the possibility that one or several substrates are similarly contaminated with other metabolites and/or inhibitors. A good example is the ATP phosphoribosyltransferase reaction, for which Bell and Koshland[17] reported partial exchange data indicative of a phosphoribosylated enzyme intermediate. Their main evidence was the occurrence of a

[16] W. A. Blätter and J. R. Knowles, *Biochemistry* **18,** 3927 (1979).
[17] R. M. Bell and D. E. Koshland, Jr., *Biochem. Biophys. Res. Commun.* **38,** 539 (1970).

PRPP ⇌ PR-ATP exchange in the absence of the complete reaction components. On learning that commercial PRPP was prepared with ATP by the Kornberg and Khorana[18] method, Brashear and Parsons[19] considered the likelihood of ATP contamination in the PRPP. These investigators succeeded in demonstrating that the earlier claim of a partial exchange reaction was entirely the consequence of ATP contamination. Even more alarming are reports that commercial ATP preparations contain significant amounts of orthovanadate[20,21] and aluminum.[22]

Another concern about the diagnostic value of exchange kinetic data regards interpretations of the relative rapidity of partial and overall exchange reactions. As noted above, the cardinal feature of ping-pong mechanisms is the ability of the enzyme to catalyze partial exchange reactions as a result of the independence of the substrate's interactions with the enzyme. This is reflected in the fact that the second substrate is obliged to await the departure of the first product from the substrate binding site before it may bind to the enzyme. Multisubstrate enzymes frequently mediate such partial reactions, but their importance in elucidating the chemical basis of the main pathway for enzymatic catalysis is sometimes unclear. One example is succinyl-CoA synthetase, for which the partial reactions are relatively slow. This property complicates interpretations of the significance of such partial reactions in the overall kinetic mechanism. Because the presence of other substrates at the active center might influence the catalytic configuration of the acid–base constellation and promote greater or more facile exchange, Bridger *et al.*[23] proposed the phenomenon of substrate synergism. By deriving appropriate rate laws for various exchanges, they concluded that the rate of a partial exchange reaction must exceed the rate of the same exchange reaction in the presence of all the other substrates, if the same catalytic steps and efficiencies are involved. When evidence to the contrary is observed, they suggested that one must consider the possibility that synergism exists.

Lueck and Fromm[24] focused their attention on the problem of substrate synergism by recognizing that misleading comparisons are often made with respect to initial velocity data and exchange rates. For the ping-pong Bi Bi mechanism, they obtained a quantitative test for substrate synergism based on the magnitudes of $R_{\text{max-A}\leftrightarrow P}$ and $R_{\text{max-B}\leftrightarrow Q}$, the maximal rates of the partial exchange reactions between the specified substrate and product, and the magnitudes of V_{mf} and V_{mr},

---

[18] A. Kornberg and H. B. Khorana, *Biochem. Prep.* **8**, 110 (1961).
[19] W. T. Brashear and S. M. Parsons, *J. Biol. Chem.* **250**, 6885 (1975).
[20] L. Josephson and L. C. Cantley, Jr., *Biochemistry* **16**, 4572 (1977).
[21] I. R. Gibbons, M. P. Cosson, J. A. Evans, B. H. Gibbons, B. Houck, K. H. Martinson, W. S. Sale, and W.-J. Y. Tang, *Proc. Natl. Acad. Sci. U.S.A.* **75**, 2220 (1978).
[22] F. C. Womack and S. P. Colowick, *Proc. Natl. Acad. Sci. U.S.A.* **76**, 5080 (1979).
[23] W. A. Bridger, W. A. Millen, and P. D. Boyer, *Biochemistry* **7**, 3608 (1968).
[24] J. D. Lueck and H. J. Fromm, *FEBS Lett.* **32**, 184 (1973).

the maximal initial reaction velocity of the overall reaction in the forward and reverse directions. Purich and Allison[25] rewrote the Fromm–Lueck expression to define the so-called synergism coefficient Q_{syn}, which equals $[(R_{max-A \leftrightarrow P})^{-1} + (R_{max-B \leftrightarrow Q})^{-1}]/[(V_{mf})^{-1} + (V_{mr})^{-1}]$. Only when Q_{syn} is substantially greater than unity can the experimenter justifiably conclude that there exists the possibility of substrate synergism. With yeast nucleoside-diphosphate kinase, Q_{syn} is about 1.3 as judged from the values reported by Garces and Cleland.[5] Because the enzyme's specific activity may vary somewhat, a value of 1.3 is quite close to unity, as expected for a ping-pong mechanism. On the other hand, for the E. coli acetate kinase reaction[26], Q_{syn} is about 32. Thus, a ping-pong mechanism is unlikely for the acetate kinase, and one might anticipate the possibility of substrate synergism; Skarstedt and Silverstein[26] proposed a so-called "activated ping-pong" mechanism for this enzyme.

Before leaving this subject, some comment on experimental protocol may be helpful. The obvious pitfall in exchange kinetic data as a criterion for catalytic participation of covalent intermediates deals with contamination. Perhaps the most direct means of evaluating contamination is radioisotope dilution with labeled A*, B*, P*, and Q*, each added separately to every component that will be used in the reaction mix. The availability of high-performance liquid chromatography (HPLC) for reversed-phase and ion-exchange separations greatly facilitates the determination of specific activity (dpm/mol). Another instructive approach, however, is to deliberately contaminate the exchange reaction mix to evaluate the effect of a very small amount of added cosubstrate on observed exchange reaction rates. The latter procedure can only be effective if the affinity of cosubstrate is not so high that the endogenous cosubstrate level may already be rate-saturating. This possibility can be evaluated by binding studies to determine the appropriate enzyme–ligand affinity constant.

Mechanistic Clues from Side Reactions

Despite the fact that specificity is a hallmark of biological catalysis, many enzymes exhibit side reactions, and systematic characterization of these processes can often provide mechanistic insights. Glutamine synthetase provides a particularly illuminating example. There was early recognition that this reaction could involve the transient formation of a covalent intermediate, and two rival hypotheses emerged: (a) that an amidophosphate intermediate formed first from ATP and ammonia, followed by attack of the γ-carboxylate of the amino acid on the phosphorus atom with the expulsion of ammonia; or (b) that a γ-glutamylphosphate compound

---

[25] D. L. Purich and R. D. Allison, *Methods Enzymol.* **64**, 1 (1980).
[26] M. T. Skarstedt and E. Silverstein, *J. Biol. Chem.* **251**, 6775 (1976).

formed by attack of the γ-carboxylate on MgATP complex, thereby expelling orthophosphate and allowing ammonia attack on the acyl-P to form glutamine. A major lead was provided by Boyer et al.,[27] who used [γ-^{18}O]glutamate to trace the fate of oxygen atoms during the overall reaction from glutamate, ATP, and ammonia. They discovered that ^{18}O-labeled orthophosphate was synthesized. Likewise, Kowalski et al.[28] showed the stoichiometric conversion of ortho [^{18}O]phosphate from used [γ-^{18}O]glutamate, with little incorporation of labeled oxygen into ADP. This finding was consistent with two possibilities: (1) a fully concerted mechanism, wherein ammonia attack at the γ-carboxylate was attended by simultaneous transfer one of the carboxylate oxygens to ATP, producing ADP and orthophosphate; or (2) the stepwise synthesis of the γ-carboxyl-P from ATP and glutamate, followed by nucleophilic attack of ammonia attended by release of orthophosphate bearing an oxygen atom once held by the γ-carboxylate. Levinthow and Meister[29] had previously examined the γ-glutamyltransferase reaction in the presence of radiocarbon-labeled glutamate; they concluded that the fact that no label appeared in either glutamine or the γ-glutamylhydroxamate argued that the amide and hydroxamate were sufficiently activated to form an unstable covalent intermediate, whereas glutamate was not. A crucial step in elucidating the actual reaction mechanism was taken by Krishnaswamy et al.,[30] who reacted the enzyme with ATP and [^{14}C]glutamic acid in ammonia-depleted solutions. This procedure demonstrated that the enzyme was capable of a side reaction in which pyrrolidone carboxylate (also known as pyroglutamate and 5-oxoproline) was synthesized. This cyclic compound that can also be formed at very low rates during the nonenzymatic dehydration of glutamate or in the nonenzymatic displacement of the amide nitrogen of glutamine by nucleophilic attack of the α-amino group on the γ-amide. The occurrence of the enzyme-catalyzed side reaction suggested that exposure to the enzyme facilitated formation of a suitably activated covalent compound corresponding to the γ-acyl-P which, on brief heating to 100°, undergoes intramolecular attack to form the cyclic amino acid. Both D- and L-glutamate were effective in the enzyme-catalyzed synthesis of the D- and L-forms of pyrrolidone carboxylate. This finding was also consistent with the ability of the enzyme to use either glutamate enantiomer. The authors also demonstrated that glutamine formed the pyrrolidone carboxylate some 600 times more slowly than in the presence of the enzyme. This observation allowed Meister[31] to conclude that an acyl-P intermediate must form in the glutamine synthetase reaction. In a subsequent study, Khedouri et al.[32]

[27] P. D. Boyer, O. J. Koeppe, and W. W. Luchsinger, *J. Am. Chem. Soc.* **78**, 356 (1956).
[28] A. Kowalski, C. Wyttenbach, L. Langer, and D. E. Koshland, Jr., *J. Biol. Chem.* **219**, 719 (1956).
[29] L. Levinthow and A. Meister, *J. Biol. Chem.* **209**, 265 (1954).
[30] P. R. Khrisnaswamy, V. Pamiljans, and A. Meister, *J. Biol. Chem.* **237**, 2932 (1962).
[31] A. Meister, "The Enzymes," 2nd Ed., Vol. 6, p. 443. Academic Press, 1962.
[32] E. Khedouri, V. Wellner, and A. Meister, *Biochemistry* **3**, 824 (1964).

chemically synthesized β-aminoglutaryl-P and demonstrated the enzymatic synthesis of β-aminoglutaramic acid (or β-glutamine) by ovine brain glutamine synthetase in the presence of ammonium ions. Likewise, they found that in the presence of hydroxylamine, β-aminoglutarylphosphate was enzymatically transformed into β-aminoglutarylhydroxamate. Moreover, when the synthetic acyl-P was incubated with ADP and enzyme, ATP was formed. None of these reactions occurred in the absence of enzyme, and β-aspartyl-P was completely ineffective in the above cited reactions. Later studies by Allison *et al.*[33] addressed the Ter-reactant initial rate kinetics of β-glutamate, ATP, and hydroxylamine as substrates for this enzyme.

Of the known side reactions catalyzed by glutamine synthetase, none is more revealing of acyl-P intermediate formation than the inhibitory properties of methionine sulfoximine (MSC).[34] There are four MSO stereoisomers, but only L-methionine-S-sulfoximine acts as an irreversible inhibitor of glutamine synthetase. Indeed, the phosphorylated inhibitor appears to mimic the transition state of the synthetase reaction. Methionine sulfoximine competes with both ammonia and glutamate for the ovine enzyme, suggesting that it occupies both pockets within the active site of the enzyme. Computer-assisted mapping of ligand interactions indicates that the sulfoximine oxygen atom of L-methionine-S-sulfoximine binds in place of the oxygen of the acyl group and that the methyl group occupies the ammonia binding site. In fact, such geometric considerations allow one to rationalize why covalent phosphorylation takes place on the sulfoximine nitrogen of MSO. A similar steric orientation cannot be achieved with either of the D-isomers of methionine sulfoximine, and while L-methionine-R-sulfoximine does bind to the enzyme, the reversed positions of the sulfoximine nitrogen and oxygen atoms do not permit efficient phosphorylation. Meister[34] also suggested (a) that the oxygen atom might be too acidic to attack ATP, and (b) that the observed binding of the methyl group in place of ammonia indicates that a hydrophobic interaction with un-ionized ammonia occurs, as opposed to ammonium ion. The latter would certainly favor nucleophilic attack on the acyl-P intermediary. Notably, inhibition by D- and L-methionine sulfone is reversible and can be easily reversed at higher glutamate concentrations; moreover, L- or D-methionine sulfone inhibition becomes more potent, but not irreversible in the presence of ATP.

Chemical Trapping of Enzyme–Substrate Covalent Compounds

Identification of a enzyme–substrate covalent intermediate by chemical trapping is a time-honored method available to the enzyme chemist. The method allows one to determine the nature of linkage formed by a particular enzyme.

---

[33] R. D. Allison, J. A. Todhunter, and D. L. Purich, *J. Biol. Chem.* **252**, 6046 (1976).
[34] A. Meister, "The Enzymes," 3rd Ed., Vol. 10, p. 699. Academic Press, 1973.

Trapping methods include (a) acid or base treatment to stabilize an otherwise fleeting species; (b) reaction with various nucleophiles, such as hydroxylamine and methoxylamine, to obtain a stable derivative; (c) treatment with such agents as sodium borohydride, sodium cyanoborohydride, or lithium aluminum hydride to obtain a chemically reduced derivative; and (d) oxidation to detect thiol ester and vicinal diol intermediates. Indeed, the efforts over the past 40 years to identify the nature of covalent intermediates have greatly reduced the universe of mechanistic possibilities. The major limitation of chemical trapping methods is that the true enzyme–substrate intermediate may be altered by the procedural conditions necessary for trapping. This may be especially true when the pH or solvent conditions are changed in the course of the trapping, such that partial denaturation of the enzyme is possible. In such cases, one might expect that a carbonium-ion intermediate stabilized in a tight ion pair with an anionic group on the enzyme might easily collapse to form a covalent compound. One can reasonably propose this possibility for sucrose phosphorylase, a enzyme for which there is considerable evidence for a covalent glucosyl-enzyme intermediate. This evidence includes the overall retention of the stereochemical configuration of the C-1 atom of the glucosyl moiety; catalysis of partial exchange reactions; the occurrence of certain hydrolytic reactions; initial velocity data supporting a modified ping-pong scheme; trapping of a glucosyl-enzyme compound on denaturation or chemical modification; differential labeling of the protein carboxyl groups in the absence and presence of substrate; and the extreme alkali sensitivity of the denatured glucosyl-enzyme intermediate.[35] Together, these chemical and kinetic properties strongly intimate the catalytic participation of a discrete glucosyl-enzyme intermediate. The actual linkage is likely to be an acylal formed by combination of the glucosyl residue and an aspartic or glutamic side-chain carboxylate. Even so, Mieyal and Abeles[35] correctly indicate that the mechanism could involve either a true covalent intermediate or a carbonium ion (possibly stabilized somewhat by formation of a strained ester or an ion pair interaction). Indeed, all of the findings may be reconciled with a oxocarbonium-ion intermediate, which under suitable trapping conditions collapses to form an acylal derivative. One might gather that the only reliable solution to the problem of distinguishing these possible pathways would come from dynamic information obtained under active catalytic conditions or by spectroscopic techniques such as nuclear magnetic resonance. For example, secondary kinetic isotope effect experiments with this enzyme might provide an indication of carbonium ion character developing in the rate-determining step(s). Although such isotope effect work can be carried out during catalysis, the absence of an appreciable k_H/k_D effect may also indicate the absence of a carbonium ion intermediate or that its formation is nested within other rate-contributing steps.

---

[35] J. J. Mieyal and R. H. Abeles, *in* "The Enzymes" (P. D. Boyer, ed.), 3rd Ed., Vol. 7, p. 515. Academic Press, 1972.

Even in the case where k_H/k_D falls in the range predicted for a carbonium ion intermediate, one cannot exclude the formation of an acylal covalent intermediate in some post-rate-determining step. Likewise, with phosphotransferases acting on acyl-P substrates or intermediates, one might anticipate the possible involvement of a metaphosphate, and the trapping of a phosphorylated enzyme compound may reflect migration and capture of this reactive species.

Analogs of Covalent Intermediates

Enzymes may promote catalysis by stabilizing transition states, by destabilizing the ground state, by enzyme-driven solvent reorganization, by means of fluctuating protein domain motions, and possibly even by vibrationally coupled atomic motions. Quantitative evaluation of the contributions of each of these phenomena is yet to be realized. In the transition-state analog approach, a major goal is to gain insight about the geometry atoms and charges in the activated complex. The able chemist designs analogs to mimic the reaction's transition state. In the Haldane–Pauling transition-state stabilization model, an enzyme strains a substrate toward the transition state at the expense of enzyme–substrate binding energy; then those agents that best reproduce the transition-state geometry and charge distribution are most apt to bind with greatest affinity. In such cases, there is no need to expend energy to reach the transition-state configuration. This approach was pioneered in Wolfenden's laboratory,[36,37] and its strengths and weaknesses have been considered at great length elsewhere.[38,39] One can also use geometrical analogs of enzyme-bound substrate–substrate covalent intermediates to probe enzyme mechanisms. Examples of the latter include the γ-glutamyl-P and aminoacyl adenylates, which appear to be important compounds in the glutamine synthetase and aminoacyl-tRNA synthetase reactions, respectively. Likewise, it is feasible to create stable covalent adducts of various coenzymes as in the case of the pyridoxal-P dependent reactions. The enhanced binding affinity is again employed as the principal criterion of the similarity of the intermediate analog to the true covalent intermediate.

The major question concerning this approach may be stated directly: To what extent does the success or failure of a particular analog to achieve tight binding indicate the participation or lack of participation of a presumed intermediate? For the sake of discussion, one might consider the conclusions of a study of the glutamine synthetases isolated from pea seeds and from *Escherichia coli* with methionine sulfoximine. With the bacterial enzyme, the analog was found to bind no more tightly than L-glutamate ($K_i = 3$ mM) and the interaction was

---

[36] R. Wolfenden, *Acc. Chem. Res.* **5**, 10 (1972).
[37] R. Wolfenden, *Methods Enzymol.* **46**, 15 (1977).
[38] V. L. Schramm, *Ann. Rev. Biochem.* **67**, 693 (1999).
[39] V. L. Schramm, *Methods Enzymol.* **308**, 301 (1999).

strictly competitive versus L-glutamate.[34,40] The binding of MSO with the pea seed enzyme was characterized by a distinctly different, two-step process: (a) a rapid, reversible step with a 10-fold greater affinity over the natural substrate; and (b) a slow, still reversible, isomerization culminating in very tight binding. In an earlier investigation,[33] both enzymes were found to bind and release substrates in a random fashion, but isotopic exchanges with partial reaction systems (indicative of ADP release from an Enz · γ-glutamylphosphate · ADP ternary complex) were only found with the pea seed enzyme experiments.[33,41] Althogether, it was concluded that the analog studies revealed differences in the active site architecture such that the bacterial enzyme fails to stabilize the acyl-P, if formed, to the extent provided by the plant enzyme.

The above example helps to draw attention to several practical problems with the use of geometrical analogs to infer information about the structure or participation of covalent intermediates of the sort represented by the γ-glutamyl-P in the glutamine synthetase reaction. First, such analogs have the collected binding interactions of two substrates—in this case, glutamate and orthophosphate. The contacts formed by the enzyme with each substrate may be largely outside the region undergoing bond-breaking and bond-making steps in catalysis, and the apparent enhancement in affinity may have little to do with the region of greatest mechanistic importance. Purich and Fromm[42] found that P^1,P^4-di(adenosine-5') tetraphosphate, while competitive with ATP and AMP sites on rabbit skeletal muscle adenylate kinase, was a rather poor analog of the Enz · MgATP^{2-} · AMP ternary complex. Later work[43] showed that the pentaphosphate bound with higher affinity, even though it is less clearly related to the likely structure of the ternary complex. On a purely geometrical basis, one might anticipate that collected substrate analogs should bind tightly, such that the total free energy released upon binding both substrates (i.e., $\Delta G = -RTK_{analog} \cong -RT\{K_{substrate-A} \cdot K_{substrate-B}\}$). However, the constellation of active site residues and the structure of the enzyme's adsorption sites for substrates may not permit efficient binding of an analog with the collected features of more than one substrate, especially if the two sites are brought closer together by means of a substrate-induced conformation change. With glutamine synthetase from *E. coli*, Todhunter and Purich,[44] for example, found that the acyl-P compound formed by ATP and glutamate in the absence of ammonia could be trapped in amounts practically stoichiometric with the enzyme. By analogy with the acetate kinase equilibrium ($K_{eq} = 10^3$ in favor of ATP and acetate), one might anticipate that much less acyl-P would be present on the enzyme if the solution

---

[40] W. B. Rowe, R. A. Ronzio, and A. Meister, *Biochemistry* **8**, 2674 (1969).
[41] F. C. Wedler and P. D. Boyer, *J. Biol. Chem.* **247**, 984 (1972).
[42] D. L. Purich and H. J. Fromm, *Biochim. Biophys. Acta* **276**, 307 (1972).
[43] G. E. Lienhard and I. I. Secemski, *J. Biol. Chem.* **248**, 1121 (1973).
[44] J. A. Todhunter and D. L. Purich, *J. Biol. Chem.* **250**, 3505 (1975).

equilibrium held. There are many examples of thermodynamically favorable internal equilibria, even though the corresponding bulk solution-phase equilibria are decidedly unfavorable.[45] The favorable conversion of Enz · ATP · glutamate to Enz · ADP · glutamyl-P may result from extremely tight binding achieved by a conformational change accompanying or following phosphoryl transfer. It is noteworthy that phosphorylation of methionine sulfoximine leads to extremely tight binding of methionine sulfoximine-P on the enzyme species with concomitant loss of exchange with MSO in the solution phase. In a sense, this special type of suicide inhibition exploits enhanced stability of the acyl-P.

Stereochemical Considerations

One of the most direct and telling methods for probing the formation of covalent intermediates is afforded by stereochemical characterization of substrates and products.[46-51] Although this approach is exhaustively described in later chapters of this volume, its mention here is necessary to achieve some degree of completeness. We shall restrict the discussion to a few limited cases where stereochemical studies were employed to discount the likelihood of covalent intermediate participation.

Chelsky and Parsons[52] examined the stereochemical course of the ATP phosphoribosyltransferase from *Salmonella typhimurium* because earlier studies pointed toward the participation of a phosphoribosyl-enzyme intermediate in its action.[17] It was once thought that this enzyme reaction occurred in two discrete steps (e.g., PRPP + Enz $\rightleftharpoons$ Enzyme complex$_1$ $\rightleftharpoons$ PR-Enz + PP$_i$; ATP + P-R-Enz $\rightleftharpoons$ Enzyme complex$_2$ $\rightleftharpoons$ PR-ATP + Enz). With S_N2 reactions, each elementary reaction step will involve inversion of configuration, here at C-1 of the ribosyl moiety. Thus, it is expected that the above mechanism should lead to overall retention of configuration as a result of suffering two inversions. If no covalent intermediate is formed in the course of the reaction cycle, then the expected configuration of the reaction product should be inverted with respect to the substrate. Starting with phosphoribosylpyrophosphate (PRPP in the above reaction sequence) the reaction was found to undergo inversion (i.e., α-PRPP + ATP = PP$_i$ + β-PR-ATP). This finding demonstrated that the above double-displacement mechanism was very unlikely and suggested a direct in-line transfer of the phosphoribosyl group.

---

[45] J. R. Knowles, *Annu. Rev. Biochem.* **49**, 877 (1980).
[46] S. L. Buchwald, D. E. Hansen, A. Hassett, and J. R. Knowles, *Methods Enzymol.* **87**, 279 (1982).
[47] P. A. Frey, J. P. Richard, H.-T. Ho, R. S. Brody, R. D. Sammons, and K.-W. Sheu, *Methods Enzymol.* **87**, 213 (1982).
[48] M. R. Webb, *Methods Enzymol.* **87**, 301 (1982).
[49] F. Eckstein, P. J. Romaniuk, and B. A. Connolly, *Methods Enzymol.* **87**, 197 (1982).
[50] H. G. Schloss, *Methods Enzymol.* **87**, 126 (1982).
[51] P. A. Frey, "The Enzymes," Vol. 20, p. 147. Academic Press, 1992.
[52] D. Chelsky and S. M. Parsons, *J. Biol. Chem.* **250**, 5669 (1975).

The utility of phosphotransfer stereochemistry can be illustrated by considering the acetate kinase reaction (acetyl-P + MgADP = acetate + MgATP). Rose et al.[53] searched for but failed to find any evidence for the participation of a phosphoryl-enzyme intermediate on the basis of kinetic and exchange studies. In the early 1970s, Spector's group at Rockefeller[54,55] provided evidence for a possible covalent intermediate: (1) enzyme phosphorylation with ATP or acetyl-P and E-P isolation by gel filtration; (2) pH stability and susceptibility to hydroxaminolysis characteristic of an acyl-P compound; (3) an inhibition by mercuric ion of the acetyl-P $\rightleftharpoons$ acetate exchange, but not the ADP $\rightleftharpoons$ ATP exchange reaction; and (4) the capacity of E-P to react with ADP or acetate, albeit slowly, to resynthesize ATP or acetyl-P. The initial rate kinetic studies were not so clear. Purich and Fromm[56] found parallel-line kinetics suggestive of a ping-pong mechanism, but Janson and Cleland[57] observed sequential kinetics by initial rate and competitive inhibition studies with exchange-inert chromium(III) nucleotide complexes. Adding to the uncertainty about the mechanism were later studies from the author's laboratory reporting the reduction of the E-P by NaB^3H_4 and recovery of α-amino-δ-hydroxyvalerate after acid hydrolysis of the enzyme reduction product,[58] but also obtaining alternative substrate kinetic data for a sequential mechanism.[12] Also, Skarstedt and Silverstein,[26] while providing evidence against a nucleotide-independent acetyl-P $\rightleftharpoons$ acetate exchange, described the overall exchange properties of the enzyme in terms of an "activated ping-pong" mechanism involving an E-P. As noted earlier, Blätter and Knowles[16] carried out stereochemical studies with [γ-(S)-^{16}O,^{17}O,^{18}O]ATP, an isotopically labeled ATP with a chiral terminal phosphoryl group. Their results with the acetate kinase reaction unambiguously demonstrated that the overall reaction proceeds with an inversion of stereochemistry, a result running counter to the participation of a phosphoryl enzyme during the catalytic round. Spector[59] later proposed a triple-displacement reaction with two discrete E-P intermediates and three inversions on the phosphoryl group. He postulated the participation of a group X on the enzyme, which forms E-X-PO_3 prior to transfer to the catalytic carboxyl group forming the acyl-P trapped in the earlier studies. He further suggested that the extra intermediate, E-X-PO_3, must be much less stable than the acyl-P, because only the latter is isolated on incubation of enzyme and ATP or acetyl-P. Blätter and Knowles[16] considered and dismissed this possibility. Such a mechanism would minimally require an adjacent displacement ("to" the phosphoryl-enzyme) and one in-line displacement ("from" the E-P). They preferred the direct in-line transfer between the two substrates.

---

[53] I. A. Rose, J. Grunberg-Manago, S. R. Korey, and S. Ochoa, *J. Biol. Chem.* **211,** 737 (1954).
[54] R. S. Anthony and L. B. Spector, *J. Biol. Chem.* **245,** 6739 (1970).
[55] R. S. Anthony and L. B. Spector, *J. Biol. Chem.* **247,** 2120 (1972).
[56] D. L. Purich and H. J. Fromm, *Arch. Biochem. Biophys.* **149,** 307 (1972).
[57] C. A. Janson and W. W. Cleland, *J. Biol. Chem.* **249,** 2572 (1974).
[58] J. A. Todhunter and D. L. Purich, *Biochem. Biophys. Res. Commun.* **60,** 273 (1974).
[59] L. B. Spector, *Proc. Natl. Acad. Sci. U.S.A.* **77,** 2626 (1980).

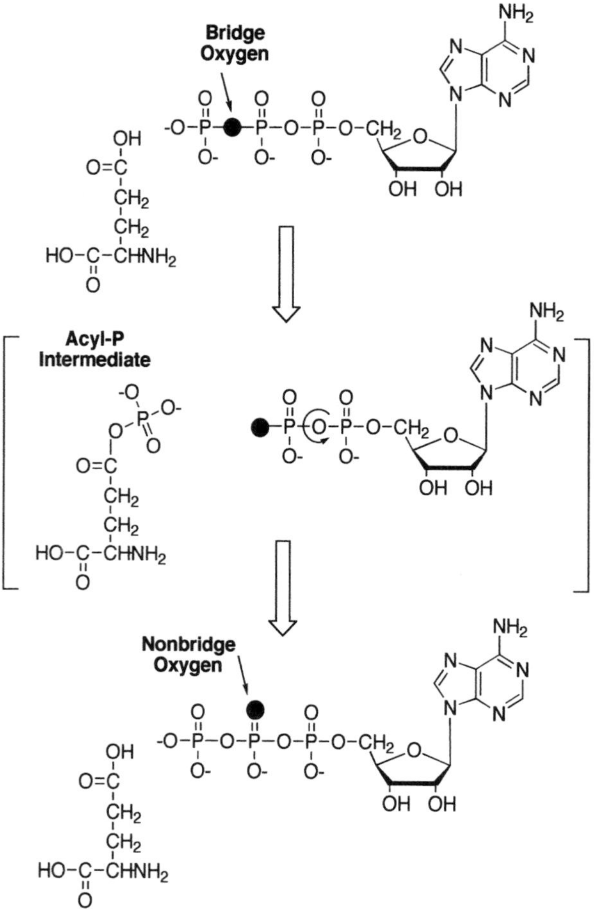

SCHEME 1

Dynamic Stereochemical Evidence

Midelfort and Rose[60] introduced an ingenious stereochemical method for detecting the reversible transfer (or positional exchange) of transfer of the terminal phosphate of ATP during enzymatic catalysis. Their isotope-scrambling technique provides direct evidence for the transient [Enz · ADP · P-X] formation from [^{18}O]ATP in enzyme reactions coupled to ATP hydrolysis. The method makes use of torsional symmetry of the newly formed group in ADP (see Scheme 1). [^{18}O]ATP

---

[60] C. F. Midelfort and I. A. Rose, *J. Biol. Chem.* **251**, 5881 (1976).

labeled in the $\beta-\gamma$ bridging oxygen is incubated with enzyme, and the occurrence of reversible cleavage of the P$_\beta$O–Pγ bond is observed by the appearance of ^{18}O in the β-nonbridge oxygens of the ATP pool. Experiments with both sheep brain and *Escherichia coli* glutamine synthetases demonstrated that cleavage of ATP to form enzyme-bound [ADP $\cdots$ P–X] requires glutamate. The exchange catalyzed by the bacterial enzyme with glutamate occurs in the absence of ammonia and is partially inhibited by added NH$_4$Cl, as would be expected if the exchange lies directly on in the catalytic pathway for glutamine synthesis. Their results provide compelling kinetic support for a two-step mechanism wherein phosphoryl transfer from ATP to glutamate, yielding a γ-glutamyl-P intermediate, precedes reaction with ammonia. Webb[61] extended the utility of this positional isotope exchange approach in nucleoside triphosphate-dependent reactions. He described methods for preparing [$\beta\gamma$-^{18}O,γ-^{18}O$_3$]ATP and [$\beta\gamma$-^{18}O,β-^{18}O$_2$]ATP, as well as an improved procedure for studying positional isotopic exchanges. Mass spectral analysis of the ^{18}O distribution in the resulting P$_i$ allowed the extent of positional isotope exchange to be determined. He also described procedures for enzymes that use GTP. The extraordinary power of positional isotope exchange (or PIX) is now firmly established.[62–64]

Burst Kinetics

The occurrence of transient formation of enzyme–substrate covalent intermediates can sometimes be demonstrated by using burst kinetic methods with reagent quantities of enzyme (see Table II). This is best accomplished with a substrate that slows down the rate of enzyme catalysis. Bursts in product formation can occur when an enzyme is first combined with its substrate(s), depending on the nature of the kinetic mechanism and the relative magnitudes of the rate constants for each step, as well as the relative concentrations of active enzyme and substrate(s). Bursts become especially apparent when the chromophoric product is released in a fast step that is followed by a much slower release of the second product. The burst phase quickly yields to the linear steady-state phase (see Fig. 2). The classical example of burst-phase kinetics is chymotrypsin-catalyzed hydrolysis of *p*-nitrophenyl ethyl carbonate or *p*-nitrophenyl acetate.[65] For either of these substrates, the product *p*-nitrophenol is released first, with the acyl-enzyme intermediate breaking down subsequently at a slower rate. Because the first product absorbs visible light so strongly, product formation is readily detectable. In this diagram,

---

[61] M. R. Webb, *Biochemistry* **19,** 4744 (1980).
[62] I. A. Rose, *Adv. Enzymol.* **50,** 361 (1979).
[63] L. S. Mullins and F. M. Raushel, *Methods Enzymol.* **249,** 398 (1994).
[64] J. J. Villafranca, *Methods Enzymol.* **177,** 390 (1989).
[65] B. S. Hartley and B. A. Kilby, *Biochem. J.* **56,** 288 (1954).

TABLE II
APPLICATION OF BURST KINETICS TO INVESTIGATION OF ENZYME-BOUND INTERMEDIATES

Enzyme	Process detected and/or characterized	Ref.
Aminoacyl-tRNA synthetase	Number of catalytically competent active sites on aminoacyl-tRNA synthetases was determined from the stoichiometry of aminoacyl-AMP formation. On mixing tRNA synthetase, cognate amino acid, [γ-^{32}P]ATP, and inorganic pyrophosphatase. Under suitable conditions, there was an initial rapid and stoichiometric "burst" (characterized by rate constant k_1) of ATP depletion as enzyme-bound aminoacyl adenylate is formed. There was then an initially linear decrease in ATP concentration as the complex undergoes hydrolysis (characterized by rate constant k_2) releasing enzyme to form further adenylate. When $k_2 < k_1$, the initial burst indicated the stoichiometry of aminoacyl-AMP formation	a
Alkaline phosphatase	Dephosphorylation step was not rate-limiting in the catalytic cycle substrates and inhibitor has no effect on phosphate release. The lack of effect of substrates on the dephosphorylation rate and on the phosphate dissociation rate indicated that the flip-flop mechanism, in which the product release is supposedly facilitated by the binding of a second molecule of substrate, cannot be valid for this enzyme	b
Dihydrofolate reductase	Kinetic mechanism was presented for the full time course of the enzymatic reaction over a wide range of substrate and enzyme concentrations. Specific rate constants were estimated by computer simulation of the full time course of single turnover, burst, and steady-state experiments using both nondeuterated and deuterated NADPH. The mechanism involves the random addition of substrates, but the substrates and enzyme are not at equilibrium prior to the chemical transformation step	c
Microtubule-kinesin ATPase	Pre-steady-state kinetics of the microtubule-kinesin ATPase were investigated by chemical-quench flow methods using the *Drosophila* kinesin motor domain. A burst of ADP product formation was observed during the first turnover of the enzyme in the acid-quench experiments that define the ATP hydrolysis transient. The observation of the burst demonstrated that product release is rate limiting even in the presence of saturating microtubule concentrations	d
β-Lactamase (E166 mutant)	Initial burst was observed (with an amplitude that was stoichiometric with enzyme concentration), suggesting rate-limiting deacylation of the acyl-enzyme intermediate. Further study revealed that in the presence of 0.5 M sodium sulfate, an agent that stabilizes the enzyme's native conformational state, the magnitude of the burst corresponded to 2 equivalents of enzyme, indicating that the mutation resulted in a change in the kinetic mechanism from the linear, acyl-enzyme pathway to one with a branch leading to an inactive form of the acyl-enzyme	e
CD38 Ectoenzyme	This enzyme catalyzes the formation of both cyclic ADP-ribose and ADP-ribose products from NAD^+ and hydrolyzes cyclic ADP-ribose to ADP-ribose. (Corresponding GDP products are formed from NGD^+.) Stopped-flow methods indicate the presence of a burst of cGDPR formation followed by the steady-state reaction rate. A lag phase, which was NGD^+ concentration-dependent, was also observed. The burst size indicates that the dimeric enzyme has a single catalytic site formed by two subunits. Product release is the overall rate-limiting step for enzyme reaction with NGD^+	f

[a] A. R. Fersht, J. S. Ashford, C. J. Bruton, R. Jakes, G. L. Koch, and B. S. Hartley, *Biochemistry* **14,** 1 (1975).
[b] J. R. Bale, C. Y. Huang, and P. B. Chock, *J. Biol. Chem.* **255,** 8431 (1980).
[c] M. H. Penner and C. Frieden, *J. Biol. Chem.* **262,** 15908 (1987).
[d] S. P. Gilbert and K. A. Johnson, *Biochemistry* **33,** 1951 (1994).
[e] W. A. Escobar, A. K. Tan, E. R. Lewis, and A. L. Fink, *Biochemistry* **33,** 7619 (1994).
[f] A. A. Sauve, C. Munshi, H. C. Lee, and V. L. Schramm, *Biochemistry* **37,** 13239 (1998).

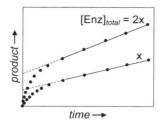

FIG. 2. Burst-phase kinetics showing that the zero-time (extrapolated) amplitude of the first-order burst and the slope of the linear, steady-state phase are proportional to the total concentration of catalytically active enzyme. The observed rate constant for the first-order phase is a direct measure of the efficiency of intermediate formation. (*Note:* When used to determine the active enzyme concentration, this technique is referred to as an active site titration.)

the instantaneous velocity (i.e., the tangent line to the curve at any instant t) is highest at the time of mixing enzyme and substrate; this rate quickly falls off to a linear rate that is characteristic of steady-state kinetic behavior. Because substrate depletion is relatively insignificant over the period shown in the figure, any change in reaction rate directly reflects the change in the concentration of free enzyme available for reaction with substrate. Extrapolation of the steady-state tangent line to $t = 0$ gives a measure of the concentration of active enzyme.

A chemically realistic representation[66] of the overall catalytic process can be written as follows:

$$E + S \rightleftharpoons E \cdot P_1 \cdot P_2 \rightleftharpoons E \cdot P_2 + P_1$$

$$E \cdot P_2 \rightarrow E + P_2$$

In this case, P_1 is the chromophoric product that is released prior to the slower conversion of $E \cdot P_2$ to form P_2, thereby completing the reaction cycle. This mechanism can be simplified to the following scheme described by two rate constants k_1 and k_2:

$$E + S \xrightarrow{k_1} E \cdot P_2 + P_1$$

$$E \cdot P_2 \xrightarrow{k_2} E + P_2$$

The magnitude of the burst equals $[E_{active}]\{k_1/(k_1 + k_2)\}$, which reduces to just the concentration of active enzyme $[E_{active}]$, if k_2 is much smaller than k_1. Burst kinetic measurements also allow one to evaluate the amount of active enzyme (i.e., $[E_{active}]/[E_{total}]$) in an enzyme preparation.

---

[66] A. R. Fersht, "Structure and Mechanism in Protein Science: A Guide to Enzyme Catalysis and Protein Folding," p. 650. W. H. Freeman, New York, 1999.

An obvious limitation in the application of the burst kinetic strategy is that one must be able to replace a naturally occurring substrate with some alternative substrate that can generate a detectable UV–visible spectral change. With enzymes displaying high substrate specificity, this technique may not be an option. Futhermore, the purist among us is always anxious to point out that the chromophoric substrate analog may not take the same catalytic path as the natural substrate. In any case, the technique is a simple and reliable indicator of the presence of a reaction intermediate.

Quenched-Flow Techniques

Once the basic chemical properties (i.e., structure, stability, and method of isolation) of a covalent enzyme–substrate compound have been established, the quenched-flow method often can be gainfully employed to learn more about the dynamics of the transient intermediates formed during enzymic catalysis.[67,68] In this procedure, the reaction is initiated by delivering two or more separate solutions, each in a syringe of appropriate volume and dimension, through a mixing device that ensures complete mixing without cavitation. One syringe typically contains a solution with reactants (with one radiolabeled) along with buffers and other required solutes, and the other syringe holds a solution of enzyme. The combined mixture of enzyme and substrate(s) is permitted to react as it flows through plastic tubing, and the reaction time is determined by the flow rate and the length and cross-sectional area of the tubing. After aging for the desired period, the sample is rapidly expelled into a quenching solution by means of a second mixing device. Quenching solutions can be prepared by using strong acid or base, metal ion chelators, or other agents known to quench the reaction's progress.

When the substrate concentration is larger than that of the enzyme, one can often detect a burst of product formation that transiently exceeds the steady-state rate. When the concentration of substrate is the limiting factor, one observes so-called single turnover kinetics. Several investigations are described in Table III to illustrate the power and versatility of rapid quench experiments.

Thermal Trapping of Enzyme–Substrate Intermediates

One of the newest techniques for exploring enzyme mechanism involves the use of subzero temperature techniques.[69–71] The basic idea is to trap each of the

---

[67] K. A. Johnson, "The Enzymes," Vol. 20, p. 1. Academic Press, 1992.
[68] K. A. Johnson, *Methods Enzymol.* **249**, 38 (1994).
[69] A. L. Fink, *Acc. Chem. Res.* **10**, 233 (1977).
[70] P. Douzou, "Cryobiochemistry." Academic Press, New York, 1977.
[71] A. L. Fink and M. H. Geeves, *Methods Enzymol.*, **63**, 336 (1979).

TABLE III
EXPERIMENTAL FINDINGS OBTAINED BY QUENCH-FLOW TECHNIQUE

Reaction/system	Experimental design and/or results	Ref.
Dynein ATPase	Effects of vanadate on the kinetics of ATP binding to and hydrolysis by *Tetrahymena* dynein were examined by quench-flow experiments which clearly demonstrated that vanadate did not alter the rate or amplitude of the pre-steady-state ATP binding or ATP hydrolysis transients, whereas the steady-state ATP hydrolysis of ATP was blocked immediately after a single turnover of ATP	a
Acetylcholine receptor	Quench-flow kinetic measurements was used to evaluate kinetic parameters governing the opening of acetylcholine receptor channels in the electric organ of *Electrophorus electricus*. Chemical kinetic measurements using carbamoylcholine (0.2–20 mM) at 12°, pH 7.0, and in the absence of a transmembrane voltage, yielded values for K_1 (dissociation constant for receptor activation), f (channel closing equilibrium constant), J (specific reaction rate for ion flux), and a_{max} (maximum inactivation rate constant) of 1 mM, 3.4, 4×10^7 M^{-1} s^{-1}, and 12 s^{-1}, respectively. Data obtained from the single-channel current measurements agreement with the values obtained by pulse-chase methods	b
Myofibrils and actomyosin	Inhibitory effect of P^3-[1-(2-nitrophenyl)ethyl]ATP (caged ATP) on the binding of Mg · ATP^{2-} to myofibrils was investigated by using the quench-flow technique to monitor single turnovers of [γ-^{32}P]ATP hydrolysis. Caged ATP behaved as a simple competitive inhibitor of ATP binding with an inhibition constant of 1.6 mM. The inhibitory effect of these ligands on ATP binding to actomyosin subfragment-1 was investigated in the same way. Inhibition constants of caged ATP and ADP were found to be 0.35 mM and 50 mM, respectively	c
Kinesin motor domain	Kinetic mechanism of bacterially expressed human kinesin ATPase motor domain K379 was investigated by transient and steady-state kinetic studies. The steps in nucleotide binding, measured with fluorescent ATP and ADP analogs, required a two-step mechanism for time-dependent changes in the fluorescence signal: an increase followed by a decrease, which indicates two isomerizations. The ATPase mechanism was fitted by a six-step reaction, and rate constants were determined.	d
*Eco*RI DNA methyltransferase	This S-adenosylmethionine-dependent enzyme transfers a methyl group to the second A in the DNA sequence GAATTC. The rate constant (41 ± 7 s^{-1}) for conversion of the central complex: Enz–DNA–S-adenosylmethionine to Enz–methylated DNA–S-adenosylhomocysteine is more than 300 times faster than k_{cat}, a finding that suggests that steps taking place after methyl transfer are rate-limiting	e
Chymotrypsin–antichymotrypsin	Serpins, serine proteinase inhibitors, form enzymatically inactive, 1 : 1 complexes (denoted E*I*) with their target proteinases, that only slowly release I*, in which the P$_1$–P$_1'$ linkage is cleaved. The serpin antichymotrypsin (ACT, I) reacts with chymotrypsin (Chtr or E) to form an E*I* complex via a three-step mechanism: E + I $\rightleftharpoons$ EI $\rightleftharpoons$ EI' $\rightleftharpoons$ E*I* in which EI', which retains the P1–P1' linkage, is formed in a partly or largely rate-determining step, depending on temperature. They extended their earlier studies through the introduction of a new assay for the formation of the postcomplex fragment, corresponding to ACT residues 359 (the P1' residue) to 398 (the C terminus), coupled with rapid quench flow kinetic analysis. We show that the E · I encounter complex of wild-type rACT and Chtr forms both E*I* and postcomplex fragment with the same rate constant, so that both species arise from EI' conversion to E*I*. These results support their earlier conclusion that the P$_1$–P$_1'$ linkage is preserved in EI' and imply that E*I* corresponds to a covalent adduct of E and I, either acyl enzyme or the tetrahedral intermediate	f

[a] E. T. Shimizu and K. A. Johnson, *J. Biol. Chem.* **258**, 13833 (1983).
[b] J. B. Udgaonkar and G. P. Hess, *Biophys. J.* **52**, 873 (1987).
[c] J. Sleep, C. Herrmann, T. Barman, and F. Travers, *Biochemistry* **33**, 6038 (1994).
[d] Y. Z. Ma and E. W. Taylor, *Biochemistry* **34**, 13233 (1995).
[e] N. O. Reich and N. J. Mashhoon, *J. Biol. Chem.* **268**, 9191 (1993).
[f] S. A. Nair and B. S. Cooperman, *J. Biol. Chem.* **273**, 17459 (1998).

intermediates and to infer the structure of all transition states along the productive catalytic pathway. One takes advantage of the temperature dependence of the reaction rate as illustrated by the Arrhenius equation ($k = Ae^{-E_{act}/RT}$). Rate reductions of 10^4 to 10^9 are attained by lowering the reaction temperature from 298 K to about 175 K. Thus, an intermediate with a lifetime in the microsecond to millisecond range may be accumulated as a result of its greatly reduced reactivity. Generally, cryosolvents are applied in such work to obviate the interpretive difficulties arising from the limited diffusivity of enzyme and substrate in the solid state. The field should now blossom as a result of the efforts expended in developing adequate cryosolvents that do not alter the structural and catalytic properties of the enzyme and in perfecting observation techniques for accumulating, stabilizing, and examining the intermediates. In principle, one may obtain kinetic, thermodynamic, and structural information about the intermediary states. When there is no spectral change in the protein or substrate attending formation of an intermediate, one may develop environmentally sensitive reporter-group methods that have been gainfully applied at normal temperatures. By changing the temperature it is possible to stabilize and accumulate a number of the reaction intermediates and to measure the first-order rate constants for interconversion. The major limitation with spectral reports (certainly with reporter groups) is that the interpretation of active-site structure and chemistry is indirect, depending largely on the experimenter's experience and intuition. In particular, the formation of acyl-enzyme, glycosyl-enzyme, and phosphoryl-enzyme intermediates must be inferred from the structural changes of the enzyme, because the substrates are not generally chromophoric. Occasionally, alternative substrates may be designed to partly surmount these difficulties. Among the significant accomplishments in cryoenzymology are X-ray diffraction studies on trapped intermediates in the crystalline state providing detailed structural information of "frozen" intermediates[72,73]; detection of tetrahedral intermediates in the reactions catalyzed by proteolytic enzymes[71]; and correlation of subzero data with fast-reaction data obtained at room temperature.[71]

As with any method, there are limitations, and the use of cryosolvents is no exception. The greatest limitation of a theoretical nature concerns the connections that may be established between the low-temperature reaction pathway and that at physiological temperature. One can usually detect breaks in Arrhenius plots even in aqueous solutions, and with enzymes these breaks may signal changes in rate-determining step, enzyme inactivation, changes in the enzyme's catalytic configuration, and/or changes in the substrate. The cryosolvent effect on the pathway may add further uncertainty to the comparison of low- and room-temperature results. The changes resulting from the relative interplay of enthalpic and entropic factors in each elementary reaction become another matter for some reservation.

[72] A. L. Fink and A. I. Ahmed, *Nature (London)* **263,** 294 (1976).
[73] T. Alber, G. A. Petsko, and D. Tsernoglou, *Nature (London)* **263,** 297 (1976).

In this regard, one might anticipate that solvation effects could greatly alter the potentially subtle balance of even internal equilibria (e.g., an enzyme-bound tight ion pair of a substrate carbocation versus a covalent enzyme–substrate compound, or an enzyme-bound imine versus carbanolamine form). It is also true that enzyme–enzyme interactions might take place at the high concentrations of enzyme required for study, and these interactions can be again altered by cryosolvent effects on such equilibria. In this respect, one might say that the greatest potential of the cryoenzymological methods is the description of mechanistic options that may influence the course of the room temperature pathway. Finally, the Boltzmann law clearly disfavors extensive accumulation of transition-state intermediates at lower temperature.

Spectral and X-Ray Methods for Detecting/Characterizing Reaction Intermediates

Spectroscopy remains one of the most powerful ways for analyzing the likely intermediates formed during catalysis. Chromophores, particularly coenzymes (e.g., FAD, NAD^+, cobalamin, pyridoxal-5-phosphate), have characteristic electronic spectra that permit the detection of intermediates in the UV/visible wavelength range. Metal ion complexes and charge-transfer complexes also produce spectral changes that have been gainfully exploited in the analysis of enzyme reaction mechanisms. During the past 40 years, electron spin resonance (ESR) and nuclear magnetic resonance (NMR) have also become especially powerful methods for deducing structural and/or distance information. Although any detailed consideration of ESR and NMR applications lies beyond the scope of this chapter, Table IV summarizes a few examples of enzyme reactions characterized by these techniques.

Although other chapters in this volume provide illuminating accounts of the power of X-ray crystallography in the investigation of enzyme intermediates, a few examples are considered here.

> *Subtilisin.* Atomic-level structural information has been especially valuable in the characterization of enzyme–inhibitor interactions. One example is the early study of Matthews *et al.*,[74] who investigated the ability of benzeneboronic acid and 2-phenylethaneboronic acid to form tightly bound enzyme adducts containing a covalent bond between the hydroxy oxygen atom of the active-site serine-221 and the inhibitor's boron atom. They found that the boron atom adopts a tetrahedral coordination, with one of the two additional boronic acid oxygen atoms lying in the "oxy-anion hole" and the other at the leaving group site.

[74] D. A. Matthews, R. A. Alden, J. J. Birktoft, S. T. Freer, and J. Kraut, *J. Biol. Chem.* **250**, 7120 (1975).

TABLE IV
DETECTION OF ENZYME-BOUND INTERMEDIATES BY ELECTRON SPIN RESONANCE OF NUCLEAR MAGNETIC RESONANCE

Enzyme	Species detected characterized	Ref.
Lysine 2,3-aminomutase	Rapid mix–freeze–quench ESR studies on the kinetic competence of a radical intermediate	a
Nitric oxide synthase	ESR spin-trapping detection of superoxide generation	b
Copper amine oxidase	ESR detection of a sequence of six intermediate species in the catalytic cycle	c
Peroxidase	ESR detection of one-electron oxidation of porphyrins to porphyrin π-cation radicals	d
Ferredoxin–ferredoxin: $NADP^+$ oxidoreductase	ESR detection of hydroxyl free radical during the reduction of molecular oxygen	e
F_1-ATPase	ESR study of asymmetry of catalytic sites in the bacterial enzyme	f
Cytochrome c oxidase	ESR detection of cyanyl and azidyl radical formation from cyanide and azide ions, respectively	g
D-Alanyl-D-alanine ligase	Rotational NMR structure of an enzyme-bound phosphorylated form of the pretransition-state analog [1(S)-aminoethyl] [2-carboxy-2(R)-methyl-1-ethyl]phosphinic acid	h
PurT GAR transformylase	NMR evidence for a proposed formyl phosphate intermediate	i
Endoglucanase	NMR study of a long-lived glycosyl-enzyme intermediate mimic produced by formate reactivation of a mutant lacking its catalytic nucleophile	j
3-Hydroxy-3-methylglutaryl-CoA	NMR study of acetyl-S-enzyme intermediate formation	k

[a] C. H. Chang, M. D. Ballinger, G. H. Reed, and P. A. Frey, *Biochemistry* **35,** 11081 (1996).
[b] J. Vasquez-Vivar, P. Martasek, N. Hogg, H. Karoui, B. S. Masters, K. A. Pritchard, Jr., and B. Kalyanaraman, *Methods Enzymol.* **301,** 169 (1999).
[c] R. Medda, A. Padiglia, J. Z. Pedersen, G. Rotilio, A. Finazzi-Agro, and G. Floris, *Biochemistry* **34,** 16375 (1995).
[d] K. M. Morehouse, H. J. Sipe, Jr., and R. P. Mason, *Arch. Biochem. Biophys.* **273,** 158 (1989).
[e] K. M. Morehouse and R. P. Mason, *J. Biol. Chem.* **263,** 1204 (1988).
[f] R. M. Losel, J. G. Wise, and P. D. Vogel, *Biochemistry* **36,** 1188 (1997).
[g] Y. R. Chen, B. E. Sturgeon, M. R. Gunther, and R. P. Mason, *J. Biol. Chem.* **274,** 24611 (1999).
[h] A. E. McDermott, F. Creuzet, R. G. Griffin, L. E. Zawadzke, Q. Z. Ye, and C. T. Walsh, *Biochemistry* **29,** 5767 (1990).
[i] A. E. Marolewski, K. M. Mattia, M. S. Warren, and S. J. Benkovic, *Biochemistry* **36,** 6709 (1997).
[j] J. L. Viladot, F. Canals, X. Batllori, and A. Planas, *Biochem. J.* **355,** 79 (2001).
[k] K. Y. Chun, D. A. Vinarov, and H. M. Miziorko, *Biochemistry* **39,** 14670 (2000).

Carboxypeptidase A. The 1.75-Å structure[75] showed that the zinc coordination number is 5, with bonding interacting with two imidazole $N^{\delta 1}$ nitrogens, the two carboxylate oxygens of glutamate-72, and a water molecule.

[75] D. C. Rees, M. Lewis, R. B. Honzatko, W. N. Lipscomb, and K. D. Hardman, *Proc. Natl. Acad. Sci. U.S.A.* **78,** 3408 (1981).

In the complex with the dipeptide glycyl-L-tyrosine, however, the water ligand is replaced by both the carbonyl oxygen and the amino nitrogen of the dipeptide. The amino nitrogen also statistically occupies a second position near glutamate-270.

Tyrosyl-tRNA synthetase. Tyrosyl adenylate and tyrosinyl adenylate were found to bind in similar conformations within a deep cleft of this aminoacylating enzyme.[76] The α-phosphate group interacted with the main-chain nitrogen of aspartate-38. The two hydroxyl groups of the ribose interacted strongly with the protein, such that the 2'-hydroxyl group bound to the carboxylate of aspartate-194 and the main-chain nitrogen of glycine-192, and the 3'-hydroxyl formed a hydrogen bond with a tightly bound water molecule.

Trypsin. The crystal and molecular structure[77] of trypsin at a transiently stable intermediate step during catalysis was determined using the substrate *p*-nitrophenyl-*p*-guanidinobenzoate under conditions where the acyl-enzyme intermediate, (guanidinobenzoyl)trypsin, was stable. The refined model provided the structural basis for the slow rate ($t_{1/2} = 12$ hr) of deacylation at 25° and pH 7.4. In addition to the rotation of the serine-195 hydroxyl away from His-157, Cβ of serine-195 was found to move 0.7 Å toward aspartate-189 at the bottom of the active site, thereby allowing the formation of energetically favorable H-bonds and an ion pair between the carboxylate of aspartate-189 and the guanidino group of the substrate. Highly ordered water molecules in the active site were no longer close enough to the scissile ester bond to serve as potential nucleophiles for hydrolysis.

β-Lactamase. The crystal structure[78] of the acyl-enzyme intermediate shows the substrate penicillin is covalently bound to O$^\gamma$ of serine-70. The deduced catalytic mechanism uses the hydroxyl group of serine-70 as the attacking nucleophile during acylation. Lysine-73 acts as a general base that abstracts a proton from serine-70 and transfers it to the thiazolidine ring nitrogen atom via serine-130. Deacylation occurs by nucleophilic attack on carbonyl by a water molecule, assisted by the general base, glutamate-166.

Triose-phosphate isomerase. The three-dimensional structure[79] of the isomerase complexed with a reactive intermediate analog, phosphoglycolohydroxamate, at 1.9-Å resolution is consistent with an acid–base mechanism

---

[76] P. Brick, T. N. Bhat, and D. M. Blow, *J. Mol. Biol.* **208**, 83 (1989).

[77] W. F. Mangel, P. T. Singer, D. M. Cyr, T. C. Umland, D. L. Toledo, R. M. Stroud, J. W. Pflugrath, and R. M. Sweet, *Biochemistry* **29**, 8351 (1990).

[78] N. C. Strynadka, H. Adachi, S. E. Jensen, K. Johns, A. Sielecki, C. Betzel, K. Sutoh, and M. N. James, *Nature* **359**, 700 (1992).

[79] R. C. Davenport, P. A. Bash, B. A. Seaton, M. Karplus, G. A. Petsko, and D. Ringe, *Biochemistry* **30**, 5821 (1991).

in which the carboxylate of glutamate-165 abstracts a proton from carbon while histidine-95 donates a proton to oxygen to form an enediol (or enediolate) intermediate. The conformation of the bound substrate stereoelectronically favors proton transfer from substrate carbon to the *syn* orbital of glutamate-165. Histidine-95 is neutral, rather than cationic, in the ground state, suggesting that it functions as an imidazole acid instead of the usual imidazolium. Lysine-12 is oriented so as to polarize the substrate oxygens by hydrogen bonding and/or electrostatic interaction, providing stabilization for the charged transition state.

4-Chlorobenzoyl-coenzyme A dehalogenase. This enzyme,[80] which catalyzes the hydrolysis of 4-chlorobenzoyl-CoA (4-CBA-CoA) to 4-hydroxybenzoyl-CoA, employs its aspartate-145 side-chain carboxylate to form a Meisenheimer-type reaction intermediate. Rapid-quench experiments using 4-[^{14}C]CBA-CoA provided evidence for the formation of a covalent enzyme intermediate during catalysis. Consumption labeled 4-CBA-CoA in the presence of a 2-fold molar excess of dehalogenase yielded a first-order rate constant of 6.5 s^{-1} for the formation of a radiolabeled enzyme intermediate. Radiolabeling reached a maximum level at 100 ms and amounted to 27% of the starting 4-[^{14}C]CBA-CoA. A single-turnover experiment conducted out in 98 g-atom %-excess ^{18}O-enriched water produced 4-HBA-CoA with 73–75 g-atom %-excess ^{16}O and 27–25 g-atom percent-excess ^{18}O at the benzoyl ring's C^4-OH. In contrast, a multiple turnover reaction carried out in 93 g-atom %-excess H$_2$ ^{18}O produced 4-HBA-CoA labeled at the C^4-OH with 89 g-atom %-excess ^{18}O and 11 g-atom %-excess ^{16}O. These data provided support for an aryl-enzyme Meisenheimer-type adduct during hydrolytic dechlorination of 4-CBA-CoA. X-ray crystallography[81] later revealed that aspartate-145 provides the side-chain carboxylate group that forms the intermediate Meisenheimer adduct and that histidine-90 serves as the general base in its subsequent hydrolysis.

Adenylosuccinate synthetase. Although early ^{18}O exchange experiments strongly intimated the formation of a 6-phosphoryl-IMP intermediate in the adenylosuccinate synthetase reaction (IMP + GTP + aspartate = adenylosuccinate + GDP + P$_i$), direct evidence of this intermediate awaited crystallographic investigations of the synthetase's structure. Poland *et al.*[82]

---

[80] G. Yang, R. Q. Liu, K. L. Taylor, H. Xiang, J. Price, and D. Dunaway-Mariano, *Biochemistry* **35,** 10879 (1996).
[81] M. M. Benning, K. L. Taylor, R.-Q. Liu, G. Yang, H. Xiang, G. Wesenberg, D. Dunaway-Mariano, and H. M. Holden, *Biochemistry* **35,** 8103 (1996).
[82] B. W. Poland, C. Bruns, H. J. Fromm, and R. B. Honzatko, *J. Biol. Chem.* **272,** 15200 (1997).

reported that enzyme crystals grown from solutions containing 6-mercapto-IMP and GTP exhibit an active-site electron density consistent with the formation of 6-thiophosphoryl-IMP and GDP. Aspartate-13 and glutamine-224 probably work in concert to stabilize the 6-thioanion of 6-mercapto-IMP, which in turn is the nucleophile in the displacement of GDP from the γ-phosphate of GTP. Once formed, 6-thiophosphoryl-IMP is stable, permitting its direct observation within the active site.

Concluding Remarks

Although the elucidation of a reaction mechanism begins with the trapping and identification of a covalent enzyme–substrate compound, the detailed examination of the catalytic mechanism requires stereochemical, kinetic, spectroscopic, and even thermodynamic data. This is obviously an ambitious undertaking, especially when one considers the fleeting nature of many intermediates. Fortunately, enzyme chemists today enjoy the availability of a robust range of complementary approaches for evaluating catalytic competence. The chapters presented in this volume provide valuable advice and experimental approaches for detecting and characterizing covalent intermediates.

[2] Rapid Mix-Quench MALDI-TOF Mass Spectrometry for Analysis of Enzymatic Systems

By JEFFREY W. GROSS and PERRY A. FREY

Introduction

Mass spectroscopy has been used as an analytical technique for several decades, but only recently, with the advent of the soft-ionization techniques matrix-assisted laser desorption/ionization (MALDI)[1,2] and electrospray ionization (ESI),[3,4] has the broader application of this discipline spread into the mainstream of biological sciences. Both MALDI and ESI allow the nondestructive ionization of large

[1] M. Karas and F. Hillenkamp, *Anal. Chem.* **60,** 2299 (1988).
[2] K. Tanaka, H. Waki, Y. Ido, S. Akita, Y. Yoshida, and T. Yoshida, *Rapid Commun. Mass Spectrom.* **2,** 151 (1988).
[3] J. B. Fenn, M. Mann, C. K. Meng, S. F. Wong, and C. M. Whitehouse, *Mass Spectrom. Rev.* **9,** 37 (1990).
[4] R. D. Smith, J. A. Loo, C. G. Edmonds, C. J. Barinaga, and H. R. Udseth, *Anal. Chem.* **62,** 882 (1990).

molecules, and thus permit the facile analysis of a wide range of biological molecules. As a result of these advances, the past decade has witnessed an explosive growth in the novel application of mass spectrometry. In this chapter, we summarize the use of MALDI–time-of-flight (TOF) mass spectrometry for the quantitative analysis of enzyme-catalyzed reactions in the pre-steady state.

Mass Shifts Associated with Enzymatic Reactions

Chemical reactions result in the rearrangement of atoms among reaction components, leading to a corresponding redistribution of mass. Because this is also true of enzymatic reactions, the changes in mass among substrates and products can be exploited to measure catalysis. The identity of mass for reaction components in enzyme-catalyzed reactions is not limited to the net reaction; many enzymatic mechanisms contain multiple intermediates. In these cases, the unique mass values of the intermediates can be exploited to investigate the chemical mechanism in the pre-steady state. To perform these analyses, one needs to be able to perform mass spectrometry on biological molecules quantitatively.

MALDI and ESI mass spectrometers provide the capacity to analyze a wide range of biological molecules. Therefore, they have great potential for investigating questions germane to mechanistic enzymology. These spectrometers can be used quantitatively, can provide isotopic resolution for molecules up to several thousand mass units, and can allow the accurate determination of mass values for molecules as large as several hundred thousand daltons. Consequently, a remarkably diverse set of steady-state enzymatic reactions have been monitored by mass spectrometry. These include glycosidase reactions,[5–7] pyruvate carboxylase,[8] carbonic anhydrase (carbonate dehydratase),[9] the acylation of peptides and proteins,[10–12] proteolysis of proteins,[13] and the depurination of RNA.[14] Although pre-steady-state mass spectrometric analysis of reactions has been less common than steady-state analysis, its value has been demonstrated in the analysis of acylation[15–17]

[5] S. Takayama, R. Martin, J. Wu, K. Laslo, G. Siuzdak, and C.-H. Wong, *J. Am. Chem. Soc.* **119,** 8146 (1997).
[6] J. Wu, S. Takayama, C.-H. Wong, and G. Suizdak, *Chem. Biol.* **4,** 653 (1997).
[7] S. A. Gerber, C. R. Scott, F. Turecek, and M. H. Gelb, *J. Am. Chem. Soc.* **121,** 1102 (1999).
[8] M. J. Kang, A. Tholey, and E. Heinzle, *Rapid Commun. Mass Spectrom.* **14,** 1972 (2000).
[9] D. B. Northrop and F. B. Simpson, *Arch. Biochem. Biophys.* **352,** 288 (1998).
[10] R. P. Brown, R. T. Alpin, and C. J. Schofield, *Biochemistry* **35,** 12421 (1996).
[11] V. A. Weller and M. D. Distefano, *J. Am. Chem. Soc.* **120,** 7975 (1998).
[12] N. L. Kelleher, C. L. Hendrickson, and C. T. Walsh, *Biochemistry* **38,** 15623 (1999).
[13] B. Bothner, R. Chavez, J. Wei, C. Strupp, Q. Phung, A. Schneemann, and G. Siuzdak, *J. Biol. Chem.* **275,** 13455 (2000).
[14] D. Fabris, *J. Am. Chem. Soc.* **122,** 8779 (2000).
[15] J. F. Krebs, G. Siuzdak, H. J. Dyson, J. D. Stewart, and S. J. Benkovic, *Biochemistry* **34,** 720 (1995).
[16] D. L. Zechel, L. Konermann, S. G. Withers, and D. J. Douglas, *Biochemistry* **37,** 7664 (1998).
[17] W.-P. Lu, Y. Sun, M. D. Bauer, S. Paule, P. M. Koenigs, and W. G. Kraft, *Biochemistry* **38,** 6537 (1999).

and phosphorylation of proteins[18] and the characterization of transiently formed intermediates.[19,20]

Mass Spectrometry

At the most fundamental level, all mass spectrometers use magnetic or electric fields to analyze ionized gaseous particles held at very low pressures. Thus prior to introduction into the spectrometer, samples must be ionized and vaporized. The methods for satisfying these two requirements distinguish the soft ionization techniques from earlier methods. Older spectrometers use predominantly electron ionization for analysis of analytes. In those techniques, samples are vaporized by heating, and then ionized by bombardment with a stream of electrons. Use of this destructive technique is limited to analysis of small, heat-stable, volatile, easily ionizable molecules.[21,22]

Early soft ionization techniques, which include chemical ionization, field ionization, and field desorption, lessen the destructiveness of the analysis, but still require thermal stability and are useful for only a limited range of mass values. Techniques using beams of atoms (fast atom bombardment) or ions (secondary ion mass spectrometry) to desorb the analyte from the sample probe and generate analyte-derived ions are widely used, but are not as effective for biological samples as MALDI and ESI.[22] In MALDI, laser-derived photons are used to excite accessory matrix molecules in a matrix–analyte crystal. This simultaneously ionizes the analyte and desorbs it from the sample probe. The photoderived desorption/ionization is effective for the nondestructive generation of gaseous ions of large proteins. ESI uses an entirely different approach to generate similar ions. In ESI, an analyte-containing solution is pumped through a high voltage nebulizing needle, resulting in the evolution of charged droplets. These are desolvated by heating, and the resulting analyte ions are drawn into the orifice of the spectrometer.

In the analysis of biomolecules, ESI and MALDI are each superior for specific applications. MALDI analysis is generally more tolerant of salts and other contaminants than ESI. This makes MALDI the more rugged and robust platform. In addition MALDI ionization generates singly charged analytes, making analysis and deconvolution of spectra more forthright. MALDI also benefits from the utilization of TOF analyzers, as compared to the quadrupole analyzers found on most ESI spectrometers. The TOF analyzer allows the analysis of analytes

---

[18] C. T. Houston, W. P. Taylor, T. S. Widlanski, and J. P. Reilly, *Anal. Chem.* **72**, 3311 (2000).
[19] A. A. Paiva, R. F. Tilton, Jr., G. P. Crooks, L. Q. Huang, and K. S. Anderson, *Biochemistry* **36**, 15472 (1997).
[20] J. W. Gross, A. D. Hegeman, M. M. Vestling, and P. A. Frey, *Biochemistry* **39**, 13633 (2000).
[21] H. M. Fales, G. W. Milne, J. J. Pisano, H. B. Brewer, M. S. Blum, J. G. MacConnell, J. Brand, and N. Law, *Recent Prog. Horm. Res.* **28**, 591 (1972).
[22] C. N. McEwen and B. S. Larsen, *in* "Mass Spectrometry of Biological Materials" (B. S. Larsen and C. N. McEwen, eds.), 2nd Ed., p. 1. Marcel Dekker, Inc., New York, 1998.

with very large mass values and provides high resolution, outstanding mass accuracy, and rapid analysis of the full mass spectrum.[23] As ESI-TOF spectrometers become more commonplace, these advantages will no longer be tightly associated with MALDI. Primary advantages of ESI over MALDI are the freedom from matrix accessory molecules and the infusion of sample as a solution. The latter point means that it is possible to infuse ongoing reactions into the spectrometer, and thus use the spectrometer to both quench and analyze the reaction.[16,19,24] A comparable assay done by MALDI must be performed as a stopped-point assay. ESI analysis of enzyme reactions will be discussed elsewhere in this volume.[25]

A conceptual hurdle in using MALDI for the analysis of enzyme catalysis is the widely held belief that MALDI is not a quantitative technique. Although in many cases MALDI is more useful for the accurate determination of analyte mass than for a quantitative analysis of analyte concentration, the nonquantitative stigma is unfounded. Over the past several years, many groups have demonstrated the use of MALDI for quantitative analysis of a diverse assortment of biomolecules including peptides,[26] proteins,[27] oligosaccharides,[28] and single-stranded DNA.[29] The primary factors for a quantitative analysis include a method to internally calibrate analyte concentration, and the deposition of a homogeneous matrix:analyte crystal on the sample target.

Comparison to Other Analyses

In many respects, analysis of enzyme catalysis by mass spectrometry is a viable alternative, and in some respects superior to traditional steady-state and pre-steady-state assays. In mass spectroscopic analyses, the separation of reaction components is based solely on the masses of the analytes; thus the need to develop chromatographic separations of reaction components is abolished. Since quantification of ligands can be performed by integration of the appropriate mass peaks in the mass spectrum, the need for radioactive or chromogenic substrates is also eliminated. Thus the bias toward the selection of substrate analogs that are easier to assay, in favor of the more relevant and, in comparison to radioactive substrates, often more easily accessible natural substrates, can be abolished. The

---

[23] D. C. Muddiman, A. I. Gusev, and D. M. Hercules, *Mass Spectrometry Rev.* **14**, 383 (1995).
[24] B. M. Kolakowski, D. A. Simmons, and L. Konermann, *Rapid Commun. Mass Spectrom.* **14**, 772 (2000).
[25] L. Konermann and D. J. Douglas, *Methods Enzymol.* **354**, [3], 2002 (this volume).
[26] H. Matsumoto, E. S. Kahn, and N. Komori, *Anal. Biochem.* **260**, 188 (1998).
[27] M. H. Allen, D. J. Grindstaff, M. L. Vestal, and R. W. Nelson, *Biochem. Soc. Trans.* **19**, 954 (1991).
[28] D. J. Harvey, *Rapid Commun. Mass Spectrom.* **7**, 614 (1993).
[29] K. Tang, S. L. Allman, R. B. Jones, and C. H. Chen, *Anal. Chem.* **65**, 2164 (1993).

only prerequisite for a mass spectrometric analysis of an enzymatic reaction is change in the mass of the ligands.

A mass spectrometric assay is amenable to the analysis of substrates enriched with stable isotopes. This labeling can be used in a variety of ways to solve mechanistic questions. By selectively enriching a substrate with stable isotopes, the reaction course for each substrate atom can be followed. This can be quite valuable for the characterization of intermediates, revealing the source of the constituent atoms. Exchange reactions can also be identified and quantified by using stable isotopes. These reactions have typically been assayed by NMR spectroscopy or using radiolabeled substrates, but could easily be analyzed using mass spectrometry. Paiva and co-workers[19] demonstrated the value and utility of stable isotopes in their identification of a reaction intermediate for 5-enolpyruvoyl-shikimate-3-phosphate synthase. They sequentially used natural abundance and 1-^{13}C-enriched phosphoenolpyruvate substrates to generate intermediates that differed by 1 mass unit, confirming that the detected analyte was substrate-derived. Other groups have also demonstrated the use of isotopically enriched substrates in mass spectral analyses. Weller and Distefano[11] used a deuterated farnesyl pyrophosphate substrate to measure the V/K isotope effect for farnesyl transfer. Dueker and co-workers[30] measured accumulation of labeled retinoids in humans following ingestion of deuterated β-carotene.

A mass spectrometric analysis of enzyme reactions does have obvious limitations. First, since the analysis is by mass alone, isomerization reactions cannot be directly observed. However, tandem mass spectrometry and other techniques based on analysis of the unique fragmentation patterns of parent ions have been used to quantify isomers.[31] Second, since the mass spectrometric analysis requires the ionization of sample, any mechanistic steps that involve protonation or deprotonation of the sample will be obscured by the analysis. Finally, mass spectrometric analyses are inhibited by the presence of salts and buffers. In most cases, these need to be removed prior to analysis.

Although mass spectrometry can be used to analyze steady-state reactions, it is particularly powerful for the investigation of the intermediates formed in the pre-steady state because it provides a method for the direct identification of the enzyme-bound intermediates. Because the identification reveals the mass of each species, the chemistry of the reaction and the identity of the intermediates can be deduced. Finally, when mass spectrometry is performed as a time-based analysis, the temporal relationship of the intermediates and the kinetics of the reaction will be also revealed. However, as is true for any pre-steady-state experiment, a mass spectrometry analysis requires that intermediates accumulate to detectable levels.

---

[30] S. R. Dueker, R. S. Mercer, A. D. Jones, and A. J. Clifford, *Anal. Chem.* **70,** 1369 (1998).
[31] G. Smith and J. A. Leary, *J. Am. Chem. Soc.* **118,** 3293 (1996).

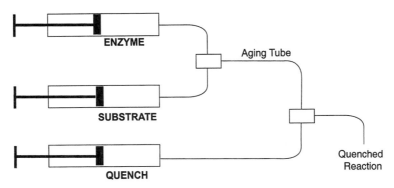

FIG. 1. Schematic diagram of rapid chemical quench apparatus. An enzymatic reaction is initiated by forcing solutions of enzyme and substrate, stored in separate syringes, into a mixing chamber, which outlets the reaction into an aging tube. The reaction is terminated by mixing with a quench solution.

Methodology

Mixing

The mass spectrometric analysis of pre-steady-state enzymatic reactions requires the mixing and quenching of reactions at times preceding a single turnover. The reaction rate dictates the type of mixing technique that must be employed. Hand mixing is effective for measuring rate constants with half-times larger than several seconds. Faster reactions will require rapid-mixing devices that mix and quench the reactions at shorter reaction times.

Rapid Mix, Chemical Quench Apparatuses

Fundamentally, all rapid-mix chemical-quench devices function similarly. Reaction components, stored in two syringes, are delivered to the mixing chamber at a precisely metered rate, sufficient to support turbulent flow in the tubes.[32] The mixed reaction then traverses an outlet (or aging) tube. At the end of the aging tube, the reaction is mixed with a quenching agent that terminates the reaction (Fig. 1). The duration of the reaction is determined by the quotient of the aging tube volume (mm^3) and the volumetric flow rate of the mixed reaction (mm^3/sec). The physical process of homogeneously mixing two solutions in the millisecond time scale has been the subject of some debate and will not be discussed here (reviewed in Ref. 33).

---

[32] H. Gutfreund, *Methods Enzymol.* **16**, 229 (1969).
[33] B. Chance, R. H. Eisenhardt, Q. H. Gibson, and K. K. Lonberg-Holm, "Rapid Mixing and Sampling Techniques in Biochemistry." Academic Press, New York, 1964.

Quench

Quenching a reaction is effected by diluting it into a chemical milieu that terminates catalysis. Often, chemical quenches are performed at the extremes of pH, in organic solvents, or using chemical denaturants.[34] Strong acids, in particular trichloroacetic acid (TCA) and perchloric acid, are quite useful for quenching because they also precipitate the enzyme. Strongly alkaline solutions can also be an effective quench, but do not cause precipitation. The drawback of acids and bases as quenching agents is the potential for the chemical degradation of analytes. Protein denaturation in guanidine hydrochloride, urea, and sodium dodecyl sulfate (SDS) or protein precipitation in organic solvents can be used as a less destructive quench. However, the efficacy of these quenching reagents must be determined for each system. The effectiveness of a quench should be evaluated under different physical conditions (i.e., temperature). If the rate of quenching is temperature dependent, then it is unlikely to be effective. In the example of dTDP-glucose-4,6-dehydratase (see Ref. 20; also described below), the efficacy of each quench was evaluated by using the transiently formed NADH chromophore as a reporter to reveal the progress of the reaction following the quench.

Stabilization

Labile intermediates may not survive the requisite delay between quenching the reaction and mass spectrometric analysis and may have to be stabilized to allow their detection. The decay of unstable components may be exacerbated by cocrystallization with matrix molecules, resulting from the generally acidic properties of those molecules. Any stabilization procedure must satisfy three criteria. First, it must enable the sample to survive the workup and analysis. Second, the identity of the derivatized analytes must be uniquely correlated to the identity of the parent molecules. Finally, the stabilization reaction must be sufficiently rapid that decay is prevented and intermediates are quantitatively retained. An example of the stabilization of a dTDP-4-ketoglucose-5,6-ene intermediate by chemical reduction with sodium borohydride is cited below.

MALDI

In MALDI-TOF, the sample of interest (analyte) and an accessory matrix molecule are cocrystallized on a stainless steel sample plate (target), then transferred into the spectrometer where they are maintained at reduced pressure. Portions of the dried sample are pulsed with a laser to excite the matrix molecules. This desorbs analyte molecules from the target and results in their ionization. Several nanoseconds after the laser burst, a high voltage electric field is applied to

---

[34] D. P. Ballou, *Methods Enzymol.* **54,** 85 (1978).

the system, accelerating the charged analytes toward the detector. Inasmuch as the analytes are ionized with either a monoanionic or monocationic charge, the acceleration in the electric field and the time-of-flight to the detector are directly proportional to the mass of the analyte.

Many factors will affect the quality of a MALDI mass spectrometry signal. These include, but are not limited to, the spectrometer, sample purity, sample concentration, counterions, matrix molecule identity, and ionizability of analyte in the selected mode (positive versus negative ion). Optimum conditions can and should be empirically determined with purified reagents.

Sample purity is by far the most important requirement for obtaining quality spectra. For most samples, minor contaminants significantly degrade the quality of results, attenuating the signals of analytes and thus lowering the signal-to-noise ratio, the threshold of detection, and the resolution. Contaminants can also induce nonuniform spotting of the sample and matrix onto the target. As discussed below in the section on homogenous sample application, this factor is critical for obtaining quantifiable MALDI results. The most problematic contaminants are salts and buffers, but other impurities such as detergents and protein molecules can also present problems. Salts can form adducts with both the analyte and the matrix, which are then detected as a series of corresponding oligomeric complexes. These adducts are most often observed with sodium and potassium. In these spectra, analyte is distributed among the free ion, sodium adducts, and potassium adducts. These diverse distributions make accurate quantification of spectra difficult.

Sample Purification

Methodology for sample purification will depend on the identity of both sample and contaminant. It is important to keep in mind that if comparisons are to be made between the abundance of two analytes, the purification must not enrich one analyte at the expense of the other. Although many methods are useful for purification of analytes, reversed-phase high-performance liquid chromatography (HPLC) is probably the most effective. This results from the high resolving power of the reversed-phase resins, the elution of analytes in volatile solutions, and the ability to manipulate small quantities of reactants. The usefulness of HPLC is exemplified by the prevalence of linked liquid chromatography–mass spectrometry (LCMS) units. Other chromatographic methods such as ion-exchange chromatography are also useful for sample purification, but even in the best circumstances result in the elution of analyte in "volatile" salts. Thus subsequent purification steps are often required. When devising a rapid mix-quench experiment, it is pragmatic to anticipate what contaminants may be present, and how they will be removed prior to the investment of expensive reagents.

Matrix Selection

The selection of a matrix for sample analysis is based on several empirical determinations; thus one should sample several matrices from the outset. The foremost requirement is the efficacy of the matrix in generating a strong analyte signal. For quantitative applications, the matrix must also be amenable to the deposition of homogenous samples. The composition of samples nonuniformly distributed in the crystal cannot be reproducibly determined. Also, in the mass spectrometric analysis of small molecules, one must verify that peaks resulting from matrix, matrix:salt adducts, or matrix oligomers do not occur at mass values coincident with those of the sample being analyzed. This final criterion may seem to be a straightforward calculation based on the mass of the matrix; however, matrix-derived peaks are abundant, occurring at unexpected mass values and cluttering most spectra at mass values below m/z 500. Table I provides a summary of the analytes and matrix molecules from several quantitative anlyses.

Quantification

Quantification of MALDI spectra cannot be performed to directly reveal absolute amounts of material in a sample. Instead, quantification is attained by determining the relative abundance of two or more samples. There are two obvious factors that will affect this comparative analysis. First, samples must be uniformly distributed within the dried crystal. Variation of sample composition within the sample introduces an unacceptable degree of randomness into the analysis (see section on homogenous sample application). Second, it is possible that two molecules may be ionized and detected with different efficiencies. Thus even if they are present in identical quantities, their signals may not be equal. Differences in detection can often be overcome by the construction of standard curves to compensate for different detection efficiencies, or by the inclusion of an internal standard against whose signal the abundance of each analyte is calculated.[13] Standard curves must be utilized with a degree of caution, as they can be nonlinear; Weinberger and co-workers[35] demonstrated a second-order correlation between sample composition and signal area. It has been demonstrated[29,36] that internal standards are more effective when they reflect the chemical properties of analytes, and thus prepurification, ionization, and detection occur with similar efficiencies. Satisfying this criterion, Gerber and co-workers[7] used a deuterated isomer of product as an internal control in their steady-state analysis of glycosidases. Similarly, Houston and

---

[35] S. R. Weinberger and K. O. Boernsen, Sixth Symposium of the Protein Society (1992).
[36] A. I. Gusev, D. C. Muddiman, A. Proctor, A. G. Sharkey, D. M. Hercules, P. Tata, and R. Venkataramanan, *Rapid Commun. Mass Spec.* **10,** 1215 (1996).

TABLE I
QUANTITATIVE MALDI ANALYSES: ANALYTES AND MATRICES

Analyte	Matrix	Matrix solution	Ref.
Oligonucleotides	2,4,6 Trihydroxyphenone, monohydrate	0.3 M in ethanol	a
Fructans (oligosaccharides)	2,4,6 Trihydroxyphenone, monohydrate	Saturated in acetone or 12.5 mg/ml in 50% (v/v) aqueous CH_3CN	b
Tacrolimus (a lipophilic macrolide)	2,5 Dihydroxybenzoic acid	20 mg/ml in 10% (v/v) aqueous ethanol augmented with 2 mg/ml fucose	c
Glyco-oxidized hemoglobin	α-Cyano-4-hydroxycinnamic acid	0.02 M in $CH_3CN : CH_3OH : H_2O$ (56 : 36 : 8)	d
Phosphopeptides, calcium calmodulin-dependent protein kinase II-derived	α-Cyano-4-hydroxycinnamic acid	Saturated solution in 50% CH_3CN/ 0.1% trifluoroacetic acid (TFA)	e
dTDP-glucose (nucleotide sugar)	α-Cyano-4-hydroxycinnamic acid	Saturated solution, in methanol: 10 mM ammonium acetate (7 : 3)	f
Stp1 protein	Ferulic acid	10 mg/ml in distilled $H_2O : CH_3CN$ (2 : 1)	g
Proteins (lysozyme, cytochrome c, hemoglobin, myoglobin, bovine serum albumin)	Sinapinic acid	Saturated solution in $CH_3CN : 0.1\%$ (v/v) aqueous TFA (30 : 70, v/v)	h
Gliadin (gluten)	Sinapinic acid	Saturated solution, in 30% aqueous CH_3CN, 0.1% TFA	i

[a] C. F. Blexzunski and C. Richert, *Rapid Commun. Mass Spectrom.* **12,** 1737 (1998).
[b] J. Wang, P. Sporns, and N. H. Low, *J. Agric. Food Chem.* **47,** 1549 (1999).
[c] A. I. Gusev, D. C. Muddiman, A. Proctor, A. G. Sharkey, D. M. Hercules, P. Tata, and R. Venkataramanan, *Rapid Commun. Mass Spectrom.* **10,** 1215 (1996).
[d] A. A. Lapolla, D. Fedele, M. Plebani, M. Garbeglio, R. Seraglia, M. D'Alpaos, C. N. Aricò, and P. Traldi, *Rapid Commun. Mass Spectrom.* **13,** 8 (1999).
[e] H. Matsumoto, E. S. Kahn, and N. Komori, *Anal. Biochem.* **260,** 188 (1998).
[f] J. W. Gross, A. D. Hegeman, M. M. Vestling, and P. A. Frey, *Biochemistry* **39,** 13633 (2000).
[g] C. T. Houston, W. P. Taylor, T. S. Widlanski, and J. P. Reilly, *Anal. Chem.* **72,** 3311 (2000).
[h] M. H. Allen, D. J. Grindstaff, M. L. Vestal, and R. W. Nelson, *Biochem. Soc. Trans.* **19,** 954 (1991).
[i] E. Camafeita, P. Alfonso, T. Mothes, and E. Méndez, *J. Mass Spectrom.* **32,** 940 (1997).

co-workers[18] used nonphosphorylated Stp1 protein as an internal standard for the calculation of phosphorylated Stp1.

Homogenous Sample Application

As stated above, a critical factor for generating quantifiable data is the ability to homogeneously spot samples onto MALDI targets. Analytes are typically spotted with matrix using the dried droplet method where small aliquots (0.5–1 μl) of analyte and matrix, each dissolved in an appropriate solvent, are sequentially deposited onto a stainless steel sample target. The samples are mixed by gently pipetting, then crystallize as the sample dries. Because crystallization is a *de facto* purification technique, this process often results in the segregation of matrix and analyte molecules and results in "hot spots," regions in the crystal where strong analyte signals are detected. These frequently occur at the edges of the crystal and result from the exclusion of analyte from the matrix crystal. Hensel and co-workers[37] provided a dramatic illustration of sample segregation by spotting the matrix 2,5-dihydroxybenzoic acid with a fluorescein isothiocyanate-labeled α-lactalbumin. Fluorescence microscopy of the air-dried samples revealed that fluorescence was concentrated in zones within the matrix crystals.

Several methods have been devised to solve the problem of sample segregation during crystallization. Premixing matrix and analyte prior to deposition onto the target can occasionally help. This guarantees uniformity in the initial drop, but may still allow segregation of components during crystallization. Fast matrix evaporation[38] reduces sample segregation during crystallization by accelerating the drying process through the use of volatile solvents and reduced pressures. Under optimized solvent conditions, a mixture of matrix and analyte can be scratched to induce their coprecipitation on the target in a homogenous polycrystalline film.[23,28] This process requires a somewhat laborious determination of solvent conditions for both the analyte and matrix, but once these are determined, it allows the facile deposition of homogenous sample. One radically different method for homogeneous sample application is the use of electrospray for the deposition (electrospray deposition, ESD[37,39]). Premixed matrix and analyte are pumped through a nebulizing needle into a strong potential field. This generates charged microdroplets that are wicked off the needle and accelerated toward the target (Fig. 2). The microdroplets have a sufficiently small volume that they are functionally dry on impacting the target, prohibiting segregation of the mixture. Disadvantages of this technique are the difficulty of obtaining of appropriate electrospray deposition hardware and the increased time associated with sample deposition.

Salt contaminants have various effects on the cocrystallization of matrix and analyte. With the assorted dried drop methods, the hygroscopic effects of the salts slow the drying process, and thus promote sample segregation. In the scratch seeding method, salts change the composition of the mixture, often preventing the rapid precipitation of sample. Electrospray deposition of homogenous samples is refractory to salt contaminants; however, Hensel and co-workers[37] report that it actually increases the salt artifacts in sample spectra, presumably because they can never be excluded from sample crystals. As in all aspects of MALDI analysis, homogenous sample deposition benefits from the rigorous removal of all contaminants in a sample.

[37] R. R. Hensel, R. C. King, and K. G. Owens, *Rapid Commun. Mass Spectrom.* **11,** 1785 (1997).
[38] O. Vorm, P. Roepstorff, and M. Mann, *Anal. Chem.* **66,** 3281 (1994).
[39] C. J. McNeal, R. D. Macfarlane, and E. L. Thurston, *Anal. Chem.* **51,** 2036 (1979).

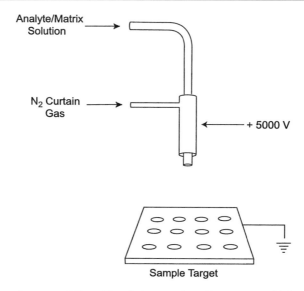

FIG. 2. Schematic representation of the electrospray deposition apparatus fabricated in-house by R. Clasen and M. Vestling (University of Wisconsin—Madison). A mixture of analyte and matrix is delivered through a nebulizing needle, held at high voltage. The resulting charged microdroplets are deposited on the sample target, held at ground. A curtain gas of N_2, delivered by the outer shell of a concentric needle, is used to focus the sample.

Integration of Sample Peaks

Quantification of analytes requires the integration of sample peaks in the spectrum. Although most spectrometers contain software that can perform this operation, it is frequently only useful for the analysis of spectra with large signal-to-noise ratios and few spurious peaks. When the spectra become more complex, overlapping peaks and the identification of baseline become problematic. In these cases it is prudent to export the raw data for peak fitting with routines that can use regression analysis algorithms to systematically fit both peaks and baseline (Peak Fit, SPSS, Inc., Chicago, IL). Once quantitative data are obtained for each time point, rate constants can be determined by fitting the data to the appropriate differential equations that refer to the chemical mechanism. Software dedicated to the analysis of enzyme mechanisms is available [Dynafit (Biokin, Ltd., www.Biokin.com); Kinsim/Fitsim[40,41]] or the differential equations can be independently derived and fitted with more universal programs [Mathematica (Wolfram Res., Inc.), Scientist (SPSS, Inc.)].

[40] B. A. Barshop, R. F. Wrenn, and C. Frieden, *Anal. Biochem.* **130,** 134 (1983).
[41] C. T. Zimmerle and C. Frieden, *Biochem. J.* **258,** 381 (1989).

Examples

To date, rapid mix-quench MALDI-MS has only been used in transient-phase analysis of two enzymatic reactions: the low molecular weight phosphatase Stp1,[18] and dTDP-glucose-4,6-dehydratase (4,6-dehydratase[20]). These two studies were quite different and demonstrated the versatility of rapid mix-quench MALDI-MS. In the case of Stp1, the chemical intermediate was a phosphorylated form of the enzyme, and the analysis monitored a burst of phosphoenzyme formation preceding and into the steady state. In the analysis of 4,6-dehydratase, the intermediates analyzed were nucleotide sugars formed in the course a single-turnover reaction converting dTDP-glucose to dTDP-4-keto-6-deoxyglucose.

Stp1

Stp1 from *Schizosaccharomyces pombe* is a small (17.5 kDa) tyrosine phosphatase (EC 3.1.3.2) that likely participates in the regulation of cell cycle control.[42] Stp1 is classified as a low molecular weight protein tyrosine phosphatase,[42,43] a subgroup of the protein tyrosine phosphatase (PTPase) superfamily.[44]

Stp1 and other low molecular weight protein tyrosine phosphatases catalyze the dephosphorylation of phosphotyrosine and phosphothreonine residues and will also accept small phosphate monoesters as substrates.[43] Thus chromogenic substrates such as *p*-nitrophenyl phosphate (*p*-NPP) have been used to simplify investigation of the chemical mechanism. Kinetic studies of low molecular weight protein tyrosine phosphatase, including Stp1, revealed a biphasic formation of product, where a burst of product formation, stoichiometric with enzyme active sites, was followed by a slower steady-state rate.[43,45] The burst phase corresponded to the rapid phosphoryl transfer from substrate to the enzyme, forming an *S*-phosphocysteinyl enzyme intermediate,[46] and the steady-state phase to the hydrolysis of the phosphoenzyme (Fig. 3). The burst kinetics of the reaction indicated that the E–P intermediate accumulated during turnover, making Stp1 a reasonable candidate for a pre-steady-state analysis.

Houston and co-workers[18] performed a series of experiments using rapid mix-quench MALDI-MS to determine the kinetics of E–P formation. A selection of small phosphate monoesters were used as substrates in the analysis. With the substrate *p*-NPP, stopped-flow experiments were performed in parallel to judge the quantitative efficacy of the rapid mix-quench MALDI-MS analysis.

---

[42] O. Mondesert, S. Moreno, and P. Russell, *J. Biol. Chem.* **269,** 27996 (1994).
[43] Z.-Y. Zhang, G. Zhou, J. M. Denu, L. Wu, X. Tang, O. Mondesert, P. Russell, E. Butch, and K.-L. Guan, *Biochemistry* **34,** 10560 (1995).
[44] Z.-Y. Zhang, *Curr. Top. Cell. Regul.* **35,** 21 (1997).
[45] Z.-Y. Zhang and R. L. VanEtten, *J. Biol. Chem.* **266,** 1516 (1991).
[46] K. L. Guan and J. E. Dixon, *J. Biol. Chem.* **266,** 17026 (1991).

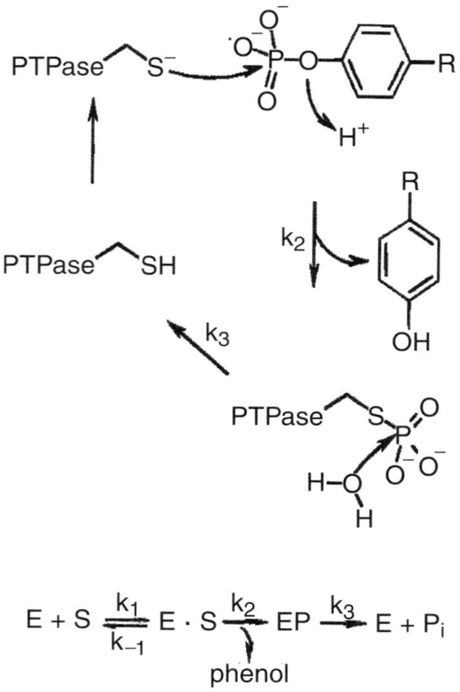

FIG. 3. Chemical and kinetic mechanism of protein tyrosine phosphatases. Substrate phosphohydrolysis involves transient formation of a phosphocysteinyl enzyme intermediate, which is detected by mass spectrometry. Reproduced with permission from C. T. Houston, W. P. Taylor, T. S. Widlanski, and J. P. Reilly, *Anal. Chem.* **72,** 3311 (2000). Copyright © 2000 American Chemical Society.

Rapid mix quench experiments were performed using a KinTek RQF-3 mixer (KinTek Corp., Austin, TX). Enzyme (13.9 μl at 120 μM in 25 mM ammonium 3,3-dimethylglutarate, 50 mM ammonium chloride, 1 mM EDTA, pH 6.5) was mixed with substrate (14.9 μM at 3–15 mM in the identical buffer), incubated for 4–1000 ms, then quenched by the addition of a fivefold volumetric excess of 0.5 M dichloroacetic acid. Quenched solutions were then mixed with a ninefold excess of the matrix ferulic acid (10 mg/ml in 2 : 1 distilled H_2O/acetonitrile), and 1 μl was spotted on the MALDI target. Samples were analyzed using a custom built MALDI-TOF mass spectrometer equipped with delayed extraction and operated in the linear mode.

Quantification of the phosphoenzyme (E–P) concentration was calculated from integrated peak area for E–P divided by the sum of nonphosphorylated and phosphorylated Stp1 peak areas, normalized to the actual concentration of Stp1 in the mixed reaction (60 μM). To minimize variation in the data collection, spectra were collected in triplicate. The authors measured the formation of E-P in the burst phase and early steady-state portion of the reaction. Consequently, the rapid mix-quench MALDI-MS results reveal both the rate of E-P formation and the level of E-P in the steady state.

Empirical determinations of experimental conditions included selection of an effective quench and an optimum matrix. The authors found that enzyme precipitation by quenching in TFA degraded the homogeneity of sample application onto the MALDI target and, in control experiments, caused unacceptable light scattering in UV/visible spectrometry. A dichloroacetic acid quench was found to terminate the reaction without precipitating the protein. The quench also denatured the enzyme and thereby prevented enzyme-catalyzed E-P phosphohydrolysis. Several matrices including ferulic acid, 3,5-dihydroxybenzoic acid, and sinapinic acid were tested. Ferulic acid was selected because it generated a superior signal.

With every substrate analyzed, the authors observed the time-dependent formation of a peak at m/z 17,470.9 [M+H] (Fig. 4). This peak was 79.1 mass units larger

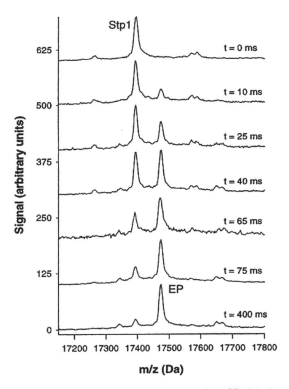

FIG. 4. MALDI-TOF mass spectra demonstrating the conversion of Stp1 (m/z = 17,392 [M+H]) to a covalent phosphoenzyme (m/z = 17,471 [M+H]) at times preceding the steady state. Reproduced with permission from C. T. Houston, W. P. Taylor, T. S. Widlanski, and J. P. Reilly, *Anal. Chem.* **72**, 3311 (2000). Copyright © 2000 American Chemical Society.

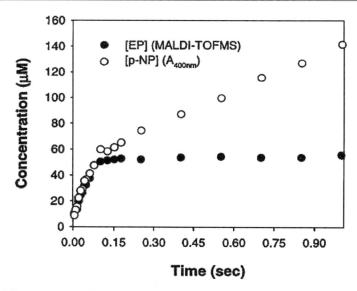

FIG. 5. Progress curves of Stp1 catalysis, demonstrating both the burst and steady-state phases of the reaction. Phosphoenzyme formation (closed circles) was assayed by MALDI-TOF MS. Product formation (*p*-nitrophenol; open circles) was assayed spectrophotometrically. Reproduced with permission from C. T. Houston, W. P. Taylor, T. S. Widlanski, and J. P. Reilly, *Anal. Chem.* **72**, 3311 (2000). Copyright © 2000 American Chemical Society.

than that of unmodified Stp1 (m/z 17,391.9 [M+H]). The 79.1 mass unit shift was consistent with the transfer of a phosphoryl group (80 mass units) to the enzyme. With the artificial substrate *p*-NPP, spectrophotometric analysis was performed in parallel with the mass spectrometric analysis. In this assay, the absorbance at 400 nm was measured in the quenched samples to quantify *p*-nitrophenol. Since the spectrophotometric analysis reported product formation, and the mass spectrometric analysis reported E-P, curves resulting from these analyses should have matched in the burst phase, but differed in the steady state. This was in fact observed (Fig. 5). Although the authors found the burst rate too rapid to accurately fit to burst kinetic equations for this mechanism,[47] a qualitative analysis reveals that the burst phases of the two assays are occurring at similar rates. With other small phosphate monoester substrates, the burst phase occurred at a slower rate and could be accurately fitted. The results revealed variability in the rate of E-P formation (k_2), but a nearly invariant rate of E-P hydrolysis (k_3). By comparing the rate of phosphoryl transfer (k_2) to the pK_a of the leaving phenoxide ion, the authors were able to conclude that the burst rate of the product formation was limited by

---

[47] M. L. Bender, F. J. Kezdy, and F. C. Wedler, *J. Chem. Ed.* **44**, 84 (1967).

the stability of the leaving group. This was consistent with kinetic isotope effect studies that led to identical conclusions.[48]

The analysis of Stp1 provides a compelling example of the value of rapid mix-quench MS for the analysis of enzymatic mechanisms. Although MALDI has been used previously to detect phosphorylation of proteins and peptides,[26,49] what distinguishes this work is the concept of performing pre-steady-state, time-resolved, quantitative analysis. This analysis demonstrates that with minimal sample preparation, and using readily available reagents, the time-dependent transfer of a phosphoryl group to the Stp1 could be measured. Since this analysis is not limited by chromophore formation, the rate of phorphoryl transfer to the enzyme can be simply and accurately determined for virtually any substrate. Thus a MALDI-MS analysis is a viable alternative to other, more well-established methods for analyzing protein phorphorylation. Other techniques such as enzyme-linked immunosorbent assay (ELISA), or the use of ^{32}P-labeled substrates could also have been employed, but these techniques have disadvantages. ^{32}P-Labeled substrates are often difficult to obtain, have a short half-life, and in many aspects are troublesome to work with. ELISA analyses can be used to detect specific phosphorylated amino acids but this is an indirect technique that requires significant sample preparation and is limited by the specificity and availability of antibodies.

dTDP-Glucose-4,6-Dehydratase

dTDP-Glucose-4,6-dehydratase (EC 4.2.1.46) catalyzes the conversion of dTDP-glucose to dTDP-4-keto-6-deoxyglucose.[50] 4,6-Dehydratase catalyzes the second step in a biosynthetic pathway for the formation of the 6-deoxy sugars, including L-rhamnose, and others that are converted into bacterial secondary metabolites.[51,52] 4,6-Dehydratase is a homodimer, with each subunit containing an irreversibly bound NAD^+ coenzyme. Conserved sequence motifs have been used to classify 4,6-dehydratase as a member of the extended short-chain dehydrogenase superfamily.[53]

The generally accepted mechanism for 4,6-dehydratase is shown in Fig. 6.[54] The first step involves the abstraction of a hydride from glucosyl C-4 and proton abstraction from the C-4 hydroxyl, forming a 4-ketoglucose intermediate and NADH. Proton abstraction from glucosyl C-5 and, in either a stepwise or concerted fashion, water elimination from carbon-6 generates a 4-ketoglucose-5,6-ene intermediate

---

[48] A. C. Hengge, Y. Zhao, and Z.-Y. Zhang, *Biochemistry* **36**, 7928 (1997).
[49] A. V. Vener, A. Harms, M. R. Sussman, and R. D. Vierstra, *J. Biol. Chem.* **276**, 6959 (2001).
[50] R. Okazaki, T. Okazaki, J. Strominger, and A. M. Michelson, *J. Biol. Chem.* **237**, 3014 (1962).
[51] J. H. Pazur and E. W. Shuey, *J. Biol. Chem.* **236**, 1780 (1961).
[52] H.-W. Liu and J. S. Thorson, *Annu. Rev. Microbiol.* **48**, 223 (1994).
[53] B. Persson, M. Krook, and H. Jörnvall, *Eur. J. Biochem.* **200**, 537 (1991).
[54] A. Melo, W. H. Elliott, and L. Glaser, *J. Biol. Chem.* **243**, 1467 (1968).

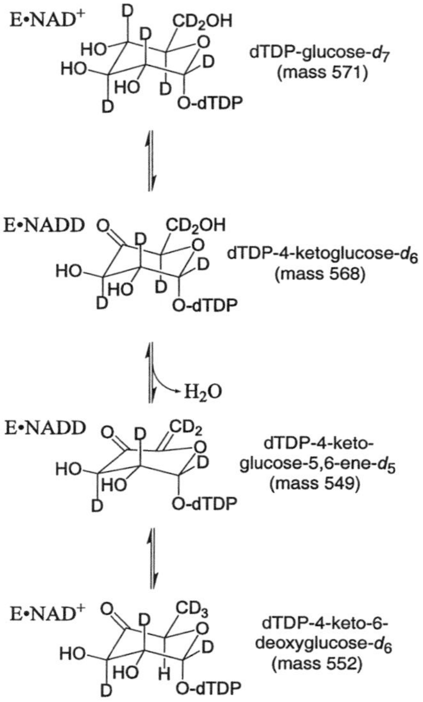

FIG. 6. Chemical mechanism for dTDP-glucose-4,6-dehydratase. Substrate is sequentially oxidized, dehydrated, and then rereduced, resulting in changes in mass for each intermediate. The chemical mechanism shown represents use of substrate synthesized from perdeuterated glucose. Reproduced with permission from J. W. Gross, A. D. Hegeman, M. M. Vestling, and P. A. Frey, *Biochemistry* **39**, 13633 (2000). Copyright © 2000 American Chemical Society.

and water. Finally, NADH reduction of the intermediate at carbon-6 in conjunction with protonation at C-5 generates the product and reforms NAD^+. NADH levels, which are below the threshold of detection in the absence of substrate, rise to account for 5–60% of the enzyme molecules in the steady state.[55,56] The formation of NADH during steady-state turnover revealed that intermediates associated with this intermediate must accumulate during turnover, making 4,6-dehydratase a reasonable candidate for a rapid mix-quench MALDI-MS analysis.

Gross and co-workers[20] used rapid mix-quench MALDI-MS to investigate the disappearance of substrate, transient formation of intermediate, and formation of

---

[55] L. Glaser and H. Zarkowsky, in "The Enzymes" (P. Boyer, ed.), Vol. 5, p. 465. Academic Press, New York, 1973.
[56] A. D. Hegeman, J. W. Gross, and P. A. Frey, submitted (2001).

product during a single turnover of the 4,6-dehydratase enzyme. These experiments were performed not only to determine the chemical identity of the intermediates, which had never been identified, but also to characterize the temporal and kinetic aspects of their formation and reactions.

Rapid mix quench experiments were performed on an Update Instruments, Inc. (Madison, WI) rapid mixer prototype interfaced with a model 745 syringe ram controller. Enzyme [600 μM in 100 mM Tris-Cl, pH 7.5, 1 mM dithiothreitol (DTT)] was mixed with an equal volume of dTDP-glucose-d_7 (400 μM in 100 mM Tris-Cl, pH 7.5, 1 mM DTT). The reactions were incubated for 10–3000 ms, then quenched by injection of 100-μl aliquots into a 10-fold volumetric excess of 6 M guanidine hydrochloride, buffered with 50 mM potassium bicinate, pH 8.0, and augmented with 50 mM sodium borohydride. Reductant was included in the quench to stabilize the labile intermediates, which could not be detected when the reductive stabilization was omitted. Protein in the quenched samples was removed by ultrafiltration, then the buffer, denaturant, and other contaminants were removed by gel filtration chromatography [TosoHaas (Montgomeryville, PA) Toyopearl HW-40S; 2.6 × 40 cm] in doubly distilled H_2O. Fractions containing the thymidine nucleotide were pooled and concentrated by lyophilization, then resuspended with water to a final concentration of approximately 1 mM. Matrix was prepared as a saturated solution of α-cyano-4-hydroxycinnamic acid in methanol: 10 mM ammonium acetate, pH 7.5 (7 : 3), then batch treated with Dowex AG 1 X-8 in the NH_4^+ form to remove nonvolatile cations. Aliquots of matrix, then sample (0.5 μl each), were spotted on the target. The solution was mixed by pipetting, then the target gently triturated to induce the homogenous coprecipitation of the matrix and sample. MALDI analysis was performed on a Bruker Reflex II spectrometer equipped with delayed extraction and operated in the negative ion, reflectron mode.

Under these conditions, the MALDI spectra revealed a time-dependent loss of an initial peak at m/z 570, the transient accumulation of a peak at m/z 569, the significant but transient accumulation of a peak at m/z 550, and the eventual formation of a solo peak at m/z 553 (Fig. 7). Controls demonstrated that the peaks at m/z 570 and 553 are, respectively, substrate and reduced product. The peak at m/z was identified as reduced dTDP-4-ketoglucose-5,6-ene-d_5, and the peak at m/z 569 was identified as dTDP-glucose-d_6 having exchanged protium for deuterium at C-5. The identification of the peak at m/z 569 as exchanged substrate and not the reduced 4-ketoglucose intermediate, which would be detected at the same value, was made by repeating the experiment and substituting a sodium borodeuteride quench. In this experiment, the mass values of the reduced dTDP-4-ketoglucose-5,6-ene-d_5 and reduced product were increased 1 mass unit, but the peak at m/z 569 and that of substrate were unchanged. This indicated that the peak at m/z 569 was not reductively stabilized, and therefore was an isotope (i.e., variant) of substrate. The identification of the dTDP 4-ketoglucose-5,6-ene intermediate was based on several results. The temporal nature of its formation

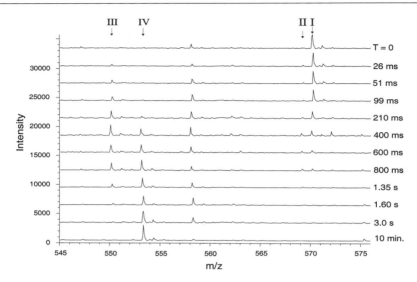

FIG. 7. MALDI-TOF ms spectra demonstrating a single turnover of dTDP-glucose 4,6-dehydratase catalysis. Substrate dTDP-glucose-d_7 (I; m/z 570 [M-H]) is converted to the dTDP-4-ketoglucose-5,6-ene-d_5 intermediate (III; detected here in its reductively stabilized form, m/z 550 [M-H]), and then to product dTDP-4-keto-6-deoxyglucose-d_6 (IV; detected in a reductively stabilized form, m/z 553 [M-H]). A C-5 protium exchanged isotope of substrate, dTDP-glucose-d_6 (II; m/z 570 [M-H]) is detected at low abundance. Reproduced with permission from J. W. Gross, A. D. Hegeman, M. M. Vestling, and P. A. Frey, *Biochemistry* **39**, 13633 (2000). Copyright © 2000 American Chemical Society.

and disappearance was consistent with it being an intermediate in the reaction. The dependence on borohydride and borodeuteride for stabilization and an appropriate shift in mass when the reductants were switched reveal the presence of a reducible group in the analyte. Most critically, any combination of natural abundance or isotopically enriched substrate and borohydride or borodeuteride resulted in the identification of the intermediate at the predicted mass value.

Spectra representing each time point were exported from the Bruker spectrometer, then fitted using a linear regression with the program Peakfit (SPSS, Inc.). Rate constants were calculated by fitting the integrated data (Fig. 8) to the mechanism shown in Fig. 9, using the program Dynafit (BioKin Ltd.). The solved rate constants are included in Fig. 9. This scheme is a variant of the mechanism shown in Fig. 6, having been modified to include the C-5 deuterium-to-protium exchange in substrate, and to exclude the dTDP-4-ketoglucose intermediate, which was not detected. The quantitative results were easily fitted to the chemical mechanism and provided statistically valid estimates of the rate constants in the reaction.

Sodium borohydride was included in the quench to reduce the keto functionality of intermediates and product. Whereas product could be detected in the absence

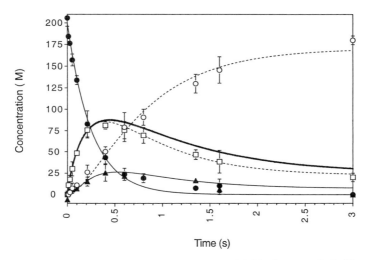

FIG. 8. Progress curves for single-turnover dTDP-glucose-4,6-dehydratase catalysis. The temporal relationship for the conversion of substrate (closed circles) to the dTDP-4-ketoglucose-5,6-ene intermediate (open squares) and then to product (open circles) is observed. Stopped flow data (collected at 350 nm; thicker solid black line), demonstrates that NADH accumulation is similar to that of intermediates. Reproduced with permission from J. W. Gross, A. D. Hegeman, M. M. Vestling, and P. A. Frey, *Biochemistry* **39,** 13633 (2000). Copyright © 2000 American Chemical Society.

of borohydride, intermediates were only detected when reductant was included. Borohydride concentrations ranging from 1 to 100 mM were tested and revealed no concentration dependence on the yield of intermediate. Similarly, a progressive delay was investigated, after the quench but before the addition of reductant, to determine if reduction was as fast as degradation. Delays of up to 3 min, the longest time tested, demonstrated no variation in the yield of intermediate. It was thus demonstrated that the rate of intermediate degradation was much slower than that of reductive stabilization.

Quenching conditions were selected to promote rapid denaturation of the enzyme, without labilizing the intermediates. Several conditions were tested, including denaturation in urea and guanidine hydrochloride and precipitation in hot organic solvents (ethanol, methanol, acetone, acetonitrile). Control experiments, analyzing the formation of the NADH chromophore, revealed that the rate of quenching was too slow in methanol, acetone and acetonitrile, and urea, but that quenching in boiling ethanol and guanidine hydrochloride was sufficiently fast. It was subsequently determined that intermediates were labile in boiling ethanol, even when reductive stabilization was included. Although the guanidine hydrochloride quench resulted in samples with high salt concentrations, it was selected because it efficiently quenched the reaction and did not contribute to analyte degradation.

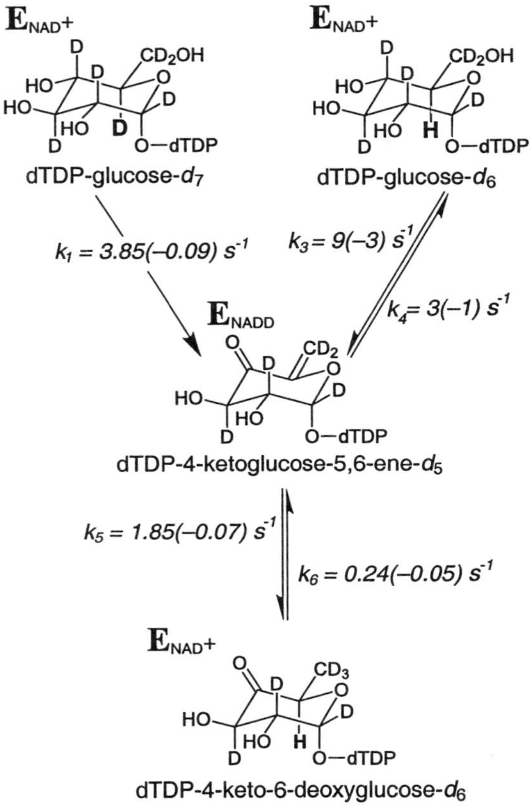

FIG. 9. Kinetic mechanism for dTDP-glucose-4,6-dehydratase, when using the substrate dTDP-glucose-d_7. Rate constants were derived by fitting the data shown in Fig. 8 to this mechanism. The dTDP-4-ketoglucose intermediate is not included in the mechanism because it did not accumulate to detectable levels. Nonetheless, it must still be retained in the chemical mechanism. Reproduced with permission from J. W. Gross, A. D. Hegeman, M. M. Vestling, and P. A. Frey, *Biochemistry* **39**, 13633 (2000). Copyright © 2000 American Chemical Society.

Several matrices were tested before α-cyano-4-hydroxycinamic acid was selected. Although other matrices generated adequate signals from purified samples of substrate, preliminary experiments demonstrated that α-cyano-4-hydroxycinnamic acid formed relatively homogeneous crystals with the analyte. With further development, the buffer conditions for matrix and analyte were determined that promoted the rapid and uniform precipitation of matrix–analyte crystals. Other matrices did not yield uniform crystals.

The 4,6-dehydratase experiments demonstrate a novel use of isotopically labeled substrate (dTDP-glucose-d_7). In initial experiments using natural-abundance substrates and the sodium borodeuteride reduction, the reduced 4-ketoglucose intermediate could not be detected. However, the predicted mass of the reduced 4-ketoglucose intermediate was 1 mass unit larger than that of substrate, and thus modest accumulation of this intermediate would be obscured by the large +1 isotope peak from the substrate. The switch to dTDP-glucose-d_7 and sodium borohydride reduction was made to isolate the reduced 4-ketoglucose intermediate at a value (m/z 569) 1 mass unit smaller than substrate (m/z 570), and thus free from overlap with isotopes of substrate. The 4-ketoglucose intermediate was nonetheless not detected, and thus its accumulation in the pre-steady state must be small (<2% of E_t).

These experiments reveal how MALDI-MS can be used to determine mechanistic information. For 4,6-dehydratase, this identification provided the first direct detection of an intermediate in this reaction, compelling evidence in support of the proposed mechanism. The detection of an enzyme-catalyzed C-5 deuterium-to-protium exchange reaction provides a new tool for the investigation of the mechanism that could be particularly useful for analysis of C-5 ^DV isotope effects, and for investigation of the dTDP 4-ketoglucose intermediate. The results demonstrated that rereduction of the 4-ketoglucose-5,6-ene intermediate is the primary contributor to overall rate limitation.

The RMQ-MS analysis provided a wealth of information not available from previous pre-steady-state analyses. These studies had been restricted to spectrophotometrically observing the NADH chromophore during the reaction. The appearance of NADH represents the sum of both intermediates and cannot be used to distinguish between the two. The RMQ-MS data demonstrated that the dTDP-4-ketoglucose-5,6-ene intermediate accumulates and that the dTDP-4-ketoglucose intermediate does not. This analysis also shows that the enzyme rapidly catalyzes a C-5 deuterium-to-protium exchange reaction concomitant with a single turnover of the enzyme, a reaction that could not be easily detected by other methods. Finally, the RMQ-MS analysis of 4,6-dehydratase has directly revealed the chemical form of the intermediates, removing all ambiguity as to the chemical course of the reaction.

[3] Pre-Steady-State Kinetics of Enzymatic Reactions Studied by Electrospray Mass Spectrometry with On-Line Rapid-Mixing Techniques

By LARS KONERMANN and DONALD J. DOUGLAS

Introduction

Steady-state kinetic experiments on enzymatic reactions provide information such as the maximum turnover number (k_{cat}) and the Michaelis constant (K_m). Studies of this kind are also useful for distinguishing between different types of enzyme inhibition. However, K_m and k_{cat} are combinations of all the rate and equilibrium constants involved in an enzymatic reaction. Therefore steady-state experiments usually provide little or no information on the actual reaction mechanism, i.e., the number of transient intermediates, their chemical structure, and the rate constants or energy barriers of individual reaction steps.[1] A proper understanding of enzyme mechanisms requires kinetic experiments in the pre-steady-state regime. Immediately after initiating an enzymatic reaction there is a short time range (usually some milliseconds to seconds, depending on the rate constants involved) during which the system approaches steady-state conditions. The mechanism of the reaction dictates the order by which short-lived intermediates become populated successively during this period. Individual rate constants can be measured, from which the energy barriers along the reaction coordinate can be estimated.[1–3] The enzyme concentration required for pre-steady-state experiments is relatively high because the enzyme is a stoichiometric "reactant." Usually this is not a problem because recombinant techniques can be used to produce sufficiently large amounts of enzyme.

Pre-steady-state kinetic experiments often involve the use of stopped-flow optical spectroscopy.[4] In these experiments, two or more reactant solutions are expelled from a set of pneumatically or stepper motor-driven syringes. The solutions meet in a mixer where the reaction is initiated. The fresh reaction mixture is rapidly transferred to an observation cell while the previous contents of the cell are flushed out. The liquid flow is abruptly halted by a stopping syringe or a valve at the exit of the cell. The kinetics are monitored by recording an optical signal

---

[1] A. Fersht, "Structure and Mechanism in Protein Science." W. H. Freeman & Co., New York, 1999.
[2] K. Hiromi, "Kinetics of Fast Enzyme Reactions: Theory and Practice." John Wiley & Sons, New York, 1979.
[3] K. S. Anderson, J. A. Sikorski, and K. A. Johnson, *Biochemistry* **27,** 7395 (1988).
[4] K. A. Johnson, in "The Enzymes" (D. S. Sigman, ed.), Vol. 20, p. 1. Academic Press, New York, 1992.

(e.g., fluorescence or absorbance) as a function of time as the reaction proceeds in the observation cell. Chemical quench-flow experiments[5] are another method for studying fast enzyme kinetics. The reaction is initiated by rapid mixing of the reactants, followed by mixing with a quenching agent such as acid, base, or an organic solvent, after a specified period of time. The quenching step abruptly stops the reaction by denaturing the enzyme and liberates noncovalently bound substrates, intermediates, and products. Various methods can be used for the off-line analysis of the quenched reaction mixture. For example the use of radiolabeled substrates allows the quantitation of individual reactive species at different reaction times by high-performance liquid chromatography (HPLC).[3] A relatively new and promising approach involves the off-line analysis of the quenched reaction mixture by matrix-assisted laser desorption/ionization (MALDI) mass spectrometry. This elegant technique is the focus of another chapter in this volume and will not be discussed here.[5a]

Stopped-flow and quench-flow methods have provided a wealth of information on enzymatic reaction mechanisms. However, these techniques have a number of disadvantages. Stopped-flow optical absorption spectroscopy is limited to reactions that are associated with chromophoric changes. Most reactive species that are involved in the enzymatic conversion of "natural" substrates are not chromophoric and hence they cannot be directly studied by stopped-flow methods (although fluorescence of the enzyme can sometimes be monitored). For this reason most studies on enzyme kinetics involve the use of synthetic, chromophoric substrates. Unfortunately, the kinetics and the reaction mechanisms observed with these artificial compounds can be different from those of the natural enzyme substrates. Chromophoric substrate analogs can pose the further disadvantage of long and difficult chemical syntheses. Usually optical spectroscopy provides little information on the chemical transformations occurring in the reaction mixture and in most cases only one reactive species can be monitored. Chemical quench-flow experiments can also have problems. The quenching agent can induce degradation or chemical modification of labile species in the reaction mixture. The use of radiolabeled substrates in quench-flow studies not only is inconvenient, it also can lead to unwanted artifacts arising from nonspecific entrapment of the radioactive label.

Here we describe the use of electrospray ionization (ESI) mass spectrometry (MS) in conjunction with on-line rapid mixing for studying the pre-steady-state kinetics of enzymatic reactions. It appears that this approach has the potential to overcome many of the limitations that are encountered with traditional stopped-flow and quench-flow methods. It seems likely that ESI MS will soon become a standard technique in the laboratories of many enzyme kineticists.

---

[5] K. A. Johnson, *Methods Enzymol.* **249**, 38 (1995).
[5a] J. W. Gross and P. A. Frey, *Methods Enzylmol.* **354**, [2], 2002 (this volume).

Electrospray Ionization Mass Spectrometry

Electrospray is one of the most widely used ionization methods in biological MS.[6] Analyte solution is pumped through a narrow capillary that is held at a high electric potential. Small, highly charged solvent droplets are emitted from the end of the ESI capillary at atmospheric pressure. These droplets rapidly shrink because of solvent evaporation, thus increasing the charge density on the droplet surface and eventually leading to the formation of offspring droplets via jet fission. Analyte gas-phase ions are generated, either as charged residues from nanometer-sized offspring droplets or by ion evaporation. This whole process of ion formation occurs on a time scale of roughly 1 ms.[7] ESI is a very gentle process that can be used for ionizing a wide range of analytes, including small organic molecules, proteins, and even noncovalent ligand–protein or protein–protein complexes.[8] The ions produced by ESI are transferred into the vacuum chamber of a mass spectrometer and analyzed according to their mass-to-charge ratio (m/z).

In enzymology, ESI MS has been used for the detection of enzyme–substrate and enzyme–inhibitor complexes[9–15] and for monitoring the concentration of reactive species at different reaction times in quench-flow experiments.[16–20] All

---

[6] J. B. Fenn, M. Mann, C. K. Meng, S. F. Wong, and C. M. Whitehouse, *Science* **246**, 64 (1989).
[7] P. Kebarle and Y. Ho, in "Electrospray Ionization Mass Spectrometry" (R. B. Cole, ed.), p. 3. John Wiley & Sons, Inc., New York, 1997.
[8] J. A. Loo, *Mass Spectrom. Rev.* **16**, 1 (1997).
[9] R. T. Aplin, J. E. Baldwin, C. J. Schofield, and S. G. Waley, *FEBS Lett.* **277**, 212 (1990).
[10] D. S. Ashton, C. R. Beddell, D. J. Cooper, B. N. Green, R. W. A. Oliver, and K. J. Welham, *FEBS Lett.* **292**, 201 (1991).
[11] R. Menard, R. Feng, A. C. Storer, V. J. Robinson, R. A. Smith, and A. Krantz, *FEBS Lett.* **295**, 27 (1991).
[12] B. Ganem, Y.-T. Li, and J. D. Henion, *J. Am. Chem. Soc.* **113**, 7818 (1991).
[13] W. B. Knight, K. M. Swiderek, T. Sakuma, J. Calaycay, J. E. Shivley, T. D. Lee, T. R. Covey, B. Shushan, B. G. Green, R. Chabin, S. Shah, R. Mumford, T. A. Dickinson, and P. R. Griffin, *Biochemistry* **32**, 2031 (1993).
[14] X. Cheng, R. Chen, J. E. Bruce, B. L. Schwartz, G. A. Anderson, S. A. Hofstadler, D. C. Gale, R. D. Smith, J. Gao, G. B. Sigal, M. Mammen, and G. M. Whitesides, *J. Am. Chem. Soc.* **117**, 8859 (1995).
[15] C. W. Duncan, H. L. Robertson, S. J. Hubbard, S. J. Gaskell, and R. J. Beynon, *J. Biol. Chem.* **274**, 1108 (1999).
[16] R. P. A. Brown, R. T. Aplin, and C. J. Schofield, *Biochemistry* **35**, 12421 (1996).
[17] F. Y. L. Hsieh, X. Tong, T. Wachs, B. Ganem, and J. Henion, *Anal. Biochem.* **229**, 20 (1995).
[18] R. J. Boerner, D. B. Kassel, S. C. Barker, B. Ellis, P. DeLacy, and W. B. Knight, *Biochemistry* **35**, 9519 (1996).
[19] W.-P. Lu, Y. Sun, M. D. Bauer, S. Paule, P. M. Koenigs, and W. G. Kraft, *Biochemistry* **38**, 6537 (1999).
[20] S. Ichiyama, T. Kurihara, Y.-F. Li, Y. Kogure, S. Tsunasawa, and N. Esaki, *J. Biol. Chem.*, **275**, 40804 (2000).

these studies were carried out on "stationary" samples, i.e., the analyte solution was assumed to undergo no chemical changes during the analysis. In this chapter we focus on recent developments that allow the direct on-line monitoring of enzyme-catalyzed reactions by ESI MS. In these experiments an enzyme/substrate mixture is injected directly into the ESI source of the mass spectrometer while the reaction proceeds in solution. With this approach it is possible to overcome many limitations of traditional kinetic methods.[21] No chemical quenching steps are required because the analysis occurs on-line and is quasi-instantaneous. Mass spectrometry does not rely on the presence of chromophores and therefore it is possible to monitor reactive species that cannot be detected optically. Radioactive labeling, artificial chromophoric substrate analogs, and coupled assays are not required. Instead, ESI MS is a sensitive and selective method for studying the reactions of enzymes with their natural, biologically significant substrates. In principle it is possible to monitor reactants, intermediates, and products simultaneously, as well as covalent and noncovalent enzyme–substrate complexes.

The enormous potential of ESI MS for the on-line monitoring of reaction mixtures in kinetic experiments was first realized by Lee et al.,[22] who used this technique for studying the steady-state kinetics of lactase. They used a reaction vessel that was closely coupled to the ion source of an ESI mass spectrometer. Reactions were initiated by manual mixing of enzyme and substrate in the vessel. Similar approaches were used for steady-state experiments on a number of other enzymes.[23–25] These elegant studies demonstrated the versatility of ESI MS as a direct method for monitoring enzyme kinetics. However, the time resolution in these experiments was relatively poor, on the order of several seconds to minutes, because manual mixing was used to initiate the reactions of interest. Although this is adequate for monitoring steady-state kinetics, it does not generally allow studies in the pre-steady-state regime. The following sections describe two different methods for the direct coupling of ESI MS with rapid mixing devices, involving continuous-flow and stopped-flow systems, respectively. The time resolution that can be achieved by these techniques is orders of magnitude better than in manual mixing experiments. The results demonstrate that the coupling of rapid mixing devices with ESI MS provides a new and powerful tool, not only for studying the pre-steady state of enzymatic reactions, but also for a wide range of other kinetic experiments in chemistry and biochemistry.

---

[21] F. B. Simpson and D. B. Northrop, in "Mass Spectrometry in Biology and Medicine" (A. L. Burlingame, S. A. Carr, and M. A. Baldwin, ed.), p. 329. Humana Press, Totowa, NJ, 2000.

[22] E. D. Lee, W. Mück, J. D. Henion, and T. R. Covey, *J. Am. Chem. Soc.* **111,** 4600 (1989).

[23] T. A. Fligge, J. Kast, K. Bruns, and M. Przybylski, *J. Am. Soc. Mass Spectrom.* **10,** 112 (1999).

[24] B. Bothner, R. Chavez, J. Wei, C. Strupp, Q. Phung, A. Schneemann, and G. Siuzdak, *J. Biol. Chem.* **275,** 13455 (2000).

[25] D. Fabris, *J. Am. Chem. Soc.* **122,** 8779 (2000).

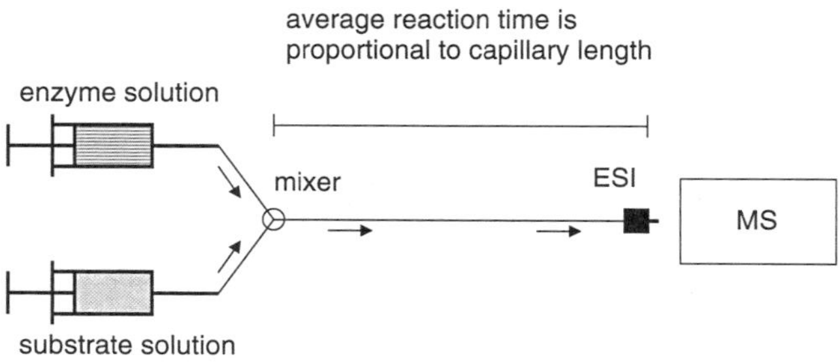

FIG. 1. Continuous-flow mixing setup for monitoring enzyme kinetics. The plungers of both syringes are advanced continuously by syringe pumps. The reaction is initiated by mixing the solutions from both syringes. The average reaction time is proportional to the length of the capillary downstream from the mixer. The end of the reaction capillary is connected to the ESI source of a mass spectrometer. Arrows indicate the direction of liquid flow. Reprinted with permission from L. Konermann, F. I. Rosell, A. G. Mauk, and D. J. Douglas, *Biochemistry* **36,** 6448 (1997). Copyright © 1997 American Chemical Society.

Pre-Steady-State Kinetics of Xylanase Monitored by Electrospray Mass Spectrometry with On-Line Continuous-Flow Mixing

Continuous-flow experiments with rapid on-line mixing are a simple and elegant method for studying biochemical reaction kinetics.[26] With optical detection a time resolution in the microsecond range can be achieved.[27] One limitation of many continuous-flow instruments is the substantial sample consumption; the original studies of Hartridge and Roughton[28] required more than 3 liters of sample for each experiment. Konermann *et al.* developed a miniaturized continuous-flow mixing device and coupled it to the ESI source of a mass spectrometer.[29] Through the use of narrow fused silica capillaries the sample consumption could be substantially reduced. This system allows extensive kinetic studies even if only a few milliliters of sample are available. The viability of this new experimental technique was amply demonstrated in a number of studies on protein folding kinetics.[29–32]

Here we describe the application of this technique to the study of the pre-steady-state kinetics of enzymatic reactions. The setup used for these experiments is shown in Fig. 1.[29] The plungers of two syringes (volume 1 ml each) are advanced simultaneously by syringe pumps. The syringes contain enzyme and substrate

---

[26] H. Gutfreund, *Methods Enzymol.* **16,** 229 (1969).
[27] M. C. R. Shastry, S. D. Luck, and H. Roder, *Biophys. J.* **74,** 2714 (1998).
[28] H. Hartridge and F. J. W. Roughton, *Proc. Roy. Soc. (London)* **A104,** 376 (1923).
[29] L. Konermann, B. A. Collings, and D. J. Douglas, *Biochemistry* **36,** 5554 (1997).
[30] L. Konermann, F. I. Rosell, A. G. Mauk, and D. J. Douglas, *Biochemistry* **36,** 6448 (1997).
[31] V. W. S. Lee, Y.-L. Chen, and L. Konermann, *Anal. Chem.* **71,** 4154 (1999).
[32] O. O. Sogbein, D. A. Simmons, and L. Konermann, *J. Am. Soc. Mass Spectrom.* **11,** 312 (2000).

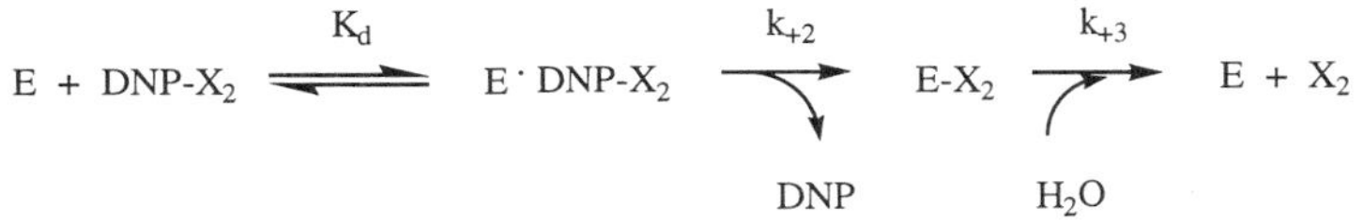

$$E + DNP\text{-}X_2 \underset{}{\overset{K_d}{\rightleftharpoons}} E\cdot DNP\text{-}X_2 \xrightarrow[DNP]{k_{+2}} E\text{-}X_2 \xrightarrow[H_2O]{k_{+3}} E + X_2$$

SCHEME 1. Mechanism for the hydrolysis of 2,5-dinitrophenyl-β-xylobioside (DNP-X_2) by *Bacillus circulans* xylanase (BCX). Reprinted with permission from D. L. Zechel, L. Konermann, S. G. Withers, and D. J. Douglas, *Biochemistry* **37**, 7664 (1998). Copyright © 1998 American Chemical Society.

solution, respectively. The reaction is initiated by mixing the solutions in a custom-made mixing "tee." This tee has a volume of 3 nl which corresponds to a mixing time (dead time) of 5 ms at a total flow rate of 30 μl/min.[29] The outlet of the tee is connected to the ESI source of a quadrupole mass spectrometer via a reaction capillary (i.d. 75 μm). The average reaction time of the solution is determined by the flow rate and the length of the reaction capillary. For the flow rate given above, 1 cm of capillary corresponds to 81 ms. Shorter reaction times can be achieved by using higher flow rates. Typically, capillary lengths between 9 mm and 2 m are used. A miniaturized ion source with an overall length of only 7 mm had to be developed for this setup in order to accommodate the shortest reaction capillaries. Ions are generated by pneumatically assisted ESI at the outlet of the reaction capillary. Within one set of kinetic experiments (i.e., for a given substrate concentration) it is preferable to change the reaction time only by varying the length of the reaction capillary while leaving the flow rate constant. This is because different flow rates can change the appearance of ESI mass spectra, even if no chemical reaction occurs.

The apparatus of Fig. 1 was used for studying the pre-steady-state kinetics of an enzymatic reaction.[33] The Y80F mutant of the extensively characterized *Bacillus circulans* xylanase (BCX, EC 3.2.1.8) was chosen as a model enzyme system. This enzyme cleaves the β-1,4-glycosidic bonds of xylan to release xylobiose. For this study, the chromophoric substrate 2,5-dinitrophenyl-β-xylobioside (DNP-X_2) was used, so the kinetics observed by ESI MS could be independently confirmed by traditional stopped-flow optical spectroscopy. The Y80F mutant of BCX has a K_m of 60 μM for DNP-X_2. The mechanism for the hydrolysis of DNP-X_2 by BCX is described in Scheme 1.[33] The free substrate and the free enzyme are in equilibrium with a noncovalent enzyme–substrate complex E.DNP-X_2 (dissociation constant K_d). Subsequently, xylobiose (X_2) is covalently bound to the enzyme. The formation of this E-X_2 complex is accompanied by the release of 2,5-dinitrophenolate (DNP) which can be monitored optically at 440 nm. The last step of the reaction is the hydrolysis of the E-X_2 complex to release X_2 and to regenerate the free enzyme. As indicated in Scheme 1, k_{+2} and k_{+3} are the rate constants for glycosylation and deglycosylation of the enzyme, respectively. Wild-type BCX

---

[33] D. L. Zechel, L. Konermann, S. G. Withers, and D. J. Douglas, *Biochemistry* **37**, 7664 (1998).

has not been amenable to pre-steady-state kinetic analysis by stopped-flow optical spectroscopy because the glycosylation step is rate limiting ($k_{+2} < k_{+3}$) with all known synthetic substrates. However, this is not the case for the Y80F mutant studied here. It shows a pre-steady-state phase which is characterized by a burst of DNP formation, indicating that deglycosylation is rate limiting ($k_{+2} > k_{+3}$).[33]

The apparatus of Fig. 1 was used for monitoring the time-resolved mass spectra of BCX Y80F at different times after mixing enzyme and substrate ("time-resolved ESI MS"). All the concentrations given below refer to the final concentrations (i.e., after the mixing step). Ammonium acetate (5 mM, pH 6.3) was used as buffer system. The enzyme concentration was 4 μM, while the initial DNP-X_2 concentrations ranged from 0.035 mM to 2.2 mM. The limited solubility of DNP-X_2 precluded experiments at higher substrate concentration. The ESI mass spectrum of the mutant BCX Y80F in 5 mM ammonium acetate buffer and in the absence of substrate displayed a charge state distribution with a characteristic maximum of 8+ (data not shown). The m/z values of the peaks in the spectrum agreed with those expected from the calculated mass of the protein (20,384 Da). A very similar spectrum was recorded 0.5 sec after mixing the enzyme with 0.11 mM DNP-X_2 (Fig. 2A).[33]

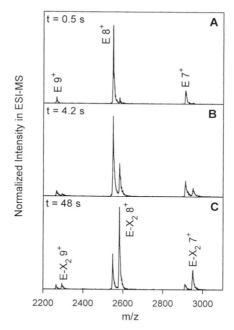

FIG. 2. ESI mass spectra recorded at 0.5 s (A), 4.2 s (B), and 48 s (C) after mixing a solution of the enzyme BCX Y80F (4 μM) with the substrate DNP-X_2 (110 μM). Peaks in the spectrum correspond to different charge states of the free enzyme (E) and the covalent xylobiosyl-enzyme intermediate (E-X_2). Reprinted with permission from D. L. Zechel, L. Konermann, S. G. Withers, and D. J. Douglas, *Biochemistry* **37**, 7664 (1998). Copyright © 1998 American Chemical Society.

However, after 4.2 sec (Fig. 2B), each of the peaks in the spectrum displayed a pronounced satellite peak. On the basis of their m/z values, these satellite peaks could be assigned to the covalent xylobiosyl-enzyme intermediate E-X_2 (20,649 Da). The spectrum recorded after 48 sec (Fig. 2C) displayed an even higher intensity for the E-X_2 peaks. Similar intensity ratios between E and E-X_2 peaks were observed for the different charge states for all reaction times and substrate concentrations. The intensity ratio of the dominant E 8+ and E-X_2 8+ peaks provides a direct measure of the relative concentrations of free enzyme (E) and covalent intermediate (E-X_2) in solution. Therefore it is possible to monitor the pre-steady-state kinetics of this reaction by measuring these peak intensities as a function of time. The results of some representative kinetic measurements are shown in Fig. 3.[33] Consistent with the kinetic scenario in which $k_{+2} > k_{+3}$ (Scheme 1), the intensity of E-X_2 shows a fast increase before it levels off and reaches a constant value as the reaction approaches the steady state regime. Higher substrate concentrations lead to an increased steady-state concentration of E-X_2, as expected. All the curves are well described by single-exponential fits with $I(t) = I_{SS}[1 - \exp(-k_{obs}t)]$, where $I(t)$ is the ion intensity as a function of time, I_{SS} is the steady-state intensity, and k_{obs} is the observed first-order rate constant.[1,2] A plot of k_{obs} versus substrate concentration is shown in Fig. 4.[33] Also shown in Fig. 4 are the values of k_{obs} that were measured by stopped-flow optical spectroscopy from the burst-phase kinetics of DNP formation.[33] The data sets are in excellent agreement. According to Scheme 1, the first-order rate constant k_{obs} is given by[1,2]

$$k_{obs} = k_{+3} + \frac{k_{+2}[S]}{K_d + [S]} \quad (1)$$

In this study the low solubility of DNP-X_2 precluded saturation of the enzyme which indicates that K_d has to be in the high millimolar range ($K_d \gg [S]$). Under these conditions, Eq. (1) simplifies to a linear relationship

$$k_{obs} = k_{+3} + \frac{k_{+2}}{K_d}[S] \quad (2)$$

The linear fits to the experimental data monitored by stopped-flow optical spectroscopy and by ESI MS are virtually identical (Fig. 4). By extrapolating these fits to a substrate concentration of zero, k_{+3} is determined to be 0.08 s^{-1}. Unfortunately, the slope in Eq. (2) cannot be used to determine the individual values of K_d and k_{+2}. However, the ratio k_{+2}/K_d of 0.65 mM^{-1}s^{-1} from these pre-steady-state experiments agrees very well with the second-order rate constant k_{cat}/K_m of 0.70 mM^{-1}s^{-1} that was determined from steady-state experiments.[1,33] Unfortunately the spectra in Fig. 2 do not show any peaks that could be assigned to noncovalent enzyme–substrate or enzyme–product complexes. Ganem et al. have shown previously that specific noncovalent complexes are only observed when the

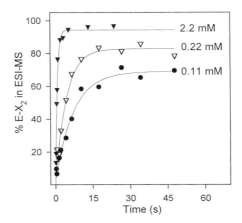

FIG. 3. Relative contribution of the 8+ ion generated from the covalent xylobiosyl-enzyme intermediate (E-X$_2$) in the ESI mass spectrum as a function of time. The substrate concentrations are as indicated. Circular and triangular symbols represent experimental data; solid lines are single-exponential fits to the data. Reprinted with permission from D. L. Zechel, L. Konermann, S. G. Withers, and D. J. Douglas, *Biochemistry* **37**, 7664 (1998). Copyright © 1998 American Chemical Society.

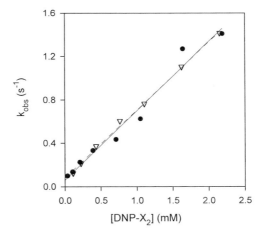

FIG. 4. First-order rate constant k_{obs} for the formation of the covalent xylobiosyl-enzyme intermediate E-X$_2$ measured by ESI MS (circles) and for the formation of DNP measured by stopped-flow optical spectroscopy (triangles), as a function of substrate concentration. The solid and dashed lines are fits to the experimental ESI and optical data, respectively, according to Eq. (2) (see text). Reprinted with permission from D. L. Zechel, L. Konermann, S. G. Withers, and D. J. Douglas, *Biochemistry* **37**, 7664 (1998). Copyright © 1998 American Chemical Society.

dissociation constant is in the micromolar range or lower.[12] As noted above, K_d in this case is much higher.

These experiments on BCX Y80F marked the first time that ESI MS in conjunction with on-line continuous flow-mixing was applied to study the pre-steady-state kinetics of an enzymatic reaction. This approach does not require radioactive labeling or chemical quenching. The chromophoric substrate DNP-X_2 was used only to validate this novel method by independently confirming the ESI MS kinetics by traditional stopped-flow optical spectroscopy. The time resolution of the current setup is on the order of some tens of milliseconds. This resolution was sufficient for studying the relatively slow pre-steady-state kinetics of BCX Y80F. However, for many enzymes a higher time resolution will be required. This should be possible by using higher liquid flow rates, a mixing chamber with a lower dead volume, and capillaries with smaller inner diameters. Instead of using fused-silica capillaries, such a continuous-flow device could be incorporated into a microfluidic chip that is directly coupled to an ESI source. Through the use of hydrodynamic focusing such a system could potentially reach a time resolution in the submillisecond range.[34] Although microfluidic chips for ESI MS are now commonly used for various analytical applications,[35] a fast kinetic mixer of this kind has not been developed yet.

Another experimental technique for studying the kinetics of enzymatic reactions has been developed by Northrop and Simpson.[36] It uses a continuous flow of carrier solution that flows past the interface of a membrane inlet mass spectrometer (MIMS). Pulses of reaction mixture are injected into the carrier stream and the reaction time is varied by using different lengths of delay line between the injection point and the interface. This setup was used for monitoring the concentration of CO_2 during the conversion of carbon dioxide to bicarbonate by carbonic anhydrase. It appears that this technique could complement the ESI-based approach described above. ESI MS allows the analysis of an extremely wide range of analytes. However, ESI cannot be used to ionize gases that are dissolved in a reaction mixture.

It is widely believed that continuous-flow systems can only produce meaningful kinetic data when they are operated in the turbulent flow regime.[1] Turbulent flow leads to continuous mixing of fast and slow liquid in a flow tube, therefore resulting in a relatively narrow distribution of solution "age" at every point along the tube, downstream from the mixer where the reaction is initiated. However, in narrow capillary devices with relatively low flow rates such as those used for the kinetic experiments by Zechel et al.,[33] the liquid flow is laminar.[37] Laminar flow is characterized by a parabolic velocity profile: the flow velocity in the center of

---

[34] J. B. Knight, A. Vishwanath, J. P. Brody, and R. H. Austin, *Phys. Rev. Lett.* **80**, 3863 (1998).
[35] R. D. Oleschuk and D. J. Harrison, *Trends Anal. Chem.* **19**, 379 (2000).
[36] D. B. Northrop and F. B. Simpson, *Arch. Biochem. Biophys.* **352**, 288 (1998).
[37] L. Konermann, *J. Phys. Chem. A* **103**, 7210 (1999).

the capillary is highest, whereas the liquid layer directly adjacent to the capillary wall is stationary. This velocity profile leads to a distribution of solution "age" at every point along the flow tube. It therefore seems somewhat surprising that reaction kinetics monitored in the laminar flow regime are in close agreement with those measured by traditional rapid mixing methods (see, e.g., Fig. 4). This apparent contradiction has been addressed in a study where computer modeling was used to simulate the liquid flow in continuous-flow systems.[37] It was shown that the distortion of reaction kinetics monitored under laminar flow conditions is surprisingly small. Radial diffusion in narrow flow tubes plays an important role since it transfers analyte between liquid layers of different velocity, thus narrowing the width of the age distribution of the analyte in the flow tube. These computer simulations have shown that the effects of laminar flow are almost undetectable for experiments such as the kinetic studies on BCX Y80F described above. The results of Ref. 37 therefore clearly indicate the feasibility of kinetic continuous-flow experiments in the laminar flow regime.

Stopped-Flow Electrospray Mass Spectrometry

Stopped-flow optical spectroscopy is one of the most widely used methods for studying the pre-steady-state kinetics of enzymatic reactions. A variety of stopped-flow instruments are commercially available. They are easy to use and provide kinetic data in simple "push-button" experiments. It was pointed out earlier in this chapter that a significant limitation of conventional stopped-flow techniques is the requirement for chromophoric substrates. ESI MS, on the other hand, does not require chromophores and allows direct monitoring of most reactive species in enzymatic assays with high sensitivity and selectivity. Therefore, there is considerable interest in the development of techniques that combine stopped-flow mixing with ESI MS.

The first steps in this direction have been described by Paiva et al.,[38] who coupled a stopped-flow device to the ESI source of an ion trap mass spectrometer. This setup was used for the detection of a short-lived tetrahedral intermediate that was formed under single turnover conditions during the reaction catalyzed by 5-enolpyruvoyl-shikimate-3-phosphate synthase. Only one single time point (corresponding to 28 ms) could be monitored by this system. Therefore a detailed kinetic study of the reaction was not possible. Ørsnes et al.[39] developed a stopped-flow mass spectrometer that is capable of monitoring the concentration of reactive species at different reaction times. They coupled a stopped-flow mixer to a membrane inlet mass spectrometer (MIMS) via a rotating ball inlet. The performance

---

[38] A. A. Paiva, R. F. Tilton, G. P. Crooks, L. Q. Huang, and K. S. Anderson, *Biochemistry* **36**, 15472 (1997).

[39] H. Ørsnes, T. Graf, and H. Degn, *Anal. Chem.* **70**, 4751 (1998).

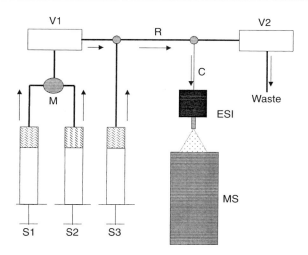

FIG. 5. Experimental setup for stopped-flow ESI MS. S1 and S2: Syringes for pulsed reactant injection; M: mixer; S3: syringe. S3 delivers solvent that pushes reaction mixture through the fused silica capillary (C) to the ESI source after the reaction tube (R) has been filled with fresh reaction mixture; MS: quadrupole mass spectrometer; V1, V2: valves. Arrows indicate the direction of liquid flow. Reprinted with permission from B. M. Kolakowski, D. A. Simmons, and L. Konermann, *Rapid Commun. Mass Spectrom.* **14,** 772 (2000). Copyright © 2000 John Wiley & Sons Ltd.

of this system was characterized by studying the concentrations of acetone and butanone in ketone–sulfite reactions. As demonstrated by Northrop and Simpson (Ref. 36, see above) such an approach involving MIMS can be very useful for monitoring dissolved gases and small organic molecules in enzyme-catalyzed reactions. However, the use of ESI MS would allow a much wider range of biochemical analytes to be monitored, most of which are not amenable to analysis by MIMS.

Kolakowski *et al.* have described a novel setup for stopped-flow ESI MS that is capable of monitoring the concentration of multiple reactive species as a function of time.[40,41] Thus far this new method has only been applied to bioorganic and protein folding reactions. Nevertheless, this novel setup will be briefly discussed in this chapter because it is anticipated that stopped-flow ESI MS will become an important tool for studying the pre-steady state of enzymatic reactions. A schematic diagram of the experimental setup is shown in Fig. 5.[40] Two 20-ml syringes S1 and S2 and the mixer M are part of a commercial stopped-flow system (Bio-Logic,

---

[40] B. M. Kolakowski, D. A. Simmons, and L. Konermann, *Rapid Commun. Mass Spectrom.* **14,** 772 (2000).
[41] B. M. Kolakowski and L. Konermann, *Anal. Biochem.* **292,** 107 (2001).

France). The plungers of S1 and S2 are operated by computer-controlled stepper motors. In its original configuration, two additional mixers and an optical observation cell are part of the instrument. These components are removed and replaced with a custom-made adapter which is installed directly downstream from M. S1 and S2 are used to deliver reactant solutions to M where the reaction is initiated. The bold lines downstream from M symbolize PEEK tubing (i.d. 0.76 mm). While the plungers of S1 and S2 are rapidly advanced, the valves V1 and V2 are open to permit rinsing of the reaction tube R with fresh reaction mixture. Subsequently both valves are closed. The timing of this valve closing has to be adjusted carefully in order to optimize the time resolution of the experiment.[41] While the reaction proceeds in R, the solvent from S3 pushes a steady "aliquot flow" of reaction mixture through a 1.5-cm-long fused silica capillary (i.d. 75 μm) into a customized ESI source.[29] Analyte gas-phase ions are generated by pneumatically assisted ESI and analyzed by a triple quadrupole mass spectrometer (PE-SCIEX, Concord, ON). Several reactants can be monitored quasi-simultaneously by operating the mass spectrometer in multiple ion mode, i.e., by rapidly switching the mass analyzer between m/z values that correspond to different analyte ions.

The performance of this apparatus can be demonstrated by using the acid-induced demetallation of chlorophyll a as a simple test reaction. This reaction is associated with a change of mass (loss of magnesium and gain of two protons) that can be easily monitored by ESI MS; it is also associated with a color change, so that the ESI MS results can be verified by traditional optical methods. Chlorophyll a (5 μM) was exposed to 14 mM HCl in methanol/water (90:10 v/v) and stopped-flow ESI MS was used to monitor the intensity of molecular ions of chlorophyll (m/z of 894) as a function of time (Fig. 6). The experimental data are well described

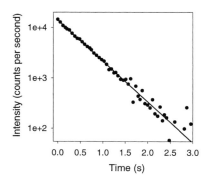

FIG. 6. Demetallation kinetics of chlorophyll a in a dilute solution of HCl monitored by stopped-flow ESI MS. The circles represent the measured count rate of chlorophyll a ions (m/z 894) as a function of time. The solid line in this semilogarithmic plot is an exponential fit with a pseudo first-order rate constant of 1.9 s^{-1}.

by an exponential fit with a pseudo first-order rate constant of 1.9 s^{-1} which is in excellent agreement with the results of optical control experiments.[41] The current time resolution that can be achieved with the stopped-flow ESI mass spectrometer shown in Fig. 5 is on the order of 100 ms. This time resolution will not be sufficient for studying the pre-steady-state kinetics of some enzymes. However, it is hoped that a substantial improvement of the time resolution will be possible by implementing a stopped-flow/ESI interface on a microfluidic chip, similar to the arrangement proposed for the continuous-flow approach described above.

Simpson and Northrop[21] have suggested that stopped-flow ESI MS could revolutionize enzyme kinetics because it will allow the direct and unambiguous determination of mechanistically important intermediates by their mass, at the same time providing detailed kinetic information. Different approaches can be used for monitoring the reaction mixture in stopped-flow ESI MS. On slow-scanning mass spectrometers such as quadrupole instruments it will be most straightforward to monitor selected ions as a function of time in multiple ion mode. It will be possible to increase the sensitivity and the selectivity of the experiment by using tandem mass spectrometry and multiple reaction monitoring (i.e., monitoring specific fragment ions of selected precursor species, instead of monitoring the precursor ions directly). By using mass spectrometers that allow a more rapid data acquisition, such as quadrupole time-of-flight instruments or quadrupole ion traps, it might be possible to monitor not only selected ion intensities, but *entire mass spectra* as a function of reaction time. This could greatly facilitate the identification of transient intermediates and the elucidation of enzymatic reaction mechanisms. Such an approach is technically demanding because it would require the acquisition of high quality mass spectra on a millisecond time scale. The feasibility of this experimental technique remains to be explored. However, spectral information can always be obtained, even with slow-scanning instruments, by using the continuous-flow methods described in this chapter. For practitioners who use quadrupole mass spectrometers it might be a good experimental strategy to (i) use continuous-flow methods to get an overview of the various species that are formed during the reaction and subsequently (ii) use stopped-flow ESI MS in order to obtain detailed and accurate kinetic data by monitoring selected ion intensities as a function of reaction time.

Concluding Remarks

The new ESI MS-based techniques described in this chapter clearly have the potential to become standard tools for studying enzymatic reactions in the pre-steady-state regime. They will also be useful for kinetic studies in other fields of chemistry and biochemistry. The current time resolution of these techniques (tens of milliseconds) is not as good as that of conventional quench-flow and stopped-flow experiments. However, it is anticipated that future developments such as the use of miniaturized flow-mixing devices will help to overcome this limitation.

Certain restrictions will be encountered when choosing solvent systems that are compatible with on-line monitoring of the reaction mixture by ESI MS. Salts, detergents, and pH buffers can interfere with the electrospray process, leading to low signal intensities and the formation of adduct species. The electrospray process is compatible with pH buffers that are based on volatile components such as acetic acid, formic acid, ammonia, or piperidine. It might be feasible to develop microdialysis systems for kinetic experiments that allow the rapid desalting of the reaction mixture immediately before it is analyzed by ESI MS.[42] At the current stage of development, experiments necessarily requiring high salt concentrations have to be carried out by other techniques such as chemical quench-flow methods in conjunction with off-line analysis by MALDI MS[5a] or other analytical methods.

Acknowledgments

This work was supported by NSERC, an NSERC-SCIEX Industrial Chair (to D.J.D.), the Canada Foundation for Innovation, the Ontario Ministry of Energy, Science and Technology, the University of British Columbia, and the University of Western Ontario. We thank students, postdoctoral researchers, and colleagues who were involved in the work described and whose names appear in the reference list. The experimental data shown in Fig. 6 were recorded by Beata M. Kolakowski.

[42] N. Xu, Y. Lin, S. A. Hofstadler, D. Matson, C. J. Call, and R. D. Smith, *Anal. Chem.* **70,** 3553 (1998).

[4] Trapping of α-Glycosidase Intermediates

By RENÉE M. MOSI and STEPHEN G. WITHERS

Introduction

α-Glycosidases and α-transglycosylases are enzymes that catalyze the transfer of glycosyl moieties from a donor sugar to an acceptor: water in the case of glycosidases and some other acceptor in the case of transglycosylases. These enzymes have been classified into families based on sequence similarities, as first described by Henrissat,[1] and more recently reviewed.[2–6] Some 2000 glycosidases have been

[1] B. Henrissat, *Biochem. J.* **280,** 309 (1991).
[2] B. Henrissat and G. Davies, *Curr. Opin. Struct. Biol.* **7,** 637 (1997).
[3] D. L. Zechel and S. G. Withers, *Acc. Chem. Res.* **33,** 11 (2000).
[4] H. Ly and S. G. Withers, *Ann. Rev. Biochem.* **68,** 487 (1999).
[5] G. Davies, M. L. Sinnott, and S. G. Withers, in "Comprehensive Biological Catalysis" (M. L. Sinnott, ed.), p. 119. Academic Press, New York, 1997.
[6] C. Rye and S. G. Withers, *Curr. Opin. Chem. Biol.* **4,** 573 (2000).

categorized into approximately 80 families, and a constantly updated listed is available (afmb.cnrs-mrs.fr/CAZY). Members within the same family have the same three-dimensional fold, and since the overall fold is more conserved than their primary sequence it turns out that several families have similar folds even though this is not evident from the sequence: these families have been assigned to clans.[7] This similarity has proved to be useful in helping to identify the residues responsible for specificity and catalysis because once a structural motif is established for one member of the family, comparisons can be made to other enzymes within the group whose structures have not yet been solved. Because of the important roles these enzymes play in the biosynthesis and biodegradation of oligosaccharides and polysaccharides, and hence their applications in industry and medicine, their mechanism of action needs to be well understood. This involves characterizing the intermediates along the reaction pathway as well as identifying the key amino acid residues involved in catalysis.

The two different stereochemical outcomes of the reactions catalyzed by transglycosylases (and glycosidases) are a result of two distinct mechanisms by which these enzymes operate. The retaining transglycosylases and glycosidases operate through a double-displacement reaction (Fig. 1) while the inverting enzymes employ a single-displacement mechanism.

The key feature of the double-displacement mechanism employed by retaining transglycosylases (and glycosidases) (Fig. 1), first proposed by Koshland in 1953, is the formation of a covalent glycosyl-enzyme intermediate.[8] The first step (glycosylation) involves the attack of the catalytic nucleophile at the anomeric center of the sugar with general acid assistance to aid in the departure of the leaving group. This generates a glycosyl-enzyme intermediate which can then undergo either transglycosylation or hydrolysis in a second step (deglycosylation), with general base assistance facilitating attack of the incoming group. Both steps proceed via transition states with substantial oxocarbenium ion character. In addition, noncovalent interactions between enzyme and substrate are believed to be responsible for much of the rate acceleration and provide considerable stabilization of the transition state.

In both steps, a pair of carboxylates function as the catalytic residues. One carboxylate performs the dual role of acid–base catalyst while the second carboxylate functions as a catalytic nucleophile. Extensive progress has been made in recent years in elucidating the identities and functions of the catalytic carboxylates in many α- and β-glycosidases and transglycosylases by means of X-ray crystallography, sequence alignment, site-directed mutagenesis, kinetic analysis of the

---

[7] B. Henrissat, I. Callebaut, S. Fabrega, P. Lehn, J. P. Mornon, and G. Davies, *Proc. Natl. Acad. Sci. U.S.A.* **92,** 7090 (1995).
[8] D. E. Koshland, *Biol. Rev.* **28,** 416 (1953).

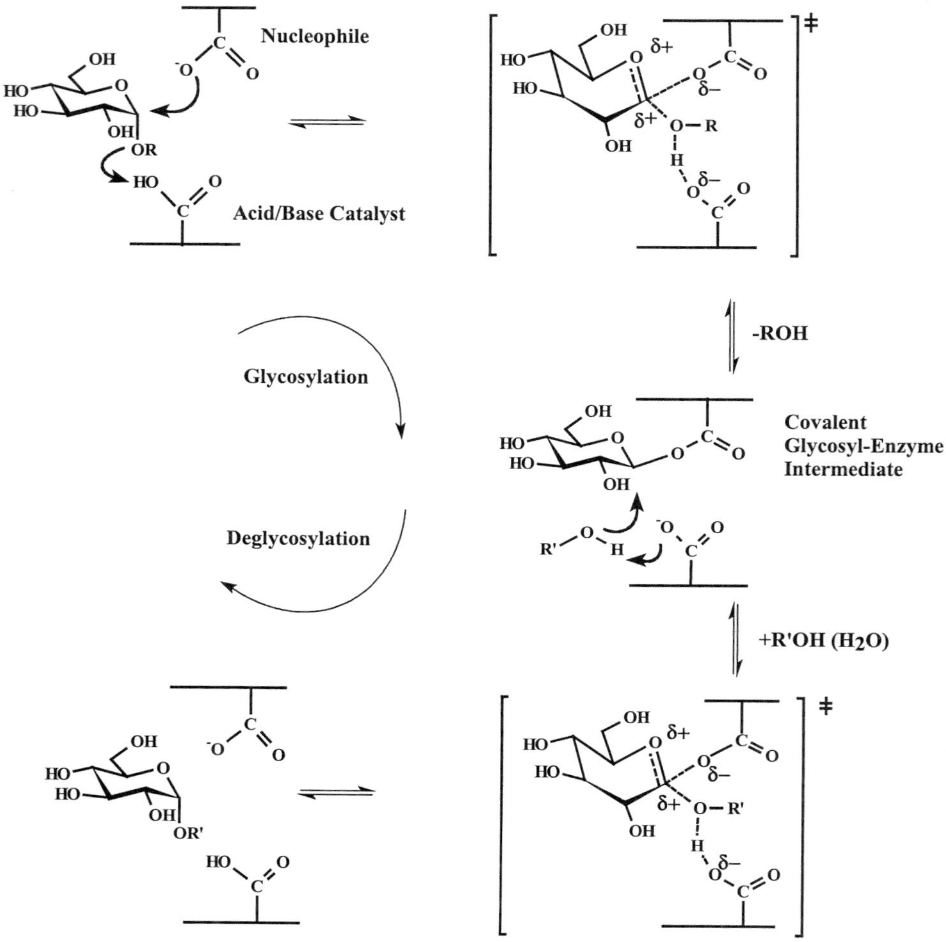

FIG. 1. The double-displacement mechanism for a retaining α-glycosyltransferase.

resulting mutant enzymes, and labeling studies (see Refs. 3–6, [9–12] for useful reviews). In all cases, these catalysts have been found to be either glutamic acid

[9] G. Davies and B. Henrissat, *Structure* **3**, 853 (1995).
[10] B. Svensson, *Plant Mol. Biol.* **25**, 141 (1994).
[11] H. M. Jespersen, E. A. MacGregor, B. Henrissat, M. R. Sierks, and B. Svensson, *J. Prot. Chem.* **12**, 791 (1993).
[12] D. L. Zechel and S. G. Withers, *in* "Comprehensive Natural Products Chemistry" (D. Barton, K. Nakanishi, and C. D. Poulter, eds.), p. 279. Elsevier, Amsterdam, 1999.

or aspartic acid residues.[13] The structures of a number of glycosidases have now been determined crystallographically (for a list of examples from each family see Refs. 6 and 12). These structures reveal that in retaining glycosidases, the average distance between the two active site carboxylic acid residues is approximately 4.5–5.5 Å, consistent with direct attack by the nucleophile. In contrast, in inverting enzymes the two carboxyl groups are separated by a greater distance, approximately 7–11 Å, presumably to create enough space for the participating water molecule.[14,15] Unfortunately, even when X-ray crystal structures are not available, it is not always easy to identify the key carboxylates unless a substrate analog or inhibitor is bound in the active site. Even then it is possible that carboxylates may be misassigned if the substrate analog or inhibitor is bound in an unproductive manner.

Nature of Glycosyl-Enzyme Intermediate and Identification of Catalytic Nucleophile

Review of Strategies to Trap Covalent Glycosyl-Enzyme Intermediates

  a. Low Temperature and Rapid Denaturation Trapping Strategies. Rapid-quench methods have been used to provide evidence for a covalent glycosyl-enzyme intermediate in retaining α-glycosidases and α-transglycosylases. In the first such report Voet and Abeles captured the glucosyl-enzyme intermediate in sucrose phosphorylase and confirmed that the glucosyl moiety was bound to the enzyme through a β-linkage.[16,17] This was achieved by incubation of the enzyme with radiolabeled sucrose, rapid acid denaturation of the protein, proteolysis, and then isolation of the radiolabeled peptides. In addition to serving as a denaturant the low pH conditions simplified the trapping by protonating the base catalyst, thereby slowing the deglycosylation step. An aspartic acid was identified as the nucleophile of *Streptococcus sobrinus* α-glucosyltransferase in a similar manner by rapidly denaturing a mixture of enzyme and radiolabeled sucrose, proteolysing and then isolating the labeled peptide.[18,19]

  Trapping the intermediate on an α-glycosidase is a more demanding task than on a transglycosylase because of the need to prevent reaction with the omnipresent substrate, water. Nonetheless, low temperature ^{13}C NMR experiments

---

[13] S. G. Withers and R. Aebersold, *Prot. Sci.* **4**, 361 (1995).
[14] J. D. McCarter and S. G. Withers, *Curr. Opin. Struct. Biol.* **4**, 885 (1994).
[15] Q. Wang, R. W. Graham, D. Trimbur, R. A. J. Warren, and S. G. Withers, *J. Am. Chem. Soc.* **116**, 11594 (1994).
[16] J. G. Voet and R. H. Abeles, *J. Biol. Chem.* **245**, 1020 (1970).
[17] J. J. Mieyal, M. Simon, and R. H. Abeles, *J. Biol. Chem.* **247**, 532 (1972).
[18] G. Mooser, S. A. Hefta, R. J. Paxton, J. E. Shively, and T. D. Lee, *J. Biol. Chem.* **266**, 8916 (1991).
[19] G. Mooser, *in* "The Enzymes" (D. S. Sigman, ed.), p. 187. Academic Press, New York, 1992.

provided evidence for the steady-state formation of a β-carboxylacetal ester covalent adduct resulting from the reaction of maltotetraose and porcine α-amylase.[20] However, actual trapping of intermediates awaited the developments described below.

    b. Inhibitors and Mechanism-Based Inactivators. An alternative to the use of extreme conditions such as low temperature and high acidity to trap intermediates involves the use of subtly modified substrate analogs and mechanism-based inactivators. By modifying the relative rates at which the glycosyl-enzyme intermediate is accumulated and broken down, it has proven possible to accumulate this intermediate in a more native-like state. In conjunction with NMR and MS methods, these inhibitors have provided strong evidence for the existence of a covalent intermediate and have allowed the identification of the nucleophilic site of attachment for many glycosidases and transglycosylases.

    i. 2-Deoxy-2-fluoroglycosides. The 2-deoxy-2-fluoroglycosides have proven to be extremely valuable tools in providing evidence for the existence of a covalent glycosyl-enzyme intermediate in several retaining glycosidases and in identifying the catalytic nucleophile. The presence of an electron-withdrawing fluorine at C-2 inductively destabilizes the oxocarbenium ion-like transition states of both the glycosylation and deglycosylation steps. The replacement of the hydroxyl group at C-2 also results in the disruption of important hydrogen bonding interactions which stabilize the transition state. Together, this results in a decrease of 10^5- to 10^8-fold in the rates of both glycosylation and deglycosylation.[21,22] However, the presence of a good leaving group [typically 2,4-dinitrophenolate (DNP) or fluoride] helps to accelerate the glycosylation step, resulting in the accumulation of the glycosyl-enzyme intermediate. This trapped glycosyl-enzyme intermediate is catalytically competent and can be turned over in the presence of a suitable acceptor to give fully active enzyme. This strategy has been used successfully to identify the catalytic nucleophiles for many β-glycosidases (see elsewhere in this volume),[22a] the first of which was Glu-358 of *Agrobacterium* β-glucosidase (Abg).[23]

    Although this strategy has worked well for β-glycosidases it has been less successful for α-retaining glycosidases or α-transglycosylases. For example, 2-deoxy-2-fluoro-α-maltosyl fluoride was found to act as a slow substrate for human pancreatic α-amylase and rabbit muscle glycogen debranching enzyme rather than as an inactivator.[24] Apparently, the deglycosylation step for an α-transglycosylase is

---

[20] B. Y. Tao, P. J. Reilly, and J. F. Robyt, *Biochim. Biophys. Acta* **995**, 214 (1989).

[21] I. P. Street, K. Rupitz, and S. G. Withers, *Biochemistry* **28**, 1581 (1989).

[22] I. P. Street, J. B. Kempton, and S. G. Withers, *Biochemistry* **31**, 9970 (1992).

[22a] J. Wicki, D. R. Rose, and S. G. Withers, *Methods Enzymol.* **354**, [5], 2002 (this volume).

[23] S. G. Withers, R. A. J. Warren, I. P. Street, K. Rupitz, J. B. Kempton, and R. Aebersold, *J. Am. Chem. Soc.* **112**, 5887 (1990).

[24] C. Braun, Ph.D. Thesis, University of British Columbia, 1995.

not slowed sufficiently by the substitution of fluorine at C-2 to allow accumulation of an intermediate, possibly indicating a fundamental difference in mechanism and charge development between β- and α-glycosidases.[24,25] Furthermore, the replacement of fluorine at C-2 makes this type of inhibitor less useful for those enzymes whose substrates are not amenable to substitution at that position (e.g., an N-acetylhexosaminidase). A solution to these problems has been developed in an effort to provide a broader approach to the labeling of both β- and α-retaining glycosidases without compromising specificity through substitution of any ring hydroxyl, as detailed below.

ii. 5-Fluoroglycosides. 5-Fluoroglycosides with a good leaving group such as fluoride might be expected to accumulate a 5-fluoroglycosyl-enzyme intermediate similar to the 2-deoxy-2-fluoroglycosyl-enzyme intermediate. A fluorine at C-5 in a pyranose ring is in an analogous position to the fluorine at C-2 as they are both adjacent to centers of developing positive charge. In the case of the 5-fluoro compound, the electron-withdrawing fluorine is adjacent to the developing positive charge of the ring oxygen, whereas in the corresponding 2-deoxy-2-fluoro case, the fluorine is closest to the developing charge at C-1. It has been suggested through modeling studies and experimentation that the greatest difference in partial charge between the ground state and the developing transition state actually occurs at O-5 rather than C-1.[26–28] Therefore, if O-5 has the greatest positive charge development in the transition state then substitution of a fluorine at C-5 rather than at C-2 would serve to destabilize the transition state to a greater extent. This is further magnified since the fluorine at C-5 replaces a hydrogen while that at C-2 replaces a significantly electronegative oxygen. Thus, on electronic grounds alone, 5-fluoroglycosides should be more effective than 2-deoxy-2-fluoroglycosides in trapping a glycosyl-enzyme intermediate. However, a substantial component of the reduction in the rates of glycosylation and deglycosylation consequent on fluorine substitution results from the disruption of transition state binding interactions with the C-2 hydroxyl group. No such contribution will likely exist at the 5-position; thus for a given enzyme, the degree to which the glycosyl-enzyme intermediate will accumulate depends on the relative importance of these binding interactions versus charge development at the transition state.[29]

Indeed, the 5-fluoro approach has proved useful in identifying the nucleophiles of several retaining α-glycosidases (Fig. 2; structures **1–6**) including yeast

---

[25] J. B. Kempton and S. G. Withers, *Biochemistry* **31**, 9961 (1992).
[26] T. Kajimoto, K. K. Liu, R. L. Pederson, Z. Zhong, Y. Ichikawa, J. A. Porco, and C. H. Wong, *J. Am. Chem. Soc.* **113**, 6187 (1991).
[27] D. A. Winkler and G. Holan, *J. Med. Chem.* **32**, 2084 (1989).
[28] M. Namchuk, J. McCarter, J. Becalski, T. Andrews, and S. G. Withers, *J. Am. Chem. Soc.* **122**, 1270 (2000).
[29] J. McCarter and S. G. Withers, *J. Am. Chem. Soc.* **118**, 241 (1996).

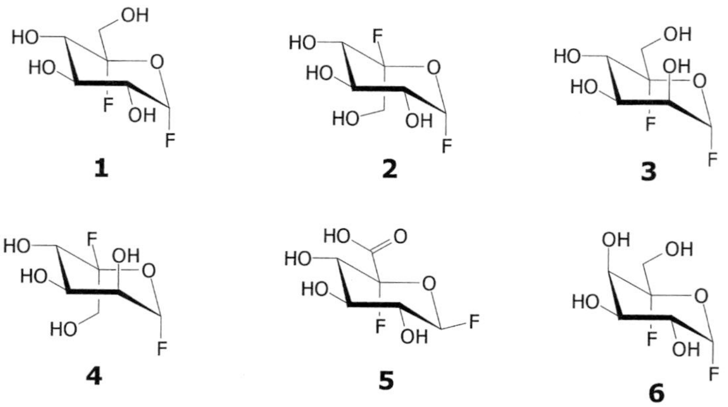

FIG. 2. Structures of 5-fluoroglycosides: 5F-α-D-GluF (**1**), 5F-β-L-IdoF (**2**), 5F-α-D-ManF (**3**), 5F-β-L-GulF (**4**), 5FβGlaF (**5**), and 5FαGalF (**6**). (*Note*: The compounds are not necessarily shown in their preferred conformation.)

α-glucosidase-Asp-214 [5-fluoro-α-D-glucosyl fluoride (5FαGluF, **1**) and 5-fluoro-β-L-idosyl fluoride (5FβIdoF, **2**)],[30] jack bean α-mannosidase-Asp [5-fluoro-α-D-mannosyl fluoride (5FαManF, **3**) and 5-fluoro-β-L-gulosyl fluoride (5FβGulF, **4**)],[30] bovine liver α-mannosidase-Asp-197 (5FαManF, **3**),[31] *Escherichia coli* β-glucuronidase (5-fluoro-β-D-glucuronosyl fluoride (5FβGlaF, **5**),[31a] and coffee bean α-galactosidase-Asp-145 (5-fluoro-α-galactosyl fluoride (5FαGalF, **6**).[32]

In many cases, full inactivation with 5-fluoro sugars was not achieved. Instead a catalytically competent intermediate was formed that was capable of normal turnover but at greatly reduced rates. If formation of the intermediate still occurs rapidly, then there will be substantial steady-state accumulation of the intermediate. This is reflected in low K_m values measured on these reagents as substrates, or low K_i values if measured as apparent competitive inhibitors. K_m or K_i values as low as 300 nM have been measured in these cases.[29] Further evidence that a relatively stable 5-fluoroglycosyl-enzyme intermediate had accumulated was obtained using electrospray mass spectrometry. An increase in mass equivalent to the covalent attachment of one 5-fluoroglycosyl moiety was obtained in most cases when samples of enzyme incubated with reagent were infused into the mass spectrometer.

---

[30] S. Howard, S. He, and S. G. Withers, *J. Biol. Chem.* **273**, 2067 (1998).
[31] S. Numao, S. He, G. Evjen, S. Howard, O. K. Tollersrud, and S. G. Withers, *FEBS Lett.* **484**, 175 (2000).
[31a] A. W. Wong, S. He, and S. G. Withers, *Can. J. Chem.* **79**, 510 (2001).
[32] H. D. Ly, S. Howard, K. Shum, S. He, A. Zhu, and S. G. Withers, *Carb. Res.* **329**, 539 (2001).

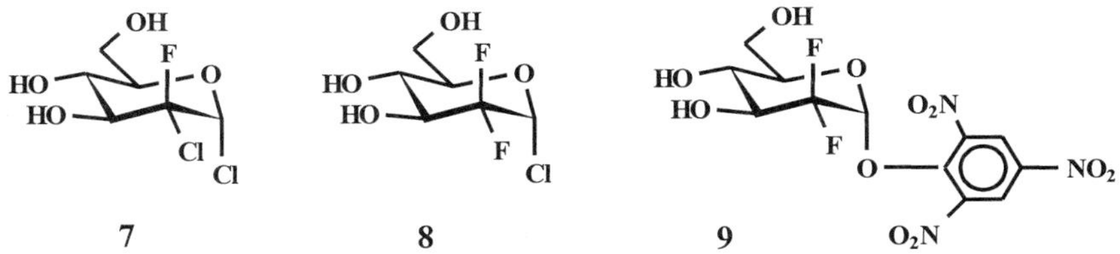

FIG. 3. Structures of 2,2-dihaloglycosides: 2-chloro-2-deoxy-2-fluoro-α-D-glucopyranosyl chloride (**7**), 2-deoxy-2,2-difluoro-α-D-*arabino*-hexopyranosyl chloride (**8**), and 2,4,6-trinitrophenyl-2-deoxy-2,2-difluoro-α-D-*arabino*-hexopyranoside (**9**).

iii. 2,2-Dihaloglycosides. In parallel with the evolution of the 5-fluoro strategy to overcome the limitations of the 2-deoxy-2-fluoro methodology, another set of mechanism-based inactivators, the 2,2-dihaloglycosides, was developed. To overcome the problem of relatively rapid turnover of the 2-fluoroglycosyl-enzyme intermediate a second halogen was substituted at C-2 (Fig. 3, compounds **7–9**). Since the presence of the second halogen would also slow the formation of the intermediate, the introduction of an extremely good leaving group, such as chloride or trinitrophenol, was necessary.

A series of 2,2-dihaloglycosides employing a chloride leaving group was synthesized and these were evaluated as potential inactivators of yeast α-glucosidase and jack bean α-mannosidase. Both 2C12FαGluCl (**7**) and 2,2FαAraCl (**8**) inactivated yeast α-glucosidase, but neither inactivated jack bean α-mannosidase, despite the similarities in their mechanisms.[33] Presumably this is because the mannosidase does not accommodate the relatively bulky equatorial fluorine substituent (versus hydrogen), whereas yeast α-glucosidase is known to be able to bind mannosides.[34] The competitive inhibitor, deoxynojirimycin, protected α-glucosidase from inactivation by both **7** and **8**, thereby providing evidence for active site directed inactivation. After a prolonged period of incubation, no reactivation of yeast α-glucosidase activity was observed (over 10 days at 37° for **7** in the presence of buffer and over 3 days at 37° for **8** in the presence of buffer or with α-methyl glucoside as a possible transglycosylation acceptor), suggesting an extremely stable, long-lived glycosyl-enzyme intermediate in both cases.

Trinitrophenolate has also been used as the leaving group in a trio of difluoroglycosides. TNP2,2FαGlu (**9**) was found to act as an inactivator of yeast α-glucosidase.[35] This inactivation was determined to be active site-directed through protection experiments with 1-deoxynojirimycin.[35] A TNP galactoside (TNP2,2FαGal) was used to inactivate the main α-galactosidase from

---

[33] J. D. McCarter, Ph.D. Thesis, University of British Columbia, 1995.
[34] S. Howard and S. G. Withers, *Biochemistry* **37**, 3858 (1998).
[35] C. Braun, G. D. Brayer, and S. G. Withers, *J. Biol. Chem.* **270**, 26778 (1995).

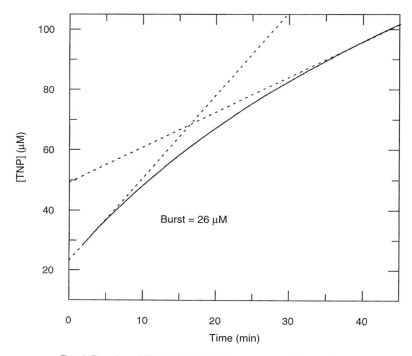

FIG. 4. Reaction of TNP2,2FαG2 with human pancreatic α-amylase.

P. chrysosporium and identify the catalytic nucleophile as Asp-130.[36] The corresponding trinitrophenyl difluoromaltoside derivative (TNP2,2FαG2) was found to inactivate human pancreatic α-amylase.[24] Evidence that an inhibitor is truly mechanism-based can be obtained by monitoring the stoichiometry of the reaction of the enzyme with inhibitor. Measurement of the mass increase in the protein provides one such measure, while monitoring of the aglycone release provides another. Reaction of TNP2,2FαG2 with a relatively large amount of α-amylase was monitored continuously using a spectrophotometer to follow the release of TNP. As seen in Fig. 4, the time course for the release of TNP was biphasic. The first phase involved a relatively rapid initial release of TNP, followed by a slower second phase, which arises from steady-state turnover of the substrate. By extrapolation of the linear phase back to time zero a burst magnitude of 26 μM was obtained. This corresponds well with the amount of α-amylase present (23 μM), thus proving that the inactivation of the enzyme is accompanied by the release of a single equivalent of aglycone (Fig. 4).

[36] D. O. Hart, S. He, C. J. Chany, S. G. Withers, P. F. Sims, M. L. Sinnott, and H. Brumer, *Biochemistry* **39**, 9826 (2000).

iv. 4-Deoxyglycosides. Another strategy that has proven useful for identifying the catalytic nucleophiles of α-transglycosylases takes advantage of the fact that transferases only transfer their sugar to the hydroxyl group of the acceptor molecule, and not to water. Thus, if the transferase of interest catalyzes the elongation of an oligosaccharide chain through the formation of an α-1,4 link, then the 4-hydroxyl group on the acceptor sugar acts as the nucleophile. Therefore, removal of the 4-hydroxyl group on the second substrate molecule will prohibit transfer and, when used in conjunction with a glycosyl donor, should result in accumulation of the intermediate, thereby allowing subsequent identification of the attachment site. The most suitable substitutions for the nonreducing 4-hydroxyl group are hydrogen (deoxy sugar) or fluorine (deoxyfluoro sugar).

This approach was adopted to trap the intermediate formed with glycogen debranching enzyme, an α-retaining glycosyltransferase.[37] This enzyme normally functions to remove branch points in glycogen through a two-step process, one of which involves the transglycosylation of a maltotriosyl moiety from the "branch chain" to the "main chain." Maltotriosyl fluoride was shown to serve as a substrate, undergoing self-condensation, presumably via the same maltotriosyl-enzyme intermediate formed in the normal mechanism. When a modified substrate, 4-deoxy-α-maltotriosyl fluoride (4DαG3F, **10,** Fig. 5), was substituted, formation of the intermediate occurred rapidly, driven by the good fluoride leaving group (see Lindhorst *et al.*[38] for the synthesis). However, in the absence of an acceptor group 4-hydroxyl on the second substrate molecule, the intermediate accumulated, thereby allowing subsequent identification of the attachment site as Asp-549 (Fig. 5).

Trapping Intermediates and Identifying Nucleophiles on Cyclodextrin Glucanotransferase

This approach was extended to cyclodextrin glucanotransferase (CGTase), an α-glycosyltransferase which works to degrade starch into a mixture of both linear and cyclic oligosaccharides through a variety of disproportionation, linearization, cyclization, and hydrolysis reactions. In each case, the reaction occurs with retention of configuration and involves three key active-site carboxylic acids, one functioning as a nucleophile (Asp-229), one as an acid–base catalyst (Glu-257), and a third (Asp-328) which appears to modulate active site pK_a values. As was the case with the glycogen debranching enzyme, CGTase catalyzes a transfer reaction with the synthetic substrate α-maltotriosyl fluoride, producing maltooligosaccharyl fluorides, hence cyclodextrins. This activity was easily assayed by following the enzyme-catalyzed release of fluoride in a time-dependent fashion

---

[37] C. Braun, T. Lindhorst, N. B. Madsen, and S. G. Withers, *Biochemistry* **35,** 5458 (1996).
[38] T. K. Lindhorst, C. Braun, and S. G. Withers, *Carbohydr. Res.* **268,** 93 (1995).

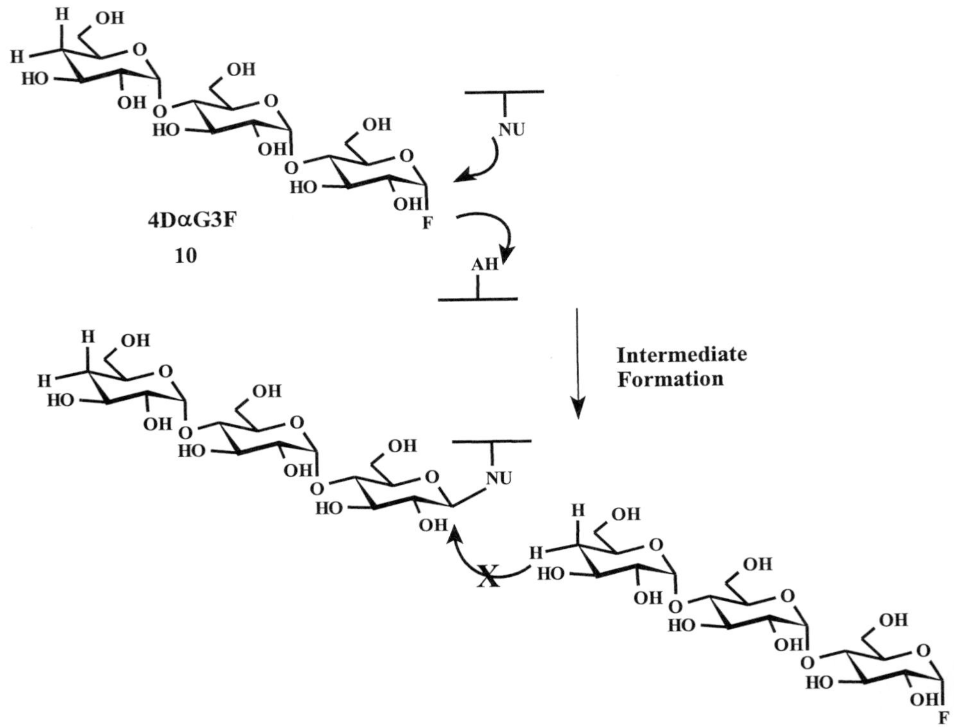

FIG. 5. General mechanism for the accumulation of a glycosyl-enzyme intermediate on a glycosyl-transferase using an incompetent substrate, 4DαG3F (**10**).

using an ion-selective fluoride electrode.[39] 4-Deoxy-α-maltotriosyl fluoride should therefore act as an incompetent substrate for CGTase through the trapping of a covalent glycosyl-enzyme intermediate, thereby allowing identification of the enzymic nucleophile.

Kinetic Analysis: Determination of Kinetic Parameters for α-Glycosyl Fluorides

The reactions of CGTase (wild type and mutants) with the glycosyl fluoride analogs have been monitored by following the release of fluoride ion using an Orion ion-selective fluoride electrode interfaced to a pH/ion-selective meter (Fisher Scientific, Toronto, Ontario, Canada). Data are collected using the program

[39] R. Mosi, S. He, J. Uitdehaag, B. Dijkstra, and S. G. Withers, *Biochemistry* **36**, 9927 (1997).

TABLE I
KINETIC PARAMETERS FOR REACTION OF *B. circulans* WILD TYPE,
Glu257Gln, AND Glu257Ala CGTASE WITH α-GLYCOSYL FLUORIDES

Enzyme	Substrate	K_m (mM)	k_{cat} (s^{-1})	k_{cat}/K_m (mM^{-1}s^{-1})
Wild type	αG3F	2.5	275	111
Wild type	4DαG3F	0.073	2.0	27.5
Wild type	4DαG2F	1.3	8.8	6.76
Glu257Gln	αG3F	0.37	1.0	2.70
Glu257Ala	αG3F	1.6	0.18	0.11
Glu257Gln	4DαG3F	0.027	0.6	22.2
Glu257Gln	4DαG2F	0.33	1.1	3.33

Terminal and values of K_m and k_{cat} are calculated from the experimental rate versus substrate concentration data, using the program GraFit.[40] In a typical experiment, an appropriate concentration of glycosyl fluoride in 50 or 100 mM citrate buffer (pH 6.0) is warmed to 30° for 10 min (total volume of 240 μl) and the reaction initiated by adding an appropriately diluted aliquot of enzyme (typically 10 μl) in citrate buffer [final concentrations: wild type (0.25 μg/ml), Glu257Gln (24 μg/ml), Asp229Asn (50 μg/ml), Glu257Ala (70 μg/ml), Asp229Ala (177 μg/ml)]. The reaction is typically followed for approximately 10 min or until 10% of the starting material is consumed. Rates are determined at 6–8 substrate concentrations ranging from 0.2 to 5 times the estimated K_m value.

α-Maltotriosyl fluoride has proved to be a good substrate for CGTase as shown in Table I ($k_{cat} = 275$ s^{-1} and $K_m = 2.5$ mM). This compares very well to a k_{cat} of 350 s^{-1} for the natural substrate starch.[41] Thus, although CGTase cannot normally use a trisaccharide as a substrate, the presence of the fluoride at the anomeric position as a good leaving group is sufficient to replace the transition state stabilization ordinarily provided by interactions between the enzyme and the oligosaccharide leaving group. As shown in Table I, 4DαG3F does indeed act as a substrate, albeit slow, indicating that the enzyme intermediate still does turn over to a significant extent. Kinetic analysis gave a K_m value of 0.073 mM, which is approximately 35 times lower than that for αG3F. Such a decrease in K_m probably reflects accumulation of a glycosyl-enzyme intermediate, resulting from a change in the relative ratios of the rate constants for the two steps of the reaction. The low k_{cat} value (reduced over 100-fold from wild type) for 4DαG3F reflects simply the hydrolytic process, transfer to sugar having been eliminated. Unfortunately, the relatively short hydrolytic half-life, $t_{1/2} = 0.5$ sec, precludes

---

[40] R. J. Leatherbarrow, Gra-Fit, Version 2.0, Erithacus Software Ltd., Staines, U.K., 1990.
[41] R. M. Knegtel, B. Strokopytov, D. Penninga, O. G. Faber, H. J. Rozeboom, K. H. Kalk, L. Dijkhuizen, and B. W. Dijkstra, *J. Biol. Chem.* **270**, 29256 (1995).

any investigation by mass spectrometry, necessitating some additional measures to increase its lifetime.

A solution to the problem of the relatively fast hydrolysis of the glycosyl-enzyme intermediate lies in a mutant of the presumed acid–base catalyst, Glu-257. Two mutants at position 257, Glu257Gln and Glu257Ala, have been previously generated by our collaborators at the University of Groningen.[41] In earlier work with cellulases and β-glucosidase,[42,43] it has been shown that by removing the acid–base catalyst and using a substrate such as a glycosyl fluoride with a good leaving group, the intermediate can be readily accumulated. This occurs because a good leaving group does not need acid catalysis; thus, despite the absence of an acid catalyst, the first step is still fast. However, because that side chain also serves as the general base catalyst for the second step, deglycosylation is slowed, allowing accumulation of the intermediate. Thus we argue that combined use of a mutant of CGTase in which the acid–base catalyst has been removed (Glu257Gln or Glu257Ala) with an incompetent substrate such as 4DαG3F should provide a means to accumulate an intermediate of longer lifetime.

As a first step, turnover of the substrate, αG3F, by Glu257Gln has been investigated. Using the fluoride electrode to monitor the time-dependent release of fluoride ion, αG3F has been found to be a substrate for Glu257Gln CGTase as shown in Table I. The k_{cat} is almost 300-fold lower than for wild-type enzyme and the K_m is reduced 7-fold; thus k_{cat}/K_m is reduced 40-fold. Similarly the K_m value for αG3F with Glu257Ala has been found to be almost 2-fold lower than that with wild-type CGTase, these general reductions in K_m likely reflecting the accumulation of an intermediate in each case. The reductions in k_{cat} relative to wild type of approximately 7500-fold for Glu257Ala and 1500-fold for Glu257Gln are comparable to those seen for similar mutations in β-glycosidases. For example, a 70- to 100-fold reduction in activity is noted upon mutating the acid–base catalyst (Glu172) in *B. circulans* β-xylanase to Gln,[44] while k_{cat} reductions in the range of 200- to 2000-fold have been measured for similar mutations in other β-retaining glycosidases.[43,44]

4DαG3F and 4DαG2F as Incompetent Substrates for Glu257Gln CGTase

Kinetic parameters for the reaction of the incompetent substrates 4DαG3F and 4DαG2F with the Glu257Gln mutant are presented in Table I. Mutation of the acid–base catalyst, Glu-257, to Gln results in reductions of 2.7-fold and 3.9-fold in K_m values for 4DαG3F and 4DαG2F, respectively, the value for 4DαG3F with Glu257Gln CGTase being 10-fold lower than that for αG3F. Such reduced K_m values are again most likely due to the accumulation of a glycosyl-enzyme intermediate. Values of k_{cat} for 4DαG3F and 4DαG2F are only 3.3-fold and 8-fold

---

[42] A. MacLeod, D. Tull, K. Rupitz, R. A. J. Warren, and S. G. Withers, *Biochemistry* **35**, 13165 (1996).

[43] Q. Wang, D. Trimbur, R. Graham, R. A. J. Warren, and S. G. Withers, *Biochemistry* **34**, 14554 (1995).

[44] S. L. Lawson, W. W. Wakarchuk, and S. G. Withers, *Biochemistry* **36**, 2257 (1997).

lower than those for the wild-type enzyme, indicating a surprisingly small apparent contribution of Glu-257 to the general base-catalyzed hydrolysis pathway. Interestingly, the k_{cat} values for αG3F and 4DαG3F with Glu257Gln are very similar, consistent with the idea that the reaction being monitored is hydrolysis and not transglycosylation in both cases. HPLC analysis of the products of the reaction of Glu257Gln CGTase with 4DαG3F and 4DαG2F showed 4DG3 and 4DG2 as the major products in each case, even after overnight incubation (results not shown). This is in contrast to the mixture of disproportionation products observed with wild-type CGTase and 4DαG3F, and consistent with the important role of the general acid catalyst in cleavage of the interglycosidic bond.

Mass Spectrometric Evidence for Covalent Enzyme–Substrate Intermediate

The k_{cat} value of 0.56 s^{-1} obtained corresponds to a half-life for the intermediate of approximately 1.1 sec, giving hope that the glycosyl enzyme species could be observed directly by electrospray ionization mass spectrometry. Mass spectrometric analyses of the protein samples were therefore carried out using a Sciex API-300 mass spectrometer interfaced with a Michrom UMA HPLC system (Michrom Bioresources, Inc., Pleasanton, CA). Intact CGTase [10–20 μg, wild type, Glu257Gln (unlabeled or labeled)] was introduced into the mass spectrometer through a microbore PLRP column (1 × 50 mm) and eluted with a gradient of 20–100% solvent B at a flow rate of 50 μl/min over 5 min (solvent A: 0.06% trifluoroacetic acid, 2% acetonitrile in water; solvent B: 0.05% trifluoroacetic acid, 90% acetonitrile in water). The MS was scanned over a range of 400–2000 Da with a step size of 0.5 Da and a dwell time of 1 ms. Upon incubation of Glu257Gln CGTase (Fig. 6A) with an excess of 4DαG3F an increase in molecular weight equivalent to a 4DG3 moiety on the intact protein [(74,513 + 471) ± 5 Da] was observed by ES-MS (Fig. 6B). The equivalent disaccharide, 4DαG2F, also acted as an incompetent substrate and similar ES-MS analysis of Glu257Gln CGTase incubated with 4DαG2F revealed a mass increase of [(74,513 + 309) ± 5 Da] consistent with the formation of a 4DG2–enzyme complex (Fig. 6C). Interestingly turnover of αG3F by Glu257Gln was slow enough that accumulations of both a G3 moiety on the intact protein {(74,513 + 487) ± 5 Da}, and a G6 moiety {(74,513 + 973) ± 5 Da} were observed by ES-MS, clearly demonstrating the transglycosylation processes occurring (Fig. 6D).

Identification of Peptide That Contains Nucleophile

Identification of the attachment site of the 4-deoxymaltotriosyl label was achieved by labeling, peptic digestion, and neutral loss tandem mass spectrometry.[45]

---

[45] S. Miao, L. Ziser, R. Aebersold, and S. G. Withers, *Biochemistry* **33,** 7027 (1994).

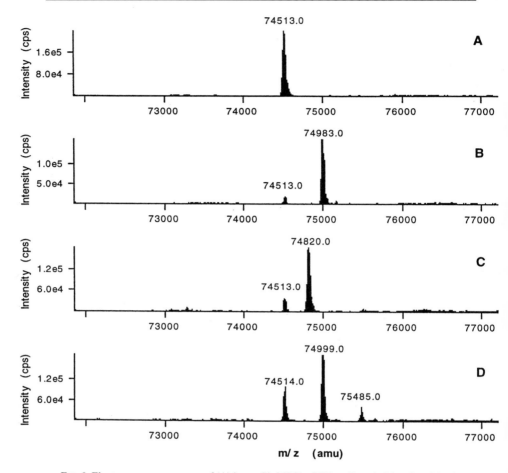

FIG. 6. Electrospray mass spectra of (A) intact Glu257Gln CGTase (3 mg/ml) incubated for 5 min with (B) 4DαG3F (62.5 mM), (C) 4DαG2F (62.5 mM), and (D) αG3F (62.5 mM). Reprinted with permission from R. Mosi, S. He, J. Uitdehaag, B. Dijkstra, and S. G. Withers, *Biochemistry* **36,** 9927 (1997). Copyright © 1997 American Chemical Society.

In this mass spectrometry technique, ions are subjected to limited fragmentation by collisions with an inert gas (nitrogen) in a collision cell located between the two mass analyzers Q1 and Q3. Because the ester linkage between the inhibitor and the peptide is one of the more labile linkages present, facile homolytic cleavage of this bond occurs, resulting in the loss of the sugar as a neutral species, with the peptide retaining its original charge. The mass spectrometer is then set so that Q1 and Q3 are scanned in a linked manner such that only those ions that lose the mass of the lost sugar moiety can pass through both quadrupoles and be detected.

Labeling of CGTase was performed by incubating Glu257Gln (10 µl, 6 mg/ml) with an excess of 4DαG3F (5 µl, 125 mM) for 10 min in 50 mM citrate buffer, pH 6.0. Proteolysis was achieved by first lowering the pH through the addition of 15 µl of 50 mM sodium phosphate buffer (pH 2) followed by immediate addition of 10 µl of pepsin (0.4 mg/ml in pH 2 buffer) and the mixture incubated at room temperature for 1 hr. ES-MS and SDS–PAGE analysis of the proteolytic digests confirmed that CGTase was completely digested under these conditions. Identical conditions were used for the labeling of Glu257Gln with 4DαG2F.

The peptides were analyzed by loading a 10 µl sample of the pepsin digest (4.5 mg/ml) onto a C_{18} column (Reliasil, 1 × 150 mm, Column Engineering, Inc., Ontario, CA) and eluting at a flow rate of 50 µl/min with a gradient of 0–60% B over 60 min. The proteolytic mixture was first examined in LC/MS mode and then in neutral loss mode. In the single quadrupole (normal LC/MS) mode, the MS conditions were as follows: the mass analyzer was scanned over the range of 300–2400 Da with a step size of 0.5 Da, a dwell time of 1 ms, an ion source voltage (ISV) of 4.8 kV, and an orifice energy (OR) of 50 V. The neutral loss spectra were obtained in the triple quadrupole mode. A scan range of 400 to 2200 amu was used with a step size of 0.5 amu and a dwell time of 1 ms. Other parameters are as follows: ISV = 5 kV, OR = 45 V.

As described above, the resulting peptide mixtures were separated by reversed phase HPLC using the ES-MS as a detector, revealing a large number of peptides (Fig. 7A). The peptide with the 4DG3 label was located by MS/MS analysis using the neutral loss experiment. The spectrometer was first scanned in the neutral loss mode searching for a peptide that loses a neutral species of 471 Da, which corresponds to the mass of the 4-deoxymaltotriosyl label. However, no significant peaks were detected. When the spectrometer was scanned for the mass loss of m/z 235.5, corresponding to the loss of a sugar from a doubly charged peptide, two strong peaks were observed: labeled peptides 1 and 2 (Fig. 7B). These two peaks, having retention times of 35.75 and 37.5 min and m/z values of 1075.5 and 1140.5 Da, respectively, were not seen in the control digest (Fig. 7C) and presumably represent two different 4-deoxymaltotriosyl-labeled peptides. Because the doubly charged labeled peptides have m/z values of 1075.5 Da (Fig. 7D) and 1140.5 Da (Fig. 7E), the singly charged unlabeled peptides must have masses of 1679 [(2 × 1075.5) − 2 − 471 + 1 H] and 1809 [(2 × 1140.5) − 2 − 471 + 1 H]. A computer generated search of all possible peptides in the Glu257Gln CGTase mutant yielded 18 possible peptides with molecular weight 1679 ± 2 Da and 34 possible peptides with molecular weight of 1809 ± 2 Da. Of these, 7 sets of peptides contained overlapping amino acid sequences. It is assumed that the two peptides arise from different modes of proteolysis of the same labeled protein since only a single oligosaccharide was initially attached. When CGTase was

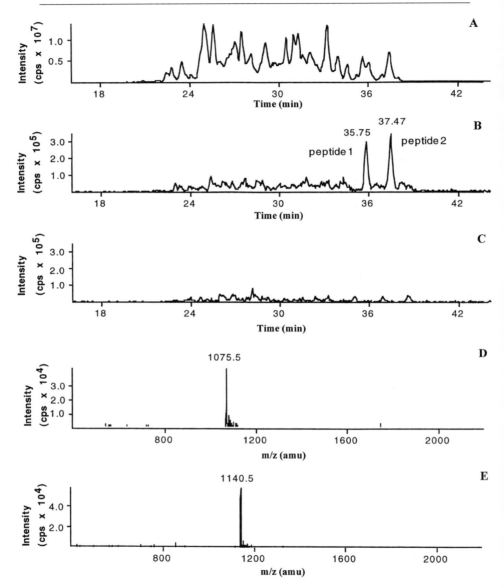

FIG. 7. Electrospray mass spectrometry experiments on Glu257Gln CGTase proteolytic digests: (A) labeled with 4DαG3F, total ion chromatogram in normal MS mode; (B) labeled with 4DαG3F, TIC in neutral loss mode; (C) unlabeled, in neutral loss mode; (D) mass spectrum of peptide 1 in (B); (E) mass spectrum of peptide 2 in (B).

incubated with 4DαG2F, digested, and analyzed by neutral loss, the same two peptides were found to be labeled (data not shown). Because of the large size of the labeled peptides, their definitive identification by MS/MS was precluded since good fragmentation was not achieved.

Identification of Catalytic Nucleophile

As shown in Fig. 7, the peptides were readily separable by HPLC and so, after purification, their sequences were determined by Edman degradation. The peptic digest (120 µl of 1.5 mg/ml) of the 4DG3-labeled CGTase was separated by HPLC using a Deltapak 300 Å C_{18} column (3.9 × 150 mm, mean particle size 15 µm) from Waters eluted with a gradient of 0–60% B over 60 min at 700 µl/min. A 13 : 1 postcolumn splitter allowed the introduction into the mass spectrometer of a portion of the peptides to be checked for purity and the fractions containing the 4DG3-labeled peptide were collected for sequencing. Prior to solid phase sequence analysis, peptides (2–5 µg in 40% CH_3CN/0.5% TFA) were coupled to arylamine-functionalized polyvinylidene fluoride membranes (Sequelon AA, Milligen/Bioresearch) using *N*-ethyl-*N'*-(3-dimethylaminopropyl) carbodiimide. Peptide sequences were determined by solid-phase Edman degradation on a Milligen/Bioresearch model 6600 protein sequenator using standard protocols with on-line HPLC analysis of the resulting phenylthiohydantoins. The sequences of ^{229}DAVKHMPFGWQKSF^{242} for peptide 1 and ^{229}DAVKHMPFGWQKSFM^{243} for peptide 2 were obtained, clearly indicating, as expected, that they are overlapping peptides. The catalytic nucleophile must be an amino acid which is common to both peptides and by analogy with other glycosidases and transglycosylases either a glutamate or aspartate.[10,13,18] The only carboxylic amino acid contained within these sequences is Asp-229 suggesting that this is indeed the catalytic nucleophile. In fact, Asp-229 is completely conserved among members of family 13.

Crystallographic Evidence for β-Linked Covalent Glycosyl-Enzyme Intermediate for CGTase

The mechanism of reaction of CGTase at the catalytic site was further probed by obtaining a crystal structure of Glu257Gln reacted with 4DαG3F.[46] Crystals of *B. circulans* 251 Glu257Gln CGTase were grown in the presence of 5% maltose as reported previously.[47] Maltose is essential for the formation of the crystals. However, as maltose would act as a good acceptor for the elongation of 4DαG3F, it was necessary to first wash the crystals repeatedly to remove all traces of maltose before exposing them to 4DαG3F. However, washing with buffer containing no

---

[46] J. C. M. Uitdehaag, R. M. Mosi, K. H. Kalk, B. A. van der Veen, L. Dijkhuizen, S. G. Withers, and B. W. Dijkstra, *Nat. Struct. Biol.* **6,** 432 (1999).

[47] C. L. Lawson, R. van Montfort, B. Strokopytov, H. J. Rozeboom, K. H. Kalk, G. E. de Vries, D. Penninga, L. Dijkhuizen, and B. W. Dijkstra, *J. Mol. Biol.* **236,** 590 (1994).

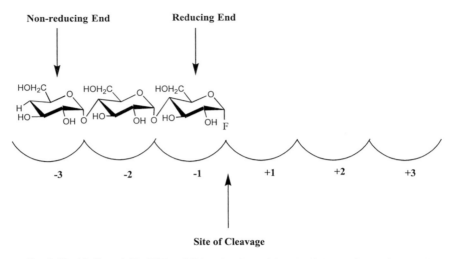

FIG. 8. The binding of 4DαG3F to CGTase showing subsites labeled according to G. J. Davies, *Biochem. J.* **321**, 557 (1997), with the site of cleavage positioned between the −1 and +1 subsites.

maltose resulted in cracking of the crystals. Therefore, the crystals were washed with 1% 4-deoxymaltose (4DG2). The crystals retained the structural stability gained from having a disaccharide moiety bound, yet the potential for elongation was eliminated. Finally, the crystal was soaked for a brief period in 125 mM 4DαG3F and frozen to 100 K for data collection.

Upon refinement to 1.8 Å, a 4DG3 moiety was observed to be bound in the −1, −2, and −3 subsites (we are using the nomenclature proposed by Davies where the site of cleavage is positioned between the −1 and +1 subsites and the nonreducing sugar is located in the −3 subsite; see Fig. 8).[48] The nucleophile, Asp-229, was observed to make a β-covalent bond to the C-1 anomeric carbon (Fig. 9). This is the first 3D structure of such a covalent intermediate for any α-1,4-retaining glycosyltransferase or α-glycosidase.[47] In addition, it is the first time that both donor and acceptor oligosaccharides have been crystallographically observed bound at the active site of CGTase.

The structure provided some particularly interesting insights into the structure and mechanism of α-transglycosylases and hydrolases, principal among which were the following: (a) The proximal sugar, located in the −1 site, is found in an undistorted $^4C_1$ conformation, there being no evidence for distortion of the type predicted by stereoelectronic theory.[49] (b) The carbonyl oxygen of the catalytic nucleophile is held close (2.7 Å) to the ring oxygen of the covalently bound sugar.

[48] G. J. Davies, *Biochem. J.* **321**, 557 (1997).
[49] P. Deslongchamps, "Stereoelectronic Effects in Organic Chemistry." Pergamon Press, Oxford, 1983.

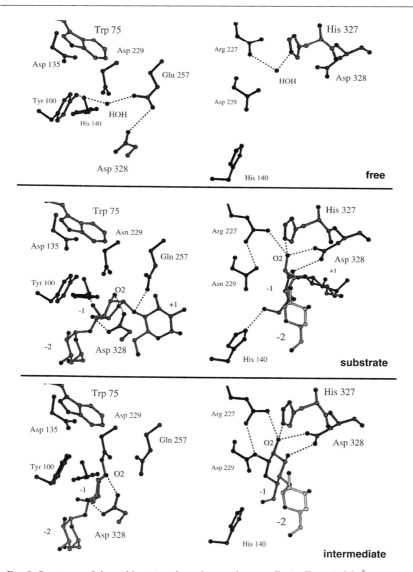

FIG. 9. Structures of the stable states along the reaction coordinate. Free: A 2.2 Å structure of uncomplexed free *B. circulans* strain 251 CGTase at 120 K and pH 7.6 [R. M. Knegtel, B. Strokopytov, D. Penninga, O. G. Faber, H. J. Rozeboom, K. H. Kalk, L. Dijkhuizen, and B. W. Dijkstra, *J. Biol. Chem.* **270,** 29256 (1995)]. Substrate: The twist in the substrate chair is seen on the left-hand side. If undistorted, it would either clash with Asp-328 or not form a hydrogen bond with His-140. Intermediate: The intermediate chair conformation is presented on the left-hand side. Reprinted with permission from J. C. M. Uitdehaag, R. M. Mosi, K. H. Kalk, B. A. van der Veen, L. Dijkhuizen, S. G. Withers, and B. W. Dijkstra, *Nat. Struct. Biol.* **6,** 432 (1999). Copyright © 1999, Nature America, Inc.

Conceivably this arrangement provides some destabilization of this intermediate, which is relieved on bond cleavage and converted to a potentially stabilizing ($O^{\delta+}-O^{\delta-}$) interaction at the transition state. (c) The fully conserved third carboxyl group at the active site of family 13 enzymes (Asp-328 in this case) is seen to form hydrogen bonds across O-2 and O-3 of the -1 site sugar, possibly providing specific stabilization of the transition state conformation of the sugar.

In conclusion, confirmation of the identity of the nucleophile of *B. circulans* 251 CGTase as Asp-229 was achieved using a unique combination of site-directed mutagenesis, an incompetent substrate, and mass spectrometry.[39] This work has also provided the first concrete evidence for a covalent glycosyl-enzyme intermediate for *B. circulans* 251 CGTase and the first 3D structure of such an intermediate for any α-glycosidase or α-glycosyltransferase. The use of 4DαG3F in crystallographic studies of the covalent intermediate bound to CGTase provided valuable information toward understanding the mechanism of reaction of CGTase at the catalytic site. Not only was the existence of a covalent intermediate confirmed but information about the conformation of the sugar in the covalently bound state was obtained. In addition, the nature of the binding of the acceptor sugar was examined and should provide an improved understanding of the processes occurring at the reaction site.

[5] Trapping Covalent Intermediates on β-Glycosidases

By JACQUELINE WICKI, DAVID R. ROSE, and STEPHEN G. WITHERS

Introduction

Glycosidases encompass a group of hydrolytic enzymes that cleave the glycosidic bond with one of two stereochemical outcomes: net retention or net inversion of anomeric configuration. Retaining glycosidases function through a mechanism involving the formation of a covalent glycosyl-enzyme intermediate. This chapter will focus on methods to trap the covalent intermediate of retaining β-glycosidases formed during catalysis. Two strategies that will be discussed involve the use of modified (fluorinated) substrates and mutation of key residues in the enzyme active site to trap the covalent intermediate. Studies featuring the trapped covalent intermediate formed on the exoglycanase from *Cellulomonas fimi* will be provided as an illustration applicable to most other retaining β-glycosidases.

More than 2000 different glycoside hydrolases have been classified into over 80 families based on amino acid sequence similarities, as originally conceived by Henrissat and currently available on an updated list (afmb.cnrs-mrs.fr/CAZY/).[1–4]

Some families have also been grouped into clans, of which there are currently 10, on the basis of overall protein fold. In addition, since the three-dimensional structures have been determined for a number of these families,[4-6] comparisons may be made to other enzymes within a group to help identify amino acid residues responsible for specificity and catalysis. Such a classification system is expected to expose evolutionary relationships and reveal common structural features shared by enzymes within the same family or clan, as well as provide details of the mechanism of action.

Hydrolysis of the glycosidic bond can occur via two distinct mechanisms, originally proposed by Koshland,[7] resulting in two possible stereochemical outcomes: inversion or retention of anomeric configuration. Both types of reaction proceed via oxocarbenium ion-like transition states and, in both cases, a pair of carboxylic acids function as catalytic residues.[8,9]

Inverting enzymes hydrolyze the glycosidic bond with net inversion of anomeric configuration and retaining enzymes do so with net retention. Inverting enzymes are believed to function by a single-step mechanism involving a general acid–base-catalyzed process in which a water molecule effects a direct displacement at the anomeric center. The two carboxyl groups involved in the single displacement reaction are typically spaced some 9–10 Å apart (O to O) in order to allow water to act as the nucleophile.[10,11]

Retaining glycosidases function through a two-step, double-displacement mechanism involving a covalent glycosyl-enzyme intermediate. A detailed mechanism for retaining β-glycosidases, which are the focus of this chapter, is shown in Fig. 1. During the glycosylation step, a nucleophilic carboxylate in the active site attacks the substrate at the anomeric carbon, displacing the leaving group in a general acid-catalyzed process, resulting in the formation of a covalently linked glycosyl-enzyme intermediate. During the deglycosylation step, general base-catalyzed attack of water at the anomeric center of the glycosyl-enzyme intermediate results in the formation of product. In contrast to inverting glycosidases, the carboxyl groups are spaced approximately 5.5 Å apart.

---

[1] B. Henrissat, *Biochem. J.* **280**, 309 (1991).
[2] B. Henrissat and A. Bairoch, *Biochem. J.* **293**, 781 (1993).
[3] B. Henrissat and A. Bairoch, *Biochem. J.* **316**, 695 (1996).
[4] B. Henrissat and G. Davies, *Curr. Opin. Struct. Biol.* **7**, 637 (1997).
[5] G. Davies, M. L. Sinnott, and S. G. Withers, *in* "Comprehensive Biological Catalysis" (M. L. Sinnott, ed.), p. 119. Academic Press, New York, 1998.
[6] G. Davies and B. Henrissat, *Structure* **3**, 853 (1995).
[7] D. E. Koshland, *Biol. Rev.* **28**, 416 (1953).
[8] D. L. Zechel and S. G. Withers, *Acc. Chem. Res.* **33**, 11 (2000).
[9] M. L. Sinnott, *Chem. Rev.* **90**, 1171 (1990).
[10] J. D. McCarter and S. G. Withers, *Curr. Opin. Struct. Biol.* **4**, 885 (1994).
[11] D. L. Zechel and S. G. Withers, *in* "Comprehensive Natural Products Chemistry" (C. D. Poulter, ed.), p. 279. Pergamon, Amsterdam, 1999.

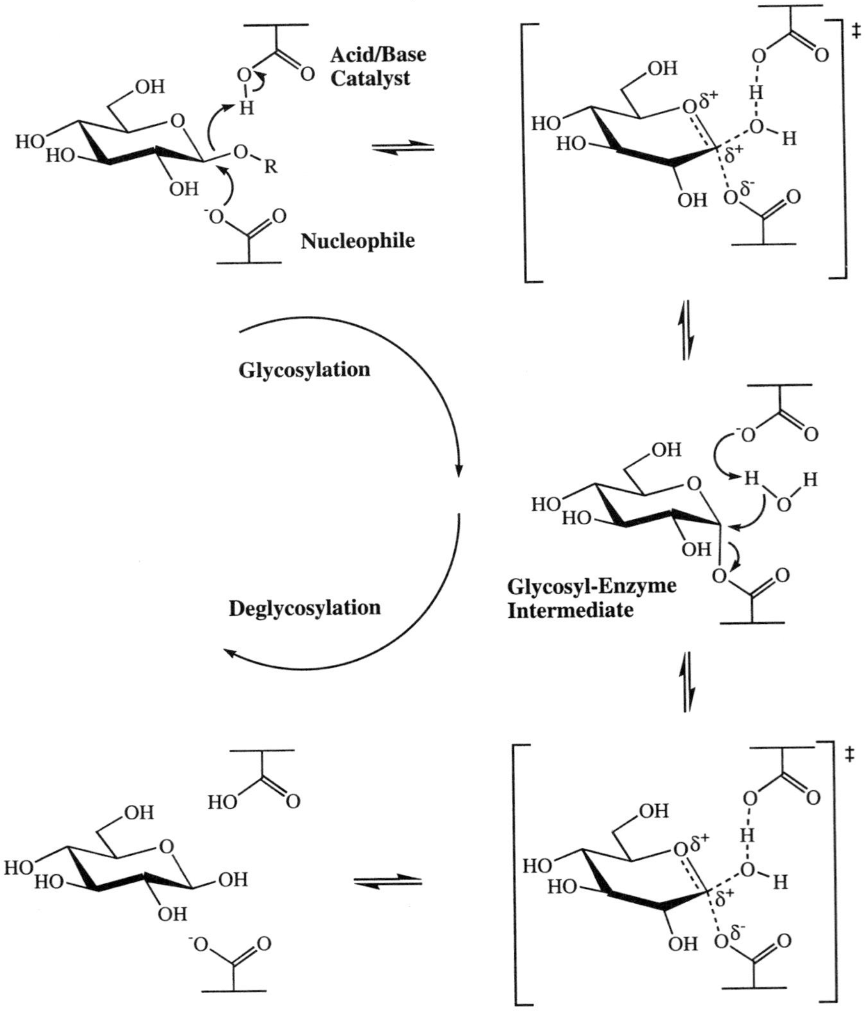

FIG. 1. Double-displacement mechanism of a retaining β-glycosidase.

Evidence for oxocarbenium ion character in the transition states for both inverting and retaining glycosidases is provided by inhibition studies with transition state analogs, as well as secondary deuterium kinetic isotope effects and Brønsted relationships.[12–14] Extensive evidence for a covalent glycosyl-enzyme intermediate exists in support of Koshland's proposed retaining mechanism including actual trapping of such an intermediate, as well as α-secondary deuterium kinetic isotope effects observed with several retaining β-glycosidases. For example, isotope effects ranging from $k_H/k_D = 1.25$ for *Escherichia coli* β-galactosidase[15] to $k_H/k_D = 1.11$ for both an *Agrobacterium* sp. β-glucosidase[13] and for *Cellulomonas fimi* exoglycanase[14] have been measured for the deglycosylation step of the double-displacement reaction. Kinetic isotope effects greater than 1 in the deglycosylation step strongly support a covalent (sp^3) intermediate flanked by sp^2-hybridized transition states, while an inverse ($k_H/k_D < 1$) measurement would suggest that the principal intermediate formed is an ion pair. Direct evidence for the covalent nature of this intermediate comes from the detection by ^{19}F NMR and electrospray mass spectrometry of stable intermediates formed from fluorinated carbohydrates such as 2-deoxy-2-fluoroglycosides. A covalent glycosyl-enzyme intermediate has also been observed crystallographically in both α- and β-glycosidases/transglycosylases.[16–19]

This chapter will focus on trapping intermediates on β-glycosidases, with an emphasis on studies of the *Cellulomonas fimi* exoglycanase (Cex). Nonetheless, most of the methods elaborated should be applicable to most β-glycosidases.

Approaches to Trapping Covalent Glycosyl-Enzyme Intermediate

Labeling Active Site Nucleophile with Fluorinated Carbohydrates

As indicated earlier, hydrolysis by retaining glycosidases proceeds by way of a glycosyl-enzyme intermediate formed and hydrolyzed through transition states with substantial oxocarbenium ion character. One approach to trapping the covalent glycosyl-enzyme intermediate has involved the synthesis of modified substrates for which the transition state of the deglycosylation step of the reaction mechanism is greatly destabilized, resulting in accumulation of the covalent species. As discussed below, 2-deoxy-2-fluoroglycosides and glycosyl fluorides have proved to be highly successful reagents that operate on this principle and thus have been used to identify the active site nucleophiles in a range of β-glycosidases.[18,20–45] 5-Fluoroglycosyl

---

[12] T. D. Heightman and A. Vasella, *Angew. Chem., Int. Ed. Engl.* **38**, 750 (1999).
[13] J. B. Kempton and S. G. Withers, *Biochemistry* **31**, 9961 (1992).
[14] D. Tull and S. G. Withers, *Biochemistry* **33**, 6363 (1994).
[15] M. L. Sinnott and I. J. Souchard, *Biochem. J.* **133**, 89 (1973).
[16] J. C. M. Uitdehaag, R. Mosi, K. H. Kalk, B. A. van der Veen, L. Dijkhuizen, S. G. Withers, and B. W. Kijkstra, *Nat. Struct. Biol.* **6**, 432 (1999).
[17] V. Notenboom, C. Birsan, M. Nitz, D. R. Rose, R. A. J. Warren, and S. G. Withers, *Nat. Struct. Biol.* **5**, 812 (1998).
[18] A. White, D. Tull, K. Johns, S. G. Withers, and D. R. Rose, *Nat. Struct. Biol.* **3**, 149 (1996).
[19] W. P. Burmeister, S. Cottaz, P. Rollin, A. Vasella, and B. Henrissat, *J. Biol. Chem.* **275**, 39385 (2000).
[20] S. Dan, I. Marton, M. Dekel, B.-A. Bravdo, S. He, S. G. Withers, and O. Shoseyov, *J. Biol. Chem.* **275**, 4973 (2000).
[21] J. C. Gebler, R. Aebersold, and S. G. Withers, *J. Biol. Chem.* **267**, 11126 (1992).

fluorides have also been found to be of great utility in the labeling of both α- and β-glycosidases.[46–50] Furthermore, structural studies on the trapped fluoroglycosyl-enzyme intermediate have provided useful insights into catalysis and the roles of specific residues in the active site of the enzyme.[17–19,51–53]

a. 2-Deoxy-2-fluoroglycosides. 2-Deoxy-2-fluoroglycosides and glycosyl fluorides function as glycosidase inactivators via the formation and accumulation of a glycosyl-enzyme intermediate which thereby blocks the active site.[39,40,54] Because the enzymatic hydrolysis of retaining glycosidases proceeds through

---

[22] S. He and S. G. Withers, *J. Biol. Chem.* **272**, 24864 (1997).
[23] L. F. Mackenzie, G. S. Brooke, J. F. Cutfield, P. A. Sullivan, and S. G. Withers, *J. Biol. Chem.* **272**, 3161 (1997).
[24] L. F. Mackenzie, G. J. Davies, M. Schülein, and S. G. Withers, *Biochemistry* **36**, 5893 (1997).
[25] L. F. Mackenzie, G. Sulzenbacher, C. Divne, T. A. Jones, H. F. Wöldike, M. Schülein, S. G. Withers, and G. J. Davies, *Biochem. J.* **335**, 409 (1998).
[26] J. D. McCarter, D. L. Burgoyne, S. Miao, S. Zhang, J. W. Callahan, and S. G. Withers, *J. Biol. Chem.* **272**, 396 (1997).
[27] J. D. McCarter, W. Yeung, J. Chow, D. Dolphin, and S. G. Withers, *J. Am. Chem. Soc.* **119**, 5792 (1997).
[28] S. Miao, J. D. McCarter, M. E. Grace, G. A. Grabowski, R. Aebersold, and S. G. Withers, *J. Biol. Chem.* **269**, 10975 (1994).
[29] S. Miao, L. Ziser, R. Aebersold, and S. G. Withers, *Biochemistry* **33**, 7027 (1994).
[30] M. Namchuk, C. Braun, J. D. McCarter, and S. G. Withers, in "Biomedical Frontiers of Fluorine Chemistry" (I. Ojima, J. R. McCarthy, and J. T. Welch, eds.), p. 279. American Chemical Society, 1996.
[31] P. Staedtler, S. Hoenig, R. Frank, S. G. Withers, and W. Hengstenberg, *Eur. J. Biochem.* **232**, 658 (1995).
[32] D. Stoll, S. He, S. G. Withers, and R. A. J. Warren, *Biochem. J.* **351**, 833 (2000).
[33] I. P. Street, J. B. Kempton, and S. G. Withers, *Biochemistry* **31**, 9970 (1992).
[34] G. Sulzenbacher, L. F. Mackenzie, K. S. Wilson, S. G. Withers, C. Dupont, and G. J. Davies, *Biochemistry* **38**, 4826 (1999).
[35] D. Tull, S. G. Withers, N. R. Gilkes, D. G. Kilburn, R. A. J. Warren, and R. Aebersold, *J. Biol. Chem.* **266**, 15621 (1991).
[36] D. Tull, S. Miao, S. G. Withers, and R. Aebersold, *Anal. Biochem.* **224**, 509 (1995).
[37] D. J. Vocadlo, L. F. Mackenzie, S. He, G. J. Zeikus, and S. G. Withers, *Biochem. J.* **335**, 449 (1998).
[38] Q. Wang, D. Tull, A. Meinke, N. R. Gilkes, R. A. J. Warren, R. Aebersold, and S. G. Withers, *J. Biol. Chem.* **266**, 14096 (1993).
[39] S. G. Withers, I. P. Street, P. Bird, and D. H. Dolphin, *J. Am. Chem. Soc.* **109**, 7530 (1987).
[40] S. G. Withers, K. Rupitz, and I. P. Street, *J. Biol. Chem.* **263**, 7929 (1988).
[41] S. G. Withers and I. P. Street, *J. Am. Chem. Soc.* **110**, 8551 (1988).
[42] S. G. Withers, R. A. J. Warren, I. P. Street, K. Rupitz, J. B. Kempton, and R. Aebersold, *J. Am. Chem. Soc.* **112**, 5887 (1990).
[43] A. W. Wong, S. He, J. H. Grubb, W. S. Sly, and S. G. Withers, *J. Biol. Chem.* **273**, 34057 (1998).
[44] D. L. Zechel, S. He, C. Dupont, and S. G. Withers, *Biochem. J.* **336**, 139 (1998).
[45] L. Ziser, I. Setyawati, and S. G. Withers, *Carbohydr. Res.* **274**, 137 (1995).
[46] J. D. McCarter and S. G. Withers, *J. Am. Chem. Soc.* **118**, 241 (1996).
[47] D. J. Vocadlo, C. Mayer, S. He, and S. G. Withers, *Biochemistry* **39**, 117 (2000).
[48] S. Howard, S. He, and S. G. Withers, *J. Biol. Chem.* **273**, 2067 (1998).
[49] H. D. Ly, S. Howard, K. Shum, S. He, A. Zhu, and S. G. Withers, *Carbohydr. Res.* **329**, 539 (2000).

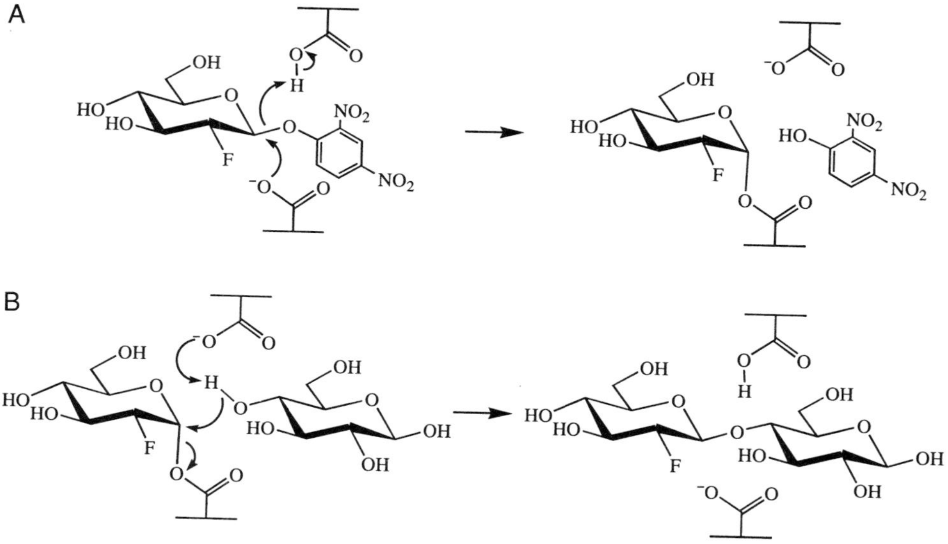

FIG. 2. Inactivation and reactivation by 2-deoxy-2-fluoroglycosides. (A) Inactivation by formation of the fluoroglycosyl-enzyme intermediate. (B) Reactivation by transglycosylation of the fluoroglycosyl-enzyme intermediate.

oxocarbenium ion-like transition states, the substitution of an electronegative fluorine atom for a hydroxyl group at C-2 adjacent to the reaction center should inductively destabilize both transition states, resulting in a decrease in rates of both formation and hydrolysis of the intermediate. In addition to the inductive destabilizing effect, replacement of the sugar 2-hydroxyl with a fluorine atom results in removal of very important hydrogen bonds that form at the transition state between the 2-hydroxyl group and the nucleophile as well as other active-site residues. The presence of a highly reactive leaving group such as 2,4-dinitrophenolate or fluoride in such a reagent will increase the rate of the glycosylation step, resulting in the accumulation of a glycosyl-enzyme intermediate according to Fig. 2A. The catalytic competence of the glycosyl-enzyme intermediate so formed may be demonstrated by measuring the turnover of the inactivated enzyme through

---

[50] S. Numao, S. He, G. Evjen, S. Howard, O. K. Tollersrud, and S. G. Withers, *FEBS Lett.* **484,** 175 (2000).

[51] G. P. Connelly, S. G. Withers, and L. P. McIntosh, *Protein Sci.* **9,** 512 (2000).

[52] G. J. Davies, L. Mackenzie, A. Varrot, M. Dauter, A. M. Brzozowski, M. Schülein, and S. G. Withers, *Biochemistry* **37,** 11707 (1998).

[53] G. Sidhu, S. G. Withers, N. T. Nguyen, L. P. McIntosh, L. Ziser, and G. D. Brayer, *Biochemistry* **38,** 5346 (1999).

[54] S. J. Williams and S. G. Withers, *Carbohydr. Res.* **327,** 27 (2000).

hydrolysis of the intermediate. Inclusion of a suitable sugar acceptor into the reactivation mixture can accelerate reactivation by transglycosylation, as shown in Fig. 2B.[33,42]

Other evidence for the mechanism of inactivation is the stoichiometric enzyme labeling performed by these compounds which has enabled active-site titrations. The stoichiometric nature of the reaction allows the direct measurement of enzyme concentration by measurement of a "burst" of either 2,4-dinitrophenolate or fluoride released on inactivation.[33] The observation by ^{19}F NMR spectroscopy of a signal corresponding to an α-glycosyl-enzyme intermediate has allowed assignment of the anomeric stereochemistry of the glycosyl-enzyme formed on the β-glucosidase from *Agrobacterium* sp. and the xylanase from *Cellulomonas fimi* by ^{19}F NMR spectroscopy.[18,41] Structures of trapped glycosyl-enzyme complexes have been solved by X-ray crystallography for a number of glycosidases as reviewed,[11] the first of which was that of the retaining exoglycanase Cex from *C. fimi*.[18]

The use of 2-deoxy-2-fluoroglycosides (Fig. 3, compounds **1–8**) has proved to be highly successful for trapping intermediates of retaining β-glycosidases, which

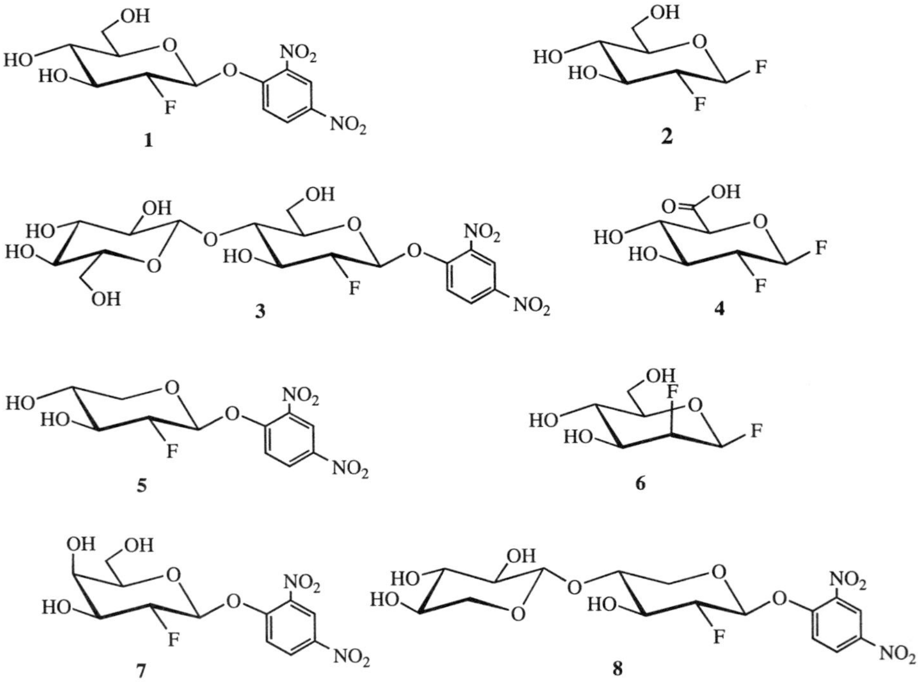

FIG. 3. Structures of 2-deoxy-2-fluoroglycosides: 2F-DNPGlc (**1**), 2F-GlcF (**2**), 2F-DNPC (**3**), 2F-GlcUAF (**4**), 2F-DNPX (**5**), 2F-ManF (**6**), 2F-DNPGal (**7**), 2F-DNPX$_2$ (**8**).

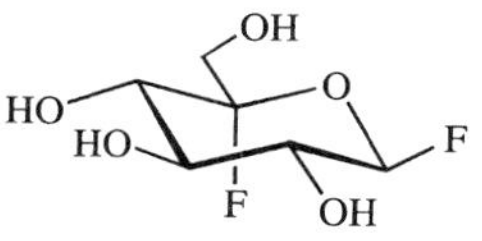

FIG. 4. Structures of 5-fluoroglycosyl fluorides: 5F-GlcF (**9**) and 5F-IdNAcF (**10**).

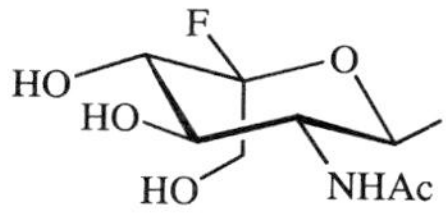

form an α-glycosyl-enzyme intermediate. The 2-fluoro sugar compounds do not appear to work with retaining α-glycosidases, which form a β-glycosyl-enzyme intermediate, but rather function as slow substrates.[55] This may be the result of the greater stability of the α-glycosyl-enzyme intermediate formed on a β-glycosidase than of the β-glycosyl-enzyme formed on an α-glycosidase. Consequently, a similar strategy that involved substitution of fluorine at C-5 in place of the hydrogen was developed as discussed below.

b. 5-Fluoroglycosides. Another related class of compounds, originally introduced by McCarter and Withers,[46] is that of the 5-fluoroglycosyl fluorides (Fig. 4, compounds **9** and **10**), which can serve to inactivate retaining glycosidases by formation of a stabilized 5-fluoroglycosyl-enzyme intermediate through a mechanism analogous to that of the 2-deoxy-2-fluoroglycosides. A sterically conservative fluorine substitution at C-5 of a glycosyl oxocarbenium ion might be expected to destabilize the transition state to a similar or greater extent than a C-2 fluorine on two counts. First, in the case of the 2-position a somewhat electronegative hydroxyl group is replaced by a more electronegative fluorine, whereas at the 5-position it is a hydrogen that is replaced by fluorine. Second, both atoms are adjacent to centers of developing positive charge, with the 5-fluoro atom adjacent to O-5 and the 2-deoxy-2-fluoro atom being adjacent to C-1. Chemical experience, along with modeling studies, suggests that the greatest difference in partial charge between the ground state glycoside and the corresponding glycosyl oxocarbenium ion is at O-5 rather than C-1; thus a fluorine substitution at C-5 is expected to destabilize the oxocarbenium ion-like transition state to a greater extent. Additionally, important transition state binding interactions between the enzyme and the usual C-2 substituent are maintained. In certain cases, functional groups such as the 2-acetamido group are crucial for catalysis. However, because of the instability of compounds in which both the 2-acetamido group and 2-fluoro groups are present simultaneously, the 5-fluoro approach serves as a viable alternative. Interestingly, it was found that the fluorinated version of the C-5 epimer of the natural substrate functioned at least as well as the 5-fluoro sugar of "correct" configuration in the case where 5-fluoro-α-L-idopyranosyl fluoride was used to inactivate a β-glycosidase.[46–48,56] Further, studies have suggested that a greater share of the positive charge is borne

---

[55] C. Braun, G. D. Brayer, and S. G. Withers, *J. Biol. Chem.* **270,** 26778 (1995).
[56] J. D. McCarter and S. G. Withers, *J. Biol. Chem.* **271,** 6889 (1996).

on the endocyclic oxygen versus the anomeric carbon for α-glycosidases than for β-glycosidases, possibly explaining why 5-fluoroglycosyl fluorides function to trap intermediates on both classes of enzyme whereas 2-fluoro sugars do not.[8,11,16]

Identification of Labeled Site by Mass Spectrometry

To identify the catalytic residues in the active sites of glycosidases, various approaches have been used such as sequence alignment, three-dimensional structural analysis, or active site labeling using mechanism-based inhibitors.[57,58] Initial approaches to the identification of the catalytic nucleophile involved synthesis and use of a radiolabeled inactivator to generate a radiolabeled enzyme, followed by reversed-phase HPLC (high-performance liquid chromatography) mapping of proteolytic digests of the inhibited enzyme compared to an unlabeled sample processed in an identical fashion. Currently, a mass spectrometric technique outlined in Fig. 5[36] is employed for the identification of covalently modified residues. This approach involves covalent labeling and proteolytic digestion of the enzyme to yield a mixture of short peptides, followed by identification and purification by reversed-phase (RP) HPLC and ESI-MS/MS (electrospray ionization–tandem mass spectrometry) and subsequent sequencing of the labeled peptide.

Pepsin turns out to be the protease of choice in such studies for several reasons. First, the ester linkage between the sugar and the peptide or denatured enzyme is labile at neutral to basic pH, but reasonably stable under the acidic conditions favored by pepsin (pH 2). Second, such low pH conditions significantly facilitate the unfolding of the glycosidase, making it susceptible to proteolytic degradation. This is extremely important since the presence of the covalently attached intermediate greatly stabilizes glycosidases toward proteolytic degradation. Thus, attempts to proteolyse labeled *Agrobacterium* sp. β-glucosidase (Abg) with trypsin or chymotrypsin at pH 7–8 were entirely unsuccessful, whereas even without prior treatment in urea, unlabeled Abg was rapidly degraded. Third, the low pH conditions may assist in slowing the deglycosylation step by protonating the general base catalyst, facilitating trapping. Finally, the low pH conditions are fully compatible with conditions used for ESI-MS analysis.

Following proteolytic digestion with pepsin, the labeled peptide is identified within an HPLC chromatogram using a tandem mass spectrometer in neutral loss scanning mode. The ester bond between inhibitor and the peptide may be cleaved by collision-induced fragmentation in the collision cell of the tandem mass spectrometer, resulting in the loss of a neutral fluoroglycosyl species and leaving the peptide with its original charge. The peptide undergoing the mass shift caused by the neutral loss can then be detected in the triple quadrupole mass spectrometer

---

[57] S. G. Withers and R. Aebersold, *Protein Sci.* **4**, 361 (1995).
[58] D. J. Vocadlo and S. G. Withers, *Methods Mol. Biol.* **146**, 203 (2000).

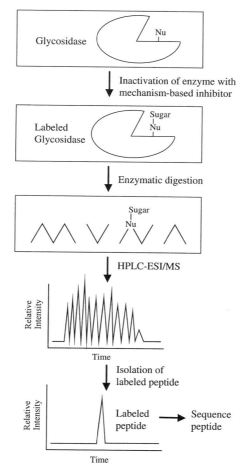

FIG. 5. Identification of catalytic active site residues by mass spectrometry. Adapted with permission from D. Tull, S. Miao, S. G. Withers, and R. Aebersold, *Anal. Biochem.* **224**, 509. Copyright © 1995 by Academic Press, Inc.

if the first and the third quadrupoles are scanned in a linked manner but offset by the mass of the eliminated neutral species. Alternatively, the labeled peptide may be identified using a comparative mapping technique whereby the labeled peptide is identified by comparison of HPLC/MS profiles of samples of enzyme (labeled and unlabeled) that have been digested and separated under identical conditions. A peptide that is unique to the labeled sample and greater in mass by the expected amount than a corresponding peptide that is exclusively present in the unlabeled sample is then an excellent candidate for the labeled peptide.

Purification of the labeled peptide and subsequent sequencing by MS/MS or through Edman degradation allows the identification of the labeled carboxyl residue as the enzyme nucleophile. The technique represents an alternative to the radioactivity-based method, which requires complex syntheses to introduce a radioisotope into the inactivator; it is also advantageous in terms of its rapid analysis time and high sensitivity, requiring only subnanomole amounts of inhibited enzyme. The 2-deoxy-2-fluoro and 5-fluoro sugar approach has proved useful in identifying the catalytic nucleophiles of more than 20 enzymes as reviewed.[11]

Enzymes whose nucleophiles have been identified in this way through the use of 2-deoxy-2-fluoro sugars include *Agrobacterium* sp. β-glucosidase[42] and *Candida albicans* exo-β-(1,3)-glucanase[23] using 2,4-dinitrophenyl-2-deoxy-2-fluoro-β-D-glucopyranoside (2F-DNPGlc, **1**); *Aspergillus niger* β-glucosidase,[20] sweet almond β-glucosidase,[22] and human glucocerebrosidase[28] using 2-deoxy-2-fluoro-β-D-glucopyranosyl fluoride (2F-GlcF, **2**); *Humicola insolens* β-1,4-endoglucanase,[25] *Streptomyces lividans* endoglucanase,[44] *Fusarium oxysporum* endoglucanase I,[24] and *Clostridium thermocellum* endo-β-1,4-glucanase[38] with 2,4-dinitrophenyl-2-deoxy-2-fluoro-β-D-cellobioside (2F-DNPC, **3**); human β-glucuronidase[43] with 2-deoxy-2-fluoro-β-D-glucopyranosyluronic acid fluoride (2F-GlucUAF, **4**); *Thermoanaerobacterium saccharolyticum* β-xylosidase[37] with 2,4-dinitrophenyl-2-deoxy-2-fluoro-β-D-xylopyranoside (2F-DNPX, **5**); *Cellulomonas fimi* β-mannosidase 2A[32] with 2-deoxy-2-fluoro-β-D-mannosyl fluoride (2F-ManF, **6**); *Escherichia coli* β-galactosidase[21] with 2,4-dinitrophenyl-2-deoxy-2-fluoro-β-D-galactopyranoside (2F-DNPGal, **7**); and *Bacillus subtilis* β-1,4-xylanase[29] with 2,4-dinitrophenyl-2-deoxy-2-fluoro-β-D-xylobioside (2F-DNPX$_2$, **8**).

The 5-fluoro sugar strategy has been employed with β-glycosidases in the following cases. 5-Fluoro-β-D-glucosyl fluoride (5F-GlcF, **9**) successfully inactivated *Agrobacterium* sp. β-glucosidase,[46] and 2-acetamido-2-deoxy-5-fluoro-α-L-idopyranosyl fluoride (5F-IdNAcF, **10**) was used to label and identify the catalytic nucleophile of *Vibrio furnisii* N-acetyl-β-D-glucosaminidase.[47] In addition, the 5-fluoroglycosyl fluorides have proved to be highly useful reagents in labeling α-glycosidases[46,48,56] as discussed elsewhere in this volume.[58a]

Trapping Glycosyl-Enzyme Intermediate through Mutagenesis

An alternative approach to trapping the glycosyl-enzyme intermediate involves mutating the enzyme rather than modifying the substrate. This approach would be preferred in order to obtain the crystal structure of an unsubstituted glycosyl-enzyme intermediate. A number of crystal structures have been obtained for the

---

[58a] R. M. Mosi and S. G. Withers, *Methods Enzymol.* **354**, [4], 2002 (this volume).

covalent glycosyl-enzyme intermediate using 2-deoxy-2-fluoroglycosides including the β-1,4-glycanase Cex from *Cellulomonas fimi* in complex with 2,4-dinitrophenyl-2-deoxy-2-fluoro-β-D-cellobioside[18] as well as 2,4-dinitrophenyl-2-deoxy-2-fluoro-β-D-xylobioside[59]; the endoglucanase Cel5A from *Humicola insolens* in complex with 2,4-dinitrophenyl-2-deoxy-2-fluoro-β-D-cellobioside (as well as the cellotrioside)[52]; the endoglucanase CelB2 from *Streptomyces lividans* in complex with 2,4-dinitrophenyl-2-deoxy-2-fluoro-β-D-cellotrioside[34]; and the β-1,4-xylanase from *Bacillus circulans*[53] and *Bacillus agaradhaerens*[60] in complex with 2,4-dinitrophenyl-2-deoxy-2-fluoro-β-D-xylobioside.

A structure with a hydroxyl, rather than a fluorine, at the 2-position is of considerable interest because of the importance of the interactions between the enzyme and substrate at this position. In order to obtain a 2-hydroxy intermediate, the glycosyl enzyme may be trapped by mutating residues that destabilize the transition state for the deglycosylation step to a greater extent than the glycosylation step of the reaction mechanism. A previous investigation that studied the role of noncovalent interactions between the enzyme and the substrate revealed the importance of hydrogen bonding interactions, particularly at the 2-position, between the enzyme and the hydroxyl groups on the substrate, in both the ground state and the transition state. Insights into the contribution of each hydroxyl group to binding and catalysis were obtained for *Agrobacterium* sp. β-glucosidase through a detailed kinetic analysis using modified substrates in which the individual hydroxyl groups on the substrate had been replaced by hydrogen or by fluorine to remove these crucial noncovalent interactions.[61] Hydrogen bonds at the 3-, 4-, and 6-positions individually were found to contribute 1–3 kcal/mol to the transition states for both glycosylation and deglycosylation, whereas in the ground state contributions were less than 1 kcal/mol. Interactions at the 2-position were by far the strongest, contributing at least 5 kcal/mol to each transition state. In fact, interactions at the 2-position are extremely important to transition state stabilization in a number of glycosidases, contributing at least 10 kcal/mol in a range of glycosidases.[18,61] The source of these interactions has been speculated to originate from the change in geometry of the substrate at the transition state, which places the 2-hydroxyl close to its hydrogen bonding partner. In addition, the increase in the acidity of the 2-hydroxyl accompanying the development of positive charge at C-1 results in a strong hydrogen bond with the carbonyl oxygen of the catalytic nucleophile. The glycosyl-enzyme intermediate can therefore be stabilized by selectively slowing

[59] V. Notenboom, C. Birsan, R. A. J. Warren, S. G. Withers, and D. R. Rose, *Biochemistry* **37**, 4751 (1998).
[60] E. Sabini, G. Sulzenbacher, M. Dauter, Z. Dauter, P. L. Jorgensen, M. Schulein, C. Dupont, G. J. Davies, and K. S. Wilson, *Chem. Biol.* **6**, 483 (1999).
[61] M. N. Namchuk and S. G. Withers, *Biochemistry* **34**, 16194 (1995).

the deglycosylation step of the enzyme mechanism through mutation of a residue involved in the stabilization of the nucleophile as it departs from the anomeric center of the sugar.[17,62-68] Additionally, removal of the acid–base catalyst may slow down the deglycosylation step as a result of the removal of base catalysis. Effects on the glycosylation step due to removal of acid catalysis may be minimized through the use of substrates with activated leaving groups. Using this strategy, the crystallographic structure of a covalent glycosyl-enzyme intermediate complex using a natural (unsubstituted) substrate was determined for the exoglycanase from *C. fimi*.

Studies on Covalent Intermediate of *Cellulomonas fimi* Exoglycanase

The β-1,4-glycanase Cex from *Cellulomonas fimi* is a 47,100 Da enzyme comprising an N-terminal catalytic domain and a C-terminal cellulose binding domain. The catalytic domain is a member of family 10 of the glycoside hydrolases, which catalyze the hydrolysis of xylan and cellulose, as well as a range of soluble aryl xylosides, xylobiosides, glucosides, and cellobiosides. Members of this family, including Cex, operate by a retaining mechanism. The acid–base catalyst was identified as Glu-127 by detailed kinetic analysis.[64] The catalytic nucleophile (Glu-233) was identified by trapping of a covalent 2-deoxy-2-fluoroglycosyl-enzyme intermediate.[35] Methods outlined below concerning the direct identification of the catalytic nucleophile[36] and crystallographic observation of the trapped intermediate[17,18] have been reported in full previously.

Labeling Active Site Nucleophile of Cex: Kinetic Analysis

The inactivation of Cex was studied using 2,4-dinitrophenyl-2-deoxy-2-fluoro-β-D-glucopyranoside (2F-DNPGlc), whose synthesis has been previously described.[42] Kinetic studies are performed at 37° in 50 mM sodium phosphate

---

[62] J. C. Gebler, D. E. Trimbur, R. A. J. Warren, R. Aebersold, M. Namchuk, and S. G. Withers, *Biochemistry* **34**, 14547 (1995).

[63] S. L. Lawson, W. W. Wakarchuk, and S. G. Withers, *Biochemistry* **36**, 2257 (1997).

[64] A. M. MacLeod, T. Lindhorst, S. G. Withers, and R. A. J. Warren, *Biochemistry* **33**, 6371 (1994).

[65] A. M. MacLeod, D. Tull, K. Rupitz, R. A. J. Warren, and S. G. Withers, *Biochemistry* **35**, 13165 (1996).

[66] R. Mosi, S. He, J. Uitdehaag, B. W. Dijkstra, and S. G. Withers, *Biochemistry* **36**, 9927 (1997).

[67] Q. Wang, D. Trimbur, R. Graham, R. A. J. Warren, and S. G. Withers, *Biochemistry* **34**, 14554 (1995).

[68] S. G. Withers, *in* "Carbohydrate Bioengineering" (S. B. Petersen, B. Svensson, and S. Pedersen, eds.), p. 97. Elsevier Science, Denmark, 1995.

buffer, pH 7.0, containing 0.1% (w/v) bovine serum albumin (BSA). A continuous spectrophotometric assay based on the hydrolysis of 2,4-dinitrophenyl-β-D-glucopyranoside (DNPGlc) is used to monitor enzyme activity by measurement of the rate of 2,4-dinitrophenolate release using a Unicam 8700 UV/Vis spectrophotometer equipped with a circulating water bath. The inactivation is monitored by incubating the enzyme in the presence of various concentrations of 2F-DNPGlc (0–4.83 mM). Residual enzyme activity is determined by addition of an aliquot (10 μl) of the inactivation mixture to a solution of DNPGlc (1.98 mM) in the above buffer and monitoring release of dinitrophenolate. Pseudo first-order rate constants for inactivation are determined by plotting the natural logarithm of the residual activity versus time and extracting the rate constant from the slope. Values for the inactivation rate constant (k_i) and the equilibrium binding constant (K_i) are determined by replotting the reciprocal of these pseudo first-order rate constants versus the reciprocal of inhibitor concentration.

The incubation of Cex with 2F-DNPGlc results in the time-dependent inactivation of the enzyme according to pseudo first-order kinetics (Fig. 6A[35]) whereas in the absence of 2F-DNPGlc the enzyme retains full activity over the time period of the inactivation. A reciprocal replot of the pseudo first-order rate constants versus inhibitor concentration (Fig. 6B) yields an inactivation rate constant of $k_i = 2.5 \times 10^{-4}$ min^{-1} and an equilibrium binding constant (K_i) of 4.5 mM. The very slow rate of inactivation is consistent with the effect of fluorine substitution at C-2 of the sugar on the rate of enzyme-catalyzed glycosyl transfer.

Evidence that inactivation occurs via stabilization and trapping of the normal intermediate in catalysis is provided by studying the reactivation of the 2-fluoroglucosyl-enzyme as follows. Enzyme (0.85 mg) is incubated with 2F-DNPGlc (13.28 mM) until 90% inactivation has occurred. Excess inactivator is then removed by extensive dialysis at 4° against large volumes of sodium phosphate buffer, pH 7.0. Aliquots of the inactivated enzyme are then incubated at 37° in the presence of phosphate buffer alone or in the presence of glucose (55 mM), β-D-glucopyranosyl benzene (55 mM), or cellobiose (55 mM). Reactivation is monitored by removal of aliquots (10 μl) over a period of 11 days for assay of enzyme activity using DNPGlc (3.08 mM) as described above. The observed first-order rate constants for reactivation are determined from the slopes of plots of the natural logarithm of the activity (full rate minus observed rate) versus time (Fig. 7[35]).

The reactivation of the enzyme follows a first-order process with a half-life of approximately 880 hr for reactivation in the absence of any reactivating ligand. Addition of glucose and glucosylbenzene to the reactivation mixtures results in some acceleration of this reactivation. The greatest increase in the rate of regeneration of the free enzyme is observed in the presence of cellobiose ($t_{1/2} = 260$ hr), a result which is consistent with the greater affinity of the Cex binding site for oligosaccharides. Presumably, the presence of a sugar facilitates turnover of the trapped glycosyl-enzyme intermediate via transglycosylation.

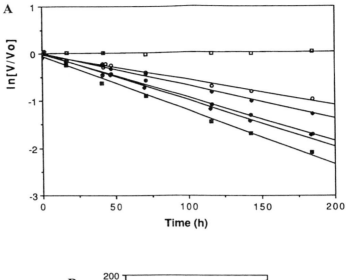
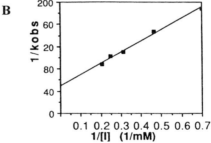

FIG. 6. Inactivation of *C. fimi* exoglycanase by 2F-DNPGlc. (A) Semilogarithmic plot of residual activity versus time at the indicated inactivator concentrations: (□) 0 mM, (○) 1.43 mM, (◇) 2.15 mM, (●) 3.22 mM, (▲) 4.03 mM, (■) 4.83 mM. (B) Replot of the first-order rate constants from (A). Reprinted with permission from D. Tull, S. G. Withers, N. R. Gilkes, D. G. Kilburn, R. A. J. Warren, and R. Aebersold, *J. Biol. Chem.* **266**, 15621. Copyright © 1991 by The American Society for Biochemistry and Molecular Biology, Inc.

Identification of Active Site Nucleophile of Cex by Mass Spectrometry

Mass spectrometric analyses of native and labeled protein samples are recorded on a PE-Sciex API III triple quadrupole mass spectrometer (Sciex, Thornhill, Ontario, Canada) equipped with an ion-spray ion source. The protein sample (10 μg, native or labeled) is introduced into the mass spectrometer through a microbore PLRP column (1 × 50 mm) equilibrated with solvent A (0.05% trifluoroacetic acid, 2% acetonitrile in water) on a Michrom HPLC system (Michrom

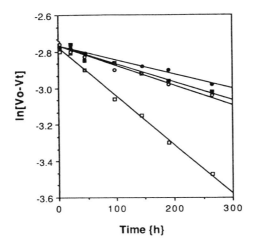

FIG. 7. Reactivation of inactivated *C. fimi* exoglycanase. Semi-logarithmic plot of activity versus time for: (●) buffer alone, (■) 55 mM glucose, (○) 55 mM glucosyl benzene, (□) 55 mM cellobiose. Reprinted with permission from D. Tull, S. G. Withers, N. R. Gilkes, D. G. Kilburn, R. A. J. Warren, and R. Aebersold, *J. Biol. Chem.* **266**, 15621. Copyright © 1991 by The American Society for Biochemistry and Molecular Biology, Inc.

BioResources Inc., Pleasanton, CA) using 20–100% solvent B (0.045% trifluoroacetic acid, 80% acetonitrile in water) over 10 min followed by 100% solvent B over 2 min. The quadrupole mass analyzer (in the single quadrupole mode) is scanned over a mass-to-charge ratio (m/z) range of 300–2400 Da with a step size of 0.5 Da and a dwell time of 1 ms per step. The ion source voltage (ISV) is set at 5 kV, and the orifice energy (OR) is 80 V. Mass spectrometric analysis reveals a mass increase of 327 Da for the labeled enzyme, corresponding to the formation of a 2-deoxy-2-fluorocellobiosyl-Cex complex.

Samples of Cex (100 μg, 9.8 mg/ml) are incubated with 2,4-dinitrophenyl-2-deoxy-2-fluoro-β-D-cellobioside, 2F-DNPC (1.0 mM), in 50 mM sodium phosphate buffer (pH 7.0) at 37° until complete inactivation is confirmed by enzyme assay. Proteolysis is performed by incubating a sample of 2-deoxy-2-fluorocellobiosyl-labeled enzyme and native enzyme with 1 : 100 pepsin (w/w, enzyme : substrate) in 50 mM sodium phosphate buffer (pH 2.0) at room temperature. Peptic proteolysis of inactivated Cex results in a mixture of peptides, which are separated by reversed-phase HPLC, using the ESI-MS as a detector. In each of the MS experiments, the proteolytic digest (10 μg) is loaded onto a C_{18} column (Reliasil, 1 × 150 mm, Column Engineering, Inc., Ontario, CA) and then eluted with a gradient of 0–60% solvent B over 20 min followed by 100% solvent B for 2 min at a flow rate of 50 μl/min. The single quadrupole mode (normal

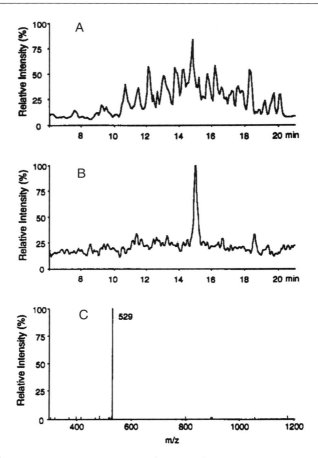

FIG. 8. Electrospray mass spectrometry experiments on Cex proteolytic digests: (A) labeled with 2F-DNPC, total ion chromatogram in normal MS mode; (B) labeled with 2F-DNPC, TIC in neutral loss mode; and (C) mass spectrum of peptide in (B). Reprinted with permission from D. Tull, S. Miao, S. G. Withers, and R. Aebersold, *Anal. Biochem.* **224,** 509, 1995. Copyright © 1995 by Academic Press, Inc.

LC/MS) MS conditions used are identical to those for analysis of the intact protein. When the spectrometer is scanned in the normal LC-MS mode, the total ion chromatogram (TIC) of the 2-deoxy-2-fluorocellobiosyl-labeled Cex digest displays a large number of peaks (Fig. 8A[36]). The 2-deoxy-2-fluorocellobiosyl-labeled peptide is then identified in a second run using the tandem mass spectrometer set up in the neutral loss scanning mode (MS/MS) (Fig. 8B). The neutral loss MS/MS spectra are obtained in the triple quadrupole neutral loss scan mode searching for the mass loss corresponding to the loss of the 2-deoxy-2-fluorocellobiosyl label from a peptide ion in the singly (327 Da) or doubly charged state (163.5). In the triple quadrupole neutral loss scan mode, the MS conditions are as follows: the mass analyzer is scanned over m/z range of 300–1200 Da with a step size of 0.5 Da and a dwell time of 1 ms per step. The ISV is set to 5 kV and the OR is 80 V. In this mode, the ions are subjected to limited fragmentation by collisions with argon in a collision cell, which are sufficient to break the ester bond between the 2-deoxy-2-fluorocellobiosyl label and the peptide, but not the peptide bonds. This results in the loss of a neutral 2-deoxy-2-fluorocellobiosyl moiety, leaving the peptide moiety with its original charge. The two quadrupoles are then scanned in a linked manner such that only ions differing in m/z by the mass corresponding to the label can pass through both quadrupoles and be detected.

When the spectrometer is initially scanned in the neutral loss mode searching for a peptide that loses the neutral 2-deoxy-2-fluorocellobiosyl moiety of m/z 327, no signal is detected. However, when the spectrometer is scanned for a mass loss of m/z 163.5, corresponding to a loss of the sugar moiety from the doubly charged peptide, a single peak is observed in the total ion chromatogram (Fig. 8B) having an m/z value of 529 Da. No such peak is observed in an equivalent chromatogram of the control digest of unlabeled Cex (not shown), suggesting that the peak identified is the peptide of interest.

The peak identified as being the 2-deoxy-2-fluorocellobiosyl peptide is collected using a splitter device and repurified. The identity of this peptide is determined through a computer-generated search of the known sequence of the protein for peptides corresponding to the calculated mass. Because the doubly charged labeled peptide observed has a m/z of 529 Da (Fig. 8C), the singly charged labeled peptide must have a molecular mass of 1056 Da [(529 × 2) − 2H]. The molecular mass of the unlabeled peptide must be 730 Da since the mass of the label is 327 Da (1056 − 327 + 1H). A computer search of the known amino acid sequence of Cex has identified nine peptides of mass 730.4 ± 0.1 Da, one of which contains the active site peptide (^{229}VRITEL^{234}) previously identified by the radioactive method using tritiated 2F-DNPG.[35] Confirmation of this identity for the peptide is obtained by Edman degradation of the peptide after further purification.

Crystallographic Observation of Covalent Fluoroglycosyl-Enzyme Intermediate

To provide insights into interactions at the active site that are crucial to substrate specificity and mechanism of action, the three-dimensional structure of a trapped fluorocellobiosyl-enzyme complex for the catalytic domain of the glycanase Cex is obtained at a resolution of 1.8 Å by X-ray diffraction. The cellulose binding domain of recombinant β-glycanase Cex is removed by proteolysis using papain and the catalytic domain (cex-cd) is crystallized as previously described.[69] A crystal of

[69] S. Bedarkar, *J. Mol. Biol.* **228,** 693 (1992).

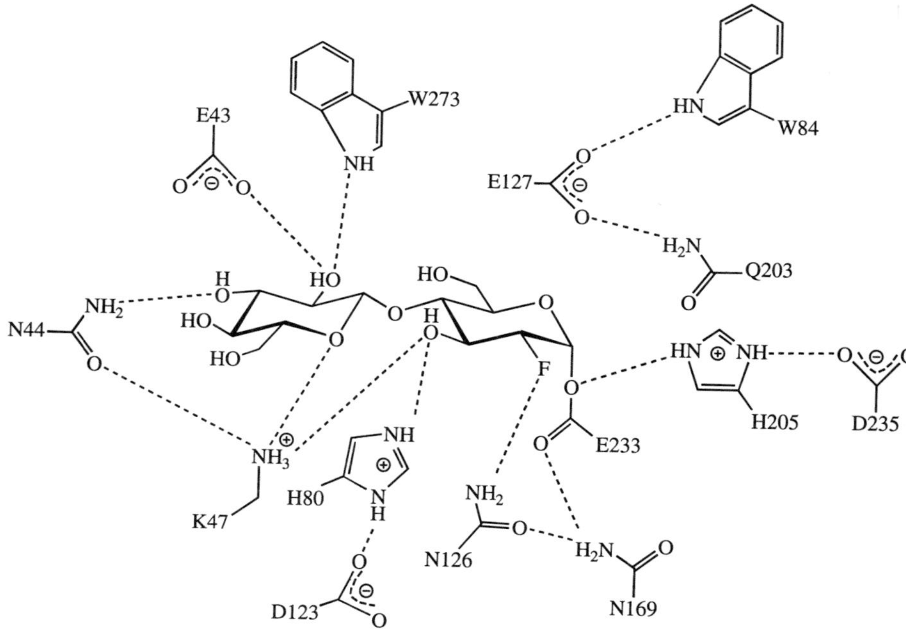

FIG. 9. Active site of the *Cellulomonas fimi* exoglycanase trapped as a covalent 2-deoxy-2-fluorocellobiosyl-enzyme intermediate. Adapted with permission from A. White, D. Tull, K. Johns, S. G. Withers, and D. R. Rose, *Nat. Struct. Biol.* **3,** 149. Copyright © 1996 Nature America, Inc.

native cex-cd is soaked for 11 hr in 0.5 mM 2,4-dinitrophenyl-2-deoxy-2-fluoro-β-D-cellobioside (2F-DNPC) in a 10 μl drop which is suspended over 1 ml of 15% polyethylene glycol 4000 and 100 mM acetate (pH 4.6). Alternatively, the 2-fluorocellobiosyl-enzyme complex may be formed prior to crystallization. X-ray diffraction data to 1.8 Å are measured on a San Diego Multiwire Systems area detector (San Diego, CA) and then reduced as previously described for the native cex-cd.[70] Rigid-body refinement is used to optimize the position of the cex-cd model, which is further refined with X-PLOR (version 3.1).[71]

Diffusion of the fluorinated mechanism-based inhibitor 2,4-dinitrophenyl-2-deoxy-2-fluoro-β-cellobioside (2F-DNPC) into crystals of the catalytic domain of Cex results in the formation of an α-D-glycopyranosyl linkage to the nucleophile Glu-233. The fluorocellobiosyl-enzyme complex identifies two binding subsites in a deep crevice at the C-terminal end of the $(\beta/\alpha)_8$-barrel motif and allows the identification of specific carbohydrate–enzyme interactions (see Fig. 9[18]). Both

---

[70] A. White, S. G. Withers, N. R. Gilkes, and D. R. Rose, *Biochemistry* **33,** 12546 (1994).
[71] A.T. Brünger, J. Kuriyan, and M. Karplus, *Science* **235,** 458 (1987).

the distal and proximal saccharides assume a standard $^4C_1$ conformation. The anomeric carbon is bonded to the conserved residue Glu-233 by a covalent linkage of 1.4 Å, which is coplanar with the endocyclic oxygen of the attached saccharide. Furthermore, the crystal structure of the enzyme identifies the key amino acid residues forming the environments around Glu-233 and therefore likely to play important roles in catalysis. The acid–base catalyst, Glu-127, is located 4.5 Å from the anomeric carbon and 5.5 Å from Glu-233. This separation from the nucleophile is consistent with the role of the acid–base catalyst in protonating the glycosidic bond during formation of the glycosyl-enzyme and the general base-catalyzed attack of water during its hydrolysis. His-205 has been found to be hydrogen bonded to both the nucleophile Glu-233 and Asp-235, forming a trio of residues that is important in the hydrolysis of the glycosyl-enzyme complex and is highly conserved within family 10 enzymes. The hydrogen bond network reveals important interactions involved in the recognition of the fluorocellobioside and a structural basis for the specificity exhibited by Cex for xylan over cellulose. The distal saccharide is involved in hydrogen bonding with residues Glu-43, Asn-44, Lys-47, and Trp-273, whereas the proximal saccharide makes hydrogen bonds with Lys-47, His-80, and Asn-126. The fluorine substituent at the 2-position of the cellobioside substrate is involved in specific interactions with the enzyme. The conserved residue, Asn-126, is located at a distance of 2.8 Å from the fluorine. The catalytic nucleophile (Glu-233) is 3.1 Å from the fluorine at C-2, despite the fact that this interaction must be destabilizing, suggesting a potentially important interaction when a hydroxyl group is present at the 2-position. Studies on related retaining β-glycosidases have shown that interactions with the hydroxyl group at the 2-position of the natural substrate contribute 10 kcal/mol to the stabilization of the oxocarbenium ion-like transition states. Site-directed mutagenesis of residue Asn-126, however, reveals an interaction worth only about 2.5 kcal/mol.[72] The remaining interaction energy is therefore likely provided by Glu-233.

In order to address and unequivocally identify the interactions involved at the 2-position, the structure of a covalent glycosyl-enzyme complex trapped using an unsubstituted (natural) substrate is preferred over that obtained using 2-deoxy-2-fluoroglycosides or glycosyl fluorides. As will be discussed below, a detailed kinetic analysis on several active-site mutants of Cex has suggested an approach to determining the three-dimensional structure of a true glycosyl-enzyme covalent intermediate formed during catalysis by a retaining glycosidase.

Crystallographic Observation of Unsubstituted Glycosyl-Enzyme Intermediate

To address the nature of interactions between the enzyme and the 2-hydroxyl of the natural substrate, the glycosyl enzyme is trapped by mutating the enzyme instead of modifying the substrate. The kinetic properties of several active-site

---

[72] V. Notenboom, personal communication (1999).

mutants of Cex (H205N, H205A, and E127A) suggest an approach. Kinetic parameters for the hydrolysis of aryl glycoside substrates of different aglycone pK_a reveal that the deglycosylation step is rate-limiting for the wild-type enzyme. Further, the low K_m values observed for substrates with good leaving groups indicate the accumulation of a glycosyl-enzyme intermediate, as does the observation of a full-sized burst of release of 1 equivalent of aglycone per equivalent of enzyme. However, turnover of the intermediate is much too fast for structural analysis. Mutation of the acid–base catalyst Glu-127 to Ala results in an active enzyme, but one with a k_{cat} value some 300-fold lower than that of the wild-type enzyme when substrates with good leaving groups are studied.[65] The covalent intermediate indeed accumulates under such conditions, as revealed by the low K_m values and mass spectral analysis; however, turnover is too fast for crystallographic analysis. Mutation of a histidine residue (His-205) which hydrogen bonds to the catalytic nucleophile Glu-233 results in 8500-fold and 4800-fold slowing of the deglycosylation step for the Asn and Ala mutants, respectively, leading to accumulation of intermediates when activated substrates are employed. However, turnover is still too fast for crystallographic analysis. A double mutant has therefore been created in which both mutations are present. In that case, a stable cellobiosyl-enzyme complex is formed when the His205Asn/Glu127Ala double mutant is reacted with 2,4-dinitrophenyl β-D-cellobioside, with no significant steady-state turnover even on lengthy incubation. The total rate reduction of 2.5×10^6-fold is consistent with the 8500-fold rate reduction expected for the His205Asn mutation coupled with the 300-fold reduction expected for the Glu127Ala mutant. Evidence for the formation of the covalent enzyme complex is obtained by the stoichiometric release of 1 equivalent of 2,4-dinitrophenolate and by the expected increase in mass (325 Da) of the mutant that corresponds to the addition of a cellobiosyl moiety.

In order to crystalize Cex it is first important to cleave off the catalytic domain using papain. Initial attempts to do this with the double mutant were unsuccessful since papain also degrades the catalytic domain. However, prior trapping of the intermediate by reaction of the double mutant with 2,4-dinitrophenyl-β-D-cellobioside stabilizes the catalytic domain, allowing proteolysis. The trapped catalytic domain is purified by gel filtration and subsequently crystallized. To ensure stability of the covalent intermediate for the duration of the experiment, the diffraction data are collected at 100 K, which allows structure determination at 1.8 Å resolution.[17] A cellobiose molecule is found covalently bound to the nucleophile Glu-233 in an α-D-glycopyranosyl linkage with the proximal sugar ring in an almost perfectly relaxed $^4C_1$ conformation, very similar to that of the wild-type enzyme in its 2-deoxy-2-fluorocellobiosyl-enzyme form. However, the orientation of the nucleophile Glu-233 has been found to be altered from that of the wild-type enzyme as a result of the His-205 mutation. Because of the repositioning of the nucleophile, the bound disaccharide is shifted approximately 1 Å with respect to that of the wild-type enzyme in its fluorocellobiosyl-enzyme form. Nevertheless, most of the hydrogen bonding interactions are maintained with the active site residues surrounding the two substrate binding sites. However, one key interaction observed between the proximal sugar 2-hydroxyl and Asn-126 is missing (4.4 Å), whereas the interaction between Asn-126 and Asn-169 is maintained at 2.9 Å. An unusually short hydrogen bond (2.4 ± 0.2 Å) between the sugar 2-hydroxyl and the carbonyl oxygen of the nucleophile, Glu233, is observed. This interaction is thought to be a common phenomenon with retaining β-glycosidases and is thought to be the primary contributor to the strong transition state interactions seen with the 2-hydroxyl. Interactions at that position with Cex have been estimated as contributing more than 10 kcal/mol to transition state stabilization.[18] Part of this interaction energy undoubtedly arises from contacts with Asn-126: kinetic analysis of mutants at that position suggests only about 2.5 kcal/mol.[72] However, a large share of the remaining 8 kcal/mol is likely due to interactions of the 2-hydroxyl with the carbonyl oxygen.

Summary

The mechanism-based inactivation and subsequent identification of the nucleophilic residue using mass spectrometry have been successfully applied and used to identify the active-site nucleophile in numerous β-glycosidases, as illustrated using *C. fimi* exoglycanase. Evidence for a covalent glycosyl-enzyme intermediate has come from X-ray crystallographic analysis of trapped complexes, the first being that of the trapped fluoroglycosyl-enzyme intermediate of Cex. The crystal structure of the trapped fluorocellobiosyl-enzyme complex for Cex has provided useful insights into catalysis and the roles of specific residues at the active site. In addition, information about the conformation of the natural sugar in the covalently bound state and the interactions at the active site was obtained using a mutant form of Cex.

Acknowledgments

We thank the Natural Sciences and Engineering Research Council of Canada (NSERC) and the Protein Engineering Network of Centres of Excellence of Canada (PENCE) for financial support.

[6] 2-Hydroxy-6-keto-nona-2,4-diene 1,9-Dioic Acid 5,6-Hydrolase: Evidence from ^{18}O Isotope Exchange for *gem*-Diol Intermediate

By TIMOTHY D. H. BUGG, SARAH M. FLEMING, THOMAS A. ROBERTSON, and G. JOHN LANGLEY

Introduction

The catalytic mechanisms for enzymatic amide (C–N bond) and ester (C–O bond) hydrolysis have been studied for many years. One important question which is central to these studies is the distinction between a nucleophilic mechanism (Fig. 1A), in which an active site nucleophile attacks the incipient carbonyl group to form a covalent acyl enzyme intermediate, and a general base mechanism (Fig. 1B), in which an active site base deprotonates a nucleophilic water molecule. If a kinetically competent acyl enzyme intermediate can be established, as has been found for the serine[1] and cysteine[2] protease families, then a nucleophilic mechanism can be inferred. If, however, there is no accumulation of an acyl enzyme intermediate, then the distinction between these two mechanisms is difficult to establish experimentally.

One method that has been used to investigate the general base mechanism is ^{18}O isotope exchange. The first step of the general base mechanism involves the reversible addition of water to the incipient carbonyl group to form a *gem*-diol (or diolate) intermediate. If the rate constants k_1 and k_{-1} are large relative to k_2 (see Fig. 1B), then in the presence of $H_2^{18}O$ there may be ^{18}O isotope exchange into the acyl substrate if in the reverse reaction there is occasional cleavage of a C–^{16}O bond rather than a C–^{18}O bond. Because the two hydroxyl groups of the *gem*-diol intermediate are diastereotopic (in the case of a peptide substrate), this circumstance requires that the enzyme exhibit a lack of stereospecificity in the reverse direction, or that it release some *gem*-diol into solution, whereupon the breakdown of *gem*-diol to acyl substrate will occur nonenzymatically. This method has been used to investigate a general base mechanism in the aspartyl proteases, which contain two catalytic aspartic acid residues in their active sites.[3,4] Porcine pepsin has been

---

[1] A. K. Balls and H. N. Wood, *J. Biol. Chem.* **219**, 245 (1956); M. L. Bender, G. R. Schonbaum, and B. Zerner, *J. Am. Chem. Soc.* **34**, 49 (1965); R. Henderson, *J. Mol. Biol.* **54**, 341 (1970).

[2] J. P. G. Malthouse, M. P. Gamesik, A. S. F. Boyd, N. E. Mackenzie, and A. I. Scott, *J. Am. Chem. Soc.* **104**, 6811 (1982).

[3] V. K. Antonov, L. M. Ginodman, L. D. Rumsh, Y. V. Kapitannikov, T. N. Barshevskaya, L. P. Yavashev, A. G. Gurova, and L. I. Volkova, *Eur. J. Biochem.* **117**, 195 (1981).

[4] L. J. Hyland, T. A. Tomaszek, G. D. Roberts, S. A. Carr, V. W. Magaard, H. L. Bryan, S. A. Fakhoury, M. L. Moore, M. D. Minnich, J. S. Culp, R. L. DesJarlais, and T. D. Meek, *Biochemistry* **30**, 8441 (1991).

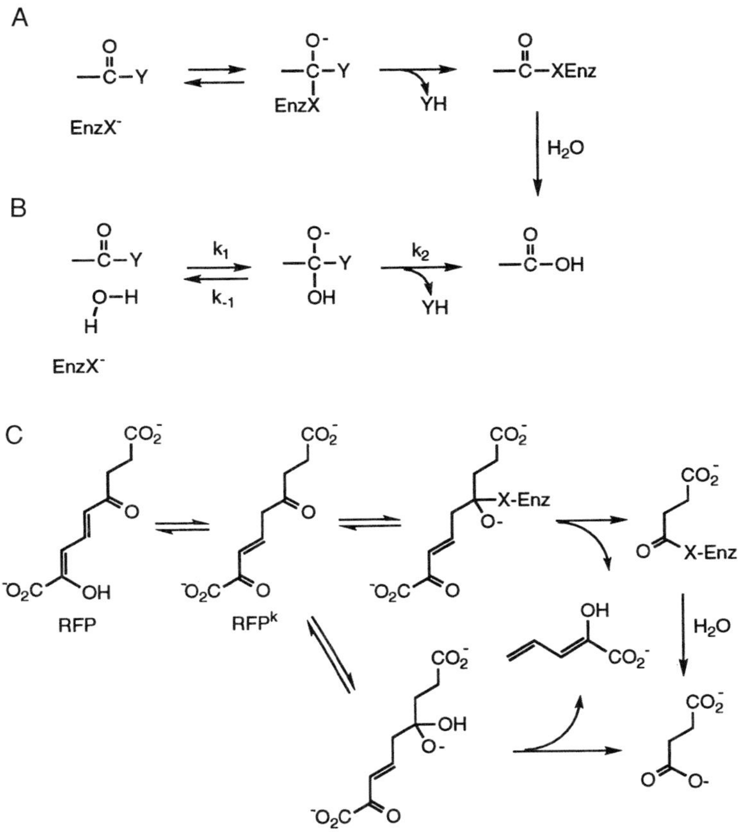

FIG. 1. Nucleophilic (A) and general base (B) mechanisms for amide/ester hydrolysis. (C) Proposed catalytic mechanisms for hydrolase MhpC, via attack of either water or an active site nucleophile upon RFP^k. (See W. W. Y. Lam and T. D. H. Bugg, *J. Chem. Soc. Chem. Commun.*, 1163 (1994); W. W. Y. Lam and T. D. H. Bugg, *Biochemistry* **36**, 12242 (1997), I. M. J. Henderson and T. D. H. Bugg, *Biochemistry* **36**, 12252 (1997).

found to catalyze the incorporation of up to 42% atom ^{18}O into the reisolated acyl substrate, and up to 65% atom ^{18}O into transpeptidation products.[3] HIV protease has been found to catalyze the incorporation of 1–11 atom % ^{18}O into reisolated peptide substrates.[4] In each case a general base mechanism has been inferred.

A small number of enzymes catalyze the hydrolytic cleavage of carbon–carbon bonds adjacent to ketones, giving a carboxylic acid product, including kynureninase from *Pseudomonas fluorescens*,[5] dioxo acid hydrolase from beef liver,[6]

---
[5] R. S. Philips and R. K. Dua, *J. Am. Chem. Soc.* **113**, 7385 (1991).
[6] H. H. Hsiang, S. S. Sim, D. J. Mahuran, and D. E. Schmidt, Jr., *Biochemistry* **11**, 2098 (1972).

and a family of C–C hydrolases involved in bacterial meta-cleavage pathways responsible for the breakdown of aromatic compounds.[7] Studies in our laboratory of the reaction catalyzed by 2-hydroxy-6-keto-nona-2,4-diene 1,9-dioic acid 5,6-hydrolase (MhpC) of *Escherichia coli* have established the overall stereochemistry of the reaction[8] and have provided evidence for an initial enol/keto tautomerization step,[9] yielding a keto intermediate (RFP^k) which contains an α,β-unsaturated ketone functional group which serves as an electron sink for the carbanion arising from C–C cleavage.

Mechanistic proposals for the C–C cleavage step of MhpC have involved attack on the C-6 carbonyl of either water or an active site nucleophile, giving rise either to a *gem*-diol intermediate or to an acyl enzyme intermediate, respectively (see Fig. 1C).[8,9] Amino acid sequence alignments imply that this family of C–C hydrolase enzymes are members of the α/β hydrolase family, containing a serine catalytic triad in the order serine–aspartate–histidine as found in the amino acid sequence.[10] The catalytic mechanisms of the α/β hydrolase family are believed to proceed via a nucleophilic mechanism,[11] although acyl enzyme intermediates have only been characterized using unnatural substrates such as *p*-nitrophenyl acetate.[12]

Attempts to trap the putative acyl enzyme intermediate of MhpC under acidic conditions using a ^{14}C-labeled substrate gave consistently low stoichiometries (0–10 mol % ^{14}C/mol subunit).[13] Pre-steady-state kinetic analysis had previously shown that under these conditions breakdown of the enzyme–product complex was rate-limiting,[9] and control incubations showed that the observed levels of ^{14}C trapping could be quantitatively accounted for by nonspecific and background nonenzymatic processes.[13] We decided therefore to use ^{18}O isotope exchange to investigate the general base mechanism for hydrolytic cleavage by this enzyme, using both the natural substrate and a nonhydrolyzable substrate analog.[13]

---

[7] T. D. H. Bugg and C. J. Winfield, *Nat. Prod. Reports* **15**, 513 (1998); C. J. Duggleby and P. A. Williams, *J. Gen. Microbiol.* **132**, 717 (1986); R. Bayly and D. D. Bernardino, *J. Bacteriol.* **134**, 30 (1978); T. Omori, K. Sugimura, H. Ishigooka, and Y. Minoda, *Agric. Biol. Chem.* **50**, 931 (1986); S. Y. K. Seah, G. Terracina, J. T. Bolin, P. Riebel, V. Snieckus, and L. D. Eltis, *J. Biol. Chem.* **273**, 22943 (1998).

[8] W. W. Y. Lam and T. D. H. Bugg, *J. Chem. Soc. Chem. Commun.*, 1163 (1994); W. W. Y. Lam and T. D. H. Bugg, *Biochemistry* **36**, 12242 (1997).

[9] I. M. J. Henderson and T. D. H. Bugg, *Biochemistry* **36**, 12252 (1997).

[10] D. L. Ollis, E. Cheah, M. Cygler, B. Dijkstra, F. Frolow, S. M. Franken, M. Harel, S. J. Remington, I. Silman, J. Schrag, J. L. Sussman, K. H. G. Verschueren, and A. Goldman, *Protein. Eng.* **5**, 197 (1992); B. Hofer, L. D. Eltis, D. N. Dowling, and K. N. Timmis, *Gene* **130**, 47 (1993); E. Diaz and K. N. Timmis, *J. Biol. Chem.* **270**, 6403 (1995).

[11] For a review of the α/β hydrolase family, see J. D. Schrag and M. Cygler, *Methods Enzymol.* **284**, 85 (1997).

[12] C. Chapus, M. Semeriva, C. Bovier-Lapierre, and P. Desnuelle, *Biochemistry* **15**, 4980 (1976).

[13] S. M. Fleming, T. A. Robertson, G. J. Langley, and T. D. H. Bugg, *Biochemistry* **39**, 1522 (2000).

Materials and Methods

General

NMR spectra are recorded on a Bruker AC300 Fourier transform spectrometer (300 MHz). Mass spectra and GC-MS are recorded on a VG 70-250 mass spectrometer. Isotopically enriched water (^{18}O, 95–98 atom %) is purchased from Promochem Ltd. (Welwyn Garden City, Hertfordshire, UK). 4-Keto-nona-1,9-dioic acid (KNDA) is prepared by treatment of 2-(2-cyanoethyl) cyclohexanone with hydrogen peroxide, according to the method of Chiusoli *et al.*,[14] in 30% yield: δ_H (300 MHz, D_2O) 2.68 (2H, t, $J = 8$ Hz), 2.43 (4H, t, $J = 8$ Hz), 2.22 (2H, t, $J = 8$ Hz), 1.4–1.5 (4H, m) ppm; δ_C (75 MHz, D_2O) 217.90, 181.27, 180.02, 44.19, 39.46, 36.05, 30.34, 26.29, 25.22 ppm; *m/z* (ES −ve ion) 237/239 $(M + Cl)^-$. All other chemicals and biochemicals are purchased from Sigma/Aldrich. 2-Hydroxy-6-keto-nona-2,4-diene 1,9-dioic acid (RFP) is prepared using *E. coli* 2,3-dihydroxyphenylpropionate 1,2-dioxygenase (MhpB) as previously described.[9]

Purification of Escherichia coli 2-Hydroxy-6-keto-nona-2,4-diene 1,9-Dioic acid 5,6-hydrolase (MhpC)

Purification buffer consists of 50 mM potassium phosphate (pH 7.0) containing 10 mM 2-mercaptoethanol. All purification steps[8,9] are carried out at 4° unless indicated otherwise.

A 5-liter culture of *Escherichia coli* W3110/pTB9[15] is grown for 16 hr at 30° with aeration in Luria broth containing 100 μg/ml ampicillin and 0.02% phenylpropionic acid. The cells are harvested by centrifugation for 10 min at 4400g and resuspended in buffer (70 ml). Cell lysis is carried out by passage through a cell disrupter (5 psi), and cell debris is removed by centrifugation for 20 min at 27,000g. Powdered ammonium sulfate is added to the crude extract to 25% saturation (144 g/liter); the suspension stirred for 1 hr and centrifuged for 20 min at 12,000g. To the supernatant is slowly added further ammonium sulfate (157 g/liter) to 50% saturation; the suspension is stirred for 1 hr and centrifuged for 20 min at 12,000g.

The resulting pellet is resuspended in buffer containing 1 M ammonium sulfate (10 ml), and the suspension loaded onto a phenyl agarose column (Sigma, 5.0 × 1.0 cm). The column is eluted at 0.5 ml/min with a gradient (200 ml) of buffer containing decreasing ammonium sulfate concentration from 1.0 to 0 M. Fractions containing dioxygenase MhpB activity eluting at 0.2–0.0 M NaCl are pooled and stored at 4°. Fractions containing MhpC activity eluting at 0.9–0.7 M NaCl are pooled and dialyzed against buffer (3 × 2 liter).

---

[14] G. P. Chiusoli, F. Minisci, and A. Quilico, *Gazz. Chim. Ital.* **87**, 100 (1957).

[15] T. D. H. Bugg, *Biochim. Biophys. Acta* **1202**, 258 (1993).

The dialyzed MhpC pool is then loaded onto a Q Sepharose FPLC column (Pharmacia, Piscataway, NJ 9 × 1.5 cm), and the column eluted at 2.0 ml/min with a gradient (200 ml) of buffer containing increasing NaCl concentration from 0 to 0.5 M NaCl. Fractions containing MhpC activity eluting at 0.3 M NaCl are pooled and dialyzed against buffer (2 liter). The resultant pool has a specific activity of 32.1 U mg^{-1} at a protein concentration of 0.2 mg ml^{-1}.

Enzyme Kinetics

MhpC is assayed by monitoring the decrease in absorbance of the substrate on addition of enzyme to a solution of RFP, either at 394 nm in 50 mM potassium phosphate buffer pH 8.0 ($\varepsilon_{394} = 15{,}600\ M^{-1}\ \text{cm}^{-1}$), or at 317 nm in 50 m$M$ sodium citrate buffer pH 5.0 ($\varepsilon_{317} = 15{,}000\ M^{-1}\ \text{cm}^{-1}$). Time-dependent inhibition of MhpC is assayed by preincubation of samples of MhpC (1 unit) with KNDA (10 mM final concentration) in 50 mM sodium citrate pH 5.0, followed by 10-fold dilution at time t into 50 mM sodium citrate pH 5.0 containing ring fission product (50 μg), and monitoring of the decrease in absorbance at 317 nm.

Incorporation of $H_2^{18}O$ into MhpC Reaction Products

Samples of 2-hydroxy-6-keto-nona-2,4-diene-1,9-dioic acid (5 μg, 23 nmol) are applied to microcentrifuge tubes by evaporation from ethanol solution. To each sample is added 110 μl of 95% $H_2^{18}O$ and 10 μl of 1 M potassium phosphate buffer (pH 8.0) prepared in $H_2^{18}O$. The samples are incubated for time intervals of 0, 120, and 300 s. Aliquots of MhpC (2 units) which have been freeze-dried and resuspended in $H_2^{18}O$ are added and left for 60 s to ensure complete conversion (reaction is visibly completed in 1–2 s as judged by the disappearance of the yellow substrate). The mixture is then acidified to pH 1 by addition of 10 μl of 2 M HCl, and the sample immediately frozen on dry ice. The samples are freeze-dried, and the resulting solids taken up in methanol (200 μl). The solutions are then treated with an ethereal solution of diazomethane. After 10 min excess diazomethane is destroyed by dropwise addition of acetic acid. Samples are then centrifuged (13,000 rpm), and the supernatant evaporated by a stream of nitrogen gas.

Samples are then analyzed by GC-MS using chemical ionization conditions and selected ion monitoring. For each sample a series of peaks are obtained for $[M+1]^+$, $[M+3]^+$, and $[M+5]^+$ ion currents for dimethyl succinate (M_r 146), retention time 8.36 min on a BPX5 25 m column with an internal diameter of 0.22 mm. The following temperature gradient is applied: 35° for 3 min; heated to 340° at 20°/min; 340° for 3 min. The observed $[M+5]^+$ intensities are corrected for the natural background isotope pattern for dimethyl succinate ($C_6H_{11}O_4$) by subtraction of 1.032% of the $[M+3]^+$ intensities. The total percentage incorporation is calculated, and the corrected $[M+5]^+$ intensities expressed as a percentage of this figure.

^{18}O Incorporation into KNDA

MhpC (5.0 mg) is freeze-dried and resuspended in 230 μl $H_2^{18}O$ (>98 atom % ^{18}O), containing ^{18}O-labeled potassium phosphate buffer pH 7.0 (50 mM) and KNDA (0.5 mg). A control incubation is also set up, containing no MhpC. At time intervals of 0.5, 15, 30, and 60 min, 50-μl aliquots are withdrawn and treated with a solution of sodium borohydride (1 mg) in ethanol (0.5 ml). Samples are then acidified by addition of formic acid (2 drops) and freeze-dried. The resulting solids are treated with an ethereal solution of diazomethane for 10 min, then excess diazomethane is destroyed by dropwise addition of acetic acid. Solutions are evaporated, then resuspended in hexane and analyzed by GC-MS (electron impact ionization). GC is carried out on a BPl 25 m column with an internal diameter of 0.22 mm. The following temperature gradient is applied: 35° for 5 min; heated to 300° at 20°/min; 300° for 5 min. Peaks corresponding to the 7-membered lactone and 5-membered lactone are observed at 1230 s and 1246 s, respectively. An authentic sample of 7-membered lactone, synthesized previously in our laboratory,[16] has been found to have a retention time of 1230 s and identical fragmentation pattern. Both peaks give m/z 201 (MH$^+$) under NH$_3$ chemical ionization conditions. The 5-membered lactone [m/z (EI) 169 (30%), 150 (20%), 115 (30%), 85 (100%), 55 (25%)] shows a major fragment ion at m/z 85 which shows clear incorporation of ^{18}O versus time. Selected ion monitoring of this peak at m/z 85 and 87 is used to determine the percentage ^{18}O incorporation.

Results

Incorporation of $H_2^{18}O$ into Enzymatic Cleavage Products

Because of the expense of $H_2^{18}O$, enzymatic conversions were carried out in the smallest practical volume, but with the highest possible percentage of ^{18}O, in order to be able to detect low levels of ^{18}O incorporation. A practical complication was that when MhpC conversions were carried out at high substrate concentrations, substrate inhibition was observed: 55% inhibition at 260 μM and 71% inhibition at 770 μM. The optimized conditions were a substrate concentration of 200 μM, in a 120 μl reaction volume, suspended in ^{18}O-containing buffer (at 95 atom % ^{18}O), and converted enzymatically to succinate by addition of MhpC (2 units, resuspended in ^{18}O-containing buffer). Enzymatic conversion could be observed visually, because of the yellow color of RFP at pH 8, and proceeded in 1–2 sec upon addition of enzyme. In order to examine the rate of nonenzymatic exchange into the carbonyl group of the substrate, samples of RFP were preincubated in ^{18}O-containing buffer for 0, 2, and 5 min prior to treatment with MhpC.

[16] J. Sanvoisin, G. J. Langley, and T. D. H. Bugg, *J. Am. Chem. Soc.* **117**, 7836 (1995).

After enzymatic conversion, samples were immediately frozen and lyophilized. The residues were derivatized using diazomethane, and the dimethyl succinate product analyzed by GC-MS. A peak corresponding to dimethyl succinate at retention time 8.36 min gave under NH_3 chemical ionization conditions a molecular ion (MH^+) at m/z 147 for ^{16}O-labeled dimethyl succinate. The ^{18}O-labeled dimethyl succinate gave an isotope peak containing m/z 147 (M + 1), 149 (M + 3), and 151 (M + 5), corresponding to the incorporation of one and two atoms of ^{18}O, respectively. In spite of the high levels of ^{18}O used (95%) in the enzymatic conversion, the total observed incorporation was in the range 70–80%, indicating that $H_2^{16}O$ had partially exchanged into the experiment through air moisture or enzyme-bound water molecules.

Since the signal for M + 5 was small (2–5% total), it was important to calculate and subtract the background isotope patterns for M + 1 and M + 3 species. The natural abundance isotope pattern for dimethyl succinate ($C_6H_{11}O_4$) is MW 147 (100.00%), 148 (7.103%), 149 (1.032%), 150 (0.061%), and 151 (0.004%). Therefore, accurate peak height intensities were measured using secondary ion monitoring, and the peak areas corresponding to 1.032% of the M + 3 peak and 0.004% of the M + 5 peak were subtracted from the M + 5 peak area, and the remaining M + 5 peak intensity expressed as a percentage of the sum of the three peak intensities. The percentage M + 5 was then corrected for the total level of ^{18}O incorporation. The data are shown in Table I.

TABLE I
INCORPORATION OF TWO ATOMS OF ^{18}O FROM $H_2^{18}O$ INTO SUCCINIC ACID BY MhpC-CATALYZED REACTION[a]

Preincubation time (sec)	m/z 147 (%)	m/z 149 (%)	m/z 151[b] (%)	Total % ^{18}O incorporation	% Incorporation of 2 atoms ^{18}O[c]
0	27.6	68.3	4.1	72.4	5.7
0	28.7	68.1	3.2	71.3	4.5
120	29.3	68.8	1.9	70.7	2.7
300	18.8	76.7	4.5	81.2	5.5
300	19.0	76.3	4.6	80.9	5.7

[a] Experiments were carried out as described in Materials and Methods, using 5 μg RFP (23 nmol) and 2 units MhpC. GC-MS analysis (under NH_3 chemical ionization) of dimethyl succinate obtained via derivatization of succinic acid product, after preincubation in 50 mM potassium phosphate pH 8.0 prior to MhpC treatment. Peak areas for M + 1 (m/z 147), M + 3 (m/z 149), and M + 5 (m/z 151) determined by selected ion monitoring of GC-MS data.

[b] The m/z 151 intensity has been corrected for the presence of natural abundance isotope peaks due to the preceding peaks.

[c] Expressed as a percentage of the total observed ^{18}O incorporation.

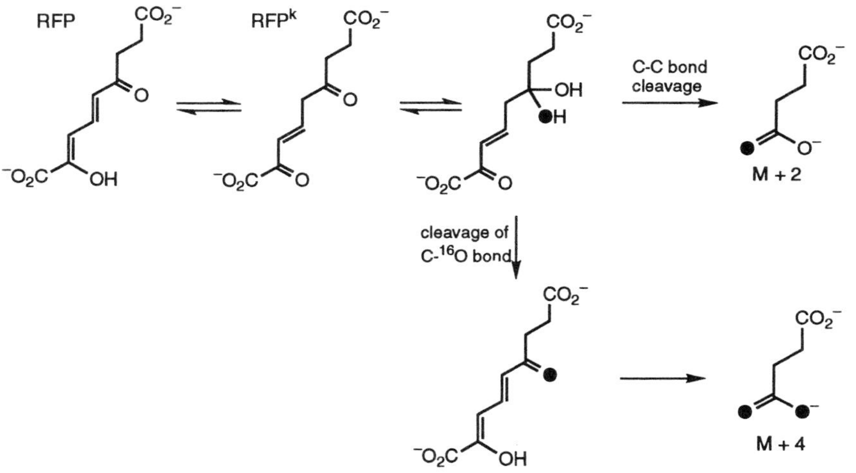

FIG. 2. Enzymatic processing of RFP by MhpC in the presence of $H_2^{18}O$, to generate singly (M + 2) and doubly (M + 4) labeled succinic acid products. See Table I for data.

In each case, after subtraction of natural abundance peaks there remained a small additional M + 4 peak corresponding to an incorporation of 5–6%. There was no significant variation in percentage incorporation versus preincubation time, indicating the nonenzymatic exchange at the C-6 ketone does not occur on this time scale. These data imply that the enzyme does catalyze a small amount of ^{18}O exchange at the acyl position during the processing of the natural substrate, consistent with the formation of a *gem*-diol intermediate (see Fig. 2). Previous kinetic analysis of the MhpC-catalyzed reaction has shown that the enzyme releases a proportion of the ketonized reaction intermediate into solution[9]; thus it is possible that cleavage of the C–^{16}O bond leading to ^{18}O exchange occurs nonenzymatically, after release of some *gem*-diol intermediate into solution.

Interaction of MhpC with Noncleavable Substrate Analog

4-Keto-nona-1,9-dioic acid (KNDA) is an analog of the MhpC substrate which contains the ketone at C-4 (C-6 of RFP) where nucleophilic attack occurs, but lacks the α,β-unsaturated ketone functional group required as an electron sink for C–C bond cleavage. This analog might therefore be susceptible to carbonyl addition at the MhpC active site, but not C–C cleavage. The analog was synthesized by a literature route[14] and was assayed against MhpC.

Addition of KNDA to assays of MhpC at 1 mM concentration gave only a small (10%) decrease in rate, implying that it binds fairly weakly to the free enzyme.

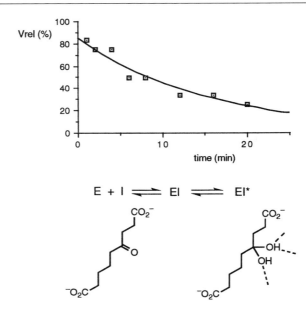

FIG. 3. Time-dependent inhibition of MhpC by KNDA (10 mM concentration). Proposed kinetic scheme for time-dependent inhibition via hydration of the C-4 ketone carbonyl at the MhpC active site.

However, preincubation of KNDA with MhpC was found to cause time-dependent inhibition of enzyme activity (see Fig. 3), implying that a more tightly binding enzyme–KNDA adduct is formed in a time-dependent fashion. Incubation of KNDA with MhpC followed by dialysis gave full recovery of enzyme activity, verifying that the observed inhibition was not irreversible. It therefore appears that KNDA acts as a slow-binding inhibitor for MhpC. In view of the evidence presented above, it is reasonable to suppose that, on binding of KNDA to the MhpC active site, the C-4 ketone is attacked by water to form a *gem*-diol intermediate, which cannot undergo C–C cleavage but would be bound more tightly by the MhpC active site (see Fig. 3). This type of inhibition is precedented by the inhibition of metalloproteases by trifluoromethylketones, believed to occur via hydration to a *gem*-diol at the protease active site.[17]

This mechanism of inhibition would predict that MhpC might catalyze the exchange of ^{18}O-labeled water into the C-4 position. A procedure was developed for the derivatization of KNDA, in order to examine the level of ^{18}O exchange from $H_2^{18}O$, as shown in Fig. 4A. Reduction of the C-4 ketone by sodium borohydride was followed by intramolecular lactonization and methylation, to give a mixture of 7-membered and 5-membered lactone products which were observed by GC-MS at

[17] M. H. Gelb, J. P. Svaren, and R. H. Abeles, *Biochemistry* **24**, 1813 (1995).

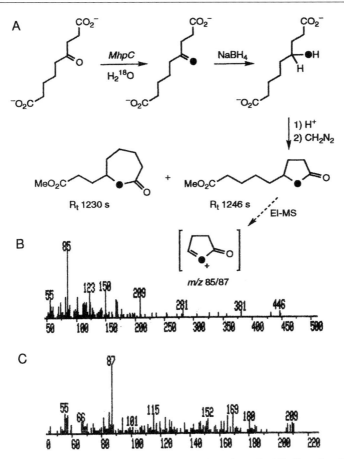

FIG. 4. (A) Derivatization scheme for KNDA used to investigate the MhpC-catalyzed exchange of ^{18}O from $H_2^{18}O$ into the C-4 carbonyl group. The GC retention times of the 5-membered and 7-membered lactone products are shown, together with the structure of the observed m/z 85/87 fragment ion. (B). GC-MS trace for unlabeled 5-membered lactone, showing m/z 85 fragment. (C) GC-MS trace for derivatized KNDA sample after 2 hr incubation with MhpC in $H_2^{18}O$, showing ^{18}O incorporation into m/z 87 peak.

1230 sec and 1246 sec, respectively, verified by comparison with authentic synthetic material. Both peaks showed m/z 201 (MH$^+$) under NH$_3$ chemical ionization conditions, and the 5-membered lactone gave a major fragment ion at m/z 85 under electron impact conditions (Figure 4B), corresponding to the 5-membered lactone portion of the molecule, which showed clear incorporation of ^{18}O versus time. Selected ion monitoring of this peak at m/z 85 and 87 was used to determine the percentage ^{18}O incorporation versus time.

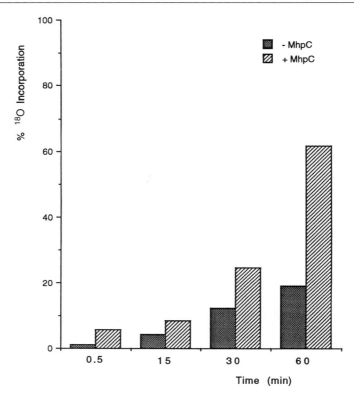

FIG. 5. Level of ^{18}O incorporation in 5-membered lactone derivative of KNDA in the presence or absence of MhpC.

A time-dependent incorporation of ^{18}O label was observed into the 5-membered lactone product arising from the incubation containing MhpC (see Fig. 5). The sample incubated for 60 min with MhpC showed 62% incorporation of one atom of ^{18}O, whereas the sample incubated in the absence of MhpC for the same length of time showed only 19% incorporation. Earlier experiments using smaller amounts of MhpC had given lower levels of ^{18}O incorporation; thus the level of ^{18}O incorporation was also dependent on the MhpC concentration.

These results imply that MhpC is able to catalyze the exchange of ^{18}O at the C-4 ketone of KNDA, consistent with the reversible formation of a *gem*-diol intermediate. Nucleophilic attack of an active-site nucleophile at the C-4 ketone would not lead to ^{18}O exchange. The observed rate of ^{18}O incorporation is approximately 0.15 μmol [^{18}O] KNDA per μmol MhpC per minute, which is comparable to the first-order rate constant for onset of time-dependent inhibition of MhpC by KNDA ($k_{obs} = 0.16$ min^{-1}). It is therefore reasonable to propose that the observed ^{18}O incorporation is caused by the same process which is responsible for time-dependent inhibition.

The rate of nonenzymatic exchange into KNDA is 3.4×10^{-5} mol dm^{-3} min^{-1} over 60 min, comparable to rates of hydration measured for ketones.[18] Ketone hydration in aqueous solution is known to be both acid- and base-catalyzed[18]; thus experiments of this kind must be designed carefully to minimize the rate of background nonenzymatic exchange.

Discussion

Attempts to trap a putative acyl intermediate in the MhpC-catalyzed reaction had previously given an upper limit of approximately 1% covalent intermediate,[13] under conditions where kinetic studies had shown that product release should be rate limiting.[9] Studies of the MhpC-catalyzed reaction in H$_2$^{18}O have shown that there is 5–6% incorporation of a second atom of ^{18}O, consistent with the existence of a *gem*-diol reaction intermediate. Further evidence for the formation of a *gem*-diol intermediate has been obtained from the slow-binding inhibition of MhpC by KNDA, a noncleavable substrate analog, and the MhpC-catalyzed exchange of ^{18}O into the ketone carbonyl of this analog. We therefore conclude that the reaction mechanism of this family of C–C hydrolases is most likely to proceed via base-catalyzed attack of water upon RFP^K to generate a *gem*-diol intermediate.

^{18}O Isotope exchange has therefore proved to be useful method in the investigation of a general base catalytic mechanism in this family of enzymes and in the aspartyl protease family.[3,4] We note that a diagnostic feature of both these classes of enzyme is their ability to catalyze their respective reactions at low pH (i.e., pH 4). Under these conditions the serine protease family is catalytically inactive, because of the protonation of the active-site histidine base; hence the acyl enzyme intermediate of those enzymes accumulates under these conditions.[1] Some lipase enzymes are also known to be active at low pH; thus it remains to be seen whether other members of the α/β-hydrolase family may also follow a general base mechanism.[11] We also note that there are members of the α/β-hydrolase family which catalyze nonhydrolytic reactions, such as cyanohydrin formation,[19] where a general base mechanism is appealing. The role of the putative serine catalytic triad also remains to be established for MhpC and its family of C–C hydrolases.

Nonhydrolyzable substrate analogs have been synthesized for other hydrolytic enzymes, but usually in the context of developing enzyme inhibitors of medicinal relevance. For example, the trifluoromethyl ketone family of protease inhibitors, whose mechanism of action depends on hydration of the ketone carbonyl, have

---

[18] M. Cohn and H. C. Urey, *J. Am. Chem. Soc.* **60,** 679 (1938); R. P. Bell, *Adv. Phys. Org. Chem.* **4,** 1 (1966); L. H. Funderburk, L. Aldwin, and W. P. Jencks, *J. Am. Chem. Soc.* **100,** 5444 (1978).

[19] U. G. Wagner, M. Hasslacher, H. Giriengl, H. Schwab, and C. Kratky, *Structure* **4,** 811 (1996).

found wide application as enzyme inhibitors.[18] We are not aware of their application for ^{18}O isotope exchange study in other enzymes, although we note that the active site hydration of a ketone analog of pepstatin to pepsin has been demonstrated using ^{13}C NMR spectroscopy.[20]

Acknowledgments

The work described in this chapter was supported by a grant (B04835) from the Biomolecular Sciences committee of BBSRC. We thank Julie Herniman (University of Southampton) for assistance with GC-MS analysis.

[20] D. H. Rich, M. S. Bernatowicz, and P. G. Schmidt, *J. Am. Chem. Soc.* **104**, 3535 (1982).

[7] Nucleoside-Diphosphate Kinase: Structural and Kinetic Analysis of Reaction Pathway and Phosphohistidine Intermediate

By JOËL JANIN and DOMINIQUE DEVILLE-BONNE

Introduction

Nucleoside-diphosphate (NDP) kinases catalyze the transfer of the γ-phosphate of a nucleoside triphosphate (N_1TP) onto the β-phosphate of a nucleoside diphosphate (N_2DP). The transfer occurs in two consecutive steps[1]:

$$N_1TP + E \rightleftharpoons N_1DP + E{\approx}P \tag{1a}$$

$$E{\approx}P + N_2DP \rightleftharpoons N_2TP + E \tag{1b}$$

The intermediate species $E{\approx}P$ is phosphorylated on a histidine which plays the part of a nucleophile. The two-step ping-pong mechanism (1) distinguishes the NDP kinase reaction from the large majority of ATP-dependent phosphorylations, where the phosphate acceptor performs a direct attack on the γ-phosphate of the donor. NDP kinase is very efficient: k_{cat} is of the order of 10^3 s^{-1}, which implies that each step takes less than 1 ms. Another peculiarity of NDP kinase is its lack of specificity: the phosphate donor N_1TP and acceptor N_2DP can be any of the nucleotides or deoxynucleotides commonly found in cells. ATP is the most likely donor in reaction (1a), and reaction (1b) can produce all other (deoxy)nucleoside

[1] R. E. J. Parks and R. P. Agarwal, *in* "The Enzymes" (P. Boyer, ed.), Vol. 8, p. 307. Academic Press, New York, 1973.

triphosphates needed for RNA and DNA synthesis. NDP kinases are ubiquitous enzymes found in all organisms and cell types, where they play a major part in maintaining the pool of intracellular NTPs. Nevertheless, deletion mutants of the *ndk* gene coding for NDP kinase in *Escherichia coli* and yeast are viable.[2,3] Thus, alternative pathways must exist for NTP synthesis, which remain to be characterized.

The first NDP kinase amino acid sequences were deduced in 1990 from those of genes cloned in the slime mold *Dictyostelium discoideum* and the bacterium *Myxococcus xanthus*.[4,5] They were closely related to the human *nonmetastatic 23* (*nm23*) gene implicated in metastasis control, and the *abnormal wing disc* (*awd*) developmental gene of *Drosophila*.[6] The human Nm23 and *Drosophila* Awd proteins were soon demonstrated to be NDP kinases. Many other sequences are now available, including several isoforms of the human Nm23 protein. All are homologous and share at least 40% identity over the 150-odd residues that constitute the mature polypeptide chain. Unless specified, we use here the sequence numbering of the human isoform B, also called Nm23-H2, which has 152 residues and forms a hexamer with a molecular mass of about 100 kDa.[7]

The X-ray structure of a *Dictyostelium* mutant NDP kinase[8] is the first of a set that includes the phosphohistidine form of the protein, dead-end complexes with ADP and other NDP substrates and a complex with ADP and beryllium fluoride mimicking the enzyme–ATP complex or with aluminum fluoride mimicking the transition state of the reaction (Table I). In parallel, ligand affinities and rate constants were obtained from solution studies for individual steps of the enzymatic reaction and also for the off-pathway hydrolysis reaction, that is the cause of a low background NTPase activity. Taken together, the biochemical and structural data provide a detailed image of the mechanism of a biologically important reaction and of a covalent intermediate which is not stable on the time scale of structural analysis.

Enzyme Structure

X-ray studies (reviewed in Ref. 9) cover the NDP kinases of the bacterium *Myxococcus xanthus,* the slime mold *Dictyostelium discoideum, Drosophila*

---

[2] H. Hama, N. Almaula, C. G. Lerner, S. Inouye, and M. Inouye, *Gene* **105**, 31 (1991).
[3] T. Fukuchi, J. Nikawa, N. Kimura, and K. Watanabe, *Gene* **129**, 141 (1993).
[4] M.-L. Lacombe, V. Wallet, H. Troll, and M. Véron, *J. Biol. Chem.* **265**, 10012 (1990).
[5] J. Munoz-Dorado, S. Inouye, and M. Inouye, *J. Biol. Chem.* **265**, 2707 (1990).
[6] J. Biggs, E. Hersperger, P. S. Steeg, L. A. Liotta, and A. Shearn, *Cell* **63**, 933 (1990).
[7] A. M. Gilles, E. Presecan, A. Vonica, and I. Lascu, *J. Biol. Chem.* **266**, 8784 (1991).
[8] C. Dumas, I. Lascu, S. Moréra, P. Glaser, P. Fourme, V. Wallet, M. L. Lacombe, M. Véron, and J. Janin, *EMBO J.* **11**, 3203 (1992).
[9] J. Janin, C. Dumas, S. Moréra, Y. Xu, P. Meyer, M. Chiadmi, and J. Cherfils, *J. Bioenerg. Biomemb.* **32**, 215 (2000).

TABLE I
NDP KINASE X-RAY STRUCTURES

Source and enzyme	Space group	Subunits per a.u.[a]	Resolution (Å)	R factor (%)[b]	PDB entry	Conditions[c]
Dictyostelium discoideum						
Free enzyme	P6322	1	1.8	0.20	1npk	a
+TDP	R32	1	2.0	0.18	1ndp	b
+ADP	R3	2	2.2	0.19	1ndc	c
+ADP, AlF3	P3121	3	2.0	0.18	1kdn	d
+ADP, BeF3	P3121	3	2.3	0.19	2bef	e
Phosphorylated	P6322	1	2.1	0.19	1nsp	f
H122G + ADP, P$_i$	P3121	3	2.5	0.20	1b4s	g
+d4T triphosphate	P3121	3	1.8	0.21	1f3f	h
Drosophila melanogaster						
Free enzyme	P3221	3	2.4	0.17	1ndl	i
Phosphorylated	P3221	3	2.2	0.18	1nsq	j
Human						
Nm23-H2	P212121	6	2.8	0.25	1nsk	k
+GDP	P212121	6	2.0	0.18	1nue	l

[a] With eukaryotic NDP kinases, the biological unit is a hexamer whatever the space group and the number of subunits par asymmetric unit (a.u.) of a particular crystal form. Additional X-ray structures are cited in J. Janin, C. Dumas, S. Moréra, Y. Xu, P. Meyer, M. Chiadmi, and J. Cherfils, *J. Bioenerg. Biomemb.* **32,** 215 (2000).

[b] The R factor measures the fit of the atomic model to the diffracted intensities. PDB is the Protein Data Bank.

[c] Crystallization conditions. Unless, specified, crystals were grown by vapor diffusion at 20° in drops containing 5 to 10 mg/ml protein in buffer and precipitant, hanging over pits containing the same buffer and precipitant. The composition of the drop and pit was: (a) drop 5% polyethylene glycol (PEG) 6000, 20 mM MgCl$_2$, 50 mM Tris pH 8.0, pit 10% PEG 6000 [S. Morera, C. Dumas, I. Lascu, M.-L. Lacombe, M. Véron, and J. Janin, *J. Mol. Biol.* **243,** 373 (1994)]; (b) drop 8% PEG 6000, 10 mM TDP, 20 mM MgCl$_2$, 50 mM Tris pH 8.0, pit 16% PEG 6000 [J. Cherfils, S. Morera, I. Lascu, M. Véron, and J. Janin, *Biochemistry* **33,** 9062 (1994)]; (c) drop 5% PEG 6000, 10 mM ADP, 20 mM MgCl$_2$, 50 mM Tris pH 8.0, pit 17.5% PEG 6000 [S. Morera, I. Lascu, C. Dumas, G. LeBras, P. Briozzo, M. Véron, and J. Janin, *Biochemistry* **33,** 459 (1994)]; (d) drop 32% PEG monomethyl ether (M_r 550), 20 mM MgCl$_2$, 25 mM NaF, 0.5 mM aluminum nitrate, 50 mM Tris pH 7.4 [Y. Xu, S. Morera, J. Janin, and J. Cherfils, *Proc. Natl. Acad. Sci. U.S.A.* **94,** 3579 (1997)]; (e) same as in (d), except that 0.5 mM BeCl$_2$ replaced aluminum nitrate; (f) same as in (a), 40 mM dipotassium phosphoramidate being added 1 hr before data collection [S. Morera, M. Chiadmi, I. Lascu, and J. Janin, *Biochemistry* **34,** 11062 (1995)]; (g) drop 21% PEG 400, 10 mM ATP, 20 mM MgCl$_2$, 50 mM Tris pH 7.5, pit 42% PEG 400 [S. J. Admiraal, B. Schneider, P. Meyer, J. Janin, M. Véron, D. Deville-Bonne, and D. I. Herschlag, *Biochemistry* **38,** 4701 (1999)]; (h) drop 19% PEG 400, 7.5 mM d4T triphosphate, 20 mM MgCl$_2$, 50 mM Tris pH 7.5, pit 38% PEG 400 [P. Meyer, B. Schneider, S. Sarfati, D. Deville-Bonne, C. Guerreiro, J. Boretto, J. Janin, M. Véron, and B. Canard, *EMBO J.* **19,** 3520 (2000)]; (i) drop 1 M ammonium sulfate, 20 mM MgCl$_2$, 30 mM Tris pH 7.2, pit 2 M ammonium sulfate [M. Chiadmi, S. Moréra, I. Lascu, C. Dumas, G. LeBras, M. Véron, and J. Janin, *Structure* **1,** 283 (1993)]; (j) same as (i), 200 mM dipotassium phosphoramidate being added 3 hr before data collection [S. Morera, M. Chiadmi, I. Lascu, and J. Janin, *Biochemistry* **34,** 11062 (1995)]; (k) drop 0.9 M ammonium sulfate, 0.1 M Tris pH 8.5, 15% (w/v) glycerol, pit 2.2 M ammonium sulfate, 0.1 M Tris pH 8.5, 20% (w/v) glycerol [P. A. Webb, O. Perisic, C. E. Mendola, J. M. Backer, and R. L. Williams, *J. Mol. Biol.* **251,** 574 (1995)]; (l) drop 9% PEG 1500, 10 mM GDP, 20 mM MgCl$_2$, 50 mM Tris pH 8.4, 1 mM dithiothreitol (DTT), pit 18% PEG 1500 [S. Morera, M.-L. Lacombe, Y. Xu, G. LeBras, and J. Janin, *Structure* **3,** 1307 (1995)].

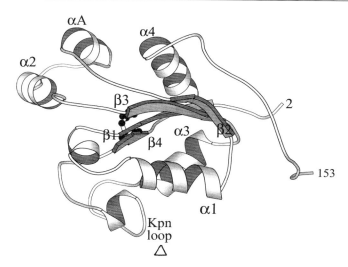

FIG. 1. The NDP kinase fold. Ribbon tracing of the *Drosophila* subunit [M. Chiadmi, S. Moréra, I. Lascu, C. Dumas, G. LeBras, M. Véron, and J. Janin, *Structure* **1**, 283 (1993)]. The product of the *awd* gene has 153 residues, one more than the human Nm23-H2 enzyme. The central four-stranded antiparallel β sheet and helices $\alpha 1-\alpha 3$ constitute the $\alpha\beta$ plait or ferredoxin motif. The active site histidine, drawn in ball-and-stick on the top face of strand $\beta 4$, becomes phosphorylated during catalysis. Like all other eukaryotic NDP kinases, the *Drosophila* Awd protein is a hexamer made of two trimers. The threefold axis (Δ) runs normal to the plane of the figure and close to the Kpn loop. The loop and the extended C-terminal segment are implicated in trimer contacts, which are weakened by the *Killer of prune* mutation, a point substitution in the Kpn loop.

melanogaster, and several human, rat, and bovine isoforms. The proteins are overexpressed in *E. coli* and purified as described in references given in Table I. The *Myxococcus* protein is a homotetramer,[10] whereas the eukaryotic proteins are hexamers. The fold of the subunit is nevertheless the same in all, as expected from the sequence similarity. It contains a four-stranded antiparallel β sheet with the topology $\beta 2\beta 3\beta 1\beta 4$, and two connecting helices $\alpha 1$ and $\alpha 3$ (Fig. 1). Strand $\beta 4$ carries the nucleophilic His-118. The four β strands and two α helices constitute a very common structural motif, which was first observed in *Pseudomonas aerogenes* ferredoxin and is sometimes called the $\beta\alpha\beta\beta\alpha\beta$ fold or $\alpha\beta$ plait. It is found in a number of nucleotide binding proteins unrelated to NDP kinase, and also in RNA and DNA binding proteins.[9,11]

[10] R. L. Williams, D. A. Oren, J. Munoz-Dorado, S. Inouye, M. Inouye, and E. Arnold, *J. Mol. Biol.* **234**, 1230 (1993).
[11] S. Morera, C. Dumas, I. Lascu, M.-L. Lacombe, M. Véron, and J. Janin, *J. Mol. Biol.* **243**, 373 (1994).

In NDP kinase, the ferredoxin fold is augmented by the insertion of a helix hairpin (αA–α2) between strands β2 and β3, and of a 22-residue loop between helix α3 and strand β4. The latter was named the Kpn loop after the *killer of prune* (*kpn*) mutation in *Drosophila*. This mutation is a point substitution in the *awd* gene, which changes the first residue of the loop and yields an active but unstable protein.[12] Beyond strand β4, the polypeptide chain continues with helix α4 and a 20-residue C-terminal segment that makes many subunit contacts in hexameric NDP kinases. The *Myxococcus* protein is a few residues shorter and its C terminus cannot make these contacts. This may explain its different quaternary structure, also shared by a number of bacterial NDP kinases.

The active site comprises the nucleophilic histidine and the substrate binding site. As shown in Fig. 2, each subunit carries a single nucleotide binding site, shaped like a slit and located between the Kpn loop and the αA–α2 helix hairpin. High temperature factors indicate that the hairpin is mobile in the absence of a ligand. It moves by 1 to 2 Å and closes the slit after the nucleotide is bound, the only significant change observed when comparing crystal structures of the free protein and the various complexes. The relatively small amplitude of the hairpin movement contrasts with the large conformation changes seen in most ATP-dependent kinases when substrates bind, well-documented in, for instance, nucleoside monophosphate kinases.[13] Nevertheless, a 2 Å displacement of the αA–α2 hairpin suffices to change the crystal packing and forbid nucleotide exchange. Except for *Myxococcus*, all NDP kinase crystals with bound nucleotides cited in Table I have been obtained by cocrystallization. They are often not isomorphous with the free enzyme crystals, but their structure is easily solved by molecular replacement.

In the complexes with ADP, GDP and TDP, all three nucleotides bind in the same way.[10,14–16] The base is sandwiched between nonpolar side chains of the αA–α2 hairpin and the Kpn loop. It makes no polar interaction with protein groups, which may explain why NDP kinases accept all common nucleotides as substrates. The eukaryotic enzymes do nevertheless have a better affinity for guanine nucleotides.[17] As shown in Fig. 2, the exocyclic amino group of guanine interacts with the C-terminal glutamate residue of an adjacent subunit of the hexamer, an interaction that cannot be made with other bases or in tetrameric NDP kinases.[15] In contrast to the base, the sugar and phosphate moieties make many polar interactions with the protein, and especially with basic residues. Two arginines

[12] I. Lascu, A. Chaffotte, B. Limbourg-Bouchon, and M. Véron, *J. Biol. Chem.* **267**, 12775 (1992).
[13] C. Vonrheim, G. Schlauderer, and G. Schulz, *Structure* **3**, 483 (1995).
[14] S. Morera, I. Lascu, C. Dumas, G. LeBras, P. Briozzo, M. Véron, and J. Janin, *Biochemistry* **33**, 459 (1994).
[15] S. Morera, M.-L. Lacombe, Y. Xu, G. LeBras, and J. Janin, *Structure* **3**, 1307 (1995).
[16] J. Cherfils, S. Morera, I. Lascu, M. Véron, and J. Janin, *Biochemistry* **33**, 9062 (1994).
[17] S. Schaertl, M. Konrad, and M. A. Geeves, *J. Biol. Chem.* **273**, 5662 (1998).

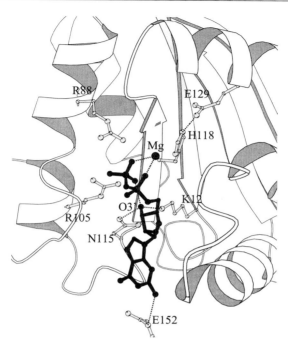

FIG. 2. The active site. Dark bonds represent GDP bound to the human Nm23-H2 enzyme [S. Morera, M.-L. Lacombe, Y. Xu, G. LeBras, and J. Janin, *Structure* **3**, 1307 (1995)]. Empty bonds are side chains interacting with the nucleotide. Strand $\beta 4$, at the center of the drawing, carries the nucleophilic His-118, held in position by a hydrogen bond to Glu-129. In GDP, the amino group of the guanine base interacts with Glu-152 at the C terminus of a neighboring subunit. The α- and β-phosphates both ligate a Mg^{2+} ion and interact with Arg-88 and Arg-105. The ribose O-3' is at the center of a hydrogen bond network (dashes) involving the Lys-12 and Asn-115 side chains, and the β-phosphate oxygen closest to His-118.

interact with the β-phosphate; a lysine (Lys-12) interacts with the 3'-OH group of the sugar and with the γ-phosphate of NTP substrates.

Two important polar interactions observed in the X-ray structures do not directly involve protein groups. One occurs within the nucleotide, the other is with a Mg^{2+} ion. The internal interaction is a short hydrogen bond between the 3'-OH of the ribose or deoxyribose and O-7, the β-phosphate oxygen closest to the nucleophilic histidine. The importance of this interaction is developed below. NDP kinase requires Mg^{2+} for activity,[18] although it can bind nucleotides in its absence.[10] Like most enzymes transferring a phosphate from nucleotides, it takes

[18] R. M. Biondi, B. Schneider, E. Passeron, and S. Passeron, *Arch. Biochem. Biophys.* **353**, 85 (1998).

as substrate a complex with a divalent ion. In the X-ray structures, Mg^{2+} is seen to ligate the phosphate groups, which provide either two (NDP) or three (NTP) oxygen ligands. No protein group is directly involved in Mg^{2+} binding and water molecules complete the octahedral coordination.

Histidine Phosphorylation

The nucleophilic histidine is located at the bottom of the nucleotide binding pocket. Its side-chain $N\varepsilon$ atom interacts with the carboxylate of a conserved glutamate (Glu-129 in the human enzyme). The interaction fixes the orientation of the imidazole group and places the $N\delta$ atom about 4.5 Å from the β-phosphate O-7 oxygen in the complexes with NDP substrates. It also implies that $N\varepsilon$ is protonated and $N\delta$ is unprotonated above the pK_a of the histidine. $N\delta$ is ready to accept a phosphate group and there is no need for a catalytic base as is needed for acylating the hydroxyl group of a serine or a tyrosine. In solution, the phosphohistidine form of the protein can be prepared by adding ATP (or any other NTP) in excess. The equilibrium constant

$$K = \frac{[E{\approx}P][NDP]}{[E][NTP]} \quad (2)$$

is in the range of 0.15 to 0.5 and approximately the same in enzymes from different sources, or for different nucleotides.[19–21] Thus, the equilibrium favors the ATP/histidine pair over the ADP/phosphohistidine pair by a factor of 2 to 6. Full phosphorylation can nevertheless easily be achieved at subsaturating ATP levels, by adding pyruvate kinase and excess phosphoenolpyruvate to convert the ADP product back to ATP.[22]

The fluorescence of a nearby tryptophan residue is a convenient reporter on the level of phosphorylation. In *Dictyostelium*, where the subunit has only one tryptophan, the fluorescence emission intensity observed at 340 nm under excitation at 295 nm is quenched by 20% on ATP addition.[20] No change in intensity is observed when ADP or a nonreactive ATP analog such as AMP–PCP is added, nor when ATP is added to a mutant protein lacking the nucleophilic histidine. Thus, the fluorescence monitors phosphorylation, not substrate binding. In the human NM23-H1 and NM23-H2 isoforms which have several tryptophan residues, a 10% quenching is also observed under the same conditions and can be used to follow phosphorylation. Moreover, the phosphorylated protein can be prepared by incubation with

---

[19] J. Lascu, R. D. Pop, H. Porumb, E. Presecan, and I. Proinov, *Eur. J. Biochem.* **135,** 497 (1983).
[20] D. Deville-Bonne, O. Sellam, F. Merola, I. Lascu, M. Desmadril, and M. Véron, *Biochemistry* **35,** 14643 (1996).
[21] B. Schneider, Y. W. Xu, O. Sellam, R. Sarfati, J. Janin, M. Véron, and D. Deville-Bonne, *J. Biol. Chem.* **273,** 11491 (1998).
[22] I. Lascu and P. Gonin, *J. Bioenerg. Biomemb.* **32,** 237 (2000).

ATP and isolated by gel filtration. Its dephosphorylation by a NDP substrate then leads to a recovery of the tryptophan fluorescence. In this way, the reaction kinetics can be followed in either direction after fast mixing in a stopped-flow apparatus:

$$E + NTP \underset{K_{NTP}}{\rightleftharpoons} E \cdot NTP \xrightarrow{k_{pho}} E{\approx}P + NDP \tag{3a}$$

$$E{\approx}P + NDP \underset{K_{NDP}}{\rightleftharpoons} E{\approx}P \cdot NDP \xrightarrow{k_{depho}} E + NTP \tag{3b}$$

In this scheme, K_{NTP} and K_{NDP} are the equilibrium dissociation constants of the two Michaelis complexes $E \cdot NTP$ and $E{\approx}P \cdot NDP$; k_{pho} and k_{depho} are the rate constants of the phosphate transfer steps. The transfer is assumed to be irreversible due to product dissociation in the absence of added NDP [reaction(3a)] or NTP [reaction (3b)], and much slower than the bimolecular steps. With natural substrates, chemical steps are very fast and the second hypothesis probably incorrect.[23] In a stopped-flow apparatus, the phosphorylation reaction can only be followed at subsaturating substrate concentrations. The kinetic data[17] yield apparent second-order rate constants in the range $5-30 \times 10^6\ M^{-1}\ s^{-1}$, which are near the diffusion limit of the bimolecular step.

With poor substrates, the chemical steps are much slower than the binding steps. Monoexponential phosphorylation and dephosphorylation kinetics are observed, with rates

$$k_{obs} = k_{pho} \frac{[NTP]}{K_{NTP} + [NTP]} \tag{4a}$$

or

$$k_{obs} = k_{depho} \frac{[NDP]}{K_{NDP} + [NDP]} \tag{4b}$$

Both the dissociation constants K_{NTP} and K_{NDP} and the rates of phosphate transfer in the two directions can be deduced from measurements done at several substrate concentrations.[21] Examples are given in Fig. 3, in which the NTP substrates are 2′,3′-dideoxy analogs of the natural nucleotides. The rate of fluorescence-detected histidine phosphorylation is seen to plateau at k_{pho} of $0.5-5\ s^{-1}$ depending on the nature of the base, the higher value being for guanine and the lower for cytosine nucleotides. Half-saturation takes place for $[NTP] = K_{NTP} \approx 1$ mM. The apparent bimolecular rate constant of the reaction is therefore $k_{pho}/K_{NTP} = 0.5-5 \times 10^3\ M^{-1}\ s^{-1}$, four orders of magnitude less than for natural nucleotides. The apparent bimolecular rate constant in the reverse direction (k_{depho}/K_{NDP}) is 2–5 times larger, in keeping with the value of equilibrium constant K in Eq. (2), which favors phosphate transfer from the histidine to NDP by this factor, which is the same for the natural nucleotides and their dideoxy analogs.[21]

---

[23] P. Gonin, Y. Xu, L. Milon, S. Dabernat, M. Morr, R. Kumar, M. L. Lacombe, J. Janin, and I. Lascu, *Biochemistry* **22**, 7265 (1999).

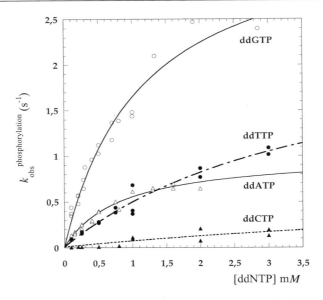

FIG. 3. Kinetics of NDP kinase phosphorylation. Fast-mixing of *Dictyostelium* NDP kinase (1 μM) with various concentration of dideoxynucleoside triphosphates is performed in a HiTech DX2 stopped-flow apparatus [B. Schneider, Y. W. Xu, O. Sellam, R. Sarfati, J. Janin, M. Véron, and D. Deville-Bonne, *J. Biol. Chem.* **273**, 11491 (1998)]. The tryptophan fluorescence intensity (excited at 295 nm for ddATP, 303 nm for other ddNTP) decays in a single exponential step. Its rate constant k_{obs} is plotted against substrate concentration and fitted to Eq. (4a). k_{obs} is less than 5 s^{-1} for ddNTP which are poor substrates, whereas the reaction with natural NTP substrates would be much faster.

The steady-state rate constants of reaction (1) are also reduced by several orders of magnitude for dideoxynucleotides relative to natural substrates, making the analogs very poor substrates of NDP kinase. They share this property with the di- and triphosphate derivatives of other nucleoside analogs lacking a 3'-OH, for instance 2',3'-dideoxy-3'-azidothymidine (AZT or zidovudine) and 2',3'-dideoxydehydrothymidine (d4T or stavudine).[21,23,24] Both thymidine analogs are major drugs in AIDS (acquired immunodeficiency syndrome) therapy. They are incorporated by HIV (human immunodeficiency virus) reverse transcriptase in place of the natural nucleotide, causing DNA chain elongation to terminate, as there is no 3'-OH on which to add the next nucleotide. Reverse transcriptase uses the triphosphate form of the analogs, which cannot pass the cell membrane and must be synthesized by cellular kinases. In part because of the low activity of NDP kinase on substrates lacking a 3'-OH, the triphosphate derivative of AZT

[24] J. Bourdais, R. Biondi, I. Lascu, S. Sarfati, C. Guerreiro, J. Janin, and M. Véron, *J. Biol. Chem.* **271**, 7887 (1996).

remains at a low level in infected cells, which severely limits the efficiency of the drug and facilitates the selection of resistant virus mutants. The design of nucleotide analogs that are better substrates of NDP kinase is therefore of great significance for antiretroviral therapies.[25]

Visualizing Phosphohistidine by X-Ray Crystallography

The NDP kinase phosphohistidine is only moderately stable to hydrolysis, which takes place in the protein at about the same rate as for free phosphohistidine. Hydrolysis is temperature dependent and faster at low pH. At 25° and pH 8.1, the half-life of the phosphorylated protein is of the order of 100 min.[26] This induces a NTPase activity at a rate $k_{hydr} \approx 10^{-4}$ s^{-1}, to be compared with $k_{cat} \approx 10^3$ s^{-1} for reaction (1). Thus, water is by seven orders of magnitude a poorer phosphate acceptor than the β-phosphate of a NDP substrate bound at the active site. Albeit very low, the NTPase activity precludes cocrystallization with NTP substrates. Direct phosphorylation of the eukaryotic enzymes in the crystal also failed, because of intermolecular contacts which prevent access of ATP to the active site. In crystals of the *Myxococcus* enzyme where the contacts allow ATP to bind, soaking yielded a mixture of species, mostly containing ADP and the free histidine.[10] Thus, we considered the possibility of using a smaller molecule as phosphate donor. We tested dipotassium phosphoramidate (K$_2$PO$_3$NH$_2$), a compound known to phosphorylate histidine as the free amino acid, and we found that it could be used to phosphorylate NDP kinase both in solution and in the crystal.[27]

Dipotassium phosphoramidate is prepared from phosphoryl chloride and ammonium hydroxide by the method of Wei and Matthews.[28] Crystals of the *Dictyostelium* and the *Drosophila* enzymes are soaked in 40 to 200 mM K$_2$PO$_3$NH$_2$ before data collection at the LURE-DCI synchrotron radiation center, a low-emittance first-generation synchrotron X-ray source where the measurements are completed in a matter of hours. With both crystals, the electron density map (Fig. 4[27]) clearly shows the presence of a phosphate group covalently attached to Nδ of His-119. No other change is observed in the protein, and the hydrogen bond between Nε and the carboxylate of Glu-129 is maintained. No other protein group is phosphorylated and, remarkably, two histidines close to the active site remain unmodified. In solution, the reaction of NDP kinase with phosphoramidate can be followed by ^{31}P NMR spectrometry and shown to yield the same product as with ATP.[26] Soaking is performed at different times and concentrations before

---

[25] P. Meyer, B. Schneider, S. Sarfati, D. Deville-Bonne, C. Guerreiro, J. Boretto, J. Janin, M. Véron, and B. Canard, *EMBO J.* **19**, 3520.
[26] A. Lecroisey, I. Lascu, A. Bominaar, M. Véron, and M. Delepierre, *Biochemistry* **34**, 12445 (1995).
[27] S. Morera, M. Chiadmi, I. Lascu, and J. Janin, *Biochemistry* **34**, 11062 (1995).
[28] Y.-F. Wei and H. R. Matthews, *Methods Enzymol.* **200**, 388 (1991).

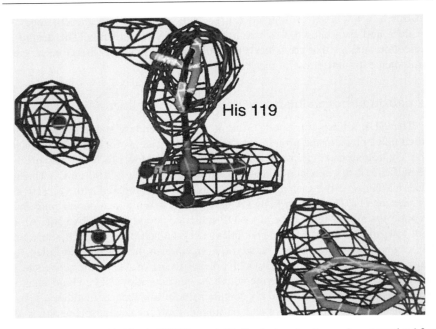

FIG. 4. The phosphohistidine in NDP kinase. A 2Fc-Fo electron density map is contoured at 1.2 σ around His-119 of the *Drosophila* enzyme, phosphorylated on Nδ by soaking crystals in dipotassium phosphoramidate before X-ray data collection. Tyr-53 and two water molecules interact with the phosphate group. [Adapted from S. Morera, M. Chiadmi, I. Lascu, and J. Janin, *Biochemistry* **34**, 11062 (1995).]

data collection, and a gross estimate of the bimolecular rate constant of the phosphorylation reaction is obtained by comparing changes in the electron density at the phosphate position. In the crystal, the rate is less than 1 M^{-1} s^{-1} instead of $\approx 10^7$ M^{-1} s^{-1} when ATP is the phosphate donor in solution. Thus, phosphoramidate is an extremely poor substrate of NDP kinase, but the specific modification of the active site histidine implies that it uses some of the catalytic machinery of the enzyme and is a useful tool for structural analysis.

Visualizing Michaelis Complexes and Transition State

The X-ray structures of complexes with ADP or other NDP substrates represent dead-end complexes, mechanistically less significant than the Michaelis complexes E · NTP and E≈P · NDP. An illustration of the E · NTP complex is obtained using the H122G mutant of the *Dictyostelium* protein, where the nucleophilic histidine (His-122 in *Dictyostelium*) is changed to Gly. The mutant cannot be phosphorylated and has no NDP kinase activity, but it binds ATP and is able to phosphorylate exogenous imidazole at a rate which is only 50-fold less than for the histidine side chain.[29] Whereas imidazole fits snugly in the cavity left by the missing side chain and is the best substrate of the H122G protein, the mutant is a general kinase that accepts as substrate the amino or hydroxyl group of a variety of aliphatic amines and alcohols.[30] It also accepts water, and therefore has a higher ATPase activity than the wild-type protein.

Cocrystallization of the mutant with ATP yielded a structure where only ADP and inorganic phosphate appeared at the active site, presumably due to hydrolysis during crystallization.[29] With d4T triphosphate, however, the electron density indicated that the nucleotide remains essentially intact, imaging a bound NTP substrate.[25] In the H122G-d4T triphosphate complex, the NDP moiety of the modified nucleotide makes the same interactions as in dead-end complexes with ADP or TDP, but the Mg^{2+} ion now coordinates all three phosphate groups. In addition, the γ-phosphate interacts with Lys-12 and the tyrosine that is seen to bind the phosphohistidine in Fig. 4. One peculiar feature of the complex is a result of the modification of the deoxyribose in d4T, which carries a double bond between C-2' and C-3'. In the bound species, a CH$\cdots$O hydrogen bond links the 3'-CH group and oxygen O-7 bridging the β- and γ-phosphates. This bond replaces the 3'-OH$\cdots$O hydrogen bond which, in a normal substrate, links the sugar to the same phosphate oxygen. The geometry of the CH$\cdots$O bond is excellent, and its contribution to catalysis significant, albeit lower than for the OH$\cdots$O bond. d4T diphosphate is phosphorylated about 10 times faster by NDP kinase than dideoxy-TDP, where the geometry of the C2'–C3' single bond prevents the formation of a CH$\cdots$O bond to O-7.[30,31]

Another image of the E·NTP complex is provided by X-ray structures of *Dictyostelium* NDP kinase where beryllium fluoride or aluminum fluoride has been added to the dead-end complex with ADP.[32] Strong electron density indicates the presence of three F^- ions in a plane halfway between the β-phosphate of ADP and the imidazole ring of the active site histidine. A regular trigonal bipyramid is formed by the three F^- ions in the equatorial plane, with $N\delta$ of the histidine and O-7 of the β-phosphate at the two apical positions (Fig. 5). The F^- ions occupy the approximate position of the γ-phosphate oxygens of d4T triphosphate bound to the H122G protein and make the same interactions with Mg^{2+} and protein groups.

[29] S. J. Admiraal, B. Schneider, P. Meyer, J. Janin, M. Véron, D. Deville-Bonne, and D. I. Herschlag, *Biochemistry* **38**, 4701 (1999).
[30] S. Admiraal, P. Meyer, B. Schneider, D. Deville-Bonne, J. Janin, and D. Herschlag, *Biochemistry* **40**, 403 (2001).
[31] B. Schneider, R. Biondi, R. Sarfati, F. Agou, C. Guerreiro, D. Deville-Bonne, and M. Véron, *Mol. Pharmacol.* **57**, 948 (2000).
[32] Y. Xu, S. Morera, J. Janin, and J. Cherfils, *Proc. Natl. Acad. Sci. U.S.A.* **94**, 3579 (1997).

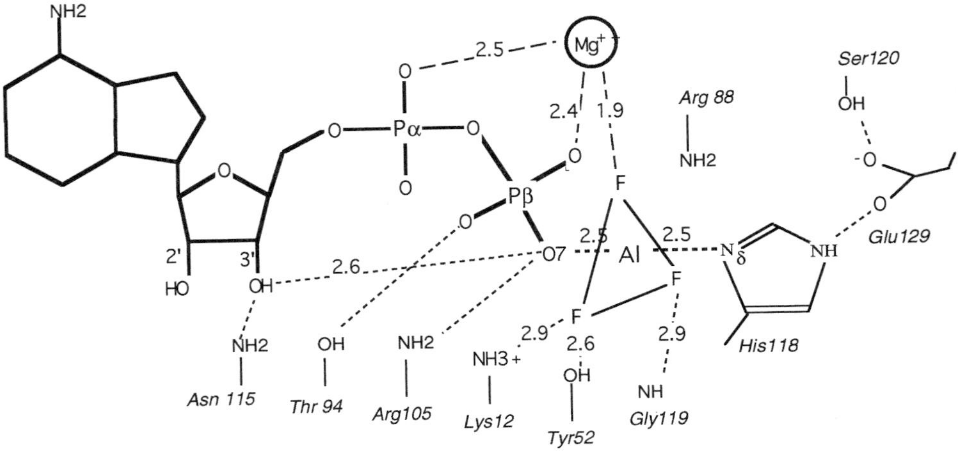

FIG. 5. Interactions of ADP–AlF$_3$ mimic the transition state. The trigonal bipyramid formed by the five ligands of Al^{3+} reproduces the geometry of the transition state for in line transfer of a γ-phosphate. Al^{3+} is at the expected position of Pγ, halfway between O7 of ADP and Nδ of the nucleophilic histidine. Distances (in Å) are from the 2.0 Å X-ray structure of the complex with *Dictyostelium* NDP kinase [Y. Xu, S. Morera, J. Janin, and J. Cherfils, *Proc. Natl. Acad. Sci. U.S.A.* **94**, 3579 (1997)]. The noncovalent Al-F bond (1.85 Å) compares well in geometry and electronic charge distribution with a P–O single bond. Protein groups interacting with the F$^-$ ions are indicated. All are known from site-directed mutagenesis to contribute to catalysis. Yet, biochemical experiments indicate that the short (2.6 Å) hydrogen bond between the 3'-OH and O-7 is the largest single contributor to stabilization of the transition state.

Whereas the two-electron Be^{2+} cation escapes observation, the density of the 10 electrons in Al^{3+} unambiguously places the trivalent cation less than 2 Å from the position of the γ-phosphorus of d4T triphosphate.

The crystallographic experiment demonstrates that beryllium fluoride binds as the BeF$_3^-$ anion and aluminum fluoride as the neutral AlF$_3$ species. BeF$_3^-$ is tetrahedral, the geometry of a phosphate group in the ground state. Thus, ADP–BeF$_3^-$ mimics ATP in the Michaelis complex. AlF$_3$ is planar and mimics the transition state of the phosphate transfer reaction. If the transfer takes place in line, the transition state should resemble a trigonal bipyramid geometry with a planar PO$_3$ group at the base and, at the apex, the leaving and entering groups. In a ATPase or a GTPase, these groups are the β-phosphate and water. In NDP kinase, the nucleophilic histidine replaces water. Beryllium fluoride and aluminum fluoride enhance the ATPase/GTPase activity of myosin, tubulin, and trimeric G proteins and they have long been proposed to be phosphate analogs.[33] X-ray structures

---

[33] J. Bigay, P. Deterre, C. Pfister, and M. Chabre, *EMBO J.* **6**, 2907 (1987).

have shown that they mimic the γ-phosphate of ATP in myosin[34] or of GTP in G proteins.[35,36]

In these structures (reviewed in Ref. 37), beryllium fluoride is present as the tetrahedral BeF_3^-, but aluminum fluoride appears as AlF_4^-, which forms a square bipyramid where Al^{3+} has six ligands instead of five in a trigonal bipyramid. The trigonal AlF_3 species seen in NDP kinase reproduces the geometry of the transition state much better than the square AlF_4^- ion. Nevertheless, five-coordinated Al^{3+} is unexpected, six ligands being the rule, and NDP kinase must select in the surrounding solution a minority species. Five-coordinated Al^{3+} uniquely fits a site designed to accommodate the γ-phosphate undergoing transfer from the NTP substrate to the nucleophilic histidine [and back in reaction (1b)]. The interactions made by ADP-AlF_3 with Mg^{2+} and protein groups described in Fig. 5 certainly contribute to catalysis, and the short (2.6 Å) hydrogen bond observed between O-3′ and O-7 implicates the 3′-OH group in the phosphate transfer, probably by stabilizing the negative charge that appears on O-7 as the bond to Pγ is broken. O-7, which is a ligand of Be/Al in the two complexes, is the leaving group in reaction (1a) and the entering group in reaction (1b).

Beryllium and aluminum fluorides were not known to inhibit ATP-dependent kinases previous to the crystallographic experiment. Indeed, we were unable to demonstrate inhibition of NDP kinase in steady-state or stopped-flow experiments, perhaps because BeF_3^- and AlF_3 compete poorly with NTP substrates. Aluminum fluoride does, however, inhibit *Dictyostelium* uridylate kinase and an X-ray structure of the complex of the enzyme with ADP, UMP, and aluminum fluoride shows that it binds AlF_3 at its active site like NDP kinase that is, forming a trigonal bipyramid with the phosphate donor and acceptor groups.[38] The K_i of 25 μM measured for Al^{3+} is an apparent value that probably indicates a very high affinity for AlF_3, as the cation is almost exclusively six-coordinated in solution.

Labile Phosphorylated Protein Derivatives

Histidine phosphorylation is an intermediate step in the mechanism of enzymes other than NDP kinase such as phosphoester phosphatases and mutases. Phosphohistidine may become a stable species in some cases. In succinyl-CoA synthetase, it is formed at the expense of ATP or GTP and used to generate succinyl phosphate, which then undergoes nucleophilic attack by the sulfur atom

---

[34] A. J. Fisher, C. A. Smith, J. B. Thoden, R. Smith, K. Sutoh, H. M. Holden, and I. Rayment, *Biochemistry* **34**, 8960 (1995).
[35] D. E. Coleman, A. M. Berghuis, E. Lee, M. E. Linder, A. G. Gilman, and S. R. Sprang, *Science* **265**, 1405 (1994).
[36] J. Sondek, D. G. Lambright, J. P. Noel, H. E. Hamm, and P. B. Sigler, *Nature* **372**, 276 (1994).
[37] D. E. Coleman and S. R. Sprang, *Methods Enzymol.* **308**, 70 (1999).
[38] I. Schlichting and J. Reinstein, *Biochemistry* **36**, 9290 (1997).

of coenzyme A to yield succinyl-CoA. Unexpectedly, the X-ray structure of the *E. coli* enzyme showed a phosphohistidine to be present at the active site even though no ATP had been added to the crystallization mixture.[39] The derivative is Nε-phosphohistidine, which is chemically more stable than the Nδ-phosphohistidine of NDP kinase.[40] Nevertheless, it must be stabilized by the protein environment. The nucleophilic histidine is in the loop located near the N terminus of two α helices, and the favorable interaction of the phosphate group with the two helix dipoles is the proposed explanation. The free protein binds inorganic phosphate at the same position and only small changes are observed between the covalent and noncovalent derivatives.[41]

Several mechanisms are known for phosphatases. Most use metal catalysis, but in others, fructose-2,6-bisphosphatase for instance, the substrate first donates its phosphate to a histidine residue. Although phosphohistidine hydrolysis must take place at least at the rate of the enzyme turnover, the intermediate can be trapped by flash-freezing in the presence of excess substrate. This led to a 2.2 Å X-ray structure of the phosphorylated rat liver enzyme.[42] In addition to the presence of N^ε-phosphohistidine, the structure shows that the fructose 6-phosphate product remains bound, suggesting that potent product inhibition of the hydrolysis reaction is the likely cause of the relatively high stability of the phosphohistidine during that experiment.

In bacteria, cellular processes often called "two-component" systems also rely on protein phosphorylation.[43] One such process is sugar import via the PTS system, an essential component of which is the histidine phosphocarrier protein (HPr). In PTS, the phosphate group of phosphoenolpyruvate is transferred first from component E1 to HPr, then to component E2 which phosphorylates the transported sugar. HPr is phosphorylated on Nδ of His-15, at the N terminus of the first α helix of this small (87 residues) protein, which has been extensively studied by NMR and X-ray crystallography. HPr is not self-phosphorylating, and its phosphohistidine is unstable as in NDP kinase. Although HPr phosphorylation by component E1 and excess phosphoenolpyruvate is easily achieved in solution, it is not practical in the crystal, and phosphorylation by phosphoramidate has not been reported. Thus, the phosphohistidine form has been studied by NMR. As there is no reporter proton to locate the phosphate group itself, NMR provides information only on the protein structure, which differs little from the free form.[44,45]

---

[39] W. T. Wolodko, M. E. Fraser, M. N. James, and W. A. Bridger, *J. Biol. Chem.* **269,** 10883 (1994).
[40] D. E. Hultquist, R. W. Moyer, and P. D. Boyer, *Biochemistry* **5,** 322 (1966).
[41] M. E. Fraser, M. James, W. A. Bridger, and W. T. Wolodko, *J. Mol. Biol.* **299,** 1325 (2000).
[42] Y. H. Lee, T. W. Olson, G. M. Ogata, D. G. Levitt, L. J. Banaszak, and A. J. Lange, *Nat. Struct. Biol.* **4,** 615 (1997).
[43] J. B. Stock, A. M. Stock, and J. M. Mottonen, *Nature* **344,** 395 (1990).
[44] N. A. J. Van Nuland, R. Boelens, R. M. Scheek, and G. T. Robillard, *J. Mol. Biol.* **246,** 180 (1995).
[45] B. E. Jones, P. Rajagopal, and R. E. Klevit, *Prot. Sci.* **6,** 2107 (1997).

Protein phosphorylation on aspartate, which is at the heart of bacterial signal transduction, also yields an unstable product. Small phosphorylating agents have been useful in structural studies of proteins involved in nitrogen response: the phosphate receiver domains of the NtrC regulatory protein from *E. coli* and the FixJ transcription activator protein from *Sinorhizobium meliloti*. NtrC has been phosphorylated by carbamoyl phosphate and studied by NMR.[46] FixJ is phosphorylated by incubating 2 hr in 20 mM acetyl phosphate and 10 mM MgCl$_2$ at pH 7.2. After excess EDTA is added, the phosphoaspartate product has been found to be stable for weeks in the absence of divalent metal ions at pH 8.0, enabling crystallization and X-ray structure determination.[47] Even more surprisingly, a stable phosphoaspartate is observed in the X-ray structure of the SpoA receiver domain of *Bacillus stearothermophilus* involved in sporulation even if no phosphorylating agent has been added and Ca^{2+} is present during crystallization.[48] The receiver domains are evolutionarily related and, unlike HPr and the phosphohistidine enzymes mentioned above, they all undergo large conformational changes upon aspartate phosphorylation.

Conclusion

Structural studies of enzymes have made major contributions to our understanding of catalytic mechanisms. Unstable chemical species such as a Michaelis complex or a covalent intermediate, which form and vanish on the millisecond to second time scale, and the even more elusive transition state, which lives a few femto- or picoseconds, can be made accessible to study. Some years ago, very fast methods of crystallographic data collection, such as Laue diffraction, seemed to be the approach of choice. In practice, their impact has been marginal and much more structural information on enzyme catalysis has come from experiments where the time scale of the chemistry has been changed to that of crystallography, rather than the converse. There are many ways to slow enzymatic catalysis: by low-temperature trapping, by removing catalytic groups through site-directed mutagenesis of the protein or chemical modification of the substrate, by replacing substrates with unreactive analogs of either the ground state or the transition state. This is essentially what we did with NDP kinase in order to image individual steps

---

[46] D. Kern, B. F. Volkman, P. Luginbühl, M. J. Nohalle, S. Kustu, and D. E. Wemmer, *Nature* **402**, 894 (1999).

[47] C. Birck, L. Mourey, P. Gouet, B. Fabry, J. Schumacher, P. Rousseau, D. Kahn, and J.-P. Samama, *Structure* **7**, 1505 (1999).

[48] R. J. Lewis, J. A. Brannigan, K. Muchova, I. Barak, and A. J. Wilkinson, *J. Mol. Biol.* **294**, 9 (1999).

[49] M. Chiadmi, S. Moréra, I. Lascu, C. Dumas, G. LeBras, M. Véron, and J. Janin, *Structure* **1**, 283 (1993).

[50] P. A. Webb, O. Perisic, C. E. Mendola, J. M. Backer, and R. L. Williams, *J. Mol. Biol.* **251**, 574 (1995).

of catalysis at the atomic scale in the crystal and correlate these images with the reaction observed in solution.

Acknowledgments

We are grateful to all our colleagues in Gif-sur-Yvette and Institut Pasteur, especially Dr. M. Véron (Paris). Pr. I. Lascu (Bordeaux), whose interest in NDP kinase stimulated this study, suggested the phosphoramidate experiment and synthesized the compound. This work was supported in part by Agence Nationale de la Recherche contre le SIDA and Association pour la Recherche contre le Cancer.

[8] Galactose-1-Phosphate Uridylyltransferase: Kinetics of Formation and Reaction of Uridylyl-Enzyme Intermediate in Wild-Type and Specifically Mutated Uridylyltransferases

By Sandaruwan Geeganage and Perry A. Frey

Hexose-1-phosphate uridylyltransferase (GalT, EC 2.7.7.12), often known as galactose-1-phosphate uridylyltransferase, is a member of the Leloir pathway of galactose metabolism and catalyzes the reaction of UDP-glucose (UDP-Glc) with galactose 1-phosphate (Gal-1-P) to produce UDP-galactose (UDP-Gal) and glucose 1-phosphate (Glc-1-P). Nucleotidyl transfer proceeds by a ping-pong kinetic mechanism, with the intermediate formation of a covalent uridylyl-enzyme (uridylyl-GalT), in which the uridylyl group is covalently bonded to a histidine residue at the active site (His-166 in the *Escherichia coli* enzyme, Fig. 1).[1–6] The detection, isolation, and characterization of the uridylyl-GalT, as well as the steady-state kinetics, have previously been presented in this series.[7]

An autosomal recessive genetic trait results from mutations in GalT and leads to defects in galactose metabolism.[8] The ensuing galactose toxicity syndrome is known as galactosemia. Patients with galactosemia show clinical symptoms such

---

[1] L.-J. Wong and P. A. Frey, *J. Biol. Chem.* **249**, 2322 (1974).
[2] L.-J. Wong and P. A. Frey, *Biochemistry* **13**, 3889 (1974).
[3] L.-J. Wong, K.-F. Sheu, S.-L. Lee, and P. A. Frey, *Biochemistry* **16**, 1010 (1977).
[4] S.-L. Yang and P. A. Frey, *Biochemistry* **18**, 2980 (1979).
[5] T. L. Field, W. S. Reznikoff, and P. A. Frey, *Biochemistry* **28**, 2094 (1989).
[6] J. Kim, F. J. Ruzicka, and P. A. Frey, *Biochemistry* **29**, 10590 (1990).
[7] P. A. Frey, L.-J. Wong, K.-F. Sheu, and S.-L. Yang, *Methods Enzymol.* **87**, 20 (1982).
[8] S. Segal, "The Metabolic Basis of Inherited Disease," 6th Ed. McGraw-Hill, New York, 1989.

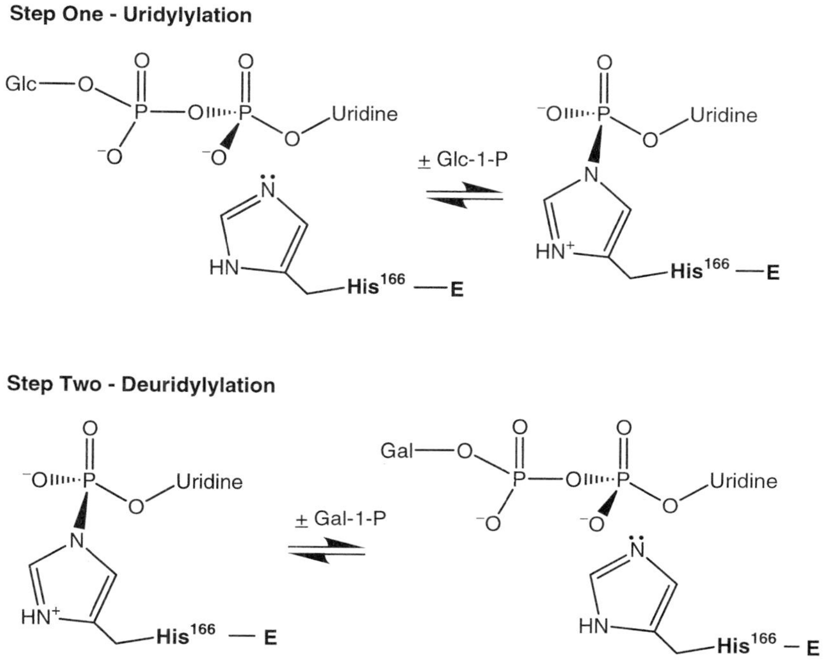

Step One - Uridylylation

Step Two - Deuridylylation

FIG. 1. Schematic representation of the ping-pong reaction mechanism of GalT. Step 1 proceeds with the attack by the histidine nucleophile on the α-phosphorus of UDP-Glc. The nucleophile is uridylylated at position $N^{\epsilon 2}$, and Glc-1-P is released. Step 2 proceeds with the attack of the β-phosphoryl group of Gal-1-P on the uridylyl-enzyme intermediate. UDP-Gal is formed, regenerating the His nucleophile.

as cataracts, jaundice, cirrhosis, brain damage, and neurodevelopmental problems. Presently, the only available therapy is to place the patients on a galactose-free diet.[8] Even then, owing to the impossibility of preventing the metabolic production of galactose, patients suffer from secondary complications later in life. These include abnormal growth, mental retardation, and premature menopause in women. One of the most common mutations found in this disease is of Gln → Arg at position 188 in the amino acid sequence of the human enzyme.[9,10]

The *E. coli* GalT, the subject of this article, is the best characterized hexose-1-P uridylyltransferase. GalT is a homodimer of molecular mass 79,288 Da containing 348 amino acids per subunit.[11,12] It is a metalloprotein, with one structural zinc ion

[9] J. K. V. Reichardt, F. Packman, and S. Woo, *Am. J. Hum. Genet.* **49,** 860 (1991).
[10] L. J. Elsas, J. L. Fridovich-Keil, and N. D. Leslie, *Int. Pediat.* **8,** 101 (1993).
[11] S. Saito, M. Ozutsumi, and K. Kurahashi, *J. Biol. Chem.* **242,** 2362 (1967).
[12] H. G. Lemaire and B. Mueller-Hill, *Nucleic Acids Res.* **14,** 7705 (1986).

and one structural iron ion per subunit.[13] These structural metal ions can be replaced with a variety of divalent metal ions, and the reconstituted enzyme displays very substantial enzymatic activity.

Since the last article on GalT in this series,[7] a number of advances have been made in study of this enzyme. A major contribution comes from the solution of several X-ray crystal structures of GalT. Two high-resolution X-ray structures are available for the wild-type enzyme, one of GalT with UDP bound to the active site and another of the covalent intermediate, uridylyl-GalT.[14,15] Two other structures of a variant of this enzyme, H166G-GalT with UDP-Gal or UDP-Glc bound to the active site have also been solved.[16] The H166G-GalT variant does not contain the nucleophilic imidazole ring of His-166 and cannot catalyze the wild-type reaction. However, in chemical rescue experiments, it readily transfers a uridylyl group to imidazole (Im) from UDP-Gal or UDP-Glc, producing UMP-imidazole (UMP-Im) and Gal-1-P or Glc-1-P.[6]

Another major advance is the establishment of transient kinetic methods for examining the kinetic competence of the uridylyl-GalT intermediate and evaluating the kinetic properties of mutated GalTs.[17] Previous steady-state kinetic studies showed that the enzyme catalyzes nucleotidyl transfer by a ping-pong kinetic mechanism and that GalT catalyzes kinetically competent isotopic exchange reactions.[2] However, the kinetic competence of the covalent intermediate was not directly established until recently.[17] Furthermore, the availability of a number of X-ray crystal structures allowed active-site residues that form potential catalytically relevant interactions to be identified. The kinetic consequences of mutations at these positions investigated by steady-state and transient kinetic methods.[18] In this chapter, we discuss the salient features of GalT X-ray crystal structures, transient-state kinetic methods, site-directed mutagenesis studies, and several other biochemical studies of GalT.

Structures of GalT and Uridylyl-GalT

Wedekind *et al.*[14,15] solved the structure of the complex between GalT and UDP at 1.8 Å resolution (1HXP) and also the structure of the uridylyl-GalT at 1.86 Å resolution (1HXQ). Neither of these structures provided information about the hexose-binding site. Thoden and co-workers generated crystals and solved two structures of H166G-GalT, one with UDP-Glc and another with UDP-Gal bound

---

[13] F. J. Ruzicka, J. E. Wedekind, J. Kim, I. Rayment, and P. A. Frey, *Biochemistry* **34**, 5610 (1995).
[14] J. E. Wedekind, P. A. Frey, and I. Rayment, *Biochemistry* **34**, 11049 (1995).
[15] J. E. Wedekind, P. A. Frey, and I. Rayment, *Biochemistry* **35**, 11560 (1996).
[16] J. B. Thoden, F. J. Ruzicka, P. A. Frey, I. Rayment, and H. M. Holden, *Biochemistry* **36**, 1212 (1997).
[17] S. Geeganage and P. A. Frey, *Biochemistry* **37**, 14500 (1998).
[18] S. Geeganage, V. W. K. Ling, and P. A. Frey, *Biochemistry* **39**, 5397 (2000).

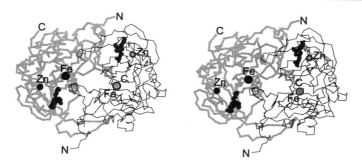

FIG. 2. The dimeric structures for the H166G-GalT is shown as α-carbon traces. UDP-Gal is shown as a ball-and-stick model. There is no significant difference between the α-carbon traces of the variant structure shown here and the native structures. Figures were made from coordinates deposited at the Protein Data Bank (1GUP-UDP-Gal·H166G-GalT; 1HXQ-uridylyl-GalT). Mol View 1.4.8 was used for generating the figures.

to the active site.[16] The structure of H166G-GalT with UDP-Gal bound is shown in Fig. 2. The wild-type GalT and the variant enzyme structures are essentially the same, except for one missing region (Ala^{36}-Gln^{43}) in the wild-type structure.

GalT exists as a homodimer, in which each subunit has 6 α helices and 13 strands in antiparallel β sheets. The subunit core is composed of nine antiparallel strands in a β sheet that forms a half-barrel structure, with the barrel axis parallel to the local dyad relating each subunit. Two amphipathic helices fill the barrel interior in a manner somewhat reminiscent of two hot dogs placed on a bun. The structural Fe ion resides on the outside of the barrel, centered on the dimeric interface.[14]

The active site resides on the external surface of the barrel rim. Loops of the barrel strands are tethered by a zinc ion, and it appears that this coordination may hold the active-site geometry intact. The region that was missing in the structures of uridylyl-GalT and the complex of GalT with UDP is ordered in the structure of the complex of UDP-Gal with H166G-GalT. This surface loop encompasses Ala^{36}-Gln^{43}, extending toward and forming a part of the active-site pocket for the neighboring subunit. The uridine-binding pocket and the diphosphate-binding region in the active site are created predominantly by residues from one subunit, and residues from the neighboring subunit form the sugar-binding pocket. Therefore, it appears that GalT must exist as a dimer to maintain catalytic activity.

Each GalT subunit has a single active site that resembles an open cleft. The active site nucleophile is His-166, and in the structure of uridylyl-GalT the $N^{\varepsilon 2}$ of this residue is covalently bonded to the UMP group. The side chain of Gln-168, which is conserved in all known GalT sequences and corresponds to Gln-188 in the human enzyme, is facing the active site adjacent to His-166. In the uridylyl-enzyme, $N^{\varepsilon 2}$ of Gln-168 lies at a distance of 3.0 Å from the bridging and nonbridging oxygens on the α-phosphorus atom (Fig. 3). In the H166G-GalT

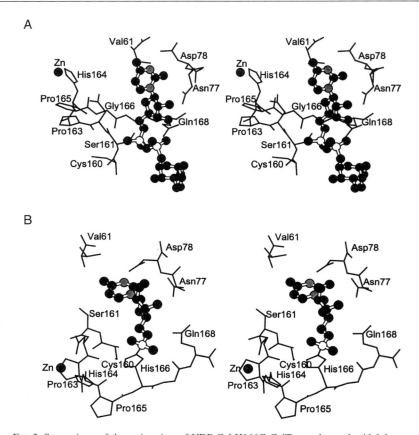

FIG. 3. Stereoviews of the active sites of UDP-Gal·H166G-GalT complex and uridylyl-enzyme. (A) The active site of H166G-GalT complexed with UDP-Gal (ball and stick) shows the interactions of Gln 168 with phosphoryl oxygens and the interaction of Ser-161 O^γ with the β-phosphoryl oxygen. The main chain amide proton of Ser-161 is pointing toward the α-phosphoryl oxygen. The vacant space left by the H166G mutation is occupied by a hydrated K^+ ion (not shown). The Cys-160 side chain is rotated away from the substrate. Val-61, Asn-77, and Asp-78 also form interactions with the uridine moiety as shown. (B) The active site of the uridylyl-enzyme show the UMP group's (ball and stick) phosphate oxygen interacting with Gln-168 side chain and Cys-160 S^γ. Ser-161 is rotated away from the substrate in this structure. Note the backbone carbonyl of His-164 is 2.5 Å away and oriented toward the N^δ of His-166. The Val-61, Asn-77, and Asp-78 interactions with the nucleotide are also shown. Figures were made from coordinates deposited at the Protein Data Bank (1GUP-UDP-Gal·H166G-GalT; 1HXQ-uridylyl-GalT). Mol View 1.4.8 was used for generating the figures.

structure, Gln-168 is within hydrogen bonding distance of nonbridging oxygen atoms of the α- and β-phospho groups of UDP-Glc or UDP-Gal.

In the structures of H166G-GalT complexed with UDP-hexoses, Ser-161 forms backbone and side-chain hydrogen bonds with the substrate β-phosphoryl group. An α-phosphoryl oxygen of the UDP-hexose lies 2.8 Å from the backbone amide nitrogen of Ser-161, and β-phosphoryl oxygen lies 2.6 Å from the Ser-161 O^γ. In the structure of uridylyl-GalT, however, the side chain of Ser-161 is rotated away from the substrate and lies 7.5 Å from the α-phosphoryl oxygens. Ser-161 is conserved in the known GalTs except in three species, where it is replaced with Thr, and in a fourth, where it is replaced with Asp.

Cys-160, which is also highly conserved, is oriented with its S^γ 3.1 Å from a nonbridging phosphate oxygen in the structure of the uridylyl-GalT. In the structures of H166G-GalT in complex with UDP-hexoses, Cys-160 S^γ lies 4.6 Å from the phosphate oxygen and is rotated 90° away from the substrate. These structural studies suggest that subtle conformational changes take place in the course of catalysis. In the binary Michaelis complexes, the Ser-161 side chain seems to be projected toward the substrate, whereas in the uridylyl-GalT form it is rotated away from the uridylyl group. Cys-160 behaves inversely.

In addition to the contacts between Gln-168, Ser-161, and Cys-160 and the phosphoryl groups of the nucleotide sugar, nucleotide and protein engage in a number of other interactions. The uridine moiety binds to the protein through interactions with the carboxylate side chain and the backbone amide proton of Asp-78. Val-61 also forms a backbone amide proton interaction with the uracil ring. The ribose 2'- and 3'-oxygens interact with the Asn-77 side chain. All these interactions are from the same subunit.

In the uridylyl-enzyme, the His-166 N^δ is 2.5 Å away from the backbone carbonyl oxygen of His-164. It appears that the His-166 $N^\delta H$ donates a hydrogen bond to that backbone carbonyl oxygen in the uridylyl-GalT.

The structures of H166G-GalT in complex with UDP-hexoses provide information about the sugar-binding site of GalT. The loop from Ala-36 to Gln-43, which was not observed in the structures of the free and uridylylated enzyme, forms part of the hexose binding pocket in the neighboring subunit. Arg-28 and Arg-31 from this loop interact with the β-phosphoryl oxygens. Similarly, most of the hydrogen bonding interactions to the hexopyranose ring are formed with residues from the neighboring subunit. Lys-311 and Gln-323 provide side-chain interactions. The structural difference between the structures with UDP-Gal and UDP-Glc bound is an elegant ~90° rotation of the side chain of Glu-317 about its C^β–C^γ bond, such that Glu-317 $O^{\varepsilon 2}$ maintains the interaction with the C4(OH) of UDP-Gal or UDP-Glc. These results confirm that UDP-Gal and UDP-Glc share a common binding site as predicted more than 20 years ago based on steady-state kinetic studies.

One other important feature revealed by the structure of H166G-GalT is the presence of potassium hexahydrate ion in the cavity created by the His→Gly mutation.[16] The ability of H166G-GalT to support a positive charge in the cavity raises the question of whether the imidazole ring of His-166 retains positive charge in the uridylyl-GalT. That is, the proton residing on the imidazole ring of His-166 may be retained upon uridylylation. Independent evidence for this will be presented in a later section.

As discussed above, GalT is a zinc and iron metalloenzyme. Zinc in its binding site lies 8 Å from the active site. His-164, which is a part of the highly conserved active-site triad $H^{164}P^{165}H^{166}$, donates one ligand of the tetrahedral coordination sphere about zinc. Active-site residues $C^{160}SNPHP^{165}$ are in a turn that wraps around the diphosphate portion of the substrate where the chemistry occurs. The nucleophilic catalyst His-166 is at the start of a β sheet following this turn. The His-164 backbone carbonyl is oriented at 2.5 Å from the $N^\delta H$ of His-166 and its oxygen forms a hydrogen bond with the $N^\delta H$ as discussed previously. Therefore, His-164 ligation to zinc appears to be important for maintaining the active-site structure. Mutation of His-164 to Asn leads to complete loss of activity. By ligating to zinc, it appears that His-164 brings structural stabilization to this region and maintains the proper alignment of active-site residues for efficient catalysis.

Cys-52 and Cys-55 are on a surface loop of the subunit, with their side chains coordinated to zinc in a rubredoxin knuckle motif. Cysteine thiolates are also involved in a network of hydrogen bonding that may provide structural stability to this region. His-115, which is somewhat buried in the structure, completes the zinc coordination. Inasmuch as all zinc coordination sites are occupied, and the zinc ion can be replaced with a wide range of divalent metal ions with retention of activity, zinc is most likely a structural metal ion in GalT.[13]

Iron is coordinated to His-296, His-298, His-281, and Glu-182 in a distorted square pyramidal geometry. His-296, His-298, and bidentate Glu-182 form the base of the pyramid and His-281 forms the axial ligand.[14] Iron is farther away from the active site and the iron can be removed selectively from a Q182A-GalT variant without affecting the GalT activity.[19]

Single Turnover Kinetics of Wild-Type GalT

The kinetic competence of uridylyl-GalT formation was not directly established in early kinetic studies.[2] One obstacle in establishing kinetic competence was the value of k_{cat} for the wild-type GalT, 780 s^{-1} at 27°. This meant that the uridylylation rate constant could be faster than 780 s^{-1}, so that the half-time for uridylylation would be less than 0.8 ms. Such a rate could not be measured by available rapid mixing methods. Therefore the uridylylation and deuridylylation

---

[19] S. Geeganage and P. A. Frey, *Biochemistry* **38**, 13398 (1999).

rate constants had to be measured at a lower temperature, 4°, where the rates are much slower.[17]

Rapid mix quench experiments have been performed by use of an Update Instruments (Madison, WI) apparatus equipped with a Model 745 Syringe-Ram Controller. Syringes, tubes, mixers, and all solutions are kept at 4° throughout the experiments. The syringe chamber is cooled to 4°, and the packed syringes are assembled in the chamber. The experiment is set up as a "three-syringe experiment" with 129 μM wild-type GalT and 10 mM 2-mercaptoethanol in 100 mM sodium bicinate buffer at pH 8.5 in the first syringe. The second syringe contains 10.2 mM UDP-Glc in sodium bicinate buffer at pH 8.5. The third syringe contains the quenching acid (0.5 M HCl). Once the syringes are assembled, the computer-controlled ram device mixes the enzyme and substrate and then quenches the reaction at times ranging from 3 ms to 100 ms. Quenched samples are collected and frozen in liquid nitrogen.

In order to measure the Glc-1-P formed, the frozen samples are thawed and the denatured enzyme removed by membrane ultrafiltration using Microcon 10 microconcentrators (Millipore Corp., Bedford, MA). The enzyme-free filtrates are collected and neutralized with 1 M Tris base. The amount of Glc-1-P produced is measured by an enzymatic fluorometric method. This method utilizes phosphoglucomutase to convert Glc-1-P into Glc-6-P, which in turn is oxidized by NADP in the presence of Glc-6-P dehydrogenase to produce NADPH.

When the transient kinetic rate constants for the deuridylylation reaction are measured, the first syringe contains 129 μM uridylyl-GalT and the second syringe contains 10.2 mM Glc-1-P or 10.2 mM Gal-1-P. Workup of time points is performed in a manner similar to that for the uridylylation reaction. The detection of UDP-Glc is by an enzymatic fluorometric method. UDP-Glc produced is dehydrogenated to UDP-glucuronate in the presence of UDP-Glc dehydrogenase and NAD$^+$ with the concomitant production of NADH, which is measured fluorimetrically. The detection of UDP-Gal is by the same method, with the addition of 0.005 unit ml^{-1} of UDP-galactose-4-epimerase to convert UDP-Gal into UDP-Glc.

Initially, time points are obtained in triplicate at a ram mixing velocity of 2.5 cm s^{-1}, and the complete time course is repeated at a mixing velocity of 5 cm s^{-1}. The rate constants are independent of mixing velocity, indicating the absence of mixing artifacts in the data. UDP-Glc concentration in the second syringe is varied from 0.2 mM to 10.2 mM, and the respective time courses are obtained and fitted to the first-order rate equation to obtain the pseudo first-order rate constants. These rate constants are plotted versus the UDP-Glc concentration to obtain the uridylylation rate constant at substrate saturation. Similar experiments have also been performed to obtain the deuridylylation rate constants.

By fitting the transient kinetic data, we have obtained rate constants of 281 ± 28 s^{-1} for the uridylylation reaction with UDP-Glc at 4° (Fig. 4). Performing the same experiments at a ram mixing velocity of 5 cm s^{-1} gives a first-order rate

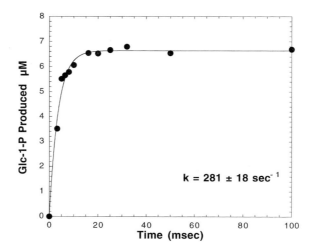

FIG. 4. Time course for wild-type enzyme uridylylation observed by transient phase kinetics. 129 μM wild-type and 10.2 mM UDP-Glc were reacted at 4° and quenched at given times using 0.5 M HCl. The amount of Glc-1-P produced at each time point is plotted as a function of time. The time points were fitted to a first-order rate equation which gave a first-order rate constant of 281 ± 18 s^{-1}. The overall turnover number for the enzyme at 4° is 62 ± 8 s^{-1}. This indicates that uridylylation of wild-type GalT is kinetically competent.

constant for uridylylation with UDP-Glc of 275 ± 28 s^{-1}. Rate constants for the deuridylylation reaction are 166 ± 13 s^{-1} with Gal-1-P and 226 ± 10 s^{-1} Glc-1-P. In order to test the kinetic competence of the uridylyl-enzyme, we have also carried out a complete steady-state kinetic analysis for the overall GalT reaction in the forward direction at 4°. We have obtained parallel-line ping-pong double-reciprocal plots for five UDP-Glc and five Gal-1-P concentrations in triplicate. The results, when fitted with the rate equation, give the following steady-state kinetic parameters: K_m Gal-1-P 0.50 ± 0.09 mM; K_m UDP-Glc 0.098 ± 0.018 mM; and k_{cat} 62 ± 8 s^{-1}. Therefore, the rate of uridylyl-GalT formation is approximately 4 times faster than the overall rate at 4°, and the rate of uridylyl-GalT reaction with Gal-1-P is approximately 3 times faster than the overall rate at 4°. The experiments prove the kinetic competence of the uridylyl-GalT as an intermediate.

At 4° neither uridylylation with UDP-Glc nor deuridylylation with Gal-1-P is strictly rate limiting in the overall reaction. The other possibilities for a rate-limiting step are product release or a conformational change following product formation. UDP-Glc binding, Gal-1-P binding, or any conformational changes before substrate binding cannot be rate limiting as such rate-limiting steps would be detected in our studies. But, our chemical quench method will not detect events following product formation such as the dissociation of the product being measured.

Therefore, by a process of elimination, only the above two possibilities remain as candidates for the rate-limiting step in GalT catalysis at 4°.

Kinetic Consequences of Active-Site Mutations

The availability of X-ray crystal structures has allowed catalytic residues in the active site to be identified. We became interested in three active-site residues that interacted with the phosphoryl oxygens of the substrate and were likely candidates for participation in transition state stabilization. The three residues, Gln-168, Cys-160, and Ser-161, are highly conserved. Our approach to determine their importance in GalT catalysis is to introduce conservative and nonconservative mutations at each position and measure the kinetic consequences in steady-state and transient phase studies.

Mutations of Cys-160, Ser-161, and Gln-168

Site-directed mutagenesis is performed as described previously using standard methods.[20] The mutated GalT proteins are expressed in *E. coli,* and cell extracts are prepared by suspending 1 g of wet cell pellet in 3.6 ml of pH 7.5 Na–HEPES 50 mM buffer and 10 mM 2-mercaptoethanol followed by sonication. Cell debris is removed by centrifugation and the cell extract prepared is used for measuring the initial protein concentration and enzyme specific activity. Protein concentration is estimated using the Warburg and Christian method. Each experiment is accompanied by a control experiment where a similar extract is prepared with an *E. coli* strain that contains the plasmid, but no GalT insert. Enzymatic activity is measured by the coupled enzyme activity assay as described previously, with the results shown in Table I.

Conservative or side-chain deletion mutations of Ser-161 or Gln-168 lead to significant effects on overall GalT activity. The mutation of Ser-161 to alanine results in a cell extract activity loss of 2×10^4-fold suggesting that the Ser-161 residue is crucial for catalysis. Mutation of Gln-168 to Asn or Gly decreases the enzyme activity in cell extracts by approximately two orders of magnitude. Such an activity loss could represent the loss of a hydrogen bond donor during the catalytic mechanism. In contrast, the cell extract activities of mutations of Cys-160 to Ala or Ser remain within a factor of 2 of the wild-type GalT activity. These observations suggest that the highly conserved Cys-160 residue does not play a crucial role in catalysis.

In order to investigate these observations further, three representative mutated forms of GalT, S161A-GalT, C160A-GalT, and Q168N-GalT, have been produced and purified. These proteins behave similarly to the wild-type GalT

---

[20] T. A. Kunkel, *Proc. Natl. Acad. Sci. U.S.A.* **82,** 488 (1985).

TABLE I
GalT ACTIVITY IN CELL EXTRACTS OF ACTIVE-SITE VARIANT GalT FORMS[a]

Form	Specific activity in cell extracts (U mg-protein^{-1})
Wild type	16.3
Gln168Asn	0.19
Gln168Gly	0.11
Gln168His	0.10
Cys160Ala	9.9
Cys160Ser	14.9
Ser161Ala	9×10^{-4}
Background	8×10^{-5}

[a] At 27°.

during purification. Each variant is subjected to CD spectroscopy, active-site titrations for uridylylation and deuridylylation, and metal analysis to ensure that they have properties identical to those of the native GalT. These control experiments suggest that any change in kinetic parameters observed is mainly due to the kinetic effect of the mutation and not to major structural perturbation, inactive enzyme populations, or loss of structural metal ions.

Steady-state kinetic analyses have been performed on the above three variant GalTs with the results summarized in Table II, which are in line with the activities observed in cell extracts. The Q168N-GalT variant shows K_m values similar to the native GalT values. The k_{cat} value is lower by 40-fold, suggesting the loss of a hydrogen bond donor due to the difference of one methylene unit between Gln and Asn.

As expected, the C160A-GalT shows only a 3-fold loss in k_{cat} value and Cys-160 can be excluded from the catalytically relevant residue list. The K_m values,

TABLE II
STEADY-STATE KINETIC PARAMETERS FOR NATIVE AND VARIANT GalT FORMS[a]

Parameter	Wild type	Q168N	C160A	S161A
Specific activity (U.mg.protein^{-1})	180	3.4	79.5	2.3×10^{-2}
Steady-state kinetics forward direction				
$k_{cat}(s^{-1})$	780 ± 19	19 ± 1	265 ± 19	$(5.4 \pm 0.3) \times 10^{-2}$
K_m UDP-Glc(mM)	0.200 ± 0.016	0.200 ± 0.016	0.082 ± 0.009	0.072 ± 0.007
K_m Gal-1-P(mM)	0.303 ± 0.033	0.293 ± 0.023	0.125 ± 0.016	0.223 ± 0.017

[a] At 27°.

however, are decreased 3-fold, indicating a tighter affinity for the substrate in the variant. Cys-160 is unlikely to be the residue that makes GalT sensitive to pCMB[3] because it shows sensitivity to pCMB similar to that of the native enzyme.[18]

The value of k_{cat} for S161A-GalT is 4 orders of magnitude lower than that of GalT in the forward direction at 27°, and the K_m values are slightly lower. Ser-161 clearly plays a crucial role in the mechanism. The importance of this residue is further revealed by the steady-state kinetic parameters for S161A-GalT at 4° and the transient-state kinetic parameters for uridylylation and deuridylylation at 4°. The K_m values do change significantly at 4° ($K_m^{\text{UDP-Glc}} = 0.12 \pm 0.01$ and $K_m^{\text{Gal-1-P}} = 0.57 \pm 0.04$ mM). The k_{cat} value is $(9.0 \pm 0.5) \times 10^{-4}$ s^{-1}. The uridylylation rate constant at 4° is $(3.70 \pm 0.02) \times 10^{-2}$ s^{-1} and the deuridylylation rate constant is $(0.50 \pm 0.05) \times 10^{-2}$ s^{-1}. The uridylylation rate is 40-fold faster than the overall rate, as expressed by k_{cat}, and the deuridylylation rate is 5-fold faster.

The most interesting comparison is between the uridylylation rate constants for native GalT and S161A-GalT. The Ser-to-Ala mutation causes the uridylylation rate constant to decrease by 7.5×10^3-fold, which accounts for an increase in the kinetic barrier for the uridylylation reaction of 4.9 kcal mol^{-1}. The deuridylylation rate with Gal-1-P is 3.3×10^4-fold slower for S161A-GalT and this relates to an increased kinetic barrier of 5.7 kcal mol^{-1}. These results indicate that Ser-161 may be playing a role in stabilizing the transition state during the uridylylation and deuridylylation. The main-chain amide hydrogen of Ser-161 forms a hydrogen bond to a nonbridging α-phosphoryl oxygen. The side chain forms a potential hydrogen bond to a nonbridging β-phosphoryl oxygen. Although these hydrogen bonds provide no chemical advantage, they could play a role in the proper orientation of phosphoryl groups in the active site and facilitate the S_N2-like transition state.

Model for Galactosemia Mutation, Q168R-GalT

When discussing the kinetic consequences of active-site mutations, one should consider the Gln-168 to Arg mutation, which is the most common mutation in human galactosemia (Gln-188 in humans). This mutation may be modeled in the bacterial enzyme. The mutation causes a 3×10^5-fold overall activity loss. Further studies showed that the Q168R-GalT contains only $(72 \pm 11)\%$ of the structural zinc and iron found in GalT. The lesser metal ion content results in a nonhomogeneous enzyme population, in which some enzyme molecules contain zinc and iron, others only zinc or only iron. Presumably owing to the low content of zinc, only $(65 \pm 1)\%$ of the active sites of Q168R-GalT undergo uridylylation and deuridylylation. The presence of structural zinc is proportional to the percentage of functional active sites in GalT.[19]

The reason for the dramatically low activity of Q168R-GalT is low kinetic reactivity in uridylylation. The uridylylation rate constant with UDP-Glc at 4° for

$$\text{E-His 166 + UDP-Glc} \xrightleftharpoons{K_{eq}} \text{E-His 166-UMP + Glc-1-P}$$

$$\text{E-His 166-UMP} \xrightarrow[\text{Slow}]{+ H_2O} \text{E-His 166}$$

SCHEME 1

this variant is $(2.2 \pm 0.4) \times 10^{-4}$ s^{-1}, and the deuridylylation rate constant with Gal-1-P is $(3.3 \pm 1.2) \times 10^{-4}$ s^{-1}. Therefore the uridylylation rate is decreased by 1×10^6-fold and the deuridylylation rate is decreased by 5×10^5-fold. The kinetic barrier for uridylylation is increased by 7.7 kcal mol^{-1} while the kinetic barrier for deuridylylation is increased by 7.3 kcal mol^{-1}. At present time it is not clear why the Gln-to-Arg point mutation in the active site causes such an enormous increase in the kinetic barrier for the two chemical steps. It is potentially significant that the mutation increases the positive charge in the active site.[17]

Hydrolysis of Uridylyl-GalT

During the course of this study, we have discovered biphasic kinetics when observing uridylylation over a long period of time (300 min). One possible explanation for this behavior is that illustrated in Scheme 1, where fast uridylyl-GalT formation is followed by slow hydrolysis, freeing the enzyme for reuridylylation and generating Glc-1-P. We isolate the uridylyl-GalT, incubate it in 100 mM sodium bicinate pH 8.5 buffer at 27°, and periodically take aliquots and measure the uridylyl content by adding Glc-1-P and the coupling enzyme system that detects UDP-Glc. The uridylyl content slowly decreases over a 300-min period, suggesting that the uridylyl group undergoes hydrolysis under these conditions. Control experiments are performed to ensure that the observed decrease in uridylyl content is not due to a loss of enzyme activity or enzyme precipitation.

UMP-imidazolide (UMP-Im) is chemically similar to the uridylyl-GalT and undergoes hydrolysis with a rate constant that is 800-fold smaller than that for the uridylyl-GalT.[21] From the pH dependence of UMP-Im hydrolysis, the pK_a for the imidazole ring is estimated to be 5.75. The monoanionic form of UMP-Im is present at pH 8.5 and does not undergo efficient hydrolysis because the unprotonated imidazole ring makes a poor leaving group. In order for uridylyl-GalT to undergo hydrolysis in the pH 8.5 range, His-166 N$^\delta$ should remain protonated. This hypothesis can be tested by examining the pH dependence for the slow hydrolysis of uridylyl-GalT.[22]

[21] F. J. Ruzicka and P. A. Frey, *Bioorganic Chem.* **21**, 238 (1993).
[22] F. J. Ruzicka, S. Geeganage, and P. A. Frey, *Biochemistry* **37**, 11385 (1998).

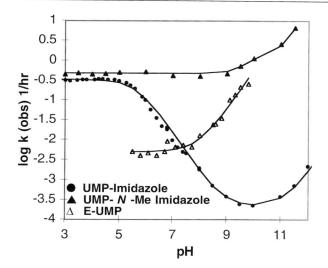

FIG. 5. pH–rate profile for the hydrolysis of the uridylyl-GalT complex. The observed first-order rate constants (k_{obs}) for the hydrolysis of E-UMP, measured at 27° as described, are plotted as open triangles versus pH. The line is calculated from the parameters obtained by fitting the data points to the equation $k_{obs} = k_1 + k_2 a_{OH}$. The other curves are the pH–rate profiles for the hydrolysis of UMP-3-MeIm and UMP-Im at 27° and were obtained by Ruzicka and Frey [F. J. Ruzicka and P. A. Frey, *Bioorganic Chem.* **21**, 238 (1993)].

Wild-type uridylyl-GalT (subunit concentration of 77.3 μM) is incubated at 27° in 0.1 M buffer and 10 mM 2-mercaptoethanol adjusted to an ionic strength of 1.0 by NaCl addition. The following buffers are used at the various pH ranges: MES for pH 5.5–7.0, MOPS for pH 6.5–8.0, and CHES for pH 8.5–9.8. Aliquots are removed periodically and added to assay mixtures containing Glc-1-P (4.5 mM), UDP-Glc dehydrogenase (0.03 units ml^{-1}), 2-mercaptoethanol (10 mM), and NAD$^+$ (1.25 mM). Uridylyl groups not hydrolyzed are converted into UDP-Glc, which can be used to reduce NAD$^+$ in the presence of UDP-Glc dehydrogenase and allows the measurement of the hydrolysis of uridylyl-GalT as a function of time. Pseudo first-order kinetics are observed for hydrolysis up to four half-lives, except when the rates are very slow. Rate constants are calculated by fitting rate data to the first-order rate equation. First-order rate constants for the hydrolysis of the uridylyl-enzyme measured as a function of pH are fitted to the equation $k_{obs} = k_1 + k_2 a_{OH}$. The rate constant for pH-independent hydrolysis is k_1, and k_2 is the rate constant for base-catalyzed hydrolysis. The pH profile in Fig. 5 is obtained along with the pH profiles for the model compounds UMP-Im and UMP-3-methylimidazolide (MeIm).

These results show that at low to neutral pHs uridylyl-GalT undergoes a pH-independent hydrolysis, and at pH values higher than 7 it shows an increase in hydrolysis rate. The profile is similar to that for the hydrolysis of UMP-3-MeIm in which the increased rate at higher pH values was attributed to cleavage by the reaction of water with the N-methylated imidazolide. The pH-rate profile for hydrolysis of the uridylyl-GalT is analogous to that for UMP-3-MeIm, but the pH-independent hydrolysis is suppressed, possibly because the enzyme protects the group from hydrolysis at lower pH values. The profile is very different from those for UMP-Im and the other UMP methylimidazolides, which display acid dependence in their pH–rate profiles.

The methyl substituent in UMP-3-MeIm is on N-3, and its presence has the electronic effect of a proton, in that it results in the imidazole ring being positively charged. Because their pK_a values lie between 5.68 and 6.4, the imidazole rings of UMP-Im and the other methylated analogs are not protonated in solutions above pH 7. The fact that the pH profile for UMP-3-MeIm looks similar to the pH profile for uridylyl-GalT slow hydrolysis suggests that the imidazole of His-166 remains protonated, and positively charged in the intermediate. The enzyme may have perturbed the pK_a for the His-166 $N^{\delta 1}$ by at least several units. A positively charged uridylyl-GalT intermediate provides a distinct mechanistic advantage. It facilitates the deuridylylation reaction by acting as a good leaving group. The protonated intermediate exists in a chemically poised and activated state for the subsequent uridylyl group transfer in the second step. Removal of the imidazole proton, which might be expected to promote uridylyl-GalT formation, would not necessarily promote the overall reaction.

Retention of the proton on the imidazole ring at pH values well above its pK_a must be made possible by stabilization through binding interactions between the uridylyl group and the active site. Some of these potential interactions were discussed under the active site structure, most importantly the interaction between His-164 backbone carbonyl and His-166 N^{δ} proton. Further biochemical and kinetic studies will be required to investigate and dissect these interactions.

[9] Kinetic Evidence for Covalent Phosphoryl-Enzyme Intermediate in Phosphotransferase Activity of Human Red Cell Pyrimidine Nucleotidases

By ADOLFO AMICI, MONICA EMANUELLI, SILVERIO RUGGIERI, NADIA RAFFAELLI, and GIULIO MAGNI

Many enzymes form covalent phosphorylated and acylated enzyme–substrate intermediates during catalysis, but their formation often goes unnoticed because of the cryptic nature of the catalyzed reaction.[1] Under favorable conditions, however, and especially when the enzyme can bind alternative reaction products, the covalent intermediate can be intercepted by group transfer to the alternative product.[2] The classical example for this behavior is the multifunctional liver microsomal glucose-6-phosphatase, which in addition to catalyzing the hydrolysis of glucose 6-phosphate to glucose and orthophosphate, also catalyzes the phosphorylation of numerous hexoses.[3] Another well-known example is the transglutaminase reaction, in which the γ-carboxamide groups of peptide-bound glutamine residues serve as acyl donors and in which water or primary amino groups within a variety of compounds function as acyl-group acceptors.[4] A characteristic feature of these branched pathway reactions is the occurrence of a modified double-displacement mechanism in which a covalently modified enzyme preserves the transfer potential of the group undergoing transfer. In fact, the transferase reaction need not exhibit ping-pong-type kinetics, because the latter is only observed in cases where one product must dissociate prior to the binding of the acceptor substrate.[2–4]

We have isolated and characterized two pyrimidine nucleotidases (PN-I and PN-II) from human erythrocytes that preferentially hydrolyze pyrimidine 5′-monophosphates and 3′-monophosphates, respectively.[5] During the course of our investigation, we observed that these enzymes also catalyze phosphoryl transfer from the pyrimidine nucleoside monophosphate donor substrate(s) to other nucleoside acceptors, including important chemotherapeutic agents such as 3′-azido-3′-deoxythymidine (AZT), cytosine-β-D-arabinofuranoside (AraC), and 5-fluoro-2′-deoxyuridine (5FdUrd). In this chapter, we illustrate how our kinetic

---

[1] D. L. Purich and R. D. Allison, "Handbook of Biochemical Kinetics," pp. 313 and 389. Academic Press, New York, 2000.
[2] R. C. Nordlie, *Methods Enzymol.* **87**, 319 (1982).
[3] K. A. Sukalski and R. C. Nordlie, *Adv. Enzymol.* **62**, 93 (1989).
[4] M. Gross and J. E. Folk, *J. Biol. Chem.* **247**, 2798 (1972).
[5] A. Amici, M. Emanuelli, G. Magni, N. Raffaelli, and S. Ruggieri, *FEBS Lett.* **419**, 263 (1997).

analysis provides convincing evidence for a covalent enzyme–substrate intermediate. We also show that the PN-I phosphotransferase activity displays higher affinity for oxynucleosides with respect to deoxynucleosides, whereas the contrary seems to be true for PN-II. Our observations that soluble pyrimidine nucleotidases are endowed with pyrimidine-specific phosphotransferase activity represent an additional way of characterizing the catalytic mechanism of these nucleotidases.

Kinetic Treatment of Transferase Activities as Means for Intercepting Covalent Intermediates

A branched pathway for nucleotidase and phosphotransferase activities can be depicted by the diagram in Fig. 1, where the primary substrate A_1P reversibly transfers its phosphoryl group by the rightmost pathway. If formed in the presence of a suitable phosphoryl acceptor substrate, the intermediate $P{\sim}E$ can suffer three fates: (a) return to yield free enzyme E plus A_1P; (b) hydrolysis to yield free E plus A_1 and P_i; or (c) combination of $P{\sim}E$ with acceptor A_2 to form A_2P. To illustrate how the absence or presence of acceptor A_2 affects the efficiency of phosphotransfer versus hydrolysis, consider the hypothetical time courses shown in Fig. 2. In the absence of acceptors A_2, hydrolysis will dominate, and A_1 and P_i will be formed in a one-to-one stoichiometry (Fig. 2a). At low A_2 concentration, the sum of P_i and A_2P will equal the A_1 formed (Fig. 2b). Finally, in the extreme case where A_2 is saturating, phosphoryl transfer to that acceptor will be the predominant pathway (Fig. 2c). Finally, Nordlie has fully considered the bisubstrate kinetic mechanisms of hydrolases that have intrinsic transferase activities. It is sufficient here to say that because group transfer requires the presence of a donor substrate A_1P and an acceptor A_2, the transferase can in principle display typical two-substrate sequential or ping-pong kinetics.[2,3]

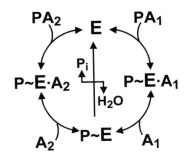

FIG. 1. Kinetic scheme for branched-mechanism hydrolase/transferase reactions. See text for details.

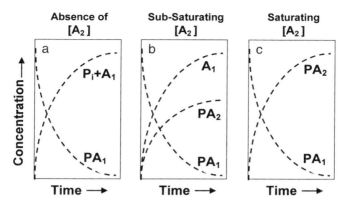

FIG. 2. Hypothetical time courses for hydrolysis, resynthesis, and group transfer catalyzed by an enzyme operating by a branched hydrolysis/transfer mechanism.

Erythrocyte Pyrimidine Nucleotidases as Examples of Branched-Mechanism Enzymes in Which Water and Pyrimidine Nucleosides Compete as Phosphoryl-Acceptor Substrates

Two pyrimidine-specific 5′-nucleotidases (PN-I and PN-II; EC 3.1.3.5) are present in the soluble fraction of human erythrocytes.[6,7] Unlike nucleotidases from other sources, PN-I and PN-II are essentially inactive toward purine nucleotides. This unique substrate restriction appears to match with the needs of the red cell metabolism, which critically depends on a limited reservoir of ATP, the adenine ring of which cannot be synthesized *de novo* in erythrocytes. The presence of a nucleotidase acting on AMP and producing diffusible adenosine should impose an inexorable drain on this metabolite.[8] It is well established that hereditary deficiency of one of these enzymes (PN-I) is an autosomal recessive condition causing hemolytic anemia characterized by marked basophilic stippling and the accumulation of high concentrations of pyrimidine nucleotides within the erythrocytes.[6,9] PN-I and PN-II operate as interconverting activities, capable of transferring the phosphate from the pyrimidine nucleoside monophosphate donor(s) to various nucleoside acceptors, including important drugs such as 3′-azido-3′-deoxythymidine (AZT),

---

[6] W. N. Valentine, K. Fink, D. E. Paglia, S. R. Harris, and W. S. Adams, *J. Clin. Invest.* **54**, 866 (1974).
[7] A. Hirono, H. Fujii, H. Natori, I. Kurokawa, and S. Miwa, *Brit. J. Haematol.* **65**, 35 (1987).
[8] D. E. Paglia, W. N. Valentine, and R. A. Brockway, *Proc. Natl. Acad. Sci. U.S.A.* **81**, 588 (1984).
[9] H. A. Simmonds, L. D. Fairbanks, G. S. Morris, D. R. Webster, and E. H. Harley, *Clin. Chim. Acta* **171**, 197 (1988).

cytosine-β-D-arabinofuranoside (AraC), and 5-fluoro-2′-deoxyuridine (5FdUrd), pyrimidine analogs widely used in chemotherapy.[5]

PN-I and PN-II Preparations and Assay of Enzymatic Activities

PN-I and PN-II have been purified to homogeneity as described elsewhere.[5,10] Nucleotidase activity is measured by a high-performance liquid chromatography (HPLC)-based assay, for which the standard reaction mixture contains 50 mM Tris–HCl (pH 7.5), 1 mM MgCl$_2$, 1 mM dithiothreitol (DTT), 1 mM substrate, and the appropriate amount of enzyme sample, in a 0.5-ml final volume. Incubations are carried out at 37° for 5–120 min. The reaction is stopped by adding 0.1 ml assay mixture to 0.05 ml ice-cold 1.2 M HClO$_4$ in an Eppendorf tube. After 10 min at 0°, the protein pellet is removed by 1 min centrifugation on microfuge. Supernatant samples (0.13 ml) are neutralized by the addition of 35 μl 1 M K$_2$CO$_3$ to form KClO$_4$ crystals that are removed by centrifugation. An appropriate aliquot of the neutralized supernatant is injected into a HPLC apparatus to evaluate the amount of produced nucleoside. The elution is performed by the isocratic separation described by Amici *et al.*[11] The HPLC is equipped with a 20 mm × 4.6 mm i.d. C$_{18}$ guard column, 5 μm particle size, equilibrated and eluted with 100 mM potassium phosphate buffer, pH 6.0, at 2 ml/min flow rate. In an alternative assay procedure, the enzyme activity is tested by measuring the amount of phosphate released, according to the ammonium molybdate–ascorbic acid complexation method of Ames.[12]

Phosphotransferase activity is measured by an HPLC-based assay of formed nucleoside monophosphate. The reaction mixture conditions are the same as for the nucleotidase assay, with the addition of an appropriate amount of nucleoside acceptor during the incubation at 37°. An appropriate aliquot of the neutralized supernatant is injected into a HPLC apparatus to evaluate the amount of produced nucleoside. The HPLC is equipped with a 250 mm × 4.6 mm i.d. C$_{18}$ column, 5-μm particle size, equilibrated with 100 mM potassium phosphate buffer pH 6.0 (buffer A) at a flow rate of 1.3 ml/min. The elution of nucleosides and nucleoside monophosphates of the mixture is achieved by the following gradient of methanol in buffer A: 9 min with no methanol, 6 min with up to 2.4% methanol, 2.5 min with up to 9% methanol, and 2.5 min with up to 20% methanol. The column is then flushed for 5 min with 20% methanol and equilibrated with buffer A for 5 min prior to the next run. Replicate determinations at different incubation times are used to calculate each enzymatic activity point. One unit of enzyme activity (U)

---

[10] A. Amici, M. Emanuelli, E. Ferretti, N. Raffaelli, S. Ruggieri, and G. Magni, *Biochem. J.* **304,** 987 (1994).

[11] A. Amici, M. Emanuelli, N. Raffaelli, S. Ruggieri, and G. Magni, *Anal. Biochem.* **216,** 171 (1994).

[12] B. N. Ames, *Methods Enzymol.* **8,** 115 (1966).

is defined as the amount of enzyme producing 1 μmol phosphate, nucleoside, or nucleoside monophosphate per minute under the standard assay conditions.

Nucleotidase Substrate Specificity

PN-I preferentially hydrolyzes, in order, 5'-UMP, 5'-CMP, 5'-AZTMP, 5'-Ara-CMP, 5'-dCMP, 5'-dTMP, 5'-dUMP. This enzyme is totally inactive toward purine nucleoside monophosphates, regardless of the position of the phosphate moiety. PN-II shows a high specificity toward pyrimidine nucleotides, being also able to dephosphorylate inosine and guanosine monophosphates, although to a lesser extent.[10] Because both PN-I and PN-II hydrolyze known antineoplastic agents, such as 5'-AZTMP, 5'-Ara-CMP, 5'-FdUMP, we suspect that these enzymes might play a relevant role in the metabolism of such pyrimidine analogs. This is a significant issue in view of the fact that the identification and/or rational design of effectors influencing PN-I and PN-II activity might prove useful for optimizing antineoplastic therapeutic protocols. Among a broad variety of compounds tested as hypothetical effectors of the nucleotidases activity, only the reaction products, i.e., the nucleoside and phosphate, are inhibitory. Product inhibition studies on PN-II have been carried out by using substrate 3'-UMP concentrations ranging from 0.1 to 1 mM, at different fixed concentrations of either phosphate (1–2 mM) or uridine (10–20 mM). Double reciprocal plots show that the inhibition exerted by phosphate appears to be competitive ($K_i = 3.5$ mM). The inhibition exerted by uridine is noncompetitive (mixed type), and the intercepts replot exhibits a nonlinear pattern approaching a plateau at high uridine concentration. Because the intercept represents the reciprocal of apparent maximal nucleotidase activity, the plateau at high inhibitor concentration represents the residual activity of the inhibited enzyme. The data obtained from the product inhibition experiments are consistent with an ordered reaction, involving the release of nucleoside as the first product and of phosphate as the second.[13]

PN-I and PN-II Associated Phosphotransferase Activity

During the study of the inhibition exerted by different nucleosides on both PN-I and PN-II, the appearance of the monophosphate of the nucleoside inhibitor has been observed in the elution pattern. In particular, when dThd is used as the inhibitor of PN-II action on 3'-dUMP, 3'-dTMP is formed. Production of 3'-dTMP depends on the presence of 3'-dUMP in the incubation mixture, and 3'-dTMP accumulates linearly with time, thus displaying the initial velocity conditions. This finding is only consistent with PN-II-associated phosphotransferase reaction,

---

[13] W. W. Cleland, in "The Enzymes" (P. D. Boyer, ed.), 3rd Ed., Vol. 2, p. 1. Academic Press, New York, 1970.

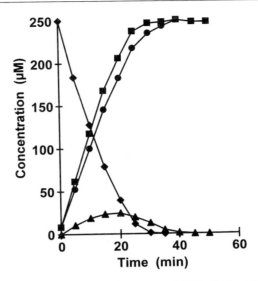

FIG. 3. Time course of PN-I activities in the presence of CMP (0.25 mM), Urd (20 mM), and 10 mU/ml of pure PN-I under the assay conditions described under Experimental. (◆), CMP; (■), Cyd; (●), phosphate and (▲), UMP were determined. Adapted, with permission, from Amici et al.[5]

wherein the phosphate from the nucleoside monophosphate substrate is enzymatically transferred to the inhibitor nucleoside acceptor. Furthermore, an identical behavior leading to 5'-UMP formation is observed for PN-I action on 5'-CMP in the presence of uridine.

This observation appears of particular interest in view of the fact that several chemotherapeutic strategies are based on the use of pyrimidine analogs, whose metabolic activation pathways may be profoundly affected by the existence of such phosphotransferase activities. This has prompted us to undertake an extensive study of the kinetic properties of the individual PN-I and PN-II enzyme activities, whose results can be exploited in the designing of novel chemotherapeutic approaches as well as in the modulation of the existing ones. Figure 3 shows the time course observed by incubating PN-I with 5'-CMP as the substrate and uridine as the inhibitor/phosphate acceptor, resulting in 5'-UMP formation. On prolonged incubation time, Fig. 3 shows that the newly formed 5'-UMP is hydrolyzed to Urd and inorganic phosphate after 40 min, whereas 5'-CMP is completely consumed after 30 min. As evidenced in the same figure, this results in a transient accumulation of 5'-UMP, the product of the phosphotransferase activity. From the time course curves it can be calculated that the sum of 5'-UMP and P_i concentration equals the consumption of the substrate, thus confirming the prediction shown by the hypothetical time course depicted in Fig. 2. Lineweaver–Burk plots of

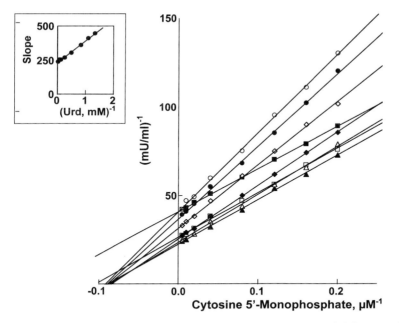

FIG. 4. Lineweaver–Burk plot of phosphotransferase activity of PN-I. CMP and Urd were used as substrates. Urd phosphorylation was measured as described in the text at different fixed concentrations of Urd (mM): (○), 0.75; (●), 0.9; (◇), 1.2; (◆), 2.0; (△), 3.75; (▲), 7.5; (□), 15; (■), 30. *Inset:* Slopes of interpolation lines are plotted against 1/[Urd]. Adapted from A. Amici, M. Emanuelli, G. Magni, N. Raffaelli, and S. Ruggieri, *FEBS Lett.* **419,** 263 (1997).

phosphotransferase activity show converging patterns with respect to both donor and acceptor substrates (Figs. 4 and 5). The evaluation of the kinetic mechanism of phosphotransferase is not straightforward, because the enzymatic protein simultaneously catalyzes both the hydrolytic and phosphotransferasic reaction. Indeed, because of the action of the intrinsic nucleotidase activity, this behavior might be consistent with both an ordered Bi Bi and a ping-pong kinetic mechanisms, as has been previously described for nucleoside phosphotransferase from barley.[14]

Phosphotransferase Substrate Specificity

Table I summarizes the results obtained with several nucleoside monophosphates tested as phosphoryl donors for various nucleoside acceptors. The relative activity is calculated as the ratio of the nucleoside monophosphate formed

---

[14] D. C. Prasher, M. C. Carr, D. H. Ives, T. C. Tsai, and P. A. Frey, *J. Biol. Chem.* **257,** 4931 (1982).

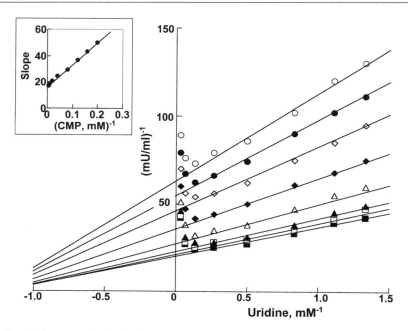

FIG. 5. Lineweaver–Burk plot of phosphotransferase activity of PN-I. CMP and Urd were used as substrates. Urd phosphorylation was measured as described in the text at different fixed concentrations of CMP (μM): (○), 5.0; (●), 6.25; (◇), 8.3; (◆), 12.5; (△), 25; (▲), 50; (□), 100; (■), 200. *Inset:* Slopes of interpolation lines are plotted against 1/[CMP]. Adapted, with permission, from A. Amici, M. Emanuelli, G. Magni, N. Raffaelli, and S. Ruggieri, *FEBS Lett.* **419**, 263 (1997).

by the phosphotransferase activity over the nucleoside liberated via the nucleotidase activity. For PN-I the data show that the apparent specificity is principally influenced by the nucleoside acceptor, as evidenced by comparing the results obtained by using a single acceptor (e.g., Cyd) and several donors, to those obtained by using a single donor (e.g., CMP) and different acceptors; in this latter case a larger spreading of the apparent specificities is observed by inspection of the data in Table I. PN-I shows also a relatively low but significant phosphotransferase activity toward AZT and AraC as the acceptors. Such an activity could play a relevant role both in the metabolism and in the pharmacokinetics of these important drugs. The phosphorylated product could be rapidly converted to the corresponding diphosphate by intracellular nucleotide kinase, widely distributed in human cells and tissues, thus protecting it from the nucleotidase activity. PN-II phosphotransferase activity is most effective toward pyrimidine nucleoside 3′-monophosphates leading to the production of nucleotides with the phosphate in the same position. PN-I exhibited K_m values for deoxynucleosides one order of magnitude higher than those for oxynucleosides, whereas the contrary appears

TABLE I
PHOSPHOTRANSFERASE ACTIVITY OF PN-I AND PN-II

PN-I				PN-II			
Donor (1 mM)	Acceptor (10 mM)	Activity (%)[a]	K_m[b]	Donor (1 mM)	Acceptor (10 mM)	Activity (%)[a]	K_m[b]
CMP	Urd	25	0.64	3'-UMP	dUrd	29	6
CMP	dUrd	1.5	—[c]	3'-dUMP	dThd	16.8	12.3
CMP	dCyd	16	7.4	3'-dUMP	Urd	2	>100
CMP	dThd	0.5	5.3	3'-dUMP	Cyd	n.d.[d]	
CMP	Ado	n.d.[d]		3'-dUMP	dCyd	n.d.	
CMP	AZT	1	—	3'-dUMP	Ado	n.d.	
CMP	AraC	4	—	3'-dUMP	dAdo	n.d.	
CMP	5FdUrd	n.d.		3'-dUMP	Ino	n.d.	
UMP	Cyd	36	0.79	3'-dUMP	dIno	n.d.	
dCMP	Cyd	43	—	3'-dUMP	AZT	n.d.	
dUMP	Cyd	15	—	3'-dUMP	AraC	n.d.	
dTMP	Cyd	30	—	3'-dUMP	5FdUrd	19.7	7.6

[a] Percentage of phosphate transferred from nucleoside monophosphate donor to nucleoside acceptor.
[b] K_m of acceptor (units, mM).
[c] Not determined.
[d] Not detectable.

to hold for PN-II.[5] The same behavior can be observed for the two nucleotidases as far as the hydrolytic activity is concerned (Table II), suggesting that the two activities, i.e., phosphotransferase and nucleotidase, are intrinsic to the catalytic capacity of the two enzymes and thus reinforcing the mechanism depicted in Fig. 1.

TABLE II
NUCLEOTIDASE ACTIVITY OF PN-I AND PN-II

PN-I				PN-II			
Substrate	V_{max}[a]	K_m[b]	V_{max}/K_m	Substrate	V_{max}[a]	K_m[b]	V_{max}/K_m
CMP	28.0	0.01	2800	3'-dUMP	110	0.20	550
UMP	29.5	0.33	89.4	3'-UMP	85	0.25	340
dCMP	15.0	0.58	25.9	3'-dTMP	117	0.30	390
dUMP	4.1	0.40	10.3	5'-dUMP	19.5	0.41	47
dTMP	12.6	1.00	12.6	5'-UMP	3.05	0.66	4.6
AZT-MP	21.0	1.20	17.5	5'-FdUMP	59.4	1.15	52

[a] Data given in U/mg.
[b] Data given in mM.

Other Enzymes Catalyzing Branched-Mechanism Reactions

Although rat liver glucose-6-phosphatase and transglutaminase are arguably the best-characterized examples of branched-mechanism catalysis, we should mention that many other enzymes exhibit this characteristic behavior. Alkaline phosphatases from many bacterial and mammalian sources form a phosphoryl-enzyme intermediate when incubated with various phosphomonoesters.[15,16] In addition to water, many other hydroxyl group-containing compounds serve as acceptor substrates. For acid phosphatases from microbial, plant, and mammalian sources, the phosphorylated enzyme is produced from numerous O- and S-substituted monoesters; water and organic alcohols compete with each other in the phosphoryl transfer reaction.[17] γ-Glutamyl-S-transferases from many sources undergo γ-glutamylation when incubated with glutathione, L-γ-glutamyl-p-nitroanilide, and L-α-methyl-L-γ-glutamyl-L-α-aminobutyrate; in this case, water and various amino groups on amino acids and peptides serve as acyl-group acceptor substrates.[18] Another well-characterized example is calf spleen NAD^+ glycohydrolase which becomes ADP-ribosylated during its hydrolytic reaction, such that the ADPR group is transferable to methanol, nicotinamide, and 3-acetylpyridine.[19] Finally, phosphoserine phosphatases[20] as well as numerous phosphoprotein phosphatases become transiently phosphorylated during their respective hydrolase reactions, and under favorable conditions, the phosphoryl group can be captured by various serine analogs as well as certain serine-containing peptides.

Concluding Remarks

The essential feature of each intermediate species in the above reactions is that the intermediate must be of sufficient reactivity to be "returned" to a respective monoester species. The additivity principle of thermodynamics allows us to write the change in the Gibbs free energy (ΔG) for the overall reaction in terms of its component or partial reactions:

$$\Delta G_{\text{overall}} = \Delta G_{\text{intermediate formation}} + \Delta G_{\text{hydrolytic bond scission}}$$

---

[15] T. W. Reid and I. B. Wilson, in "The Enzymes" (P. D. Boyer, ed.), 3rd Ed., Vol. 4, p. 373. Academic Press, New York, 1971.

[16] H. N. Fernley, in "The Enzymes" (P. D. Boyer, ed.), 3rd Ed., Vol. 4, p. 417. Academic Press, New York, 1971.

[17] V. P. Hollander, in "The Enzymes" (P. D. Boyer, ed.), 3rd Ed., Vol. 4, p. 449. Academic Press, New York, 1971.

[18] A. M. Karkowsky, M. V. Bergamini, and M. Orlowski, *J. Biol. Chem.* **251**, 4736 (1976).

[19] F. Schuber, P. Travo, and M. Pascal, *Eur. J. Biochem.* **69**, 593 (1976).

[20] W. L. Byrne, in "The Enzymes" (P. D. Boyer, H. Lardy, and K. Myrbäck, eds.), 2nd Ed., Vol. 5, p. 373. Academic Press, New York, 1961.

The occurrence of phosphotransferase, acyltransferase, and ADPR transferase activities in the above suggests that the Gibbs free energy of the overall hydrolytic reaction is dissipated in steps, such that initial formation of the phosphoryl-, acyl-, or ADPR-containing intermediate can remain reversible. Whether this stepwise process offers a practical advantage during catalysis has yet to be convincingly demonstrated, but these group transfer reactions pose limitations on the types of metabolites that may undergo unanticipated modification reactions on exposure to a suitably activated enzyme–substrate covalent compound.

Despite the fact that the putative phosphoryl-enzyme intermediate remains to be chemically characterized, we have presented conclusive evidence that human red cell pyrimidine nucleotidases operate by so-called branched mechanisms, with water and pyrimidine nucleosides competing as phosphoryl acceptors. Beyond the interesting catalytic features of PN-I and PN-II, the spectrum of substrates with which one can observe phosphotransfer activity may also prove to be invaluable in comprehending the action of antineoplastic agents and other unphosphorylated prodrugs. Although most cells presumably possess the ability to prevent formation of undesirable metabolic by-products, our discovery of these transfer reactions may represent opportunities for therapeutic intervention. This outcome would be especially favored in cases where the enzyme displays little selectivity for the acceptor substrate. The fact that 3'-azido-3'-deoxythymidine (AZT), cytosine-β-D-arabinofuranoside (AraC), and 5-fluoro-2'-deoxyuridine (5FdUrd) are suitable phosphoryl-acceptor substrates indicates a potentially important pathway that some cells may possess for converting prodrug species into their likely therapeutic form. Presumably, one might even be able to increase the susceptibility of certain cell types through the expression and/or transfection of prodrug-activating enzymes in target tissues.

[10] Characterization of $\alpha(2\rightarrow 6)$-Sialyltransferase Reaction Intermediates: Use of Alternative Substrates to Unmask Kinetic Isotope Effects

By BENJAMIN A. HORENSTEIN and MICHAEL BRUNER

Glycosyltransferases transform oligosaccharides and polysaccharides by facilitating attack of a saccharide hydroxyl group at the anomeric carbon of a sugar nucleotide. Because mechanistic work on glycosyltransferases utilizing glycosyl phosphates has not advanced to the level of corresponding work on glycosidases, our laboratory has established an ongoing interest in the mechanism(s) of sialyltransferase catalysis. These enzymes catalyze N-acetylneuraminic acid

(NeuAc) transfer from cytidine monophosphate glycoside of N-acetylneuraminic acid (CMP-NeuAc) with overall inversion of configuration to acceptor hydroxyl groups at or near the nonreducing termini of oligosaccharide chains of glycoproteins and glycolipids. An interesting feature of sialyltransferase catalysis is the unique structure and reactivity of the donor substrate CMP-NeuAc, with its carboxylate group immediately adjacent to the anomeric carbon. Acid-catalyzed solvolysis of CMP-NeuAc proceeds by an extremely late transition state, with no significant nucleophilic participation exerted by solvent or the carboxylate group.[1,2] After this transition state, the formation of a short-lived oxocarbenium ion intermediate appears to be stabilized by intramolecular ion-pairing with the carboxylate group of the substrate.[3,4]

Kinetic isotope effects (KIEs) for CMP-NeuAc solvolysis are considerably different from the kinetic isotope effects observed with the sialyltransferase-catalyzed transfer.[5] Although the β-secondary dideuterium KIE for solvolysis is 1.276, a much smaller KIE value of 1.022 is observed in the sialyltransferase reaction. The primary ^{14}C-KIE was also diminished, decreasing from 1.030 for solvolysis to 1.000 for the sialyltransferase-catalyzed reaction when CMP-NeuAc is the substrate. These findings were consistent with two possibilities, namely that the enzymatic mechanism differs considerably from its nonenzymatic counterpart or that additional kinetic complexity of the enzymatic reaction masks the true intrinsic value of the KIE. The present chapter illustrates how one can address these rival explanations by employing a poorer alternative substrate to unmask intrinsic kinetic isotope effects and to reveal significant mechanistic details about sialyltransferase catalysis.[6]

Experimental Procedures

Details about commercial sources of substrates, isotopically labeled metabolites and sialyltransferase are presented elsewhere.[1,5,6] High-performance liquid chromatography (HPLC) using a Pharmacia (Piscataway, NJ) Mono Q HR10/10 anion-exchange resin is monitored by UV absorbance at 260 nm. Radioactivity is determined using a Packard 1600 TR liquid scintillation counter (Packard, Meriden, CT), with data analysis carried out on a personal computer. An Orion Sure-Flow pH probe (Orion, Beverly, MA) is satisfactory for pH measurements.

[1] B. A. Horenstein and M. Bruner, *J. Am. Chem. Soc.* **118**, 10371 (1996).
[2] B. A. Horenstein, *J. Am. Chem. Soc.* **119**, 1101 (1997).
[3] B. A. Horenstein and M. Bruner, *J. Am. Chem. Soc.* **120**, 1357 (1998).
[4] B. A. Horenstein, in "Transition-State Modeling for Catalysis," ACS Symp. Ser. #721 (D. G. Truhlar and K. Morokuma, eds.). American Chemical Society, Washington, DC, 1999.
[5] M. Bruner and B. A. Horenstein, *Biochemistry* **37**, 289 (1998).
[6] M. Bruner and B. A. Horenstein, *Biochemistry* **39**, 2261 (2000).

In solvent isotope effect studies, the pD of the D_2O mixture was determined by using the following correction[7]: pD = pH meter reading + 0.4.

Kinetic Isotope Effect Methods

The methods described here have been used to measure KIEs for sialyltransferase with CMP-NeuAc[5] and uridine monophosphate glycoside of N-acetylneuraminic acid (UMP-NeuAc)[6] as the glycosyl donor. To prepare the appropriate UMP-NeuAc isotopomer, the corresponding CMP-NeuAc isotopomer is deaminated with nitrous acid. CMP-NeuAc isotopomer (2–10 μCi/reaction; 16–50 μCi/μmol specific activity for ^{14}C and 10 mCi/μmol specific activity for ^{3}H) is dissolved in 50–500 μl 1 N $NaNO_2$, adjusted to pH 4–5 with 1 N HCl (10–50 μl), and allowed to react at 4° for 48 hr, using manual addition of 1 N HCl to maintain the pH. Evolution of N_2 from the diazonium salt occurs on acidification; deamination is monitored by anion-exchange HPLC (100 mM NH_4HCO_3, 15% (v/v) methanol, pH 8.0, 2 ml/min). Under these conditions, CMP-NeuAc has a retention time of 13.5 min, and UMP-NeuAc has a retention time of 17.5 min. UMP-NeuAc is purified from the reaction mixture by anion-exchange chromatography [50 mM NH_4HCO_3, 15% (v/v) methanol, pH 8.0, 2 ml/min]. Collected UMP-NeuAc fractions are desalted with Amberlite IR120-H$^+$ cation-exchange resin, as described for CMP-NeuAc.[1]

Kinetic isotope effects[5,6] on sialyltransferase reactions are measured by the dual-label competitive method,[8,9] employing approximately 100,000 cpm of each ^{3}H- and ^{14}C-labeled UMP-NeuAc isotopomers. Typically, one prepares a master mixture containing a given ^{3}H/^{14}C-labeled UMP-NeuAc isotopomeric pair and 20 mM N-acetyllactosamine (LacNAc) in 40 mM cacodylate buffer [0.2 mg/ml bovine serum albumin (BSA), 0.2% Triton CF-54, pH 7.0]. Aliquots are withdrawn from this master to make up individual reaction samples and to measure the reference ^{3}H/^{14}C ratio at time t_0, when no product has formed. Reactions are initiated by the addition of enzyme to 50–100 μl reaction mixtures to give 40–60%[10] conversion in <45 min at 37°. Unreacted substrate is isolated by anion-exchange HPLC [100 mM NH_4HCO_3, 15% (v/v) methanol, pH 8.0, 2 ml/min], and 2-ml fractions are collected directly into 20-ml glass scintillation vials. Care is exercised to collect the entire UMP-NeuAc peak. The percent conversion is determined from the ratios of the UMP-NeuAc and UMP peaks in the HPLC chromatogram. The initial ^{3}H/^{14}C ratio should be obtained in triplicate by injecting aliquots of the master mixture

---

[7] P. Salomaa, L. L. Schaleger, and F. A. Long, *J. Am. Chem. Soc.* **86**, 1 (1964).
[8] F. W. Dahlquist, T. Rand-Meir, and M. A. Raftery, *Biochemistry* **8**, 4214 (1969).
[9] D. L. Parkin, in "Enzyme Mechanisms from Isotope Effects" (P. F. Cook, ed.), p. 269. CRC Press, Boca Raton, FL, 1991.
[10] R. G. Duggleby and D. B. Northrop, *Bioorg. Chem.* **17**, 177 (1989).

of labeled UMP-NeuAc on the Mono Q column and recollecting the entire UMP-NeuAc peak. The ^{3}H/^{14}C ratios for collected fractions are determined by dual-channel liquid scintillation counting (channel A, 0–12 keV; channel B, 12–80 keV) with each tube being counted for 10 min, and all tubes are cycled through the counter for 6–10 successive sets of radioactivity measurements. The internal ^{133}Ba source is used to estimate the quench parameter for each tube, which typically varies by ±1%. Triplicate samples of [^{14}C]CMP-NeuAc are used to determine the ratio of ^{14}C counts in channels A and B (A : B^{14}). Because ^{3}H is only detected in channel A, the ^{3}H/^{14}C ratio can be calculated with Eq. (1). Control experiments established that the HPLC does not introduce artifactual isotopic fractionation.[1] The observed KIE, which can be calculated using Eqs. (2) or (3), is then corrected for fractional conversion with Eq. (4):

$$^3\text{H}/^{14}\text{C} = \frac{[\text{cpm B}^*(\text{A} : \text{B}^{14})]}{[\text{cpmB} + \text{B}^*(\text{A} : \text{B}^{14})]} \quad (1)$$

$$^3\text{H-KIE}_{\text{observed}} = (^3\text{H}/^{14}\text{C})_0/(^3\text{H}/^{14}\text{C})_t \quad (2)$$

$$^{14}\text{C-KIE}_{\text{observed}} = (^{14}\text{C}/^3\text{H})_0/(^{14}\text{C}/^3\text{H})_t \quad (3)$$

$$\text{KIE}_{\text{corrected}} = \ln(1 - f)/\ln[(1 - f)\text{KIE}_{\text{observed}}] \quad (4)$$

where f is the fraction of reaction.[11] The reported value and error of a KIE represents the mean and standard deviation of three separate KIE reactions taken over 6–10 cycles through the liquid scintillation counter.

Solvent Isotope Effect Measurements

To measure the solvent deuterium isotope effect on V_{max} with UMP-NeuAc, a stock solution containing 4 mM [9-^{3}H]UMP-NeuAc (40,000 cpm/reaction, 0.69 μCi/mol) and 40 mM LacNAc is prepared in 250 mM Bis–Tris propane buffer, 0.2 mg/ml BSA, 0.2% Triton CF-54, pH 7.8. Half of the reaction mixture is retained for the reaction in H$_2$O, and the other half is concentrated to dryness on a rotary evaporator equipped with a mechanical vacuum pump. The residue is dissolved and twice reconcentrated in D$_2$O before being dissolved in 99.9% D$_2$O and brought to the original volume. Reaction samples (60 μl) are preincubated at 37° for 1 min before addition of 5 μl sialyltransferase (1.2 mU). Aliquots (20 μl) are removed after 6, 12, and 18 min, and the product is quantified with Dowex mini-columns.[12] The solvent deuterium isotope effect on $V/K_{\text{UMP-NeuAc}}$ is measured by the same procedure and in the same buffer system, but in these experiments, the final concentration of UMP-NeuAc is 0.3 M. Solvent deuterium isotope effects

---

[11] J. Bigeleisen and M. Wolfsberg, *Adv. Phys. Chem.* **1**, 15 (1958).
[12] J. Weinstein, U. Souza-e-Silva, and J. C. Paulson, *J. Biol. Chem.* **257**, 13845 (1982).

with CMP-NeuAc are measured by the same procedure described above, but are carried out at pH 6.5 in 40 mM sodium cacodylate buffer containing 16 mM LacNAc, 0.2 mg/ml BSA, and 0.2% Triton CF-54. In the V_{max} experiments, the concentration of CMP-NeuAc is 500 μM; for V/K, the concentration is 11 μM. Control experiments for viscosity effects of D$_2$O at 37° are made by taking initial velocity measurements under V_{max} and V/K conditions in the presence and absence of 9% (v/v) glycerol.[13,14]

(2→6)-Sialyltransferase Catalysis with Alternative Substrate UMP-NeuAc

Despite their broad acceptor-substrate specificity, (2→6)-sialyltransferases prefer to sialylate terminal Gal-β(1→4)-GlcNAc residues. Minor modifications in the NeuAc ring of the donor substrate CMP-NeuAc are tolerated, but there is an evident preference for an unmodified cytidine nucleotide that is marked.[15-17] To simplify pH–rate profiles and to probe for kinetic isotope effects that are potentially masked when CMP-NeuAc is the donor substrate, our laboratory has employed a nonnatural sugar-nucleotide UMP-NeuAc. The kinetic parameters for (2→6)-sialyltransferase using UMP-NeuAc and LacNAc as the donor–acceptor substrate pair are estimated at pH 7.5 and 37°. The K_m for UMP-NeuAc is 1.2 ± 0.1 mM, the K_m for LacNAc is 4.9 ± 0.3 mM, and the V_{max} is 1.8 μmol/(min · mg) with a corresponding k_{cat} value of 1.2 s^{-1}. Relative to CMP-NeuAc, UMP-NeuAc is a very weakly bound substrate for sialyltransferase, and V_{max} is somewhat reduced. With CMP-NeuAc as the donor substrate, the pH profiles obtained are complex (Fig. 1A). On the acid limb, both the V and V/K profiles show a limiting slope of 2 and an apparent pK_a near 5.3. The V_{max} profile reaches an optimum near pH 6, and then decreases with an apparently nonintegral slope toward alkaline pH, whereas the V/K profile reaches a maximum limiting plateau after the acid-side pK_a of ~5.3. The pH profiles with the slow substrate UMP-NeuAc (Fig. 1B) are quite different from those discussed above. The bell-shaped data are well fit to a model which requires a single ionizable group in its unprotonated form, and another group in its protonated form. For the V_{max} plot, the pK_a values are 5.5 and 9.0, whereas for $V/K_{UMP-NeuAc}$, pK_a values are 6.2 and 8.9.

---

[13] I. Kirshenbaum, "Physical Properties and Analysis of Heavy Water," p. 33. McGraw-Hill, New York, 1951.
[14] C. S. Miner and N. N. Dalton, "Glycerol," ACS Monograph Series, p. 240. Reinhold Publishing Corp., New York, 1953.
[15] H. H. Higa, J. C. Paulson, and J. Weinstein, *J. Biol. Chem.* **260,** 8838 (1985).
[16] H. J. Gross, U. Rose, J. M. Krause, J. C. Paulson, K. Schmid, R. E. Feeney, and R. Brossmer, *Biochemistry* **28,** 7386 (1989).
[17] R. G. Kleineidam, T. Schmelter, R. T. Schwartz, and R. Schauer, *Glycoconjugate J.* **14,** 57 (1997).

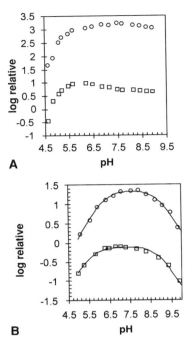

FIG. 1. (A) pH–log rate profiles with CMP-NeuAc as donor substrate. The circles represent the experimental k_{cat}/K_m data, and the squares represent the k_{cat} data. Units are M^{-1} s^{-1} for k_{cat}/K_m and s^{-1} for k_{cat}. (B) pH–log rate profiles with UMP-NeuAc as donor substrate. The circles represent the experimental k_{cat}/K_m data, and the squares represent the k_{cat} data. Units are M^{-1} s^{-1} for k_{cat}/K_m and s^{-1} for k_{cat}, and the solid lines represent the fit of the data to a Michaelis-type pH function.

The kinetic isotope effect methodology used in this study utilizes radiolabeled substrates, so the isotope effects are on the kinetic parameter V/K.[18] Figure 2 presents the structure of UMP-NeuAc and identifies the positions with isotopic labels. The V/K KIEs for the enzyme-catalyzed reaction with UMP-NeuAc are measured at pH 7 and 37° and are presented in Table I. For comparison, the KIEs for the enzyme reaction with the natural donor substrate CMP-NeuAc and solvolysis of both CMP-NeuAc and UMP-NeuAc are provided.

A β-dideuterio KIE of 1.218 ± 0.010 is measured with UMP-NeuAc as the donor substrate and LacNAc as the acceptor substrate. The magnitude of this effect is close to that measured for the acid-catalyzed hydrolysis of CMP-NeuAc and much greater than that measured for the enzyme reaction with the natural donor substrate CMP-NeuAc. A value of 1.28 ± 0.01 for the β-dideuterio isotope effect

---

[18] H. Simon and D. Palm, *Angew. Chem., Int. Ed. Engl.* **5,** 920 (1966).

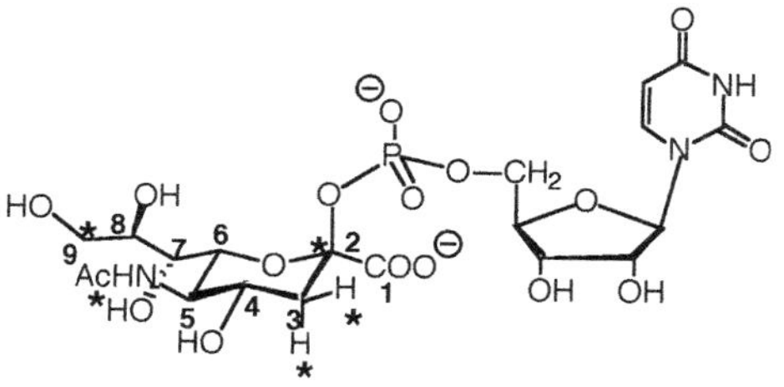

FIG. 2. UMP-NeuAc isotopomers. Asterisks are included to locate labeled atoms, and the numbers identify the carbon atoms for the NeuAc residue.

on the acid-catalyzed hydrolysis reaction of UMP-NeuAc has also been measured, which is identical within experimental error to that measured for CMP-NeuAc. A primary ^{14}C-KIE of 1.028 ± 0.010 has been measured on the enzyme-catalyzed reaction with UMP-NeuAc, which is almost identical to that observed for the solvolysis reaction of CMP-NeuAc. A large ^{3}H inverse binding isotope effect of 0.944 ± 0.010 is measured at C9 of the C6–C9 glycerol tail of NeuAc. Finally, a control KIE of 1.005 ± 0.005 is measured for ^{3}H and ^{14}C labels positioned in the N-acetyl group of UMP-NeuAc. This control shows that the N-acetyl remote labels do not experience KIEs.

The effects of solvent deuterium substitution on V_{max} and V/K for the altered donor substrate (UMP-NeuAc) have been measured for sialyltransferase.[6] In both

TABLE I
OBSERVED KINETIC ISOTOPE EFFECTS FOR SIALYLTRANSFERASE[a]

Isotopomeric pair	Type of KIE	KIE$_{UMP-NeuAc}$	KIE$_{CMP-NeuAc}$	Solvolysis[b]
[1-^{3}H-N-acetyl; 3,3′-^{2}H$_2$]	β-Secondary	1.218 ± 0.010	1.022 ± 0.007	1.276 ± 0.008
[1-^{14}C-N-acetyl]				1.28 ± 0.01^c
[2-^{14}C], [1-^{3}H-N-acetyl]	Primary ^{14}C	1.028 ± 0.010	1.000 ± 0.004	1.030 ± 0.005
[9-^{3}H], [1-^{14}C-N-acetyl]	Binding	0.944 ± 0.010	0.984 ± 0.007	NA
[2-^{14}C], [9-^{3}H]	Primary ^{14}C and binding	1.102 ± 0.012	NA	NA
[1-^{3}H-N-acetyl], [1-^{14}C-N-acetyl]	Control	1.005 ± 0.005	1.003 ± 0.004	1.002 ± 0.010

[a] With CMP-NeuAc and UMP-NeuAc [M. Bruner and B. A. Horenstein, *Biochemistry* **37**, 289 (1998); *Biochemistry* **39**, 2261 (2000)]. The isotope effects obtained with CMP-NeuAc and the isotope effects for solvolysis are included for comparison.

[b] KIEs for solvolysis of CMP-NeuAc [B. A. Horenstein and M. Bruner, *J. Am. Chem. Soc.* **118**, 10371 (1996)].

[c] KIE for solvolysis of UMP-NeuAc [M. Bruner and B. A. Horenstein, *Biochemistry* **39**, 2261 (2000)].

cases, effects close to unity are measured. In contrast, CMP-NeuAc shows a solvent isotope effect of 1.3 on V/K and 2.6 on V_{max}. The isotope effects are apparently not related to the difference in viscosity of H_2O and D_2O[13] since control experiments with glycerol as an added viscogen failed to show measurable differences in velocity for V/K or V_{max}. The results for the substrate dependence of the solvent isotope effects are in counterpoint to those for the -^{2}H and primary ^{14}C isotope effects. The observed solvent isotope effects are higher with the "fast" substrate CMP-NeuAc than they are with UMP-NeuAc. The KIEs for the isotope-labeled NeuAc are suppressed with CMP-NeuAc but are maximal with the "slow" substrate UMP-NeuAc.

Mechanistic Inferences Regarding Sialyltransferase Catalysis

The very small KIEs observed for sialyltransferase action with CMP-NeuAc suggest that kinetic barriers other than that for the chemical step are partially rate-limiting. Separate isotope trapping experiments yielded a commitment factor of 1.0, but even after accounting for this commitment to catalysis,[19] the corrected KIEs are still small,[5] suggesting that additional slow steps after substrate binding remain to be explained. With the exception of the oxo group in the pyrimidine ring, the alternative substrate UMP-NeuAc resembles CMP-NeuAc. UMP-NeuAc binds more weakly to the sialyltransferase than CMP-NeuAc, as indicated by its 30-fold higher K_m. The k_{cat} for UMP-NeuAc is also five times lower than that for CMP-NeuAc. These findings suggested that the increased barrier for the chemical step and the weaker binding of UMP-NeuAc should provide conditions that are more favorable for full expression of the kinetic isotope effects. The magnitude of the isotope effects measured with UMP-NeuAc (Table I) supports this inference. The $\alpha(2\rightarrow 6)$ β-^{2}H isotope effect (1.22) is nearly as large as the 1.28 β-^{2}H isotope effects measured for solvolysis of either CMP-NeuAc or UMP-NeuAc. Further, the primary ^{14}C isotope effects for the sialyltransferase UMP-NeuAc reaction and solvolysis of CMP-NeuAc are identical (1.030). The β-^{2}H KIEs for the solvolysis reaction present a rough upper limit for the magnitude of the enzymatic KIE, and the observed similarity between the two sets of KIEs argues that the enzyme KIEs measured with UMP-NeuAc are intrinsic ones, or very nearly so.

Structural Characteristics of Transition State

Our working model for sialyltransferase catalysis starts with a preprotonation of a nonbridging phosphate oxygen before cleavage of the glycosidic bond to

---

[19] D. B. Northrop, *in* "Isotope Effects on Enzyme Catalyzed Reactions" (W. W. Cleland, M. H. O'Leary, and D. B. Northrop, eds.). University Park Press, Baltimore, 1977.

CMP. The structure of the transition state that follows is identified on the basis of deuterium and ^{14}C KIEs.[20] The β-dideuterium KIE of 1.218 ± 0.010 is nearly as large as the KIE for solvolysis of CMP- or UMP-NeuAc (Table I). The solvolysis reaction transition state is purely dissociative, and by the comparable size of the KIEs, the sialyltransferase transition state must be similar. This is characteristic of a transition state with oxocarbenium ion character.[21] The primary ^{14}C KIE of 1.028 ± 0.010 with UMP-NeuAc supports this explanation. This effect is in the range seen for dissociative type mechanisms[22-24] and is identical to the KIE measured for solvolysis of CMP-NeuAc.[1] This small primary carbon KIE rules out significant nucleophilic participation in concert with loss of the UMP leaving group. The β-^{2}H KIEs show that the transition state for glycosidic bond cleavage has substantial positive charge. The lack of significant participation by the acceptor oligosaccharide would allow for the possibility that after the transition state a very short-lived oxocarbenium ion intermediate is formed; this is precedented for β-galactosidase.[25] In solution, the NeuAc oxocarbenium ion species is very short-lived,[3] so if it were generated in the active site, it is likely that it would be readily trapped by the acceptor in a kinetically invisible step. Another aspect of the transition state involves a binding interaction at C-9 of the glycerol tail on the NeuAc residue. This follows from the substantial inverse ^{3}H isotope effect of 0.944 that indicates that the hydrogen at C-9 is in a tighter vibrational environment in the transition state than in the ground state. It is not yet clear what specific molecular interactions are responsible for the isotope effect. The apparent binding and recognition of this side chain by sialyltransferase may be an important component of inhibitor design for sialyltransferases.

Concluding Remarks

Enzyme action on a preferred substrate is often characterized by rate processes that lead to high catalytic efficiency, but also prevent a thorough kinetic analysis of the reaction mechanism. This limitation frequently can be obviated by use of alternative substrates that exhibit lower rates of catalysis and often unmask kinetic features that are indiscernible with the natural substrate. In the case of $\alpha(2\rightarrow 6)$-sialyltransferase, the alternative donor substrate UMP-NeuAc provides the opportunity to probe the mechanism in greater detail than achieved with CMP-NeuAc.

---

[20] V. L. Schramm, *Methods Enzymol.* **308**, 301 (1999).
[21] L. Melander and W. H. Saunders, "Reaction Rates of Isotopic Molecules," Chapter 6. Kreiger, Malabar, FL, 1980.
[22] R. K. Goiten, D. Chelsky, and S. M. Parsons, *J. Biol. Chem.* **253**, 2963 (1978).
[23] A. J. Bennett and M. L. Sinnott, *J. Am. Chem. Soc.* **108**, 7287 (1986).
[24] L. Melander and W. H. Saunders, *in* "Reaction Rates of Isotopic Molecules," p. 242. Kreiger, Malabar, FL, 1980.
[25] J. P. Richard, R. E. Huber, C. Heo, T. L. Amyes, and S. Lin, *Biochemistry* **35**, 12387 (1996).

The specificity of CMP-NeuAc synthase for CTP posed a synthetic challenge for the preparation of UMP-NeuAc isotopomers by direct enzymatic means from UTP and labeled NeuAc. This was overcome by use of diazotization and subsequent hydrolysis to convert radioisotopically labeled CMP-NeuAc isotopomers directly to the corresponding UMP-NeuAc isotopomers, thereby greatly facilitating our analysis of kinetic isotope effects.

Acknowledgments

Support was provided by the National Science Foundation (CAREER award Grant MCB-9501866) and the University of Florida Division of Sponsored Research.

[11] Use of Sodium Borohydride to Detect Acyl-Phosphate Linkages in Enzyme Reactions

By DANIEL L. PURICH

The nature of the leaving group (or exiphile) that is expelled from a substrate during a nucleophilic displacement greatly influences both the kinetics and thermodynamics of the overall reaction.[1] A common theme in nucleophilic catalysis is that the strongest acids make the best leaving groups. Therefore, loss of an oxygen atom from a carboxyl group as a hydroxide ion is particularly disfavored. Formation of acyl-phosphate intermediates enhance nucleophilic substitution on the carboxyl carbonyl. In these mixed anhydrides, the unlikely OH⁻ leaving group of the carboxyl is replaced by phosphate, to the effect that reactivity is enhanced enormously. It is therefore not surprising that acyl phosphates are the chief intermediates in reactions involving carboxyl-group activation.[2–4] There are several classical methods for detecting the presence of acyl-P compounds in enzyme reactions, including the well-known susceptibility to attack by hydroxylamine. Even so, borohydride reduction has become a standard method[5] for demonstrating acyl-P formation in an enzyme-catalyzed reaction, and this chapter describes protocols for detecting both covalently and noncovalently bound acylphosphate compounds.

---

[1] W. P. Jencks, "Catalysis in Chemistry and Enzymology." McGraw-Hill, New York, 1969.
[2] D. L. Purich, *Adv. Enzymol.* **72**, 9 (1998).
[3] R. M. Bell and D. E. Koshland, Jr., *Science* **172**, 1253 (1971).
[4] R. D. Allison and D. L. Purich, *Methods Enzymol.* **354**, [33], 2002 (this volume).
[5] C. Degani and P. D. Boyer, *J. Biol. Chem.* **248**, 8222 (1973).

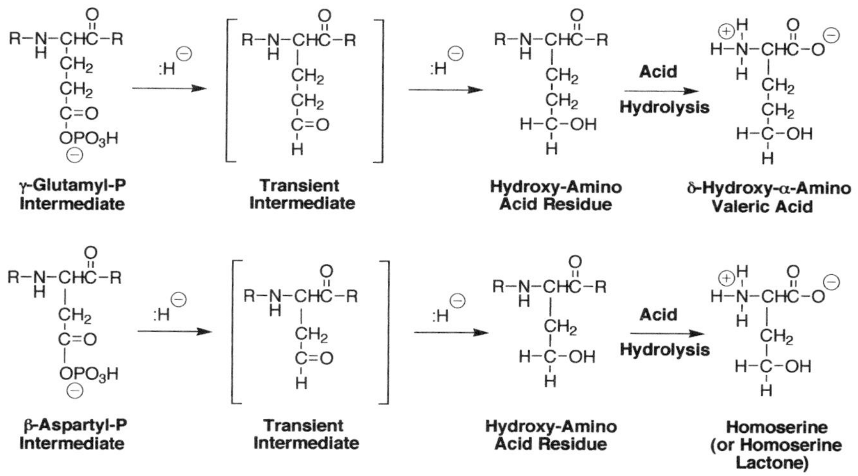

FIG. 1. Reductive dephosphorylation of an activated carboxyl group of γ-glutamyl-P (or γ-glutamyladenylate or γ-glutamylthiol ester) results in the formation of α-amino-δ-hydroxyvalerate. Reductive dephosphorylation of an activated carboxyl group of β-aspartyl-P (or β-aspartyladenylate or β-aspartylthiol ester), respectively, and homoserine. (Aldehydes, ketones, and imines are also reduced by sodium borohydride or sodium cyanoborohydride.)

Borohydride Reduction of Acyl Phosphates

The method exploits the fact that an unmodified carboxyl group is insufficiently activated to undergo sodium borohydride reduction (Fig. 1). For enzymes that are covalently modified by phosphorylation on the β- and γ-carboxyls of aspartate and glutamate residues, one employs [^{3}H]NaBH$_4$ to form respectively tritiated homoserine (or its lactone) and α-amino-δ-hydroxyvaleric acid.[5,6] In the event that the C-terminal amino is converted to an acyl-P compound, one then detects the corresponding [^{3}H]amino alcohol (e.g., ethanolamine in the case of a C-terminal glycyl residue[7]). Partial proteolytic fragmentation generates tritium-containing peptides whose sequences can be deduced through the use of mass spectrometry. More frequently, the phosphorylated enzyme is subjected to complete hydrolysis, and the liberated amino alcohol is purified by chromatography, relying on radioactivity measurements for detection and quantification. If, as in the case of glutamine synthetase, the acyl-P is noncovalently bound to the enzyme, then the acyl-P reduction can be accomplished either in the presence of the enzyme or immediately after the acyl-P is liberated from the enzyme by the addition of a denaturing agent.[8]

---

[6] J. A. Todhunter and D. L. Purich, *Biochem. Biophys. Res. Commun.* **60**, 273 (1974).

[7] C. M. Pickart and I. A. Rose, *J. Biol. Chem.* **261**, 10210 (1986).

[8] J. A. Todhunter and D. L. Purich, *J. Biol. Chem.* **250**, 3505 (1975).

Reduction of Phosphorylated Enzyme Compounds

Preliminary experiments using a ^{32}P-labeled phosphoryl donor should be carried out to establish optimal reaction conditions for enzyme phosphorylation. The observed stoichiometry of phosphoryl transfer establishes the maximal yield expected upon borohydride reduction. Ideally, companion experiments should be likewise conducted with a ^{14}C/^{3}H-labeled phosphoryl donor to confirm that the presumed phosphorylation is not the trivial consequence of noncovalent binding of the phosphoryl donor.

Two equivalents of hydride ion are required to convert an acyl-P to the corresponding alcohol, thereby increasing the sensitivity of acyl-P detection. Reduction experiments therefore require 0.5–4 mg protein, corresponding to 10–80 nmol covalently bound phosphoryl groups. Sodium borohydride reduction reactions are carried out essentially as described by Degani and Boyer,[5] who used a 2.5-mM final concentration of [^{3}H]NaBH$_4$. Commercially available borohydride typically has a radiospecific activity of 2×10^{11} Bq/mmol, and the reagent is used without mixing with carrier borohydride.

To a solution containing 1–3 mg phosphorylated protein (experimental) and an identical separate sample of unphosphorylated protein (control), each dissolved in 0.3 ml dimethyl sulfoxide (DMSO), add 0.2 ml [^{3}H]NaBH$_4$ (12.5 mM). Allow the reaction to proceed for 15–30 min and terminate the reduction by addition of 5 ml perchloric acid (0.44 M), followed by low-speed centrifugation to obtain the protein-containing pellet. The resulting precipitate is washed three times by resuspension and brief centrifugation. Protein samples are then transferred to vials containing 6 N HCl, sealed, and placed in a flask directly above refluxing toluene (108–110°) for 22 hr. Resulting hydrolyzates are evaporated to dryness under vacuum at room temperature and then redissolved in 0.5 ml water. Exchangeable tritium label is removed by a second evaporation to dryness, readdition of 0.5 ml water, and a third evaporation to dryness. The sample is finally dissolved in 0.1–0.3 ml water and neutralized by addition of pyridine.

Warning: Radioactive tritium gas is produced on acidification of sodium borohydride, and all experimental procedures should be conducted in a fume hood. One also must be mindful that dimethyl sulfoxide undergoes borohydride reduction to form dimethyl sulfide. Because DMSO is a highly efficient drug-delivery vehicle that readily transports radiolabeled compounds through skin, one should always use acid-resistant rubber gloves. Avoid using the far more permeable latex gloves that are widely available in most clinical and biochemical laboratories.

Reduction of Noncovalently Bound Acyl-Phosphate Intermediates

Because radiolabeled carboxyl group-containing substrate is used in these experiments, the efficiency of reduction can be improved by using 30 mM unlabeled sodium borohydride. The data presented in Fig. 2 demonstrate the utility of the

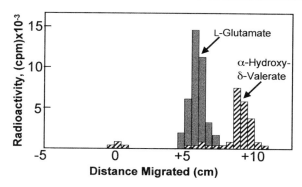

FIG. 2. High-voltage electropherograms of sodium borohydride-treated glutamine synthetase reaction mixtures containing enzyme, ATP, glutamate, and metal ion. Glutamine synthetase (0.35 mg) was incubated for 5 min at 37° in a 1.0-ml reaction mixture containing 0.5 mM ATP, 0.1 mM 1-[^{14}C]glutamate (radiospecific activity, 11.3 Ci/mol), 1.0 mM magnesium ion, and adjusted to pH 7.5 with dilute KOH. Prior to borohydride reduction the samples were deproteinized by addition of 1.5 volumes of cold dimethyl sulfoxide and subsequent centrifugation at 8000g for 5 min at 37°. Reduction commenced with the addition of 5 mg of solid NaBH$_4$ to each tube; 12 hr later 5 volumes of water were added, and the resulting solution was repeatedly freeze-dried to remove the dimethyl sulfoxide. The residue, which was taken up in 0.5 ml water and treated with 0.05 ml concentrated formic acid to destroy the remaining borohydride, was adjusted to pH 8.5 and applied to a Sephadex G-25 column (0.3 × 2 cm). Borate remained bound to the column provided a flow rate not exceeding 0.1 ml/min was used, and the removal of the borate was monitored by the alizarin method. Pooled radioactive samples were again freeze-dried and taken up in 0.05 ml water, and 0.1 ml of alcohol was added to precipitate the protein which was then removed by centrifugation. Aliquots (0.01 ml) were spotted on Whatman (Clifton, NJ) No. 1 paper and electrophoresis was carried out for 30 min at 2000 V (15°) in 7% formic acid after a desalting period of 5 min at 500 V. Strips of the dried electropherograms were treated with ninhydrin to localize the standards, and radioactivity of 5-mm sections was determined by direct counting of the paper in a toluene-based scintillation cocktail with an overall efficiency of about 40%. Of each sample, approximately 14 nCi of radioactivity was spotted and electrophoresis was carried out as described above. (AHV, α-Amino-δ-hydroxyvalerate.) Radioactivity distribution from the electrophoretic origin for the Mg^{2+}-supported system. [Reproduced with permission of the American Society for Biochemistry and Molecular Biology from J. A. Todhunter and D. L. Purich, *J. Biol. Chem.* **250**, 3505 (1975).]

borohydride reduction method[5] in establishing the identity of an acyl-P intermediate in the glutamine synthetase reaction. In the reaction catalyzed by *Escherichia coli* glutamine synthetase,[8] ammonium ion interferes with the accumulation of the γ-glutamyl-P intermediate when enzyme is incubated with L-glutamate and ATP. Therefore, before use, all enzyme-containing solutions must be subjected to extensive dialysis against ammonia-free buffers (treated with Permutit) to remove traces of ammonium ions.[9,10] L-[U-^{14}C]Glutamate (specific radioactivity 265 mCi/mmol)

---

[9] P. R. Krishnaswamy, V. Pamiljans, and A. Meister, *J. Biol. Chem.* **237**, 2932 (1962).
[10] J.-F. Collet, V. Stroobant, and E. van Schaftingen, *Methods Enzymol.* **354**, [12], 2002 (this volume).

should be purified by ascending paper chromatography. Glutamine synthetase (0.35 mg) is incubated for 5 min at 37° in a 1.0-ml reaction mixture containing 0.5 mM ATP, 0.1 mM 1-[^{14}C]glutamate (radiospecific activity, 11.3 Ci/mol), 1.0 mM magnesium ion, and adjusted to pH 7.5 with dilute KOH. Reactions are quenched by addition of 1.5 ml ice-cold dimethyl sulfoxide, and after cooling on ice for 1 min, 5 mg of solid NaBH$_4$ is added to each tube. (Dimethyl sulfoxide is the solvent of choice for borohydride reductions. Use of alcohol-containing solvents should be avoided, especially in the early steps, because additional reduction products are formed from contaminating aldehydes.) Twelve hours later, 5 volumes of water are added and the resulting solution is freeze-dried repeatedly to remove the dimethyl sulfoxide. The residue, which is taken up in 0.5 ml water and treated with 0.05 ml concentrated formic acid to destroy the remaining borohydride, is adjusted to pH 8.5 and applied to a Sephadex G-25 column (0.3 × 2 cm). Borate remains bound to the column if the flow rate does not exceed 0.1 ml/min, and the removal of the borate is easily monitored fluorescently by the alizarin method. (To a dried spot of the effluent of the Sephadex column on Whatman No. 1 paper is added a drop of a 0.1% alcoholic solution of alizarin, and the spot is inspected under a near-ultraviolet lamp to detect the fluorescent adduct. The limit of detection by this approach is approximately 10 nmol.) The pooled radioactive samples

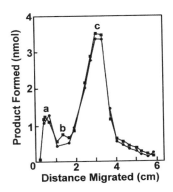

FIG. 3. Plot of the radioactivity distribution of (a) glutamate, (b) 5-oxoproline (OxPro), and (c) α-amino-δ-hydroxy-valerate (AHV) as a function of distance migrated from the origin of the thin layer chromatogram. Reaction conditions are those described in the legend to Fig. 2. Aliquots of the reaction mix containing Mg^{2+} (filled squares) and and Mn^{2+} (filled circles) were spotted, and the chromatograms were developed for 8 hr in an ascending system containing a collidine–lutidine–water solvent. Strips of the plastic-backed thin-layer sheets were cut and sprayed with ninhydrin to localize the position of authentic standards. Radioactivity was determined by carefully scraping the thin layer away from the plastic backing with a razor blade and counting in a toluene-based scintillation system. Counts have been converted to nmol glutamate, 5-oxoproline, and α-amino-δ-hydroxyvalerate formed under the reaction conditions described. Represented are the reactions carried out in the presence of magnesium ion (■) or manganese ion (●). [Reproduced with the permission of the American Society for Biochemistry and Molecular Biology from J. A. Todhunter and D. L. Purich, *J. Biol. Chem.* **250**, 3505 (1975).]

TABLE I
FORMATION OF α-AMINO-δ-HYDROXYVALERATE AND 5-OXYPROLINE[a]

Reaction conditions	Sample number				
	1	2	3	4	5
ATP (0.5 mM)	+	+	+	−	−
Magnesium ion (1 mM)	+	+	−	+	−
Manganous ion (1 mM)	−	0	+	−	+
1-[^{14}C]Glutamate (0.1 mM)	+	+	+	+	+
Deproteinized before NaBH$_4$ treatment?	No	Yes	Yes	No	No
AHVa produced (nmol)	14.5	14.7	15.4	None	None
5-Oxoproline formed (nmol)	1.2	1.0	1.1	None	None

[a] In the presence of *Escherichia coli* glutamine synthetase. Enzyme (7 nmol) was incubated in the absence of ammonium ions with ATP, glutamate, and metal ion in the quantities indicated [J. A. Todhunter and D. L. Purich, *J. Biol. Chem.* **250**, 3505 (1975)]. The 1.0-ml reaction mixtures were incubated at 30° for 5 min and processed as described in the legend to Fig. 2. [Reproduced with the permission of the American Society for Biochemistry and Molecular Biology from J. A. Todhunter and D. L. Purich, *J. Biol. Chem.* **250**, 3505 (1975).]

are again freeze-dried and taken up in 0.05 ml water, and alcohol (0.1 ml) is added to precipitate the protein which is removed by centrifugation.

The data presented in Fig. 3 and Table I established the formation of a γ-glutamyl-P intermediate in the glutamine synthetase reaction. In these experiments, 20 μM enzyme yields 14 μM α-amino-δ-hydroxyvalerate detected an observation that accords with earlier findings on 5-oxoproline formation.[9] Suitable control experiments were carried out to confirm that the free unactivated amino acid is inert to the borohydride reduction. In the absence of ATP or enzyme, radiolabeled glutamate should give no labeled α-amino-δ-hydroxyvalerate. That essentially all of the acyl-P is trapped as α-amino-δ-hydroxyvalerate can also be determined by confirming that 5-oxoproline formation is negligible (see Table I).

Separation and Detection of Amino-Alcohol Reduction Products

Paper electrophoresis is accomplished in a Savant high-voltage electrophorator at 10° using a 7% (v/v) formic acid : water buffer.[8] Ascending thin-layer chromatography of deproteinized samples is carried out at room temperature using a solvent of the following composition: 2,6-lutidine : collidine : water (1 v : 1 v : 1 v) with 2% diethylamine.[8] Amino acids were located by ninhydrin spray reagent. The radioactivity content of unsprayed samples can be determined by liquid scintillation counting, once the sample is scraped free from the plastic backing of the thin-layer chromatogram. To further verify that the radioactive compound is radiolabeled δ-hydroxy-α-aminovalerate or radiolabeled homoserine, one can also separately combine the radioactive protein hydrolyzate (or a partially purified compound) with saturated solutions of and α-amino-δ-hydroxyvalerate or homoserine to evaluate their repetitive recovery after multiple recrystallizations.

One may also utilize a high-sensitivity amino acid analyzer to detect and quantify hydroxyamino acids formed by borohydride reduction. Likewise, mass spectroscopic detection of peptide hydroxyamino acid-containing residues is highly effective.[10]

Complications

Enzymes frequently contain disulfide cross-links (or oxidized dithiothreitol), minor amounts of imine and ketimine species, bound reducing sugars, and even esters formed from peptide bonds by N-to-O acyl shift reactions. Therefore, enzyme inactivation upon exposure to borohydride is insufficient proof of acyl-P formation. As first observed by Degani and Boyer,[5] only a fraction of the tritium radioactivity associated with the reduced enzyme arises strictly from reduction of the acyl-P residue. However, because acyl-P reduction generates a stable and structurally defined reaction product, and because there are good methods for detecting and quantifying these reduction products, there is relatively little ambiguity to the approach. The notable exception to this statement is that β-aspartyl- and γ-glutamylthiol esters and adenylates should produce homoserine and α-amino-δ-hydroxyvalerate, respectively, on borohydride reduction. Knowledge of other properties of the enzyme reaction, however, should indicate the likelihood of their formation.

Finally, destruction of hydroxyamino acids during acid hydrolysis is a widely acknowledged problem in protein chemistry, and in some cases, one may underestimate the actual extent of acyl-P formation. One method for detecting this problem is to include nonradioactive homoserine and/or α-amino-δ-hydroxyvalerate as internal standards during the acid hydrolysis step. It may also be advisable to prepare 12-, 18-, and 24-hr hydrolyzates, thereby permitting extrapolation to obtain the time $(t) = 0$ values for homoserine and/or α-amino-δ-hydroxyvalerate.

Use of Borohydride in Other Acyl Phosphate-Forming Processes

Ubiquitin Carboxyl-Terminal Hydrolase

Ubiquitin carboxyl-terminal hydrolase (Enz) catalyzes the hydrolysis, at the ubiquitin-carboxyl terminus, of a wide variety of C-terminal ubiquitin derivatives. Pickart and Rose[7] proposed that the hydrolase mechanism exploits nucleophilic catalysis with an acyl-ubiquitin-Enz intermediate. The enzyme is inactivated by millimolar concentrations of either $NaBH_4$ or hydroxylamine, provided that ubiquitin is present. The borohydride-inactivated enzyme, tritium-containing one-to-one complex of enzyme and ubiquitin, is stable at neutral pH in 5 M urea and can be isolated by gel filtration. Acid treatment yields ubiquitin-carboxyl-terminal aldehyde, and further reduction of the ubiquitin released with [^{3}H]NaBH$_4$ and complete acid hydrolysis yielded tritium-labeled ethanolamine, the expected product if the carboxyl-terminal glycine residue of ubiquitin formed an ester or thiol ester adduct. Hydrolase inactivation by hydroxylamine of occurs once during the enzymatic hydrolysis of 1200 molecules of ubiquitin-hydroxamate, and this hydrolysis/inactivation ratio was constant over the 10–50 mM range of hydroxylamine, showing that forms of Enz-ubiquitin with which hydroxylamine and water react are not in rapid equilibrium. The inactive enzyme may be an acylhydroxamate formed from an Enz-ubiquitin mixed anhydride generated from the Enz-ubiquitin-thiol ester.

Myelin Proteolipid Protein

Reductive cleavage with sodium borohydride of myelin proteolipid protein-acylated *in vitro* or *in vivo* yielded [^{3}H]hexadecanol, thereby identifying at least one of the acyl linkages as a thiolester with palmitate.[11] Myelin proteolipid protein was acylated nonenzymatically on incubation with with acyl-CoA as the fatty acid donor.

Phosphomannomutase

When incubated with their respective substrates, phosphomannomutase and L-3-phosphoserine phosphatase become phosphorylated, and the chemical characteristics suggested the formation of an acyl phosphate.[12] In phosphomannomutase, the phosphorylated residue identified by mass spectrometry after borohydride reduction of the phosphoenzyme and trypsin digestion is the first aspartate in a conserved DVDGT motif. Replacement of either aspartate of this motif by asparagine or glutamate resulted in complete inactivation of the enzyme. L-3-Phosphoserine phosphatase also belongs to the phosphotransferase family with an amino-terminal DXDX(T/V) motif serving as an intermediate phosphoryl acceptor. Of seven other phosphomutases or phosphatases sharing a similar DXDX(T/V) motif β-phosphoglucomutase forms a phosphoenzyme with acyl-P characteristics.

Nucleoside Phosphotransferase

This carrot enzyme[13] formed *N*-phosphorylhydroxylamine when substrates were hydrolyzed in the presence of hydroxylamine. If the enzyme was denatured

---

[11] N. W. Ross and P. E. Braun, *J. Neurosci. Res.* **21,** 35 (1988).
[12] J. F. Collet, V. Stroobant, M. Pirard, G. Delpierre, and E. Van Schaftingen, *J. Biol. Chem.* **273,** 14107 (1998).
[13] B. Stelte and H. Witzel, *Eur. J. Biochem.* **155,** 121 (1986).

after brief incubation with its substrates, reduction with [^{3}H]NaCNBH$_3$ and subsequent hydrolysis with 6 M HCl, radiolabeled homoserine could be detected, providing the first experiment evidence for an acyl-P intermediate.

Prothymosin α

The thymosin precursor prothymosin α is a glutamate-rich, 100-residue nuclear polypeptide that is covalently linked to a small rRNA of about 20 nucleotides. Prothymosin α is thought to mediate changes in chromatin compaction by inducing chromatin fiber unfolding through its interaction with histone H1. The protein is phosphorylated, and the U-shaped pH profile for its hydrolysis is a signature for acyl-P compounds.[14] Through the use of [^{3}H]NaBH$_4$, the authors demonstrate the presence of stoichiometric amounts of glutamyl phosphate.

UDP-N-Acetyl-Muramate L-Alanine Ligase (MurC) and
UDP-N-Acetylmuramoyl-L-Alanine:D-Glutamate Ligase (MurD)

The mechanism of the Mur synthetase reactions involved in peptidoglycan biosynthesis is thought to involve an acyl-P intermediate.[15] Reduction by sodium borohydride and subsequent acid hydrolysis yield the δ-hydroxy-α-amino valerate, thereby firmly establishing that *Escherichia coli* MurC and MurD enzymes form acyl phosphates.

CheY Chemotaxis Regulator

This regulatory protein, which controls the direction of flagellar rotation during bacterial chemotaxis, is phosphorylated on aspartate-57, based on reductive cleavage by sodium [^{3}H]borohydride and high-performance tandem mass spectrometry of radiolabeled proteolytic peptides.[16] The phosphorylation is transient and has a much shorter half-life than that expected for a typical acyl-P intermediate, suggesting that the sequence and conformation of the protein are designed to achieve a rapid hydrolysis.

Vanadate-Sensitive Membrane ATPase

This *Streptococcus faecalis* enzyme forms an acyl-P intermediate as part of its reaction cycle.[17] The phosphorylated amino acid residue was identified by reducing the purified reconstituted phosphoenzyme with [^{3}H]borohydride, followed by acid

---

[14] M. W. Trumbore, R. Wang, S. A. Enkemann, and S. L. Berger, *J. Biol. Chem.* **272**, 26394 (1997).
[15] A. Bouhss, S. Dementin, J. van Heijenoort, C. Parquet, and D. Blanot, *FEBS Lett.* **453**, 15 (1999).
[16] D. A. Sanders, B. L. Gillece-Castro, A. M. Stock, A. L. Burlingame, and D. E. Koshland, Jr., *J. Biol. Chem.* **264**, 21770 (1989).
[17] P. Furst and M. Solioz, *J. Biol. Chem.* **260**, 50 (1985).

hydrolysis of the protein and quantitative amino acid analysis. Tritiated homoserine was found to be the resulting reaction product, generated through the reduction of a β-aspartyl phosphate residue.

Collagen α_2 Chains

Purified components of chicken bone collagen contain approximately 4 organic-phosphorus atoms per mole collagen, mainly in the α_2 chains. Cohen-Solal *et al.*[18] used $NaB[^3H]H_4$ reduction to demonstrate that chicken bone collagen contains γ-glutamyl phosphate which was detected as α-amino-δ-hydroxyvalerate in the α_2 chains. Tritiated was not detected in dephosphorylated α_2 chains. The collagen α_2 chain is thus the first structural protein found to contain an acyl-P.

[18] L. Cohen-Solal, M. Cohen-Solal, and M. J. Glimcher, *Proc. Natl. Acad. Sci. U.S.A.* **76,** 4327 (1979).

[12] Evidence for Phosphotransferases Phosphorylated on Aspartate Residue in N-Terminal DXDX(T/V) Motif

By JEAN-FRANÇOIS COLLET, VINCENT STROOBANT, and EMILE VAN SCHAFTINGEN

The reaction mechanism of many phosphatases or phosphotransferases involves the formation of an intermediate phosphoenzyme. Four types of amino acid residues have been shown to act as phosphate acceptor in catalytic sites: serine, histidine, cysteine,[1,2] and aspartate. The group of enzymes forming a phosphoaspartate intermediate comprises P-type ATPases,[3] bacterial protein phosphatases acting in signal transduction,[4] and a newly identified class of phosphotransferases that is characterized by a conserved DXDXT/V motif close to the N terminus.[5,6]

Sequence comparisons indicate that this last group comprises at least 18 different enzymes (Table I). These are either phosphomutases (eukaryotic phosphomannomutase and β-phosphoglucomutase) or phosphatases acting mostly on low

[1] J. B. Vincent, M. W. Crowder, and B. A. Averill, *Trends Biochem. Sci.* **17,** 105 (1992).
[2] Y. Y. Wo, M. M. Zhou, P. Stevis, J. P. Davis, Z. Y. Zhang, and R. L. Van Etten, *Biochemistry* **31,** 1712 (1992).
[3] R. L. Post and S. Kume, *J. Biol. Chem.* **248,** 6993 (1973).
[4] D. A. Sanders, B. L. Gillece-Castro, A. M. Stock, A. L. Burlingame, and D. E. Koshland, *J. Biol. Chem.* **264,** 21770 (1989).
[5] J. F. Collet, V. Stroobant, M. Pirard, G. Delpierre, and E. Van Schaftingen, *J. Biol. Chem.* **273,** 14107 (1998).
[6] M. C. Thaller, S. Schippa, and G. M. Rossolini, *Protein Sci.* **7,** 1647 (1998).

TABLE I
DEMONSTRATED OR PRESUMPTIVE PHOSPHORYLATION SITES IN ENZYMES OF DXDX(T/V) FAMILY[a]

Name of enzyme	Species	Accession number		Motif 1		Phosphoenzyme
DXDX(T/V) family						
Phosphomannomutase	Homo sapiens	Q92871	14	V**LCLFD**V**DGT**LTPARQ	29	Localized
β-Phosphoglucomutase	Lactococcus lactis	P71447	3	KA**VLFDLDGV**T**D**TAE	18	Acyl phosphate
Phosphoserine phosphatase	Homo sapiens	NP_004568	15	**DA**V**CFD**V**DS**TVIREEG	30	Localized
Phosphoglycolate phosphatase	Escherichia coli	P32662	8	RGVA**FDLDGT**LVDSAP	23	Acyl phosphate[b]
2-Deoxyglucose-6-phosphate phosphatase	Saccharomyces cerevisiae	P38774	7	D**LCLFDLDGT**VSTTV	22	
Glycerol-3-phosphate phosphatase	Escherichia coli	P41277	12	NAA**LFD**V**DGT**IIISQP	27	
Histidinol-phosphate phosphatase	Saccharomyces cerevisiae	P06987	4	K**YLFIDRDGT**LISEPP	19	
Trehalose-6-phosphate phosphatase	Saccharomyces cerevisiae	P31688	572	R**LFLFD**Y**DGT**LTPIVK	587	
p-Nitrophenylphosphate phosphatase	Saccharomyces cerevisiae	P19881	25	**DTFLFDCDGVL**WLGSQ	40	
Bacterial acid phosphatases (class B)	Escherichia coli	P32695	67	MAVG**FDIDD**T**V**LFSSP	82	
Plant acid phosphatases (class C)	Lycopersicon esculentum	P27061	107	**DV**W**IFDVDETLL**SNLP	122	
3X11A protein (chondrocyte-specific hypothetical phosphatase)	Gallus gallus	CAA07090	27	Y**LL**V**FDFDGT**IINESS	42	
Polynucleotide 5'-kinase/3'-phosphatase	Homo sapiens	AF126486	166	K**VAGFDLDGT**LITTRS	181	
Cytosolic purine nucleotidase	Homo sapiens	NP_036361	47	KC**FGFD**M**D**Y**T**LAVYKS	62	
C-Terminal domain (CTD)-phosphatase	Homo sapiens	AAD42088	183	**L**V**LM**V**DLDQT**LIHTE	198	
Homoserine kinase (ThrH)	Pseudomonas aeruginosa	CAA07580	2	EIAC**LDLEG**V**L**VPEIW	16	
Sucrose-6-phosphate phosphohydrolase	Zea mays	AF283564	10	**L**MI**VSDLDHT**MVDHHD	25	
Psr1p/Psr2p (plasma membrane phosphatases)	Saccharomyces cerevisiae	NP_013091	258	KC**LILDLDET**L**V**HSSF	273	
P-ATPases						
Ca²⁺ (SERCA1)	Homo sapiens	014983	346	SVICS**D**K**TGT**LTTNQM	361	Localized
Phosphonatases						
	Pseudomonas aeruginosa	AAC45742	10	QAA**ILDWAGT**V**D**FGS	25	
Haloacid dehalogenases						
	Xanthobacter autotrophicus	AAA27590	3	KAVV**FD**AY**GT**LFDVQS	18	Localized

[a] Residues in boldface type are strictly conserved residues in each subfamily and the phosphorylated residue is underlined in the last column. The type of evidence for a phosphoenzyme is indicated in the last column. *Localized* means that the phosphorylated aspartate has been experimentally localized in the sequence. *Acyl phosphate* means that experimental evidence has been obtained for a phosphoenzyme with the characteristic lability of an acylphosphate bond.

[b] Demonstrated for the enzyme from spinach leaves. In the case of haloacid dehalogenase, the underlined residue is implicated in the formation of an acyl ester intermediate.

molecular weight phosphate esters (e.g., phosphoserine phosphatase and phosphoglycolate phosphatase). One of the identified enzymes, however, is a protein serine phosphatase acting on the C-terminal domain of the larger subunit of RNA-polymerase II.[7,8] The list in Table I also includes an enzyme, homologous to phosphoserine phosphatase, that acts *in vivo* as a homoserine kinase, most likely by using L-3-phosphoserine as a phosphate donor[9]; in this enzyme, the second aspartate of the conserved motif is replaced by a glutamate.

Formation of an acyl-phosphate intermediate was unambiguously shown for the following enzymes: spinach phosphoglycolate phosphatase,[10] human phosphoserine phosphatase,[11] human phosphomannomutases 1 and 2,[12,13] and *Lactococcus lactis* β-phosphoglucomutase.[5] The position of the phosphorylated residue was identified as the first aspartate in the DXDXT/V motif in human phosphomannomutase 1 and in human phosphoserine phosphatase.[5,14] Results of site-directed mutagenesis of the two aspartates in phosphomannomutase 1 and in phosphoserine phosphatase are compatible with this conclusion.[5]

In most of these enzymes, the DXDXT/V motif is located close to the N terminus. An exception such as yeast trehalose-6-phosphate phosphatase is due to the fact that this enzyme is bifunctional,[15] with a trehalose-6-phosphate synthase domain N-terminal to the phosphatase domain. Other exceptions may correspond to proteins with several domains.

Iterated sequence comparisons and position-specific iterated BLAST (PSI-BLAST) searches[16] starting from haloacid dehalogenase have shown that this enzyme shares three statistically significant motifs with the new class of phosphomonoesterases/phosphomutases and P-type ATPases.[17,18] The first of these motifs (DXDXT in phosphomonoesterases/phosphomutases; DKTGT in ATPases; and DXYGT in dehalogenases) contains an absolutely conserved aspartate, which covalently binds phosphate in the first two classes of enzymes and an α-hydroxy acid

[7] H. Cho, T. K. Kim, H. Mancebo, W. S. Lane, O. Flores, and D. Reinberg, *Genes Dev.* **13**, 1540 (1999).
[8] M. S. Kobor, J. Archambault, W. Lester, F. C. Holstege, O. Gileadi, D. B. Jansma, E. G. Jennings, F. Kouyoumdjian, A. R. Davidson, R. A. Young, and J. Greenblatt, *J. Mol. Cell* **4**, 55 (1999).
[9] J. C. Patte, C. Clepet, M. Bally, F. Borne, V. Mejean, and M. Foglino, *Microbiology* **145**, 845 (1999).
[10] S. N. Seal and Z. B. Rose, *J. Biol. Chem.* **262**, 13496 (1987).
[11] J. F. Collet, I. Gerin, M. H. Rider, M. Veiga-da-Cunha, and E. Van Schaftingen, *FEBS Lett.* **408**, 281 (1997).
[12] M. Pirard, J. F. Collet, G. Matthijs, and E. Van Schaftingen, *FEBS Lett.* **411**, 251 (1997).
[13] M. Pirard, Y. Achouri, J. F. Collet, E. Schollen, G. Matthijs, and E. Van Schaftingen, *Biochem. J.* **339**, 201 (1999).
[14] J. F. Collet, V. Stroobant, and E. Van Schaftingen, *J. Biol. Chem.* **274**, 33985 (1999).
[15] A. Vandercammen, J. François, and H. G. Hers, *Eur. J. Biochem.* **182**, 613 (1989).
[16] S. F. Altschul, T. L. Madden, A. A. Schaffer, J. Zhang, Z. Zhang, W. Miller, and D. J. Lipmann, *Nucleic Acids Res.* **25**, 3389 (1997).
[17] E. V. Koonin and R. L. Tatusov, *J. Mol. Biol.* **244**, 125 (1994).
[18] L. Aravind, M. Y. Galperin, and E. V. Koonin, *Trends Biochem. Sci.* **23**, 127 (1998).

in haloacid dehalogenases.[19] The second motif (not shown) contains a strictly conserved serine or threonine and the third motif, a strictly conserved lysine residue followed, at some distance, by lesser conserved residues and a strictly conserved aspartate. This group of enzymes appears to be also related to phosphonoacetaldehyde phosphatases, with which they share the first motif (DWAGT), and possibly also part of the third motif.[20]

The three-dimensional structures of haloacid dehalogenases[21–23] and of sarcoplasmic Ca^{2+}-ATPase[24] have been solved. In both structures, the residues of the three conserved motifs are located in the catalytic pocket, where they occupy almost identical positions. This is most probably also the case for the phosphotransferases with the DXDX(T/V) motif.

Identification of Phosphorylated Residue

It is good practice to start by a preliminary characterization of the stability of the phosphate bond, because this will have an impact on the methods used for the identification of the phosphorylated residue. Several methods can be used to separate the denatured phosphorylated protein from a low molecular weight substrate, including centrifugation, adsorption on paper with several washings in 5% (w/v) trichloroacetic acid,[25] or adsorption on phosphocellulose papers followed by washings in dilute phosphoric acid.[26] The filtration technique proposed below is completed in less than 1 min and has therefore the advantage of minimizing the time during which hydrolysis of the phosphate ester can occur.

The phosphoaspartate residues of denatured phosphoserine phosphatase and phosphomannomutase have been found to be: (1) resistant to acid (1 M HCl) at 0° but not at elevated temperatures ($\approx$50% hydrolysis in 60 min at 37°; 100% hydrolysis in 15 min at 100°); (2) extremely labile to alkali (100% hydrolysis in less than 10 min at 0° in 1 M NaOH); labile to NH_2OH (100% hydrolysis in 10 min at 20° in 0.2 M NH_2OH, 0.1 M sodium acetate, pH 5.5). These properties are in sharp contrast to those of phosphohistidine residues, which are acid-labile but resistant to alkali, and of phosphoserine residues, which are resistant to acid

[19] J. Q. Liu, T. Kurihara, M. Miyagi, N. Esaki, and K. Soda, *J. Biol. Chem.* **270**, 18309 (1995).
[20] A. S. Baker, M. J. Ciocci, W. W. Metcalf, J. Kim, P. C. Babbitt, B. L. Wanner, B. M. Martin, and D. Dunaway-Mariano, *Biochemistry* **37**, 9305 (1998).
[21] T. Hisano, Y. Hata, T. Fujii, J. Q. Liu, T. Kurihara, N. Esaki, and K. Soda, *J. Biol. Chem.* **271**, 20322 (1996).
[22] I. S. Ridder, H. J. Rozeboom, K. H. Kalk, D. B. Janssen, and B. W. Dijkstra, *J. Biol. Chem.* **272**, 33015 (1997).
[23] Y. F. Li, Y. Hata, T. Fujii, T. Hisano, M. Nishihara, T. Kurihara, and N. Esaki, *J. Biol. Chem.* **273**, 15035 (1998).
[24] C. Toyoshima, M. Nakasako, H. Nomura, and H. Ogawa, *Nature (London)* **405**, 647 (2000).
[25] J. D. Corbin and E. M. Reimann, *Methods Enzymol.* **38**, 287 (1974).
[26] R. Roskoski, *Methods Enzymol.* **99**, 3 (1983).

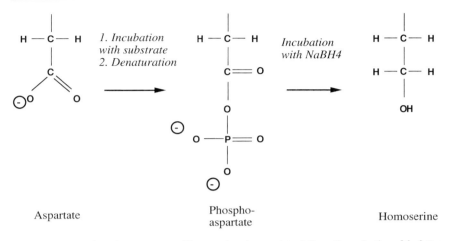

FIG. 1. Conversion of an aspartate residue to a phosphoaspartate, followed by reduction of the latter to homoserine.

and labile to alkali at elevated temperatures.[27] They are also distinct from those of phosphocysteines, which are maximally unstable at pH ≈ 3.0.[28]

Another useful step is the demonstration that a polypeptide chain with the expected size is labeled with ^{32}P. SDS–PAGE at room temperature is not suitable to demonstrate this because this procedure results in a virtually complete hydrolysis of the acylphosphate bond. Rapid SDS–PAGE (a few hours) in the cold as detailed below works, provided that the samples are not heated before loading.[12] Because of the relative stability of phosphoaspartates at acidic pH, another possibility is to use an acid-denaturing gel system.[29] However, SDS–PAGE at low temperature has the advantage that it uses commonly available reagents.

Phosphoaspartate and phosphoglutamate residues are too labile to resist the procedure commonly used to identify phosphorylated residues, which involves digestion of the phosphorylated protein with trypsin, separation of the resulting peptides by high-performance liquid chromatography (HPLC), and sequencing of the phosphorylated peptide. The problem of the lability can be circumvented (Fig. 1) by reducing the carboxyl-phosphate group of the denatured protein to a hydroxymethyl group with sodium borohydride.[30] This reduction is specific, as free carboxyl groups do not react. If tritiated borohydride is used, radioactivity screening allows one to identify the fractions containing the modified peptide.

[27] M. Weller, "Protein Phosphorylation." Pion, London, 1979.
[28] K. L. Guan and J. E. Dixon, *J. Biol. Chem.* **266,** 17026 (1991).
[29] A. Amory, F. Foury, and A. Goffeau, *J. Biol. Chem.* **255,** 9353 (1980).
[30] C. Degani and P. Boyer, *J. Biol. Chem.* **248,** 8222 (1973).

Furthermore, the replacement of a carboxyl group by a hydroxymethyl group results in a 14 Da decrease of the molecular mass (provided ^{3}H is a minor component compared to ^{1}H in the borohydride used) and allows the localization of the modified residue by tandem mass spectrometry.[31]

This identification of the modified peptide in the mass spectrum is facilitated by the fact that the reduced peptide (containing a homoserine residue) and the nonreduced peptide (with a free aspartate residue) are eluted in the same fraction. Inspection of the mass spectrum indicates therefore the presence of two ions that differ by 14 amu. If the sequence of the protein is known, the m/z value of the nonreduced ion allows one to predict the identity of the nonreduced peptide, which can be readily confirmed by fragmentation. Fragmentation of the reduced peptide is expected to show (1) ions with identical m/z values as fragments from the nonreduced peptide and (2) ions with m/z values lower by 14 amu. This allows one to identify the position of the reduced amino acid. A flow chart summarizing these various steps is shown in Fig. 2 (left-hand side).

In carrying out such experiments, it should be kept in mind that the incorporation of radioactivity from tritiated borohydride is not specific for the acyl-phosphate bond. This is indicated by the facts that (1) there is background incorporation of radioactivity in all peptides, possibly due to partial reduction of the peptide bonds to secondary amines, and (2) some peaks of radioactivity are found that do not correspond to the reduction of an acyl phosphate, but to the reduction of another bond, probably a peptide bond which is particularly reactive to borohydride. In the case of phosphomannomutase two such peaks were identified with both the nonphosphorylated and the phosphorylated enzyme.[5] Mass spectrometry (MS) analysis indicated the presence in the corresponding fractions of two peptides with m/z differing by 14 amu (indicating the replacement of an oxygen atom by two hydrogens). Reduction of an acyl-phosphate bond could be excluded by fragmentation of the ions, which allowed the identification of the peptides, as well as the approximate localization of the reduced bond.

As a consequence of this lack of specificity, identification of the radioactive fractions containing the phosphorylated/reduced peptide in the HPLC eluate may be difficult or impossible in the case of enzymes with low phosphorylation stoichiometry. This was the case for human phosphoserine phosphatase, which has a phosphorylation stoichiometry of 0.07 as compared to 0.50 for phosphomannomutase.[5] In that case, a "candidate peptide" approach was used (Fig. 2 right-hand side): the mass of the peptide containing the potential phosphorylation site could be predicted from the expected digestion pattern of phosphoserine-phosphatase. The ion with the appropriate m/z ratio, as well as that with a mass lower by 14 amu (corresponding to the phosphorylated/reduced peptide), was then directly searched in the spectrum of the digest (without any HPLC separation) and fragmented. This

---

[31] L. M. Cagen and H. C. Friedmann, *J. Biol. Chem.* **247**, 3382 (1972).

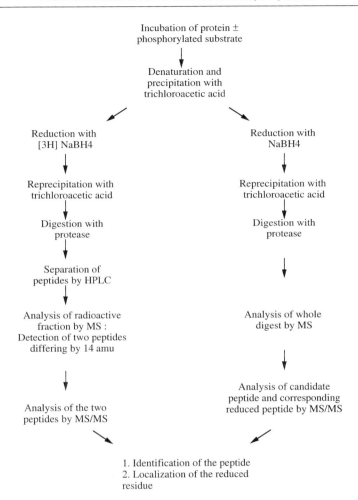

FIG. 2. Flow chart showing the steps in the identification of a phosphorylated aspartate. The right-hand side corresponds to the "candidate peptide" approach. Reprinted with permission from J. F. Collet, V. Stroobant, M. Pirard, G. Delpierre, and E. Van Schaftingen, *J. Biol. Chem.* **273**, 14107 (1998).

allowed confirmation of the identity of the peptides and localization of the position of the phosphorylated aspartate.

Protein Preparations

An impure preparation may be enough for preliminary characterization of the phosphoenzyme intermediate, provided the enzyme under study is sufficiently concentrated and there is no incorporation of radioactivity in contaminating proteins, as assessed by SDS–PAGE. For the identification of the phosphorylated residue, the enzyme should be purified (close) to homogeneity. Prior knowledge of its sequence considerably facilitates the interpretation of MS data. In the case of human phosphomannomutase 1[5,12] and phosphoserine phosphatase,[11,14] recombinant purified proteins are used.

Preparation of Radiolabeled Substrates

L-[^{32}P]Phosphoserine is synthesized by incubating 1 mM L-serine with 10 μM ^{32}P-labeled inorganic pyrophosphate (0.1 mCi), 2 mM magnesium acetate, 1 mM dithiothreitol, 25 mM Tris-HCl, pH 7.1, and 0.03 U pyrophosphate: L-serine phosphotransferase[31] for 40 min at 30° in a final volume of 3 ml. It is purified by chromatography on Dowex AG1 X8 formate and elution with 1 M formic acid.[32] ^{32}P-Labeled mannose 6-phosphate and mannose 1,6-bisphosphate are prepared as previously described.[12,13]

Incorporation of ^{32}P into Protein

The protein is incubated with ^{32}P-labeled substrate of sufficiently high specific activity in the presence of a buffer and of 0.1 to 1 mg/ml albumin, to facilitate protein precipitation upon addition of trichloroacetic acid. The incubation is carried out for a few seconds to a few minutes at 0° to minimize enzymatic hydrolysis of the substrate. The reaction is arrested by addition of $\approx$3 volumes of 5% ice-cold trichloroacetic acid and the whole mixture is filtered under vacuum on a polyether sulfone membrane (Supor 200 from Gelman, Ann Arbor, MI; 2.5 cm diameter; 0.2 μm pores; prerinsed with 5 ml 5% trichloroacetic acid). The filter is rinsed with 10 ml 5% trichloroacetic acid, dried, and counted in the presence of scintillation fluid (e.g., Hisafe 2 from Packard, Meriden, CT).

In the case of phosphoserine phosphatase, the purified enzyme (25 μg) is typically incubated in the presence 50 mM HEPES, pH 7.5, 1 mM dithiothreitol (DTT), 1 μM phosphoserine, 60,000 cpm of [^{32}P]phosphoserine, 5 mM MgCl$_2$, 20 mM serine (to slow down the hydrolysis of the substrate), and 1 mg/ml albumin in a final volume of 100 μl.[11] In the case of phosphomannomutase 1, the incubation mixture (100 μl) typically contains 5 μg homogeneous enzyme, 25 mM HEPES, pH 7.1, 5 mM MgCl$_2$, 0.1 mg/ml albumin, 100,000 cpm mannose 1,6-bis [^{32}P]phosphate, and 1 μM mannose-1,6-bisphosphate.[13] In the case of phosphomannomutase, mannose 6 [^{32}P]phosphate can also be used in combination with

---

[32] L. F. Borkenhagen and E. P. Kennedy, *J. Biol. Chem.* **234**, 849 (1959).

cold mannose 1,6-bisphosphate[12] instead of mannose 1,6-bis [^{32}P]phosphate, since the latter is readily formed by the enzyme through an exchange reaction.

Variation in the incubation time allows one to optimize the conditions. The apparent dissociation constant and stoichiometry of phosphorylation can be determined by varying the concentration of cold substrate.

Sensitivity of Phosphoenzyme to Acid, Alkali, and Hydroxylamine

To determine the sensitivity of the phosphoenzyme to acid, alkali, and hydroxylamine, the protein is incubated with its radiolabeled substrate as described above. After the addition of 3 volumes of cold trichloroacetic acid, the mixture is centrifuged for 10 min at 10,000g and 4°. The pellet is washed with cold 10 mM HCl and dried under vacuum. It is then resuspended in 100 μl 1 M HCl, 1 M NaOH, or 0.2 M hydroxylamine (in 25 mM sodium acetate, pH 5.5) and incubated for various times at different temperatures. The incubation is arrested by addition of 3 volumes of cold 5% trichloroacetic acid and the mixture is filtered as described above.

SDS–PAGE Analysis

To avoid hydrolysis of the acyl-phosphate bond, slab gel SDS–PAGE[33] is carried out at low temperature by placing the electrophoresis apparatus in crushed ice. No SDS is added in the lower tank buffer to prevent precipitation of the detergent due to the low temperature. Labeled enzyme prepared as described above is precipitated with trichloroacetic acid. After centrifugation and removal of the supernatant, the pellet is dissolved in 50 μl of a mixture containing 2.5% (w/v) SDS, 12% (w/v) sucrose, 12 mM dithiothreitol, 0.02% (w/v) Coomassie Blue, and 50 mM Tris-HCl, pH 6.8. The samples are then loaded onto a 10% (w/v) SDS–PAGE gel, which is run for 5 hr at 0° under a voltage corresponding to 20 V/cm. The gel is then dried and autoradiographed.

Identification of Phosphorylated Residue by Reduction with [^{3}H]NaBH$_4$, Separation of Peptides by HPLC and Mass Spectrometry[5]

Human recombinant phosphomannomutase 1 (50 μg) is incubated at 0° in a mixture (100 μl) containing 20 mM HEPES, pH 7.1, 1 μM mannose 6-phosphate, 0 (control) or 20 μM mannose 1,6-bisphosphate, 1 mM dithiothreitol, and 2 mM

---

[33] U. K. Laemmli, *Nature* **227**, 680 (1970).

MgCl$_2$. The reaction is stopped after 5 sec by the addition of 500 μl of 10% ice-cold trichloroacetic acid. After centrifugation (20 min at 10,000g and 4°), the pellet is resuspended in 500 μl of ice-cold 5% trichloroacetic acid and recentrifuged for 10 min at 4°. The pellet is washed a second time with 10 mM ice-cold HCl, centrifuged for 10 min at 4° and dried under vacuum. After resuspension of the pellet in 57 μl of dimethyl sulfoxide (DMSO), 10 mCi of [^{3}H]NaBH$_4$ (5 mCi/μmol, in 10 μl DMSO) and 2.7 μmol cold NaBH$_4$, in 33 μl DMSO, are successively added. After 10 min of incubation at 20°, 900 μl of ice-cold 0.44 M HClO$_4$ is added to stop the reaction and precipitate protein; the sample is placed on ice for 30 min and then centrifuged for 20 min at 4°. The resulting pellet is washed with 500 μl of cold acetone and dried under vacuum. The protein is then resuspended in 200 μl of a mixture containing 0.1 M Tris–HCl, pH 8.6, 2 M urea, and 1 μg of trypsin and the mixture incubated at 37° for 18 hr. The reaction is stopped by the addition of 8 μl of 50% trifluoroacetic acid and the mixture is applied onto a C$_{18}$ reversed-phase HPLC column from Alltech (Deerfield, IL) (1 × 250 mm). Elution is accomplished with a linear aqueous 0–80% acetonitrile gradient in 0.5% trifluoroacetic acid, over 70 min at a flow rate of 80 μl/min. Samples of the fractions are mixed with liquid scintillation cocktail and counted for radioactivity. Inspection of the elution profiles corresponding to the preparations of control or phosphorylated enzyme allows the identification of a radioactive peak that is specific for the phosphorylated form of phosphomannomutase, besides two other peaks that are common to the nonphosphorylated and phosphorylated proteins.

Radioactive fractions are analyzed by mass spectrometry. Mass spectrometry is performed on a Finnigan LCQ ion trap equipped with an electrospray source. The samples are introduced directly into the source at a flow of 2 μl/min. The spray is obtained by applying a potential difference of 5 kV and with the help of a sheet gas (N$_2$). The capillary temperature is 250°. The LCQ is operated under manual control in the Tune plus view with default parameters and Automatic Gain Control active. Two positive ions with m/z 1044 and 1030 have been identified. From the sequence of phosphomannomutase 1,[34] only two peptides (DVDGTLT PAR and LLSKQTIQN) have calculated monoisotopic masses in agreement with the measured mass of 1044.2 amu and only the first one is compatible with the cleavage specificity of trypsin and chymotrypsin, a common contaminant of trypsin preparations. It is indeed preceded by a phenylalanine and ends with an arginine. Analysis of this 1044 m/z peptide by tandem mass spectrometry (Fig. 3) has confirmed the identity of the peptide, the most abundant ions obtained (m/z 929, 830, and 715) corresponding to fragments y9, y8, and y7. (By convention, y ions are the C-terminal fragments resulting from cleavage of the peptide bond.[35])

---

[34] G. Matthijs, E. Schollen, M. Pirard, M. L. Budarf, E. Van Schaftingen, and J. J. Cassiman, *Genomics* **40**, 41 (1997).

[35] K. Biemann, *Biomed. Environ. Mass Spectrom.* **16**, 99 (1988).

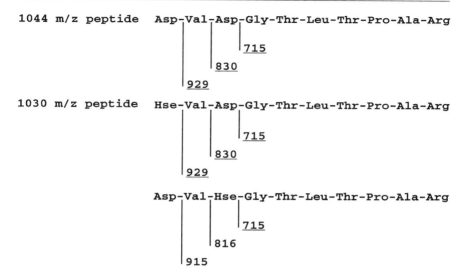

FIG. 3. Main fragments derived from the *m/z* 1044 and 1030 peptides of human phosphomannomutase 1. The observed fragments are underlined. The fact that no 915 and 816 *m/z* ions were observed among the fragments of the reduced peptide indicates that the homoserine (Hse) residue is in the first, not the third, position. Hence, the phosphorylated residue is the first aspartate. Reprinted with permission from J. F. Collet, V. Stroobant, and E. Van Schaftinger, *J. Biol. Chem.* **274**, 33985 (1999).

Fragmentation of the *m/z* 1030 peptide yielded fragments of the same size as with the *m/z* 1044 peptide, indicating that the two peptides differed from each other by the replacement of an aspartate residue by a homoserine in N-terminal position. This allowed us to conclude that the phosphorylated residue in phosphomannomutase 1 is Asp-19. This residue is the first in the completely conserved DVDGT motif present in eukaryotic phosphomannomutases.

Identification of Phosphorylated Aspartate without Separation of Peptides by HPLC Following "Candidate Peptide Approach"

Phosphorylated phosphoserine phosphatase is prepared in the following manner[14]: 500 μg of the human recombinant enzyme (prepared as in Ref. 10) is incubated at 0° in a mixture (500 μl) containing 25 m*M* HEPES, pH 7.5, 20 m*M* serine, 10 m*M* phosphoserine, 1 m*M* dithiothreitol, and 5 m*M* MgCl$_2$; a control is incubated in the absence of phosphoserine. The reaction is stopped after 5 sec by the addition of 1 ml of 10% ice-cold trichloroacetic acid. After 10 min on ice, the mixture is heated for 10 min at 30° to prevent renaturation of the enzyme (this treatment does not significantly hydrolyze the phosphoenzyme), placed on ice for 5 min, and then centrifuged for 20 min at 10,000*g* and 4°. The pellet is washed in

200 μl of ice-cold 10 mM HCl, centrifuged for 10 min at 4°, dried under vacuum, and used for reduction with borohydride.

For the reduction with borohydride, the protein pellets corresponding to phosphorylated and control enzymes are resuspended in 67 μl of DMSO and mixed with 2.7 μmol of NaBH$_4$ in 33 μl DMSO. After 10 min of incubation at 30°, 1 ml of ice-cold 0.44 M HClO$_4$ is added; the samples are placed on ice for 30 min, then centrifuged for 20 min at 10,000g and 4°. The resulting pellets are washed with 200 μl of ice-cold 10 mM HCl and dried under vacuum. The protein is resuspended in 50 μl of 50 mM Tris/HCl pH 8.5 containing 1% octylglucoside and 3 μg of trypsin, and the mixture is incubated overnight at 37°. The reaction is stopped by the addition of 2 μl of 50% trifluoroacetic acid and the mixture is analyzed by mass spectrometry. Mass spectral analyses are performed as indicated above.

Prediction of the digestion pattern indicates that the positive ion corresponding to the phosphorylatable peptide (DVDSTVIR) would have an m/z of 904 if not reduced and of 890 after reduction of phosphoaspartate to homoserine. An m/z 904 ion is indeed present in the mass spectrum of the two digests, as well as a small amount of the m/z 890 ion (not shown). On fragmentation, this m/z 904 ion yields two major ions with m/z 789 and 575, corresponding to fragments y7 (VDSTVIR) and y5 (STVIR) (not shown). We have also fragmented the minor m/z 890 ion. In the case of the phosphorylated/reduced enzyme preparation, two major fragments of m/z 789 and 575 are also observed, showing that this peptide is identical to the m/z 904 ion, except for reduction of the first aspartate to homoserine. By contrast, in the fragmentation spectrum of the m/z 890 ion originating from the control enzyme, the peak at m/z 789 is much smaller and no m/z 575 ion is present, indicating that the reduced peptide is not present in the tryptic/chymotryptic digest.

Acknowledgments

J.F.C. is Chargé de Recherches of the Belgian FNRS. This work was supported by the Actions de Recherche Concertées of the Communauté Française and by the Belgian Federal Service for Scientific, Technical, and Cultural Affairs.

[13] MurC and MurD Synthetases of Peptidoglycan Biosynthesis: Borohydride Trapping of Acyl-Phosphate Intermediates

By AHMED BOUHSS, SÉBASTIEN DEMENTIN, JEAN VAN HEIJENOORT, CLAUDINE PARQUET, *and* DIDIER BLANOT

Introduction

The biosynthesis of bacterial peptidoglycan is a complex two-stage process. The first stage consists in the formation of the disaccharide–pentapeptide monomer unit, whereas the second stage concerns the polymerization and maturation reactions.[1] The peptide moiety of the monomer unit is assembled by a series of cytoplasmic synthetases responsible for the successive additions of L-alanine, D-glutamic acid, *meso*-diaminopimelic acid or L-lysine, and a dipeptide (generally D-alanyl-D-alanine) to UDP-*N*-acetylmuramic acid (UDP-MurNAc); these enzymes are referred to as MurC, MurD, MurE, and MurF, respectively.[2,3] A fifth synthetase, which catalyzes the addition of tripeptide L-Ala-γ-D-Glu-*meso*-diaminopimelate to UDP-MurNAc in the recycling process of peptidoglycan, has been identified and designated as Mpl.[4] All these enzymes catalyze the formation of an amide or peptide bond with concomitant cleavage of ATP into ADP and inorganic phosphate:

$$R-COOH + H_2N-R' + ATP \rightleftharpoons R-CO-NH-R' + ADP + P_i$$

The enzyme-catalyzed amide bond formation driven by ATP hydrolysis to ADP and P_i is not a frequently encountered biochemical reaction. Aside from the Mur synthetases considered here, only a few other enzymes performing this type of reaction have been found and extensively studied: glutamine synthetase, γ-glutamylcysteine synthetase, glutathione synthetase, folylpoly-γ-L-glutamate synthetase, D-alanine : D-alanine ligase (an enzyme also involved in peptidoglycan biosynthesis), etc. It has generally been thought that their reaction mechanism consists in the activation of the carboxylic substrate R–COOH by ATP into an acyl phosphate; this species then undergoes the nucleophilic attack by the amine substrate R'–NH$_2$ to yield a tetrahedral intermediate, which breaks down into amide (or peptide) and inorganic phosphate (Scheme 1). This mechanism has been

---

[1] J. van Heijenoort, *in* "*Escherichia coli* and *Salmonella*" (F. C. Neidhardt, ed.), p. 1025. American Society for Microbiology, Washington, D.C., 1996.
[2] T. D. H. Bugg and C. T. Walsh, *Nat. Prod. Rep.* **9,** 199 (1992).
[3] J. van Heijenoort, *Cell. Mol. Life Sci.* **54,** 300 (1998).
[4] D. Mengin-Lecreulx, J. van Heijenoort, and J. T. Park, *J. Bacteriol.* **178,** 5347 (1996).

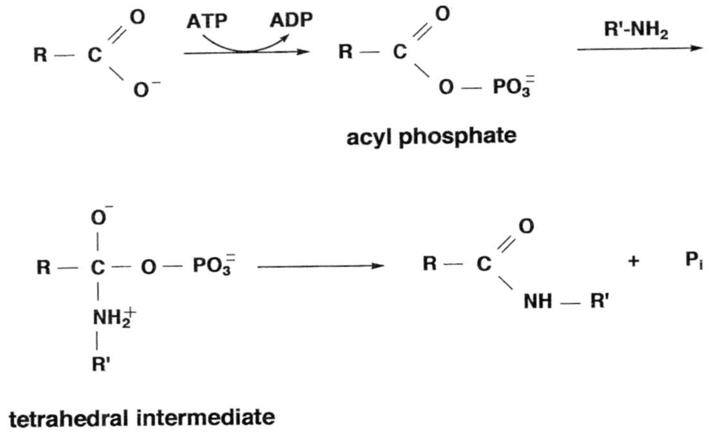

$$R-C\underset{O^-}{\overset{O}{\diagup}} \xrightarrow[ADP]{ATP} R-C\underset{O-PO_3^=}{\overset{O}{\diagup}} \xrightarrow{R'-NH_2}$$

acyl phosphate

$$\underset{\underset{R'}{\overset{NH_2^+}{|}}}{\overset{O^-}{\underset{|}{R-C-O-PO_3^=}}} \longrightarrow R-C\underset{NH-R'}{\overset{O}{\diagup}} + P_i$$

tetrahedral intermediate

SCHEME 1

demonstrated by different techniques for glutamine synthetase,[5] γ-glutamylcysteine synthetase,[6] glutathione synthetase,[7,8] or D-alanine : D-alanine ligase.[9,10] A possible analogy between the reaction mechanism of these amide- or peptide-forming enzymes and MurE (L-lysine-adding enzyme) from *Staphylococcus aureus* was proposed 40 years ago by Ito and Strominger.[11,12] Work aimed at verifying this mechanism for the Mur synthetases has been performed.[2,13–21] However, the problem of the occurrence of the aforementioned intermediates in a reaction pathway

---

[5] A. Meister, *Methods Enzymol.* **113**, 185 (1985).
[6] G. F. Seelig and A. Meister, *Methods Enzymol.* **113**, 379 (1985).
[7] A. Meister, *Methods Enzymol.* **113**, 393 (1985).
[8] J. Hiratake, H. Kato, and J. Oda, *J. Am. Chem. Soc.* **116**, 12059 (1994).
[9] L. S. Mullins, L. E. Zawadzke, C. T. Walsh, and F. M. Raushel, *J. Biol. Chem.* **265**, 8993 (1990).
[10] C. Fan, P. C. Moews, C. T. Walsh, and J. R. Knox, *Science* **266**, 439 (1994).
[11] E. Ito and J. L. Strominger, *J. Biol. Chem.* **235**, PC7 (1960).
[12] E. Ito and J. L. Strominger, *J. Biol. Chem.* **239**, 210 (1964).
[13] P. J. Falk, K. M. Ervin, K. S. Volk, and H. T. Ho, *Biochemistry* **35**, 1417 (1996).
[14] D. Liger, A. Masson, D. Blanot, J. van Heijenoort, and C. Parquet, *Microb. Drug Resist.* **2**, 25 (1996).
[15] M. E. Tanner, S. Vaganay, J. van Heijenoort, and D. Blanot, *J. Org. Chem.* **61**, 1756 (1996).
[16] S. Vaganay, M. E. Tanner, J. van Heijenoort, and D. Blanot, *Microb. Drug Resist.* **2**, 51 (1996).
[17] J. J. Emanuele, H. Jin, B. L. Jacobson, C. Y. Chang, H. M. Einspahr, and J. J. Villafranca, *Prot. Sci.* **5**, 2566 (1996).
[18] J. J. Emanuele, H. Jin, J. Yanchunas, and J. J. Villafranca, *Biochemistry* **36**, 7264 (1997).
[19] D. J. Miller, S. M. Hammond, D. Anderluzzi, and T. D. H. Bugg, *J. Chem. Soc. Perkin Trans. 1*, 131 (1998).
[20] L. D. Gegnas, S. T. Waddell, R. M. Chabin, S. Reddy, and K. K. Wong, *Bioorg. Med. Chem. Lett.* **8**, 1643 (1998).
[21] B. Zeng, K. K. Wong, D. L. Pompliano, and M. E. Tanner, *J. Org. Chem.* **63**, 10081 (1998).

is difficult to tackle by a chemical approach owing to their instability. As far as the acyl phosphates are concerned, we use a chemical trapping method, the *in situ* sodium borohydride reduction, to firmly establish their formation by MurC and MurD from *Escherichia coli*.[22]

Materials and Methods

Materials

UDP-*N*-acetyl[^{14}C]glucosamine (9.29 GBq/mmol) is purchased from Amersham France (Les Ulis, France), and L-alaninol from Sigma (St. Louis, MO). UDP-MurNAc and 2-amino-2-deoxy-3-*O*-[(*S*)-1-(hydroxymethyl)ethyl]-D-glucose are synthesized according to Blanot *et al*.[23] and Jeanloz and Walker,[24] respectively. UDP-MurNAc-L-Ala and UDP-MurNAc-L-[^{14}C]Ala (5.44 GBq/mmol) are prepared by addition of cold or radioactive L-alanine to UDP-MurNAc, catalyzed by MurC.[25,26] MurC is purified from *E. coli* JM83(pAM1005) by the procedure already described,[27] except that phosphate is replaced by HEPES throughout the purification[28]; the enzyme is stored in 20 mM K$^+$–HEPES, 1 mM EDTA, 2 mM dithiothreitol (DTT), 15% (v/v) glycerol, pH 7.2. MurD is purified from *E. coli* JM83(pMLD58)[25]; it is dialyzed against 20 mM Na$^+$–HEPES, 1 mM DTT, pH 7.5.

Cloning and Overexpression of murA and murB Genes

In order to synthesize radioactive UDP-MurNAc, crude extracts containing the highly overproduced MurA or MurB activity are prepared. For this purpose, the *murA* and *murB* genes from *E. coli* are cloned using procedures different from those already described.[29–31] Two oligonucleotides, corresponding to the 5' and 3' ends of the *murA* sequence and containing the *Nco*I and *Pst*I sites, are used for the PCR amplification of the *murA* gene from λ521 phage DNA of Kohara's library.[32]

---

[22] A. Bouhss, S. Dementin, J. van Heijenoort, C. Parquet, and D. Blanot, *FEBS Lett.* **453**, 15 (1999).
[23] D. Blanot, G. Auger, D. Liger, and J. van Heijenoort, *Carbohydr. Res.* **252**, 107 (1994).
[24] R. W. Jeanloz and E. Walker, *Carbohydr. Res.* **4**, 504 (1967).
[25] G. Auger, L. Martin, J. Bertrand, P. Ferrari, E. Fanchon, S. Vaganay, Y. Pétillot, J. van Heijenoort, D. Blanot, and O. Dideberg, *Prot. Express. Purif.* **13**, 23 (1998).
[26] F. Pratviel-Sosa, F. Acher, F. Trigalo, D. Blanot, R. Azerad, and J. van Heijenoort, *FEMS Microbiol. Lett.* **115**, 223 (1994).
[27] D. Liger, A. Masson, D. Blanot, J. van Heijenoort, and C. Parquet, *Eur. J. Biochem.* **230**, 80 (1995).
[28] G. Auger and D. Blanot, unpublished results (1997).
[29] M. J. Pucci, L. F. Discotto, and T. J. Dougherty, *J. Bacteriol.* **174**, 1690 (1992).
[30] J. L. Marquardt, D. A. Siegele, R. Kolter, and C. T. Walsh, *J. Bacteriol.* **174**, 5748 (1992).
[31] T. E. Benson, J. L. Marquardt, A. C. Marquardt, F. A. Etzkorn, and C. T. Walsh, *Biochemistry* **32**, 2024 (1993).
[32] Y. Kohara, K. Akiyama, and K. Isono, *Cell* **50**, 495 (1987).

The PCR fragment is digested with NcoI and PstI and ligated into pTrc99A cut with the same enzymes. After transformation of DH5α cells by the resulting plasmid, the transformants are selected by their resistance to fosfomycin (120 μg/ml) on isopropylthiogalactoside (IPTG) (1 mM) plates, then cultivated on 2YT medium (50 ml) in the presence of ampicillin (100 μg/ml). The overexpression of the protein is induced by addition of IPTG (1 mM) at an $A_{600\,nm}$ of 0.5. Cells are collected by centrifugation, suspended in 50 mM Tris-HCl, pH 8.0, containing 1 mM DTT, and disrupted by sonication. After centrifugation, the crude extract contains 31 mg protein with a specific activity of 0.75 μmol/min/mg. SDS–PAGE analysis shows that the MurA protein has a molecular mass of ∼45 kDa and represents ∼50% of the material colored by Coomassie Brilliant Blue.

Concerning murB, the PCR is performed with a 5′-end oligonucleotide starting at the second codon of the gene sequence and a 3′-end oligonucleotide carrying a PstI site. The λ534 phage DNA of Kohara's library is used as a template for amplification. Before ligation, the PCR product is treated with T4 DNA polymerase and PstI enzyme, and vector pTrc99A is submitted successively to NcoI restriction, T4 DNA polymerase to fill the protuberant NcoI ends, and PstI restriction. The recombinant plasmid is used to transform a thermosensitive murB strain.[33] The growth of the transformant and the preparation of the crude extract are carried out as described in the preceding paragraph. The extract contains 25 mg protein with a specific activity of 0.91 μmol/min/mg. SDS–PAGE analysis shows that the MurB protein has a molecular mass of ∼36 kDa and represents ∼50% of the material colored by Coomassie Brilliant Blue.

Enzymatic Synthesis of UDP-[^{14}C]MurNAc

UDP-N-acetyl[^{14}C]glucosamine (19.9 nmol, 185 kBq) is dissolved in 50 mM Tris-HCl, 5 mM DTT, 20 mM KCl, pH 8.0, containing 10 mM phospho*enol*pyruvate (volume: 20 μl). A mixture of MurA- and MurB-containing extracts (3.5 μg protein each) is added. NADPH (32 nmol) is added in 8-nmol portions every 5 min in order to avoid the inhibition of MurB by this substrate. After 1 hr at 37°, the reaction is stopped by addition of 50 mM ammonium formate, pH 3.5 (180 μl). The product is purified by HPLC on Nucleosil 5C$_{18}$ (250 × 4.6 mm; Alltech France, Templemars, France) in the same formate buffer at 0.5 ml/min. The overall yield calculated from the radioactivity is 74%.

Sodium Borohydride Reduction

MurC (0.57 nmol) is incubated for 60 min at 37° in a 20-μl reaction mixture containing 0.1 M Tris-HCl, pH 8.6, 20 mM MgCl$_2$, 5 mM ATP and

---

[33] P. Doublet, J. van Heijenoort, and D. Mengin-Lecreulx, in "Bacterial Growth and Lysis: Metabolism and Structure of the Bacterial Sacculus" (M. A. De Pedro, J. V. Höltje, and W. Löffelhardt, eds.), p. 139. Plenum Press, New York, 1993.

UDP-[^{14}C]MurNAc (2.86 nmol, 26.6 kBq). A control, in which the enzyme has been previously inactivated for 5 min at 100°, is also performed. Reactions are quenched by addition of water (980 μl) and dimethyl sulfoxide (1.5 ml), and after cooling on ice for 5 min, NaBH$_4$ (5 mg) is added to each tube. After 16 hr at room temperature, 5 volumes of water are added and the resulting solutions are lyophilized repeatedly to remove dimethyl sulfoxide. The residues are taken up in water (500 μl) and treated with formic acid (50 μl) for 60 min to destroy the remaining borohydride. The pH is adjusted to 8.5 and the solutions are applied to Sephadex G-25 columns (50 × 5 mm) in order to remove borate ions. Elution is carried out at a flow rate of 60 μl/min. The radioactive fractions are pooled, lyophilized, and hydrolyzed (6 M HCl, 95°, 16 hr). The hydrolyzates are evaporated and taken up with water, and the radioactive compounds are separated by high-voltage paper electrophoresis. Essentially the same procedure is followed with MurD (0.60 nmol) and UDP-MurNAc-L-[^{14}C]Ala (3.83 nmol, 21.2 kBq), except that the concentration of MgCl$_2$ is 5 mM.

High-Voltage Electrophoresis

It is performed on 3469 filter paper (Schleicher & Schüll, Dassel, Germany) in water/pyridine/acetic acid 971 : 6 : 23 (v/v; pH 4.0) for 1 hr at 40 V/cm, using an LT36 apparatus (Savant Instruments, Hicksville, NY). Radioactive spots on electropherograms are detected by autoradiography (type R2 films, 3M, St. Paul, MN) and quantified with a radioactivity scanner (Tracemaster LB285, EG&G Berthold/Wallac, Evry, France).

Results

Todhunter and Purich[34] used the borohydride reduction method, introduced by Degani and Boyer,[35] to identify the γ-glutamyl phosphate intermediate of the *E. coli* glutamine synthetase reaction. In their experiment, L-[^{14}C]glutamate was transformed into the easily characterized alcohol derivative α-amino-δ-hydroxy [^{14}C]valerate. Although attempts to repeat this procedure with sheep brain glutamine synthetase and *E. coli* γ-glutamate kinase turned out to be unsuccessful,[36] we tried to adapt it to MurC and MurD. Labeled substrate UDP-[^{14}C]MurNAc or UDP-MurNAc-L-[^{14}C]Ala was incubated with ATP and the respective enzyme in order to form the acyl phosphate, which was then reduced by sodium borohydride according to the published procedure.[34] The resulting material was submitted to acid hydrolysis, supposedly yielding radioactive 2-amino-2-deoxy-3-O-[(S)-1-hydroxymethyl)ethyl]-D-glucose (reduced muramic acid) for MurC, and

---

[34] J. A. Todhunter and D. L. Purich, *J. Biol. Chem.* **250**, 3505 (1975).
[35] C. Degani and P. D. Boyer, *J. Biol. Chem.* **248**, 8222 (1973).
[36] A. P. Seddon, K. Y. Zhao, and A. Meister, *J. Biol. Chem.* **264**, 11326 (1989).

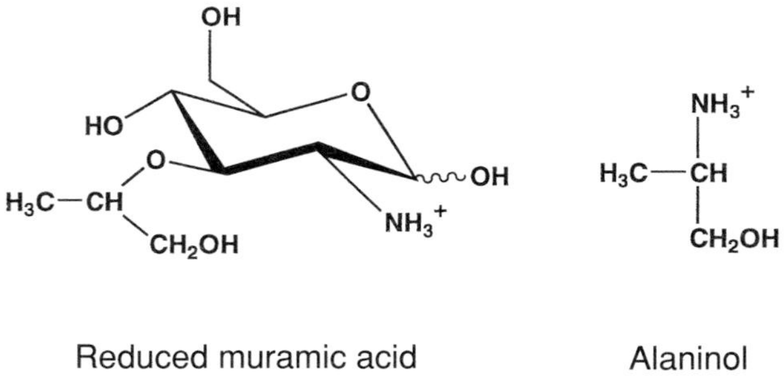

Reduced muramic acid   Alaninol

SCHEME 2

radioactive L-alaninol for MurD (Scheme 2). The radioactive amino alcohols were indeed clearly identified by comparison of their electrophoretic migration with those of unlabeled reference compounds revealed with ninhydrin (since the presence of salts in the hydrolyzates modified their migration, the reference compounds were included in the samples to be spotted) (Fig. 1).

The electropherogram of the reduced, hydrolyzed MurC reaction mixture displayed three radioactive spots (Fig. 1A): muramic acid (-2.7 cm; 80% of the radioactivity spotted), an unidentified compound (-13 cm; 10%), and reduced muramic acid (-15 cm; 10%). Since only the two former spots were detected in the control with boiled MurC, it can be deduced that part of the substrate had been transformed into the alcohol derivative via the acyl phosphate. The intermediary spot, which was also present in an acid hydrolyzate of UDP-[^{14}C]MurNAc alone (not shown), is undoubtedly a product of decomposition of muramic acid, which is known to be partially destroyed on acid hydrolysis.[37]

With MurD, three radioactive spots were also detected (Fig. 1B): alanine (-3.5 cm; 74% of the radioactivity spotted), an unidentified compound (-24 cm; 21%), and alaninol (-31 cm; 5%). However, only the alanine spot was present in the control with boiled MurD. Therefore, the procedure gave rise to the appearance of two radioactive products, one of them being alaninol. Because the alaninol is stable under the hydrolysis conditions used, the possibility that the unknown spot was a degradation product is unlikely. Owing to the minute amounts available, no further attempts were made to identify the unknown compound.

Taking into account the radioactivity of the spots of reduced muramic acid and alaninol and assuming a quantitative yield for reduction, it was calculated that the amounts of acyl phosphate were 0.50 and 0.32 mol/mol enzyme for MurC and MurD, respectively. Interestingly enough, the value for MurD was nearly equal to the burst amplitude observed in the kinetics of synthesis of adenosine 5'-tetraphosphate by MurD (0.31 mol/mol enzyme); this nucleotide has been

---

[37] M. Zaoral, J. Ježek, V. Krchňák, and R. Straka, *Coll. Czech. Chem. Commun.* **45,** 1424 (1980).

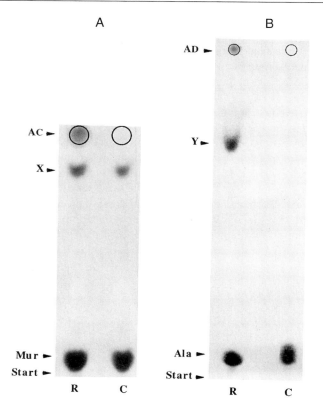

FIG. 1. High-voltage electrophoretic analysis of the sodium borohydride reduction experiments with MurC (A) and MurD (B). R, Reaction; C, control with previously boiled enzyme; Mur, muramic acid; AC, reduced muramic acid; AD, alaninol; X, degradation product of muramic acid; Y, unidentified compound. Circle, ninhydrin spot of the reference compound (reduced muramic acid or alaninol, 50 nmol) added to the sample prior to electrophoresis. The direction of migration of the radioactive compounds is that of the cathode. For reasons of convenience, (A) and (B) are not at the same scale (see migration distances in Results). From A. Bouhss, S. Dementin, J. van Heijenoort, C. Parquet, and D. Blanot, *FEBS Lett.* **453**, 15 (1999).

shown to originate from UDP-MurNAc-L-alanyl phosphate in the absence of D-glutamic acid.[22]

Conclusion

Acyl phosphates can be identified by their chemical trapping into stable derivatives.[34,38] In the present work, the alcohol derivatives originating from the

[38] S. G. Powers and A. Meister, *Proc. Natl. Acad. Sci. U.S.A.* **73**, 3020 (1976).

chemical reduction of the acyl phosphates for MurC and MurD were detected, thereby firmly establishing the formation of such compounds. The possibility that these acyl phosphates are off-pathway, dead-end adducts is ruled out by the observation that isotope exchange reactions occur when the amino acid substrate (L-alanine for MurC, D-glutamic acid for MurD) is present.[13,16]

The comparison of 20 Mur synthetases from various bacterial species has revealed seven invariant amino acids and the same ATP-binding sequence at the same position in the alignment.[39,40] The Mur synthetases thus appear to be a well-defined class of closely functionally related proteins originating presumably from a common ancestor. Moreover, the conservation of a constant backbone length between certain invariants suggests common 3D structural motifs.[40] Therefore, it is reasonable to assume that they all share the same reaction mechanism, which involves an acyl phosphate and a tetrahedral compound as reaction intermediates (Scheme 1). Crystallographic[41,42] and site-directed mutagenesis[43] studies of MurD have allowed the assignment of a role in acyl phosphate formation to some invariant residues.

Acknowledgments

This work was supported by grants from the Centre National de la Recherche Scientifique and the Action Concertée Coordonnée Sciences du Vivant No. V. Financial support from Hoechst Marion Roussel to one of us (A.B.) is gratefully acknowledged. We thank Christophe Dini for the synthesis of reduced muramic acid, Geneviève Auger for the gift of the enzyme preparations, and Dominique Mengin-Lecreulx for helpful discussions.

---

[39] S. S. Eveland, D. L. Pompliano, and M. S. Anderson, *Biochemistry* **36,** 6223 (1997).

[40] A. Bouhss, D. Mengin-Lecreulx, D. Blanot, J. van Heijenoort, and C. Parquet, *Biochemistry* **36,** 11556 (1997).

[41] J. A. Bertrand, G. Auger, E. Fanchon, L. Martin, D. Blanot, J. van Heijenoort, and O. Dideberg, *EMBO J.* **16,** 3416 (1997).

[42] J. A. Bertrand, G. Auger, L. Martin, E. Fanchon, D. Blanot, D. Le Beller, J. van Heijenoort, and O. Dideberg, *J. Mol. Biol.* **289,** 579 (1999).

[43] A. Bouhss, S. Dementin, C. Parquet, D. Mengin-Lecreulx, J. A. Bertrand, D. Le Beller, O. Dideberg, J. van Heijenoort, and D. Blanot, *Biochemistry* **38,** 12240 (1999).

[14] Transaldolase B: Trapping of Schiff Base Intermediate between Dihydroxyacetone and ε-Amino Group of Active-Site Lysine Residue by Borohydride Reduction

By GUNTER SCHNEIDER and GEORG A. SPRENGER

Introduction

Class I aldolases form a family of enzymes not related in amino acid sequence that catalyze a large variety of aldol cleavage/condensation reactions. Catalysis by class I aldolases occurs via a covalent Schiff base intermediate, formed between a catalytic lysine residue and the substrate.[1,2] In several cases, reduction of this intermediate by borohydride and subsequent peptide sequencing has been used successfully to identify the nature of the amino acid that is covalently modified.[3,4]

The ubiquitous enzyme transaldolase (EC 2.2.1.2) is a member of this aldolase family, and it catalyzes the reversible transfer of a dihydroxyacetone moiety, derived from fructose 6-phosphate to erythrose 4-phosphate yielding sedoheptulose 7-phosphate and glyceraldehyde 3-phosphate.[1] Transaldolase from *Escherichia coli* consists of a single domain and forms a homodimer in solution[5,6] and in the crystal.[6,7] Like other class I aldolases, the subunit folds into an $(\alpha/\beta)_8$ barrel and the active site is located at the carboxyl end of the β strands.[7,8] A lysine residue (Lys-132 in the *E. coli* transaldolase sequence) is invariant in all transaldolase sequences known. This active-site residue is located at the end of β strand 4 of the $(\alpha/\beta)_8$ barrel and was proposed to form the covalent Schiff base intermediate during catalysis. A unique feature of transaldolase is the relative stability of the enzyme–Schiff base complex,[1] which prompted us to use this enzyme as a suitable model system to trap the intermediate for subsequent structure determination by X-ray crystallography.[9]

We describe here protocols used for the purification of recombinant transaldolase B from *Escherichia coli*, stoichiometric reduction of the Schiff-base

---

[1] O. Tsolas and B. L. Horecker, in "The Enzymes" (P. D. Boyer, ed.), Vol. 7, p. 259. Academic Press, New York, 1972.
[2] T. Gefflaut, C. Blonski, J. Perie, and M. Willson, *Prog. Biophys. Mol. Biol.* **63**, 301 (1995).
[3] C. Y. Lai, C. Chen, and O. Tsolas, *Arch. Biochem. Biophys.* **121**, 790 (1967).
[4] C. Y. Lai and T. Oshima, *Arch. Biochem. Biophys.* **144**, 363 (1971).
[5] G. A. Sprenger, U. Schörken, G. Sprenger, and H. Sahm, *J. Bacteriol.* **177**, 5930 (1995).
[6] U. Schörken, J. Jia, H. Sahm, G. A. Sprenger, and G. Schneider, *FEBS Lett.* **441**, 247 (1998).
[7] J. Jia, W. Huang, U. Schörken, H. Sahm, G. A. Sprenger, Y. Lindqvist, and G. Schneider, *Structure* **4**, 715 (1996).
[8] S. Thorell, P. Gergely, K. Banki, A. Perl, and G. Schneider, *FEBS Lett.* **475**, 205 (2000).
[9] J. Jia, U. Schörken, Y. Lindqvist, G. A. Sprenger, and G. Schneider, *Prot. Sci.* **6**, 119 (1997).

intermediate, and subsequent crystallization and structure determination of the enzyme–dihydroxyacetone complex.

Purification of Recombinant Transaldolase B from *Escherichia coli*

Materials

The plasmid pGS451 encodes a recombinant transaldolase B from *Escherichia coli*.[6] The *Escherichia coli* strain DH5 is used as host for pGS451.

Method

Expression of plasmid pGS451 in *E. coli* and purification of the enzyme are carried out using procedures described previously.[5] Cells are grown overnight in LB medium (400 ml) at 37° and collected by centrifugation. After washing once with glycylglycine buffer (pH 8.5, containing 1 mM dithiothreitol), the cells are broken by ultrasonic treatment consisting of eight 30-sec bursts at 40 W with cooling in an ethanol–ice bath. In order to remove cell debris, the extract is spun down in a centrifuge at 20,000g. Crystalline ammonium sulfate is added slowly to the cell extract to a final concentration of 277 g/liter (45% saturation). After 1 hr of stirring the extract is centrifuged and the precipitate is discarded. The concentration of ammonium sulfate is slowly raised to 561 g/liter (80% saturation). After centrifugation the precipitate is dissolved in glycylglycine buffer (pH 8.5, containing 1 mM dithiothreitol) and subjected to another round of ammonium sulfate precipitation. Concentration intervals are 0–351 g/liter (55% saturation) and 351–472 g/liter ammonium sulfate (70% saturation), respectively. The precipitate from the 70% fraction is dissolved in glycylglycine buffer and desalted by ultracentrifugation. The protein solution is then applied to a Q-Sepharose FF anion-exchange column (XK 26/20; 26 by 200 mm). Elution is carried out at a flow rate of 0.5 ml/min using a linear NaCl gradient (100–150 mM). Active fractions are collected, dialyzed against buffer, and applied to a column packed with EMD-DEAE-650 tentacle material. Transaldolase is again eluted in a linear NaCl gradient (0–400 mM). Purified transaldolase is analyzed for homogeneity by SDS–PAGE.

Activity Measurements

For all activity measurements the formation of glyceraldehyde 3-phosphate from fructose 6-phosphate is monitored by the decrease of NADH at 340 nm with the auxiliary enzymes triose-phosphate isomerase and glycerol-3-phosphate dehydrogenase as described before.[1,5] The specific activity of purified recombinant *E. coli* transaldolase B is 60 U/mg.

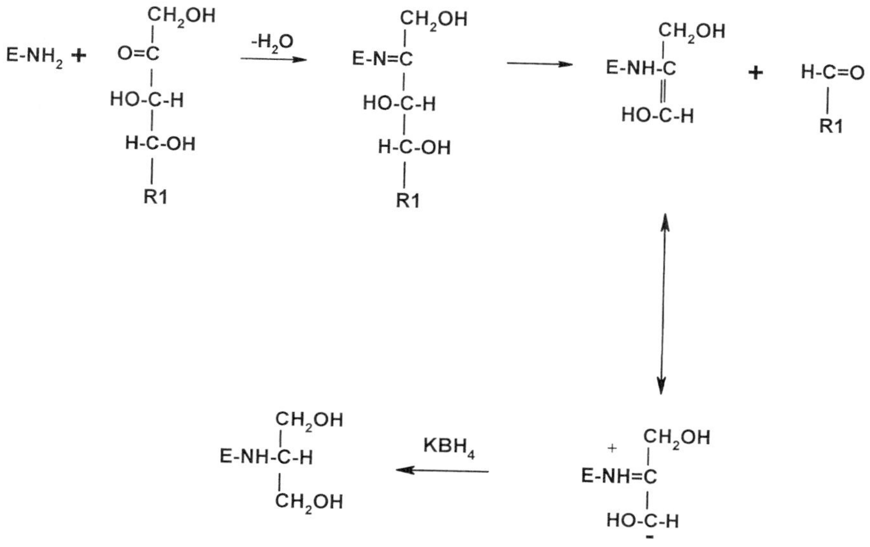

FIG. 1. Reaction scheme for trapping of the reduced transaldolase–dihydroxyacetone complex.

Trapping of Schiff Base Intermediate by Borohydride Reduction

Strategy

Addition of the donor substrate fructose 6-phosphate to transaldolase results in the formation of the enzyme–dihydroxyacetone complex with concomitant release of glyceraldehyde 3-phosphate (Fig. 1). In the absence of an acceptor substrate, the Schiff-base complex can be reduced by borohydride to yield N^6-β-glyceryl lysine.[3] Trapping of this compound is carried out in order to obtain a crystal structure of this complex, which requires a complete conversion of the free enzyme into the enzyme–intermediate complex for subsequent crystallization. This is achieved by two modifications of previous procedures: (i) a high molar excess of fructose 6-phosphate during incubation and borohydride reduction and (ii) enzymatic removal of the product glyceraldehyde 3-phosphate in order to drive the equilibrium further toward complete complex formation (Fig. 1).

Methods

Purified transaldolase (10 mg) is incubated with fructose 6-phosphate (200-fold molar excess) in 40 ml glycylglycine buffer (20 mM, pH 8.0) at 25°. The solution further contains 200 μl of a mixture of triose-phosphate isomerase/glycerol-3-phosphate dehydrogenase (Boehringer, Mannheim) and a 40-fold excess of

NADH. After incubation for 30 min, 40 ml of ice-cold triethanolamine buffer (pH 5.8) is added, the solution is cooled to 0°, and the pH is adjusted to pH 6.0 with 5 M acetic acid. Portions of 1 ml of a 3 M KBH$_4$ solution are then added under continuous stirring at 0°. After each addition, the pH is adjusted to pH 6. The loss of activity is equivalent to the amount of stable dihydroxyacetone–transaldolase complex formed. Aliquots are therefore removed for determination of enzymatic activity in order to monitor progress of the reaction. When the reaction is completed to about 95%, the solution is concentrated to 5 ml and the buffer is changed to 50 mM glycylglycine containg 1 mM dithiothreitol. This enzyme solution is then used for crystallization experiments.

Crystallization

Crystals of the transaldolase–dihydroxyacetone complex are obtained by a combination of micro- and macroseeding techniques using the hanging drop method as described previously for the wild-type enzyme.[10] Protein samples (20 mg/ml) are mixed with mother liquid containing 15% (w/w) polyethylene glycol 6000 and 0.1 M sodium citrate buffer at pH 4.0 at 20°. Crystals grown in this way are severely twinned. In order to obtain single crystals, the twinned crystals are crushed into small fragments and diluted with mother liquor. This diluted solution is used as microseeds and transferred to hanging drops which are set up as usual and have been preequilibrated for 5 days, but with lower content of polyethylene glycol 6000. After 2 days, small, well-shaped crystals appear in the drops. These crystals are used as macroseeds which after washing for 10 min in mother liquid are transferred to new droplets. These drops are set up as above and preequilibrated for 5 days, but with the addition of 15% polyethylene glycol 6000 as precipitant. After 10–14 days, single crystals of a size of up to 0.4 × 0.5 × 0.7 mm can be harvested. These crystals belong to space group $P2_12_12_1$, with cell dimensions $a = 69.2$ Å, $b = 91.9$ Å, and $c = 130.9$ Å.

Structure of Enzyme–Dihydroxyacetone Complex

The crystal structure of the complex of the borohydride-trapped reaction intermediate and transaldolase has been determined to 2.2 Å resolution using difference Fourier methods. Initial phases are obtained from the model of the free enzyme, determined at 1.9 Å resolution.[7] The structure of the complex is refined to an R-factor of 20.4% (R$_{free}$ 24.4%). Details of the structure analysis are given elsewhere.[9]

The structure analysis provides the first crystallographic evidence for Schiff-base formation in a class I aldolase. The overall structure of the complex is similar to the structure of the free enzyme, and no large conformational changes occur on

---

[10] J. Jia, Y. Lindqvist, G. Schneider, U. Schörken, H. Sahm, and G. A. Sprenger, *Acta Crystallogr.* **D52**, 192 (1996).

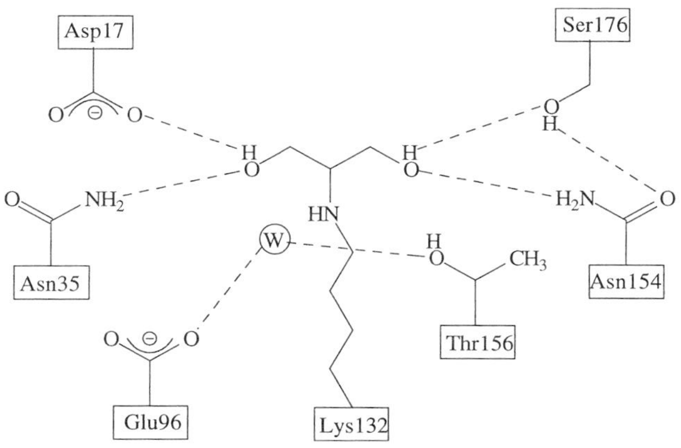

[14] SCHIFF BASE INTERMEDIATE IN TRANSALDOLASE B 201

FIG. 2. Schematic view of the borohydride-reduced Schiff base intermediate in the active site of transaldolase. The dihydroxyacetone moiety covalently linked to Lys-132 and its interactions with conserved active site residues are shown. Hydrogen bonds are indicated with dashed lines [from J. Jia, U. Schörken, Y. Lindqvist, G. A. Sprenger, and G. Schneider, *Prot. Sci.* **6**, 119 (1997) with permission of Cold Spring Harbor Laboratory Press].

formation of the enzyme–intermediate complex. The electron density map clearly shows that the C-2 carbon atom of dihydroxyacetone is covalently linked to the ε-amino group of Lys-132, thus confirming that this invariant residue in transaldolases is directly involved in Schiff-base formation. The dihydroxyacetone moiety interacts through a number of hydrogen bonds with groups on the enzyme (Fig. 2). The C-1 hydroxyl group forms hydrogen bonds with Asn-154 and Ser-176, and the C-3 hydroxyl group is within hydrogen bonding distance to the side chains of Asp-17 and Asn-35. Close to the ε-amino group of Lys-132 a water molecule is found which forms hydrogen bonds to the side chains of Glu-96 and Thr-156. This water molecule has been proposed to be component of a proton relay system in the active site, from Glu-96 to the Schiff base intermediate, thus actively participating in catalysis. Site-directed mutagenesis of active-site residues surrounding the Schiff base intermediate confirms the critical roles of these amino acids in binding of substrate/intermediate and/or in catalysis.[11–13]

Acknowledgments

This work was supported by grants from the Swedish Natural Science Research Council and the Deutsche Forschungsgemeinschaft through SFB 380/Teilprojekt B21.

[11] T. Miosga, I. Schaaff-Gerstenschläger, E. Franken, and F. Zimmermann, *Yeast* **9**, 1241 (1993).
[12] K. Banki and A. Perl, *FEBS Lett.* **378**, 161 (1996).
[13] U. Schörken, S. Thorell, M. Schürmann, J. Jia, G. A. Sprenger, and G. Schneider, *Eur. J. Biochem.*, **268**, 2408 (2001).

[15] T4 Endonuclease V: Use of NMR and Borohydride Trapping to Provide Evidence for Covalent Enzyme–Substrate Imine Intermediate

By M. L. DODSON, ANDREW J. KURTZ, and R. STEPHEN LLOYD

Introduction

Bailly and Verly[1] obtained evidence that the base excision repair (BER)[2] glycosylase, *Escherichia coli* endonuclease III, catalyzed DNA strand breakage by a β-elimination reaction. Grossman *et al.*[3] followed soon after with a general proposal that the strand breakage caused by those BER glycosylases that were capable of producing DNA strand breaks, utilized this mechanism. Kim and Linn[4] confirmed and extended these observations. The experimental methodologies used in these studies characterized the nature of the ends of the product DNA strands, and the experimental data were consistent with the proposed β-elimination mechanism. These interpretations implied a Schiff base (imine) intermediate in the DNA strand breakage reaction and a mechanistic parallel with the aldolase reaction.[5] In this class of enzyme mechanisms, formation of the protonated imine labilizes the bond from the adjacent carbon to carbon or hydrogen.[6]

Our laboratory sought to directly test these proposals and to identify the enzyme amine involved in Schiff base formation, specifically for the BER glycosylase T4-PDG.[7] The chemical characteristics of the Schiff base intermediate suggested several experimental approaches to answering these questions. Since the unprotonated form of the enzyme amine was required for Schiff base formation,[6] the relatively high pK_a values of solvent-exposed lysine ε-amino groups would disfavor their involvement in experiments carried out at neutral pH. An ε-amino group with a substantially lowered pK_a (for example, as might occur if the amine were deeply embedded in an appropriate active site) or an α-amino group would display

---

[1] V. Bailly and W. G. Verly, *Biochem. J.* **242**, 565 (1987).
[2] BER, Base excision repair; SDS, sodium dodecyl sulfate; FPG, formamidopyrimidine-DNA glycosylase; NMR, nuclear magnetic resonance.
[3] L. Grossman, P. R. Caron, S. J. Mazur, and E. Y. Oh, *FASEB J.* **2**, 2696 (1988).
[4] J. Kim and S. Linn, *Nucleic Acids Res.* **16**, 1135 (1988).
[5] R. Venkataraman and E. Racker, *J. Biol. Chem.* **236**, 1876 (1961).
[6] R. E. Feeney, G. Blankenhorn, and H. B. F. Dixon, *Adv. Prot. Chem.* **29**, 135 (1975).
[7] Formerly, this enzyme was designated T4 endonuclease V. Since it has been demonstrated that this enzyme, and other BER glycosylases which catalyze DNA backbone scissions, proceed by way of β-elimination mechanisms, i.e., they cleave a carbon–oxygen bond, they should more properly be called lyases. The term endonuclease has come to imply an enzyme which cleaves a nucleic acid chain via a phosphotransfer mechanism, i.e., they break an oxygen–phosphorus bond.

properties consistent with efficient Schiff base formation. Reductive alkylation of amines, specifically reductive methylation, has been shown to proceed through a Schiff base intermediate,[8] so that the rate of methylation of a particular amine would correlate with the concentration of the unprotonated form of that amine, and, therefore, with its pK_a. Of course, this kind of argument could not prove that the most readily methylated enzyme amine was located in the active site, with putative involvement in the mechanism, but would strongly suggest that possibility.

Schrock and Lloyd[9] observed that ratios up to approximately 0.8 methyl group per T4-PDG enzyme molecule correlated on an approximately one-to-one basis with enzyme inactivation. This level of methylation did not preclude enzyme–substrate binding, although higher levels of methylation did do so (4.4 methyl groups per enzyme). This strongly implied that the most readily methylated amine was directly involved in the T4-PDG mechanism. [^{14}C]Formaldehyde incorporation via reductive methylation, followed by proteolytic peptide separation and amino acid compositional analysis, identified the N-terminal α-amino group as the most readily methylated species.

Jentoft and Dearborn[8] demonstrated that reductive mono- and dimethylation of protein amino groups had only a minor perturbing effect on their pK_a values, and that both primary and secondary amines could be methylated. Jentoft et al.[10] used [^{13}C]formaldehyde to methylate ribonuclease A and were able to assign resonances in the ^{13}C NMR spectrum by their chemical shifts and apparent pK_a values in NMR titration experiments. They assigned one resonance to a particular lysine residue (K41) whose involvement in the enzyme mechanism had been suggested previously. Jentoft et al.[11] showed that active site ligands were inhibitors of K41 methylation. These studies constituted an early demonstration of the efficacy of amino group reductive methylation coupled with biochemical and NMR experiments in elucidating the involvement of specific residues in enzyme catalysis.

Dodson et al.[12] used [^{13}C]formaldehyde to methylate T4-PDG at levels of 1 and 3.5 methyl groups per enzyme molecule. At 1 methyl per enzyme, a single ^{13}C NMR resonance predominated, whose chemical shift and titration behavior allowed an assignment to the N-terminal α-amino group. The change in chemical shift of the dominant species after SDS denaturation occurred between pH 7 and 9.3, whereas the expected pK_a of an α-amino group in denaturing conditions was 8.[13,14] Jentoft et al.[11] found the pK_a of the dimethyl α-amino terminus of methylated ribonuclease A to be approximately 6.6, a value similar to that found for

---

[8] N. Jentoft and D. G. Dearborn, *J. Biol. Chem.* **254**, 4359 (1979).

[9] R. D. Schrock and R. S. Lloyd, *J. Biol. Chem.* **266**, 17631 (1991).

[10] J. E. Jentoft, N. Jentoft, T. A. Gerken, and D. G. Dearborn, *J. Biol. Chem.* **254**, 4366 (1979).

[11] J. E. Jentoft, T. A. Gerken, N. Jentoft, and D. G. Dearborn, *J. Biol. Chem.* **256**, 231 (1981).

[12] M. L. Dodson, R. D. Schrock, and R. S. Lloyd, *Biochemistry* **32**, 8284 (1993).

[13] G. R. Stark, *Methods Enzymol.* **11**, 590 (1967).

[14] C. Tanford and J. D. Havenstein, *J. Am. Chem. Soc.* **78**, 5287 (1956).

the unmethylated α-amino group of ribonuclease A. The differences between these two studies could be rationalized as differences between the titration behaviors of native versus denatured proteins. Additional NMR resonances were observed in the spectra of enzymes modified at 3.5 methyl groups per enzyme molecule, and these were assigned to mono- and dimethyl ε-amino groups based on their titration behavior. The chemical characteristics of the Schiff base intermediate also implied that it should readily undergo nucleophilic addition, e.g., by cyanide. Since the formation of the Schiff base during enzyme catalysis involved functional groups on both the substrate and the enzyme, cyanide inhibition should be dependent on the presence of the substrate. That is, preincubation of the enzyme with cyanide, followed by its dilution or removal, should not result in inhibition. This prediction was supported by the experimental data.

The indirect identification of the N-terminal α-amino group of T4-PDG as the catalytic nucleophile, based on its methylation behavior, was decisively confirmed by the direct isolation of a covalent enzyme–substrate product after reduction of the Schiff base intermediate.[12] Dodson et al. took advantage of the location of a methionine residue near the N terminus of the T4-PDG amino acid sequence. The N-terminal peptide resulting from cyanogen bromide cleavage at this methionine contained no lysine residues, and it was sufficiently small to be easily distinguished from the larger C-terminal portion. Demonstration that the N-terminal peptide was covalently attached to the DNA after cyanogen bromide treatment of the reduced Schiff base product completed the proof that the active-site nucleophile was the N-terminal α-amino group and directly demonstrated the involvement of an imine intermediate in the reaction. Zharkov et al.[15] have used mass spectrometry analysis in a similar experimental design to identify the N-terminal proline of the Escherichia coli FPG protein as the active-site nucleophile.

The ability of a protonated imine to labilize the bond from the adjacent carbon to carbon or hydrogen[6] suggested to Dodson et al.[16] that the key difference between those BER glycosylases causing DNA strand breaks and those that did not might be the type of the active-site nucleophile. The pure glycosylases (those not catalyzing DNA strand breakage) would use a water molecule, possibly activated by interaction with an acidic group, whereas the glycosylases causing DNA strand breaks would use an amine nucleophile. The β-elimination DNA cleavage reaction would be a consequence of the chemistry of the Schiff base intermediate. To date, this prediction has survived all tests, although sometimes with unexpected confounding observations.[17] Some of these glycosylases catalyze a δ- as well as a β-elimination reaction. The δ-elimination has been shown to require a prior

---

[15] D. O. Zharkov, R. A. Rieger, C. R. Iden, and A. P. Grollman, *J. Biol. Chem.* **272**, 5335 (1997).
[16] M. L. Dodson, M. L. Michaels, and R. S. Lloyd, *J. Biol. Chem.* **269**, 32709 (1994).
[17] Y. Guan, R. C. Manuel, A. S. Arvai, S. S. Parikh, C. D. Mol, J. H. Miller, R. S. Lloyd, and J. A. Tainer, *Nat. Struct. Biol.* **5**, 1058 (1998).

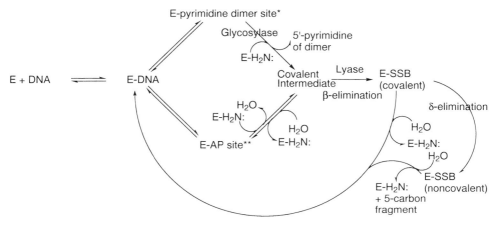

FIG. 1. The T4-PDG reaction mechanism. The * designates an enzyme–pyrimidine dimer substrate complex. The ** identifies an enzyme–abasic site substrate complex. The "Covalent Intermediate" is the Schiff base. The initial lyase (β-elimination) step produces an α,β-unsaturated Schiff base product. Its hydrolysis generates a DNA product with one side of the single strand break terminated with a 5'-phosphate and the other with an α,β-unsaturated aldehyde. A δ-elimination event may occur before hydrolysis. This results in terminal phosphates on both side of the resulting single nucleotide gap after hydrolysis of the covalent intermediate. The balance among the rate constants for these possibilities accounts for the elaborate product spectra displayed by these types of enzymes.

β-elimination.[18] These enzymes have come to be called BER glycosylase–abasic site lyases. Figure 1 shows a proposed scheme for the reaction of T4-PDG at a pyrimidine dimer or an abasic substrate site and illustrates the concepts discussed. To recapitulate: the DNA backbone breakage associated with the lyase-capable BER glycosylases and their elaborate product spectra can be satisfactorily rationalized by the chemistry of their enzyme–substrate Schiff base intermediate.

Methods

The experimental design will, to a first approximation, dictate the choice of reducing agent: NaBH$_4$ or NaCNBH$_3$. Two broad classes of experiments can be distinguished. In the first, the objective is to accumulate as much of the reduced Schiff base product as possible, or to have the reducing agent present throughout the course of the experiment. This implies that NaCNBH$_3$ would likely be used. We term these experiments "integration" experiments. In the second design, a transient Schiff base intermediate is to be "quenched" and stabilized for further analysis. This design implies that NaBH$_4$, as a stronger reducing agent, would

---

[18] M. Bhagwat and J. A. Gerlt, *Biochemistry* **35,** 659 (1996).

likely be the better choice. Note, however, that even $NaBH_4$ does not reduce Schiff base intermediates sufficiently quickly to be used as a quencher in rapid kinetic analyses.[19] We call this class "snapshot" experiments. The pH of the reaction is another key variable. $NaBH_4$ is unstable in acidic conditions, whereas we have successfully used $NaCNBH_3$ at pH 5. Because $NaBH_4$ is a good reducing agent for aldehydes and ketones, as well as for Schiff bases, formaldehyde used in labeling experiments or aldehydic substrates would be directly reduced to alcohols as competing side reactions. $NaCNBH_3$ is a much weaker reducing agent that does not reduce aldehydes or ketones at neutral pH but readily reduces Schiff bases.

We prepare aqueous solutions of $NaCNBH_3$ on the day of use. $NaBH_4$ solutions are prepared immediately before use. $NaCNBH_3$ solutions can be buffered at the normal enzyme reaction pH, whereas we use $NaBH_4$ either unbuffered (the pH of a 100 mM solution is approximately pH 9), or buffered at pH 9 with CHES.[19]

Although inhibitors may be present in solutions of $NaCNBH_3$ or $NaBH_4$, in our experience no purification has been found to be necessary. When Jentoft and Dearborn[8] used recrystallized $NaCNBH_3$, the rate of the reaction was found to be relatively insensitive to the $NaCNBH_3$ concentration over a range of 10 to 250 mM. This suggested that Schiff base reduction was not the rate-limiting step in the reaction and must be much faster than Schiff base formation. The differences between these observations and those of McCullough et al.[19] are likely due to the concentration of unprotonated lysine species being very low in the Jentoft and Dearborn experiments. Recall that the unprotonated form of the amine is required for Schiff base formation.[6] As might be expected, exogenous amines, such as Tris, in the reaction are potent inhibitors of protein labeling by Schiff base reduction. A final side reaction considered by Jentoft and Dearborn[8] was the possible effect of $NaCNBH_3$ on disulfide linkages, because $NaBH_4$ readily reduces disulfides. This was not found to occur at a rate sufficient to pose a problem.

Figure 2 shows one application of peptide–DNA cross-linking with $NaCNBH_3$. This experiment was part of an investigation of small peptide models of the BER glycosylase–lyase reaction. We have similarly used this reaction to prepare protein–DNA cross-links as substrates for other DNA repair enzyme systems.

Future Applications

This methodology could be more widely applied in the investigation of enzymes utilizing Schiff base intermediates. For example, [^{13}C]formaldehyde-labeled active site amino groups could be used as spectroscopic (NMR) probes to study structural aspects of substrate binding. Similarly, the BER glycosylases use a "base flipping" mechanism to achieve access to the scissile base–sugar bond. ^{13}C-Labeled active site amines could be used as probes of this process.

---

[19] A. K. McCullough, A. Sanchez, M. L. Dodson, P. Marapaka, J.-S. Taylor, and R. S. Lloyd, Biochemistry **40**, 561 (2001).

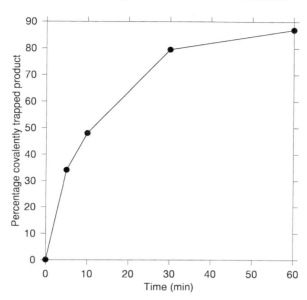

FIG. 2. A 17 μl reaction mixture contained 1 unit of uracil DNA glycosylase and 0.2 pmol 5'-^{32}P-labeled duplex oligonucleotide DNA (with a uracil residue replacing one thymine) in 50 mM HEPES (pH 6.8) and 5 mM NaCl. After incubation at 37° for 75 min to ensure complete removal of the uracil, the mixture was diluted to 170 μl with water. Several 17 μl aliquots were removed. These (except for appropriate controls) received 2 μl of 250 mM NaCNBH$_3$ (in 1 M sodium acetate, pH 5) and 1 μl of the peptide KFHEKHHSHRGY dissolved in 10 mM HCl. The reaction mixtures were incubated at 37°. At selected time points an individual reaction tube was removed, and 20 μl of 95% (v/v) formamide, 20 mM EDTA, 0.5% (v/v) SDS, 0.02% (v/v) bromphenol blue, and 0.02% (v/v) xylene cyanole were added. The products were separated on a 20% (v/v) acrylamide gel containing 7 M urea and quantitated with a phosphorimager (Molecular Dynamics, Amersham Biosciences, Sunnyvale, CA).

Jentoft et al.[11] showed that the ^{13}C-methylated ribonuclese A K41 ε-amino group had a shortened T_1 (spin lattice relaxation time), 0.4 versus 0.8–0.9 sec, and smaller nuclear Overhauser enhancement, 2.3 versus 3.0, when compared to solvent-exposed lysine residues. This was correlated with reduced mobility of the side chain as had been predicted from X-ray diffraction studies. This type of investigation could be more widely used.

Finally, the ability to reductively stabilize Schiff base reaction intermediates could be more widely used to investigate the structures of components along the enzyme reaction pathway, since these structures should be only slightly perturbed from that of the normal unreduced Schiff base species.

[16] Detection of Covalent Tetrahedral Adducts by Differential Isotope Shift ^{13}C NMR: Acetyl-Enzyme Reaction Intermediate Formed by 3-Hydroxy-3-Methylglutaryl-CoA Synthase

By HENRY M. MIZIORKO and DMITRIY A. VINAROV

Introduction

The substitution of heavy stable isotopes (e.g., D for H, ^{18}O for ^{16}O) in metabolites that contain NMR observable nuclei (e.g., ^{13}C, ^{15}N, ^{31}P) produces small shifts in the signals of these observable nuclei. Examples of the detection of such isotope-induced upfield shifts include the work of Pfeffer et al.[1] on deuterium labeled carbohydrates, the report of Risley and Van Etten[2] on ^{18}O-labeled carbohydrates, the documentation by Cohn and Hu[3] of ^{18}O shifts of ^{31}P signals in nucleotides, and the experiments of Van Etten and Risley[4] on ^{18}O shifts for ^{15}N-nitrite. These small shifts in NMR signal are usually reported for the atom to which the heavy isotope is covalently attached. However, in the case of deuterium labeling of carbohydrate, ^{13}C NMR shifts have been detected[1,5] for carbons with two or three intervening bonds.

Mechanistic studies involving use of ^{18}O-induced isotope shifts of resonances in ^{13}C-enriched and ^{31}P-containing metabolites have been reviewed earlier in this series by Risley and Van Etten[6] as well as by Villafranca.[7] Exchange reactions have been demonstrated and measurements of exchange rates have been reported. Similarly, rates of hydrolysis reactions have been studied and cleavage positions of chemical bonds have been elucidated. Advantages invoked for the differential isotope shift approach include the ability to continuously monitor for development of the isotope-induced shift and elimination, in many cases, of the need for subsequent sample derivatization prior to analysis. An obvious limitation in the case of ^{13}C or ^{15}N NMR experiments involves sensitivity limitations that translate into the need to isotopically label with ^{13}C or ^{15}N the compound into which the heavy isotope will be introduced.

The work alluded to above has relied on analysis of metabolites and other small molecules; ample amounts of such samples can be generated using only catalytic

[1] P. E. Pfeffer, K. M. Valentine, and F. W. Parrish, *J. Am. Chem. Soc.* **101**, 1265 (1979).
[2] J. M. Risley and R. L. Van Etten, *Biochemistry* **21**, 6360 (1982).
[3] M. Cohn and A. Hu, *Proc. Natl. Acad. Sci. U.S.A.* **75**, 200 (1978).
[4] R. L. Van Etten and J. M. Risley, *J. Am. Chem. Soc.* **103**, 5633 (1981).
[5] J. L. Jurlina and J. B. Stothers, *J. Am. Chem. Soc.* **104**, 4609 (1982).
[6] J. M. Risley and R. L. Van Etten, *Methods Enzymol.* **177**, 376 (1989).
[7] J. J. Villafranca, *Methods Enzymol.* **177**, 390 (1989).

amounts of enzyme. Reports of isotope-induced shifts for protein–metabolite complexes or adducts have been much less common even though recombinant DNA technology has made many enzymes available at levels that would support such experiments. The infrequent use of the isotope shift approach may be due, in part, to difficulty in resolving the NMR signal due to unbound ligand from that corresponding to the protein-bound species. Nonetheless, Schmidt et al.[8] have documented deuterium and ^{18}O shifts for interaction of pepsin with a ^{13}C-enriched tight binding inhibitor, pepstatin. The covalent acetyl-enzyme reaction intermediate formed on incubation of 3-hydroxy-3-methylglutaryl (HMG)-CoA synthase with acetyl-CoA was detected by Vinarov et al.[9] using a ^{13}C NMR approach. Even though the native protein is reasonably large (homodimer of 58 kDa subunits), the [^{13}C]acetyl peak is easily detected and is well resolved from excess [^{13}C]acetyl-CoA substrate, as well as from any free [^{13}C]acetic acid produced on slow, abortive hydrolysis of the reaction intermediate. These observations prompted Vinarov and Miziorko[10] and Chun et al.[11] to further investigate the reaction intermediate in a series of ^{13}C NMR experiments which include the differential isotope shift work on which this article is focused. In contrast with other techniques, the isotope shift approach affords a reasonably straightforward characterization of the reaction intermediate.

Experimental Design

Spectroscopic Equipment and Measurements of ^{13}C NMR Spectra

The ^{13}C NMR measurements that we have reported for substrates, product, and reaction intermediates involved in the HMG-CoA synthase reaction are performed using a 5-mm QNP probe on a Bruker AC-300 spectrometer operating at 75.468 MHz for ^{13}C. Earlier observations of ^{18}O shifts were accomplished by Risley and Van Etten[12] at lower fields, indicating that access to medium/high field strength instrumentation is not a strict requirement for these experiments. However, the higher field strengths of modern spectrometers and probes result in notably improved sensitivity for ^{13}C. This can translate into shorter signal acquisition periods that would make it more straightforward to follow the kinetics of the development of isotope-induced shifts. Moreover, the improved sensitivity afforded by modern medium and high field instruments makes possible measurements on samples that are much lower in protein concentration than the samples which we used for measurements on the HMG-CoA synthase reaction intermediate.

[8] P. G. Schmidt, M. W. Holladay, F. G. Salituro, and D. H. Rich, *Biochem. Biophys. Res. Commun.* **129**, 597 (1985).
[9] D. A. Vinarov, C. Narasimhan, and H. M. Miziorko, *J. Am. Chem. Soc.* **121**, 270 (1999).
[10] D. A. Vinarov and H. M. Miziorko, *Biochemistry* **39**, 3360 (2000).
[11] K. Y. Chun, D. A. Vinarov, and H. M. Miziorko, *Biochemistry* **39**, 14670 (2000).
[12] J. M. Risley and R. L. Van Etten, *J. Am. Chem. Soc.* **101**, 252 (1979).

All spectra are recorded at 21°. Chemical shifts are referenced to tetramethylsilane (TMS). A sweep width of 16,000 Hz is used and 16K data points collected. Signal acquisition employs a 35-degree pulse angle and a 2-sec delay between transients. A typical spectrum of ^{13}C-enriched acetyl-CoA, measured in samples with a 2 : 1 substrate/enzyme site ratio, requires 1.5–5 hr of data collection (1500–5000 transients). FID data are processed with a Gaussian lineshape analysis subroutine for center frequency determination using a convergence limit of 0.001. For spectra shown in the figures, the collected data have been zero-filled to 64K points and then processed with 5 Hz line broadening to improve signal-to-noise.

Synthesis of [1-^{13}C]- and [1,2-^{13}C]Acetyl-CoA

[1,1-^{13}C]- and [1,1,2,2-^{13}C]acetic anhydride (99% enrichment) are purchased from Isotec (Miamisburg, OH). Synthesis employs the anhydride method of Simon and Shemin.[13] Briefly, a solution of 100 μmol of CoA in 1 ml of nitrogen-purged water is adjusted to pH 8.0 with LiOH prior to addition of 200 μmol of [1,1-^{13}C]- or [1,1,2,2-^{13}C]acetic anhydride. The mixture is periodically vortexed over 5 min, after which the reaction is complete, as demonstrated by a negative nitroprusside test. The pH is readjusted to 3.0 with HCl. [1-^{13}C]Acetyl-CoA or [1,2-^{13}C]acetyl-CoA are precipitated with chilled methanol/acetone (1 : 3) three times and recovered with a yield of 95%. The purity of the product is evident from a proton-decoupled ^{13}C NMR spectrum, which shows a resonance (doublet in the case of [1,2-^{13}C]acetyl-CoA) at 204 ppm corresponding to C-1 of acetyl-CoA and a doublet in the case of [1,2-^{13}C]acetyl-CoA at 33 ppm corresponding to C-2 of acetyl-CoA.

Synthesis of [1,3-^{13}C]Acetoacetyl-CoA

[1,3-^{13}C]Acetoacetyl-CoA is prepared following the method of Hersh and Jencks[14] as modified by Miziorko and Lane.[15] Ethyl [1,3-^{13}C]acetoacetate (99% enrichment, Isotec) is hydrolyzed with LiOH (1 : 2 ratio, micromoles) in a final volume of 1 ml at 30° for 5 hr. After the hydrolysis, pH is adjusted to 7.2. Unlabeled acetoacetyl-CoA (Li$^+$ salt), EDTA, and one unit of porcine heart succinyl-CoA transferase (Sigma, St. Louis, MO) are added to the hydrolyzed product in a final volume of 2 ml. The exchange is allowed to proceed for 30 min at 30°. The reaction mixture is cooled on ice and brought to pH 2 with HCl. The mixture is extracted with a threefold excess of ether three times to remove free acetoacetate. The aqueous layer is taken to near dryness and cooled. The sample was dissolved in chilled methanol prior to addition of 3 volumes of cold acetone to precipitate [1,3-^{13}C]acetoacetyl-CoA as the Li$^+$ salt. The purity of the nucleotide product has been evaluated by proton-decoupled ^{13}C NMR, which indicates only

---

[13] E. J. Simon and D. Shemin, *J. Am. Chem. Soc.* **75,** 2520 (1953).
[14] L. B. Hersh and W. P. Jencks, *J. Biol. Chem.* **242,** 3468 (1967).
[15] H. M. Miziorko and M. D. Lane, *J. Biol. Chem.* **252,** 1414 (1977).

two resonances at 198.5 and 208.9 ppm attributable to C-1 thioester carbonyl and C-3 keto carbons, respectively.

Synthesis of [1,3,5-^{13}C]HMG-CoA

[1,3,5-^{13}C]HMG-CoA is enzymatically synthesized by condensing [1,3-^{13}C]acetoacetyl-CoA with [1-^{13}C]acetyl-CoA in the presence of catalytic amounts of purified recombinant avian HMG-CoA synthase, produced as described by Misra *et al.*[16] [1,3-^{13}C]Acetoacetyl-CoA (13 μmol) and [1-^{13}C]acetyl-CoA (12 μmol) are incubated with 1 mg of HMG-CoA synthase (4 U/mg) at 30° in 100 mM Tris-Cl, pH 8.2, containing 0.10 mM EDTA. After 3 hr, additional [1,3-^{13}C]-acetoacetyl-CoA (1.0 μmol) and HMG-CoA synthase (0.2 mg) are added to drive the reaction to completion. After 6 hr, the reaction monitored by high-performance liquid chromatography (HPLC) is found to be >90% complete. The product is purified by DEAE-Sephadex chromatography (1 × 45 cm column; 1 liter gradient ranging from 20 to 300 mM LiCl in 3 mM HCl). The fractions containing the product are brought to near dryness by rotary evaporation and the nucleotide is recovered by precipitation from cold methanol/acetone (1 : 3). Proton-decoupled ^{13}C NMR evaluation of the isolated product indicates resonances at 73.5, 182.9, and 202.9 ppm, attributable to C-3 hydroxyl, C-5 carboxyl, and C-1 thioester carbonyl carbons, respectively.

Preparation of [^{13}C]Acetyl-Enzyme Reaction Intermediates

HMG-CoA synthase and β-ketothiolase proteins are buffer exchanged into 10 mM sodium phosphate pH 7.0 using Centricon-25 membrane cones (Amicon, Danvers, MA). Both proteins and metabolites used for preparation of NMR samples are treated with Chelex resin to remove metal contaminants that could contribute to NMR line broadening. After concentration to an enzyme site concentration of about 1 mM, paired samples are prepared and each is separately lyophilized. Each protein sample is dissolved (without significant loss of activity) in an appropriate volume of either deionized water supplemented with 10% D$_2$O (MSD Isotopes, Dorval, Quebec, Canada) for internal lock or H$_2$^{18}O (>95% ^{18}O; Mound-Monsanto) supplemented with 10% D$_2$O for internal field lock, prior to addition of [^{13}C]acetyl-CoA and measurement of the ^{13}C NMR spectrum. Two paired, independently exchanged samples are run consecutively, allowing calculation of a standard error for the differential isotope shift.

Experimental Results

Differential isotope ^{13}C NMR shifts measured upon substitution of deuterium for hydrogen can be of considerable magnitude (>100 ppb[1]), reflecting the large fractional increment in mass resulting from this isotopic substitution. Replacement

---
[16] I. Misra, C. Narasimhan, and H. M. Miziorko, *J. Biol. Chem.* **268**, 12129 (1993).

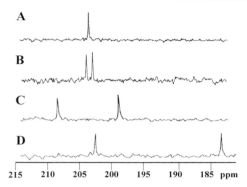

FIG. 1. NMR spectra of ^{13}C-labeled acyl-CoA derivatives. Proton-decoupled ^{13}C NMR spectra (180–220 ppm, referenced to TMS) were measured at 21° using a Bruker-AC300 spectrometer operating at 75.469 MHz for ^{13}C. Samples, buffered with 10 mM potassium phosphate (pH 7.0, 20% D$_2$O for internal lock), contained: (A) 2.0 mM [1-^{13}C] acetyl-CoA; (B) 2.0 mM [1,2-^{13}C] acetyl-CoA; (C) 2.0 mM [1,3-^{13}C] acetoacetyl-CoA; (D) 2.0 mM [1,3,5-^{13}C] HMG-CoA. Spectra were zero-filled to 64K points and processed with 5 Hz line broadening to improve signal-to-noise. Spectra A–D were obtained with 2000 transients each, which required approximately 1.5 hr of acquisition time. Reprinted with permission from D. A. Vinarov and H. M. Miziorko, *Biochemistry* **39**, 3360 (2000). Copyright © 2000 American Chemical Society.

of ^{16}O by ^{18}O results in smaller upfield shift of ^{13}C NMR peaks (<60 ppb). Multiply substituted carbons will reflect the additive nature of the individual isotope-induced shifts. The magnitude of any individual shift will reflect bond order,[17] e.g., the ^{13}C shift in ^{18}O-labeled alcohols will be smaller than that measured in ^{18}O-labeled ketones. The modest magnitude of the ^{18}O-induced ^{13}C NMR shifts demands careful sample preparation. The use of appropriate standards and control samples will help validate the interpretation of any shifts measured for the experimental samples. Our work on the acetyl-enzyme intermediate formed by HMG-CoA synthase illustrates these points.

Utility of ^{13}C-Enriched Metabolites as Calibration Standards for Magnitude of ^{18}O-Induced Shifts and for Uncatalyzed Control Samples

Figure 1 shows the 180–215 ppm region of the proton-decoupled ^{13}C NMR spectra of acetyl-CoA, acetoacetyl-CoA, and HMG-CoA with ^{13}C enrichment at various positions. The ^{13}C NMR spectra of [1-^{13}C]- and [1,2-^{13}C]acetyl-CoA show a resonance (a doublet in the case of [1,2-^{13}C]acetyl-CoA) at 204 ppm corresponding to C-1 (Figs. 1A, 1B) and, in the case of [1,2-^{13}C]acetyl-CoA, a doublet at 33 ppm corresponding to C-2 (not shown). As explained below, the doubly labeled metabolite affords a useful internal control in experiments on the HMG-CoA

---

[17] J. M. Risley and R. L. Van Etten, *J. Am. Chem. Soc.* **102**, 4609 (1980).

synthase reaction intermediate. The ^{13}C NMR spectrum of [1,3-^{13}C]acetoacetyl-CoA shows two resonances at 198.5 and 208.9 ppm attributable to C-1 thioester carbonyl and C-3 keto carbons, respectively (Fig. 1C). These metabolites, substrates for HMG-CoA synthase, are precursors for enzymatic synthesis of [1,3,5-^{13}C]HMG-CoA, which contains a C-1 thioester carbonyl that resonates at 202.9 ppm, a C-3 hydroxyl carbon resonating at 73.5 ppm (not shown), and a C-5 carboxyl resonating at 182.9 ppm (Fig. 1D). This C-5 position is useful in the context of a calibration standard. In order to confirm the magnitude of ^{13}C resonance shift on incorporation of an ^{18}O substituent, 5-^{18}O-enriched [1,3,5-^{13}C]HMG-CoA was synthesized by reacting catalytic amounts of HMG-CoA synthase with its ^{13}C-enriched substrates (2 mM each) in H$_2$^{18}O. This incorporates one ^{18}O into the free carboxylate of the HMG-CoA which should exhibit a C–O bond order of 1.5 due to resonance averaging of one single and one double bond. Proton-decoupled ^{13}C NMR spectra of [1,3,5-^{13}C]HMG-CoA synthesized in H$_2$^{18}O versus H$_2$^{16}O are shown in Fig. 2A. Substitution of H$_2$^{18}O for H$_2$^{16}O resulted in an upfield shift of 0.026 ± 0.005 ppm for C-5 (182.25 ppm) of [1,3,5-^{13}C]HMG-CoA (Table I). Resonances corresponding to C-1 (202.9 ppm) and C-3 (73.5) of HMG-CoA showed insignificant upfield shifts of 0.008 ± 0.001 ppm for C-1 thioester carbon and 0.004 ± 0.001 ppm for C-5 carboxyl carbon (Table I). Because no shift should be observed for peaks due to C-1 or C-3 of HMG-CoA (no solvent oxygen will label these positions under the conditions employed for HMG-CoA synthesis), these values provide a useful indication of the magnitude of experimental error.

Because acquisition of ^{13}C NMR spectra for enzyme intermediates or enzyme–metabolite complexes may require an extended period of time, it becomes necessary to ensure that nonenzymatic exchange mechanisms do not compromise the interpretation of any ^{18}O-induced shift that is detected. In order to evaluate such a contribution in experiments on the HMG-CoA synthase reaction intermediate, proton-decoupled ^{13}C NMR spectra of 2 mM [1,2-^{13}C]acetyl-CoA in buffer/H$_2$^{18}O versus buffer/H$_2$^{16}O were measured, as shown in Fig. 2B. Substitution of H$_2$^{18}O for H$_2$^{16}O resulted in insignificant upfield shifts of 0.005 ± 0.001 ppm for the C-1 thioester carbon and 0.006 ± 0.001 ppm for the C-2 methyl carbon of acetyl-CoA, measured after 5 hr of signal accumulation. These results (Table I) demonstrate that no uncatalyzed exchange labeling occurs over the time course employed in a reaction intermediate experiment.

Detection of a Specific ^{18}O-Induced Shift in an Acetyl-S–Enzyme Reaction Intermediate

The ability of the acetyl-S enzyme of HMG-CoA synthase to reversibly form/collapse a tetrahedral intermediate was tested by reconstituting freeze-dried enzyme in H$_2$^{18}O versus H$_2$^{16}O and adding [1,2-^{13}C]acetyl-CoA, appropriately

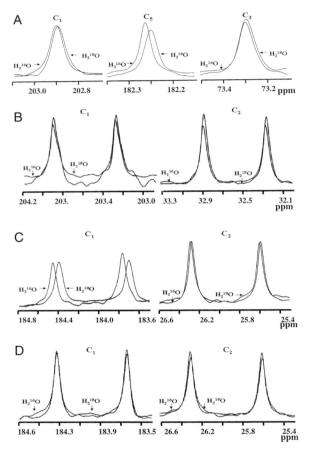

FIG. 2. ^{18}O-induced shifts of ^{13}C NMR signals of ^{13}C-enriched metabolites and acyl-enzyme reaction intermediates. Samples contained: (A) [1,3,5-^{13}C] HMG-CoA synthesized in $H_2^{18}O$ versus $H_2^{16}O$; (B) 2 mM [1,2-^{13}C] acetyl-CoA in buffer/$H_2^{18}O$ versus buffer/$H_2^{16}O$; (C) HMG-CoA synthase (1 mM) with [1,2-^{13}C] acetyl-CoA (2 mM) in $H_2^{18}O$ versus $H_2^{16}O$; (D) β-ketothiolase (1 mM) with [1,2-^{13}C] acetyl-CoA (2 mM) in $H_2^{18}O$ versus $H_2^{16}O$. Reprinted with permission from D. A. Vinarov and H. M. Miziorko, *Biochemistry* **39**, 3360 (2000). Copyright © 2000 American Chemical Society.

dissolved in $H_2^{18}O$ or $H_2^{16}O$ (Fig. 2C). Use of doubly labeled [1,2-^{13}C]acetyl-CoA affords an internal control, since the 26 ppm C-2 methyl peak can be observed in $H_2^{18}O$ and $H_2^{16}O$ experiments to verify that no significant upfield shifts are observed for this carbon, which contains no C–O bond. For production of the acetyl-enzyme reaction intermediate of HMG-CoA synthase it is possible to use a modest excess of [1,2-^{13}C]acetyl-CoA over enzyme sites without compromising

TABLE I
^{18}O-INDUCED ISOTOPE SHIFTS OF ^{13}C RESONANCES OF HMG-CoA,
Ac-CoA, ACETYL-S-HMG-CoA SYNTHASE, AND ACETYL-S-THIOLASE
C378G REACTION INTERMEDIATES[a]

Sample	^{13}C signal	$\Delta[\delta_{(H2\,16O)} - \delta_{(H2\,18O)}]$ (ppm)
HMG-CoA in buffer	C-1	0.008 ± 0.001[b]
	C-3	0.004 ± 0.001
	C-5	0.026 ± 0.005
Ac-CoA in buffer	C-1	0.005 ± 0.001
	C-2	0.006 ± 0.001
Acetyl-S-HMG-CoA synthase	C-1	0.055 ± 0.005
	C-2	0.009 ± 0.002
Acetyl-S-thiolase C378G	C-1	0.004 ± 0.001
	C-2	0.006 ± 0.001

[a] Reprinted with permission from D. A. Vinarov and H. M. Miziorko, *Biochemistry* **39**, 3360 (2000). Copyright © 2000 American Chemical Society.
[b] Standard error for differential isotope shift values was calculated from two independently exchanged and consecutively run samples.

the ^{13}C NMR experiment. Vinarov et al.[9] established that transfer of the thioesterified acetyl moiety from CoASH to C-129 of HMG-CoA synthase results in a 20 ppm upfield shift of the ^{13}C resonance attributable to C-1 of the acetyl group. Thus, [1,2-^{13}C]acetyl-CoA in excess of enzyme sites produces a 204 ppm resonance well removed from the signal which is scrutinized for the appearance of an ^{18}O-induced shift. For the resonance corresponding to the C-1 thioester carbon of acetyl-S–enzyme (184.3 ppm), the substitution of $H_2^{18}O$ for $H_2^{16}O$ resulted in an upfield shift of 0.055 ± 0.005 ppm, as measured after 3 hr of signal accumulation (Fig. 2C). The magnitude of the shift (Table I) does not increase on further signal accumulation. In contrast to the observations on C-1, the substitution of $H_2^{18}O$ for $H_2^{16}O$ had very little effect on the resonance corresponding to the C-2 methyl carbon (26 ppm), as expected. After 3 hr of signal accumulation the observed upfield shift was 0.009 ± 0.002 ppm (Fig. 2C; Table I). This value remains unchanged on further signal accumulation and reflects experimental error.

Although the observed ^{18}O-induced isotope shift seems compelling in contrast to the results obtained for the nonenzymatic control incubation of [1,2-^{13}C]acetyl-CoA in buffer, another persuasive negative control experiment is afforded by a parallel experiment on another acetyl-enzyme intermediate. In its interconversion of acetoacetyl-CoA and acetyl-CoA, β-ketothiolase also forms an acetyl-S-enzyme.[18] Vinarov et al.[9] demonstrated that this intermediate accumulates as an

[18] M. A. Palmer, E. Differding, S. F. Williams, C. T. Walsh, A. J. Sinskey, and S. Masamune, *J. Biol. Chem.* **266**, 8369 (1991).

NMR detectable species when the C378G variant is employed for the experiment, since this general base mutant is unable to form acetoacetyl-CoA condensation product from acetyl precursor. Reaction of C378G β-ketothiolase (1 mM) with [1,2-^{13}C]acetyl-CoA (2 mM) in $H_2^{18}O$ versus $H_2^{16}O$ resulted in the ^{13}C NMR spectra shown in Fig. 2D. For the resonance corresponding to the C-1 thioester carbon (184.3 ppm), the substitution of $H_2^{18}O$ for $H_2^{16}O$ resulted in an upfield shift of 0.004 ± 0.001 ppm, measured after 10 hr of signal accumulation (Table I). The magnitude of the shift did not increase upon further signal accumulation. The substitution of $H_2^{18}O$ for $H_2^{16}O$ had a similarly minimal effect on the resonance corresponding to the C2 methyl carbon (26 ppm; Fig. 2D). After 10 hr of signal accumulation, the observed upfield shift was 0.006 ± 0.001 ppm (Table I) and remained unchanged on further signal accumulation. The observed upfield shift for the C2 methyl carbon reflects experimental error.

The results for β-ketothiolase, which does not catalyze thioester hydrolysis during its normal catalytic cycle, contrast strikingly with the ^{18}O-induced shift observed for the C-1 carbon of HMG-CoA synthase's acetyl-S-enzyme reaction intermediate. Access of ^{18}O enriched solvent to the active site seems assured when lyophilized protein is dissolved in $H_2^{18}O$, as in the case of the experiments outlined above. Thus, the negative observations for β-ketothiolase cannot be explained by limited solvent access. In experiments where concentrated aqueous protein samples are diluted or buffer exchanged (e.g., by centrifugal gel filtration) to introduce $H_2^{18}O$, equilibrium exchange is usually assumed. Although this may normally be a good assumption, there are instances when a preformed enzyme–substrate/analog complex or reaction intermediate is so stable that equilibrium solvent exchange is not assured. Figure 3 illustrates a test of the accessibility of the active site of HMG-CoA synthase to a solvent-borne paramagnetic perturbant, Mn^{2+}. In aqueous solution, this cation causes paramagnetic broadening of the thioester carbonyl signal of [^{13}C]acetyl-CoA to undetectable levels and visibly broadens the C2 methyl group signal (Fig. 3A versus 3B). In contrast, when mixed with a preformed [^{13}C]acetyl-S-enzyme sample which contains excess [^{13}C]acetyl-CoA, only the unbound species exhibits paramagnetic broadening (Fig. 3C–3F). No substantial signal diminution of the reaction intermediate's acetyl group is observed, even after a 5 hr incubation. In such a case, the ability of isotopically enriched solvent to access the ^{13}C-labeled probe should not be assumed.

Investigation of Ability of Acylation-Impaired HMG-CoA Synthase Mutants to Produce ^{18}O-Induced Isotope Shifts

Chun et al.[11] have demonstrated that elimination of carboxyl side chains at D99, D159, and D203 results in diminutions in k_{cat} of 2, 3, and 4 orders of magnitude, respectively. A mutation at any of these sites markedly impairs the rate at which the acetyl-S–enzyme reaction intermediate forms. These slow acetylation

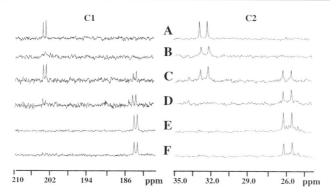

FIG. 3. Effect of Mn^{2+} on ^{13}C resonances of free $[^{13}C]$acetyl-CoA and $[^{13}C]$acetyl-S-enzyme reaction intermediate. Samples contained: (A) 3 mM [1,2-^{13}C]acetyl-CoA in buffer; (B) 3 mM [1,2-^{13}C]acetyl-CoA in buffer in the presence of 0.5 mM Mn^{2+}; (C) HMG-CoA synthase (1 mM) with [1,2-^{13}C]acetyl-CoA (3 mM) (spectra were acquired after 1 hr of signal accumulation); (D) HMG-CoA synthase (1 mM) with [1,2-^{13}C]acetyl-CoA (3 mM) in the presence of 0.5 mM Mn^{2+} (spectra were acquired after 1 hr of signal accumulation); (E) HMG-CoA synthase (1 mM) with [1,2-^{13}C]acetyl-CoA (3 mM) (spectra were acquired after 3 hr of continuous signal accumulation); (F) HMG-CoA synthase (1 mM) with [1,2-^{13}C]acetyl-CoA (3 mM) in the presence of 0.5 mM Mn^{2+} (spectra were acquired after 5 hr of continuous signal accumulation). Reprinted with permission from D. A. Vinarov and H. M. Miziorko, *Biochemistry* **39**, 3360 (2000). Copyright © 2000 American Chemical Society.

mutants were, therefore, tested to determine whether the functions that D99, D159, or D203 support in formation of acetyl-S-enzyme or other steps in the overall reaction were also important to formation/collapse of a diol-containing tetrahedral adduct, which should (if performed in $H_2{}^{18}O$) result in a detectable isotope shift. The active site integrity of mutants D99A, D159A, and D203A is validated by their formation of [^{13}C]acetyl-S-enzyme intermediates which exhibit substantial upfield NMR shifts for the C-1 and C-2 acetyl resonances that match those observed for wild-type enzyme (Fig. 4). Although any *precise* measurement of the kinetics of exchange is precluded by the extended time required to acquire a ^{13}C NMR spectrum, Fig. 4 demonstrates that this approach does indeed discriminate between the functions of these residues. The ability of wild-type HMG-CoA synthase's acetyl-S-enzyme to reversibly form/collapse a tetrahedral intermediate is demonstrated by reconstituting freeze-dried enzyme in $H_2{}^{18}O$ versus $H_2{}^{16}O$ and adding [1,2-^{13}C]acetyl-CoA, appropriately dissolved in $H_2{}^{18}O$ or $H_2{}^{16}O$ (Fig. 4A). As indicated above, use of doubly labeled [1,2-^{13}C]acetyl-CoA affords an internal control, since the 26 ppm C-2 methyl peak can be observed in $H_2{}^{18}O$ and $H_2{}^{16}O$ experiments to verify that no significant upfield shifts are observed for this carbon, which contains no C–O bond. For the resonance corresponding to the C-1 thioester carbon of acetyl-S-enzyme (184.3 ppm), the substitution of $H_2{}^{18}O$ for $H_2{}^{16}O$ resulted in an upfield shift of 0.055 ± 0.005 ppm, as measured after 3 hr of

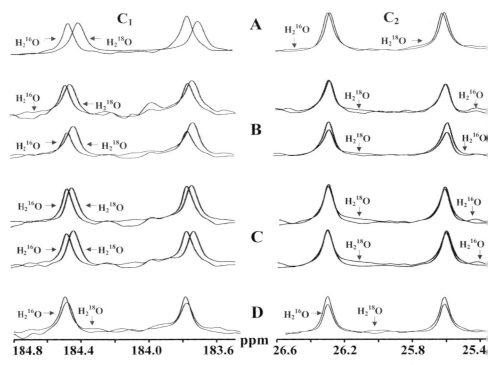

FIG. 4. ^{18}O-induced shift of the ^{13}C NMR signal due to the thioester carbonyl of the [^{13}C]acetyl-S–enzyme reaction intermediate. Samples contained: (A) wild-type enzyme (1 mM) with [1,2-^{13}C]acetyl-CoA (2 mM) in $H_2^{18}O$ versus $H_2^{16}O$; (B) D159A (1 mM) with [1,2-^{13}C]acetyl-CoA (2 mM) in $H_2^{18}O$ versus $H_2^{16}O$; (C) D203A (1 mM) with [1,2-^{13}C]acetyl-CoA (2 mM) in $H_2^{18}O$ versus $H_2^{16}O$; (D) D99A (1 mM) with [1,2-^{13}C]acetyl-CoA (2 mM) in $H_2^{18}O$ versus $H_2^{16}O$. Spectra on left-hand side depict signals due to the C-1 thioester carbonyl; spectra on right-hand side depict signals due to the C-2 methyl group. The spectrum for sample A was measured after 3 hr. The spectra for samples B and C were measured after 5 hr (upper trace) and 10 hr (lower trace). The spectrum for sample D was measured after 10 hr. Reprinted with permission from K. Y. Chun, D. A. Vinarov, and H. M. Miziorko, *Biochemistry* **39,** 14670 (2000). Copyright © 2000 American Chemical Society.

signal accumulation. The magnitude of the shift did not increase on further signal accumulation. In contrast to the observations on C-1, the substitution of $H_2^{18}O$ for $H_2^{16}O$ had very little effect on the resonance corresponding to the C-2 methyl carbon (26 ppm), as expected. After 3 hr of signal accumulation the observed upfield shift was 0.009 ± 0.002 ppm (Fig. 4A), a value which remained unchanged upon further signal accumulation and reflects experimental error. Reaction of D159A and D203A HMG-CoA synthases ([enzyme sites] = 1 mM in each sample) with [1,2-^{13}C] acetyl-CoA (2 mM) in $H_2^{18}O$ versus $H_2^{16}O$ resulted in the ^{13}C NMR spectra shown in Figs. 4B and 4C. For the resonance corresponding to the C-1

TABLE II
^{18}O-INDUCED ISOTOPE SHIFTS OF ^{13}C RESONANCES OF AC-COA, ACETYL-S-WILD-TYPE, ACETYL-S-D159A, ACETYL-S-D203A, AND ACETYL-S-D99A HMG-COA SYNTHASE REACTION INTERMEDIATES[a]

Sample	Linewidth (Hz)		$\Delta[\delta_{(H2\ 16O)} - \delta_{(H2\ 18O)}]$ (ppm)
Ac-CoA in buffer	C-1	7.9	0.005 ± 0.001[b]
	C-2	6.1	0.006 ± 0.001
Acetyl-S-WT synthase	C-1	17.0	0.055 ± 0.005
	C-2	15.8	0.009 ± 0.002
Acetyl-S-D159A synthase	C-1	16.9	$0.027 \pm 0.004 \Rightarrow 0.040 \pm 0.004$[c]
	C-2	16.2	$0.003 \pm 0.001 \Rightarrow 0.003 \pm 0.001$[c]
Acetyl-S-D203A synthase	C-1	16.9	$0.020 \pm 0.003 \Rightarrow 0.035 \pm 0.003$[c]
	C-2	16.6	$0.003 \pm 0.001 \Rightarrow 0.003 \pm 0.001$[c]
Acetyl-S-D99A synthase	C-1	16.2	0.010 ± 0.002
	C-2	16.4	0.002 ± 0.001

[a] Reprinted with permission from K. Y. Chun, D. A. Vinarov, and H. M. Miziorko, *Biochemistry* **39**, 14670 (2000). Copyright © 2000 American Chemical Society.
[b] Standard error for differential isotope shift values was calculated from two independently exchanged and consecutively run samples.
[c] For D159A and D203A enzymes, isotope shifts measured after 5 and 10 hr are listed. Measurement reported for D99A enzyme was determined after a 10 hr incubation of protein with acetyl-CoA.

thioester carbonyl (184.3 ppm) of acetyl-S-D159A and acetyl-S-D203A, the substitution of $H_2^{18}O$ for $H_2^{16}O$ resulted in upfield shifts of 0.027 ± 0.004 ppm and 0.020 ± 0.003, respectively, measured after 5 hr of signal accumulation. The magnitude of the shifts increased to 0.040 ± 0.004 and 0.035 ± 0.003, respectively, after 10 hr of signal accumulation (Table II) and remained unchanged on further signal accumulation. The substitution of $H_2^{18}O$ for $H_2^{16}O$ had a minimal effect on the resonances corresponding to the C-2 methyl carbon (26 ppm) of acetyl-S-enzyme species produced using D159A and D203A. After 5 and 10 hr of signal accumulation, both observed upfield shifts were 0.003 ± 0.001 ppm; these values remained unchanged on further signal accumulation and reflect the magnitude of experimental error in these experiments.

Reaction of D99A HMG-CoA synthase (1 mM) with [1,2-^{13}C] acetyl-CoA (2 mM) in $H_2^{18}O$ versus $H_2^{16}O$ resulted in the ^{13}C NMR spectra shown in Fig. 4D. In contrast with the results obtained using wild-type, D159A, and D203A enzymes, the substitution of $H_2^{18}O$ for $H_2^{16}O$ resulted in a minimal upfield shift of 0.010 ± 0.002 ppm for the resonance corresponding to the C-1 thioester carbonyl (184.3 ppm) of acetyl-S-D99A, measured after 10 hr of signal accumulation. The magnitude of the shift (Table II) remained unchanged on further signal accumulation. The substitution of $H_2^{18}O$ for $H_2^{16}O$ had a negligible effect on the resonance corresponding to the C-2 methyl carbon (26 ppm) of acetyl-S-D99A. After 10 hr

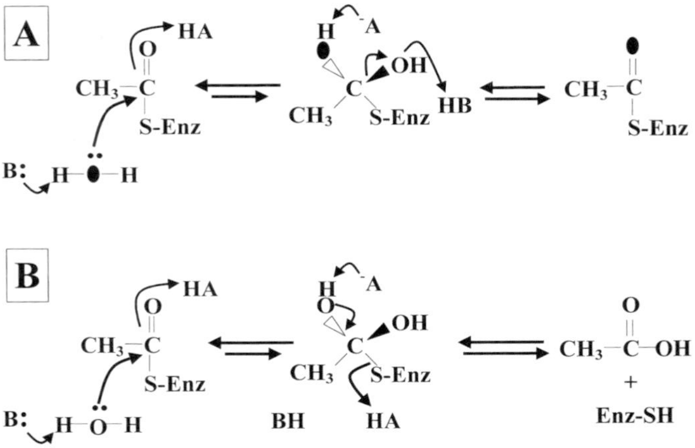

FIG. 5. Mechanism for ^{18}O exchange (A) and hydrolysis (B) of the acetyl-S-enzyme reaction intermediate of HMG-CoA synthase. Reprinted with permission from D. A. Vinarov and H. M. Miziorko, *Biochemistry* **39**, 3360 (2000). Copyright © 2000 American Chemical Society.

of signal accumulation, the observed upfield shift was 0.002 ± 0.001 ppm, a value which remained unchanged on further signal accumulation. Although these three mutants vary in kinetics of ^{18}O exchange, they all exhibit reduced (or negligible) exchange rates, as expected if they are impaired in the acetyl-S-enzyme formation step which is a prerequisite to hydration and formation of the diol-containing tetrahedral adduct (Fig. 5) required to account for the observed solvent exchange reaction.

Interpretation of Isotope Shift Data and Evaluation of Utility of Differential Isotope Shift Approach

The issue of whether an enzymatic reaction proceeds through a tetrahedral carbon intermediate can, in principle, be addressed by both deuterium and ^{18}O shift approaches. As indicated in Fig. 5, the addition of water across the thioester carbonyl of an acyl-S–enzyme intermediate produces (albeit transiently, in most cases) a diol intermediate. If the reaction is performed in D$_2$O, a deuterium shift of the ^{13}C signal should be observed. Indeed, an analogous experiment on a ^{13}C-labeled pepstatin–pepsin adduct[8] showed a convincing upfield shift of a 99 ppm ^{13}C NMR peak attributed to a diol-containing tetrahedral adduct. In the case of HMG-CoA synthase, the acetyl C-1 resonance shifts from 204 ppm (for acetyl-CoA) to 184 ppm for the acetyl-enzyme reaction intermediate, suggesting that the carbon retains substantial carbonyl (rather than thiohemiacetal) character. Nonetheless, Vinarov and Miziorko[10] showed that preparation of the reaction intermediate in D$_2$O results

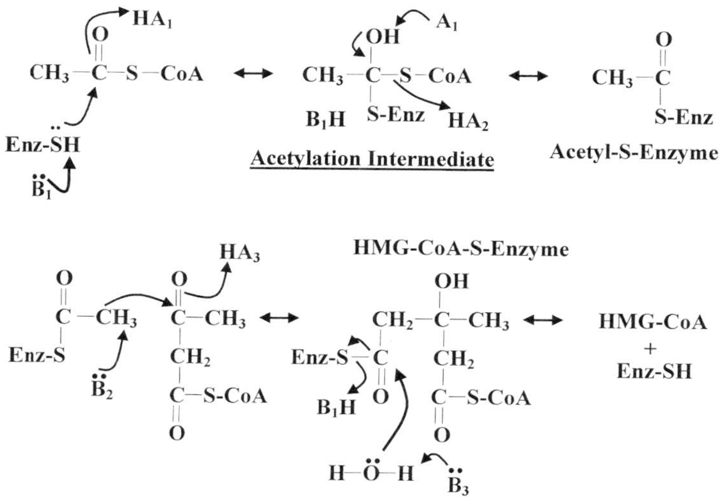

FIG. 6. The mechanism for formation of HMG-CoA synthase's acetyl-S-enzyme reaction intermediate and its subsequent condensation with acetoacetyl-CoA to transiently form an enzyme-S-HMG-S-CoA species prior to hydrolytic release of product. Reprinted with permission from K. Y. Chun, D. A. Vinarov, and H. M. Miziorko, *Biochemistry* **39,** 14670 (2000). Copyright © 2000 American Chemical Society.

in a significant time-dependent upfield shift. This is presumably attributable to hydrogen bonding that involves solvent or exchangeable amino acid side chains. Thus, in this situation, the deuterium-based differential isotope shift approach does not cleanly settle the issue of the existence of a tetrahedral adduct. Fortunately, the ^{18}O isotope shift approach cleanly resolves the issue.

In contrast to the dissipation of the acetyl-S-enzyme reaction intermediate of HMG-CoA synthase by the abortive hydrolysis partial reaction (Fig. 5B), formation of the acetyl-S-enzyme involves no solvent oxygen (Fig. 6). Thus, strictly speaking, involvement of a tetrahedral adduct has only been demonstrated for acetyl group hydrolysis (Fig. 5). Nonetheless, the homology between hydrolysis and the terminal reaction step that releases product HMG-CoA is clear (Fig. 6). Although the formation of a tetrahedral adduct during production of acetyl-S-enzyme remains to be directly demonstrated, the hypothesis can be tested by extension of the isotope shift approach to selected HMG-CoA synthase mutants. The diminution in catalytic efficiency of D99A, D159A, and D203A HMG-CoA synthase mutants varies considerably, reflecting the different and potentially multiple roles for these residues. However, each of these three variants exhibits retarded kinetics of acetyl-S-enzyme formation. Thus, the hypothesis that formation of a tetrahedral adduct occurs as a necessary prelude to production of the acetyl-S-enzyme reaction intermediate is convincingly supported by the observation that substantial differential ^{18}O shifts of the acetyl-S-enzyme's C-1 thioester carbonyl resonance do occur for D159 and D203, but only after extended incubation times (Table II). The more exaggerated deficiency of D99A in catalysis of solvent oxygen exchange (Fig. 4; Table II) is interesting and invites speculation concerning the function of this side chain. If acyl transfer were to occur via a mechanism which requires attacking and leaving groups to be positioned on opposite faces of the carbonyl carbon, the related process of solvent exchange into acetyl-S-enzyme would have to accommodate the same spatial requirements. Failure of D99A to support solvent exchange (or efficient acetyl-S-enzyme formation) may simply derive from a perturbation of the precise active site positioning that is required. Since the diminution in k_{cat} is modest (10^2-fold), an indirect role for this residue in positioning either reactants or active site residues more directly involved in reaction chemistry merits consideration.

A series of kinetic isotope effect experiments by Hengge and Hess[19] focused initially on uncatalyzed acyl transfer reactions but were later extended by Hess et al.[20] to enzymatic acyl transfers. The results were interpreted as arguing for a concerted process in which any tetrahedral species might exist in the transition state but not as a stable intermediate. Although the isotope effects are clearly demonstrable, the interpretation of those data in the context of stepwise (stable tetrahedral intermediate) versus concerted mechanisms relies on assumptions concerning intermediate partitioning. In contrast, the ^{18}O isotope shift approach seems more direct. The isotopic shift reported for a pepstatin–pepsin adduct by Schmidt et al.,[8] as well as the chemical shift documented for that adduct, argues strongly for participation of a tetrahedral species in that reaction. However, since the pepsin–inhibitor complex cannot turn over, it remained unclear whether this species mimics an intermediate or the short-lived transition state. Thus, the pepsin study does not resolve the issue raised by the kinetic isotope effect approach. In the case of HMG-CoA synthase, the evidence presented by Miziorko and Lane[15] indicates that the acetyl-S-enzyme species is a stable, bona fide reaction intermediate. Chun et al.[11] showed that the formation of this species represents the rate-limiting step in the reaction. No large steady-state population of a diol-containing intermediate is detectable (indeed, the 184 ppm ^{13}C NMR peak argues for the carbonyl nature of acetyl C-1). The marked isotope shift (0.055 ppm) which is measured is clearly indicative of a two-bond effect. Such an effect could only be explained by *repetitive* addition of solvent water and collapse to reform the dominant acetyl-enzyme species. Although the different experimental models employed in the kinetic isotope effect studies on oxoester cleavage and our differential ^{18}O isotope shift studies on acyl-S-enzymes make precise comparisons difficult, it seems that, in certain reactions, tetrahedral intermediates do not invariably partition forward to

---

[19] A. C. Hengge and R. A. Hess, *J. Am. Chem. Soc.* **116**, 11256 (1994).
[20] R. A. Hess, A. C. Hengge, and W. W. Cleland, *J. Am. Chem. Soc.* **120**, 2703 (1998).

product. Undoubtedly, kinetic isotope effect approaches remain valuable experimental tools. However, the directness of the differential isotope shift approach and the straightforward interpretation of the experimental data qualify this technique as an attractive and useful mechanistic tool.

Acknowledgments

Studies on HMG-CoA synthase in the authors' laboratory have been supported by NIH DK-21491. D.A.V. was the recipient of a postdoctoral fellowship from the American Heart Association. Dr. Kelly Chun provided the acylation-deficient mutant forms of HMG-CoA synthase used for some of the ^{18}O-induced shift experiments described in this account.

[17] Detection of Intermediates in Reactions Catalyzed by PLP-Dependent Enzymes: O-Acetylserine Sulfhydrylase and Serine–Glyoxalate Aminotransferase

By WILLIAM E. KARSTEN and PAUL F. COOK

Background

A prerequisite to the detection of an intermediate in an enzyme-catalyzed reaction rests in one's ability to identify the intermediate via its physical and chemical properties. There is an extensive literature[1] on the spectral properties of intermediates in pyridoxal 5'-phosphate (PLP) dependent enzyme-catalyzed reactions that greatly simplify the task of intermediate detection. A brief overview of the general properties of PLP-dependent enzyme reactions in general, with respect to common intermediates along the reaction pathway, is informative.

All PLP-dependent enzymes catalyze their reaction via the initial formation of an external Schiff base intermediate as shown in Fig. 1. The internal and external Schiff bases exist in several different tautomeric forms. The two predominant forms, the enolimine and ketoenamine tautomers, are pictured at the top of the figure, along with ranges of their respective λ_{max} values. The ketoenamine form alone is shown in the case of the external Schiff base. The external Schiff base is formed via the intermediacy of *geminal*-diamine intermediates, two of which are shown in Fig. 1. Once the external Schiff base is formed, the reaction pathway is determined by the kind of reaction catalyzed, and this is determined by the identity of the

---

[1] Y. V. Morozov, in "Vitamin B$_6$ Pyridoxal Phosphate: Chemical, Biochemical, and Medical Aspects, Part A" (D. Dolphin, R. Poulson, and O. Avramovic, eds.), p. 131. John Wiley & Sons, New York, 1986.

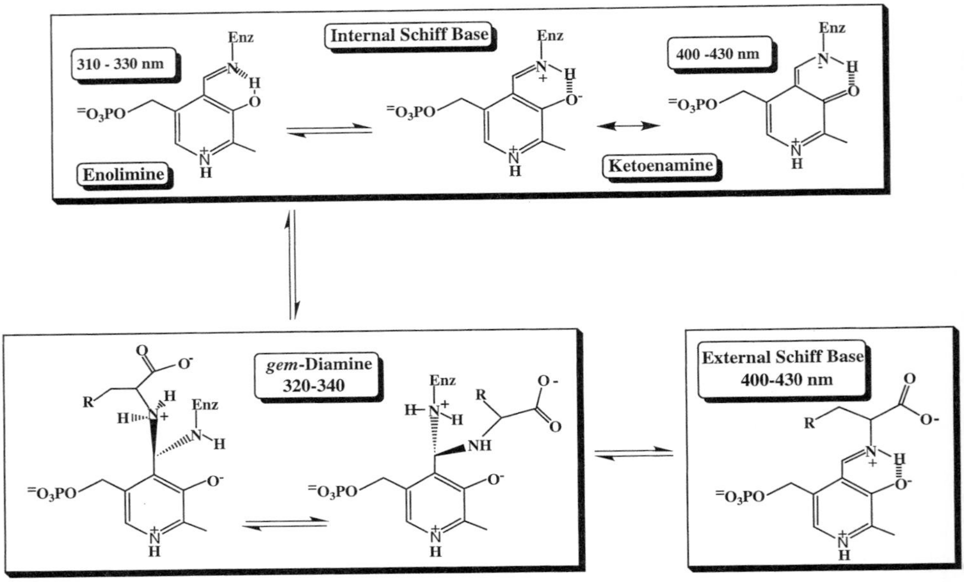

FIG. 1. Species involved in transimination to generate the external Schiff base. An amino acid is added to the internal Schiff base to generate the *geminal*-diamine intermediates. The most commonly observed tautomers and resonance forms are shown. Wavelength ranges are provided for the intermediates shown.

functional group on the α-carbon that is orthogonal to the PLP ring. In the case of the reaction types considered in this article, β-replacement and aminotransferase, the α-proton will be orthogonal to the ring for both reaction types.

Additional intermediates that would be expected in the case of a β-replacement reaction are shown in Fig. 2. Abstraction of the α-proton can generate a quinonoid intermediate, which can then eliminate the functional group in the β-position to generate an α-aminoacrylate Schiff base. The second half-reaction is then the reverse of the first half of the reaction, i.e., it would likely proceed via the same intermediates. Thus, all of the intermediates can be distinguished via their UV–visible spectral properties.

In the case of an aminotransferase reaction, a quinonoid intermediate is also formed, followed by protonation at C-4' to generate a ketimine intermediate (Fig. 3). Hydrolysis of the ketimine intermediate proceeds via a carbinolamine intermediate, which collapses to give pyridoxamine 5'-phosphate (PMP). The second half-reaction is then the reverse of the first half of the reaction, i.e., it would likely proceed via the same intermediates. Unlike the case with the β-replacement

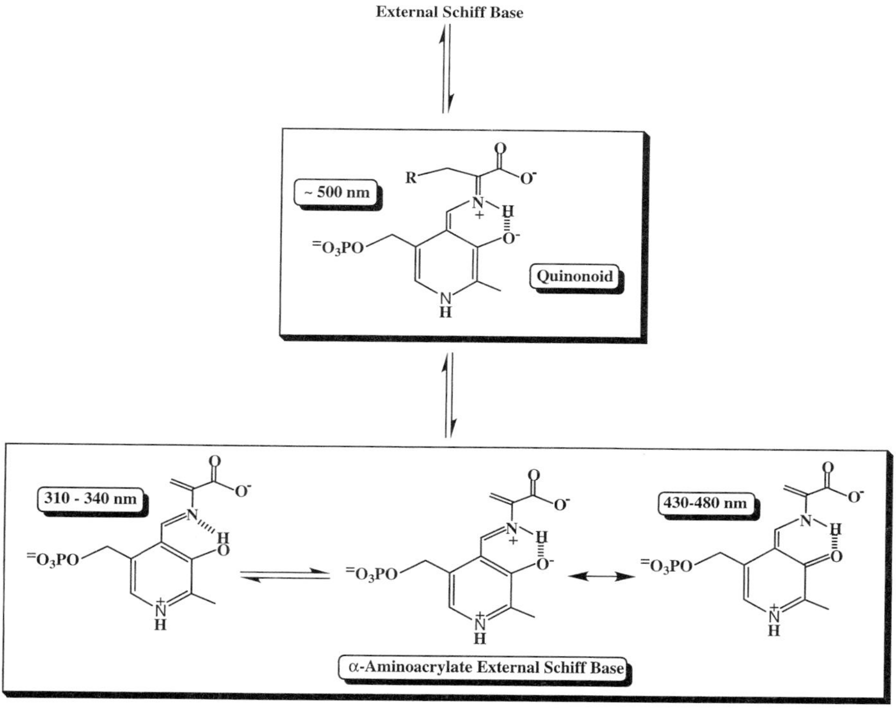

FIG. 2. Species that may be involved in formation of the α-aminoacrylate intermediate from the external Schiff base in a β-substitution reaction.

reaction, several of the intermediates absorb in the 325–340 nm range, but a number can still be distinguished from one another based on their UV–visible spectral properties.

Enzymes

O-Acetylserine sulfhydrylase (Cysteine synthase, EC 4.2.99.8; OASS) catalyzes the final step in the *de novo* synthesis of L-cysteine in enteric bacteria, i.e., the conversion of O-acetyl-L-serine (OAS) and bisulfide to L-cysteine and acetate (Fig. 4).[2] The *cysK* gene, encoding for OASS-A, has been cloned and sequenced from *Salmonella typhimurium*, and the enzyme overexpressed using the wild-type

[2] M. A. Becker, N. M. Kredich, and G. M. Tomkins, *J. Biol. Chem.* **244**, 2418 (1969).

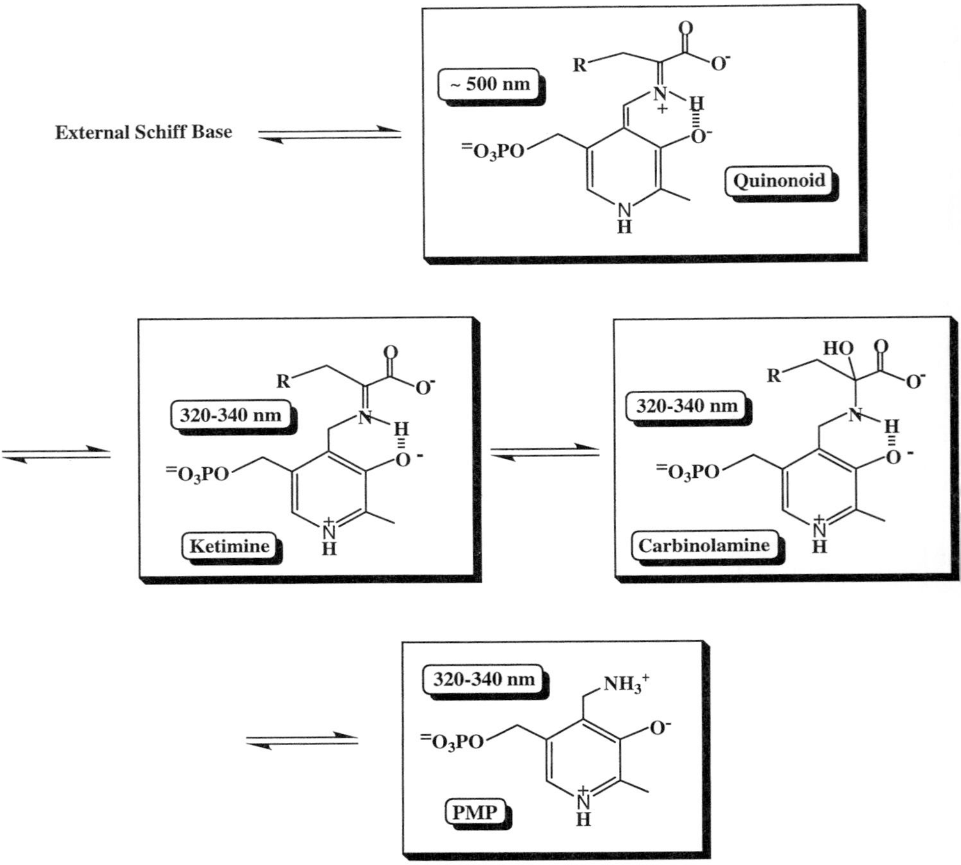

FIG. 3. Species involved in formation of the pyridoxamine 5'-phosphate intermediate in an aminotransferase reaction.

cysK promoter.[3] There are two isozymes of OASS, A and B, in enteric bacteria (those from *Salmonella typhimurium* are best studied), and of these the A isozyme has been the best studied. The A and B isozymes are thought to be expressed under aerobic and anaerobic conditions, respectively.[4] O-Acetylserine sulfhydrylase A is a homodimeric enzyme with a molecular weight of 68,900[5] with 1 mol of

[3] N. M. Kredich, in "*Escherichia coli* and *Salmonella*" (F. C. Neidhardt, ed.), 2nd Ed., p. 514. ASM Press, Washington, DC, 1996.
[4] T. Nakamura, H. Kon, H. Iwahashi, and Y. Eguchi, *J. Bacteriol.* **156**, 656 (1983).
[5] C. R. Byrne, R. S. Monroe, K. A. Ward, and N. M. Kredich, *J. Bacteriol.* **190**, 3150 (1988).

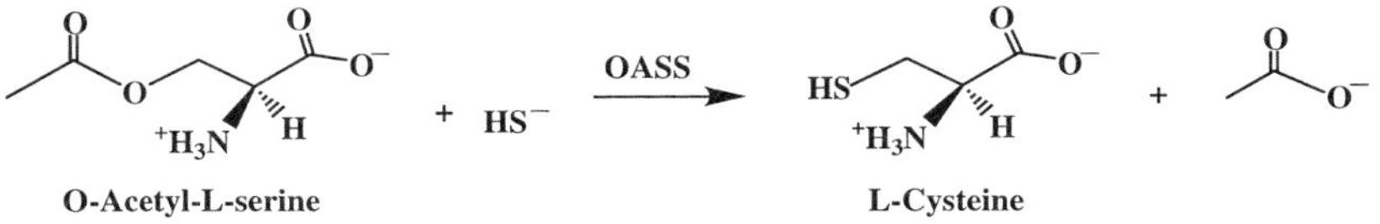

FIG. 4. Reaction catalyzed by *O*-acetylserine sulfhydrylase.

PLP bound in Schiff base linkage to lysine-41 in each of the monomers.[6–8] A rapid purification procedure has been developed for OASS-A.[9] Overnight fermentation in 12- to 15-liter batches of minimal medium gives 0.75–1 g of enzyme that is >95% pure. The OASS catalyzes its reaction via a ping-pong Bi Bi kinetic mechanism.[10,11]

Serine–glyoxalate aminotransferase (EC 2.6.1.45; SGAT) catalyzes the interconversion of L-serine and glyoxalate to glycine and hydroxypuruvate in the methylotrophic bacterium *Hyphomicrobium methylovorum* GM2 (Fig. 5). The formation of hydroxypyruvate is a precursor to phosphoenolpyruvate and is part of the C-1 assimilation pathway in methylotrophs.[12] The gene encoding SGAT has been sequenced and cloned into the expression vector pKK223-3.[13] Serine–glyoxalate aminotransferase is a homotetrameric enzyme with a molecular weight of 175,520. Based on multiple sequence alignments,[14] the active site PLP is covalently bound in Schiff base linkage to lysine-196 in each of the subunits. A purification procedure has been developed for the overexpressing strain.[15,16] Overnight growth in 12- to 15-liter batches at 34° in LB medium supplemented with 1% glucose gives 50–70 mg of >90% pure enzyme. The enzyme catalyzes its reaction via a ping-pong Bi Bi kinetic mechanism.[16]

Because both OASS and SGAT catalyze their reaction via a ping-pong mechanism, some intermediates can be detected via simple equilibrium spectral studies.

[6] N. M. Kredich and G. M. Tomkins, *J. Biol. Chem.* **241**, 4955 (1966).
[7] V. Rege, C.-H. Tai, N. M. Kredich, W. E. Karsten, K. D. Schnackerz, and P. F. Cook, *Biochemistry* **35**, 13485 (1996).
[8] P. Burkhard, G. S. J. Rao, E. Hohenester, P. F. Cook, and J. N. Jansonius, *J. Mol. Biol.* **283**, 111 (1998).
[9] S. Hara, M. A. Payne, K. D. Schnackerz, and P. F. Cook, *Protein Expression Purif.* **1**, 70 (1990).
[10] P. F. Cook and R. T. Wedding, *J. Biol. Chem.* **251**, 2023 (1976).
[11] C.-H. Tai, S. R. Nalabolu, T. M. Jacobson, D. E. Minter, and P. F. Cook, *Biochemistry* **32**, 6433 (1993).
[12] Y. Izumi, T. Yoshida, T. Hagishita, Y. Tanaka, T. Mitsunaga, T. Ohshiro, T. Tanabe, H. Miyata, C. Yokoyama, J. D. Goldberg, and D. Brick, in "Microbial Growth on C1 Compounds" (M. E. Lidstrom and F. R. Tabita, eds.), p. 25. Kluwer Academic, Dordrecht, the Netherlands, 1996.
[13] T. Hagishita, T. Yoshida, Y. Izumi, and T. Mitsunaga, *Eur. J. Biochem.* **241**, 1 (1996).
[14] P. K. Mehta, T. I. Hale, and P. Christen, *Eur. J. Biochem.* **214**, 549 (1993).
[15] Y. Izumi, T. Yoshida, and H. Yamada, *Agric. Biol. Chem.* **54**, 1573 (1990).
[16] W. E. Karsten, T. Ohshiro, Y. Izumi, and P. F. Cook, *Arch. Biochem. Biophys.* **388**, 267 (2001).

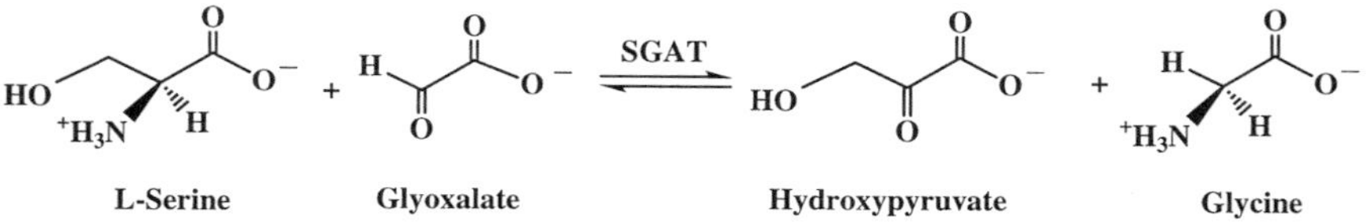

FIG. 5. Reaction catalyzed by serine–glyoxalate aminotransferase.

Minimally, the F form of the enzyme that is produced as a result of the first half-reaction can be observed on addition of the first substrate, OAS in the case of OASS, and L-serine in the case of SGAT.

Steady-State UV–Visible Spectroscopy

Experimental

Equilibrium spectral studies can be carried out using any scanning UV-visible spectophotometer, but are most easily measured using a diode array spectrophotometer. The diode array spectrophotometer has the advantage of collecting data at all wavelengths from 200 to 800 nm in less than 1 sec. The only caveat to using this instrument is that it utilizes a relatively high intensity xenon lamp to generate a beam of white light. If one anticipates photosensitive intermediates, a more conventional dual-beam spectrophotometer should be used. In most cases an HP-8452a or HP-8453 diode array spectrophotometer (Hewlett-Packard Co., Palo Alto, CA) is used to obtain spectral information for OASS and SGAT. A scanning instrument can be used successfully for steady-state experiments providing the spectral changes being observed are fast relative to the rate at which data can be collected. One can test for the rapidity of the changes measured by collecting spectra as a function of time. Buffers or substrates that absorb in the wavelength range being monitored for enzyme spectral changes should be avoided if possible.

Spectra for both enzymes are measured using cuvettes 1 cm in path length with a 1 ml total volume containing 100 mM HEPES, pH 7. A reference scan of absorbance versus wavelength is collected first and subtracted from all subsequent scans with enzyme and reactant(s). Spectra are measured using a final concentration of 30 μM and 45 μM active sites for OASS and SGAT, respectively. For spectra taken in the presence of substrates, stock substrate solutions are used so that the volume of substrate solution added to the cuvette is minimal (2 μl), and thus volume correction is negligible. This method is preferable to obtaining separate spectra for each reactant concentration. For spectral pH titrations the buffers used are 100 mM MES, HEPES, and CHES adjusted to the desired pH with 10 M KOH. Sample blanks of buffer alone are taken before addition of enzyme at each pH value. After the spectrum is recorded the pH of the cuvette solution is measured using an Accumet Basic pH meter (Fisher Scientific, Houston, TX).

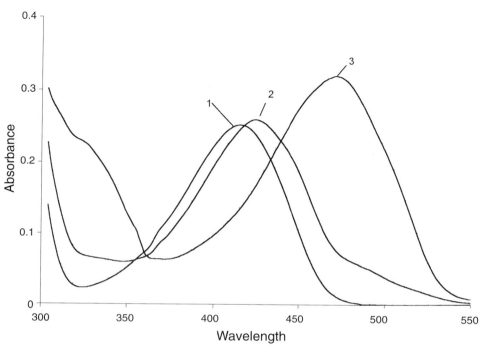

FIG. 6. Spectra of OASS in the absence and presence of OAS and L-cysteine. Enzyme is present at a concentration of 30 μM active sites. (1) OASS alone at pH 6.5, 100 mM MES. (2) Enzyme plus 1 mM OAS, pH 6.5, 100 mM MES. (3) Enzyme plus 1 mM L-cysteine, pH 9.5, 100 mM CHES.

O-Acetylserine Sulfhydrylase

The spectrum of OASS in the absence of reactants exhibits a visible λ_{max} of 412 nm indicating the ketoenamine tautomer of the internal Schiff base predominates (Fig. 6).[17] The spectrum is pH independent over the pH range from pH 5.5 to 10.85. Addition of 1 mM OAS (pH 7) to a solution containing OASS results in a decrease in the absorbance at 412 nm and the appearance of two new bands at 330 and 470 nm. The new bands reflect the enolimine (330 nm) and ketoenamine (470 nm) tautomers of the α-aminoacrylate external Schiff base (Fig. 6).[17,18] The α-aminoacrylate intermediate is stable for hours and can be studied in a manner similar to the free enzyme.

---

[17] P. F. Cook, S. Hara, S. Nalabolu, and K. D. Schnackerz, *Biochemistry* **31,** 2298 (1992).
[18] K. D. Schnackerz, C.-H. Tai, J. W. Simmons III, T. M. Jacobson, G. S. J. Rao, and P. F. Cook, *Biochemistry* **34,** 12152 (1995).

Substrate analogs can sometimes be used to generate intermediates that would occur along the enzyme's reaction pathway. Addition of the product, L-cysteine, at a concentration of 1 mM (pH 9.0) to a solution containing OASS results in a slight red shift in the 412 nm absorbance to 418 nm, consistent with the formation of the ketoenamine tautomer of the external Schiff base, (Fig. 6).[18]

Thus, OASS in the absence of reactants exists as an internal Schiff base with PLP predominately as the ketoenamine tautomer. Formation of the external Schiff base results in a red shift in the spectrum to 418 nm and may, depending on the nature of the amino acid substrate, exist as a mixture of ketoenamine and enolimine tautomers. Formation of the α-aminoacrylate external Schiff base on elimination of acetate results in a mixture of the ketoenamine and enolimine tautomers (Figs. 1 and 2).

The above observed spectral changes can be utilized for a variety of different titrations. The changes observed on formation of the external Schiff base of L-cysteine or L-serine can be used to estimate their dissociation constants.[18] A plot of the change in absorbance versus L-cysteine concentration gives the K_d for the cysteine external Schiff base. A repeat of the experiment as a function of pH provides information on the optimum protonation state of cysteine and enzyme functional groups required for formation of the external Schiff base.[18] The pH-independent K_d is 120 μM, achieved under conditions where L-cysteine is monoanionic. The inability to demonstrate formation of an α-aminoacrylate external Schiff base suggests the overall OASS reaction is irreversible. Similar experiments have also been carried out with L-serine.

Formation of the α-aminoacrylate external Schiff base by addition of OAS to OASS is reversible, and the amount of the external Schiff base formed is thus dependent on the concentration of acetate. The equilibrium constant for the first half-reaction can thus be quantitated by recording spectra in the presence of a stoichiometric amount of OAS and increasing concentrations of acetate. At the concentration of acetate where half of the enzyme is converted to the α-aminoacrylate Schiff base, the concentrations of the internal and external Schiff bases will be equal and the two will be equal to the remaining OAS concentration. The equilibrium constant is thus calculated as the ratio of the acetate added to the OAS concentration remaining; K_{eq} is 15,000 at pH 7.[19] The equilibrium constant for the first half-reaction is pH dependent and should have a slope of 1 when plotted as log K_{eq} versus pH. This can be confirmed by repeating the above experiment at different pH values.

Serine–Glyoxalate Aminotransferase

In the absence of substrates SGAT exhibits two absorbance maxima at 330 and 413 nm corresponding to the enolimine and ketoenamine tautomers of the internal

---

[19] C.-H. Tai, S. R. Nalabolu, T. M. Jacobson, J. W. Simmons III, and P. F. Cook, *Biochemistry* **34**, 12311 (1995).

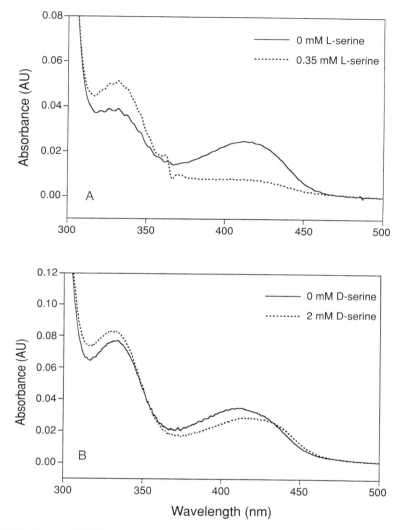

FIG. 7. Spectra of SGAT in the absence and presence of L-serine (A) and D-serine (B). Enzyme is present at a concentration of 45 μM active sites. Adapted from W. E. Karsten, T. Ohshiro, Y. Izumi, and P. F. Cook, *Arch. Biochem. Biophys.* **388**, 267 (2001).

Schiff base with the enolimine tautomer predominating (Fig. 7).[16] At high pH, the 413 nm peak decreases concomitant with a slight increase in the 330 nm peak. The 413 nm absorbance does not disappear at high pH nor is there a shift in the 330 nm absorbance to higher wavelength at high pH as would be expected if the pH-dependent changes in the absorbance spectra were the result of deprotonation of the internal Schiff base. The likely explanation is the deprotonation of a group with a pK of about 8.2 in or near the active site that shifts the tautomeric equilibrium in favor of the enolimine. The tautomeric equilibrium between the enolimine and ketoenamine tautomers is also shifted by the addition of the keto acid substrates hydroxypyruvate or glyoxalate, but the addition of the keto acid substrate induces a shift in favor of the ketoenamine tautomer.

Addition of L-serine to the internal Schiff base results in an increase in absorbance at 330 nm and a decrease at 413 nm indicating formation of the pyridoxamine enzyme (Fig. 3). If this experiment is carried out as a function of hydroxypyruvate, the equilibrium constant for the first half-reaction is estimated as 0.08; the K_{eq} is pH-independent. If a keto acid such as glyoxalate is added to the pyridoxamine form of the enzyme, the internal Schiff base form of the enzyme is regenerated. Addition of D-serine to the internal Schiff base results in a red shift in the λ_{max} from 413 nm to 421 nm, with no substantial change at 330 nm, indicative of formation of the external Schiff base (Fig. 7).

The above experiments provide information concerning intermediates that may build up in the steady state and allow characterization of the intermediates spectrally and with respect to their thermodynamics. Transient kinetic studies allow a detection of the buildup and decay of intermediates as the reaction occurs, and this aspect will be discussed below.

Stopped-Flow Absorbance and Fluorescence

Experimental

Experiments are generally carried out using an Olis RSM (On-Line Instrument Systems, Inc., Bogart, GA) in the rapid-scanning or single wavelength absorbance modes. Some of the earlier studies were carried out with a Durrum D-110 rapid-mixing stopped-flow (Dionex Corp., Sunnyvale, CA) and a Princeton Applied Research (Oak Ridge, TN) multichannel analyzer with a 1214 photodiode array detector.[20,21] The OLIS RSM in the scanning mode collects up to 1000 scans per second and has a dead time of about 2 ms. The stopped-flow contains two loading syringes and a stopping syringe. One of the loading syringes typically contains enzyme, while the other contains substrate [or other ligand(s)]. For both enzymes, OASS and SGAT, the wavelength range used is 300 to 520 nm. Once data have been collected, any wavelength can be selected, and the time course can be fitted using the appropriate rate expression and the software provided by Olis. The software also provides errors and goodness of fit parameters to aid in determining the kinetic model that best fits the data. In addition, global analysis software is provided with the instrument so that the complete spectral time course can be fitted to kinetic

---

[20] M. F. Dunn, S. A. Bernhard, D. Anderson, A. Copeland, R. G. Morris, and J. P. Roque, *Biochemistry* **29**, 2346 (1979).

[21] S. C. Koerber, A. K. H. MacGibbon, H. Dietrich, M. Zeppezauer, and M. F. Dunn, *Biochemistry* **22**, 3424 (1983).

models. The latter is, of course, greatly aided by the investigator's knowledge of the system.

Enzyme samples used in the stopped-flow experiments are first dialyzed against an appropriate buffer (HEPES or K_2HPO_4 at pH 7 for SGAT and MES at pH 6.5 for OASS) and adjusted to the desired pH. The pH is maintained lower in the case of OASS, because the substrate OAS is base labile. Typically a 2–3 mg/ml solution of SGAT or 700 μg/ml solution of OASS is loaded into one injection syringe and substrate at the same pH into the other. The substrate concentration in the injection syringes is twice the desired concentration of enzyme and substrate upon mixing. For experiments carried out in D_2O the enzyme is concentrated to less than one-tenth the original volume and resuspended in buffered D_2O to the original volume and the process is then repeated. Substrates are first dissolved in D_2O, and the pD is adjusted with KOD, and the solution is lyophilized, and finally redissolved in D_2O.

O-Acetylserine Sulfhydrylase

Mixing of OASS with 1 mM OAS at pH 6.5 gives a first-order appearance ($k \geq 1000$ s^{-1}) of the OAS external Schiff base with OAS based on the shift in λ_{max} from 412 to 418-nm (first few spectra in Fig. 8), followed by a first-order decay of the 418-nm species to the α-aminoacrylate external Schiff base absorbing maximally at 470 nm (Fig. 8).[22] The α-aminoacrylate intermediate can be preformed in one syringe by adding stoichiometric amounts of OAS and OASS. Mixing the α-aminoacrylate intermediate with 50 μM SH$^-$ gives the internal Schiff base spectrum in the dead time of the instrument; results are identical using 5 μM SH$^-$. Data suggest a diffusion-limited second half-reaction.

The first-order rate constant ($1/\tau$) for formation of the α-aminoacrylate intermediate is a hyperbolic function of OAS concentration. Data indicate a preequilibrium formation of the OAS external Schiff base ($K_{ESB} = 0.2$ mM^{-1}) followed by rate-limiting deprotonation of the α-carbon to generate the α-aminoacrylate external Schiff base (Fig. 2). The rate constant for the α-aminoacrylate external Schiff base formation is 400 s^{-1}. The rate constant for α-proton abstraction can be compared to a steady-state V/E_t value of 240 s^{-1}, suggesting that deprotonation of the α-carbon of OAS is also rate limiting for the overall synthesis of L-cysteine. To corroborate these data, the experiment was repeated with OAS-2-D giving a primary deuterium kinetic isotope effect of 2.4.

A quinonoid intermediate (Fig. 2) was not observed in the above experiments, nor would one be expected if it were an intermediate since it would occur after the rate-limiting α-proton abstraction. However, since protonation of the α-carbon should be rate limiting in the direction of OAS formation from the α-aminoacrylate intermediate, a quinonoid intermediate would be expected in the pre-steady state as

---

[22] E. U. Woehl, C.-H. Tai, M. F. Dunn, and P. F. Cook, *Biochemistry* **35**, 4776 (1996).

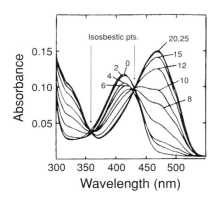

FIG. 8. Rapid-scanning stopped-flow data obtained on mixing 10 μM OASS active sites with 1 mM OAS at pH 6.5, 100 mM MES. Spectra shown are measured at the following times after mixing: 0; 2, 0.017 sec; 4, 0.034 sec; 6, 0.06 sec; 8, 0.084 sec; 10, 0.13 sec; 12, 0.23 sec; 15, 0.38 sec; 20, 0.85 sec; 25, 1.71 sec. Note the lack of single isosbestic points, corroborating the rapid buildup of the external Schiff base by spectrum 6, followed by a slow appearance of the α-aminoacrylate intermediate absorbing maximally at 470 nm. From E. U. Woehl, C.-H. Tai, M. F. Dunn, and P. F. Cook, *Biochemistry* **35**, 4776 (1996).

acetate is mixed with the α-aminoacrylate external Schiff base. When this experiment was carried out, no quinonoid intermediate was observed, consistent with the $anti$-E_2 mechanism proposed for formation of the α-aminoacrylate external Schiff base from OAS.[23]

The irreversibility of the second half-reaction, i.e., starting with L-cysteine and acetate, was always puzzling considering a number of other PLP-dependent enzymes including the closely related β subunit of tryptophan synthase catalyzed the synthesis of tryptophan from L-cysteine (in place of L-serine) and indole. Mixing L-cysteine with OASS in the stopped flow gave a rapid appearance of the L-cysteine external Schiff base absorbing at 418 nm, followed by a relatively rapid appearance of the α-aminoacrylate external Schiff base, absorbing maximally at 470 nm. The α-aminoacrylate external Schiff base then decayed to give a new species absorbing at 418 nm. Data were interpreted in terms of the thiol of a second molecule of L-cysteine attacking the α-aminoacrylate external Schiff base to generate the lanthionine external Schiff base.[22] To probe these changes more carefully, fluorescent properties of the enzyme were utilized as discussed below.

Excitation of OASS at 280 nm gives two emission bands, at 337 and 498 nm.[24] The 337-nm band reflects intrinsic tryptophan fluorescence, whereas the 498-nm band reflects triplet-to-singlet energy transfer from tryptophan to the internal Schiff

---

[23] C.-H. Tai and P. F. Cook, *Acc. Chem. Res.* **34**, 49 (2001).
[24] G. D. McClure and P. F. Cook, *Biochemistry* **33**, 1674 (1994).

base.[24-26] Formation of the lanthionine external Schiff base (addition of L-cysteine) gives a large enhancement in the intensity of the 498-nm fluorescence band, resulting from a change in the orientation of the PLP relative to tryptophan in the external versus internal Schiff base.[24-27] The fluorescence enhancement that occurs on formation of the external Schiff base was used as a more sensitive probe of the reaction of L-cysteine with OASS. The OASS was excited at 284 nm, and the fluorescence of the 498-nm band was monitored as an indicator of formation and decay of the external Schiff base with L-cysteine. Data clearly indicated a rapid ($\sim$1000 s^{-1}) formation of the cysteine external Schiff base followed by a slower decay of the external Schiff base to give the α-aminoacrylate external Schiff base. Fitting individual time courses using a model that allows for a rapid preequilibrium formation of the external Schiff base followed by a slow decay to give the α-aminoacrylate intermediate gives a K_{ESB} of 6.7 mM^{-1} and a first-order rate constant for intermediate formation of 110 s^{-1}.[22]

Serine–Glyoxalate Aminotransferase

The rapid-scanning stopped-flow experiments with SGAT have allowed detection of intermediates, and an estimate of rate constants and isotope effects for the two half-reactions catalyzed by the enzyme. Mixing enzyme with a saturating concentration of L-serine or L-serine-2-d leads to the first-order disappearance of the 413-nm peak reflecting the internal Schiff base, concomitant with an increase in the absorbance at 330 nm representing pyridoxamine phosphate. No other intermediates along the pathway of the first half-reaction are observed. A fit of the data at 413 or 330 nm using a single exponential kinetic model gives a rate constant of about 34 s^{-1}. The rate constant can be compared to a V/E_t value of 38 s^{-1}. The lack of an observable isotope effect indicates that abstraction of the α-proton to generate the quinonoid intermediate does not limit the overall rate of the first half-reaction. On the other hand, if the experiment is carried out in D$_2$O, the rate constant decreases to about 17 s^{-1}, giving a solvent deuterium isotope effect of about 2. Thus, a solvent-sensitive step(s) is(are) at least partially rate limiting in the first half-reaction; likely hydrolysis of the ketimine intermediate. The inability to detect the buildup of an intermediate in the first half-reaction is due to a low concentration of the intermediate, and/or to overlap between the spectral properties of the intermediate and the final product, the pyridoxamine form of the enzyme.

The pyridoxamine form of the enzyme can be generated by preincubation with L-serine in one syringe and then mixed with a keto acid substrate such as glyoxalate, hydroxypyruvate, or the alternative substrate ketomalonate. Rapid mixing with

---

[25] G. B. Strambini, P. Cioni, and P. F. Cook, *Biochemistry* **35**, 8392 (1996).
[26] S. Benci, S. Vaccari, A. Mozzarelli, and P. F. Cook, *Biochemistry* **36**, 15419 (1997).
[27] S. Benci, S. Bettati, S. Vaccari, G. Schianchi, A. Mozzarelli, and P. F. Cook, *Photochem. Photobiol.* **48**, 17 (1999).

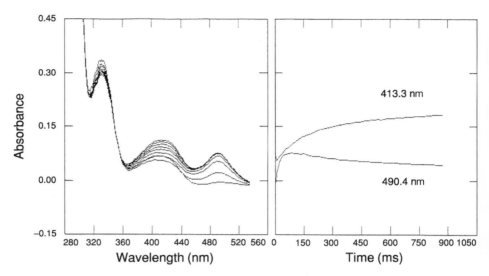

FIG. 9. Rapid-scanning stopped-flow data obtained on mixing 22.8 μM SGAT active sites in the pyridoxamine form in D_2O with ketomalonate at pD 7.5, 100 mM HEPES. Spectra shown are obtained every 16 ms after mixing. The bottom spectrum is obtained immediately after mixing and increasing time is from bottom to top. Note that equilibrium is established between the internal aldimine absorbing maximally at 413 nm and the quinonoid intermediate absorbing maximally at 490 nm. The panel on the right shows a time course for appearance and disappearance of the quinonoid intermediate at 490 nm and a time course for the appearance of the internal aldimine at 413 nm with the same rate constant as the disappearance of the 490 nm band.

either of the two natural substrates results in a decrease in absorbance at 330 nm and an increase in absorbance at 413 nm associated with formation of internal Schiff base, and thus no intermediates are detected. A fit of these data using a single exponential with hydroxypyruvate and glyoxalate yields rate constants of 55 s^{-1} and 25 s^{-1}, respectively. In contrast, when ketomalonate is used as the keto acid substrate a rapid increase in absorbance at 490 nm is followed by a slow decrease in the 490 nm absorbance, and a concomitant increase at 413 nm. The increase in absorbance at 490-nm is too rapid to allow estimation of a rate constant, whereas the slow decay gives a rate constant of 25 s^{-1}. The 490-nm peak is indicative of the rapid formation of a quinonoid intermediate generated on deprotonation at C-4' of the ketimine intermediate, followed by a slower decay to give the external Schiff base, which is rapidly converted to the internal Schiff base (Fig. 3). In D_2O the quinonoid decays with a rate constant of about 5.4 s^{-1} (Fig. 9), about 4.5 times slower than in H_2O, indicating that protonation at C_α to form the external Schiff base is rate limiting with ketomalonate. In agreement with these data, the appearance of the internal Schiff base mirrors the disappearance of the quinonoid intermediate. Data show the importance of utilizing a number of substrates that are utilized with varying efficiency to visualize intermediates along the reaction pathway.

As a note added in closing, one can also deconvolute rapid-scanning time courses to generate spectra for intermediates. The resulting spectra are useful in assigning intermediates, but there are certain caveats that must be taken into account. One must, of course have sufficient data to define the spectrum. It is relatively easy to produce the spectrum from a simple model such as A goes to B, but for more complex models, such as A goes to B, which goes to C, one must be aware of how close isosbestic points are relative to one another. If they are too close, one can obtain nonsensical spectra. In the case of the above experiments, spectral deconvolution was not discussed because spectra can usually be easily enough generated from equilibrium studies.

Acknowledgment

This work was supported by a grant to P.F.C. from the National Science Foundation (MCB 9729609) and funds for P.F.C. from the Grayce B. Kerr Endowment to the University of Oklahoma.

[18] Protein Tyrosine Phosphatases: X-Ray Crystallographic Observation of Cysteinyl-Phosphate Reaction Intermediate

By DAVID BARFORD

Physiology and Structure of Protein Tyrosine Phosphatases

The formation of phosphoryl-enzyme intermediates is a necessary component of numerous enzymatic reactions that involve phosphoryl-transfer processes. These include, for example, the dephosphorylation of phosphoamino acids by members of the protein tyrosine phosphatase (PTP) superfamily. The opposing activities of the protein phosphatases and protein kinases regulate the overall level of intracellular protein phosphorylation, an essential process of the signal transduction pathways triggered by hormones, mitogens, and oncogenes that regulate growth, differentiation, and proliferation. The PTPs are a large and diverse family of enzymes that comprise tyrosine-specific transmembrane receptor-like PTPs and soluble cytosolic proteins as well as the subfamily of dual-specificity phosphatases (DSPs). The DSPs are related in terms of structure and catalytic mechanism to the tyrosine-specific phosphatases but differ in their substrate selectivity,

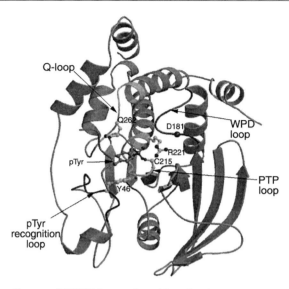

FIG. 1. Ribbon diagram of PTP1B in complex with a phosphotyrosine residue. The four loops critical for catalysis and substrate recognition (PTP loop, WPD loop, pTyr recognition loop, and the Q-loop) are indicated. The catalytic site residues, Cys-215, Arg-221, and Gln-262 are labeled and the Cα atom of Asp-181 is indicated.

being capable of dephosphorylating all three phosphoamino acids.[1,2] Genetic and biochemical studies have demonstrated the importance of the PTPs in regulating signal transduction processes. For example, PTP1B has been implicated as the phosphatase that negatively regulates signaling from the insulin receptor kinase,[3] whereas the dual-specificity phosphatases Cdc25 and the mitogen-activated protein (MAP) kinase phosphatases regulate cell cycle progression and MAP kinase signaling, respectively. PTPs are defined by the presence of a conserved PTP signature motif: $C(X)_5R(S/T)$, containing the catalytically essential Cys and Arg residues. Architecturally, the common characteristic of the PTPs is a catalytic domain consisting of a central β sheet surrounded by α helices[4,5] (Fig. 1). The catalytic site is

[1] D. Barford, A. K. Das, and M.-P. Egloff, *Annu. Rev. Biophys. Biomol. Struct.* **27**, 133 (1998).
[2] N. K. Tonks and B. G. Neel, *Curr. Opin. Cell Biol.* **13**, 182 (2001).
[3] M. Elchebly, P. Payette, E. Michaliszyn, W. Cromlish, S. Collins, A. L. Loy, D. Normandin, A. Cheng, J. Himms-Hagen, C. C. Chan, C. Ramachandran, M. J. Gresser, M. L. Tremblay, and B. P. Kennedy, *Science* **283**, 1544 (1999).
[4] D. Barford, A. J. Flint, and N. K. Tonks, *Science* **263**, 1397 (1994).
[5] J. A. Stuckey, H. L. Schubert, E. B. Fauman, Z. Y. Zhang, J. E. Dixon, and M. A. Saper, *Nature* **370**, 571 (1994).

at the center of the molecule, the base of which is formed by residues of the PTP signature motif that forms a loop connecting the C terminus of the central β strand of the β sheet with the N terminus of an α helix. The structure of the PTP signature motif is that of a cradle, ideal for the coordination of the phosphate group of the phosphoamino acid substrate. Main-chain amide groups and the guanidinium side chain of the invariant Arg residue from the motif donate a total of eight hydrogen bonds to the terminal oxygen atoms of the substrate phosphate group. This positions the phosphorus atom of the phosphate group 3.7 Å from the Sγ atom of the catalytic Cys for in-line nucleophilic attack onto the substrate. In the case of the tyrosine-specific PTPs, three other loops bearing invariant residues form the sides of the cleft and contribute to catalysis and substrate recognition. Engagement of phosphopeptides by PTP1B promotes a major conformational change of one of these loops (the WPD loop) consisting of residues 179–187 that shift by as much as 8 Å to close over the phenyl ring of pTyr and allow the side chain of Asp-181 to act as a general acid in the catalytic reaction.[6] The Arg-221 side chain reorients to optimize salt bridge interactions with the phosphate bound to the catalytic site. This shift is coupled to motion of the WPD loop via a hydrogen bond between the NH_2 group of Arg-221 and the carbonyl oxygen atom of Pro-180, and hydrophobic interactions between the aliphatic moiety of Arg-221 and the side chain of Trp-179. These interactions and the hydrophobic packing between Phe-182 and the phenyl ring of the pTyr substrate stabilize the closed, catalytically competent conformation of the loop. The second loop, the pTyr recognition loop, is unique to the tyrosine-specific PTPs and defines the selectivity of these enzymes for pTyr-containing peptides and proteins by creating, in combination with the Phe-182 of the WPD loop, a deep catalytic site cleft that matches the length of a pTyr residue, but which excludes the shorter phosphoserine and phosphothreonine side chains. Finally, the Q-loop contains two invariant glutamine residues. The results of structural studies described here and kinetic experiments reported elsewhere[7,8] indicate that the role of one of these glutamines (Gln-262 of PTP1B) is to coordinate a water molecule for hydrolysis of the catalytic cysteinyl-phosphate intermediate. In addition, Gln–262 determines the specificity of the nature of the nucleophilic attack onto the cysteinyl-phosphate intermediate.[8]

Insights into the catalytic mechanism of PTPs have been obtained from combined structural and kinetic studies of these enzymes.[1,9,10] The phosphotyrosine dephosphorylation reaction commences with nucleophilic attack by the Sγ atom of the catalytic cysteine (Cys-215 of PTP1B) onto the pTyr phosphorus atom

---

[6] Z. Jia, D. Barford, A. J. Flint, and N. K. Tonks, *Science* **268**, 1754 (1995).
[7] A. D. B. Pannifer, A. J. Flint, N. K. Tonks, and D. Barford, *J. Biol. Chem.* **273**, 10454 (1998).
[8] Y. Zhao, L. Wu, S. J. Noh, K.-L. Guan, and Z.-Y. Zhang, *J. Biol. Chem.* **273**, 5484 (1998).
[9] J. M. Denu and J. E. Dixon, *Curr. Opin. Chem. Biol.* **2**, 633 (1998).
[10] Z.-Y. Zhang, *Crit. Rev. Biochem. Mol. Biol.* **33**, 1 (1998).

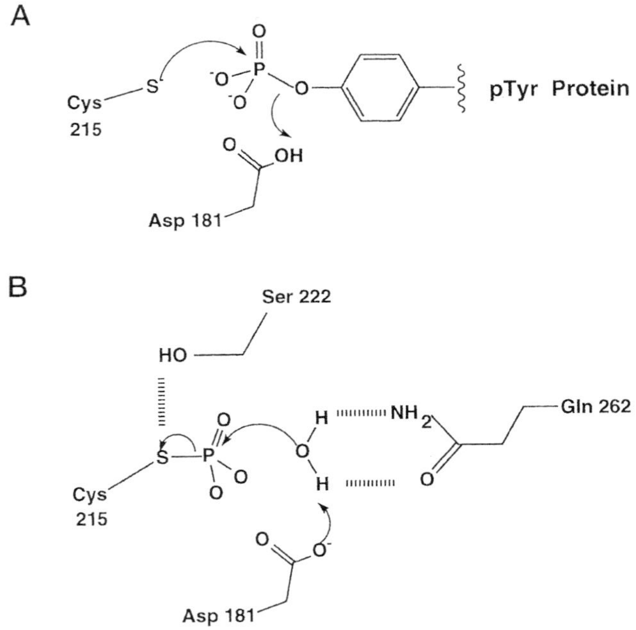

FIG. 2. Schematic of the pTyr dephosphorylation reaction catalyzed by PTP1B. (A) Formation of the cysteinyl-phosphate intermediate. (B) Hydrolysis of the cysteinyl-phosphate intermediate. The amide group of Gln-262 coordinates the water molecule.

(Fig. 2). Cleavage of the scissile P–O bond is facilitated by protonation of the phenolic oxygen by Asp-181 of the WPD loop, with the consequent formation of a cysteinyl-phosphate intermediate. The tyrosine residue diffuses out of the catalytic site and subsequently an enzyme-activated water molecule hydrolyzes the transient phosphoryl-enzyme intermediate. Numerous biochemical data support the reaction mechanism delineated above. Cysteinyl-phosphate intermediates have been trapped by rapid denaturation of PTPs during catalytic turnover, and the existence of the cysteinyl-phosphate intermediate has been monitored using ^{32}P NMR.[11–13] The presence of a two-step reaction, involving an enzyme intermediate, is also indicated from the results of rapid-reaction kinetic analysis.[14] These

[11] K. L. Guan and J. E. Dixon, *J. Biol. Chem.* **266**, 17026 (1991).
[12] H. Cho, R. Krishnaraj, E. Kitas, W. Bannworth, C. T. Walsh, and K. S. Anderson, *J. Am. Chem. Soc.* **114**, 7296 (1992).
[13] J. M. Denu, D. L. Lohse, J. Vijayalakshmi, M. A. Saper, and J. E. Dixon, *Proc. Natl. Acad. Sci. U.S.A.* **93**, 2493 (1996).
[14] Z.-Y. Zhang, *J. Biol. Chem.* **270**, 11199 (1995).

studies demonstrate that catalytic turnover is characterized by a rapid burst phase, corresponding to formation of a phosphoryl-enzyme intermediate and substrate hydrolysis, followed by a slower steady-state rate, corresponding to phosphate release from the catalytic site. Finally, substituting a Ser for the catalytic Cys residue abolishes catalytic activity without impairing the ability of the enzyme to bind substrate. Indeed, this property of catalytically inactive mutants of PTPs has allowed the generation of stable enzyme–substrate Michaelis–Menten complexes for protein crystallography[6,15,16] and has been utilized as a tool for generating substrate trapping mutants for the identification of *in vivo* substrates of PTPs.[17] The nucleophilicity of the Cys residue results from its close proximity to the main-chain amide groups and a hydrogen bond with the side chain of Ser-222 of the PTP signature motif, resulting in an unusually low pK_a of 4.5–5.5.[18,19] The catalytic Asp residue, Asp-181 of PTP1B, contributes to the basic limb of the pH activity profile, and its substitution by Asn lowers k_{cat} by 10^5-fold, suggestive of a role as a general acid catalyst.[20,21] These results imply that Asp-181 is necessary for the first step of the reaction, namely cleavage of the pTyr P–O bond and intermediate formation, a notion consistent with the findings that Asp-181 mutants of PTP1B allow phosphorylated substrates to form stable complexes with the enzyme *in vivo*.[21]

Dissecting Catalytic Reaction Coordinate

We will first consider some of the theoretical and practical parameters for optimizing accumulation of the phosphoryl-enzyme intermediate in solution because this is a prerequisite for trapping the phosphoryl enzyme in the crystalline state. An enzyme-catalyzed reaction is a multistep event, with each step governed by discrete kinetic rate constants. In the case of protein tyrosine phosphatases, and for most enzymatic reactions that are not limited by diffusion rates, the rate-limiting step is that of chemical conversion of the substrate into product. Enzymatic reactions, which proceed via an enzyme intermediate, involve two chemical steps, namely, formation and breakdown of the intermediate. To trap the enzyme intermediate, the kinetic rate constant governing the rate of cysteinyl-phosphate hydrolysis should be significantly smaller than all other rate constants. Assuming that substrate is

---

[15] A. Salmeen, J. N. Andersen, M. P. Myers, N. K. Tonks, and D. Barford, *Mol. Cell* **6**, 1401 (2000).
[16] H. Song, N. Hanlon, N. R. Brown, N. E. M. Noble, L. N. Johnson, and D. Barford, *Mol. Cell* **7**, 615 (2001).
[17] H. Sun, C. H. Charles, L. F. Lau, and N. K. Tonks, *Cell* **75**, 487 (1993).
[18] Z.-Y. Zhang and J. E. Dixon, *Biochemistry* **32**, 9340 (1993).
[19] J. M. Denu, G. Zhou, Y. Guo, and J. E. Dixon, *Biochemistry* **34**, 3396 (1995).
[20] Z.-Y. Zhang, Y. Wang, and J. E. Dixon, *Proc. Natl. Acad. Sci. U.S.A.* **91**, 1624 (1994).
[21] A. J. Flint, T. Tiganis, D. Barford, and N. K. Tonks, *Proc. Natl. Acad. Sci. U.S.A.* **94**, 1680 (1997).

in excess and the reaction is irreversible, the reaction catalyzed by PTPs is shown schematically as:

Enzyme + substrate → enzyme–substrate complex $\xrightarrow{k_1}$ phosphoryl-enzyme intermediate $\xrightarrow{k_2}$ enzyme–phosphate complex → enzyme + phosphate

The rate of cysteinyl-phosphate hydrolysis (k_2) is always substrate *independent*, although, under certain conditions, the rate of the first step of this reaction, formation of a cysteinyl-phosphate intermediate (k_1), is substrate *dependent*. However, the rates of both steps are dependent on the pH of the reaction. Finally, the relative rates of these two reactions can also be altered by site-directed mutagenesis of the protein. We will examine conditions that favor the formation of the cysteinyl-phosphate intermediate in solution, a process that is achieved by reducing k_2, the rate of cysteinyl-phosphate hydrolysis, and maximizing k_1, the rate of cysteinyl-phosphate formation.

Buffer pH and Choice of Substrate

Previous kinetic studies performed by Denu, Dixon, Zhang, and their colleagues on the *Yersinia* PTP and rat PTP1, and later extended to the dual-specificity phosphatases, have defined many of the chemical steps of the reaction process[9,10] (Fig. 2). Attack by the catalytic site Cys residue onto the phosphate of the phosphoamino acid substrate is accompanied by proton transfer from the general acid (Asp-181 in PTP1B and equivalent residues in other PTPs and DSPs) onto the phosphate-ester oxygen atom, an event that prevents the accumulation of negative charge on the leaving group during cleavage of the scissile bond. Analysis of the dependence of k_{cat} and k_{cat}/K_m as a function of pH indicates a bell-shaped profile with a pH maxima from pH 5.5 to 6.5 for PTP1B.[20] The basic limb of the plot (i.e., at a pH above the pH optimum) results from the deprotonation of the general acid. This would suggest that at higher pH where Asp-181 becomes deprotonated, the first step of the reaction is rate limiting. Numerous factors contribute to the acidic limb of the pH–reactivity profile. In the case of PTP1B, the acidic limb results from the deprotonation of Glu-115,[20] a structural residue that defines the conformation of Arg-221 that is critical for substrate binding and transition state stabilization.[1,22] For the DSP, VHR, the acidic limb results from deprotonation of the catalytic Cys residue and is also due to deprotonation of the substrate *p*-nitrophenol (*p*-NPP). At the pH optima of PTPs, a number of experimental results suggest that cysteinyl-phosphate formation is not rate limiting and that k_1 exceeds k_2. First, rapid reaction kinetic analysis of PTPs indicates a burst phase,

---

[22] Z.-Y. Zhang, Y. Wang, L. Wu, E. B. Fauman, J. A. Stuckey, H. L. Schubert, M. A. Saper, and J. E. Dixon, *Biochemistry* **33**, 15266 (1994).

corresponding to phosphoryl-enzyme formation, followed by a slower steady-state phase, corresponding to the rate-limiting step of the reaction, described by k_{cat}, and attributable to either phosphoryl-enzyme hydrolysis or phosphate release. Moreover, in uncatalyzed reactions, the rate of scissile P–O bond cleavage decreases with an increase in the pK_a of the leaving group, a phenomenon known as the Bronsted effect. Analysis of the rates of dephosphorylation of a number of artificial PTP substrates at the pH optima for PTPs showed that the rates of scissile bond cleavage, as determined by pre-steady-state kinetic analysis and measurements of k_{cat}/K_m, was independent of the pK_a of the leaving group of the substrate. These data suggest that proton transfer from the general acid catalyst effectively compensates the accumulation of negative charge on the leaving group during the first step of the reaction, and that this step is not rate limiting in the catalytic reaction. In contrast to the situation with the wild-type enzyme at optimal pH values, mutant forms of the enzymes with Ala or Asn substitutions for the catalytic Asp residue demonstrate reaction rates that are dependent on the pK_a value of the leaving group of the substrate. This would suggest that at pH values higher than the pH optimum, where the general acid will be deprotonated and therefore a poor catalyst, formation of the cysteinyl-phosphate intermediate becomes rate limiting. The pK_a of tyrosine, the leaving group of phosphotyrosine in the reactions catalyzed by PTPs, is 10.0, compared with a pK_a of 5.1 for p-nitrophenol, the leaving group of p-nitrophenol phosphate. Fjeld and colleagues report the use of aryl-phosphate substrates such as 3-O-methylfluorescein phosphate with a pK_a of 4.6.[23] The conditions used to crystallize the protein may not match the pH optima for the reaction catalyzed by the enzyme. When the pH of the crystallization buffer is higher than the pH optima, a condition which may slow the scissile P–O bond cleavage and subsequently cysteinyl-phosphate accumulation using the physiological substrate, use of artificial substrates with leaving groups with low pK_a values may compensate for the higher pH values used.

Mutagenesis of Catalytic Site

Stabilizing the lifetime of the enzyme intermediate, to favor both its accumulation and hence ease of structural analysis, may also be possible by altering the enzyme catalytic site structure to specifically reduce the rate constant k_2 for hydrolysis of the enzyme intermediate relative to that of k_1, the rate of phosphoryl-enzyme formation. This approach requires that at least one catalytic site residue be involved in promoting enzyme-intermediate hydrolysis, but not be required for its formation. In the case of PTP1B, a structure-based approach was applied that identified Gln-262 as one such residue. Kinetic studies had indicated that Gln-262, a residue that is invariant among all PTPs and conserved among some of the DSPs,

[23] C. C. Fjeld, A. E. Rice, Y. Kim, K. R. Gee, and J. M. Denu, *J. Biol. Chem.* **275,** 6749 (2000).

for example KAP,[16] contributes to the enzymatic reaction mechanism.[21] Substituting an alanine for Gln at this position reduces the steady-state rate constant k_{cat}, a parameter that is a measure of the rate of dephosphorylation, by 83-fold.[21] In contrast, the pseudo second-order rate constant k_{cat}/K_m, which describes the rate of the first step of the reaction, is reduced by only 7-fold. Interestingly, K_m for this mutant is 12-fold lower than for the wild type. These data indicated that Gln-262 plays a role to promote cysteinyl-phosphate hydrolysis but makes little or no contribution to determining the rate of cysteinyl-phosphate formation. The structure of the PTP1B-phosphopeptide complex provides a structural explanation for these findings.[6,15] Relative to the apoenzyme, in the PTP1B-phosphopeptide Michaelis–Menten complex, the side chain of Gln-262 rotates out of the catalytic site to avoid a steric clash with the phosphoamino acid substrate, and the side chain of Gln-262 plays no role in promoting the pre-steady-state dephosphorylation of the pTyr residue. However, a role for Gln-262 in hydrolyzing the cysteinyl-phosphate intermediate was suggested from the structure of PTP1B in complex with the PTP inhibitor orthovanadate.[7] A covalent bond is formed between the Sγ atom of the nucleophilic Cys-215 residue and the vanadium atom, so that the vanadate ion forms a pentavalent transition-state intermediate. The three equatorial oxygen atoms of vanadate form some eight hydrogen bonds to main-chain NH groups of the PTP signature motif and guanidinium side chain of Arg-221, whereas the apical oxygen atom forms hydrogen bonds with the side chains of Asp-181 and Gln-262. The apical oxygen atom most closely resembles the position of an attacking nucleophilic water molecule during the hydrolysis of the cysteinyl-phosphate intermediate. Thus, the hydrogen bonds to Asp-181 and Gln-262 suggest that these residues may play a role in positioning and/or activating a water molecule in this step. Hence, by combining structural and kinetic analysis, modification of enzyme catalytic sites may be employed to stabilize an enzyme intermediate. This analysis would predict that substituting Ala for Gln-262 would reduce k_2 but not k_1, hence allowing the cysteinyl-phosphate intermediate to accumulate.

Once consideration has been given to the buffer pH, substrate, and enzyme parameters that favor accumulation of the enzyme intermediate, a useful step prior to performing high resolution structural studies is to monitor the formation of the enzyme intermediate by means of appropriate spectroscopic analysis. In the case of PTPs, Denu and colleagues accomplished this by use of ^{32}P NMR.[13] These studies using a S222A mutant of PTP1B indicated the formation and steady-state existence of a cysteinyl-phosphate intermediate with a resonance peak discrete from those of the substrate and product.

Protein Crystallography

The arrangement of enzyme molecules in an ordered three-dimensional lattice that is required for protein crystallography alters the kinetic parameters of the enzymatic reaction in numerous ways. First, the buffer conditions used to obtain suitable protein crystals may not match those optimal for the enzyme-catalyzed reaction. For example, the pH used for protein crystallization is unlikely to match the pH optimum of the reaction. Such a situation would be particularly problematic if the pH of the crystallization solution reduced k_1 relative to k_2, leading to a change of the rate-limiting step and reduced or loss of accumulation of the phosphorylenzyme intermediate. This problem can be resolved by incubating preexisting crystals grown under ideal crystallization conditions in buffers of different pH more appropriate for the enzymatic reaction. In many instances crystals are resistant to buffer changes of one to two pH units. In addition, as discussed above, when general acid catalysis is not functioning efficiently, for example at pH values higher than the pH optimum, or when substitutions of the general acid occur, the rate of cysteinyl-phosphate formation may be accelerated by use of phospho substrates which have leaving groups with low pK_a values. Other properties of the crystallization buffer, such as the ionic strength and the nature of crystallization additives, may interfere with substrate binding to the catalytic site. Preincubation of preexisting crystals in alternative buffers may be required to maximize substrate–enzyme interactions.

The most fundamental problems associated with the study of transient enzyme intermediates using protein crystallography are twofold. The first is the time required to capture structural information describing the intermediate. Conventional monochromatic data collection methods, requiring a three-dimensional data set, even using third-generation synchrotron sources, require periods of minutes to hours. The rediscovery of the Laue method utilizing polychromatic radiation increases data collection rates to the second and millisecond time scale, but suffers from loss of low-resolution data, the requirement for crystals with mosaicities of less than 0.1°, and extreme sensitivity to crystal disorder.[24] Second, although solvent channels comprise some 30–70% of the volume of a protein crystal, the ordered arrangement of protein molecules within the crystal restricts the free diffusion of substrate. Empirical findings suggest that the uniform diffusion of ligands within crystals occurs within 10–20 sec. Since this time is longer than the catalytic cycle of most enzyme reactions, a mixed population of free enzyme, enzyme–substrate, and enzyme–product molecules develops during the initial stages of adding substrate to a protein crystal. The consequence is that the enzymatic reaction coordinate will not be synchronized in all molecules of the protein so that when the rates of diffusion are slower than the k_{cat}, which is usually the situation for wild-type enzymes, substrate turnover by enzyme molecules on the surface of the crystal will have occurred before substrate binding to molecules at the center of the crystal. The most elegant technique developed to circumvent this problem and to synchronize reactions in all molecules in the crystal is the use of inert "caged" substrates capable of diffusing into the crystal and of being converted to substrate

[24] J. Hajdu and I. Andersson, *Ann. Rev. Biophys. Biomol. Struct.* **22**, 467 (1993).

by flash photolysis by means of an intense laser light source. Unfortunately, despite the elegance of this approach in theory, various practical problems limit its use. First, chemical synthesis of caged compounds of substrates is a specialized process and therefore for most instances the appropriate caged substrate will not be readily available. Second, the efficiency of photolysis of caged compounds is often very low and numerous bursts of a laser light are required to liberate sufficient quantities of the substrate from the caged compound. The main reason for this is that protein crystals absorb much of the incident light required to release the substrate with the consequence that the protein crystal absorbs heat, increasing the risk of crystal disorder, and moreover, the time elapsed in releasing the substrate may exceed the catalytic turnover of the enzyme.

A method for studying the enzyme intermediate that we have employed is to determine the structure of the enzyme, as it exists in the steady-state phase of the catalytic reaction pathway. Assuming that all kinetic rate constants are negligible compared with the rate of cysteinyl-phosphate hydrolysis, then during the steady-state phase, the enzyme molecules will exist predominantly as the cysteinyl-phosphate intermediate. This state will exist for as long as the substrate concentration is high enough to saturate the catalytic site and as long as substrate binding is not inhibited by product formation. Given that appropriate parameters have been satisfied, such a condition may persist for minutes to hours, providing ample time to trap the enzyme intermediate at 100K for structural analysis. For this approach to be successful, it is necessary that the enzyme be capable of catalytic turnover within the crystal. Such a condition requires that as well as being capable of binding substrate, any protein conformational changes that are associated with the reaction mechanism can be accommodated within the crystal lattice. In addition, it is necessary that the rate of diffusion of substrate molecules to the catalytic sites of the enzyme molecules in the crystal exceed the rate at which all phosphoryl-enzyme molecules are hydrolyzed. However, because we are studying the steady state, it is likely that molecules on the surface of the crystal will have catalyzed numerous enzymatic turnovers before enzyme molecules at the center of the protein crystal have achieved steady state. By incubating crystals in a huge molar excess of substrate, the prospect of substrate depletion at the crystal surface will be minimized. All enzyme molecules present at steady state will be indistinguishable structurally.

Diffusion times are dependent on the size of the substrate, its free diffusion coefficient, the pore size of the crystals, the crystal dimensions, buffer composition, and temperature. Importantly, rates of mass transport of substrates into the crystal will be proportional to the concentration of substrates in the mother liquor surrounding the crystal. Therefore, higher substrate concentrations will lead to more rapid diffusion of the substrate into the crystal and saturation of enzyme catalytic sites, and a choice of substrate that increases diffusion rates should be considered.

Method for Trapping Cysteinyl-Phosphate Intermediate of PTP1B

Our approach to trapping the phosphocysteine intermediate of PTP1B is the following. By assuming that the rate of substrate diffusion into the crystal exceeds the rate of cysteinyl-phosphate hydrolysis, and with an excess of substrate added to the crystal incubation buffer, a steady-state accumulation of intermediate should be obtained, which may be captured by flash-freezing a crystal in a stream of nitrogen gas at 100K. The cysteinyl-phosphate intermediate is preserved in the frozen crystal, which is suitable for X-ray analysis that can be performed at leisure. PTP1B crystals are grown at pH 7.5 and 4°.[25] However, at this pH, the rate of PTP-catalyzed hydrolysis of p-NPP is 10% of its maximal rate observed between pH 5.5 and 6.5.[14,20] The reduction in rate is attributable to the deprotonation of Asp-181, the presumed acid catalyst in the p-NPP hydrolysis reaction, which would imply that the rate-limiting step becomes that of enzyme phosphorylation and hence a phosphoryl-enzyme intermediate would not accumulate under these conditions. Thus, to optimize conditions for enzyme phosphorylation, we equilibrate crystals at pH 6.5 prior to the start of the kinetic experiment. To start the experiment, we incubate small crystals in a large molar excess of substrate ($\sim$1000-fold) for 12 min at high concentrations (25 mM) to saturate the enzyme catalytic site to achieve V_{max} for enzyme phosphorylation. The K_m for p-NPP is 1 mM.[14] Small crystals are used to optimize uniform distribution of the substrate in the crystal. The substrate p-NPP is preferable to pTyr for two reasons. First, p-NPP is 80% of the mass of pTyr and therefore will diffuse more rapidly into the protein crystal, and second, the pK_a of the leaving group of p-NPP (p-nitrophenol) is 5.1 compared with that of pTyr (tyrosine), which is 10, thus enhancing the formation of the cysteinyl-phosphate intermediate. Favorable features of PTP1B crystals for this study are: (1) high solvent content of 60%, which facilitates substrate diffusion, (2) the ability of small substrates such as p-NPP and pTyr to bind to the catalytic site and promote closure of the WPD loop, and (3) the fact that PTP1B crystals are grown at 4°, hence prolonging the lifetime of the enzyme intermediate. The wild-type enzyme k_{cat} is 40 s^{-1}, whereas that of the Q262A mutant is $\sim$0.5 s^{-1} at 25°.[20,21] At 4° this corresponds to one enzymatic reaction in $\sim$10 sec.

We can calculate the concentration of enzyme molecules in the crystal from knowledge of the unit cell dimensions and number of molecules in each unit cell. For PTP1B crystals, the concentration of PTP1B is 14 mM, a value similar to that of other protein crystals. From consideration of the enzyme concentration in the crystal, estimated substrate diffusion times into the crystal, and the length of an enzyme catalytic cycle, one can estimate the appropriate substrate concentration that will be required to generate a uniform population of enzymes in the crystal that exist in the phosphoryl-enzyme steady state. The condition can be stated

---

[25] D. Barford, J. C. Keller, A. J. Flint, and N. K. Tonks, *J. Mol. Biol.* **239**, 726 (1994).

by the simple formula t_e [substrate] > t_s [enzyme], where t_e and t_s are the catalytic turnover and substrate diffusion times, respectively. For the PTP1B reaction, by assuming that t_e and t_s are each about 10 sec, hence with a substrate concentration of 25 mM and an enzyme concentration of 14 mM, we should achieve uniform steady-state accumulation of the phosphoryl-enzyme intermediate.

Experimental Procedures

Buffers

Buffer A: 20 mM imidazole (pH 7.5) 1.0 mM EDTA, 0.5 mM EGTA, 2.0 mM benzamidine hydrochloride, 0.5 mM phenylmethylsulfonyl fluoride (PMSF), 3.0 mM dithiothreitol (DTT), 10% (v/v) glycerol

Buffer B: 25 mM NaH$_2$PO$_4$ (pH 6.5), 1.0 mM EDTA, 0.5 mM EGTA, 2.0 mM benzamidine hydrochloride, 0.5 mM PMSF, 3.0 mM dithiothreitol, 10% (v/v) glycerol

Buffer C: 25 mM Tris-HCl (pH 7.5), 1.0 mM EDTA, 0.5 mM EGTA, 2.0 mM benzamidine hydrochloride, 0.5 mM PMSF, 3.0 mM dithiothreitol

Cloning and Purification of PTP1B (Q262A)

A 1.0-kb fragment encoding the 37 kDa form of PTP1B (amino acid residues 1–321) is cloned downstream of the phage T7 RNA polymerase promoter at the NcoI site of the pET-19b (Novagen, Madison, WI) vector, thus providing translational initiation at Met1 of PTP1B. Site-directed mutagenesis of the Gln-262 residue of PTP1B to alanine is performed by means of the U-labeled template method of Kunkel.[21,26] LB broth (4 × 1 liter) with 100 µg/ml of ampicillin is inoculated with 40 ml of an overnight culture of BL21(DE3) cell transformed with the pET expression vector containing the cDNA encoding the N-terminal 321 amino acids of PTP1B. The cultures are grown at 37° until the absorbance at 600 nm reaches 0.6 (approximately 2.5 hr) at which time the cells are induced with 0.6 mM isopropylthio-β-D-galactopyranoside (IPTC). After three hours the cells are harvested and stored at −70°.

All purification steps are carried out at 4° and performed as described by Barford et al.[25] The frozen cell pellet is thawed on ice and the cells are resuspended in 80 ml of buffer A with 2 µg/ml each of leupeptin, pepstatin, aprotinin, and DNase A and lysed by one pass through a French press at 1000 to 1500 psi. The lysate is cleared by centrifugation at 20,000g for 45 min and diluted with 100 ml of buffer A. The protein solution is loaded onto a 100 ml Q Sepharose fast flow anion-exchange column (Amersham Pharmacia Biotech, Piscataway, NJ) at a flow

---

[26] T. A. Kunkel, *Proc. Natl. Acad. Sci. U.S.A.* **82**, 488 (1985).

rate of 1 ml/min. The column is washed in 800 ml of buffer A. The position of peak fractions is assessed by Coomassie blue staining of SDS–polyacrylamide gels. PTP1B elutes at approximately 0.3 M NaCl. Peak fractions are pooled and dialyzed for 16 hr against buffer B. Dialyzed PTP1B is loaded onto an HR 10/10 Mono S cation-exchange column (Amersham Pharmacia Biotech) previously equilibrated in buffer B and then washed in 100 ml of buffer B. The protein is eluted by applying a linear gradient from 0 to 0.5 M NaCl in 150 ml of buffer B. PTP1B elutes as a single peak at ~0.2 M NaCl. Peak fractions are pooled and brought to 1.4 M ammonium sulfate by addition of 10 mM Tris-HCl (pH 7.5) containing 4.0 M ammonium sulfate. The protein solution is diluted with 1 volume of buffer C + 1.4 M ammonium sulfate before being loaded onto a phenyl-TSK column (TosoHaas) previously equilibrated in buffer C + 1.4 M ammonium sulfate. The column is washed with 20 ml of buffer C + 1.4 M ammonium sulfate and PTP1B is eluted by applying a linear gradient of 30 ml of buffer C with 10% (v/v) glycerol. The protein at >99% homogeneity is dialyzed against a buffer containing 10 mM Tris-HCl (pH 7.5), 25 mM NaCl, 0.5 mM EDTA, 3 mM DTT and concentrated to 10 mg/ml (assuming an extinction coefficient of 1.2 per mg/ml of PTP1B). The yield of the protein is approximately 60–80 mg.

Crystallizations

Crystals are grown using the vapor diffusion method at 4°. PTP1B (4 μl at 10 mg/ml) is mixed with an equal volume of precipitating solution, 0.1 M HEPES (pH 7.5), 0.2 M magnesium acetate, and 12% to 16% (w/v) polyethylene glycol 8000 and applied to a glass coverslip inverted over a well solution containing 0.7 ml of the precipitating solution in tissue culture trays (Linbro, Hampton Research, Laguna Niguel, CA). Crystals appear within 1–2 days.

Preparation of Crystals

Selection of Crystals. To ensure that substrate diffusion into the crystal is as uniform as possible, a compromise with the crystal size is made. Small crystals allow substrate to diffuse uniformly within the lattice more rapidly than larger crystals; however, there are attendant disadvantages associated with lower diffraction power of small crystals. This can be remedied by the use of a synchrotron source to obtain data of sufficient resolution and quality. All steps are performed at 4°. To prepare crystals of the PTP1B–cysteinyl phosphate complex, Q262A PTP1B crystals that have been grown at pH 7.5 are incubated for 5 min in a buffer of 100 mM Bis–Tris (pH 7.0), 200 mM magnesium acetate, 18% (w/v) polyethylene glycol 8000 and subsequently for another 5 min in the same buffer at pH 6.5. To prepare for cryofreezing, and for trapping the intermediate, crystals are incubated in this buffer with 5% and then 10% (v/v) methylpentanediol each for 5 min before being transferred to 3 ml of 100 mM Bis–Tris (pH 6.5), 200 mM magnesium acetate, 18% (w/v) polyethylene glycol 8000, 25 mM p-nitrophenol phosphate, and 15% (v/v) methylpentanediol for 12 min. Incubations are performed in 3 cm diameter polystyrene petri dishes. After the incubation, crystals are frozen and either used for data collection immediately or stored in liquid nitrogen dewars. All crystals are frozen using the loop method.[27] Briefly, rayon loops approximately 0.5 mm in diameter are glued to a hollow metal pin. The pin is then fixed in a top hat comprising a hollow cylindrical body accommodating the pin at one end and having a flattened base at the other (designed at the Laboratory of Molecular Biophysics, University of Oxford, and available from Oxford Cryosystems, Oxford, UK). The crystal is captured in the loop by agitating the crystal to the top of the cryoprotectant buffer and picking it, and the surrounding buffer surface film, up with the loop. Microbridges (Hampton Research) are ideal for capturing crystals in this manner. The top hat is then rapidly mounted on a magnetic goniometer via the flattened base and using an Oxford Cryosystems cryostream at 100K the crystal is frozen. The precise arrangement of the goniometer differs depending on whether the data are to be collected immediately or the crystals are to be stored. If the crystals are used immediately, they are simply mounted on the goniometer with the pin extending along the phi axis of the goniometer with the crystal 0.5–1 cm from the nozzle of the cryostream. If the crystals are to be stored for later data collection, a special detachable extended arc goniometer head designed at the Laboratory of Molecular Biophysics (Oxford Cryosystems) is used. This comprises a standard goniometer with a detachable metal arc containing a track along which the magnetic pad can travel. The arc extends from the data collection position of the goniometer (with the pin axis extending along phi) through about 90° so that the magnetic pad can be oriented anywhere between two orthogonal positions. To freeze the crystal, the magnetic pad is slid along the arc to orient the pad to face vertically downward, the top hat with the crystal is mounted in this position, with the pin pointing vertically downward, and the crystal is frozen using a cryostream. A cryovial containing liquid nitrogen is then brought up rapidly underneath the top hat to immerse the crystal and the top hat is dislodged into the cryovial and stored on aluminum canes in liquid nitrogen dewars. To collect data, the stored frozen crystal is quickly transferred from the cryovial to the magnetic pad of the goniometer with the magnetic pad in the same vertical orientation on the arc as before and maintained at 100K with a cryostream. The pad is then run along the arc through 90° to orient the pin along phi. The dimensions of the arc are such that the loop is at the center of rotation and does not undergo any translation, which would remove the crystal from the jet of nitrogen gas. The arc is then removed and data collection can begin using conventional approaches, the nature of which is

[27] E. Garman, *Acta Crystallogr. D* **55**, 1641 (1999).

specific to the area detector and X-ray source. Crystals may also be frozen directly into liquid nitrogen and later mounted onto the extended arc goniometer head for data collection.

Summary of Results

By replacing Gln-262 with Ala, we prolonged the lifetime of the cysteinyl-phosphate enzyme intermediate, allowing us to trap and visualize it using X-ray crystallography.[7] The structure (1) revealed that the WPD loop adopts a closed conformation that sequesters both the cysteinyl-phosphate intermediate and the nucleophilic water molecule, (2) suggests roles for Asp-181 and Gln-262 in catalyzing the enzyme dephosphorylation reaction, and (3) explains why phosphoryl transfer occurs to a water molecule and not to other phosphoryl acceptors.

[19] GTP:GTP Guanylyltransferase: Trapping Procedures for Detecting and Characterizing Chemical Nature of Enzyme–Nucleotide Phosphoramidate Reaction Intermediate

By JARED L. CARTWRIGHT and ALEXANDER G. MCLENNAN

Introduction

Enzyme–nucleotide phosphoramidates are involved as intermediates in a variety of reactions, including those catalyzed by mRNA-capping enzymes, DNA and RNA ligases,[1,2] and by proteins of the GAFH superfamily, including galactose-1-phosphate uridylyltransferase.[3] The chemical nature of the enzyme–nucleotide phosphoramidate intermediate employed by the unique GTP:GTP guanylyltransferase from yolk platelets of *Artemia franciscana* cysts to synthesize diguanosine tetraphosphate (Gp_4G) has been investigated. Using this example, the following article aims to provide detailed practical advice to the investigator who wishes to characterize the chemical nature of an enzyme–nucleotide phosphoramidate reaction intermediate.

Diguanosine $5',5'''-P^1,P^4$-tetraphosphate (Gp_4G) and diguanosine $5',5'''-P^1,P^3$-triphosphate (Gp_3G) are major nucleotides in the brine shrimp *Artemia*

---
[1] K. Mizumoto and Y. Kaziro, *Prog. Nucleic Acid Res. Mol. Biol.* **34**, 1 (1987).
[2] P. Cong and S. Shuman, *J. Biol. Chem.* **268**, 7256 (1993).
[3] J. E. Wedekind, P. A. Frey, and I. Rayment, *Biochemistry* **35**, 11560 (1996).

franciscana and related Branchiopoda including *Daphnia magna,* but are found only at submicromolar concentrations in other organisms.[4-6] The enzyme responsible for the synthesis of Gp_4G in *Artemia,* the 480-kDa GTP:GTP guanylyltransferase (Gp_4G synthetase), catalyzes the reversible reaction $GTP + GTP \rightleftharpoons Gp_4G + PP_i$.[7] It can also synthesize Gp_3G according to the scheme: $Gp_4G + GDP \rightarrow Gp_3G + GTP$.[8] Purification of this unique enzyme from *Artemia* cysts showed the enzyme activity to be associated with a 110-kDa polypeptide and the reaction mechanism to involve a covalent enzyme–nucleotide intermediate.[9,10] The procedures described here were used to show that this polypeptide forms a histidyl-GMP reaction intermediate via the $N\varepsilon 2$ ring nitrogen of an enzyme histidine residue. This information suggests an evolutionary origin for this unique enzyme.

Experimental Procedures

The formation and trapping of the enzyme–guanylate reaction intermediate are described, followed by experiments to establish the identity of the bound nucleotide and to determine that it represents a genuine reaction intermediate. The chemical stability of the intermediate yields information about the nature of the covalent linkage and may even suggest the identity of the amino acid involved. Identification of this amino acid is achieved by labeling the enzyme with $[\alpha\text{-}^{32}P]GTP$ followed by isolation of the labeled phosphoamino acid by periodate treatment and alkaline hydrolysis. Comparison of the product to phosphoamino acid standards by thin-layer and ion-exchange chromatography completes the characterization of the trapped reaction intermediate.

Preparation of Enzyme–Guanylate Intermediate

Gp_4G synthetase is purified as previously described.[9,10] The formation of the enzyme–guanylate reaction intermediate is optimized by setting up the labeling reaction described below under various conditions. The conditions quoted have been found to be optimal. The final protein pellets are analyzed by SDS–PAGE and incorporation of ^{32}P measured by autoradiography.

---

[4] A. Warner, *in* "Regulation of Macromolecular Synthesis by Low Molecular Weight Mediators" (G. Koch and D. Richter, eds.), p. 161. Academic Press, New York, 1979.
[5] A. Warner, *in* "Ap$_4$A and Other Dinucleoside Polyphosphates" (A. G. McLennan, ed.), p. 275. CRC Press, Boca Raton, FL, 1992.
[6] T. G. Oikawa and M. Smith, *Biochemistry* **5**, 1517 (1966).
[7] A. H. Warner, P. C. Beers, and F. L. Huang, *Can. J. Biochem.* **52**, 231 (1974).
[8] A. H. Warner and F. L. Huang, *Can. J. Biochem.* **52**, 241 (1974).
[9] J. J. Liu and A. G. McLennan, *J. Biol. Chem.* **269**, 11787 (1994).
[10] J. L. Cartwright and A. G. McLennan, *Arch. Biochem. Biophys.* **361**, 101 (1999).

1. Set up the following reaction mix in 100 μl at 20° and incubate for 10 min.[10]

0.2 M Triethanolamine hydrochloride (pH 6.0)	10 μl
0.2 M Magnesium acetate	10 μl
0.1 M 2-Mercaptoethanol	10 μl
[α-^{32}P]GTP (10 mCi/ml)	0.5 μl
Purified Gp$_4$G synthetase (50 μg/ml)	20 μl
H$_2$O (double distilled)	49.5 μl

2. Terminate the reaction by the addition of 11.1 μl of 100% (w/v) trichloroacetic acid (TCA) and incubate at 4° for 30 min.

3. Recover the precipitated protein and wash at 4° with 1 ml 10% (w/v) TCA by centrifugation at 16,000g for 5 min.

4. Wash the pellet three times with 1 ml cold acetone, hold at −10° for 10 min, and then centrifuge as before.

5. Allow the pellet to air dry for further analysis.

Figure 1 shows clearly that when a sample of such a pellet is solubilized in sample buffer and analyzed by SDS–PAGE and autoradiography, labeling of

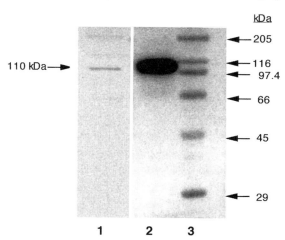

FIG. 1. SDS–polyacrylamide gel electrophoresis and autoradiography of purified Gp$_4$G synthetase. Lane 1: the purified Gp$_4$G synthetase (1 μg) was analyzed by SDS–PAGE in a 10% gel and stained with Coomassie blue. Enzyme–guanylate intermediate was prepared as described under Experimental Procedures. After electrophoresis and staining, the gel was dried and autoradiographed for 30 min at room temperature and the autoradiograph realigned over the gel. Lane 2: labeled enzyme–guanylate intermediate (1 μg); lane 3: protein standards (8 μg): rabbit muscle myosin, 205 kDa; *Escherichia coli* β-galactosidase, 116 kDa; rabbit muscle phosphorylase B, 97.4 kDa; bovine serum albumin, 66 kDa; hen ovalbumin, 45 kDa; and bovine erythrocyte carbonic anhydrase, 29 kDa.

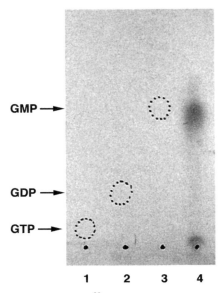

FIG. 2. Autoradiograph of TLC-separated [^{32}P]guanylyl moiety following acid hydrolysis. Samples (5 μl) of GMP, GDP, GTP, and enzyme–guanylate intermediate (prepared as described under Experimental Procedures and dissolved in 100 μl of 50 mM sodium carbonate, pH 10.5, 1% SDS) were treated with 0.15 ml of 0.15 M HCl and incubated at 37° for 20 min. Samples were applied directly to a polyethyleneimine cellulose plate and chromatographed in dioxane/35% ammonia/doubly distilled H$_2$O (6 : 1 : 4 v/v). Positions of the authentic standards were visualized under UV light (254 nm) and marked on the autoradiograph. Lane 1: GTP; lane 2: GDP; lane 3: GMP; lane 4: [^{32}P]guanylyl moiety.

the 110-kDa band can clearly be seen (lane 1), confirming that this polypeptide contains the catalytic site of the purified Gp$_4$G synthetase (lane 2).

The labeled nucleotide moiety was confirmed as GMP by acid hydrolysis of the intermediate (see later for chemical stability) and thin-layer chromatography on polyethyleneimine cellulose (Fig. 2). That it represents a true reaction intermediate was confirmed by the ability of the isolated complex to generate Gp$_4$G, Gp$_4$I, and Gp$_4$X on the addition of GTP, ITP, and XTP, respectively (results not shown).[9,10]

Chemical Stability of Enzyme–Guanylate Intermediate

Characterization of the covalent enzyme–GMP intermediate by analysis of its chemical stability reveals information about the nature of the linkage between the labeled moiety and the polypeptide chain.

1. Dissolve the enzyme–guanylate intermediate, prepared as described above, in 50 μl of 10 mM Tris-HCl (pH 7.5), 1% SDS, and store on ice.

TABLE I
CHEMICAL STABILITY OF ENZYME–GUANYLATE REACTION INTERMEDIATE
OF Gp_4G SYNTHETASE[a]

Treatment	cpm	% Unhydrolyzed
No additions or incubation	24,049	100.0
0.15 M Tris-HCl, pH 7.5, 5 min, 85°	19,160	79.7
0.15 M HCl, 5 min, 85°	233	1.0
0.15 M NaOH, 5 min, 85°	24,682	102.6
3.86 M Hydroxylamine, pH 4.75, 20 min, 37°	174	0.7
0.2 M Hydroxylamine, pH 7.5, 20 min, 37°	601	2.5
4 M Sodium acetate, pH 7.5, 20 min, 37°	23,084	96.0

[a] Aliquots of enzyme–GMP intermediate were subjected to various chemical treatments and the TCA-precipitable counts remaining after treatment were measured.

2. Dispense the labeled mixture into 5-μl aliquots, each containing approximately 20,000 cpm, and treat with 0.15 ml of each of the reagents listed in Table I and incubate as specified.

3. Following treatment, place the samples on ice and add 12 μl of 10 mg/ml denatured salmon sperm DNA and 2 ml of cold 5% (w/v) TCA and incubate for 30 min.

4. Collect the acid-insoluble material on glass microfiber filters (Whatman, Clifton, NJ, GF/C) and measure the radioactivity in 2 ml Optiphase (PerkinElmer, Boston, MA) by liquid scintillation spectrometry.

Analysis of the chemical stability of the enzyme–guanylate intermediate of Gp_4G synthetase toward acid, base, and hydroxylamine is performed as described above. Characteristically, nucleotidyl-(5' $\rightarrow$ N)phosphoramidate bonds are stable to basic and neutral conditions, but are hydrolyzed by acid treatment. If the nucleoside moiety is guanosine, it can be expected that a higher degree of hydrolysis will result because the phosphoamide nitrogen is protonated near to neutrality. Sensitivity to acidic, but not neutral, hydroxylamine is indicative of an amide-type bond and tends to exclude an ester or anhydride-type linkage involving, for example, serine or threonine. Importantly, among the phosphoramidates, nucleotide-imidazolides have the greatest reactivity and readily accept and transfer phosphate.[11] The results in Table I confirm that the linkage is a phosphoramidate with the sensitivity to neutral hydroxylamine suggesting the possibility of a histidyl-GMP.[12]

---

[11] Z. A. Shabarova, *Prog. Nucleic Acid Res. Mol. Biol.* **10**, 145 (1970).
[12] Y.-F. Wei and H. R. Matthews, *Methods Enzymol.* **200**, 388 (1991).

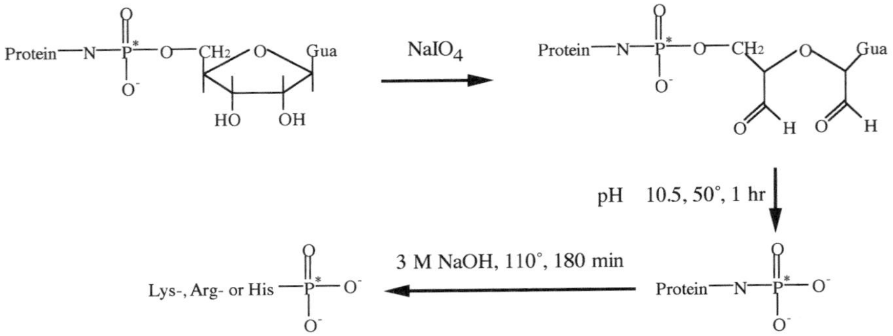

FIG. 3. Scheme outlining the generation of [^{32}P]phosphoamino acid from the [^{32}P]guanylyl-enzyme reaction intermediate. P* represent the ^{32}P-labeled phosphate.

Preparation of Authentic Phosphoamino Acids

In order to identify precisely the amino acid involved in the reaction intermediate, authentic phosphoamino standards are required. Of the possible N-phosphoamino acids—N-ω-phosphoarginine, N-ε-phospholysine, $N^{δ1}$-phosphohistidine, and $N^{ε2}$-phosphohistidine—only N-ω-phosphoarginine is sufficiently stable to be sold commercially; thus synthesis of phosphohistidines and N-ε-phospholysine must be carried out in the laboratory. N-ω-Phosphoarginine was obtained from Sigma (St. Louis, MO). N-ε-Phospholysine was synthesized[13] and phosphohistidines synthesized[14] and purified[12] as described. 1-Phosphohistidine and 3-phosphohistidine in the nomenclature of Hultquist et al.[14] correspond to $N^{δ1}$-phosphohistidine and $N^{ε2}$-phosphohistidine, respectively. The phosphoamino acids should be used shortly after preparation.

Characterization of Guanylylated Amino Acid Residue

Identification of the guanylylated amino acid involves elimination of the nucleosidyl moiety with periodate followed by hydrolysis of the resulting [^{32}P]phosphoprotein and chromatographic identification of the labeled amino acid residue. The experimental scheme is outlined in Fig. 3.

1. Dissolve the enzyme–guanylate intermediate, prepared as described above, in 100 μl of 50 mM sodium carbonate (pH 10.5) containing 1% sodium dodecyl sulfate (SDS).

---

[13] J. M. Fujitaki, A. W. Steiner, S. E. Nichols, E. R. Helander, Y. C. Lin, and R. A. Smith, *Prep. Biochem.* **10**, 205 (1980).

[14] D. E. Hultquist, R. W. Moyer, and P. D. Boyer, *Biochemistry* **5**, 322 (1966).

2. Add 11 μl of 100 mM NaIO$_4$ to give a final concentration of approximately 10 mM and incubate the solution for 30 min at room temperature in the dark.[15]

3. Quench the periodate by the addition of 40 μl of 0.18 M ethylene glycol.

4. Adjust the pH to 11.0 with 1 M NaOH and heat the solution at 50° for 1 hr to eliminate the cleaved nucleoside.

5. Transfer the solution to a 5-ml glass test tube and adjust the volume to 200 μl with doubly distilled H$_2$O. Add 86 μl of 10 M KOH and place the test tube inside a tightly sealed 100-ml Duran flask containing 10 ml of 3 M KOH (necessary to prevent sample evaporation). Hydrolyze the protein at 110° for 180 min in dilute solution (5 μg protein/ml) to minimize randomization of the ^{32}P-labeled phosphoryl group among the basic endogenous amino acids in the sample.[15]

6. Neutralize the hydrolyzate with 10% (v/v) HClO$_4$ and remove the insoluble precipitate by centrifugation at 16,000g for 5 min at room temperature.

Thin-Layer Chromatography

7. Mix a 3-μl sample of the supernatant with carrier standard phosphoamino acids (N-ω-phosphoarginine, N-ε-phospholysine, and $N^{\varepsilon 2}$-phosphohistidine) and analyze on a silica gel TLC plate along with samples of the individual phosphoamino acids (3 μl of 15 mg/ml). Perform ascending chromatography with a solvent mixture of absolute ethanol : 25% ammonia solution (3.5 : 1.6 v/v) for 90 min at room temperature.[16]

8. Visualize the phosphoamino acid markers by spraying the dried plate with 0.3% (w/v) ninhydrin solution in n-butanol containing 3% (v/v) acetic acid, followed by heating for 10 min at 60°.

9. Detect the ^{32}P by autoradiography overnight at −70°.

The results in Fig. 4 (lanes 1 and 5) show that the label derived from the enzyme–guanylate separates into three spots on TLC. The label remaining at the origin corresponds to free phosphate generated by secondary hydrolysis of the phosphoamino acid[16] (and confirmed in our own experiments), whereas the principal migrating spot corresponds to the phosphohistidine standard (lane 3, Fig. 4). The third, minor spot migrating ahead of the phosphohistidine is unidentified but probably represents a small acid-soluble oligopeptide(s) resulting from incomplete hydrolysis of the labeled protein under the minimal alkaline hydrolysis conditions used to minimize hydrolysis of the phosphoamino acids.[15] No significant amount of label comigrated with phosphoarginine (lane 2, Fig. 4) or phospholysine (lane 4, Fig. 4).

---

[15] S.-L. L. Yang and P. A. Frey, *Biochemistry* **18**, 2980 (1979).

[16] Z. Tuháčková and J. Krivánek, *Biochem. Biophys. Res. Commun.* **218**, 61 (1996).

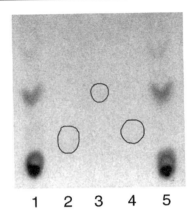

FIG. 4. Identification of [^{32}P]phosphoamino acid by thin-layer chromatography and autoradiography. Samples of ^{32}P-labeled, neutralized protein hydrolyzate were mixed with unlabeled standard phosphoamino acids and analyzed by TLC along with the individual phosphoamino acid markers as described under Experimental Procedures. The TLC plate was then stained with ninhydrin and autoradiographed and the positions of the stained standards were marked on the autoradiograph. Lane 1: protein hydrolyzate; lane 2: N-ω-phosphoarginine; lane 3: $N^{\varepsilon 2}$-phosphohistidine; lane 4: N-ε-phospholysine; lane 5, protein hydrolyzate.

Because the TLC system used does not distinguish between $N^{\delta 1}$-phosphohistidine and $N^{\varepsilon 2}$-phosphohistidine, a sample of unneutralized [^{32}P]phosphoamino acid hydrolyzate is mixed with equal amounts of $N^{\delta 1}$-phosphohistidine, $N^{\varepsilon 2}$-phosphohistidine, and $N^{\delta 1}$, $N^{\varepsilon 2}$-diphosphohistidine and subjected to anion-exchange chromatography.

High-Performance Liquid Chromatography

10. Mix a further sample (10 μl) of the unneutralized protein hydrolyzate with freshly prepared phosphohistidines[12,14] in a final volume of 100 μl and then load on to a 1 ml Mono Q HR5/5 anion-exchange column (Pharmacia, Piscataway, NJ) equilibrated in 0.05 M KHCO$_3$ (pH 8.5).

11. Elute the column at 1 ml/min with a 35-ml linear gradient from 0.05 to 0.5 M KHCO$_3$. If on-line detection facilities are available, the elution of histidines can be monitored at 235 nm simultaneously with the elution and detection of ^{32}P label. Alternatively, the detection procedures in steps 12–16 can be followed.

12. Collect fractions (0.5 ml) and determine their ^{32}P content by Cerenkov counting.

13. Recover the fractions and detect the histidine-containing compounds by a modification of the Pauly colorimetric test.[17] First, hydrolyze the phosphohistidines

---

[17] H. T. MacPherson, *Biochem. J.* **36**, 59 (1942).

by adding 111 μl of 18 M H_2SO_4 and heat to 65° for 15 min. Allow to cool to room temperature.

14. Neutralize each fraction by the addition of 10 M NaOH and then adjust the volume to 1 ml with doubly distilled H_2O.

15. To the neutralized fractions, add 100 μl of 1% sulfanilic acid in 10% HCl and 100 μl of 5% $NaNO_2$, mix, and incubate at room temperature for 30 min.

16. Develop the color by the addition of 300 μl of 20% Na_2CO_3 and 1 ml of 75% (v/v) ethanol. Measure the absorbance at 480 nm.

In the case of the enzyme–GMP intermediate of Gp_4G synthetase, peaks were identified by comparison with standards and with published results.[12,15] As before, the major radioactive peak (peak 2, Fig. 5) corresponds to free phosphate while the principal amino acid-associated radioactive peak (peak 4, Fig. 5) cochromatographs with carrier $N^{\varepsilon 2}$-phosphohistidine (Fig. 5). The later eluting, minor radioactive peaks (peaks 6 and 7, Fig. 5) again correspond to [^{32}P]phosphopeptides.[15] Phospholysine elutes between the mono- and diphosphohistidine peaks, whereas phosphoarginine does not bind to the column. Together, these results indicate that

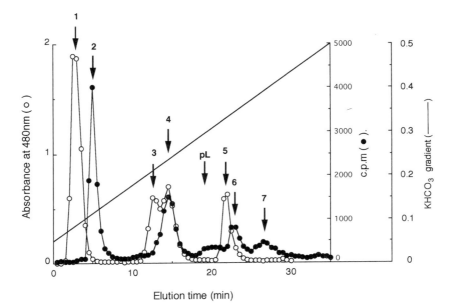

FIG. 5. Anion-exchange chromatography of ^{32}P-labeled protein hydrolyzate mixed with unlabeled phosphohistidine isomers. See Experimental Procedures for details. Peaks are identified as follows: 1: histidine; 2: [^{32}P]P_i; 3: $N^{\delta 1}$-phosphohistidine; 4: $N^{\varepsilon 2}$-phosphohistidine; 5: $N^{\delta 1}$, $N^{\varepsilon 2}$-diphosphohistidine; 6: unknown; 7: unknown. pL, elution position of phospholysine.

the α-phosphorus of the GMP of the enzyme–guanylate reaction intermediate is linked to $N^{\varepsilon 2}$ of an enzyme histidyl residue.

Discussion

Several enzymes employ nucleosidyl phosphoramidate reaction intermediates, including DNA and RNA ligases, mRNA capping enzymes,[1,2] adenosylcobinamide-phosphate guanylylytransferase,[18] and members of the recently described GAFH superfamily,[19] which includes galactose-1-phosphate uridylyltransferase (GalT),[3] the *FHIT* tumor suppressor diadenosine polyphosphate hydrolase,[20] and diadenosine polyphosphate phosphorylases.[21,22] The intermediate in the ligases and capping enzymes involves an enzyme lysyl residue while the others employ a histidyl residue. In the absence of sequence and structural information, as is the case with the *Artemia* Gp$_4$G synthetase, characterization of the reaction intermediate can help in the tentative allocation of an enzyme to a particular protein family. Because the linkage in GalT is also to Nε2 and because other members of the GAFH superfamily have dinucleoside polyphosphates as substrates, the *Artemia* Gp$_4$G synthetase may ultimately prove to be a new member of this superfamily.

[18] M. G. Thomas, T. B. Thompson, I. Rayment, and J. C. Escalante-Semerena, *J. Biol. Chem.* **275**, 27576 (2000).
[19] K. Huebner, P. N. Garrison, L. D. Barnes, and C. M. Croce, *Annu. Rev. Genet.* **32**, 7 (1998).
[20] L. D. Barnes, P. N. Garrison, Z. Siprashvili, A. Guranowski, A. K. Robinson, S. W. Ingram, C. M. Croce, M. Ohta, and K. Huebner, *Biochemistry* **35**, 11529 (1996).
[21] Y. Huang, P. N. Garrison, and L. D. Barnes, *Biochem. J.* **312**, 925 (1995).
[22] J. W. Booth and G. Guidotti, *J. Biol. Chem.* **270**, 19337 (1995).

[20] γ-Glutamyl Thioester Intermediate in Glutaminase Reaction Catalyzed by *Escherichia coli* Asparagine Synthetase B

By HOLLY G. SCHNIZER, SUSAN K. BOEHLEIN, JON D. STEWART, NIGEL G. J. RICHARDS, and SHELDON M. SCHUSTER

In humans, asparagine is biosynthesized from aspartic acid by asparagine synthetase (AS) (EC 6.3.5.4) via an ATP-dependent reaction in which glutamine is employed as a source of nitrogen.[1] Several lines of evidence suggest that inhibitors of AS, a glutamine-dependent N-terminal nucleophile (Ntn) amidotransferase,[2]

[1] N. G. J. Richards and S. M. Schuster, *Adv. Enzymol. Relat. Areas Mol. Biol.* **72**, 145 (1998).
[2] H. Zalkin and J. L. Smith, *Adv. Enzymol. Relat. Areas Mol. Biol.* **72**, 87 (1998).

[20] γ-GLUTAMYL THIOESTER INTERMEDIATE IN GLUTAMINASE REACTION 261

will have potential clinical utility in the treatment of leukemia and solid tumors.[3–5] Unfortunately, large-scale screening studies employing a range of substrate and product analogs have failed to identify compounds that act as potent and selective AS inhibitors.[6–9] The use of rational strategies[10,11] to discover inhibitors of human AS has also been precluded by a lack of information on the structure and catalytic mechanism of the enzyme.[1] In seeking a fuller understanding of AS catalysis, we have focused on the bacterial glutamine-dependent asparagine synthetase B (AS-B) (EC 6.3.5.4), encoded by the *asnB* gene in *Escherichia coli*,[12,13] primarily because this enzyme can be obtained in large amounts using standard expression protocols. Like the human enzyme, AS-B catalyzes the ATP-dependent conversion of aspartate to asparagine using glutamine or ammonia as a nitrogen source. In the absence of aspartate, AS-B catalyzes the hydrolysis of glutamine to glutamate and ammonia, a reaction that can be stimulated by the addition of ATP[14,15] (Scheme 1). The crystal structure of AS-B has revealed that the enzyme has two active sites, each catalyzing one of the two half-reactions that are combined to effect the overall transformation.[16] In the C-terminal domain, ATP and aspartate react to form a β-aspartyl-AMP intermediate,[17] and glutamine is hydrolyzed in the other active site in a reaction for which the thiol side chain of the conserved, N-terminal cysteine (Cys-1) is required.[18,19] The catalytic importance of

[3] R. Chakrabarti and S. M. Schuster, *J. Ped. Haemotol./Oncol.* **4**, 597 (1997).
[4] J. D. Broome, *J. Exp. Med.* **127**, 1055 (1968).
[5] G. F. Sanz, M. A. Sanz, F. J. Rafecas, J. A. Martinez, G. Martin-Aragon, and M. L. Marty, *Cancer Treat. Rep.* **70**, 1321 (1986).
[6] D. A. Cooney, J. S. Driscoll, H. A. Milman, H. N. Jayaram, and R. D. Davis, *Cancer Treat. Rep.* **60**, 1493 (1976).
[7] H. N. Jayaram, D. A. Cooney, J. A. Ryan, G. Neil, R. L. Dion, and V. H. Bono, *Cancer Chemother. Rep.* **59**, 481 (1975).
[8] H. N. Jayaram, D. A. Cooney, H. A. Milman, E. R. Homan, and R. J. Rosenbluth, *Biochem. Pharmacol.* **25**, 1571 (1976).
[9] D. A. Cooney, M. T. Jones, H. A. Milman, D. M. Young, and H. N. Jayaram, *Int. J. Biochem.* **11**, 519 (1980).
[10] R. S. Bohacek, C. McMartin, and W. C. Guida, *Med. Res. Dev.* **16**, 3 (1996).
[11] H.-J. Boehm and G. Klebe, *Angew. Chem. Int. Ed. Engl.* **35**, 2588 (1996).
[12] S. K. Boehlein, N. G. J. Richards, and S. M. Schuster, *J. Biol. Chem.* **269**, 7450 (1994).
[13] M. A. Scofield, W. S. Lewis, and S. M. Schuster, *J. Biol. Chem.* **265**, 12895 (1990).
[14] S. K. Boehlein, N. G. J. Richards, E. S. Walworth, and S. M. Schuster, *J. Biol. Chem.* **269**, 26789 (1994).
[15] S. K. Boehlein, S. M. Schuster, and N. G. J. Richards, *Biochemistry* **35**, 3031 (1996).
[16] T. M. Larsen, S. K. Boehlein, S. M. Schuster, N. G. J. Richards, J. B. Thoden, H. M. Holden, and I. Rayment, *Biochemistry* **38**, 16146 (1999).
[17] S. K. Boehlein, J. D. Stewart, E. S. Walworth, R. Thirumoorthy, N. G. J. Richards, and S. M. Schuster, *Biochemistry* **37**, 13230 (1998).
[18] G. Van Heeke and S. M. Schuster, *J. Biol. Chem.* **264**, 19475 (1989).
[19] S. Sheng, D. A. Moraga-Amador, G. Van Heeke, R. D. Allison, N. G. J. Richards, and S. M. Schuster, *J. Biol. Chem.* **268**, 16771 (1993).

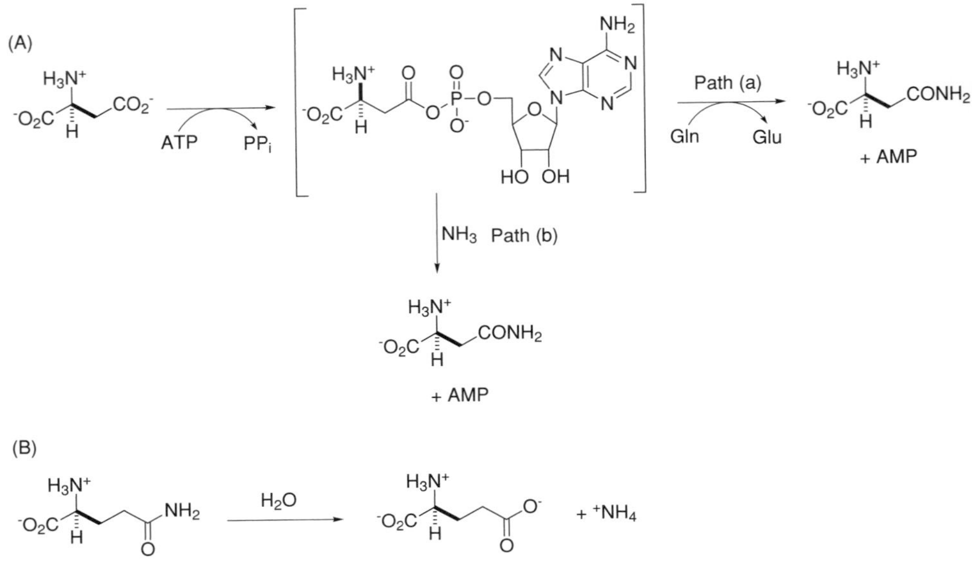

SCHEME 1

this cysteine places AS in the family of the Ntn amidotransferases,[20] which also includes glutamine 5'-phosphoribosyl-1-pyrophosphate amidotransferase (GPATase [EC 2.4.2.14]),[21] glutamine fructose-6-phosphate amidotransferase (GFAT) (EC 2.4.1.90),[22] NAD$^+$ synthetase (EC 6.3.5.1),[23] and glutamate synthase (EC 1.4.1.14).[24] Ammonia that is released in the N-terminal domain by cleavage of glutamine to glutamate is then thought to pass through an intramolecular channel[16,25–28] to the other active site containing the β-aspartyl-AMP intermediate resulting in the formation of asparagine and AMP. This ammonia transfer event therefore couples the two half-reactions catalyzed by the enzyme.

In view of the essential role of Cys-1 in catalyzing the glutamine-dependent activity of AS, GPATase, and GFAT, it was initially proposed that amide bond

---

[20] J. L. Smith, *Biochem. Soc. Trans.* **23**, 894 (1995).
[21] J. Y. Tso, H. Zalkin, M. van Cleemput, C. Yanofsky, and J. M. Smith, *J. Biol. Chem.* **257**, 3525 (1982).
[22] M. A. Badet-Denisot, L. René, and B. Badet, *Bull. Chim. Soc. Fr.* **130**, 249 (1993).
[23] M. Rizzi, M. Bolognesi, and A. Coda, *Structure Fold. Des.* **6**, 1129 (1998).
[24] M. A. Vanoni, L. Nuzzi, M. Rescigno, G. Zanetti, and B. Curti, *Eur. J. Biochem.* **202**, 181 (1991).
[25] F. M. Raushel, J. B. Thoden, and H. M. Holden, *Biochemistry* **38**, 7891 (1999).
[26] E. W. Miles, S. Rhee, and D. R. Davies, *J. Biol. Chem.* **274**, 12193 (1999).
[27] F. Massière and M.-A. Badet-Denisot, *Cell. Mol. Life Sci.* **54**, 205 (1998).
[28] J. L. Smith, *Curr. Opin. Struct. Biol.* **8**, 686 (1998).

hydrolysis proceeded in a similar manner to thiol proteinases,[29] such as the cathepsins,[30] papain,[31] and caspases.[32] In the latter enzymes, protonation of the leaving group nitrogen is a key step in facilitating subsequent C–N bond cleavage in the catalytic mechanism[33] and these enzymes therefore possess an active site cysteine/histidine dyad.[29] Hence the His-159 side chain is specifically oriented in papain so as to promote breakdown of the initial tetrahedral intermediate and to stabilize the thiolate of Cys-25.[34] Extensive mutagenesis studies using recombinant AS-B, however, failed to identify catalytically important histidine residues,[12] and the crystal structures of AS,[16] GPATase,[35] and GFAT[36] appear to confirm this observation. The mechanistic relationship between amide hydrolysis in Ntn amidotransferases and thiol proteases[29,31] is therefore intriguing, given the subtle differences in catalytic functional groups used by these two families of enzymes.[37–39] Thioesters are proposed to be intermediates in thiol-catalyzed amide hydrolysis, and evidence for their participation in glutamine breakdown by Class I amidotransferases has been reported.[40–43] We now summarize conditions for isolating a covalent adduct that is formed when AS-B is incubated with glutamine, and present evidence that this molecular species corresponds to a γ-glutamyl thioester intermediate formed by attack of the Cys-1 side chain on the amide of the substrate.[44]

---

[29] K. Brocklehurst, F. Willenbrock, and E. Salih, in "Hydrolytic Enzymes" (A. Neuberger and K. Brocklehurst, eds.), p. 39. Elsevier, Amsterdam, 1987.

[30] D. Musil, D. Zucic, D. Turk, R. A. Engh, I. Mayr, R. Huber, T. Popovic, V. Turk, T. Towatari, N. Katunuma, and W. Bode, *EMBO J.* **10**, 2321 (1991).

[31] A. C. Storer and R. Ménard, *Methods Enzymol.* **244**, 486 (1994).

[32] N. A. Thornberry and S. M. Molineaux, *Protein Sci.* **4**, 3 (1995).

[33] M. H. O'Leary, M. Urberg, and A. P. Young, *Biochemistry* **13**, 2077 (1974).

[34] I. G. Kamphuis, H. M. Kalk, M. B. A. Swarte, and J. Drenth, *J. Mol. Biol.* **179**, 233 (1984).

[35] J. M. Krahn, J. H. Kim, M. R. Burns, R. J. Parry, H. Zalkin, and J. L. Smith, *Biochemistry* **36**, 11061 (1997).

[36] M. N. Isupov, G. Obmolova, S. Butterworth, M.-A. Badet-Denisot, B. Badet, I. Polikarpov, J. A. Littlechild, and A. Teplyakov, *Structure* **4**, 801 (1996).

[37] S. K. Boehlein, J. G. Rosa-Rodriguez, S. M. Schuster, and N. G. J. Richards, *J. Am. Chem. Soc.* **119**, 5785 (1997).

[38] J. A. Brannigan, G. Dodson, H. J. Duggleby, P. C. E. Moody, J. L. Smith, D. R. Tomchick, and A. G. Murzin, *Nature* **378**, 416 (1995).

[39] M. Peräkylä and P. A. Kollman, *J. Am. Chem. Soc.* **119**, 1189 (1997).

[40] M. G. Chaparian and D. R. Evans, *J. Biol. Chem.* **266**, 3387 (1991).

[41] B. Roux and C. T. Walsh, *Biochemistry* **31**, 6904 (1992).

[42] C. J. Lusty, *FEBS Lett.* **314**, 135 (1992).

[43] J. B. Thoden, S. G. Miran, J. C. Phillips, A. J. Howard, F. M. Raushel, and H. M. Holden, *Biochemistry* **37**, 8825 (1998).

[44] H. G. Schnizer, S. K. Boehlein, J. D. Stewart, N. G. J. Richards, and S. M. Schuster, *Biochemistry* **38**, 3677 (1999).

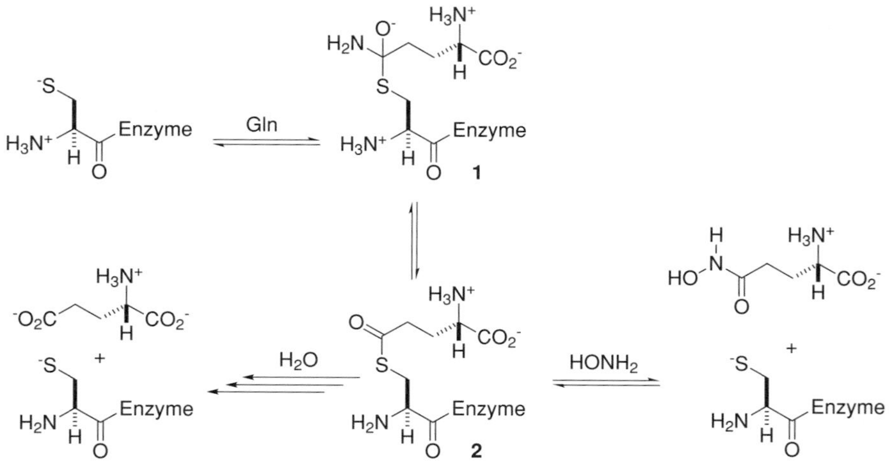

SCHEME 2

Reactions Catalyzed by Asparagine Synthetase

Biosynthetic and Partial Reactions

Asparagine biosynthesis (Scheme 1A) proceeds by initial reaction of aspartate and ATP to yield a β-aspartyl-AMP intermediate (βAspAMP). In the presence of glutamine, ammonia released in the N-terminal active site reacts with βAspAMP to yield asparagine and AMP [path (a)]. Ammonia can also be employed as an alternative nitrogen source for asparagine synthesis [path (b)]. AS-B also catalyzes the hydrolysis of glutamine to glutamate in the absence of aspartate (Scheme 1B).

Minimal Mechanism for Formation of γ-Glutamyl Thioester Intermediate in AS-B Glutaminase Activity and Its Partitioning in Presence of Hydroxylamine

Reaction of the Cys-1 thiolate with glutamine (Scheme 2) yields tetrahedral intermediate **1** that then loses ammonia, in a reaction for which the N-terminal amine group acts as a general acid, to give a thioester derivative **2**. Reaction with water gives glutamate and regenerates the free enzyme. The thioester **2** can also be trapped by hydroxylamine to yield γ-glutamylhydroxamate.

Methods

Partitioning γ-Glutamyl Thioester Intermediate to γ-Glutamylhydroxamate

Hydroxylamine has been widely used to provide indirect evidence for the formation of acyl-enzyme intermediates in amide hydrolysis reactions catalyzed by serine proteases[45,46] and the Class I glutamine-dependent amidotransferases carbamoyl-phosphate synthetase–aspartate transcarbamoylase–dihydroorotase (CAD)[40] (EC 3.5.2.3) and carbamoyl-phosphate synthetase (CPS) (EC 6.3.5.5).[42] Reaction mixtures containing wild-type AS-B (9.3 μg), 50 mM glutamine, 100 mM Bis–Tris, Tris–HCl (pH 8), and variable concentrations of hydroxylamine are incubated at 37° for 45 min (300 μl total volume). The reaction is terminated by the addition of 100 μl of 16% trichloroacetic acid (TCA). An aliquot (50 μl) is removed, and the amount of glutamate formed is measured using glutamate dehydrogenase in the presence of NAD^+.[47] In this end-point assay, coupling reagent (300 mM glycine, 250 mM hydrazine, pH 9, 1 mM ADP, 1.6 mM NAD^+, 2.2 units of glutamate dehydrogenase) is added to the aliquot and the resulting mixture incubated for 10 min at room temperature. The absorbance at 340 nm is then measured and the glutamate concentration determined by comparison to a standard curve. γ-Glutamyl-NHOH formation is assayed by adding a solution containing 80% TCA, 6 N HCl, and 10% $FeCl_3$ in 0.02 N HCl to the remaining reaction solution (500 μl final volume). After centrifugation of the samples to remove particulates, the absorbance at 540 nm is determined, which is associated with the complex formed between Fe^{3+} and γ-glutamyl-NHOH. A standard curve, obtained using solutions of authentic γ-glutamyl-NHOH, is used to quantitate the amount of product formed in these reactions.

Isolation of Putative γ-Glutamyl Thioester Formed by Incubation of AS-B with [^{14}C]Glutamine

Wild-type AS-B (0.74 nmol) or the C1A AS-B mutant (0.74 nmol) is incubated with 1 mM L-[U-^{14}C]-glutamine (SA 22 000 dpm/nmol) in 100 mM Tris-HCl (pH 8.0) at room temperature (100 μl total reaction volume). After 30 sec, the reaction is quenched using 8% TCA (1 ml) and 100 μl BSA (10 mg/ml) is added. After 2 min in the quench solution the samples are filtered under vacuum on a 2.5-cm nitrocellulose filter (0.45 μm porosity), and the filter is washed with 1 N HCl (50 ml). the filter is transferred to 5 ml of scintillation fluid (ScintiVers II*) and the ^{14}C activity measured on a Beckman LS6000IC scintillation counter. This procedure is repeated in the absence of enzyme to determine the nonspecific binding of L-glutamine to the filter.

Stability of Putative γ-Glutamyl Thioester Formed by Incubation of AS-B with [^{14}C]Glutamine

Wild-type AS-B (4.8 nmol) is incubated at room temperature for 30 sec with 1.5 mM L-[U-^{14}C]glutamine (SA 25 633 dpm/nmol) in 100 mM Bis–Tris and

---

[45] M. Caplow and W. P. Jencks, *J. Biol. Chem.* **239**, 1640 (1964).
[46] R. M. Epand and I. B. Wilson, *J. Biol. Chem.* **238**, 1718 (1963).
[47] E. Bernt and H. U. Bergmeyer, in "Methods of Enzymatic Analysis" (H. U. Bergmeyer, ed.), p. 1704. Academic Press, New York, 1974.

Tris–HCl (pH 8) (100 μl total volume). The enzyme-catalyzed reaction is terminated by the addition of 22 μl of 0.5 M sodium acetate (pH 4) containing 5% sodium dodecyl sulfate (SDS). The reaction solution is divided into two portions, and protein is separated from free L-glutamine on a Sephadex G-50 Fine spin column equilibrated in either (a) 0.1 M sodium phosphate (pH 12) containing 1% SDS, or (b) 0.1 M sodium phosphate (pH 2) containing 1% SDS. The amount of radiolabeled enzyme that passes through the column is determined by liquid scintillation counting, and the total protein concentration is measured using a standard Bio-Rad (Hercules, CA) assay kit. The percentage of radiolabeled enzyme is calculated as an average of values from six separate experiments.

Base Lability of Covalent Adduct Isolated by Gel Filtration

Wild-type AS-B (4.8 nmol) is incubated for 30 sec at room temperature with 1.5 mM L-[U-^{14}C]glutamine (SA 25 633 dpm/nmol) in 100 mM Bis–Tris and Tris–HCl (pH 8). The enzyme-catalyzed reaction is terminated by adding 22 μl of 0.5 M sodium acetate (pH 4) in 5% SDS. To determine the stability of the putative γ-glutamyl thioester intermediate at pH 2 and 12, protein is separated from free glutamine[48] using a Sephadex G-50 Fine spin column equilibrated in 0.1 M sodium phosphate (either at pH 2 or pH 12) containing 1% SDS. The amount of radiolabeled enzyme in 70 μl effluent is then determined by scintillation counting, and the protein concentration measured using a standard assay kit (Bio-Rad). Average values, together with the standard error, are derived from six independent experiments.

Deacylation Rate of Thioester Intermediate

The γ-glutamyl enzyme adduct is formed at 5° by incubating 0.74 nmol of wild-type AS-B with 1.5 mM L-[U-^{14}C]glutamine (SA 20,000 dpm/nmol) in a 100 μl reaction containing 100 mM Bis–Tris and Tris (pH 8.0). After the reaction is incubated for 1 min, 100 μl of 200 mM nonradioactive glutamine is added to each sample to dilute unreacted L-[U-^{14}C]glutamine. Samples are quenched at various time intervals after dilution, as described above, and the amount of radioactivity associated with the protein is determined by scintillation counting.

Glutaminase Assays

Initial rates of L-glutamine hydrolysis are determined by a modified procedure that employs glutamate dehydrogenase in the presence of NAD$^+$ to measure glutamate concentration.[47] Reaction mixtures (100 μl) contain 100 mM Bis–Tris and Tris–HCl (pH 8.0), 8 mM MgCl$_2$, and various concentrations of glutamine.

[48] H. S. Penefsky, *Methods Enzymol.* **56**, 527 (1979).

Glutamine hydrolysis is initiated by the addition of wild-type AS-B at various temperatures (4–40°). Spontaneous breakdown of the substrate to form glutamate under these conditions is monitored in a control mixture lacking the enzyme. The reactions are terminated by the addition of 20 μl 1 N acetic acid. A coupling reagent (300 mM glycine, 250 mM hydrazine, pH 9, 1 mM ADP, 1.6 mM NAD$^+$, 2.2 units of glutamate dehydrogenase) is then added and the resulting mixture incubated for 10 min at room temperature. The absorbance at 340 nm is measured, and the concentration of glutamate is determined by comparison to a standard curve. Initial velocities are measured at nine different substrate concentrations, and each velocity is an average of three measurements. The stability of the wild-type AS-B is confirmed by incubation in buffer at the appropriate temperature for various amounts of time, followed by an assay of its activity at pH 8. Initial velocity conditions are confirmed at each pH by linear plots of velocity versus time and velocity versus enzyme concentration. The pH of the solution is adjusted to pH 8 at each temperature to account for the change in buffer pK_a.

Properties of Thiol Ester Intermediate

In order to establish that the adduct formed by incubation of L-glutamine with wild-type AS-B represented a true intermediate in the AS-B-catalyzed glutaminase reaction, quantitative kinetic measurements were undertaken by using filter binding assays, similar to those employed in other successful attempts to isolate thioacylenzyme intermediates.[41] Thus, AS-B was mixed with L-[U-^{14}C]glutamine for 30 sec and then denatured with TCA to yield a precipitate that was filtered and washed with 1 N HCl. Radioactivity remaining on the filter was determined by liquid scintillation counting. Approximately 37.5% of wild-type AS-B became radiolabeled under these experimental conditions (Table I). In control experiments, no radiolabeled adduct was observed when the C1A AS-B mutant, in which an alanine residue substitutes Cys-1, was incubated with L-[U-^{14}C]glutamine in place of the wild-type enzyme.[44] The site-specific C1A mutant retains the ability to synthesize asparagine when ammonia is used as a nitrogen source but lacks both glutamine-dependent activities of wild-type AS-B. In addition, the C1A AS-B

TABLE I
ISOLATION OF COVALENT ADDUCT FORMED BY INCUBATION OF AS-B AND RADIOLABELED GLUTAMINE USING FILTER-BINDING ASSAY

Incubation conditions	Radioactivity (dpm$_{ave}$)	Fraction of enzyme labeled (%)
Wild-type AS-B + [U-^{14}C]Gln	6500 ± 183	37.5
Wild-type AS-B + DON + [U-^{14}C]Gln	190 ± 55	<1
C1A AS mutant + [U-^{14}C]Gln	300 ± 115	<1
[U-^{14}C]Gln only	400 ± 187	—

mutant binds L-glutamine with a higher affinity than wild-type AS-B.[12] In a second set of control experiments, wild-type AS-B was pretreated with 6-diazo-5-oxo-L-norleucine (DON),[49] an irreversible inhibitor of the glutamine-dependent activity of class II amidotransferases.[50] DON inactivation of wild-type AS-B can also be blocked by L-glutamine, suggesting that this reagent forms a covalent bond specifically with the side chain of Cys-1, a hypothesis that has been confirmed crystallographically for *Escherichia coli* GPATase.[35,51] The DON-modified AS-B completely retained ammonia-dependent synthetase activity, but did not yield a radiolabeled covalent adduct when incubated with L-[U-^{14}C]glutamine under standard conditions (Table I). The thiolate side chain of Cys-1 therefore appears necessary for formation of radiolabeled protein when AS-B is incubated with radiolabeled glutamine.

Two observations provided additional support for the hypothesis that the radiolabeled protein adduct isolated in these experiments was the γ-glutamyl thioester **2**. First, examination of the effect of glutamine concentration on the amounts of the covalent adduct formed under steady-state conditions revealed a saturation curve in which the maximum molar ratio of adduct to AS-B was 0.37 (Fig. 1). Second, the glutamine concentration required to reach half-saturation was 0.14 mM, which is similar to the apparent K_m for glutamine of 0.17 mM (data not shown) observed for wild-type AS-B glutaminase activity.[9] Additional evidence for the idea that the covalent adduct isolated in these assays was indeed the γ-glutamyl thioester intermediate was provided by the pH dependence of the stability of the radiolabeled adduct, which was investigated using a gel filtration protocol. Briefly, wild-type AS-B was incubated with L-[U-^{14}C]glutamine under standard conditions, and then the resulting adduct was subjected to gel filtration chromatography through columns equilibrated at either low or high pH. Approximately $36 \pm 4\%$ and $6 \pm 2\%$ of radiolabeled wild-type AS-B was obtained after gel filtration at pH 2 and 12, respectively, demonstrating the putative intermediate was acid-stable and base-labile. This is consistent with the pH dependence observed for thioester stability. The reliability of the gel filtration procedure was further indicated by the observation that the amount of radiolabeled enzyme isolated from the column at pH 2 is identical to that observed at saturating concentrations of L-glutamine in the filter binding measurements.

Attempts to measure the rate constant for forming the putative thioester were unsuccessful. Even at 5°, formation of the intermediate seems to occur within 10 sec. The rate constant associated with loss of bound radioactivity from AS-B was determined by incubating the preformed complex with a large excess of unlabeled

---

[49] R. E. Handschumacher, C. J. Bates, P. K. Chang, A. T. Andrews, and G. A. Fischer, *Science* **161**, 62 (1968).

[50] H. Zalkin, *Adv. Enzymol. Relat. Areas Mol. Biol.* **66**, 203 (1993).

[51] J. H. Kim, J. M. Krahn, D. R. Tomchick, J. L. Smith, and H. Zalkin, *J. Biol. Chem.* **271**, 15549 (1996).

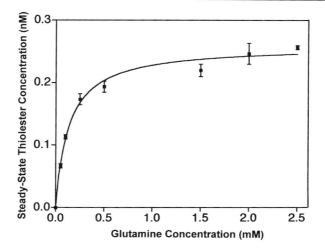

FIG. 1. Effect of glutamine on the steady-state concentration of the putative γ-glutamyl thioester intermediate. The steady-state concentration of the putative thioester intermediate was measured at 5° by incubating wild-type AS-B (0.74 nmol) with various concentrations of [^{14}C]glutamine in 100 mM Tris-HCl (pH 8) (100 μl total volume). After 30 sec, the reaction was quenched and radiolabeled enzyme was trapped using the filter binding protocol described in text. The glutamine concentration at half-saturation was determined from nonlinear regression analysis. Reprinted with permission from H. G. Schnizer, S. K. Boehlein, J. D. Stewart, N. G. J. Richards, and S. M. Schuster, *Biochemistry* **38**, 3677 (1999). Copyright © 1999 American Chemical Society.

glutamine. Under these conditions, the disappearance of radiolabeled protein could be modeled using a single exponential, giving a rate constant of 0.16 ± 0.02 s^{-1} (Fig. 2). This value is 4 times higher than the k_{cat} value of 0.043 ± 0.007 s^{-1} measured for AS-B catalyzed glutamine hydrolysis.

Concluding Remarks

Glutamine-dependent amidotransferases have been classified into three classes on the basis of conserved residues in the deduced amino acid sequences of their glutamine-hydrolyzing domains, or subunits.[20,50] The family of Class I amidotransferases, which includes carbamoyl-phosphate synthetase (CPS)[25,52] and guanosine-5'-monophosphate synthetase (GMPS),[53] is characterized by a conserved Cys-His dyad that is essential in catalyzing the conversion of glutamine to glutamate.[42,53,54] The conserved glutamate in the glutaminase active site of these

[52] J. B. Thoden, H. M. Holden, G. Wesenberg, F. M. Raushel, and I. Rayment, *Biochemistry* **36**, 6305 (1997).
[53] J. J. G. Tesmer, T. J. Klem, M. L. Deras, V. J. Davisson, and J. L. Smith, *Nat. Struct. Biol.* **3**, 74 (1996).
[54] X. Huang and F. M. Raushel, *Biochemistry* **38**, 15909 (1999).

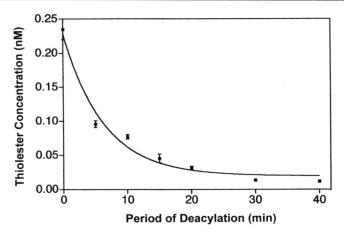

FIG. 2. Rate of deacylation of the thioester intermediate formed during glutamine hydrolysis catalyzed by wild-type asparagine synthetase B. Reprinted with permission from H. G. Schnizer, S. K. Boehlein, J. D. Stewart, N. G. J. Richards, and S. M. Schuster, *Biochemistry* **38,** 3677 (1999). Copyright © 1999 American Chemical Society.

enzymes has been shown to have no catalytic role, at least in CPS,[54] suggesting that Class I amidotransferases catalyze glutamine hydrolysis in a manner akin to thiol proteases such as papain (EC 3.4.22.2).[31] Hence, initial attack of the thiolate on the amide yields a tetrahedral intermediate, the existence of which has been demonstrated in kinetic isotope effect (KIE) studies,[55] that collapses to give a γ-glutamyl thioester and ammonia. Subsequent hydrolysis of the reactive thioester regenerates the enzyme and produces glutamate. The active-site histidine facilitates this reaction by acting as a general acid/base catalyst. There is ample evidence to support such a mechanism, including isolation of the thioester in CAD,[40] the PabA subunit of *Escherichia coli* *p*-aminobenzoic acid synthase (EC 4.13),[41] and CPS.[42,56] In more recent work, using the HN CPS mutant, the thioester intermediate formed in CPS has been directly observed by X-ray crystallography.[57]

In contrast, only the N-terminal, conserved cysteine residue is required for glutamine-dependent activity in Ntn amidotransferases.[18,58–60] A family of amidotransferases sharing a third type of glutamine-hydrolyzing domain that

---

[55] M. A. Rishavy, W. W. Cleland, and C. J. Lusty, *Biochemistry* **39,** 7309 (2000).
[56] V. P. Wellner, P. M. Anderson, and A. Meister, *Biochemistry* **12,** 2061 (1973).
[57] J. B. Thoden, X. Y. Huang, F. M. Raushel, and H. M. Holden, *Biochemistry* **38,** 16158 (1999).
[58] B. Mei and H. Zalkin, *J. Biol. Chem.* **264,** 16613 (1989).
[59] N. Kucharczyk, M.-A. Denisot, F. Le Goffic, and B. Badet, *Biochemistry* **29,** 3668 (1990).
[60] M. A. Vanoni, D. E. Edmondson, M. Rescigno, G. Zanetti, and B. Curti, *Biochemistry* **30,** 11478 (1991).

contains an amidase sequence motif[61] has also been identified, of which glutamine Glu-tRNA amidotransferase is the first member.[62,63] Although the mechanistic relationship between Class I amidotransferases and thiol proteases is now firmly established, the situation remains less clear for Ntn amidotransferases given the absence of an active site histidine that can participate in general acid/base catalysis. In addition, Ntn amidotransferases exhibit significant differences in their ability (i) to hydrolyze glutamine in the absence of other substrates and (ii) to employ ammonia as an alternative nitrogen source.[1] It has therefore been proposed that the N-terminal amino group of Cys-1 performs the role of an active-site histidine (Scheme 2). Hence, nucleophilic attack of the Cys-1 thiolate on the amide of glutamine yields a tetrahedral intermediate **1** that can then break down to give a thioester **2** and ammonia. C–N bond cleavage is thought to be catalyzed by proton transfer from the N-terminal amine, although unambiguous kinetic evidence supporting this hypothesis has yet to be reported. In addition, heavy-atom KIE determinations reveal subtle differences in the detailed kinetics of the underlying mechanisms of amide hydrolysis employed by AS-B and papain.[64]

In conclusion, the stoichiometry, saturation behavior, and rates of formation and breakdown of the covalent intermediate isolated in these experiments are consistent with its participation in the normal mechanism of AS-B catalyzed glutamine hydrolysis. The covalent adduct isolated when AS-B is incubated with radiolabeled glutamine is the putative γ-glutamyl thioester intermediate **2**. The fact that the rates of formation and breakdown of the adduct were at least 4.7- and 4-fold greater, respectively, than the steady-state k_{cat} value for AS-B glutaminase activity suggests that one or more steps after thioester formation are largely rate-limiting. The observation of nearly identical k_{cat} values for the enzyme-catalyzed hydrolyses of glutamine and more reactive substrate analogs, such as γ-glutamyl hydroxamate and glutamate γ-methyl ester, further supports this proposal.[15] Although the rates at which these substrates form the covalent γ-glutamyl thioester intermediate **2** may differ, the overall turnover rate is limited by subsequent events that are common to all three mechanisms.

Acknowledgment

This work was supported by Grant CA 28725 from the National Cancer Institute of the National Institutes of Health, DHHS.

---

[61] J. F. Mayaux, E. Cerbelaud, F. Soubrier, P. Yeh, F. Blanche, and D. Petre, *J. Bacteriol.* **173**, 6694 (1991).

[62] A. W. Curnow, K. W. Hong, R. Yuan, S. I. Kim, O. Martens, W. Winkler, T. M. Henkin, and D. Soll, *Proc. Natl. Acad. Sci. U.S.A.* **94**, 11819 (1997).

[63] D. Ahn, Y. C. Kim, Y. Ishino, M. W. Chen, and D. Soll, *J. Biol. Chem.* **265**, 8059 (1990).

[64] P. W. Stoker, M. H. O'Leary, S. K. Boehlein, S. M. Schuster, and N. G. J. Richards, *Biochemistry* **35**, 3024 (1996).

[21] γ-Glutamyltranspeptidase and γ-Glutamyl Peptide Ligases: Fluorophosphonate and Phosphonodifluoromethyl Ketone Analogs as Probes of Tetrahedral Transition State and γ-Glutamyl-Phosphate Intermediate

By JUN HIRATAKE, MAKOTO INOUE, and KANZO SAKATA

Introduction

L-Glutamine is an important metabolic nitrogen source in plants and bacteria,[1] and a tripeptide glutathione (γ-L-Glu-L-Cys-Gly) is a universal biological antioxidant and detoxifying agent.[2] These physiologically important molecules contain a unique γ-glutamylamide bond which is not observed in normal proteins and peptides. The γ-glutamylamide bond is neither formed by the ribosomal peptidyltransferases nor cleaved by the conventional proteinases, but is formed and cleaved by specific γ-glutamylamide ligases and hydrolases. Glutamine synthetase (GS, EC 6.3.1.2, glutamate–ammonia ligase) and γ-glutamylcysteine synthetase (γ-GCS, EC 6.3.2.2, glutamate–cysteine ligase) are typical ligases for the γ-glutamylamide synthesis, and γ-glutamyltranspeptidase (GGT, EC 2.3.2.2, γ-glutamyltransferase) is a specific amidohydrolase cleaving the γ-glutamylamide bond. Thus, GS catalyzes the synthesis of L-glutamine from L-glutamic acid and NH_3 with concomitant hydrolysis of ATP.[1] Similarly, γ-GCS catalyzes the formation of γ-L-glutamyl-L-cysteine, a glutathione precursor, from L-glutamic acid and L-cysteine with ATP as a coupling agent.[3] On the other hand, GGT catalyzes the transfer of the γ-glutamyl group of glutathione and its derivatives to water or to amino acids and peptides to give L-glutamic acid (hydrolysis) or γ-glutamyl peptides (transpeptidation).[4]

The reaction mechanisms of these enzymes represent a typical chemistry employed by the γ-glutamyl group transferring enzymes; the activation of the γ-carboxy group is achieved by phosphorylation for the synthesis, and a γ-glutamyl-enzyme intermediate is involved in the cleavage of the γ-glutamylamide bond. Thus, the reactions catalyzed by GS and γ-GCS are thought to proceed by the initial phosphorylation of L-glutamic acid to form the γ-glutamyl-P as a common intermediate. The activated carbonyl is then attacked by the substrate amine (NH_3 and L-cysteine for GS and γ-GCS, respectively) to yield the respective

[1] D. L. Purich, *Adv. Enzymol. Relat. Areas Mol. Biol.* **72**, 9 (1998).
[2] O. W. Griffith, *Free Rad. Biol. Med.* **27**, 922 (1999).
[3] O. W. Griffith and R. T. Mulcahy, *Adv. Enzymol. Relat. Areas Mol. Biol.* **73**, 209 (1999).
[4] N. Taniguchi and Y. Ikeda, *Adv. Enzymol. Relat. Areas Mol. Biol.* **72**, 239 (1998).

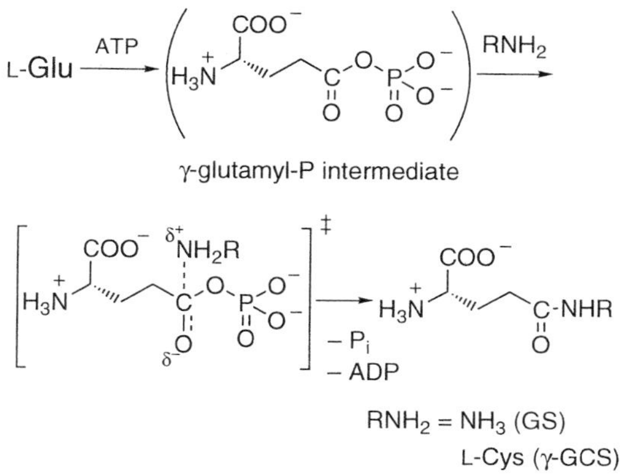

[21] γ-GLUTAMYLTRANSPEPTIDASE AND γ-GLUTAMYL PEPTIDE LIGASES

SCHEME 1

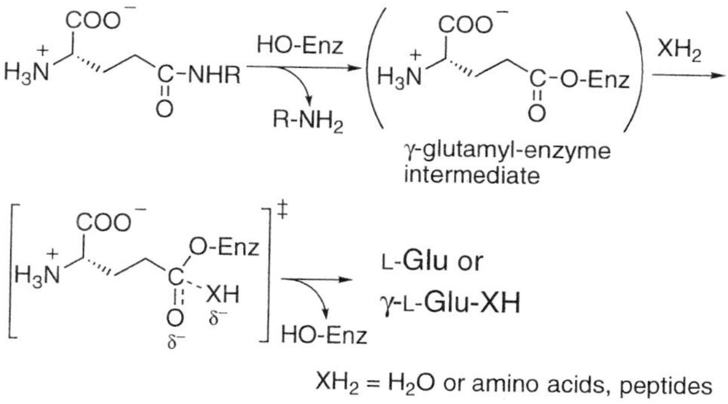

SCHEME 2

γ-glutamylamide (Scheme 1).[1,3] On the other hand, GGT is proposed to catalyze the reaction through the initial formation of a γ-glutamyl-enzyme intermediate by the attack of a catalytic nucleophile of the enzyme on the γ-carbonyl of the substrate. This intermediate is then attacked by water or by amino acids and peptides to complete the hydrolysis or the γ-glutamyl transfer, respectively (Scheme 2).[4] Like the chemical nature of the nucleophilic substitutions on a carbonyl carbon,[5] the anionic and tetrahedral intermediate or transition state flanking the trigonal acyl intermediate is also expected to be catalytically important in the enzymatic formation and cleavage of the γ-glutamylamide bond. Thus the rate-determining

---

[5] R. H. DeWofe and R. C. Newcomb, *J. Org. Chem.* **36**, 3870 (1971).

breakdown or formation of the tetrahedral intermediate should be facilitated in the enzymatic formation and cleavage of the γ-glutamylamide bond. Since these transient species and reactive intermediates are not directly accessible, the use of stable analogs is an effective way to approach such key structures. This chapter describes the synthesis and the use of the monofluorophosphonate **1** and the phosphonodifluoromethyl ketone **2** as analogs of the tetrahedral transition state and the γ-glutamyl-phosphate intermediate to probe the reaction mechanisms of GS, γ-GCS, and GGT.

γ-Glutamyltranspeptidase: Monofluorophosphonate Analog as Mechanism-Based Affinity Labeling Agent

Monofluorophosphonate 1 as Stable Affinity Label

GGT cleaves the γ-glutamyl bond of glutathione and other γ-glutamyl compounds such as S-substituted glutathiones and γ-glutamylamides to transfer the γ-glutamyl group either to water (hydrolysis) or to amino acids and peptides (transpeptidation). This enzyme plays a central role in glutathione degradation and in the regulation of intracellular glutathione level by the cooperative action with other enzymes such as glutathione-synthesizing enzymes.[6] From a mechanistic point of view, GGT is proposed to catalyze the cleavage of the γ-glutamylamide bond by an acylation–deacylation double-displacement mechanism involving an anionic and tetrahedral transition state as observed with classical serine hydrolases. Thus, as shown in Scheme 2, the reaction catalyzed by GGT is thought to proceed via a γ-glutamyl-enzyme intermediate where a nucleophilic residue in the small subunit is γ-glutamylated. A hydroxy nucleophile such as Ser and Thr has been proposed to be the catalytic residue, but the γ-glutamyl-enzyme intermediate has never been isolated, and the identification of the catalytic nucleophile was not achieved either by chemical modification or by site-directed mutagenesis.[4]

A major problem in the chemical modification is that classical inhibitors of glutamine-dependent enzymes failed to detect the catalytic nucleophile unambiguously. For example, L-(αS, $5S$)-α-amino-3-chloro-4,5-dihydro-5-isoxazoleacetic

---

[6] C. O. Harding, P. Williams, E. Wagner, D. S. Chang, K. Wild, R. E. Colwell, and J. A. Wolff, *J. Biol. Chem.* **272**, 12560 (1997).

[21] γ-GLUTAMYLTRANSPEPTIDASE AND γ-GLUTAMYL PEPTIDE LIGASES   275

$$\underset{\text{acivicin}}{\text{H}_3\overset{+}{\text{N}}-\overset{\text{COO}^-}{\underset{}{\text{CH}}}-\text{CH}_2-\underset{\underset{\text{N}}{|}}{\underset{\text{O}}{\underset{||}{\text{C}}}}-\text{Cl}} \qquad \underset{\substack{\text{6-diazo-5-oxo-L-norleucine}\\\text{(DON)}}}{\text{H}_3\overset{+}{\text{N}}-\overset{\text{COO}^-}{\underset{}{\text{CH}}}-\text{CH}_2-\underset{\underset{||}{\text{O}}}{\text{C}}-\text{CH}=\overset{+}{\text{N}}=\overset{-}{\text{N}}}$$

acid (acivicin) and 6-diazo-5-oxo-L-norleucine (DON) inactivate GGT by forming a covalent bond to the small subunit. The labeling studies on several GGTs with [^{14}C]acivicin identified Thr-523 (rat kidney enzyme),[7] Ser-405 (pig kidney), and Ser-406 (human kidney)[8] in the small subunit as the labeled site. However, site-directed mutagenesis revealed that none of these amino acid residues was essential for catalysis or for the inactivation by acivicin.[8] This misleading result probably comes from the chemical nature of acivicin: that it is bound to the enzyme by an unstable hydroximic ester bond[9] and is prone to migrate by transesterification to any nucleophilic hydroxy residue(s) nearby from the initial modification site. Therefore an ideal reagent to trap the catalytic nucleophile of GGT is a compound that will react rapidly and selectively with a hydroxy nucleophile in a mechanism-based manner to afford a stable adduct devoid of postmodificational migration. From a chemical and mechanistic point of view, 2-amino-4-(fluorophosphono)butanoic acid (**1**),[10] a γ-monofluorophosphono derivative of glutamic acid, meets the requirements. Compound **1** binds covalently to the catalytic nucleophile in a mechanism-based manner, forming an anionic and tetrahedral transition-state-like adduct in the enzyme active site (Scheme 3). A negative charge on the phosphonate oxygen stabilizes the phosphonic acid monoester bond between the reagent and the catalytic nucleophile, thereby preventing the postmodificational transesterification during the analysis of the inactivated enzyme. In addition, the negative charge increases the hydrolytic stability of the fluorophosphonate **1** to ease the handling and storage of the reagent. The detailed synthesis and the use of this reagent for the labeling study are described with *Escherichia coli* GGT as an example.[10]

Synthesis

The monofluorophosphonate **1** is synthesized in four steps from commercially available racemic 2-amino-4-phosphonobutanoic acid (**3**) (Scheme 4).[10,11] A key is the monofluorination of phosphonic acid and the purification of the resulting

---

[7] E. Stole, A. P. Seddon, D. Wellner, and A. Meister, *Proc. Natl. Acad. Sci. U.S.A.* **87**, 1706 (1990).
[8] T. K. Smith, Y. Ikeda, J. Fujii, N. Taniguchi, and A. Meister, *Proc. Natl. Acad. Sci. U.S.A.* **92**, 2360 (1995).
[9] E. Stole, T. K. Smith, J. M. Manning, and A. Meister, *J. Biol. Chem.* **269**, 21435 (1994).
[10] M. Inoue, J. Hiratake, H. Suzuki, H. Kumagai, and K. Sakata, *Biochemistry* **39**, 7764 (2000).
[11] M. Inoue, J. Hiratake, and K. Sakata, *Biosci. Biotechnol. Biochem.* **63**, 2248 (1999).

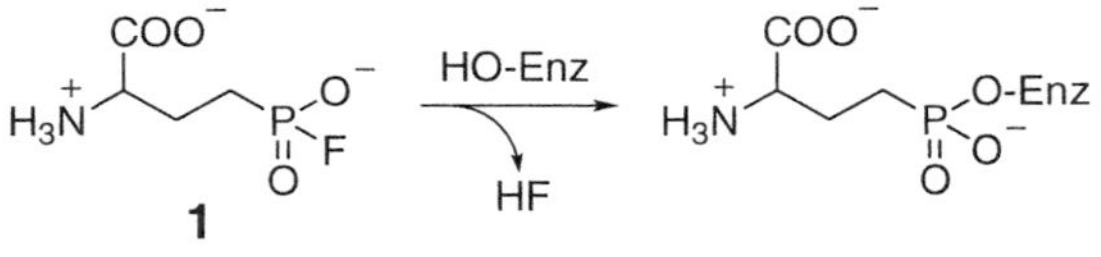

SCHEME 3

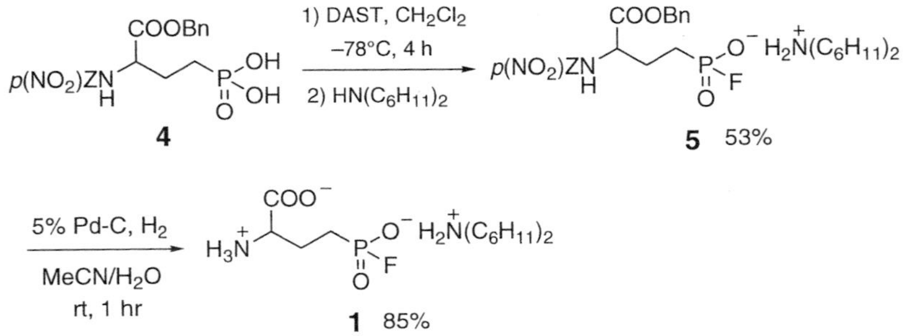

SCHEME 4

monofluorophosphonate as dicyclohexylammonium salt. The detailed synthetic procedures for the key steps are as follows.[11a]

Benzyl 2-(N-4-nitrobenzyloxycarbonylamino)-4-(fluorophosphono)butanoate Dicyclohexylammonium Salt (5). Benzyl 2-(*N*-4-nitrobenzyloxycarbonylamino)-4-phosphonobutanoate (**4**) (2.99 g, 6.61 mmol) prepared from **3** is suspended in dry CH_2Cl_2 (40 ml). Diethylaminosulfur trifluoride (DAST, 1.75 ml, 13.2 mmol) is added to the suspension at $-78°$, and the mixture is stirred for 3.5 hr at $-78°$. The reaction is quenched by adding H_2O (3 ml). The solvent is evaporated, and the residue is passed through a short column of Dowex 50W × 8 (NH_4^+ form, eluted with H_2O) to remove diethylamine. The elute is concentrated and is purified by a medium pressure reversed-phase column chromatography (Yamazen Co., Osaka, Japan) equipped with an ODS column [Daisogel-SP-120-ODS-B-40/60 μm (Daiso Co., Ltd., Osaka, Japan)]. The column is eluted with a linear gradient from 10 to 20% CH_3CN in H_2O, and the fractions containing the product are lyophilized to give a hygroscopic solid (2.23 g). The product contains $\sim$5% of the phosphonic acid **4**, which is formed by hydrolysis during purification. The crude product is dissolved in acetone, and dicyclohexylamine (0.995 g, 5.48 mmol) is added. The volatile is evaporated, and the residue is recrystallized

---

[11a] Adapted with permission from M. Inoue, J. Hiratake, H. Suzuki, H. Kumagai, and K. Sakata, *Biochemistry* **39**, 7764. Copyright © 2000 American Chemical Society.

from $CH_2Cl_2/(C_2H_5)_2O$ (5:1) to afford pure **5** as a nonhygroscopic solid (2.23 g, 53%): mp 128–130°; IR (KBr) ν_{max} 3200, 3000–2600, 1720, 1600, 1520, 1340, 1220 cm^{-1}; ^1H NMR (200 MHz, acetone-d_6, TMS as internal standard) δ_H 1.0–2.3 (m, 24H, PCH_2CH_2 and cyclohexyl), 3.0 (m, 2H, cyclohexyl), 4.4 (m, 1H, α-H), 5.19 (s, 2H, benzyl), 5.25 (m, 2H, benzyl), 7.4 (m, 5H, aromatic), 7.64 (d, J = 8.6 Hz, 2H, aromatic), 7.7 (br, 1H, NH), 8.21 (d, J = 8.6 Hz, 2H, aromatic), 9.4 (br s, 2H, cyclohexyl-NH_2); ^{31}P NMR (81.0 MHz, acetone-d_6, 85% H_3PO_4 as external standard) δ_P 21.90 (d, $^1J_{P-F}$ = 969 Hz); ^{19}F NMR (282 MHz, acetone-d_6, TFA as external standard) δ_F 17.67 (d, $^1J_{F-P}$ = 968 Hz). Analysis: Calculated for $C_{31}H_{43}FN_3O_8P$: C, 58.57; H, 6.82; N, 6.61. Found: C, 58.58; H, 6.81; N, 6.65.

*2-Amino-4-(fluorophosphono)butanoic Acid Dicyclohexylammonium Salt (**1**)*. A solution of **5** (1.71 g, 2.69 mmol) in CH_3CN/H_2O (9:1, 20 ml) is purged with H_2 gas in the presence of 5% Pd-C (100 mg) for 1 hr at room temperature. The resulting mixture is filtered, and the filtrate is evaporated. The residue is dissolved in H_2O and washed with $(C_2H_5)_2O$. The aqueous layer is lyophilized to give a colorless solid. The crude product contains ~10% of the phosphonic acid **3**. The monofluorophosphonate **1** is isolated by flash column chromatography on neutral silica gel [silica gel 60N (Kanto Chemical Co., Inc., Tokyo, Japan), 21:5, CH_3CN/H_2O]. The lyophilized solid is dissolved in CH_2Cl_2 and is filtered through Celite-545 to remove silica gel. Evaporation of the filtrate gives pure **1** as a colorless solid (0.840 g, 85%): mp 152–155° (dec); IR (KBr) ν_{max} 3700–2300, 1620, 1220 cm^{-1}; ^1H NMR (D_2O) δ_H 1.0–2.3 (m, 24H, PCH_2CH_2 and cyclohexyl), 3.2 (m, 2H, cyclohexyl), 3.78 (t, J = 6.0 Hz, 1H, α-H); ^{31}P NMR (D_2O) δ_P 26.98 (d, $^1J_{P-F}$ = 982 Hz); ^{19}F NMR (D_2O) δ_F 12.59 (d, $^1J_{F-P}$ = 980 Hz); HRMS (FAB, glycerol) calculated for $C_{16}H_{33}FN_2O_4P$ (MH^+) 367.2162, found 367.2173.

*Chemical Stability of **1***

The monofluorophosphonate **1** is highly stable in the solid state and can be stored at room temperature with exclusion of moisture for several months without appreciable decomposition. Compound **1** is relatively stable in an acidic aqueous solution, but is easily hydrolyzed in alkaline to neutral media to the phosphonic acid **3**. For example, no hydrolysis is observed over 1 week at 4° in acetate buffer (pH 5), but the half-life of the hydrolysis is 21.6 hr in 1 M sodium succinate buffer (pH 5.5) at 37°, under which conditions *Escherichia coli* GGT exhibits the maximal hydrolase activity toward γ-glutamyl-*p*-nitroanilide. Under alkaline conditions (>pH 7.5) where the enzyme exhibits much higher transpeptidase activity with Gly-Gly as an acceptor peptide, compound **1** is hydrolyzed with a half-life of 20 min in 1 M Tris-HCl (pH 7.5) at 37°. As described below, compound **1** inactivates the enzyme quickly at pH 5.5 to deprive the enzyme of both the hydrolase and transpeptidase activities.

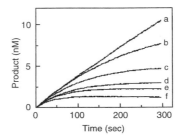

FIG. 1. Time-dependent inhibition of *E. coli* GGT by **1**. The enzymatic reaction was carried out in the presence of the following concentrations of **1**: (a) 0, (b) 0.17, (c) 0.33, (d) 0.5, (e) 0.67, and (f) 1.0 μM. Reprinted with permission from *Biochemistry* **39**, 7764 (2000). Copyright © 2000 American Chemical Society.

Inactivation of Escherichia coli GGT by 1

The inactivation of *E. coli* GGT by **1** followed by ion-spray mass spectrometric analysis illustrates how this reagent is used to trap the catalytic nucleophile. The experiments and methods described here are general approaches that can be applied to other GGTs.

The hydrolase activity of GGT is measured in 100 mM succinate buffer (pH 5.5) using 7-(γ-L-glutamylamino)-4-methylcoumarin[12] as the substrate at 37° without acceptor amino acids. To the preincubated mixture of **1** and 0.2 μM 7-(γ-L-glutamylamino)-4-methylcoumarin in 0.1 M sodium succinate buffer (pH 5.5) at 37° is added the *E. coli* GGT to a final concentration of 0.26 nM to initiate the reaction. The enzyme is rapidly inactivated in a time-dependent manner (Fig. 1). The inactivation is competitive with respect to the substrate. The time-dependent inhibition of the enzyme is monitored until the product release reaches a plateau with various concentrations of **1**. The curves of the time-dependent loss of the enzyme activity are fit to the first-order rate equation [Eq. (1)]:

$$[P] = [P]_\infty [1 - \exp(-k_{obs}t)] \qquad (1)$$

where [P] and [P]$_\infty$ are the concentrations of product formed at time t and at the time approaching infinity, respectively,[13] to determine the pseudo first-order rate constant for the inactivation (k_{obs}) at each inhibitor concentration.

A plot of k_{obs} against the inhibitor concentration [I] often gives a saturation kinetic as a result of reversible binding of inhibitor prior to inactivation,[14] but in this particular case, the plot gives a straight line up to 1.0 μM of **1**. Saturation kinetics might be observed at higher inhibitor concentrations, but rapid enzyme

---

[12] G. D. Smith, J. L. Ding, and T. J. Peters, *Anal. Biochem.* **100**, 136 (1979).
[13] C. L. Tsou, *Adv. Enzymol. Relat. Areas Mol. Biol.* **61**, 381 (1988).
[14] J. F. Morrison and C. T. Walsh, *Adv. Enzymol. Relat. Areas Mol. Biol.* **61**, 201 (1988).

inactivation does not allow the initial reaction rate to be measured by a conventional assay method. Actually the second-order rate constant for the inactivation of the enzyme (k_{on}) is determined as $4.83 \times 10^4 \, M^{-1}s^{-1}$ according to Eq. (2) derived from the following kinetic mechanism:

$$E + S \underset{}{\overset{K_m}{\rightleftharpoons}} ES \longrightarrow E + P$$
$$\downarrow k_{on} \, [I]$$
$$E\text{-}I$$
$$k_{obs} = k_{on} \, [I]/(1 + [S]/K_m) \tag{2}$$

where [S] and K_m are 0.2 and 0.3 μM, respectively.

The value of k_{on} is 3 orders of magnitude larger than the inactivation rate of mammalian kidney GGTs (rat, pig, and human) by acivicin[8,15] and DON[16] and is only 37-fold less than the second-order rate constant (k_{cat}/K_m) for the enzymatic hydrolysis of the substrate [7-(γ-glutamylamino)-4-methylcoumarin] ($1.79 \times 10^6 \, M^{-1}s^{-1}$). Furthermore, the observed inactivation rate is comparable to or even higher than that of typical serine proteinases by neutral fluorophosphono ester analogs of amino acids.[17,18] This is worth noting, because phosphonic acid monoesters are usually much less reactive toward nucleophilic substitution than neutral phosphonic diesters because of the negative charge.[19] The extremely high inactivation rate of *E. coli* GGT by the monofluorophosphonate **1** suggests that the compound **1** recruits the catalytic power of the enzyme to modify the catalytic nucleophile and serves as a mechanism-based inactivator. On the other hand, the modification of GGT by acivicin and DON seems to have relied mostly on their reactive groups rather than on the enzyme catalysis, thereby forming a covalent bond with a nearby nucleophile much more slowly with less fidelity to the catalytic nucleophile. This is in contrast to rapid and specific inactivation of cysteine proteinases by peptidyl diazoketone analogs, where the catalytic cysteine is alkylated with extremely high inactivation rate in an enzyme-activated manner.[20] In this regard, glutamine amidohydrolases,[21] another class of γ-glutamylamide cleaving enzymes, are modified effectively and selectively by a diazoketone derivative DON, because their catalytic mechanism involves Cys as the catalytic nucleophile. Conversely, the phosphonylating agents such as **1** are not expected to be good modification agents for these enzymes because of the unstable nature of the phosphorus–sulfur bond.

[15] Y. Ikeda, J. Fujii, M. E. Anderson, N. Taniguchi, and A. Meister, *J. Biol. Chem.* **270**, 22223 (1995).
[16] M. Inoue, S. Horiuchi, and Y. Morino, *Eur. J. Biochem.* **73**, 335 (1977).
[17] L. A. Lamden and P. A. Bartlett, *Biochim. Biophys. Res. Commun.* **112**, 1085 (1983).
[18] P. A. Bartlett and L. A. Lamden, *Bioorg. Chem.* **14**, 356 (1986).
[19] E. J. Behrman, M. J. Biallas, H. J. Brass, J. O. Edwards, and M. Isaks, *J. Org. Chem.* **35**, 3063 (1970).
[20] E. Shaw, *Methods Enzymol.* **244**, 649 (1994).
[21] H. Zalkin and J. L. Smith, *Adv. Enzymol. Relat. Areas Mol. Biol.* **72**, 87 (1998).

Although the enzyme is modified by **1** under acidic conditions where the hydrolase activity is maximized and the transpeptidase activity is diminished, the modified enzyme loses both activities. No activities are regained after a prolonged dialysis, indicating that the inhibitor is attached covalently to the catalytic nucleophile of *E. coli* GGT.

Ion-Spray Mass Spectrometric Analysis of Modified GGT

The inactivated enzyme is analyzed by ion-spray mass spectrometry according to the following procedure: (1) measurement of the overall mass increase of the modified enzyme to determine the number of the inhibitor attached to the enzyme, (2) proteolytic digestion of the modified enzyme followed by LC/MS analysis to identify the modified peptide fragment, and (3) MS/MS analysis of the modified peptide to identify the modified amino acid.

The *E. coli* GGT is composed of two subunits with a molecular mass of 20,010 and 39,196 Da for the small and large subunit, respectively. The small and large subunits isolated from the native enzyme give a major peak with a molecular mass of 20,014 and 39,209 Da, respectively (Figs. 2A,2C). On the other hand, the small subunit from the inactivated enzyme gives a molecular mass of 20,178, while the mass of the large subunit remains unchanged (39,209 Da) (Figs. 2B,2D). The observed mass increase of the small subunit (164 Da) agrees well with the calculated mass increase (165 Da) by phosphonylation with **1**, indicating that one molecule of the small subunit is phosphonylated by one molecule of **1**.

The unmodified and phosphonylated small subunits are digested by lysyl endopeptidase (LysC), and the proteolytic digests are analyzed by reversed-phase high-performance liquid chromatography (HPLC) using the ion-spray mass spectrometer as the detector. The peptide fragment containing the phosphonylation site is identified by comparing the HPLC elution profile (Fig. 3) and by measuring the molecular mass of each peptide fragment with a mass increase of 165 Da as a marker. By comparing the observed molecular mass of each peptide with the predicted mass calculated from the amino acid sequence, the phosphonylated peptide can be identified as the N-terminal nanopeptide, Thr^{391}-Lys^{399} (Fig. 4). The phosphonylation of this peptide is supported not only by mass difference ($[M+H]^+ = 1050.0$ and 1215.2 for unmodified and modified peptide, respectively), but also by the formation of two characteristic peaks (m/z 184.1 and 517.1) in the mass spectrum of the modified peptide (Fig. 4B). These peaks are formed during the ionization process by the β-elimination of the phosphonic acid **3** (183.0 Da) from the phosphonylated peptide Thr^{391}-Lys^{399} (1213.5 Da) and correspond to the $(M+H)^+$ ion of the phosphonic acid **3** and the $(M-183+2H)^{2+}$ ion of the elimination product (calculated mass of 1030.5 Da), respectively. β-Elimination is a well-known reaction with O-phosphorylated or sulfonylated serine and

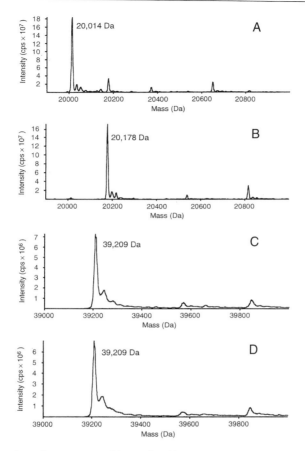

FIG. 2. Reconstructed mass spectra of the small and large subunit. (A) Small subunit from the unmodified enzyme. (B) Small subunit from the modified enzyme. (C) Large subunit from the unmodified enzyme. (D) Large subunit from the modified enzyme. Reprinted with permission from *Biochemistry* **39**, 7764 (2000). Copyright © 2000 American Chemical Society.

threonine derivatives[22]; similar fragment ions are reported for the mass spectra of phosphorylated proteins, where the phosphothreonine and phoshoserine residues undergo the elimination of phosphonic acid (98.0 Da) to produce the (y−98) product ions. This is often taken as a proof that a threonine or serine residue is phosphorylated.[23]

[22] D. Samuel and B. L. Silver, *J. Chem. Soc.,* 289 (1963).
[23] D. M. Payne, A. J. Rossomando, P. Martino, A. K. Erickson, J. H. Her, J. Shabanowitz, D. F. Hunt, M. J. Weber, and T. W. Sturgill, *EMBO J.* **10,** 885 (1991).

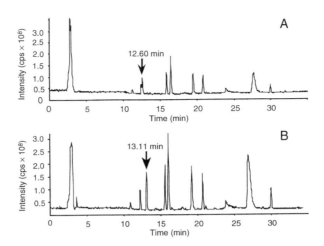

FIG. 3. HPLC elution profiles of the proteolytic digests of the small subunit. Proteolytic digests were loaded onto a ODS column, and the column was eluted with 2% acetonitrile containing 0.1% formic acid for 5 min, followed by linear gradient of 2–82% acetonitrile containing 0.1% formic acid over 40 min at a flow rate of 1 ml/min. (A) Digest from the unmodified enzyme. (B) Digest from the modified enzyme. The peak with different elution time between (A) and (B) is indicated by an arrow. Reprinted with permission from *Biochemistry* **39**, 7764 (2000). Copyright © 2000 American Chemical Society.

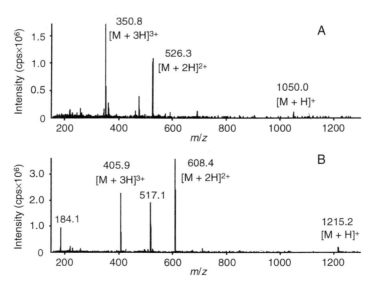

FIG. 4. Mass spectra of the isolated peptides of the proteolytic digest. (A) The peptide eluted at 12.60 min from the unmodified enzyme. (B) The peptide eluted at 13.11 min from the modified enzyme. Reprinted with permission from *Biochemistry* **39**, 7764 (2000). Copyright © 2000 American Chemical Society.

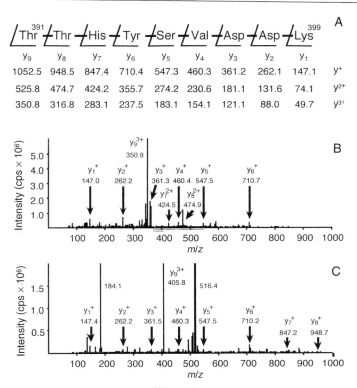

FIG. 5. MS/MS analysis of the peptide Thr^{391}-Lys^{399}. (A) Predicted monoisotopic masses for product ions of type y derived from the sequence shown above. (B) CID spectrum of the m/z 350.7 precursor ion from the unmodified enzyme. (C) CID spectrum of the m/z 405.9 precursor ion from the modified enzyme. Reprinted with permission from *Biochemistry* **39**, 7764 (2000). Copyright © 2000 American Chemical Society.

Finally, the phosphonylated amino acid is identified by the MS/MS analysis of a precursor ion of the phosphonylated peptide Thr^{391}-Lys^{399}. The prominent $(M+3H)^{3+}$ ions at m/z 350.7 (unmodified peptide) and 405.9 (phosphonylated peptide) are subjected to collision-induced decay (CID). The unmodified peptide precursor ion (m/z 350.7) gives a CID spectrum with all possible product ions of type y (y_1–y_8, Fig. 5B).

On the other hand, the phosphonylated peptide precursor ion (m/z 405.9) gives the same product ions y_1–y_8 with no mass increase of 165 Da. The mass increase of 165 Da is observed only in the precursor ion y_9, indicating that the phosphonylation site is the N-terminal Thr-391 (Fig. 5A,C). The marker ions corresponding to the β-elimination products (m/z 184.1 and 516.4) are also observed in the CID spectrum of the phosphonylated precursor ion (Fig. 5C). The β-elimination is a good marker also in the type y fragment ions; the phosphonylated peptide Thr^{391}-Lys^{399} gives the (y−183) fragment ions only from the precursor ion (y$_9$), not from other product ions y$_1$–y$_8$, supporting the identification of the N-terminal Thr as the phosphonylation site.

Discussion

The rapid and specific modification with the tetrahedral phosphonylating agent **1** identified the N-terminal Thr-391 as the catalytic nucleophile of *E. coli* GGT. This is good evidence that *E. coli* GGT is a member of the N-terminal nucleophile (Ntn) hydrolases, a recently recognized hydrolase family.[24] Other GGTs such as mammalian enzymes share the same chemical and structural features: the reaction proceeds via a γ-glutamylenzyme intermediate, and the N-terminal Thr is conserved among all the GGTs for which the primary sequences are known.[4] As far as a hydroxy nucleophile is involved in the γ-glutamyl transfer reaction, the monofluorophosphonate **1** is expected to serve as an effective affinity labeling agent for other GGTs. One potential problem, however, is that most GGTs exhibit maximum transpeptidase activity at pH 8–9[15,16] where the monofluorophosphonate **1** is unstable. This problem was resolved by lowering the pH of the modification experiment for the *E. coli* enzyme, but another way is to use a stable derivative of **1**. The stability of **1** can be increased by replacing the fluorine atom by less acidic leaving groups such as *p*-nitrophenoxy and phenoxy groups. Since the substrate specificity of GGT is rather broad and γ-glutamyl-*p*-nitroanilide is a good substrate, the replacement of the fluorine atom by a substituted phenoxy group is a reasonable modification to increase the stability without sacrificing the electrophilic nature. The replacement of a fluorine atom by a phenoxy group is successfully used for increasing the hydrolytic stability of the phosphonyl analogs of amino acids and peptides for serine hydrolase inhibitors: the phenyl esters serve as stable and specific phosphonylating agents of serine proteinases, although the inactivation rate is decreased.[18,25]

The rapid and selective inactivation of GGT by the monofluorophosphonate **1** is reminiscent of the specific inactivation of class C β-lactamases by phosphonic acid monoesters **6** and **7**.[26] This is in marked contrast to the fact that mechanistically related classical serine proteinases are inactivated by neutral phosphonic acid diesters such as **8**,[18] but not by anionic phosphonic monoesters **6** and **7**. The susceptible nature of the *E. coli* GGT to the anionic phosphonylating agent might suggest that this enzyme is more related to the class C β-lactamases than to the serine hydrolases in terms of the catalytic mechanism or the active site environment. In fact, unlike

---

[24] J. A. Brannigan, G. Dodson, H. J. Duggleby, P. C. E. Moody, J. L. Smith, D. R. Tomchick, and A. G. Murzin, *Nature* **378**, 416 (1995).
[25] N. S. Sampson and P. A. Bartlett, *Biochemistry* **30**, 2255 (1991).
[26] J. Rahil and R. F. Pratt, *Biochemistry* **31**, 5869 (1992).

the classical serine hydrolases with a Ser-His-Asp (Glu) catalytic triad, the Ntn hydrolases are proposed to have a different catalytic machinery of "nucleophile/ α-amine" catalytic dyad, where the free α-amino group of the catalytic residue is used as a base.[24,27] A similar active site environment is observed with class C β-lactamases[28] where a Ser nucleophile is interacting with a Lys ε-amino group to form a "Ser/Lys" catalytic dyad. In class C β-lactamases, an increased electrostatic positive potential in the catalytic site is suggested to account for the susceptibility to the anionic phosphonate monoesters; the positive potential is likely to allow the anionic phosphonate inactivators to phosphonylate the catalytic nucleophile effectively.[28] If the active site environment or the electrostatic potential is associated with the unique catalytic machinery of the "nucleophile/amine" catalytic dyad, then the anionic monofluorophosphonate-based inhibitors might be effective for other N-terminal nucleophile hydrolases such as prokaryotic proteosomes with a "hydroxy nucleophile–amine" catalytic dyad.[29,30]

γ-Glutamyl Peptide Ligases: Phosphonodifluoromethyl Ketone and Monofluorophosphonate Analogs as Probes

Design of Phosphonodifluoromethyl Ketone 2

The intermediacy of γ-glutamyl-P as an activated form of glutamic acid has been well evidenced in the reaction catalyzed by GS[1] and γ-GCS.[3] Stable analogs of this intermediate are therefore potential inhibitors and can be used as probes to understand the reaction mechanisms of the γ-glutamyl peptide ligases. 4-(Phosphonoacetyl)-L-α-aminobutyrate (PALAB) is such a compound designed

---

[27] M. Paetzel and R. E. Dalbey, *Trends Biochem. Sci.* **22**, 28 (1997).
[28] E. Lobkovsky, E. M. Billings, P. C. Moews, J. Rahil, R. F. Pratt, and J. R. Knox, *Biochemistry* **33**, 6762 (1994).
[29] M. Bogyo, J. S. McMaster, M. Gaczynska, D. Tortorella, A. L. Goldberg, and H. Ploegh, *Proc. Natl. Acad. Sci. U.S.A.* **94**, 6629 (1997).
[30] J. Adams, M. Behnke, S. Chen, A. A. Cruickshank, L. R. Dick, L. Grenier, J. M. Klunder, Y.-T. Ma, L. Plamondon, and R. L. Stein, *Bioorg. Med. Chem. Lett.* **8**, 333 (1998).

$$\text{H}_3\overset{+}{\text{N}}\underset{}{\overset{\text{COO}^-}{\diagup}}\text{C}\underset{\text{O}}{\overset{}{\diagdown}}\text{CH}_2\diagdown\text{P}\underset{\text{O}}{\overset{\text{OH}}{\diagup}}\text{O}^-$$

4-(phosphonoacetyl)-L-α-aminobutyrate (PALAB)

to mimic the γ-glutamyl-P by substituting the bridging oxygen with CH_2.[31] PALAB is a relatively good analog of the reaction intermediate and inhibits the pea seed GS with an initial K_i of 0.59 mM and causes a time-dependent loss of the enzyme activity. However, the *E. coli* enzyme is inhibited only moderately (K_i of 4.8 mM) with no time-dependent loss of the enzyme activity.

α,α-Difluorination of the bridging methylene is an attractive way of improving the properties of phosphonates as phosphate mimics by lowing the phosphonate pK_a and by introducing the hydrogen-bonding interactions which are lost in simple methylene analogs. Although the nature of analogy between O and CF_2 linkages is still in controversy,[32] the (α,α-difluoromethyl)phosphonates have been used successfully as hydrolytically stable and isopolar mimics of ATP,[33] O-phosphoserine, O-phosphothreonine,[34] and O-phosphotyrosine[35] analogs with significant biological activities.

In the reaction catalyzed by the γ-glutamyl peptide ligases, the tetrahedral transition-state geometry is reached by nucleophilic attack of the substrate amine on the γ-glutamyl-P intermediate (Scheme 1). It is well known that fluorinated peptidyl ketones are potent inhibitors of proteinases because of the formation of stable hydrates and hemiketals which mimic the tetrahedral transition states.[36] In this context, the phosphonodifluoromethyl ketone **2** is particularly interesting as a potential inhibitor of γ-glutamyl peptide ligases, because the highly electrophilic carbonyl next to CF_2 is expected to undergo the nucleophilic attack of the substrate amine (NH_3 and L-Cys for GS and γ-GCS, respectively) to form a transition-state-like adduct in the enzyme active site (Scheme 5).

Synthesis and Characterization of 2

The synthesis of the phosphonodifluoromethyl ketone **2** is outlined in Scheme 6.[11] Among several methods for PCF_2-carbon bond formation, the substitution of simple methyl esters by lithium(difluoromethyl)phosphonate[37] is

---

[31] F. C. Wedler, B. R. Horn, and W. G. Roby, *Arch. Biochem. Biophys.* **202**, 482 (1980).
[32] G. R. J. Thatcher and A. S. Campbell, *J. Org. Chem.* **58**, 2272 (1993).
[33] G. M. Blackburn, D. E. Kent, and F. Kolkmann, *J. Chem. Soc. Perkin Trans.* **1**, 1119 (1984).
[34] D. B. Berkowitz, M.-J. Eggen, Q. Shen, and R. K. Shoemaker, *J. Org. Chem.* **61**, 4666 (1996).
[35] T. R. Burke, Jr., M. S. Smyth, A. Otaka, M. Nomizu, P. P. Roller, G. Wolf, R. Case, and S. E. Shoelson, *Biochemistry* **33**, 6490 (1994).
[36] M. H. Gelb, J. P. Svaren, and R. H. Abeles, *Biochemistry* **24**, 1813 (1985).
[37] M. Obayashi, E. Ito, K. Matsui, and K. Kondo, *Tetrahedron Lett.* **23**, 2323 (1982).

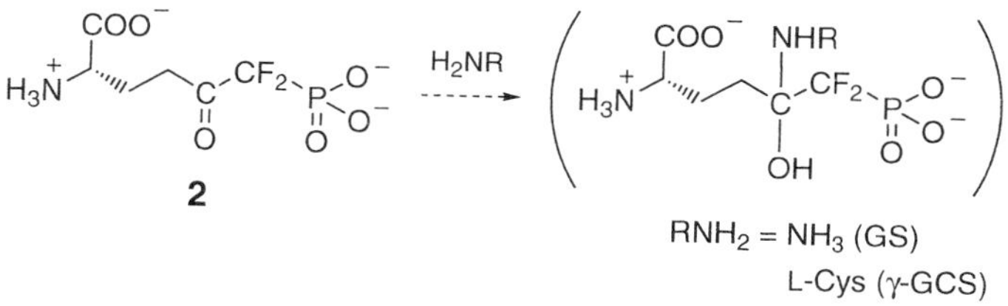

[21] γ-GLUTAMYLTRANSPEPTIDASE AND γ-GLUTAMYL PEPTIDE LIGASES   287

SCHEME 5

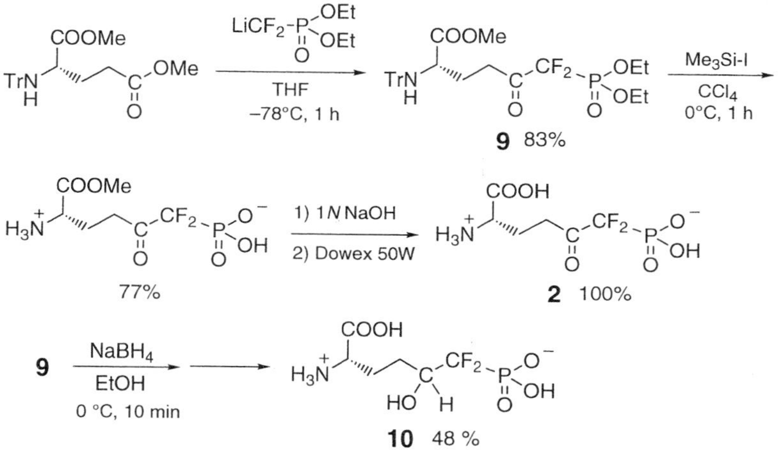

SCHEME 6

conveniently carried out with amino acid derivatives.[34] Thus, the reaction with N-trityl-L-glutamic acid dimethyl ester gives the phosphonodifluoromethyl ketone **9** in 83% yield. The sterically hindered N-trityl protecting group effectively drives the reaction to the γ-carbonyl group of glutamic acid to yield the γ-phosphonomethylated product selectively.[38] Deprotection by iodotrimethylsilane (TMS-I) followed by alkaline hydrolysis gives the product ketone **2** as crystals. The corresponding alcohol **10** is synthesized by NaBH$_4$ reduction of the ketone **9** followed by deprotection and purification by the same procedure. The synthesis of the ketone **2** is as follows.

Methyl (R)-6-(Diethoxyphosphoryl)-6,6-difluoro-5-oxo-2-(N-tritylamino)hexanoate (9). Diethyl difluoromethylphosphonate (8.92 g, 47.4 mmol) in dry tetrahydrofuran (THF, 50 ml) is added to a solution of lithium diisopropylamide (LDA) (47.4 mmol) in dry THF (100 ml) at −78°, and the mixture is stirred for 20 min.

---

[38] D. E. Rudisill and J. P. Whitten, *Synthesis*, 851 (1994).

To this solution is added a solution of N-trityl-L-glutamic acid dimethyl ester (11.0 g, 26.3 mmol) in dry THF (30 ml) at $-78°$. After 1 hr, the reaction mixture is quenched by adding acetic acid (5.68 g, 94.9 mmol) and warmed to room temperature. The mixture is partitioned into ethyl acetate (300 ml)–H_2O (300 ml), and the organic layer is washed with H_2O (300 ml) and saturated NaCl (300 ml), dried over Na_2SO_4, and evaporated. Medium pressure column chromatography (Silica Gel 60, 14 : 25, ethyl acetate/hexanes) affords **3** as a colorless gum (12.5 g, 83%): $[\alpha]_D^{26} + 29.8°$ (c 1.25, ethanol); IR (NaCl) v_{max} 3300, 2990, 1740, 1280, 1170, 1020 cm^{-1}; ^{1}H NMR (200 MHz, CDCl$_3$) δ_H 1.38 (t, J = 7.0 Hz, 6H, POCH$_2$CH_3 × 2), 2.1 (m, 2H, CH_2CH$_2$C=O), 2.7 (br, 1H, NH), 2.9 (m, 2H, CH$_2$CH_2C=O), 3.16 (s, 3H, COOCH_3), 3.42 (m, 1H, α-H), 4.3 (quintet, J = 7.4 Hz, 4H, POCH_2CH$_3$ × 2), 7.1–7.3 (m, 9H, aromatic), 7.48 (d, J = 8.4 Hz, 6H, aromatic); ^{19}F NMR (188 MHz, CDCl$_3$) δ_F −42.07 (d, $^2J_{F-P}$ = 96.7 Hz); ^{31}P NMR (81.0 MHz, CDCl$_3$) δ_P 3.84 (t, $^2J_{P-F}$ = 96.7 Hz); ^{13}C NMR (99.5 MHz, CDCl$_3$) δ_C 16.25 (d, $^3J_{C-P}$ = 5.0 Hz, POCH$_2$$CH_3$), 28.03 (s, CH$_2CH_2$C=O), 33.39 (s, CH$_2$$CH_2$C=O), 51.51 (s, CHCOOCH$_3$), 54.74 (s, COOCH$_3$), 65.26 (d, $^2J_{C-P}$ = 6.6 Hz, POCH$_2$CH$_3$), 70.99 (s, trityl), 113.02 (dt, $^1J_{C-P}$ = 195 Hz, $^1J_{C-F}$ = 273 Hz, CF$_2$), 126.14, 127.51, 128.40 and 145.31 (s, aromatic), 174.20 (s, COOCH$_3$), 197.75 (dt, $^2J_{C-P}$ = 14.0 Hz, $^2J_{C-F}$ = 23.9 Hz, C=O). Analysis: Calculated for C$_{30}$H$_{34}$F$_2$NO$_6$P: C, 62.82; H, 5.97; N, 2.44. Found: C, 62.75; H, 6.07; N, 2.45.

(R)-2-Amino-6,6-difluoro-5-oxo-6-phosphonohexanoic Acid (2). To an ice cooled solution of **9** (4.00 g, 6.97 mmol) in dry CCl$_4$ (50 ml) is added TMS-I (5.58 g, 27.9 mmol). After stirring for 1 hr at 0°, the reaction mixture is evaporated. The residue is partitioned with (C$_2$H$_5$)$_2$O (80 ml)/H$_2$O (80 ml), and the aqueous layer is washed with (C$_2$H$_5$)$_2$O (80 ml × 4). Purification by medium pressure reversed-phase column chromatography (ODS, Daisogel, 100% H$_2$O) followed by lyophilization affords the methyl ester as a pale yellow solid (1.47 g, 77%).

The methyl ester (1.10 g, 3.99 mmol) is hydrolyzed in 1 N NaOH (12.0 ml, 12.0 mmol) for 1 hr at 0°. The reaction mixture is passed through Dowex 50W× 8 (H$^+$ form) and lyophilized to give a colorless hygroscopic solid (1.06 g, 100%). Recrystallization from acetone/H$_2$O affords **2** as nonhygroscopic colorless crystals (0.439 g, 42%): mp 141–143° (dec); $[\alpha]_D^{26}$ +11.6° (c 1.05, H$_2$O); IR (KBr) v_{max} 3300–2000, 1730, 1710, 1620, 1500, 1250 cm^{-1}; ^{1}H NMR (D$_2$O) (mixture of keto and hydrated forms) δ_H 1.9–3.2 (m, 4H, CH_2CH_2CO), 4.1 (m, 1H, α-H); ^{19}F NMR (D$_2$O) δ_F −46.68 (d, $^2J_{F-P}$ = 90.9 Hz) and −44.14 (t, $^2J_{F-P}$ = 86.2 Hz); ^{31}P NMR (D$_2$O) (mixture of keto and hydrated forms) δ_P 1.01 (t, $^2J_{P-F}$ = 85.9 Hz) and 3.70 (t, $^2J_{P-F}$ = 90.8 Hz); ^{13}C NMR (99.5 MHz, D$_2$O) (mixture of keto and hydrated forms) δ_C 25.34 and 25.55 (s, CHCH$_2$CO), 32.3 and 35.8 (m, CH$_2$$CH_2$CO), 54.45 and 55.35 (s, CHCOOH), 97.2 [m, C(OH)$_2$], 118.01 (dt, $^1J_{C-P}$ = 176 Hz, $^1J_{C-F}$ = 269 Hz) and 120.68 (dt, $^1J_{C-P}$ = 184 Hz, $^1J_{C-F}$ = 268 Hz) (CF$_2$), 173.82 and 174.15 (s, COOH), 204.5 (m, C=O). Analysis: Calculated for C$_6$H$_{10}$F$_2$NO$_6$P: C, 27.60; H, 3.86; N, 5.36. Found: C, 27.28; H, 3.79; N; 5.31.

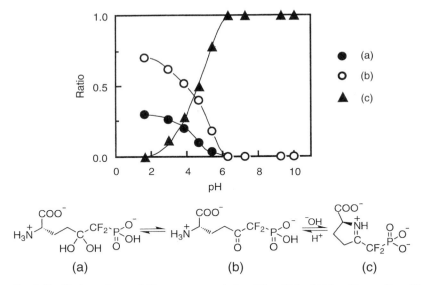

FIG. 6. Equilibrium mixture of **2** in an aqueous solution. [**2**] = 80.4 mM. The pH was adjusted by adding K_2CO_3. The ratios of the hydrated form (a) ●, the keto form (b) ○, and the cyclic imine form (c) ▲ were calculated from the integral of the ^{19}F NMR peak areas. Reprinted with permission from M. Inoue, J. Hiratake, and K. Sakata, *Biosci. Biotechnol. Biochem.* **63**, 2248 (1999).

Compound **2** is isolated as the keto form (^{13}C NMR and elemental analysis). In an aqueous solution, however, **2** exists as an equilibrium mixture of the following three components whose ratios (determined by ^{19}F NMR) are dependent on the pH of the solution (Fig. 6): the hydrated form (a) and the keto form (b) in acidic media, and the cyclic imine form (c) in alkaline solution.[11] The ease of addition of water and amino group is consistent with the increased electrophilicity of the carbonyl group flanked by CF_2.

Enzyme Inhibition by 2

The inhibitory activity of compound **2** toward *E. coli* GS and γ-GCS has been examined.[11] The enzyme activity of γ-GCS and GS is determined by measuring ADP formation by a pyruvate kinase–lactate dehydrogenase coupled enzyme assay.[39]

Under physiological conditions (pH 7.5), in which **2** exists almost exclusively in the cyclic imine form (c), neither enzyme is inhibited with up to 6.1 mM of **2**.

---

[39] S. E. Brolin, *in* "Methods of Enzymatic Analysis," 3rd Ed. (H. U. Bergmeyer and M. Graßl, eds.), Vol. 3, p. 540. VCH Verlag, Weinheim, 1983.

However, moderate inhibition is observed in acidic media in which the hydrated form (a) and the keto form (b) are contained as minor components. Thus, γ-GCS and GS are inhibited by 47% at pH 5.5 and 54% at pH 6.0, respectively, with 6.3 mM of **2**. The effective concentrations (IC$_{50}$) of the keto form (b) are 1 and 0.3 mM at pH 5.5 and 6.0, respectively, as calculated from the pH–ratio profile (Fig. 6). The keto form (b) seems to be responsible for the inhibition, because the corresponding alcohol **10** does not inhibit either enzyme up to 6.9 mM. Thus, the extent of inhibition by **2** is significantly higher than that reported with the inhibition of *E. coli* GS by PALAB. To see if the enzyme inactivation occurs according to Scheme 5, the effect of the substrate amine on the enzyme inhibition has been examined. γ-GCS and GS are incubated at pH 6.0 with **2** (6.3 mM) and ADP (1 mM) in the presence of L-Cys or L-2-aminobutyric acid (150 mM) and NH$_4$Cl (50 mM) for 2 hr at 37°, respectively, and the remaining activity is measured. The remaining activities of γ-GCS and GS are 93 and 107%, respectively. Thus the enzyme inactivation or tight-binding inhibition is not observed in the presence of the corresponding substrate amine. Although the equilibrium is not favorable for the keto form of **2**, the expected inhibition scheme with respect to the ketone **2** and the substrate amine does not seem to be operative with these enzymes.

Enzyme Inhibition by Monofluorophosphonate **1**

The *E. coli* γ-GCS and GS are likely to catalyze the reaction with a tetrahedral intermediate as the most stabilized one. This notion is supported by the inhibition studies. For example, *E. coli* GS is inhibited much more potently by tetrahedral phosphinothricin[40] and methionine sulfoximine[41] than by trigonal PALAB. The

phosphinothricin   methionine sulfoximine

ligases with the same kinetic mechanism such as γ-GCS, glutathione synthetase, and D-Ala : D-Ala ligase are also inhibited strongly by the tetrahedral inhibitors: buthionine sulfoximine,[3] the phosphinate **11** and the sulfoximine **12**[42] (for γ-GCS), the phosphinate **13**[43] (for glutathione synthetase), and the phosphinates **14**[44] (D-Ala : D-Ala ligase). According to the inhibition studies, the enzymes are strongly inhibited by mechanism-based phosphorylation of the P–O oxygen or

---

[40] J. A. Colanduoni and J. J. Villafranca, *Bioorg. Chem.* **14**, 163 (1986).
[41] W. B. Rowe, R. A. Ronzio, and A. Meister, *Biochemistry* **8**, 2674 (1969).
[42] N. Tokutake, J. Hiratake, M. Katoh, T. Irie, H. Kato, and J. Oda, *Bioorg. Med. Chem.* **6**, 1935 (1998).
[43] J. Hiratake, H. Kato, and J. Oda, *J. Am. Chem. Soc.* **116**, 12059 (1994).
[44] K. Duncan and C. T. Walsh, *Biochemistry* **27**, 3709 (1988).

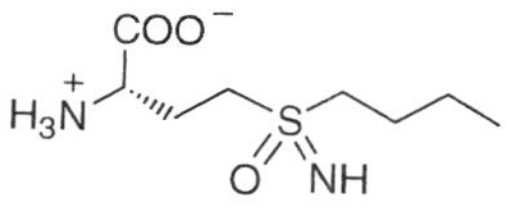

buthionine sulfoximine

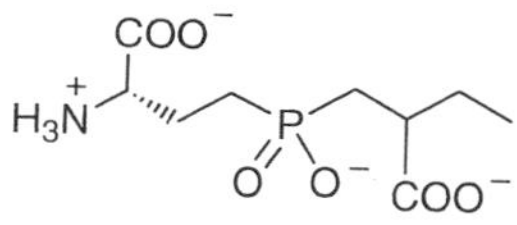

11

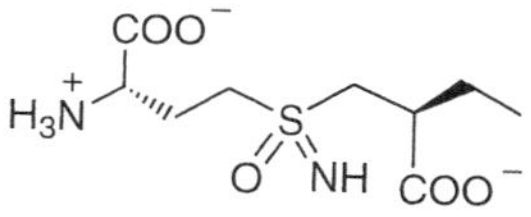

12

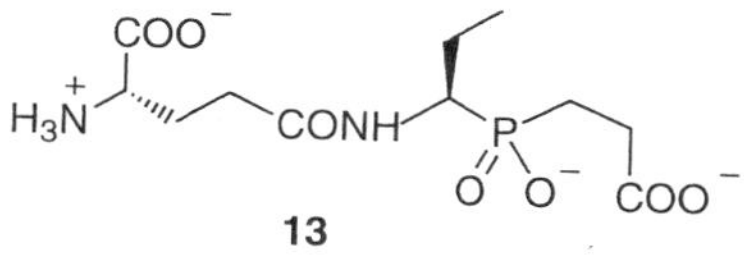

13

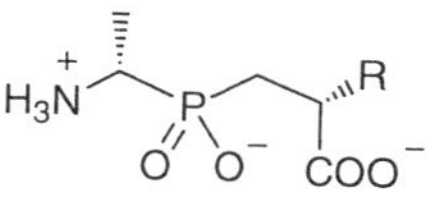

14 R = CH$_3$, C$_7$H$_{15}$

S=NH nitrogen of the inhibitor in the enzyme active site when ATP is present. These results suggest that the most stabilized intermediate is the tetrahedral transition-state-like adduct formed by the nucleophilic addition of the substrate amine to the γ-glutamyl-P intermediate (Scheme 1). In this regard, the tetrahedral monofluorophosphonate **1** is a potential inhibitor of the γ-glutamyl peptide ligases, in the hope that the substrate amine may attack the highly electrophilic phosphorus to substitute the fluorine, followed by phosphorylation of the P–O oxygen with ATP in a mechanism-based manner (Scheme 7). The inhibition of *E. coli* GS and γ-GCS by **1** has been examined.

The monofluorophosphonate **1** inhibits γ-GCS and GS with a K_i of 83 and 59 μM, respectively. The inhibition is reversible and competitive with respect to L-Glu. To examine the enzyme inactivation or tight-binding inhibition, γ-GCS and GS are incubated for 20 min with sufficient concentrations of **1** and the substrate amine (L-Cys and NH$_3$ for γ-GCS and GS, respectively) in the presence of ATP, and the remaining activity is measured. No inactivation is observed with γ-GCS (102% of the remaining activity), but the residual activity of GS is 59% relative to the control. The extent of the inactivation depends on the preincubation time (Fig. 7). The loss of enzyme activity is recovered by dialysis, but the partial inactivation of GS is unique to the monofluorophosphonate **1**; no inactivation is observed with the simple phosphonic acid **3** under the same conditions. Furthermore, the lack of either NH$_3$ or ATP fails to inactivate the enzyme, and the replacement

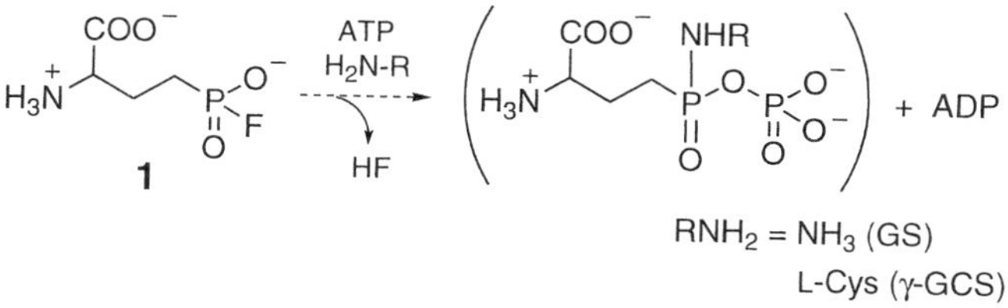

RNH$_2$ = NH$_3$ (GS)
L-Cys (γ-GCS)

SCHEME 7

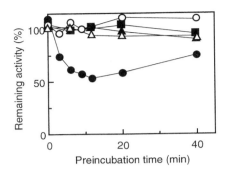

FIG. 7. Partial inactivation of GS by **1**. GS was incubated at 37° with **1** (200 μM) in the presence of ATP (6 mM) and NH$_4$Cl (50 mM), ●; NH$_4$Cl alone, ▲; ATP alone, △; and AMP-PCP (6 mM) and NH$_4$Cl, ○ in 100 mM HEPES–KOH (pH 7.5). GS was preincubated with **3** (300 μM), ATP (6 mM), and NH$_4$Cl (50 mM), ■. The remaining activity was measured after a certain period of preincubation (abscissa) followed by a 100-fold dilution. Reprinted with permission from M. Inoue, J. Hiratake, and K. Sakata, *Biosci. Biotechnol. Biochem.* **63,** 2248 (1999).

of ATP by 5′-adenylylmethylene diphosphonate (AMP-PCP), a nonhydrolyzable ATP analog, does not cause enzyme inactivation (Fig. 7). This mode of partial inactivation of GS is consistent with the initially expected inhibition scheme, although direct evidence for the formation of a transition-state mimic is not available at present.

Discussion

Wedler *et al.* reported in their studies using PALAB and 3-(phosphonoacetylamido)-L-alanine (PA$_2$LA) as γ-glutamyl-P analogs that the putative γ-glutamyl-P intermediate was not a most recognized species in the course of the reaction catalyzed by *E. coli* GS.[31,45] Although the intermediacy of γ-glutamyl-P is well documented in the glutamine synthetase reaction,[1] PA$_2$LA and PALAB were found to bind to the *E. coli* enzyme no more tightly than the substrate L-Glu. Interestingly, the same compounds served as strong inhibitors of the pea seed enzyme, causing a time-dependent inhibition with an overall inhibition constant of 10^{-6} M.[31] The large difference between these enzymes is interpreted by the relative importance or the degree of stabilization of this intermediate in the course of each enzymatic reaction. Thus, the *E. coli* enzyme is likely to form the γ-glutamyl-P intermediate only transiently as a less stabilized intermediate than that with a tetrahedral geometry at the γ-carbonyl of glutamic acid, whereas the pea seed enzyme recognizes the γ-glutamyl-P as a highly stabilized intermediate.[45] The relatively poor

---

[45] F. C. Wedler and B. R. Horn, *J. Biol. Chem.* **251,** 7530 (1976).

3-(phosphonoacetylamido)-
L-alanine (PA₂LA)

[Structure: H₃N⁺–CH(COO⁻)–NH–C(=O)–CH₂–P(=O)(OH)(O⁻) with stereochemistry at α-carbon]

inhibitory activity of compound **2** toward *E. coli* γ-GCS and GS is not inconsistent with this notion, if these enzymes form the γ-glutamyl-P intermediate transiently. This assumption seems reasonable if one considers that both enzymes catalyze the reaction by a sequential mechanism and that in many cases γ-glutamyl-P is not kinetically competent in the reaction pathway catalyzed by GS[46] and γ-GCS.[47,48]

Although the *E. coli* γ-GCS and GS are not inhibited effectively by **2**, the phosphonomethyl ketones and the related carbonyl compounds are still reasonably good analogs of the putative acyl phosphate and the equivalent intermediates for other ATP-dependent ligases. Table I shows the structure of such compounds and their inhibitory activities toward the corresponding ligases. A key structural feature is that the hydrolytically labile acyl phosphate or acyl adenylate anhydride is replaced by isosteric and hydrolytically stable phosphonoacetyl, *N*-acylphosphoramidate, and *N*-acylsulfamoyl linkages. In terms of inhibitory activity, the replacement of the anhydride bridging oxygen with NH is better than or at least equally effective as the substitution by CH_2, probably because of the isosteric and isopolar nature of nitrogen with oxygen. However, the *N*-acylphosphoramidate linkage is relatively unstable under acidic conditions.[49] The replacement of CH_2 by CF_2 in phosphonoacetyl derivatives does not seem to have a significant effect on the inhibitory activity.

The extent of enzyme inhibition or the relative affinity of the inhibitor is most likely to reflect the relative importance of the trigonal versus tetrahedral geometry in the reaction intermediate for each enzymatic reaction. Mechanistically related ADP-forming ligases such as GS, γ-GCS, glutathione synthetase, and D-Ala : D-Ala ligase are all inhibited strongly by tetrahedral phosphinate- or sulfoximine-based inhibitors. On the other hand, the trigonal aminoacyl sulfamoyladenosines (Table I), which are isosteric and isopolar analogs of the aminoacyl adenylate intermediates, serve as extremely potent inhibitors of the corresponding aminoacyl-tRNA synthetases. These results are corresponding to the kinetic mechanism of aminoacyl-tRNA synthetases; each enzyme shows the typical two-step mechanism involving the aminoacyl adenylate as a stable and kinetically viable

---

[46] T. D. Meek and J. J. Villafranca, *Biochemistry* **19**, 5513 (1980).
[47] B. Yip and F. B. Rudolph, *J. Biol. Chem.* **251**, 3563 (1976).
[48] V. B. Schandle and F. B. Rudolph, *J. Biol. Chem.* **256**, 7590 (1981).
[49] J. Robles, E. Pedroso, and A. Grandas, *J. Org. Chem.* **60**, 4856 (1995).

TABLE I
ATP-Dependent Ligases and Acyl-Phosphate Intermediate Analogs

Acyl-P intermediate analogs	Enzyme	K_i [M]	Refs.
PALA	Aspartate transcarbamoylase	2.7×10^{-8} 1×10^{-9}	a b
DIKEP	Aspartate transcarbamoylase	1×10^{-8}	b
(CF$_2$ phosphonate, N-H)	Aspartate transcarbamoylase	$IC_{50} = 5\ \mu M$	c
(CF$_2$ phosphonate, CH)	Aspartate transcarbamoylase	No inhibition	c
(H$_3$N$^+$, CH$_2$ phosphonate)	D-Ala:D-Ala ligase	5.1×10^{-4}	d
(H$_3$N$^+$, NH phosphoramidate)	D-Ala:D-Ala ligase	5.0×10^{-5}	d
(phosphonate-biotin)	Acetyl CoA carboxylase	8.4×10^{-3}	e
(aspartyl-AMP analog)	Asparagine synthetase B (E. coli)	8.0×10^{-5}	f

Acyl-P intermediate analogs	Enzyme	K_i [M]	Refs.
(structure)	Alanyl-tRNA synthetase (*E. coli*)	Strong inhibition	g
(structure)	Seryl-tRNA synthetase (*E. coli*)	10^{-9}	h
(structure)	Prolyl-tRNA synthetase (*E. coli*) (human)	4.3×10^{-9} 6.0×10^{-10}	i i

[a] K. D. Collins and G. R. Stark, *J. Biol. Chem.* **246**, 6599 (1971).
[b] E. A. Swyryd, S. S. Seaver, and G. R. Stark, *J. Biol. Chem.* **249**, 6945 (1974).
[c] S. D. Lindell and R. M. Turner, *Tetrahedron Lett.* **31**, 5381 (1990).
[d] P. K. Chakravarty, W. J. Greenlee, W. H. Parsons, A. A. Patchett, P. Combs, A. Roth, R. D. Busch, and T. N. Mellin, *J. Med. Chem.* **32**, 1886 (1989).
[e] D. R. Amspacher, C. Z. Blanchard, F. R. Fronczek, M. C. Saraiva, G. L. Waldrop, and R. M. Strongin, *Organic Lett.* **1**, 99 (1999).
[f] D. C. Pike and L. Beevers, *Biochim. Biophys. Acta* **708**, 203 (1982).
[g] H. Ueda, Y. Shoku, N. Hayashi, J. Mitsunaga, Y. In, M. Doi, M. Inoue, and T. Ishida, *Biochim. Biophys. Acta* **1080**, 126 (1991).
[h] H. Belrhali, A. Yaremchuk, M. Tukalo, K. Larsen, C. Berthet-Colominas, R. Leberman, B. Beijer, B. Sproat, J. Als-Nielsen, G. Grübel, J.-F. Legrand, M. Lehmann, and S. Cusack, *Science* **263**, 1432 (1994).
[i] D. Heacock, C. J. Forsyth, K. Shiba, and K. Musier-Forsyth, *Bioorg. Chem.* **24**, 273 (1996).

intermediate.[50] In this sense, the intermediate analogs such as phosphonomethyl and phosphonodifluoromethyl ketones are expected to be more potent inhibitors for the ligases with a kinetic mechanism similar to that of aminoacyl-tRNA synthetases, where a trigonal acyl phosphate or an equivalent intermediate is formed as a significantly stabilized species.

[50] A. Fersht, "Structure and Mechanism in Protein Science," p. 235. Freeman, New York, 1999.

[22] Stoichiometric Redox Titrations of Complex Metalloenzymes

By PAUL A. LINDAHL

Introduction

In principle, performing redox titrations of metalloenzymes with numerous redox centers is no different from doing so on proteins containing single redox centers or on low molecular weight compounds. However, in practice, titrating complex metalloenzymes is substantially more difficult, especially if a comparable level of precision and accuracy are desired. As the result of our efforts to perform such titrations on the NiFe hydrogenase from *Desulfovibrio gigas*[1–3] and the acetyl-coenzyme A synthase (also known as carbon monoxide dehydrogenase) from *Clostridium thermoaceticum*[4–6] we have developed what can be considered a method applicable to many enzymes. The purpose of this article is to describe this method and point out difficulties and avoidable pitfalls.

Summary of Method

Quantified aliquots of a redox titrant are reacted with a known amount of enzyme while redox-dependent spectroscopic properties of the enzyme are monitored. Spectral changes are plotted against the number of redox equivalents per mole added. Theoretical titration curves, based on various candidate descriptions of the complete redox/spectroscopic system in the enzyme, are generated and fitted to the data. The simplest description that generates the best-fit simulations to the data is concluded to be the redox system present in the enzyme.

Problems Addressed by Method

A number of metalloenzymes have the following profile: they have been purified to apparent homogeneity but their atomic level structures have not been determined. Metal analysis reveals that they contain numerous metal ions, although the *exact* number is not known. (This uncertainty arises because routine metal and protein concentration determinations have a combined relative uncertainty of 20%

---

[1] D. P. Barondeau, L. M. Roberts, and P. A. Lindahl, *J. Am. Chem. Soc.* **116**, 3442 (1994).
[2] L. M. Roberts and P. A. Lindahl, *Biochemistry* **33**, 14339 (1994).
[3] L. M. Roberts and P. A. Lindahl, *J. Am. Chem. Soc.* **117**, 2565 (1995).
[4] W. Shin, P. R. Stafford, and P. A. Lindahl, *Biochemistry* **31**, 6003 (1992).
[5] W. K. Russell and P. A. Lindahl, *Biochemistry* **37**, 10016 (1998).
[6] D. M. Fraser and P. A. Lindahl, *Biochemistry* **38**, 15697 (1999).

or more.) Spectroscopic studies of the oxidized and reduced forms of the enzyme reveal that the metal centers are redox active. The exact nature of the spectroscopic features is unknown and spectral assignments are controversial. For example, it may not be clear whether features arise from well-behaved isolated redox centers, *coupled* redox centers, or spin-state mixtures. Despite these uncertainties, a center has been assigned to (or defined by) each spectroscopic feature observed. The intensity of each feature has been monitored as a function of solution potential (via potentiometric titrations), and the Nernst equation has been fitted to the resulting titration curves to determine the apparent reduction potentials for each putative center. These putative centers and their apparent redox processes form the foundation of proposed catalytic mechanisms.

For such enzymes, perplexing questions inevitably arise. Might the enzyme contain additional (mechanistically important) redox centers for which there is no spectroscopic evidence? Given that the spectroscopic features have uncertain origins and might arise from coupled redox centers, to what process or processes do the reduction potentials assigned to these features actually refer? Is it possible that only a small fraction of enzyme molecules in a given preparation (say those that are catalytically *inactive*) actually exhibit the monitored spectroscopic properties, while active molecules are spectroscopically silent? The purpose of the method described in this chapter is to address these questions and to provide a more solid foundation on which catalytic mechanisms may be built.

Requirements

The method can be applied to enzymes that catalyze redox reactions. Consider a hypothetical enzyme that catalyzes reaction (1).

$$\text{Substrate} + \text{Oxidant} \rightleftharpoons \text{Product} + \text{Reductant} \tag{1}$$

Substrate and Product compose a physiological redox couple (e.g., CO_2/CO or H^+/H_2) with reduction potential $E^0{}_{PS}$. The Ox/Red redox couple (with $E^0{}_{OR}$) may be another physiological substrate (e.g., O_2/H_2O) or a nonphysiological species used in *in vitro* assays (e.g., oxidized/reduced methyl viologen). The enzyme can be reduced by substrate and reductant, and oxidized by product and oxidant, as shown in reactions (2) and (3).

$$\text{Substrate} + E_{ox} \rightleftharpoons \text{Product} + E_{red} \tag{2}$$

$$E_{red} + \text{Oxidant} \rightleftharpoons E_{ox} + \text{Reductant} \tag{3}$$

This hypothetical enzyme employs three redox centers in catalysis, called A, B, and C. E_{ox} represents the state $\{A_{ox}, B_{ox}, C_{ox}\}$ while E_{red} represents $\{A_{red}, B_{red}, C_{red}\}$. There may be other intermediate redox states of the enzyme in which some centers are reduced and others oxidized.

For the method described here to be effective, some redox centers must have spectroscopic features that are sensitive to their redox state and can be monitored during titration. However, the total number of redox centers need not be known. In favorable cases (see below), spectroscopically silent redox centers may be identified from their effect on the titration curves of spectroscopically active centers. Another requirement is that the enzyme can be isolated in fairly large amounts (~1 μmol). Also required are an anaerobic glove box, titration cuvettes and syringes, a spectrometer, an accurate protein determination method, an accurate metal analysis method, and a personal computer with software for generating theoretical titration curves and fitting the data. The most important requirement, of course, is a bright meticulous energetic graduate student to perform all the experiments.

Details of Method

Ascertaining Thermodynamic Reduction Potentials of Observable Redox Sites

The analysis component of this method uses the reduction potentials of as many redox processes as possible (as input parameters in simulations). If these values are unknown, they should be determined if possible, as the ability to fix them in fitting routines increases the reliability of the best-fit values for those parameters that cannot be fixed. Reduction potentials are typically determined by potentiometric titrations in which the intensity of a spectroscopic feature is monitored as a function of solution potential, and the Nernst equation is fitted to the resulting curve.[7,8] In such titrations, sufficient reductant or oxidant is added to the enzyme solution in the presence of redox mediators to achieve a desired potential, but the amounts of titrant required to achieve such potentials are not quantified.

Selecting Potential Region to Be Titrated

The region of interest typically includes potentials where mechanistically relevant redox events occur. Because the redox centers in our hypothetical enzyme are reducible by substrate and reductant, the lowest relevant potential will be about 0.1 V lower than the lower of the two reduction potentials (E^0_{PS} or E^0_{OR}). At potentials lower than this, any reduction processes occurring will be mechanistically irrelevant. At the other extreme, mechanistically relevant redox centers are oxidizable by product and oxidant, so the highest relevant potential will be about 0.1 V higher than the higher of the two reduction potentials. Thus, the approximate potential range will be $|E^0_{PS} - E^0_{OR}| + 0.2$ V, centered at the average of the two potentials, $\{|E^0_{PS} + E^0_{OR}\}/2$. Ideally, titrations should be performed in both oxidative and reductive directions so that reversibility can be assessed.

[7] W. G. Zumft, L. E. Mortenson, and G. Palmer, *Eur. J. Biochem.* **46**, 525 (1974).
[8] N. A. Stombaugh, J. E. Sundquist, R. H. Burris, and W. H. Orme-Johnson, *Biochemistry* **15**, 2633 (1976).

Selecting Reductant or Oxidant to Be Used

Substrates and products of the enzyme-catalyzed reaction (in this case Substrate and Product, as well as Oxidant and Reductant) should be used as redox titrants. Besides having appropriate redox potentials, electron transfer kinetics will almost certainly be rapid (since oxidation and reduction of the enzyme by products and substrates are part of the catalytic mechanism). Analysis of the titration curves assumes equilibrium, and rapid kinetics makes achieving this more likely. Also, including a fixed amount of one member of the redox pair and increasing amounts of the other (e.g., a fixed known amount of product as increasing amounts of substrate are added) will allow the solution potential to be stable and reliably calculated using the Nernst equation. Finally, redox intermediates generated using substrate and product will be mechanistically relevant, whereas those generated using nonphysiological redox agents may depend on the nature of the redox agent and be of questionable mechanistic importance. In any event, redox agents should be calibrated, in terms of number of equivalents per mole, by titration against a primary standard such as potassium ferricyanide.

Considering Requirements of Spectroscopic Method to Be Employed

Identify any constraints, limitations, or restrictions arising from the spectroscopic method to be employed. We have used electron paramagnetic resonance (EPR) and electronic absorption spectroscopy in our studies. For the EPR studies, protein aliquots are transferred into a series of EPR tubes. Increasing volumes of the redox agent are added and samples are frozen after an incubation period. In our particular experiment, the reductants used are gases (CO, H_2) which can partition into the gas phase of the reaction vessel. We have avoided such complications by designing small EPR cuvettes that can be filled completely with solution (leaving no gas phase). Standard Wilmad 5-mm EPR tubes are cut 4 cm from the bottom, and small holes opposite each other are drilled into the walls, ~1 mm from the top.[9] Small quartz chips are added and used for mixing. Rubber septa which have been intentionally punctured ~6 times are used to seal the cuvettes. Then cuvettes are filled with enzyme solution until liquid emanates from the punctures. The reaction is initiated by injecting a known volume of CO- or H_2-saturated buffer. After samples are mixed and frozen, a nylon rod modified to lock into the holes is attached and the assembly is inserted into the EPR spectrometer. In the UV–Vis titrations, the protein is transferred into a single small-volume anaerobic cuvette and precise volumes of a standardized redox agent are added incrementally, followed by the collection of a spectrum. Again, no gas phase is present. Different spectroscopic methods and cuvettes will undoubtedly require different experimental setups.

---

[9] W. Shin and P. A. Lindahl, *Biochemistry* **31,** 12870 (1992).

Considering Idiosyncrasies of Enzyme to Be Titrated

Every enzyme has a unique set of idiosyncrasies, and the titration protocol should be tailor-made with this in mind. Some proteins are denatured/inactivated by repeated freeze–thawing cycles, or by storage under certain conditions or temperatures. Others become unstable when diluted, or precipitate when concentrated. Others may exhibit different behavior in the presence of different substrates or products. Some enzymes may be robust toward reuse while others must be discarded after a single use. Such unpleasantries may not be emphasized or described clearly in the literature, and some investigative probing may be required to identify them. If no such information can be found, testing the stability of the system under the conditions of the titrations is critical to the design, execution, and interpretation of the experiment. Unfortunately the varied and unpredictable nature of proteins precludes generating a step-by-step procedure to follow. Rather, the specific protocol must arise from a dedication to achieving valid and interpretable results. For example, if protein precipitation is noticed during the titration, do not ignore it—find conditions to eliminate or minimize it. If no such conditions can be found, determine the extent of precipitation and include that in simulating titration curves.

Using Rigorously Anaerobic Techniques

The presence of oxygen in air makes it imperative to perform quantitative redox titrations anaerobically, even if the enzyme is stable in oxygen. Ideally, such titrations can be performed in an inert-atmosphere glove box with a solid-phase oxygen-binding catalyst through which the atmosphere is continuously circulated. Glove boxes from Vacuum/Atmospheres Inc. (Hawthorne, CA) are capable of maintaining O_2 pressures below 1 ppm, as monitored continuously by Teledyne O_2 analyzers. Titrations should be performed in reaction tubes or spectroscopic cuvettes sealed with rubber septa.

The time required to make materials anaerobic is far longer than naively assumed. Metals and glasses require the least time (minutes to hours), whereas plastic, rubber, and paper products require very long degassing times. Rubber septa should degas for *weeks or months* in the box prior to use. Teflon stir bars should be avoided completely as the O_2 impregnated in these materials is very difficult to remove. If they must be used, store them in an alkaline solution of dithionite/methyl viologen in the box for weeks or months prior to use. Glass-covered bar magnets (or tiny glass chips that agitate solutions on manual mixing) should be used instead. Teflon tips of gas-tight syringes are also a source of O_2. Such syringes, as well as plastic pipetters, should be stored in the box for weeks or months before use. Buffers used in titrations should be thoroughly degassed by repeatedly evacuating and refilling containers holding such solutions with O_2-scrubbed argon using a rigorously anaerobic Schlenk vacuum manifold. Such buffers should be subsequently stored in the box for days or weeks prior to use (to allow residual O_2 outgassing). Alternatively, buffers can be further sparged with box atmosphere using a bubbling apparatus (permanently stored in the box) made from vinyl tubing and an aquarium pump.

Performing Titration Using Many Data Points at Closely Spaced Intervals

Redox titrations analyzed by spectroscopies such as EPR are sample-intensive. As a result, relatively few (4–5) widely separated (0.5 to 1 equivalent/mol) data points are commonly obtained. Unfortunately, obtaining many data points (10–20) separated by small (0.1–0.25 equivalent/mol) intervals becomes increasingly important for enzymes with complex redox properties, as a high density of points will reveal the shape or structure of the titration curve. This structure can be important for fitting models to the data and for evaluating the presence of spectroscopically silent redox centers, heterogeneity, and spin coupling between redox centers.

Performing Multiple Titrations

Cells from which protein is obtained often grow at variable rates and final cell densities, despite efforts to employ uniform growth conditions and harvesting times. Harvesting large quantities of cells may require hours and result in heterogeneous product, because cells harvested at the beginning of the process may differ physiologically from those harvested at the end. Protein isolation is typically an arduous task fraught with unanticipated problems. Thus, despite considerable effort to make isolation and purification procedures uniform, there can be noticeable differences in catalytic, redox, and spectroscopic properties of the resulting product. For this reason, it is important to titrate samples from numerous (at least three) batches.

Thoroughly Characterizing Protein Samples

Another significant source of error in these titrations is in determining protein concentration. The most accurate determination method, quantitative amino acid analysis, should be employed if possible. It may be best to perform the experiment first, and then sacrifice the samples for protein and metal analyses at the end of the experiment. Pure protein samples should be used in these titrations, as impurities distort the enzyme concentration measurement and generate uncertainties in the titration curves. The level of impurities can be easily underestimated if small amounts of variable molecular weight species are present and if the gel used for densitometry analysis is underloaded, or if the percentage of polyacrylamide used does not sufficiently resolve all protein bands. If impurities cannot be removed, they should be quantified by densitometry analysis of stained SDS–polyacrylamide electrophoretic gels and used in calculating the concentration of the enzyme of interest. It is also important to quantify the metal content and the spectroscopic signal intensities of the samples. Considered together, these analytical determinations will indicate the concentration of the enzyme to be titrated, as well as relative uncertainties. Expect uncertainties between 10% and 25%.

Proposing Candidate Redox Systems

Once titration curves have been obtained and samples have been characterized, candidate redox/spectroscopic systems are constructed. Such systems consist of explicit sets of chemical reactions that describe all redox processes occurring in the enzyme during the titration, as well as descriptions of how those reactions give rise to the spectroscopic features monitored during the titration.

For example, our hypothetical enzyme is EPR silent when oxidized, but exhibits three EPR signals (A, B, and C) when reduced. One signal (C) has unusual features that are difficult to interpret while the other two (A and B) are typical $S = \frac{1}{2}$ spin systems, although signal A quantifies to only 0.3 spin/mol of enzyme. The experimentally ascertained reductive titration curves reveal that A develops first, followed by B, and then C. Curiously, as C develops, B disappears.

This behavior suggests numerous possibilities. For example, only 30% of enzyme molecules may contain a redox-active form of A. Alternatively, 100% of molecules may have a redox-active center A but the spin intensity may be low for other (physical/spectroscopic) reasons. Center B may be reduced by two electrons in one-electron steps, and the semireduced form may yield the observed signal. Alternatively, B may be reduced by just one electron, but the reduced form may be spin-coupled to the reduced form of center C. Two possible redox systems (among many) describing this behavior are shown below:

Possibility 1 ($n = 2$ B center and no redox heterogeneity in A)

$$A_{ox} + 1e^- \rightleftharpoons A_{red} \quad \text{and } [I_A] = P_A[A_{red}]$$
$$B_{ox} + 1e^- \rightleftharpoons B_{red1} \quad \text{and } [I_B] = P_B[B_{red1}]$$
$$B_{red1} + 1e^- \rightleftharpoons B_{red2}$$
$$C_{ox} + 1e^- \rightleftharpoons C_{red} \quad \text{and } [I_C] = P_C[C_{red}]$$

Possibility 2 (redox heterogeneity in A and spin-coupling between B and C)

$$0.3\{A_{ox} + 1e^- \rightleftharpoons A_{red}\} \quad \text{and } [I_A] = P_A(0.3)[A_{red}]$$
$$(0.7)\{A'_{ox} + 2e^- \rightleftharpoons A'_{red}\}$$
$$B_{ox} + 1e^- \rightleftharpoons B_{red} \quad \text{and } [I_B] = P_B[B_{red}]$$
$$C_{ox} + 1e^- \rightleftharpoons C_{red} \quad \text{and } [I_C] = P_C[C_{red}][B_{red}]$$

In these examples, only half-cell redox couples are shown. The parameter "I" refers to the quantified spin intensity of a given signal (given in concentration units).

P refers to a proportionality factor between the intensity of the signal and the concentration of the center.

Numerous redox/spectroscopic systems should be proposed, ideally exhausting all realistic possibilities. For example, if the enzyme has been proposed to contain an additional redox center for which there is no spectroscopic evidence, models should be constructed to include/exclude such a center. If the enzyme is thought to suffer from redox heterogeneity, one model may assume this while another might assume that all centers present in a population of enzyme molecules are redox active. The presence or absence of spin-coupled redox centers can also be specified by such systems. Unfortunately, it is likely that some model systems will be indistinguishable using the titration data obtained.

Simulating Titration Curves Using Models

A titration curve is a plot of the proportion of a given center in a detected state, measured at equilibrium, as a function of the number of redox equivalents added. For a single center and corresponding reaction, such curves can be simulated if the equilibrium constant K_{eq} for the corresponding equilibrium expression is known. For example, at any point in the reduction of A_{ox} ($n = 1$) by S ($n = 2$), forming A_{red} and P (meaning the total concentration of S added up to that point, called $[S]_{init}$), the extent of reaction x can be determined by solving the expression:

$$K_{eq} = (\{[A_{red}]_{init} + 2x\}^2 \{[P]_{init} + x\})/(\{[A_{ox}]_{init} - 2x\}^2 \{[S]_{init} - x\}) \quad (4)$$

A plot of $[A_{red}]/[A_{tot}]$ versus $[S]_{init}$ should simulate the experimental titration curve (assuming the correct K_{eq}). The equilibrium constant for a redox reaction can be obtained from the thermodynamic standard reduction potentials for the two half-cell couples ($E^0_{A_{ox}/A_{red}}$ and $E^0_{S/P}$).

$$K_{eq} = \exp(nF/RT)\{E^0_{A_{ox}/A_{red}} - E^0_{S/P}\} \quad (5)$$

In Eq. (5), n equals the number of electrons transferred in the reaction (2 in this case), and F, R, and T have their standard designations.

For a system of many reactions, each of which involves the same titrated redox agent, the equilibrium expressions must be solved simultaneously. This can be achieved by solving a series of equations (for the mathematically adept) or by an iterative cycling method. In the latter method, the equilibrium expression associated with the first reaction in the system is solved in the manner just described, using initial concentrations and increasing initial amounts of titrant. Then the second expression is solved, using the final equilibrium concentrations obtained from the first reaction as initial concentrations. This process is iterated until all equilibrium expressions are simultaneously satisfied. Although this technique is not the most elegant, computing power is plentiful and it allows redox systems of virtually any complexity to be handled without difficulty by chemists or biochemists unable to obtain analytical solutions for large numbers of simultaneous relationships.

Comparing Simulations of Candidate Redox/Spectroscopic Systems to Experimental Curves

To compare the simulated titration curves to experimental curves, the concentration of various redox centers must be related to the intensity of the spectroscopic feature monitored during the titration. EPR signals can be quantified by double integration, and the values obtained should equal the concentration of the redox center in its EPR-active state. However, for the systems we have studied (and we suspect for many others), it was not certain that this relationship held. Thus, it may be necessary to introduce a heterogeneity proportionality constant as done above for the hypothetical redox system 2.

For electronic absorption spectra, titration curves can be plotted as the relative change in absorbance at a wavelength. If the wavelength is sensitive to the redox status of a particular redox transition (e.g., $C_{ox} + e^- \rightarrow C_{red}$) and one state (e.g., C_{ox}) is titrated stoichiometrically to the other (C_{red}), then

$$[Abs_{pt} - Abs_{beg}]/[Abs_{end} - Abs_{beg}] = [C_{red}]/[C_{tot}] \quad (6)$$

where Abs_{pt} is the absorbance at any point in the titration, and Abs_{beg} and Abs_{end} are the beginning and ending absorbances, respectively.

The next step is to compare simulated and experimental titration curves. Assume that the experimental curve consists of a series of 10 points, designated $[X_{dat}, Y_{dat}]_i$ ($i = 1\text{--}10$). Let each simulation curve consist of a series of 100 points, designated $[X_{sim}, Y_{sim}]_j$ ($j = 1\text{--}100$). For each data point, the length to each simulation point, normalized to the overall data length in the X and Y dimensions, is:

$$L_{ij} = \sqrt{\{[(X_{dat\text{-}i} - X_{sim\text{-}j})/(X_{dat\text{-}10} - X_{dat\text{-}1})]^2 + [(Y_{dat\text{-}i} - Y_{sim\text{-}j})/(Y_{dat\text{-}10} - Y_{dat\text{-}1})]^2\}} \quad (7)$$

Because of normalization, differences in either the X or Y directions are weighted equally. Of the 100 L_{ij} values obtained for each data point, the smallest values ($L_{ij\text{-min}}$) are summed.

$$Q_{fit} = \sum L_{ij\text{-min}} (\text{sum from } i = 1 \text{ to } 10) \quad (8)$$

Q_{fit} is the quality-of-fit parameter describing how well simulated curves fit experimental data. Simulations can be generated and assessed using different input parameters (K_{eq} or E^0 values) and different models. Simulations are judged relative to that yielding the best fit to the data, by comparing $Q_{best\text{-}fit}$ values (lower Q_{fit} is $Q_{best\text{-}fit}$).

At least two candidate redox systems should be fitted to the experimental titration curves. The *control* system should include all known or undisputed redox events while the *test* should be identical to the control but include disputed events.

The critical issue is whether the test fits significantly better than the control. We have used $(Q_{\text{best-fit}})_{\text{control}}/(Q_{\text{best-fit}})_{\text{test}}$ ratios >1.5 as our criterion for significance.

Uncertainties in best-fit values should be determined. One parameter should be varied systematically to either side of the best-fit value as all other parameters are fixed at their best-fit values. The upper and lower values for which $Q_{\text{fit}}/Q_{\text{best-fit}}$ ratios >1.5 are taken to be the $\pm$relative uncertainties for that parameter.

Examples

NiFe Hydrogenase from Desulfovibrio gigas

NiFe hydrogenases catalyze the reversible reduction of molecular hydrogen to protons. The enzyme from *Desulfovibrio gigas* contains a NiFe active site cluster, a proximal $[Fe_4S_4]^{2+/1+}$ cluster, an $[Fe_3S_4]^{1+/0}$ cluster, and a distal $[Fe_4S_4]^{2+/1+}$ cluster.[10] The active site is stable in four redox states. The most oxidized state, Ni-AB, exhibits either of two $S = \frac{1}{2}$ EPR signals. Both have g values typical of Ni^{3+} ions and intensities that quantify to ~ 0.6 spin/mol protein. On reduction, these signals disappear forming the silent-intermediate Ni-SI state. Further reduction causes development of the Ni-C state. Ni-C is $S = \frac{1}{2}$ and also exhibits an EPR signal. Further reduction leads to the EPR-silent Ni-R state.

The issue addressed through redox titrations was the electronic nature of the Ni-C state. This state had been designated as Ni^{3+}, a Ni^{3+} hydride, a Ni^{3+} molecular H_2 complex, a protonated Ni^{1+} complex, and a Ni^{1+} hydride. We noticed that these designations could be separated into three groups according to whether they were 0, 2, or 4 electrons more reduced than the Ni-AB state. Our strategy was to quantify the number of electrons required to reduce the active site from Ni-B to Ni-C.

One difficulty was that Ni-C had been reported to be an unstable transient species that could not be titrated under equilibrium conditions. However, we demonstrated that this state was quite stable in the strict absence of O_2.[1] The presence of Fe-S clusters also complicated our study, as some of these centers underwent redox at potentials in the region where the Ni-B to Ni-C transition occurred. Thus, quantitative stoichiometric titrations were performed in the oxidative and reductive directions, followed by both EPR and UV–Vis spectroscopies, using H_2 and thionin as redox titrants (Fig. 1).

Another complication was that the reason for the low-spin intensity of the Ni-B and Ni-C signals was not understood, and so whether the EPR-silent Ni centers underwent redox reactions was unknown. We developed and fitted three models to the experimental titration curves, in which Ni-C was assumed to be 0, 2, and 4 electrons more reduced than Ni-AB. Also, we allowed the EPR-silent Ni ions to be either redox active or inactive. Most of the reduction potentials for the reactions

---

[10] A. Volbeda, M. H. Charon, C. Piras, E. C. Hatchikian, M. Frey, and J. C. Fontecilla-Camps, *Nature* **6515,** 580 (1995).

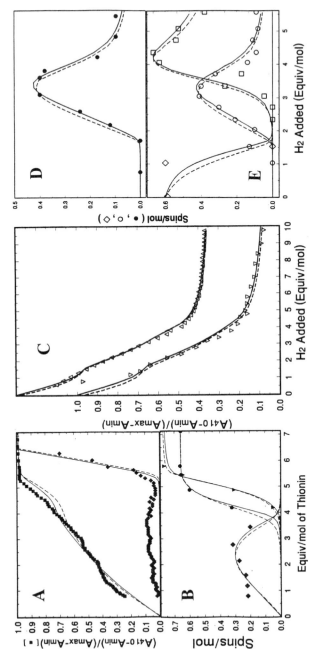

FIG. 1. Stoichiometric titrations of NiFe hydrogenase. (A) Oxidative titration using thionin as oxidant and monitored by electronic absorption spectroscopy at 410 (squares) and 600 (diamonds) nm. Solid and dashed lines are simulations generated from models described [D. P. Barondeau, L. M. Roberts, and P. A. Lindahl, *J. Am. Chem. Soc.* **116,** 3442 (1994); L. M. Roberts and P. A. Lindahl, *Biochemistry* **33,** 14339 (1994)]. Increased absorbance at 410 nm was due predominantly to oxidation of Fe-S clusters, whereas that at 610 nm was due to oxidized thionin and thus served as an endpoint marker. (B) Same as (A), except monitored by EPR spectroscopy. Diamonds, Ni-C; circles, Ni-B; triangles, the $g = 2.02$ signal from the oxidized $[Fe_3S_4]^{1+}$ cluster. (C) Reductive titration using H_2 as reductant and monitored by electronic absorption spectroscopy at 410 nm [L. M. Roberts and P. A. Lindahl, *J. Am. Chem. Soc.* **117,** 2565 (1995)]. Data (triangles) from two titrations are shown. (D and E) Same as (C) except monitored by EPR. Open and closed circles, Ni-C from two titrations; diamonds, Ni-B; squares, the $g = 2.21$ signal. Simulations in solid and dashed lines assume that the $g = 2.21$ signal arises from spin coupling between Ni-C and $[Fe_{4b}]^{1+}$. Reprinted with permission from the American Chemical Society.

in the redox systems had been measured, and these values were fixed. The best-fit model assumed that Ni-C was two electrons more reduced than Ni-AB and that the EPR-silent Ni ions were redox inactive.

The enzyme exhibits an unusual EPR signal thought to reflect spin coupling between Ni-C and one of the Fe_4S_4 clusters. Whether this was correct, and which cluster was involved ($[Fe_{4a}]^{2+/1+}$ or $[Fe_{4b}]^{2+/1+}$), was unknown. Our approach was to monitor the intensity of the so-called $g = 2.21$ signal in these titrations and simulate the resulting plots using models assuming that the signal intensity was proportional to either the product of $[Ni-C][Fe_{4a}]^{1+}$ or $[Ni-C][Fe_{4b}]^{1+}$. One of the clusters ($[Fe_{4b}]^{2+/1+}$) has a lower reduction potential than the other ($[Fe_{4a}]^{2+/1+}$) and so the two simulations differed. The model that assumed coupling between Ni-C and $[Fe_{4b}]^{1+}$ fit significantly better (Fig. 1E, squares), thereby identifying the source of this unusual signal and the cluster which is located nearer to the active site.

Acetyl-CoA Synthase/Carbon Monoxide Dehydrogenase from Clostridium thermoaceticum

This enzyme is an $\alpha_2\beta_2$ tetramer that catalyzes the reversible reduction of CO_2 to CO and the synthesis of acetyl-CoA from CO, a methyl group and coenzyme A. Each α subunit contains a Ni-X-Fe_4S_4 cluster known as the A-cluster, while each β subunit contains an Fe_4S_4 B-cluster and another Ni-X-Fe_4S_4 cluster called the C-cluster. The A-cluster is the active site for acetyl-CoA synthesis, the C-cluster is the active site for CO_2/CO redox catalysis, and the B-cluster functions in electron transfer. These clusters exhibit extensive redox chemistry. The oxidized A-cluster (A_{ox}) can be reduced by one electron and bound with CO to yield an EPR-active $S = 1/2$ state (A_{red}-CO). The oxidized B-cluster (B_{ox}) can be reduced by 1 electron to yield B_{red}. The oxidized $S = 0$ C-cluster (C_{ox}) can be reduced by one electron to the $S = 1/2$ C_{red1} state. C_{red1} can be reduced further to the EPR-silent C_{int} state, and C_{int} can be reduced even further to an $S = 1/2$ state called C_{red2}.

Two issues warranted use of quantitative stoichiometric redox titrations. First, a number of groups had proposed additional, albeit EPR silent, redox centers. Second, the EPR signal intensities from the A-, B-, and C-clusters were notoriously low (typically 0.2, 0.6, and 0.4 spins/$\alpha\beta$, respectively), and it had been suggested that these centers suffered from redox and spin-state heterogeneity. Both issues could be examined by stoichiometric titrations.

One idiosyncratic aspect of the enzyme is that it slowly but spontaneously oxidizes if reduced fully, and spontaneously reduces if oxidized fully.[4] Thus, the enzyme tends toward an intermediate state of reduction. We decided to make this intermediate state the starting point for these titrations. Fortunately, the rate of spontaneous redox is far slower than the rate at which the physiological redox

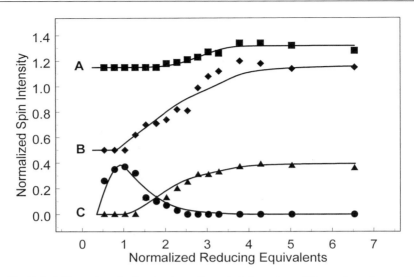

FIG. 2. Stoichiometric reductive titrations of acetyl-CoA synthase. CO was the reductant (in a background of CO_2) and EPR was used to monitor the titrations. (A) NiFeC signal (squares) from the A_{red}-CO state; (B) $g_{av} = 1.94$ signal (diamonds) from the B_{red} state; (C) $g_{av} = 1.82$ (circles) and 1.86 (triangles) signals from the C_{red1} and C_{red2} states, respectively. Solid lines are simulations, as described [D. M. Fraser and P. A. Lindahl, *Biochemistry* **38**, 15697 (1999)]. Curves A and B have been offset; both start at a spin intensity of 0. Reprinted with permission from the American Chemical Society.

centers in the enzyme react. Thus, samples were given enough time for the centers to react and reach equilibrium but not long enough for other nonphysiological processes to progress appreciably. The reaction used CO and thionin as reductant and oxidant, respectively. The titrations were monitored by EPR spectroscopy (we had performed an earlier, less sophisticated study using UV–Vis) and are shown in Fig. 2. The solid lines in Fig. 2 are best-fit simulations assuming no additional redox center (within the potential region probed) and the presence of redox heterogeneity, as previously proposed.

Advantages and Limitations of Method

The stoichiometric redox titration method described in this chapter has a number of worthy characteristics. It can provide reliable evidence for a spectroscopically silent redox center in an enzyme, as long as the reduction potential of the center is within the potential range probed. Such centers will consume redox equivalents and thus alter the shape of the titration curves of spectroscopically active centers. Spectroscopically observable stable redox states are commonly viewed as being more important than those which cannot be easily observed, but this need not be the case. Thus, the results of stoichiometric titration studies may be critical for formulating the correct mechanism of catalysis. Related to this is the ability of these titrations to demonstrate the presence of redox heterogeneity. Such heterogeneity is likely to be more common in metalloenzymes than is presently acknowledged. Although it is tempting to view heterogeneity as a nuisance, the as-yet unexplained phenomena causing this may provide insight into the mechanism of metal cluster assembly and function. Quantitative stoichiometric redox titrations can also be used for spin-coupling assignments, as long as the spin-coupled system yields a monitorable spectroscopic feature.

The major disadvantage of the method is the large amounts of money, time, and effort required to employ it, in contrast to what some regard as the unglamorous results that are obtained. Moreover, the method must be executed carefully using equipment (i.e., glove boxes) that is not routinely available. Large quantities of enzyme are required, and analysis requires some computer programming. Finally, despite best efforts, some uncertainty in the results and analysis will inevitably remain. This ultimately translates into some persistent uncertainty regarding the redox/spectroscopic system present in the enzyme.

Despite these limitations, the results of such titrations provide a fairly solid foundation on which an accurate mechanistic structure can be built. Complex metalloenzymes that have not been subjected to quantitative stoichiometric redox titrations lack this foundation. As a result, the catalytic mechanisms of such enzymes will inevitably remain less certain and more controversial than need be the case. In short, the researcher needs to make the effort, perform the titrations, and test the mechanism.

Acknowledgments

I thank the National Institutes of Health (GM46441) for supporting the development of the method described here, and all the bright meticulous and energetic graduate and undergraduate students who contributed to this development, including Woonsup Shin, Phillip Stafford, Melvin Roberts, David Barondeau, William Russell, and Daniel Fraser.

[23] Urate Oxidase: Single-Turnover Stopped-Flow Techniques for Detecting Two Discrete Enzyme-Bound Intermediates

By PETER A. TIPTON

Introduction

Urate oxidase catalyzes the formation of 5-hydroxyisourate from urate. It is a true oxidase; although the product is hydroxylated the oxygen in the hydroxyl group derives from solvent, and 1 equivalent of dioxygen is reduced to H_2O_2 concomitant with urate oxidation. In leguminous plants urate oxidase is a constituent of the ureide pathway by which fixed nitrogen is converted to allantoin and allantoate, the metabolites that serve to transport nitrogen from the roots to the stems, leaves, and seed pods of the plant. In mammals urate oxidase facilitates the excretion of excess nitrogen by virtue of the greater solubility of allantoin relative to urate. Urate oxidase is also found in a wide variety of microorganisms, and some bacteria can grow on urate as their sole carbon and nitrogen source.

Mechanistic interest in urate oxidase derives largely from the fact that no cofactor participates in the catalytic reaction. Transition metals such as copper and iron and the organic cofactors flavin and pterin play key roles in mediating reactions between triplet state dioxygen and spin-paired organic substrates in most oxidase and oxygenase reactions. Indeed, urate oxidase was long believed to be a copper-dependent enzyme, but studies have shown that the enzyme from soybean root nodules does not contain a cofactor and is active in the absence of copper,[1] and the X-ray crystal structure of urate oxidase from *Aspergillus flavus* revealed the absence of any cofactors at the active site.[2]

A potential mechanism for the urate oxidase reaction is presented in Fig. 1. Similarities with the well-studied mechanism for oxidation of reduced flavin are evident. Stopped-flow spectroscopy has played a key role in the determination of the mechanisms of flavin-dependent enzymes and was utilized to evaluate the urate oxidase reaction as well.[3] The goals of these studies were both qualitative and quantitative. In particular, the mechanism illustrated in Fig. 1 proposes the formation of three discrete intermediates in the course of the catalytic reaction, and stopped-flow studies were conducted in order to determine whether evidence could be adduced for their existence. An alternative mechanism for the urate oxidase reaction has

---

[1] K. Kahn and P. A. Tipton, *Biochemistry* **36**, 4731 (1997).
[2] N. Colloc'h, M. El Hajji, B. Bachet, G. L'Hermite, M. Schiltz, T. Prange, B. Castro, and J. Mornon, *Nat. Struct. Biol.* **4**, 947 (1997).
[3] K. Kahn and P. A. Tipton, *Biochemistry* **37**, 11651 (1998).

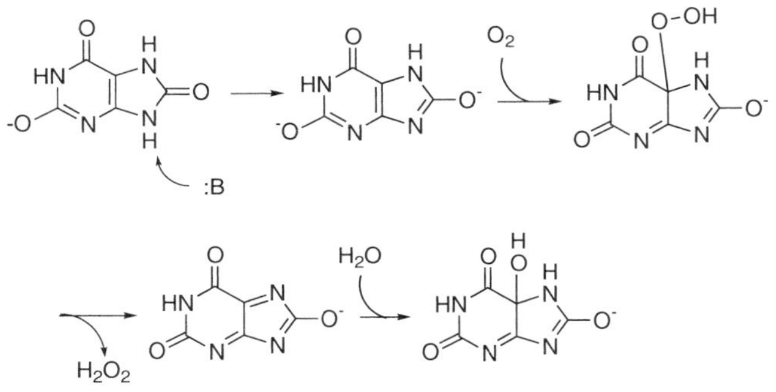

FIG. 1. Proposed chemical mechanism for the reaction catalyzed by urate oxidase.

been proposed that suggests that oxidation of urate occurs without formation of any discrete intermediates.[2] Stopped-flow spectroscopy offered the means to evaluate whether intermediates form at all, and if so, to determine the time dependence of their formation and decay. Experiments were conducted using recombinant soybean root nodule urate oxidase. Transient-state kinetic studies of urate oxidases from other organisms have not been conducted. However, amino acid sequence conservation among urate oxidases from species as diverse as *Bacillus subtilis* and primates is extensive, and it is likely that the essential features of the catalytic mechanism are conserved as well.

Stopped-Flow Spectroscopic Studies

Urate oxidase reactions are carried out in air-saturated solutions under single-turnover conditions. Soybean root nodule urate oxidase is relatively insoluble and active site concentrations greater than 30 μM cannot be achieved routinely. At neutral pH the absorbance maximum of urate is at 292 nm, so all time course data are collected after the spectrophotometer has been blanked against solutions containing enzyme alone. Although the absolute magnitude of the absorbance changes that are observed are small they are quite reproducible, and, as described below, can be interpreted in a coherent manner.

The time course for the reaction at 306 nm is illustrated in Fig. 2. Four kinetically distinct phases are evident. The initial phase, in which the absorbance increases, is observable only at reduced temperature, and even at 15° ends within 20 ms. The second phase, characterized by decreasing absorbance, lasts until approximately 90 ms. The absorbance increases again in the third phase, which lasts for approximately 300 ms, and then the absorbance decreases slowly. The final phase of the reaction requires several minutes to come to completion. In order to ensure that the experiments are conducted under single turnover conditions the ratio of enzyme to substrate is varied, and the time required to reach the absorbance

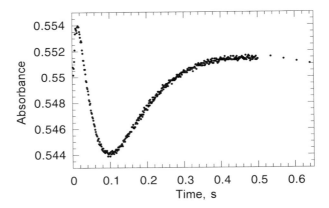

FIG. 2. Time course of the urate oxidase reaction under single turnover conditions, monitored by absorbance at 306 nm at 15°. Reprinted with permission from K. Kahn and P. A. Tipton, *Biochemistry* **37**, 11651 (1998). Copyright © 1998 American Chemical Society.

minimum corresponding to the end of the second kinetic phase is monitored. Under first-order conditions the time after mixing at which this minimum is reached is independent of enzyme concentration; a two- to threefold excess of urate oxidase over urate has been found to be sufficient to meet this criterion.

The data in Fig. 2 demonstrate that spectroscopically distinct intermediates form during the urate oxidase reaction. The primary challenge of further experiments has been to obtain unique spectra for those intermediates. Technical complications stem from the facts that the absorbance spectra of the substrates, intermediates, and products overlap with each other and the absorbance spectrum of the enzyme, and the product of the enzymatic reaction is itself unstable.

In order to resolve the spectra of the species that form during the urate oxidase reaction and to determine their kinetic behavior, a series of stopped-flow experiments has been conducted with detection at several discrete wavelengths (Fig. 3). These data are collected at 20° and so the initial phase of the reaction seen in Fig. 2 is not observed. Analysis of this family of data proceeds through several stages. The initial data analysis is conducted using singular value decomposition (SVD) methods.[4] The advantage of SVD is that the analysis is model-free and therefore allows a relatively objective evaluation of the number of species in solution that are resolved by the data. The family of data shown in Fig. 3 is decomposed into pairs of basis spectra and amplitude vectors and the significance factors associated with each pair of components. The magnitude of the significance factors, the signal-to-noise ratio in the output components, and the number of pairs of output components that are required to reconstruct the experimental data in a satisfactory

---

[4] E. R. Henry and J. Hofrichter, *Methods Enzymol.* **210**, 129 (1994).

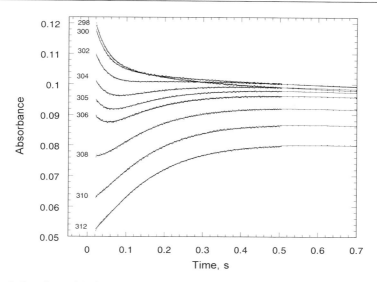

FIG. 3. Experimental (points) and calculated (line) time courses of the urate oxidase reaction, monitored at the wavelengths indicated. Data were collected under single turnover conditions at 20°. The lines were calculated as described in the text. Reprinted with permission from K. Kahn and P. A. Tipton, *Biochemistry* **37**, 11651 (1998). Copyright © 1998 American Chemical Society.

fashion are evaluated. It has been concluded that the data support a model in which four species, sequentially converted from one to another, contribute to the absorbance of the samples ($X_1 \rightarrow X_2 \rightarrow X_3 \rightarrow X_4$). Preliminary estimates for the rate constants governing the interconversion of the species have been derived by fitting the amplitude vectors from the SVD analysis to the equation describing the sum of three first-order processes.

Further analysis has been conducted using global data fitting[5] with the DYNAFIT software package.[6] Global fitting is used to extract the spectra of the individual components of the reaction mixture and to refine the estimates for the rate constants for their interconversion. Consideration of the lifetimes of the species X_3 and X_4 suggests that they are hydroxyisourate (HIU), the true product of the urate oxidase reaction, and 2-oxo-4-carboxy-4-hydroxy-5-ureidoimidazoline (OHCU), the product of hydrolysis of HIU, respectively. The analysis of the stopped-flow data is simplified by independent determination of the spectra of HIU and OHCU.[3] Thus, the data are fitted to the model $X_1 \rightarrow X_2 \rightarrow HIU \rightarrow OHCU$. The global analysis yields the rate constants for the interconversion of the four species in the model

[5] J. M. Beechem, *Methods Enzymol.* **210**, 37 (1992).
[6] P. Kuzmic, *Anal. Biochem.* **237**, 260 (1996).

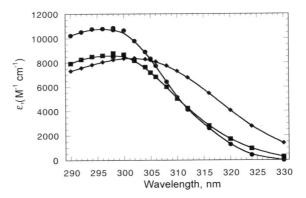

FIG. 4. Calculated absorbance spectra of INT 1 (●), INT II (■), and 5-hydroxyisourate (◆). Reprinted with permission from K. Kahn and P. A. Tipton, *Biochemistry* **37,** 11651 (1998). Copyright © 1998 American Chemical Society.

and their absorbance spectra. The comparison between the experimental data and the fit is shown in Fig. 3, and the calculated absorbance spectra are in Fig. 4.

The agreement between experimental data and the fit is quite good; however, a good fit can always be obtained if enough parameters are put into the model. Therefore, it is important to evaluate whether four species is the minimum number required to give a good fit. The quality of the fit is evaluated quantitatively by the parameter χ^2, the weighted sum of the squared deviations between experimental and fitted values, divided by the number of data points; the better the fit, the closer χ^2 is to unity. The F-test is used to determine whether the values of χ^2 given by two fits are significantly different. The simplest model that gives a small value for χ^2 is selected as the appropriate model. The value for χ^2 for the fit shown in Fig. 3 was 0.90; when the data were fitted to a model for the interconversion of three species χ^2 was 9.86. The data from a single wavelength, 304 nm, were fitted to the four species model and a value for χ^2 of 0.495 was obtained; when these data were fitted to a model with five species the value for χ^2 was 0.489, and it was concluded that there was no reasonable basis for accepting the more complex model.

An interesting mathematical complication in the interpretation of stopped-flow spectroscopic data has been pointed out by Alcock *et al.*[7]: namely, that multiple solutions exist for the equation that describes the absorption of a solution arising from the sequential conversion of reactant to intermediate and intermediate to product if the absorbance spectrum of the intermediate is not known *a priori*. For the case in which S→I→P the quality of the fit obtained when k_1 is assigned to the first process and k_2 is assigned to the second process is exactly the same as when k_1 is assigned to the second process and k_2 is assigned to the first process. The difference

[7] N. W. Alcock, D. J. Benton, and P. Moore, *Trans. Faraday Soc.* **66,** 2210 (1970).

between the two cases lies in the absorbance spectrum of I that each solution yields. The three-step process that the stopped-flow data for the urate oxidase reaction requires has multiple solutions; however, the slowest step in the sequence, which occurs with a rate constant of 0.05 s^{-1}, can be assigned with confidence to the conversion of HIU to OHCU. Two solutions that are equally good mathematically are left: assigning a rate constant of 32 s^{-1} to the first step and a rate constant of 6.6 s^{-1} to the second step, or the alternative assignment. The former solution yields the calculated spectrum for intermediate II that is shown in Fig. 4; the latter solution has been discarded because it yields a calculated spectrum for intermediate II with a maximum extinction coefficient of approximately 4000 M^{-1}cm^{-1}, which was judged to be chemically unreasonable.

An assessment of the validity of the values of the calculated rate constants is provided by an analysis of the time course of the reaction when it is monitored by stopped-flow fluorescence spectroscopy. 5-Hydroxyisourate is fluorescent, but its fluorescence is efficiently quenched by urate oxidase. Urate is not fluorescent under the conditions of our experiments, so fluorescence detection provides a means to monitor only HIU that has been released from the enzyme. The time course is shown in Fig. 5. The line drawn through the experimental points is calculated using a four-step model, $X_1 \rightarrow X_2 \rightarrow [HIU] \rightarrow HIU^* \rightarrow OHCU$, where [HIU] is enzyme-bound HIU and HIU* is free HIU and is the only fluorescent species in solution. The values for the first, second, and fourth rate constants are fixed at the values derived from the analysis of the stopped-flow absorbance data. A value of 32 s^{-1} is determined for the third rate constant, which is the product release step. The value of

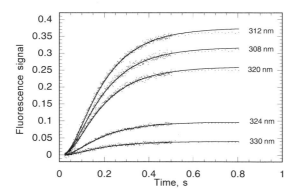

FIG. 5. Time course of the urate oxidase reaction under single turnover conditions at 20°, monitored by fluorescence. Fluorescence was monitored at 380–400 nm using the excitation wavelengths indicated in the figure. The points are experimental and the solid lines are the calculated time courses derived from the four-step model described in the text. Reprinted with permission from K. Kahn and P. A. Tipton, *Biochemistry* **37**, 11651 (1998). Copyright © 1998 American Chemical Society.

χ^2 that is derived from this analysis, 0.32, is significantly better than that derived from a three-step model (χ^2, 2.37). The close match between the experimental points and the calculated time course suggests that the values for the rate constants that are derived from the absorbance data are reasonably accurate. Note that this analysis does not speak to the assignment of the first two rate constants to specific processes.

What are X_1 and X_2? On the basis of the calculated absorbance spectra X_1 has been assigned to the urate dianion and X_2 to the urate hydroperoxide. In aqueous solution the urate dianion has an absorption maximum of 295 nm; the absorption maximum of the urate monoanion is at 292 nm. However, environmental effects can influence the absorption spectrum of urate and thus complicate the assignment of the spectrum of an enzyme-bound species. The crystal structure of *Aspergillus flavus* urate oxidase reveals that an arginine residue is hydrogen-bonded to the C-6 carbonyl and N-7 of the ligand at the active site. Donation of a hydrogen bond would be expected to cause a blue shift in the absorption maximum of urate, so although the absorption maxima of the urate monoanion and dianion are not very different, the calculated spectrum of X_1 seems more consistent with assignment to the urate dianion rather than to a red-shifted urate monoanion.

The second intermediate, X_2, has an absorption maximum at 296–298 nm. Based on the reaction mechanism illustrated in Fig. 1 urate hydroperoxide and dehydrourate are the most likely candidates for X_2. Dehydrourate has been observed as a fleeting intermediate in the electrochemical oxidation of urate, but its UV–Vis absorption spectrum has not been measured. Methylated model compounds have been prepared, however, and are characterized by absorption maxima at 290 and 350–360 nm.[8] Methylation shifts the absorption maxima of purines only slightly, and since X_2 does not exhibit absorbance above 330 nm it is unlikely to be dehydrourate. Authentic samples of urate hydroperoxide are not available, either. However, hydroxy and hydroperoxy adducts of pterin and flavin compounds have been characterized[9,10] and it has been observed that changing the substituent from hydroxy to hydroperoxy changes the absorption maximum by only 2–10 nm. The absorption maximum of 5-hydroxyisourate is at 302 nm so it is reasonable to assign X_2 to 5-hydroperoxyisourate (urate hydroperoxide).

The mechanism in Fig. 1 proposes that three intermediates form on the pathway to 5-hydroxyisourate formation, yet the spectroscopic data provide evidence for the existence of only two intermediates. Why is dehydrourate not observed? As mentioned above, model studies have provided evidence for the existence of dehydrourate as an intermediate in the electrochemical oxidation of urate.[11] Its

---

[8] A. Palkovic and M. Poje, *Tetrahedron Lett.* **31**, 6101 (1990).
[9] C. Kemal and T. C. Bruice, *Proc. Natl. Acad. Sci. U.S.A.* **73**, 995 (1976).
[10] G. Moad, C. L. Luthy, P. A. Benkovic, and S. J. Benkovic, *J. Am. Chem. Soc.* **101**, 6068 (1979).
[11] A. Brajter-Toth and G. Dryhurst, *J. Electroanal. Chem.* **122**, 205 (1981).

lifetime in aqueous solution has been estimated to be ~20 ms, so even if urate oxidase does not catalyze its hydration it would not be expected to accumulate to detectable levels during enzyme turnover. *Ab initio* calculations which indicate that dehydrourate is almost 10 kcal/mol less stable than urate hydroperoxide[12] are also consistent with the notion that dehydrourate would be undetectable in stopped-flow spectroscopy experiments. In light of these considerations it is rather gratifying that the analysis of the stopped-flow absorbance data does not suggest that dehydrourate is detected.

Supporting Evidence

In the absence of authentic samples of all of the proposed intermediates one cannot presume that the stopped-flow absorbance data are conclusive. Several independent lines of investigation have provided evidence for the existence of the proposed intermediates, however.

Evidence for the formation of the urate dianion is provided by steady-state kinetic studies.[1] Determination of the pH dependence of inhibition of the urate oxidase reaction by 9-methylurate and xanthine established that the inhibitors bound to the enzyme as monoanions. By analogy one can assume that urate also binds as the monoanion. The pH dependence of V_{max} and V/K revealed that the enzyme has a residue which must be unprotonated for reaction to occur. Putting these two pieces of information together suggests that the role of the catalytic residue is to abstract a proton from the urate monoanion bound at the active site to generate the urate dianion.

Urate hydroperoxide has been detected indirectly through two experiments. Rapid-mixing chemical quench experiments showed that hydrogen peroxide was present in samples generated by quenching reaction mixtures with ethanol; a burst of hydrogen peroxide lasting approximately 100 ms was observed. The hydrogen peroxide in the burst phase of the reaction is presumed to arise from the decomposition of urate hydroperoxide after it has been released from the enzyme active site upon mixing with ethanol. The observed burst of hydrogen peroxide evolution corresponds to the second kinetic phase evident in Fig. 2, which was assigned to urate hydroperoxide formation.

The second line of experimentation that provides evidence for urate hydroperoxide formation arose from studies of the effects of thiols on the urate oxidase reaction. A full explanation of these studies has been published.[13] When the urate oxidase reaction is conducted in the presence of dithiothreitol (DTT) or cysteine no hydrogen peroxide formation can be detected but the thiol reagent is oxidized. One pathway for thiol oxidation is by reduction of 5-hydroxyisourate to urate; this

---

[12] K. Kahn, *Bioorg. Chem.* **27**, 351 (1999).
[13] A. D. Sarma and P. A. Tipton, *J. Am. Chem. Soc.* **122**, 11252 (2000).

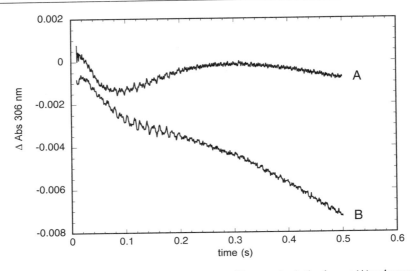

FIG. 6. Stopped-flow absorbance traces of the urate oxidase reaction in the absence (A) and presence of 20 mM DTT (B). The transients have been displaced by an arbitrary amount for visual clarity. Reprinted with permission from A. D. Sarma and P. A. Tipton, *J. Am. Chem. Soc.* **122**, 11252 (2000). Copyright © 2000 American Chemical Society.

does not explain why hydrogen peroxide formation is curtailed, however. It was postulated that the thiol also reacts with the urate hydroperoxide at the enzyme active site to reduce it directly to 5-hydroxyisourate. Stopped-flow absorbance measurements demonstrated reaction between DTT and an enzyme-bound species (Fig. 6). 5-Hydroxyisourate produced by reduction of urate hydroperoxide should contain an oxygen atom derived from molecular oxygen rather than from solvent water. Thus, the reaction was conducted in $H_2^{18}O$ and it was demonstrated by mass spectrometry that the allantoin that was formed by the decomposition of 5-hydroxyisourate did not contain ^{18}O.

Isotope labeling studies provide evidence for the intermediacy of dehydrourate in the urate oxidase reaction. In the absence of thiol reagents when the reaction is conducted in $H_2^{18}O$ the hydroxyl group at C-5 contains ^{18}O, suggesting that product formation occurs through hydration of dehydrourate.[14] It has been established that hydration occurs at the enzyme active site, not after release of dehydrourate from the enzyme, by showing that the 5-hydroxyisourate produced by urate oxidase is optically active.[15]

---

[14] K. Kahn, P. Serfozo, and P. A. Tipton, *J. Am. Chem. Soc.* **119**, 5435 (1997).
[15] A. D. Sarma, P. Serfozo, K. Kahn, and P. A. Tipton, *J. Biol. Chem.* **274**, 33863 (1999).

Mechanistic Rationale

Urate oxidase is quite unusual in its lack of a cofactor to participate in the catalytic reaction with O_2. However, consideration of the chemical mechanism reveals that nature has found a very elegant solution for urate oxidation. The barrier to direct reaction between O_2 and spin-paired organic species arises from the fact that O_2 is a ground-state triplet and its reaction with spin-paired species is a forbidden, i.e., low probability or high energy, process. One solution to this problem is for the reaction to proceed via single electron transfer steps, and it is believed that this is how the flavin cofactor reacts with oxygen. A single electron is transferred from reduced flavin to O_2 and then radical recombination rapidly occurs to form the flavin hydroperoxide. This reaction pathway is energetically accessible because flavin can form relatively stable one-electron oxidized species by delocalizing the unpaired electron throughout the isoalloxazine ring system. Similarly, urate has been shown to undergo one-electron oxidation in a facile manner under physiological conditions.[16] Thus, the basic prerequisite for reaction with O_2 is fulfilled in the urate oxidase reaction by the inherent chemical properties of the substrate itself.

The role of the enzyme must be to facilitate the reaction between urate and O_2, and what may appear to be the most pedestrian step in the mechanism, the proton transfer to generate the urate dianion, may in fact be the key step in the overall transformation. *Ab initio* calculations show that the ionization potential decreases as one and then two protons are removed from uric acid, so that in the gas phase one-electron oxidation of the urate dianion is a spontaneous process.[12] Therefore, urate oxidase enhances the reactivity of urate by proton abstraction; the resulting dianion is energetic enough to undergo direct reaction with O_2. The remaining steps in the reaction presumably flow smoothly through involvement of active site residues that act as general acid and general base catalysts.

Acknowledgments

Work in the author's laboratory on urate oxidase has been supported by grants from the USDA (94-37305-0578, 96-35305-3542, and 98-35305-6548).

[16] K. R. Maples and R. P. Mason, *J. Biol. Chem.* **263**, 1709 (1988).

[24] Nitric Oxide Synthase: Use of Stopped-Flow Spectroscopy and Rapid-Quench Methods in Single-Turnover Conditions to Examine Formation and Reactions of Heme–O_2 Intermediate in Early Catalysis

By Chin-Chuan Wei, Zhi-Qiang Wang, *and* Dennis J. Stuehr

Introduction

Nitric oxide synthases (NOS, EC 1.14.13.39) catalyze a two-step oxidation of L-arginine (Arg) to nitric oxide (NO) and citrulline (Fig. 1). Arg first undergoes N-hydroxylation to generate N^{ω}-hydroxy-L-arginine (NOHA) as a bound intermediate. NOHA is further oxidized to citrulline and NO. Both steps represent a mixed-function oxidation that consumes NADPH and O_2 and generates water as a coproduct.[1–4]

In active form, all NOS are homodimers whose subunits are comprised of an N-terminal oxygenase domain and C-terminal reductase domain. Two oxygenase domains form the dimer interface, and each contains a catalytic site consisting of cysteine-ligated iron protoporphyrin IX (heme), $6R$-tetrahydrobiopterin (H_4B), and the Arg binding site. The attached reductase domains contain flavin mononucleotide (FMN), flavin adenine dinucleotide (FAD), and an NADPH binding site. Reductase domains transfer NADPH-derived electrons to the oxygenase domain for O_2 activation during NO synthesis. A calmodulin binding sequence is located in each subunit and its occupancy by calmodulin triggers electron transfer between the reductase and oxygenase domains.[1–4]

Several aspects of NOS make rapid-mixing single-turnover studies particularly well suited to investigating the reaction mechanism: the heme binds and activates O_2 twice in two experimentally separable catalytic cycles to generate NO from Arg. The intermediate product (NOHA) is commercially available and stable. The NOS oxygenase domain dimer (NOSoxy) can be overexpressed in an active form independent of the reductase domain. Several catalytic steps occur within the millisecond to second time range, enabling study with conventional rapid mixing equipment. Some of the electron transfer and binding events are accompanied by

---

[1] D. J. Stuehr and S. Ghosh, *in* "Handbook of Experimental Pharmacology" (B. Mayer, ed.), Vol. 143, p. 33. Springer-Verlag, Berlin, 1999.
[2] M. A. Marletta, A. R. Hurshman, and K. M. Rusche, *Curr. Opin. Chem. Biol.* **2**, 656 (1998).
[3] B. Hemmens and B. Mayer, *Methods Mol. Biol.* **100**, 1 (1998).
[4] B. R. Crane, A. S. Arvai, D. K. Ghosh, C. Wu, E. D. Getzoff, D. J. Stuehr, and J. A. Trainer, *Science* **279**, 2121 (1998).

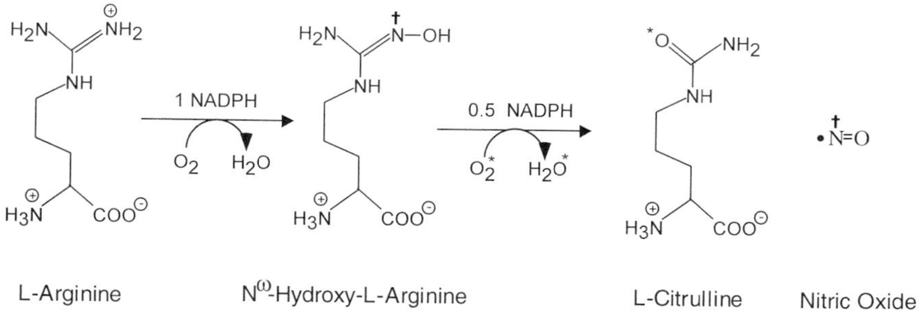

FIG. 1. Stepwise NO synthesis by NOS. Symbols (†,*) trace the sources of nitrogen and oxygen in the product.

spectral transitions with good extinction. Finally, steady-state measurements of NOS can be affected by buildup of enzyme heme–NO complexes, which complicate or preclude investigation of the reaction mechanism.

The following chapter describes how one may study NOS catalysis in a single-turnover setting using stopped-flow and stop-quench techniques. More general reviews of these methods[5,6] and of NOS biochemistry[1–3] are available.

Enzyme Materials

Experiments can be performed either with NOSoxy or full-length NOS enzymes, although NOSoxy proteins are more typically expressed in sufficient quantities for single-turnover studies. NOSoxy proteins with a six-histidine (His$_6$) affinity tag at their C terminus, or full-length NOS containing an N-terminal six-histidine tag, can be expressed and prepared in the presence or absence of substrate and H$_4$B as previously described.[7–10] Concentrated enzymes are stored at $-70°$. Enzyme concentration can be determined from the absorbance of the ferrous–CO complex generated in the presence of Arg and H$_4$B by using an A_{444}–A_{500} extinction coefficient of 76 mM^{-1} cm^{-1}. To minimize autoxidation, H$_4$B solutions should be prepared and stored in the presence of excess dithiothreitol (DTT).

---

[5] P. S. Brzovic and M. F. Dunn, *Methods Enzymol.* **246**, 168 (1995).
[6] K. A. Johnson, *Methods Enzymol.* **249**, 38 (1995).
[7] I. Rodriguez-Crespo and P. R. Ortiz de Montellano, *Arch. Biochem. Biophys.* **336**, 151 (1996).
[8] R. Gachhui, D. K. Ghosh, C. Wu, J. Parkinson, B. R. Crane, and D. J. Stuehr, *Biochemistry* **36**, 5097 (1997).
[9] K. McMillan and B. S. S. Masters, *Biochemistry* **34**, 3586 (1995).
[10] A. Leber, B. Hemmens, B. Klosch, W. Goessler, G. Raber, B. Mayer, and K. Schmidt, *J. Biol. Chem.* **274**, 37658 (1999).

Preparation of Reduced (Ferrous) NOS Proteins

Ferrous NOSoxy or full-length NOS are prepared by titrating the ferric proteins with a dithionite solution. A suitable dithionite solution is prepared by placing 4 mg dithionite in a 10-ml double-necked flask that has one neck sealed with a septum and the other attached through a ground glass joint to a vacuum gas train. A gastight syringe is used to transfer 4 ml of buffer (i.e., 50 mM HEPES, pH 7.5) that has been purged with catalyst-deoxygenated N_2 into the flask. This solution is mixed and flushed with catalyst-deoxygenated N_2. The dithionite concentration can be determined spectrally by titrating an anaerobic ferricyanide solution of known concentration, using an extinction coefficient of 1.2 mM^{-1} cm^{-1} at 421 nm.

Concentrated ferric NOSoxy protein is placed in a septum-sealed quartz cuvette that is connected to a vacuum gas train and is made anaerobic by several cycles of evacuation and purging with deoxygenated N_2. Initial evacuation times should be short to minimize bubble formation, and subsequent evacuation times can be increased. A buffer (i.e., 50 mM HEPES, pH 7.5) which contains H_4B, Arg, and DTT and has been previously degassed and purged with N_2 is transferred into the anaerobic cuvette with a gas-tight syringe. Typical final concentrations for NOSoxy, H_4B, Arg, and DTT are 20 μM, 0.2 mM, 0.2 mM, and 0.6 mM, respectively. When using full-length NOS, EDTA or EGTA should be added to the enzyme solution to prevent calmodulin binding to NOS if one is interested in observing just a single turnover by the ferrous heme.[11,12] The cuvette solution is further evacuated and flushed with N_2 three times. The absorption of ferric NOSoxy is then recorded in a UV–Vis spectrophotometer. The enzyme is reduced by sequential addition of 2–5 μl of dithionite solution prepared as described above. Heme reduction is monitored by UV–Vis scanning (see example below). The reduced enzyme solution is transferred into the driver syringe of various rapid mixing instruments using a gas-tight syringe.

Rapid-Scanning Stopped-Flow Spectrometer

A variety of stopped-flow spectrophotometers are commercially available. We have used a model SF-51 stopped-flow apparatus from Hi-Tech Ltd. (Salisbury, UK) equipped with a temperature control unit and a Hi-Tech MG-6000 rapid-scan diode array detector that is designed to collect a complete spectrum (280 nm to 700 nm) within 1 ms. Ninety-six consecutive spectra are obtained within predetermined time frames following each mixing. The detector is calibrated to five principal peak absorption wavelengths of a HY1 holmium oxide calibration filter: 362, 420, 446, 460, and 536 nm. An anaerobic chamber with septum ports covers the sample loading reservoirs, and a continuous stream of nitrogen is flowed through the chamber during experiments.

---

[11] H. M. Abu-Soud, A. Presta, B. Mayer, and D. J. Stuehr, *Biochemistry* **36**, 10811 (1997).
[12] N. Bec, A. C. Gorren, C. Voelker, B. Mayer, and R. Lange, *J. Biol. Chem.* **273**, 13502 (1998).

Two drive syringes must be flushed with desired buffer prior to loading reaction solutions. The drive syringe and internal tubing that will contain anaerobic reduced enzyme solution is flushed with a weak dithionite solution to remove traces of oxygen followed by flushing with nitrogen-purged buffer several times. Around 1 ml of ferrous enzyme solution is transferred from the anaerobic cuvette to the drive syringe reservoir using a gas-tight syringe and then taken into the drive syringe immediately. An air (or O_2)-saturated buffer is loaded quickly into the other drive syringe. The drive syringes are bathed continuously in circulating water at the desired temperature and are incubated for 10 min after receiving the reaction solutions to achieve thermal equilibrium prior to starting the mixing experiments.

Rapid-Quench Analysis

Several chemical quench instruments are commercially available. We have used a Hi-Tech RFQ-63 rapid-quench instrument that is equipped with a constant temperature bath. The instrument can be calibrated using alkaline hydrolysis of 2,3-dinitrophenyl acetate as described in the manufacturer's manual. The dead time of mixing plus quenching was determined to be less than 5 ms by this method. For studying experimental reaction times shorter than 200 ms, a continuous mixing mode is used, whereas for longer reaction times an interrupt mode (push–delay–push) is used. Sample dilution that occurs in each mixing mode must be determined in advance, and this can be done spectroscopically using bromphenol blue. Final dilution of the enzyme solution typically ranges between three- and fourfold in the Hi-Tech instrument.

Before loading anaerobic enzyme sample into the driver syringe, a procedure to remove traces of O_2 is performed using a weak dithionite solution as described above for the rapid-scanning stopped-flow instrument. The requirement for an O_2-free condition is very important in the drive syringe that will contain reduced enzyme because traces of O_2 will enable it to oxidize substrate prior to initiating the reaction. Anaerobic ferrous NOSoxy containing substrate, H_4B, and DTT is loaded into one syringe and O_2-containing solution is loaded into the other as with the stopped-flow instrument. Solutions are incubated for 10 min in the driver syringes prior to starting the experiment to allow for thermal equilibration.

In a typical mixing experiment, a solution containing anaerobic reduced enzyme and a substoichiometric amount of [^{14}C]Arg is rapidly mixed at 10° with an air or O_2-saturated solution containing 20 mM unlabeled Arg to ensure the enzyme is limited to a single turnover.[13] The quench syringe contains 0.5 or 1 N HCl in 50% (v/v) 2-propanol plus 100 μM Arg and 100 μM NOHA (these last two additives are present to minimize ^{14}C-labeled product loss during sample workup and analysis). Quenched reaction samples are collected from the instrument and can be stored at −70°.

---

[13] T. Iwanaga, T. Yamazaki, and S. Kominami, *Biochemistry* **38**, 16629 (1999).

Workup and Separation of ^{14}C-Labeled Reaction Products by High-Performance Liquid Chromatography

Rapid-quench samples are vortexed for 30 min and then centrifuged at 10,000g for 10 min at room temperature. Vigorous vortexing is essential to completely denature the protein and release all bound amino acids into solution.[13] An aliquot (100 μl) of each sample is injected onto a Nucleosil 250/3 100-5 SA cation-exchange column (Macherrey Nagel, Duran, Germany) that has been equilibrated with 50 mM sodium acetate, pH 6.5, at a flow rate of 0.5 ml/min. High-performance liquid chromatography (HPLC) column fractions (250 μl) are collected directly into scintillation vials to measure ^{14}C radioactivity. Elution times for Arg, NOHA, and citrulline can be determined using nonlabeled compounds run under the same HPLC conditions, with detection by thin-layer chromatography (TLC) of HPLC fraction aliquots on silica plates. Using a 4 : 1 : 1 (v/v) ethanol : acetic acid : water solvent system, the R_f values for Arg, NOHA, and citrulline on silica gel are 0.38, 0.42, and 0.52.[14]

Data Analysis

Diode array rapid-scanning spectral data obtained in stopped-flow experiments can be analyzed using commercially available global analysis programs such as SpecFit (Spectrum Software Associates, Chapel Hill, NC) and/or can be analyzed as absorbance change at single wavelengths typically with software provided by the instrument manufacturer. In NOS experiments the analysis was limited to absorbance change between 350 and 700 nm because DTT and dithionite contribute absorbance below this range and make calculation unreliable. For SpecFit global analysis the data are fit to different reaction models (i.e., A to B, A to B to C, etc.) to obtain the best fit regarding the calculated number of species, their individual spectra, their concentrations versus time, and the rate constants for each transition. Analysis at specific wavelengths can be done to illustrate, compare, or confirm results obtained from global analysis.

The kinetics of product formation in rapid quench experiments can be determined by dividing the counts associated with each individual product peak by the total counts in each HPLC sample trace for each time point, and plotting these values as a function of time. Rates can be determined by fitting the curves to single or multiple exponential equations using commercially available software such as DeltaGraph (Delta Point, Inc., Monterey, CA).

Examples

The following selected stopped-flow and rapid quench experiments were done under single-turnover conditions using mouse inducible NOSoxy (iNOSoxy) and

[14] D. J. Stuehr, N. S. Kwon, C. F. Nathan, O. W. Griffith, P. L. Feldman, and J. Wiseman, *J. Biol. Chem.* **266**, 6259 (1991).

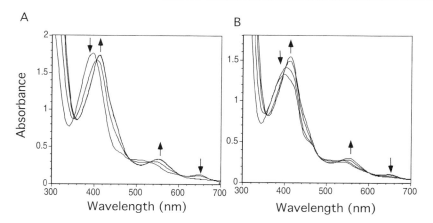

FIG. 2. Anaerobic dithionite titration of ferric iNOSoxy. Representative spectra of a 20 μM ferric iNOSoxy containing 0.2 mM H$_4$B and 0.6 mM DDT (A) in the presence of 0.2 mM Arg or (B) in the absence of Arg before and after consecutive addition of dithionite solution. The arrows indicate the direction of spectral change during reduction.

rat neuronal NOSoxy (nNOSoxy). These can serve to guide investigations with different NOS or various NOS mutants.

Reductive Titration of iNOSoxy

Figure 2 contains typical spectral data obtained during titration of anaerobic ferric iNOSoxy (approximately 20 μM) with dithionite solution. The starting protein contains H$_4$B plus Arg (Fig. 2A) or H$_4$B alone (Fig. 2B). In Fig. 2A the ferric heme is in a predominantly high spin state with Soret peak absorbance at 390 nm. As dithionite solution is added the Soret peak is red shifted and absorption at 556 nm is increased, while absorbance at 650 nm is decreased. Note that the position and shape of the Soret peak in the ferric enzyme differs depending on the presence of substrate and H$_4$B: the enzyme containing H$_4$B alone has a broader Soret absorbance with maximum near 400 nm (Fig. 2B), whereas one that contains neither H$_4$B nor Arg has a Soret absorbance near 417 nm because its heme is in a predominantly low spin state (not shown). However, when complete heme reduction is achieved the Soret peak is at 412 nm in all cases and absorbance peaks at 556 and 650 nm reach their maximum and minimum, respectively. Monitoring absorption changes at 556 nm or 650 nm becomes more practical when higher protein concentrations are used. Adding excess dithionite (up to 3 times the heme concentration) does not alter the kinetic results or product yields in single-turnover experiments with NOSoxy proteins.[12,15] However, it can complicate the spectral analysis because of increased absorbance between 350 and 400 nm.

[15] S. Boggs, L. Huang, and D. J. Stuehr, *Biochemistry* **39,** 2332 (2000).

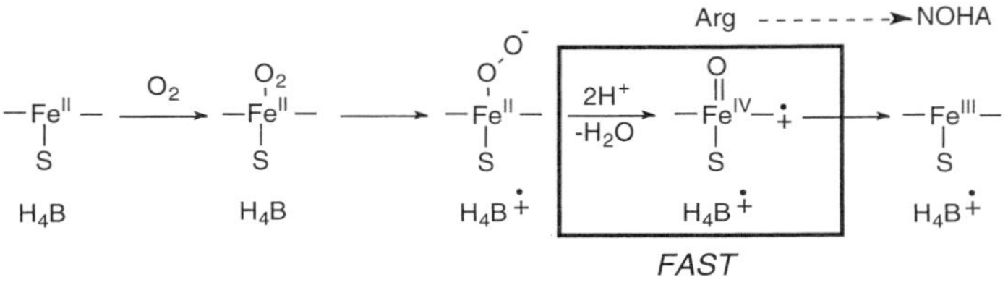

$$-\text{Fe}^{\text{II}}- \xrightarrow{O_2} -\overset{O_2}{\underset{S}{\text{Fe}^{\text{II}}}}- \longrightarrow -\overset{\overset{O}{\overset{|}{O^-}}}{\underset{S}{\text{Fe}^{\text{II}}}}- \boxed{\xrightarrow{\underset{-H_2O}{2H^+}} -\overset{\overset{O}{\parallel}}{\underset{S}{\text{Fe}^{\text{IV}}}}-\overset{\cdot}{+} \xrightarrow{\text{Arg} \dashrightarrow \text{NOHA}}} -\underset{S}{\text{Fe}^{\text{III}}}-$$

H₄B   H₄B   H₄B $\overset{\cdot}{+}$ H₄B $\overset{\cdot}{+}$ H₄B $\overset{\cdot}{+}$

FAST

FIG. 3. Model for Arg hydroxylation by NOSoxy in the single-turnover reaction. Oxygen binds to the ferrous protein to form the $\text{Fe}^{\text{II}}\text{O}_2$ species, which receives an electron from H_4B to form a putative heme–peroxo intermediate. Subsequent steps generate a putative heme–oxo intermediate that may hydroxylate Arg. This leaves ferric NOSoxy containing NOHA and the H_4B radical at the end of the single turnover.

Reaction of Ferrous iNOSoxy with Oxygen and Arg

The first reaction cycle in NO synthesis involves O_2 binding to ferrous NOS, electron transfer from bound H_4B to the heme $\text{Fe}^{\text{II}}\text{O}_2$ intermediate to form the ultimate heme-based oxidant, and Arg hydroxylation, to generate ferric enzyme, NOHA, water, and bound H_4B radical as products (Fig. 3). The kinetics and/or magnitude of some individual steps have been examined in single-turnover reactions using various NOSoxy or full-length NOS enzymes.[11–13,15–23] However, to accurately determine kinetic or quantitative relationships among these steps it is best to perform measurements with the same enzyme preparation under identical reaction conditions whenever possible.

Stopped-flow spectrophotometry can follow the kinetics and magnitude of NOS heme transitions during the single-turnover reaction. As an example, a reaction was initiated by mixing ferrous iNOSoxy containing Arg and H_4B with an air-saturated buffer in the stopped flow instrument at 10° (see also Ref. 20). In this circumstance stopped-flow analysis has the potential to identify any of three intermediates in addition to the beginning ferrous and ending ferric enzymes. These are the $\text{Fe}^{\text{II}}\text{O}_2$, heme–peroxo, and heme–oxo intermediates (see Fig. 3). Ninety-six spectra were recorded within 0.28 sec, which encompasses the time required for a full reaction at 10°. Figure 4 shows 12 of these scans to highlight the spectral evolution versus time. Changes in position and magnitude of the Soret and visible bands

---

[16] H. M. Abu-Soud, K. Ichimori, A. Presta, and D. J. Stuehr, *J. Biol. Chem.* **275**, 17349 (2000).

[17] H. Sato, I. Sagami, S. Daff, and T. Shimizu, *Biochem. Biophys. Res. Commun.* **253**, 845 (1998).

[18] M. Couture, D. J. Stuehr, and D. L. Rousseau, *J. Biol. Chem.* **275**, 3201 (2000).

[19] A. R. Hurshman, C. Krebs, D. E. Edmondson, B. H. Huynh, and M. A. Marletta, *Biochemistry* **38**, 15689 (1999).

[20] C.-C. Wei, Z.-Q. Wang, Q. Wang, A. L. Meade, C. Hemann, R. Hille, and D. J. Stuehr, *J. Biol. Chem.* **276**, 315 (2001).

[21] T. Iwanaga, T. Yamazaki, and S. Kominami, *Biochemistry* **39**, 15150 (2000).

[22] P. P. Schmidt, R. Lange, A. C. Gorren, E. R. Werner, B. Mayer, and K. K. Andersson, *J. Biol. Inorg. Chem.* **6**, 151 (2001).

[23] H. M. Abu-Soud, R. Gachhui, F. M. Raushel, and D. J. Stuehr, *J. Biol. Chem.* **272**, 17349 (1997).

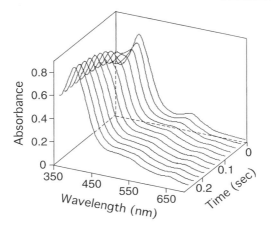

FIG. 4. Rapid-scanning stopped-flow spectra recorded during reaction of ferrous iNOSoxy with O_2 and Arg. Ferrous iNOSoxy saturated with Arg and H_4B was rapid-mixed with air-saturated buffer at 10°. Ninety-six consecutive spectra were recorded within 0.28 sec. Twelve of these spectra are displayed.

are apparent. SpecFit analysis of the data using different reaction models showed that only one fit well, namely the A → B → C transition. Thus, the analysis detects only a single intermediate (species B) during the reaction. The spectrum for each species derived from SpecFit analysis is shown in Fig. 5. Species A has absorption maxima at 412 nm and 557 nm and is consistent with the initial ferrous iNOSoxy

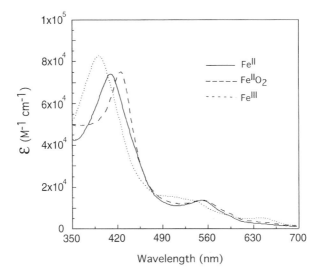

FIG. 5. Spectra of ferrous, heme-dioxy, and ferric iNOSoxy species present during the Arg reaction as calculated from rapid-scanning stopped-flow data.

TABLE I
LIGHT ABSORBANCE SPECTRAL FEATURES OF NOS $Fe^{II}O_2$ INTERMEDIATE UNDER SELECTED CONDITIONS

Condition	Peak positions (nm)		
	iNOS	nNOS	eNOS
$+H_4B$, $+Arg$	427, 551, 595 (sh)[a]	427, 557, 585 (sh)[b]	427, 560, 590 (sh)[c]
		427, 560[d]	
		427[e]	
		430[f]	
$+H_4B$, $+NOHA$		429, 557, 585 (sh)[b]	
$+H_4B$, $+NMA$[g]		419, 553, 585 (sh)[h]	
$+H_4B$, $-Arg$		427, 557, 585 (sh)[b]	
		416[i]	
$-H_4B$, $-Arg$		420[f]	
		415[i]	

[a] Mouse iNOSoxy at 10° [C.-C. Wei, Z.-Q. Wang, Q. Wang, A. L. Meade, C. Hemann, R. Hille, and D. J. Stuehr, *J. Biol. Chem.* **276**, 315 (2001)].

[b] Rat nNOSoxy at 10° [S. Boggs, L. Huang, and D. J. Stuehr, *Biochemistry* **39**, 2332 (2000)].

[c] Bovine full-length eNOS at 10° [H. M. Abu-Soud, K. Ichimori, A. Presta, and D. J. Stuehr, *J. Biol. Chem.* **275**, 17349 (2000)].

[d] Rat nNOSoxy at 10° [H. Abu-Soud, R. Gachhui, F. M. Roushel, and D. J. Stuehr, *J. Biol. Chem.* **272**, 17349 (1997)].

[e] Rat nNOSoxy at 25° [M. Couture, D. J. Stuehr, and D. L. Rousseau, *J. Biol. Chem.* **275**, 3201 (2000)].

[f] Rat nNOS full-length at 10° [H. Sato, I. Sagami, S. Daff, and T. Shimizu, *Biochem. Biophys. Res. Commun.* **253**, 845 (1998)].

[g] NMA, N^{ω}-methyl-L-arginine.

[h] Rat nNOSoxy at $-30°$ [A. P. Ledbetter, K. McMillan, L. J. Roman, B. S. Masters, J. H. Dawson, and M. Sono, *Biochemistry* **38**, 8014 (1999)].

[i] Rat nNOSoxy at $-30°$ [N. Bec, A. C. Gorren, C. Voelker, B. Mayer, and R. Large, *J. Biol. Chem.* **273**, 13502 (1998); A. C. Gorren, N. Bec, A. Schrammel, E. R. Werber, R. Large, and B. Mayer, *Biochemistry* **39**, 11763 (2000)].

species (compare with spectrum obtained by dithionite reduction in Fig. 2). Species B has a Soret peak at 427 nm and visible spectrum features that identify it as the $Fe^{II}O_2$ intermediate of iNOSoxy. A spectral intermediate with identical features has been characterized in nNOSoxy[15,18,23] and in full-length eNOS[16] or nNOS[17] (Table I). Interestingly, a nNOSoxy $Fe^{II}O_2$ species with somewhat different spectral features has been reported in conventional mixing experiments that were conducted in ethylene glycol at subfreezing temperature[12,24,25] (Table I). Species C has a Soret

[24] A. P. Ledbetter, K. McMillan, L. J. Roman, B. S. Masters, J. H. Dawson, and M. Sono, *Biochemistry* **38**, 8014 (1999).

[25] A. C. Gorren, N. Bec, A. Schrammel, E. R. Werber, R. Lange, and B. Mayer, *Biochemistry* **39**, 11763 (2000).

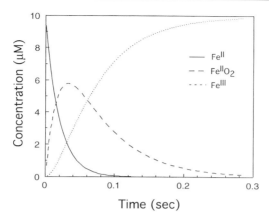

FIG. 6. Concentration of iNOSoxy species present during Arg hydroxylation in the single-turnover reaction. Time courses were calculated from spectral data using the SpecFit program.

peak at 390 nm and visible absorbance at 650 nm, consistent with the ending ferric iNOSoxy species. Thus, the stopped-flow analysis detected an $Fe^{II}O_2$ intermediate during the Arg single-turnover reaction of ferrous iNOSoxy, but did not detect putative heme–peroxo or heme–oxo intermediates that are expected to form downstream from $Fe^{II}O_2$ in the reaction (Fig. 3). This leaves three possibilities: (a) these latter two intermediates did not form in the iNOSoxy reaction. (b) They formed but did not accumulate during the reaction. (c) They formed and accumulated but have spectral features not easily distinguished from the $Fe^{II}O_2$ intermediate. The fact that the H_4B radical also forms during the reaction,[19,20,26] and that its formation is kinetically coupled to $Fe^{II}O_2$ disappearance and Arg hydroxylation,[20] argues against the first possibility. Regarding the third possibility, the spectra of heme–oxo and heme–peroxo intermediates in related heme–thiolate enzymes are distinct from the spectrum of the iNOSoxy $Fe^{II}O_2$ intermediate.[12,27–29] This suggests they could be detectable in iNOSoxy, but may not have accumulated sufficiently during the single-turnover reaction at 10°. Interestingly, downstream intermediates such as the heme–oxo species may accumulate when NOS single-turnover reactions are run at subfreezing temperatures.[12,15] In any case, in the present example we are constrained to follow the kinetics and reactions of the $Fe^{II}O_2$ intermediate.

Figure 6 shows how the concentrations of the three species change with time during the single turnover as determined by SpecFit analysis. Ferrous iNOSoxy

[26] N. Bec, A. C. Gorren, B. Mayer, P. P. Schmidt, K. K. Andersson, and R. Lange, *J. Inorg. Biochem.* **81**, 207 (2000).
[27] D. E. Benson, K. S. Suslick, and S. G. Sligar, *Biochemistry* **36**, 5104 (1997).
[28] C. Larroque, R. Lange, L. Maurin, A. Bienvenue, and J. E. van Lier, *Arch. Biochem. Biophys.* **282**, 198 (1990).
[29] M. M. Palcic, R. Rutter, T. Araiso, L. P. Hager, and H. B. Dunford, *Biochem. Biophys. Res. Commun.* **94**, 1123 (1980).

reacts with O_2 with concurrent buildup of the $Fe^{II}O_2$ intermediate. The $Fe^{II}O_2$ species reaches a maximum concentration of 6 μM at 30 ms, which represents 60% of the total iNOSoxy present in the reaction. Its conversion to ferric iNOSoxy is complete within 300 ms. Both $Fe^{II}O_2$ formation and disappearance are monophasic transitions by the analysis. The calculated formation rate of the $Fe^{II}O_2$ intermediate was 52.7 ± 2.2 s^{-1} and the rate of its transition to ferric iNOSoxy was 12.5 ± 0.2 s^{-1} (Table II). It is important to note that O_2 binding to ferrous heme is a bimolecular process; therefore the observed rate of $Fe^{II}O_2$ formation will depend on the concentrations of O_2 and ferrous iNOSoxy present at the time of mixing.

TABLE II
KINETIC VALUES FOR STEPS OCCURRING IN NOS SINGLE-TURNOVER REACTIONS

Step	iNOS	nNOS	eNOS
O_2 binding to heme[a] kon ($M^{-1} s^{-1}$)			
+H_4B, −Arg		7.8×10^5	3.1×10^5
+H_4B, +Arg		9.0×10^5	3.4×10^5
−H_4B, +Arg		1.1×10^6	
−H_4B, −Arg		1.4×10^6	
$Fe^{II}O_2$ disappearance (s^{-1})			
+H_4B, −Arg	9.5^b	10^c	
+H_4B, +Arg	12.5^d	74(36%) 6.1(29%) 0.17(35%)[e] 10^c, 20^g	3.0^f
−H_4B, +Arg	0.4	0.1^c	2.8^d
−H_4B, −Arg	0.3^e	0.1 2.3^c 60(28%) 5.5(111%) 0.048(61%)[e]	
Arg hydroxylation (s^{-1})			
+H_4B, +Arg	9.2^d		
H_4B radical formation (s^{-1})			
+Arg	11^d, 15^h		

[a] Data obtained with rat nNOSoxy at 10° [H. M. Abu-Soud, R. Gachhui, F. M. Raushel, and D. J. Stuehr, *J. Biol. Chem.* **272**, 17349 (1997)] or bovine full-length eNOS at 10° [H. M. Abu-Soud, K. Ichimori, A. Presta, and D. J. Stuehr, *J. Biol. Chem.* **275**, 17349 (2000)].

[b] C. C. Wei and D. J. Stuehr, unpublished data, 2000.

[c] Rat nNOSoxy at 10° [H. M. Abu-Soud, R. Gachhui, F. M. Raushel, and D. J. Stuehr, *J. Biol. Chem.* **272**, 17349 (1997)].

[d] Mouse iNOS at 10° [C.-C. Wei, Z.-Q. Wang, Q. Wang, A. L. Meade, C. Hemann, R. Hille, and D. J. Stuehr, *J. Biol. Chem.* **276**, 315 (2001)].

[e] Rat full-length nNOS at 10° [H. Sato, I. Sagami, S. Daff, and T. Shimizu, *Biochem. Biophys. Res. Commun.* **253**, 845 (1998)].

[f] Bovine full-length eNOS at 10° [H. M. Abu-Soud, K. Ichimori, A. Presta, and D. J. Stuehr, *J. Biol. Chem.* **275**, 17349 (2000)].

[g] Rat nNOSoxy at 10° [S. Boggs, L. Huang, and D. J. Stuehr, *Biochemistry* **39**, 2332 (2000)].

[h] Mouse iNOSoxy at 4° [A. R. Hurshman, C. Krebs, D. E. Edmondson, B. H. Hugnh, and M. A. Marletta, *Biochemistry* **38**, 15689 (1999)].

Second-order rate constants for O_2 binding to ferrous nNOSoxy and ferrous eNOS have been determined under several different conditions (Table II).[16,23] In the same reports, conversion of the $Fe^{II}O_2$ intermediate to ferric enzyme was shown to be independent of O_2 concentration, which is consistent with the reaction mechanism in Fig. 3.

Independent analysis of collected spectral traces can confirm results of the SpecFit analysis. For example, Fig. 7A overlays several spectral traces collected during conversion of the $Fe^{II}O_2$ intermediate to ferric iNOSoxy during the Arg single-turnover reaction. A relatively smooth Soret peak transition from 427 nm to 395 nm is evident with an isosbestic point at 410 nm. Three other isosbestic points occur at 480, 541, and 603 nm. Plots describing absorbance change at two single wavelengths (395 or 427 nm) versus time are shown in Fig. 7B. The spectral change that occurs within the first 20 ms is due to O_2 binding to ferrous heme and is not considered in the analysis. Absorbance change beyond 20 ms at these two wavelengths fit best to a single exponential equation, indicating both spectral transitions are monophasic. Residual analysis provides a goodness of fit for this assignment (Fig. 7C). The rates of spectral change at 395 and 427 nm (10.8 and 10.5 s^{-1}, respectively) are similar to one another and to the value derived from SpecFit analysis (12.5 s^{-1}, Table II). This confirms that no spectrally distinct species accumulate during conversion of the $Fe^{II}O_2$ intermediate to ferric enzyme in the Arg single-turnover reaction. A similar stopped-flow kinetic study has been done for the nNOSoxy $Fe^{II}O_2$ intermediate[15] (Table II).

Influence of Substrate, H_4B, and Their Analogs

In theory, the rate of O_2 binding, spectral characteristics of the $Fe^{II}O_2$ intermediate, and its subsequent rate of disappearance could all be affected by the chemical nature of the bound substrate and pterin cofactor or by amino acid substitutions in the protein itself. Effects are not easily predicted and should be determined for each structurally distinct substrate, pterin analog, or NOS protein. The rate of O_2 binding to NOS ferrous heme is not greatly affected by bound Arg or H_4B.[16,23] However, the spectral characteristics of the $Fe^{II}O_2$ intermediate (i.e., peak positions) may be altered in some cases when these molecules are absent or when structural analogs are used (Table I).[23,25] The rate of $Fe^{II}O_2$ disappearance is not greatly changed by bound Arg, but is significantly increased by bound H_4B(Table II).[16,17,20,23] Faster rates of $Fe^{II}O_2$ disappearance have been linked to an ability of H_4B to reduce the $Fe^{II}O_2$ intermediate.[20] Presumably, this generates heme–oxy species that are more reactive (or unstable) than the $Fe^{II}O_2$ intermediate itself (see Fig. 3).

Kinetics of Arg Hydroxylation

Arg hydroxylation by iNOSoxy can occur as soon as the $Fe^{II}O_2$ intermediate receives an electron and forms the ultimate heme-based oxidant (Fig. 3). Light

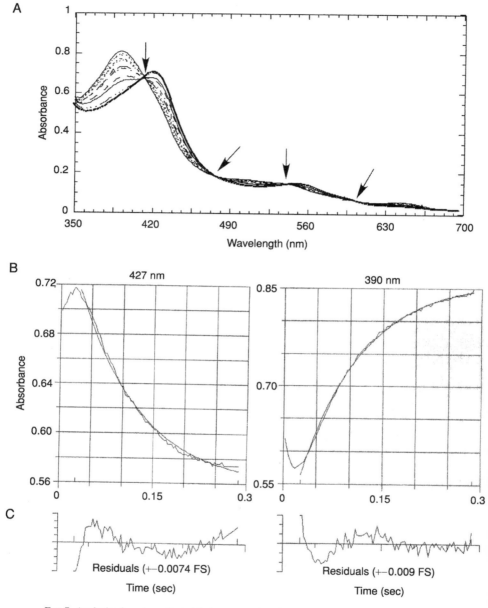

FIG. 7. Analysis of spectra collected during conversion of the heme–dioxy species to ferric iNOSoxy in the Arg single-turnover reaction. (A) Overlay of 12 spectral traces collected from 30 ms to 1.8 sec after mixing. The initial spectrum is designated by a heavy black line. The arrows indicate points in isosbestic the spectra. (B) Absorbance change versus time at 427 and 395 nm. Smooth line in each trace is the line of best fit according to a single exponential equation. (C) Residual analysis of fitted curve versus actual absorbance trace at each wavelength.

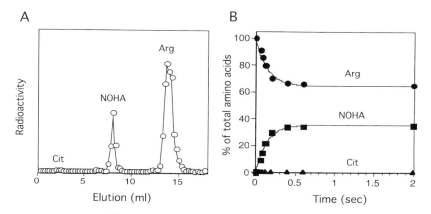

FIG. 8. Conversion of [^{14}C]Arg to products by iNOSoxy in the single-turnover reaction. Anaerobic ferrous iNOSoxy containing [^{14}C]Arg and H$_4$B was rapid-mixed with air-saturated buffer at 10° to start the reaction, and the reaction was quenched at various times. (A) The amino acid composition of the reaction after 125 ms as determined by HPLC and liquid scintillation counting analysis. (B) The relative amounts of [^{14}C]Arg, [^{14}C]NOHA, and [^{14}C]citrulline versus time during the reaction.

absorbance characteristics of Arg and NOHA are too similar to allow their direct monitoring by stopped-flow spectroscopy during the single-turnover reaction. Instead, a rapid-quench apparatus can be used in combination with HPLC to determine the kinetics of Arg hydroxylation. As in the example experiment described above, anaerobic ferrous iNOSoxy containing H$_4$B and Arg is rapid mixed with an air-saturated buffer to start the reaction, except that here the enzyme contains [^{14}C]Arg instead of unlabeled Arg. The reaction is quenched at various times, followed by sample workup, HPLC separation, and scintillation counting of column fractions (see earlier section for details). Typical results are shown in Fig. 8. Each reaction contained 0.05 μCi [^{14}C]Arg (2 μM, approximately 111,000 dpm). About 40% of total [^{14}C]Arg was converted to [^{14}C]NOHA under the reaction conditions (iNOSoxy was present at 3 μM). Further conversion of [^{14}C]NOHA product to citrulline was not observed (Fig. 8). This is consistent with a single catalytic turnover, and with the fact that the quenching step, sample workup, and HPLC analysis are done under acidic conditions which inhibits hydrolysis of NOHA to citrulline.[14] Rates of NOHA formation or Arg disappearance can both be determined and are equivalent in the example. However, such equivalence in rate would not necessarily hold for substrate analogs that can stabilize an intermediate enzyme species or form intermediates that react covalently with the NOS protein.[30–33]

[30] N. C. Gerber and P. R. Ortiz de Montellano, *J. Biol. Chem.* **270**, 17791 (1995).
[31] M. A. Marletta, *J. Med. Chem.* **37**, 1899 (1994).
[32] R. Bryk and D. J. Wolff, *Biochemistry* **37**, 4844 (1998).
[33] S. Jianmongkol, J. L. Vuletich, A. T. Bender, D. R. Demady, and Y. Osawa, *J. Biol. Chem.* **275**, 13370 (2000).

The rate of NOHA formation by iNOSoxy in this example was 9 s^{-1}, which is close to the rate of Fe^{II}O$_2$ disappearance (10.5 to 12.5 s^{-1}, Table II). This indicates the Fe^{II}O$_2$ spectral intermediate fulfills the criteria of kinetic competence. That is, it forms quickly and then disappears at a rate that is as fast or slightly faster than product formation in the single-turnover reaction. Disappearance of the Fe^{II}O$_2$ intermediate can therefore be said to be kinetically coupled to Arg hydroxylation in iNOSoxy. The observation that Arg hydroxylation and Fe^{II}O$_2$ conversion to ferric enzyme occur at similar speeds is also consistent with the spectral analysis suggesting that putative heme–peroxo and heme–oxo intermediates do not accumulate during the single-turnover reaction. Instead, Arg would have to react with these species as quickly or more quickly than they are formed through Fe^{II}O$_2$ reduction (see Fig. 3). Accumulation of the Fe^{II}O$_2$ species as a sole intermediate, together with the relative kinetics of Fe^{II}O$_2$ disappearance and Arg hydroxylation, allows one to propose that Fe^{II}O$_2$ reduction is the rate-limiting step in the single-turnover reaction.

H$_4$B Radical Formation during Arg Hydroxylation

The rate of H$_4$B radical formation during Arg hydroxylation can be measured in the single-turnover experiment using a rapid-quench apparatus that shoots the mixed reaction solution into a freezing bath instead of mixing it with a chemical quench solution. This has been done for iNOSoxy in single-turnover reactions[19,20] and the method is not described here. One study showed that H$_4$B radical formation begins after the Fe^{II}O$_2$ complex forms,[20] and then proceeds at a rate that is essentially equivalent to disappearance of the Fe^{II}O$_2$ intermediate and the rate of Arg hydroxylation (Table II). Thus, oxidation of H$_4$B to its radical is kinetically coupled to reduction of the Fe^{II}O$_2$ intermediate and formation of a heme-based oxidant that quickly reacts with Arg. The kinetic relationships between Fe^{II}O$_2$ disappearance, H$_4$B radical formation, and Arg hydroxylation are consistent with the slow step being reduction of the Fe^{II}O$_2$ intermediate in the single-turnover reaction.[20]

Reaction of Ferrous iNOSoxy with Oxygen and NOHA

The second reaction cycle in NO synthesis involves O$_2$ binding to ferrous NOS to form the Fe^{II}O$_2$ intermediate, followed by oxidation of NOHA to generate ferric enzyme, NO, citrulline, and water as products (Fig. 9). As drawn in Fig. 9, the reaction may also involve electron transfer from H$_4$B to the Fe^{II}O$_2$ intermediate. However, this has not been conclusively demonstrated, because in contrast to the Arg reaction the H$_4$B radical does not accumulate during the NOHA single-turnover reaction.[19] As shown in Fig. 9, this could be because the H$_4$B radical acts as an electron sink in a later point in the reaction, so as to allow NOS to

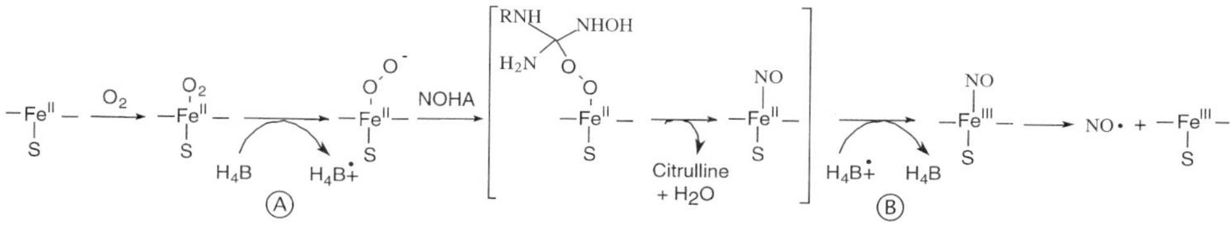

FIG. 9. A model for NOHA oxidation by NOSoxy in the single turnover reaction. Oxygen binds to the ferrous protein to form the $Fe^{II}O_2$ species, which upon receiving an electron from H_4B forms a putative heme–peroxo intermediate that reacts with NOHA. Rearrangement could generate citrulline and a ferrous heme–NO species. One of the heme intermediates formed by reaction with NOHA could transfer an electron back to the H_4B radical, generating the ferric heme–NO species, which dissociates into free NO and ferric enzyme.

generate NO instead of NO^- (see Ref. 34 for further discussion). On the other hand, the H_4B radical may simply not form in this reaction cycle.[19] The kinetics and/or magnitude of certain steps involved in NOHA oxidation have been examined in single-turnover reactions using various NOSoxy and full-length NOS proteins.[11–13,15,19,25,26]

Stopped-flow spectrophotometry can follow the kinetics and magnitude of heme transitions during the reaction. As for Arg, a reaction was initiated by mixing ferrous nNOSoxy containing NOHA and H_4B with an air-saturated buffer in the stopped-flow instrument at 10°. Here the same concepts hold as in the Arg reaction for identifying heme-based reaction intermediates, although some of the proposed intermediates may be unique to the NOHA reaction (see Fig. 9). Ninety-six spectra were recorded within 0.72 sec, which encompasses the time required for a full reaction at 10°. Spectra were also collected within selected shorter time frames to more closely study individual transitions.[15] SpecFit analysis of the data showed that in this case the best fit was the reaction model A → B → C → D transition. Each of the three transitions was monophasic. The analysis indicates that two intermediate species were detectable during the reaction. Their calculated spectra are shown in Fig. 10. Species A and D represent the beginning ferrous and ending ferric nNOSoxy, respectively, whereas species B has features identical to the $Fe^{II}O_2$ intermediate formed in the Arg reaction. Species C is unique to the NOHA reaction and has absorbance maxima at 440, 547, and 580 nm, identical to the spectrum of the ferric heme–NO complex of nNOS.[35] Thus, the single-turnover experiment reveals that newly formed NO binds to the ferric heme iron before leaving the enzyme. Rates for the three transitions as calculated by SpecFit

---

[34] S. Adak, Q. Wang, and D. J. Stuehr, *J. Biol. Chem.* **275**, 33554 (2000).
[35] J. Wang, D. L. Rousseau, H. M. Abu-Soud, and D. J. Stuehr, *Proc. Natl. Acad. Sci. U.S.A.* **91**, 10512 (1994).

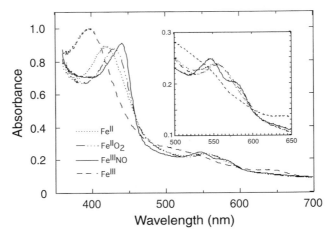

FIG. 10. Spectra of ferrous, heme–dioxy, ferric heme–NO, and ferric nNOSoxy species present during the NOHA reaction as calculated from the rapid-scanning stopped-flow data.

analysis are (in s^{-1}):

$$Fe^{III} \xrightarrow{113\pm20} Fe^{II}O_2 \xrightarrow{26\pm2} Fe^{III}NO \xrightarrow{5.4\pm0.4} Fe^{III}$$
$$\quad A \qquad\qquad\quad B \qquad\qquad\quad C \qquad\qquad\quad D$$

Figure 11 shows the calculated concentration versus time plots for all four species. The kinetic data imply that $Fe^{II}O_2$ disappearance is kinetically coupled

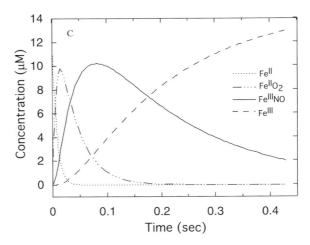

FIG. 11. Concentration of nNOSoxy species present during NOHA oxidation in the single-turnover reaction. Time courses were calculated from spectral data using the SpecFit program.

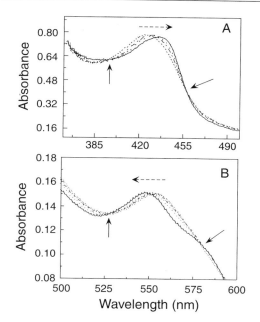

FIG. 12. Spectra collected during conversion of the heme–dioxy species to the ferric heme–NO species in the NOHA single turnover reaction. Overlay of four spectra recorded within 7.2–37.4 ms after mixing. (A) Wavelength range 360 nm to 500 nm; (B) Wavelength range 500–700 nm. The solid arrows indicate isosbestic points; the dashed arrows indicate direction of spectral change versus time for the major absorbance peak in each panel.

to formation of the ferric heme–NO species, which subsequently dissociates into ferric enzyme and free NO. This concept has fundamental implications for NOS function during steady-state catalysis.[36] Near-quantitative (about 75%) buildup of the heme–NO intermediate is consistent with the calculated transition rates, and with flash photolysis studies that show that the NOS heme displays a high trapping efficiency for NO released into the heme pocket.[37,38] Thus, stopped-flow spectroscopy under single-turnover conditions provides a direct demonstration that a ferric heme–NO species is an "immediate product" of NOS catalysis.

NO synthesis from NOHA occurs during conversion of the $Fe^{II}O_2$ intermediate to the ferric heme–NO species in the single-turnover reaction. Unfortunately, other heme intermediates that may form during this transition (see Fig. 9) were not indicated by SpecFit analysis. Actual spectral traces recorded during the $Fe^{II}O_2$ to $Fe^{III}NO$ transition are overlayed in Fig. 12 and reveal isosbestic points as marked.

[36] J. Santolini, S. Adak, C. M. L. Curran, and D. J. Stuehr, *J. Biol. Chem.* **276,** 1233 (2001).
[37] J. S. Scheele, E. Bruner, V. G. Kharitonov, P. Martasek, L. J. Roman, B. S. S. Masters, V. S. Sharma, and D. Magde, *J. Biol. Chem.* **274,** 13105 (1999).
[38] M. Negrerie, V. Berka, M. H. Vos, U. Liebl, J.-C. Lambry, and J.-L. Martin, *J. Biol. Chem.* **274,** 24694 (1999).

Single-wavelength analysis indicates the transition kinetics are best described as monophasic,[15] and therefore also suggest no buildup of intermediates. Thus, the presumed heme intermediates do not form, react more quickly than they form (i.e., do not accumulate), or have spectral features that make their appearance hard to distinguish. Conceivably, H_4B radical formation and its subsequent reduction back to H_4B would also occur within this time frame (see Fig. 9), but this has yet to be demonstrated.

Concluding Remarks

Stopped-flow spectroscopy is useful for investigating reaction kinetics of the NOS $Fe^{II}O_2$ intermediate. This method can be combined with other techniques such as rapid chemical quenching and rapid freezing to investigate temporal and quantitative links between $Fe^{II}O_2$ formation and disappearance, Arg or NOHA oxidation, and H_4B radical formation in NOS. Application of these three methods within a single-turnover setting is further defining the NOS reaction mechanism.[12,13,15–26] A future challenge will be to create reaction conditions or NOS mutant proteins that stabilize heme intermediates other than the $Fe^{II}O_2$ species, such that they build to detectable levels during the NOS single-turnover reaction.

[25] Myeloperoxidase: Kinetic Evidence for Formation of Enzyme-Bound Chlorinating Intermediate

By H. BRIAN DUNFORD and LEAH A. MARQUEZ-CURTIS

Introduction

Myeloperoxidase is a heme-containing enzyme which exhibits typical peroxidase activity, forming compound I from hydrogen peroxide.[1] An oxygen atom from hydrogen peroxide converts the heme iron(III) to a ferryl group $Fe^{IV}=O$. The additional oxidizing equivalent is stored on the porphyrin ring as a π-cation radical. Compound I can be reduced in two steps, via compound II, by hydrogen atom donors such as ascorbate so that the native enzyme is regenerated. Compound II retains the $Fe^{IV}=O$ group but the porphyrin cation radical is reduced so the normal porphyrin structure is regenerated.

Myeloperoxidase and chloroperoxidase from *Caldariomyces fumago* are unique enzymes because their compounds I possess sufficient oxidization potential

---

[1] H. B. Dunford, "Heme Peroxidases." Wiley, New York, 1999.

to oxidize chloride ion in a two-electron reaction.

$$MPO + H_2O_2 \longrightarrow MPO\text{-}I + H_2O$$
$$MPO\text{-}I + Cl^- + H^+ \rightarrow \rightarrow \rightarrow MPO + HOCl$$

where MPO is native myeloperoxidase and MPO-I is its compound I. Hypochlorous acid has been widely believed to be the killing agent of bacteria ingested by neutrophils.

The mechanism by which chloride ion is oxidized by myeloperoxidase and a substrate, the amino acid taurine, is chlorinated is the topic of this article.[2] We show that the only plausible mechanism is one in which oxidized chlorine, bound to the enzyme, chlorinates taurine before hypochlorous acid can be released into the medium.

Taurine is the common name for 2-aminoethanesulfonic acid. The pK_a values are 8.7 for the amino group and 1.5 for the sulfonic acid group, so the predominant form of taurine at the pH of our experiments, pH 4.7, is

$$^-O\text{-}\underset{\underset{O}{\|}}{\overset{\overset{O}{\|}}{S}}\text{-}CH_2CH_2NH_3^+$$

Kinetics of Taurine Chlorination by Myeloperoxidase

Preliminary Experiments

In order to obtain accurate kinetic data on the chlorination of taurine by myeloperoxidase, it is necessary to establish whether taurine can also act as a reducing substrate. Rapid spectral scan experiments have been conducted, in which MPO is mixed with hydrogen peroxide in the presence and absence of taurine. Taurine has no effect on the formation and decay of MPO-I. Also, taurine does not react with preformed MPO-II. Therefore taurine is an ideal substrate to study chlorination reactions catalyzed by myeloperoxidase.

Optical spectra were obtained for taurine $TauNH_2$, monochlorinated taurine $TauNHCl$, and dichlorinated taurine $TauNCl_2$. A significant difference in absorbance was observed at 252 nm between taurine and its monochlorinated derivate, confirming earlier published results.[3] A smaller but significant absorbance of the more slowly formed dichlorinated product was observed at this wavelength compared to that of the monochlorinated derivate. The effect of formation of dichlorotaurine was eliminated in the subsequent steady-state experiments by using only initial rate data.

---

[2] L. A. Marquez and H. B. Dunford, *J. Biol. Chem.* **269**, 7950 (1994).
[3] E. L. Thomas, M. M. Jefferson, and M. B. Grisham, *Biochemistry* **21**, 6299 (1982).

Experiments were conducted to determine the optimum pH for a detailed kinetic study. It was concluded that pH 4.7, which is within the pH range inside the phagolysosome, was ideal. All results reported here are for pH 4.7.

In order to compare the enzymatic chlorination rate to that of free HOCl, the nonenzymatic reaction was studied:

$$\text{TauNH}_2 + \text{HOCl} \longrightarrow \text{TauNHCl} + \text{H}_2\text{O} \qquad (1)$$

These experiments were conducted so as to obtain the most accurate value of the second-order rate constant as follows. Various excess concentrations of taurine were reacted with a fixed concentration of freshly prepared HOCl. The resultant pseudo first-order rate constants were plotted against taurine concentration and the second-order rate constant obtained from the linear slope. The result was $k_{\text{app}} = 3.5 \times 10^3 \ M^{-1}\text{s}^{-1}$.

Steady-State Experiments

A Photal stopped flow apparatus was used for the steady-state kinetic measurements. A stopped flow provides a 1000-fold decrease in the dead time in comparison with measurements made on a conventional spectrophotometer, and in our view should always be used for steady-state experiments.

Two most probable mechanisms were considered. The first assumes that free HOCl is the chlorinating intermediate.

Mechanism I

$$\text{MPO} + \text{H}_2\text{O}_2 \xrightarrow{k_1} \text{MPO-I} + \text{H}_2\text{O} \qquad (2)$$

$$\text{MPO-I} + \text{Cl}^- \xrightarrow[\text{H}^+]{k'_2} \text{MPO} + \text{HOCl} \qquad (3)$$

$$\text{HOCl} + \text{TauNH}_2 \xrightarrow{k'_3} \text{TauNHCl} + \text{H}_2\text{O} \qquad (4)$$

This leads to the rate equations:

$$v = k'_3[\text{HOCl}][\text{TauNH}_2] \qquad (5)$$

$$\frac{v}{[\text{E}]_{\text{tot}}} = \frac{1}{\dfrac{1}{k_1[\text{H}_2\text{O}_2]} + \dfrac{1}{k'_2[\text{Cl}^-]}} \qquad (6)$$

which can be rearranged into the form

$$\frac{v}{[\text{E}]_{\text{tot}}} = \frac{k'_2[\text{Cl}^-][\text{H}_2\text{O}_2]}{\dfrac{k'_2}{k_1}[\text{Cl}^-] + [\text{H}_2\text{O}_2]} \qquad (7)$$

$$\frac{v}{[\text{E}]_{\text{tot}}} = \frac{k_{\text{cat}}[\text{H}_2\text{O}_2]}{K_m + [\text{H}_2\text{O}_2]} \qquad (8)$$

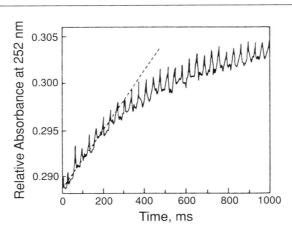

FIG. 1. Typical time course of the enzymatic formation of taurine monochloramine. The chlorination rate was determined from the slope of the initial linear portion of the trace. Reaction conditions were as follows: 0.1 μM MPO, 7 mM taurine, 0.03 M KCl, 2 mM H$_2$O$_2$ in 0.1 M phosphate buffer, pH 4.7. From H. B. Dunford and L. A. Marquez-Curtis, *J. Biol. Chem.* **269**, 7950 (1994).

where

$$k_{cat} = k'_2[\text{Cl}^-] \tag{9}$$

$$K_m = \frac{k'_2}{k_1}[\text{Cl}^-] \tag{10}$$

Experimental measurements of initial velocity v have been obtained as illustrated in Fig. 1. A rectangular hyperbola is obtained for a series of measurements at different [H$_2$O$_2$], keeping [Cl$^-$] constant. Then a second hyperbola is obtained for a different fixed [Cl$^-$], by varying [H$_2$O$_2$], and so on. Thus the family of rectangular hyperbolas is obtained, shown in Fig. 2. The hyperbolas, each for a different value of [Cl$^-$], yielded a family of values of k_{cat} and K_m.

The prediction from mechanism I is very clear: plots of both k_{cat} and K_m versus [Cl$^-$] should be linear.

Mechanism II. In this mechanism it is assumed that the chlorinating species is MPO-I–Cl$^-$, in which the two oxidizing equivalents of MPO-I have been transferred to a chlorine species contained within the enzyme.

$$\text{MPO} + \text{H}_2\text{O}_2 \xrightarrow{k_1} \text{MPO-I} + \text{H}_2\text{O} \tag{2}$$

$$\text{MPO-I} + \text{Cl}^- \xrightarrow{k_2} \text{MPO-I–Cl}^- \tag{11}$$

$$\text{MPO-I–Cl}^- + \text{TauNH}_2 \xrightarrow[\text{H}^+]{k_3} \text{MPO} + \text{TauNHCl} \tag{12}$$

One of the latter two reactions must liberate a water molecule in order to have balanced overall equations. We defer further discussion of mass balance until later.

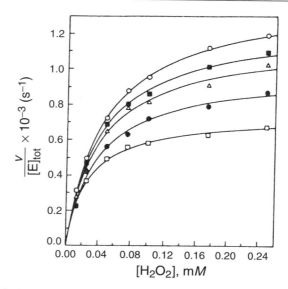

FIG. 2. Chloride dependence of chlorination rate of taurine. Reaction conditions: 0.1 μM MPO, 5 mM taurine in 0.1 M phosphate buffer, pH 4.7. KCl concentrations in mM are: 0.50 □, 0.75 ●, 1.0 △, 1.5 ■, and 2.0 ○. The data were fitted using Eq. (8). From H. B. Dunford and L. A. Marquez-Curtis, *J. Biol. Chem.* **269**, 7950 (1994).

Mechanism II leads to the following equations:

$$v = k_3[\text{MPO-I-Cl}^-][\text{TauNH}_2] \tag{13}$$

$$\frac{v}{[\text{E}]_{\text{tot}}} = \frac{1}{\dfrac{1}{k_1[\text{H}_2\text{O}_2]} + \dfrac{1}{k_2[\text{Cl}^-]} + \dfrac{1}{k_3[\text{TauNH}_2]}} \tag{14}$$

Thus the equation for $v/[\text{E}]_0$ from mechanism II has an additional term involving [TauNH$_2$], not predicted by mechanism I. The latter equation can be rearranged into the form

$$\frac{v}{[\text{E}]_{\text{tot}}} = \frac{\dfrac{k_3[\text{TauNH}_2][\text{Cl}^-]}{\frac{k_3}{k_1}[\text{TauNH}_2]+[\text{Cl}^-]}[\text{H}_2\text{O}_2]}{\dfrac{\frac{k_3}{k_1}[\text{TauNH}_2][\text{Cl}^-]}{\frac{k_3}{k_2}[\text{TauNH}_2]+[\text{Cl}^-]}+[\text{H}_2\text{O}_2]} \tag{15}$$

$$\frac{v}{[\text{E}]_{\text{tot}}} = \frac{k_{\text{cat}}[\text{H}_2\text{O}_2]}{K_m + [\text{H}_2\text{O}_2]} \tag{8}$$

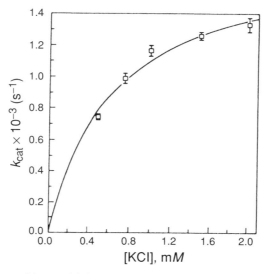

FIG. 3. Dependence of k_{cat} on chloride concentration. The vertical bars are standard deviations from the computer fit of the curves from Fig. 2. The data were fitted to Eq. (16), in which [TauNH$_2$] is constant. From H. B. Dunford and L. A. Marquez-Curtis, *J. Biol. Chem.* **269,** 7950 (1994).

where

$$k_{cat} = \frac{k_3[\text{TauNH}_2][\text{Cl}^-]}{\frac{k_3}{k_2}[\text{TauNH}_2] + [\text{Cl}^-]} \quad (16)$$

$$K_m = \frac{\frac{k_3}{k_1}[\text{TauNH}_2][\text{Cl}^-]}{\frac{k_3}{k_2}[\text{TauNH}_2] + [\text{Cl}^-]} \quad (17)$$

The prediction of mechanism II is very clear: plots of both k_{cat} and K_m at constant [TauNH$_2$] versus [Cl$^-$] should be rectangular hyperbolas. The results are shown in Figs. 3 and 4. They disprove mechanism I and show that mechanism II fits the experimental data. In the case of taurine, at least, the chlorination reaction catalyzed by myeloperoxidase does not involve free hypochlorous acid. Further evidence for the validity of mechanism II was obtained by showing that variation of [TauNH$_2$] caused changes in values of k_{cat} and K_m, a result predicted by mechanism II and contrary to the predictions of mechanism I (Fig. 5).

Additional Data. Analysis of the parameters obtained from the experimental data yielded the following results for mechanism II: $k_1 = (3.3 \pm 0.2) \times 10^7 \, M^{-1}\text{s}^{-1}$, $k_2 = (2.8 \pm 1.2) \times 10^6 \, M^{-1}\text{s}^{-1}$, and $k_3 = (4.4 \pm 0.2) \times 10^5 \, M^{-1}\text{s}^{-1}$. Thus the rate constant k_3 for chlorination of taurine by the myeloperoxide compound I-chlorinating intermediate is over 100 times larger than that for chlorination by free hypochlorous acid.

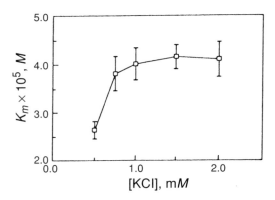

FIG. 4. Dependence of K_m on chloride concentration. The vertical bars are standard deviations from the computer fit of the curves from Fig. 2. From H. B. Dunford and L. A. Marquez-Curtis, *J. Biol. Chem.* **269**, 7950 (1994).

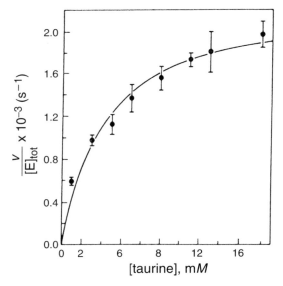

FIG. 5. Dependence of chlorination rate on taurine concentration. Reaction conditions were as follows: 0.1 μM MPO, 0.5 mM H_2O_2, 5 mM KCl in 0.1 M phosphate buffer, pH 4.7. Vertical bars represent standard deviations from the calculated initial rates. From H. B. Dunford and L. A. Marquez-Curtis, *J. Biol. Chem.* **269**, 7950 (1994).

Possible Complications

1. The formation of compound I is slightly reversible.[4–6]

$$\text{MPO} + \text{H}_2\text{O}_2 \rightleftharpoons \text{MPO-I} + \text{H}_2\text{O} \tag{18}$$

However with an excess of H_2O_2 the back reaction is negligible. Its neglect caused no error in the taurine kinetic analysis.

2. Chloride ion binds weakly to native myeloperoxidase.[2,7–9]

$$\text{MPO} + \text{Cl}^- \rightleftharpoons \text{MPO-Cl}^- \tag{19}$$

The binding constant has been accurately measured as a function of pH. In the original paper describing the taurine chlorination, a small accurate correction was made to both mechanisms I and II. This correction was neglected here as an unnecessary complication to the story of the striking difference between mechanisms I and II.

3. Hydrogen peroxide reacts as a reductant of compound I.[6,10]

$$\text{MPO-I} + \text{H}_2\text{O}_2 \longrightarrow \text{MPO-II} + \text{HO}_2\cdot \tag{20}$$

where MPO-II is myeloperoxidase compound II and $\text{HO}_2\cdot$ is the hydroperoxyl radical. The rate constant for the reduction of MPO-I by H_2O_2 is $(8.2 \pm 0.2) \times 10^4\ M^{-1}\text{s}^{-1}$, so the reaction would be unable to compete with the formation of the myeloperoxidase chlorinating intermediate from compound I and Cl^- which has an approximately 40-fold faster rate constant.

4. Hypochlorous acid reacts with native myeloperoxidase to form compound I.[11,12]

$$\text{MPO} + \text{HOCl} \longrightarrow \text{MPO-I} + \text{H}^+ + \text{Cl}^- \tag{21}$$

However, hypochlorous acid is never released from the enzyme in its rapid chlorination of taurine.

---

[4] B. G. J. M. Bolscher and R. Wever, *Biochim. Biophys. Acta* **788,** 1 (1984).

[5] J. K. Hurst, in "Peroxidases in Chemistry and Biology," Vol. I (J. Everse, K. E. Everse, and M. B. Grisham, eds.), p. 37. CRC Press, Boca Raton, FL, 1991.

[6] L. A. Marquez, J. T. Huang, and H. B. Dunford, *Biochemistry* **33,** 1447 (1994).

[7] A. R. J. Bakkenist, J. E. G. De Boer, H. Plat, and R. Wever, *Biochim. Biophys. Acta* **613,** 337 (1980).

[8] P. C. Andrews and N. I. Krinsky, *J. Biol. Chem.* **257,** 13240 (1982).

[9] M. Ikeda-Saito, *J. Biol. Chem.* **260,** 11688 (1985).

[10] H. Hoogland, H. L. Dekker, C. van Riel, A. van Kuilenburg, A. O. Muijsers, and R. Wever, *Biochim. Biophys. Acta* **955,** 337 (1988).

[11] R. Floris and R. Wever, *Eur. J. Biochem.* **207,** 697 (1992).

[12] P. G. Furtmüller, U. Burner, W. Jantschko, G. Regelsberger, and C. Obinger, *Redox Report* **5,** 173 (2000).

5. Myeloperoxidase acts as a catalase, so that compound I formation is followed by the reaction

$$\text{MPO-I} + \text{H}_2\text{O}_2 \longrightarrow \text{MPO} + \text{O}_2 + \text{H}_2\text{O} \tag{22}$$

Catalase activity was reported for myeloperoxidase by Kjell Agner, in the paper describing the first isolation and purification of the enzyme in 1941.[13] He reported that the catalatic activity of myeloperoxidase was only 1/10,000 of that observed with catalase. True catalase activity does not occur with myeloperoxidase. The reaction that is observed when preformed compound I is mixed with hydrogen peroxide is reduction of compound I to compound II and oxidation of hydrogen peroxide to hydroperoxyl radical. Although there is some oxygen evolution observed in the absence of oxidizable substrates other than hydrogen peroxide, it must be caused by subsequent reactions of superoxide. Superoxide formation is not occurring in the presence of taurine.

6. The chlorinating reagent is Cl_2. This requires either of the following reactions to occur.

$$\text{HOCl} + \text{Cl}^- + \text{H}^+ \longrightarrow \text{Cl}_2 + \text{H}_2\text{O} \tag{23}$$

$$\text{MPO-I-Cl}^- + \text{Cl}^- + 2\text{H}^+ \longrightarrow \text{Cl}_2 + \text{H}_2\text{O} + \text{MPO} \tag{24}$$

The former reaction cannot occur because free HOCl is not released from the enzyme. The latter reaction leads to a steady-state mechanism in which k_{cat} and K_m are predicted to be linearly dependent on chloride ion concentration and independent of taurine concentration, as was the case for mechanism I, and is contrary to the experimental evidence.

Thus of all these possible complications, only the weak chloride ion binding to the native enzyme is of any importance, and it was accurately corrected for in the original analysis.

Discussion

The value of the rate constant for myeloperoxidase compound I formation, k_1, obtained in this steady-state kinetic study, $(3.3 \pm 0.2) \times 10^7 \, M^{-1}s^{-1}$, is in good agreement with values obtained by transient-state kinetics: $(2.3 \pm 0.1) \times 10^7 M^{-1}s^{-1}$ for human myeloperoxidase,[14] and $(1.8 \pm 0.1) \times 10^7 M^{-1}s^{-1}$ for beef spleen myeloperoxidase.[15] There is now considerable evidence that the kinetic behaviors of the beef spleen and human enzymes are identical.

---

[13] K. Agner, *Acta Physiol. Scand.* **II** (Suppl. VIII), 1 (1941).
[14] P. G. Furtmüller, U. Burner, and C. Obinger, *Biochemistry* **37**, 17923 (1998).
[15] L. A. Marquez and H. B. Dunford, *J. Biol. Chem.* **270**, 30434 (1995).

TABLE I
APPARENT SECOND-ORDER RATE CONSTANTS
OBTAINED FOR REACTION OF MPO-I WITH Cl^{-a}

$[Cl^-]$ Range	k_{app} ($M^{-1}s^{-1}$)
0–100 μM	$(1.59 \pm 0.06) \times 10^6$
100 μM–1.5 mM	$(1.92 \pm 0.08) \times 10^5$
1.5–6.0 mM	$(4.6 \pm 0.7) \times 10^4$

^{a} At pH 7.0, 25°. From Ref. 16.

Attempts to measure the rate of reaction of myeloperoxidase compound I with chloride, using transient state kinetics, were hindered because of the instability of compound I. This instability was overcome by using the sequential-mixing stopped flow technique in which an optimum yield of compound I was obtained in the aging loop and then reacted rapidly with chloride ion. The following value was reported: $(4.7 \pm 0.1) \times 10^6\ M^{-1}s^{-1}$ for the beef spleen enzyme.[15] This is in good agreement with the steady-state result,[2] reported here from mechanism II for the monochlorination of taurine: $(2.8 \pm 1.2) \times 10^6\ M^{-1}s^{-1}$, the rate constant for formation of MPO-I–Cl^-.

Reaction of Myeloperoxidase with Hydrogen Peroxide and Chloride Ion

The two values,[2,15] quoted above, for the rate constant for the formation of the myeloperoxidase chlorinating intermediate from compound I and chloride ion give an average value of $(3.8 \pm 1.0) \times 10^6\ M^{-1}s^{-1}$. Both values, at 25°, were obtained by the University of Alberta group, Edmonton. A very different rate constant was published by the University of Agricultural Sciences group, Vienna, for the rate of reaction of myeloperoxidase compound I with chloride ion at 15°[14]: $(2.5 \pm 0.3) \times 10^4\ M^{-1}s^{-1}$. In order to resolve the apparent discrepancy, the two groups collaborated in an intensive transient-state study of the reaction of myeloperoxidase with hydrogen peroxide and chloride ion. The results showed that when myeloperoxidase compound I was preformed in the aging loop of a sequential-mixing stopped flow apparatus, and subsequently reacted with chloride ion, three distinctly different regions were observed. These were dependent on chloride ion concentration, with the higher the value of $[Cl^-]$ the slower the reaction. In each region linear plots of the pseudo first-order rate constant versus $[Cl^-]$ were observed. Thus three different values of the second-order rate constant k_{app} were obtained as shown in Table I.[16]

[16] P. G. Furtmüller, C. Obinger, Y. Hsuanyu, and H. B. Dunford, *Eur. J. Biochem.* **267**, 5858 (2000).

The results were explained by the following mechanism:

$$\text{MPO-I} + \text{Cl}^- \underset{k_{-2}}{\overset{k_2}{\rightleftharpoons}} \text{MPO-I–Cl}^- \quad (25)$$

$$\text{MPO-I–Cl}^- \xrightarrow[\text{H}^+]{k_4} \text{MPO} + \text{HOCl} \quad (26)$$

$$\text{MPO} + \text{H}_2\text{O}_2 \underset{k_{-1}}{\overset{k_1}{\rightleftharpoons}} \text{MPO-I} + \text{H}_2\text{O} \quad (27)$$

Results for the lowest range of [Cl$^-$] are governed solely by MPO-I–Cl$^-$ formation and so $k_{app} = k_2$. For the intermediate range of [Cl$^-$] sufficient MPO-I–Cl$^-$ has formed so that the back reaction, dissociation of MPO-I–Cl$^-$ to MPO-I and Cl$^-$, becomes important. Thus the value of the apparent second-order rate constant k_{app} is decreased. Finally, for the largest range of [Cl$^-$], dissociation of MPO-I–Cl$^-$ to MPO and HOCl also occurs, the excess H$_2$O$_2$ which is present recycles the enzyme and k_{app} is further decreased. Rate constants obtained from a global analysis of results for the entire range of [Cl$^-$], according to the above three-step mechanism, yielded the following results: $k_2 = (2.2 \pm 0.3) \times 10^6\ M^{-1}\text{s}^{-1}$, $k_{-2} = (1.9 \pm 0.7) \times 10^5\ \text{s}^{-1}$, and $k_4 = (5.2 \pm 0.2) \times 10^4\ \text{s}^{-1}$. Literature values, well established for k_1 and k_{-1}, were used in the analysis. The results show that in the absence of substrate to be chlorinated, an equilibrium concentration of MPO-I–Cl$^-$ is approached: it dissociates to its starting reactants faster than it liberates HOCl. The presence of taurine reacts with MPO-I–Cl$^-$ before it can undergo either dissociation reaction.

Thus taurine is an excellent trapping reagent, not only for free HOCl,[17] but for the chlorinating reagent bound to myeloperoxidase compound I. There is well-documented evidence that HOCl reacts rapidly with primary amines and amino side chains of proteins. It is likely the neutral amino form of taurine and all other amines that react with the strong electrophilic reagents such as HOCl and MPO-I–Cl$^-$.

What could be the structure of the chlorinating intermediate? Our suggestion is that whatever species is first generated by MPO-I and Cl$^-$ reacts rapidly with an amino acid side chain within the enzyme to form RNHCl. The internal enzyme reaction could involve HOCl:

$$\text{RNH}_2 + \text{HOCl} \longrightarrow \text{RNHCl} + \text{H}_2\text{O} \quad (28)$$

or it might possibly be

$$\text{RNH}_2 + \text{Cl}^+ \longrightarrow \text{RNHCl} + \text{H}^+ \quad (29)$$

---

[17] E. L. Thomas and D. B. Learn, in "Peroxidases in Chemistry and Biology," Vol. I (J. Everse, K. E. Everse, and M. B. Grisham, eds.), p. 83. CRC Press, Boca Raton, FL, 1991.

In the sequential-mixing transient-state experiments on the reaction of myeloperoxidase with hydrogen peroxide and chloride ion, measurements were made at 430 nm, where preformed MPO-I appeared to be reduced to form native enzyme, so that an increase in absorbance was observed. A chlorinating intermediate that is not bound to the iron, and that now contains the two oxidizing equivalents which were on the oxoheme group of MPO-I, would be expected to have a spectrum similar to that of native MPO. There was no indication of formation of a species with a spectrum uniquely different from that of native enzyme, which would have been expected if an Fe–O–Cl species with a finite lifetime were formed.

X-ray crystallographic work, in which bromide was substituted for chloride in order to provide higher electron density, identified four binding sites for chloride on native myeloperoxidase.[18] One of these could be the site of the two-electron oxidation of chloride by MPO-I, followed immediately by chlorination of a nearby amine or imidazole side chain where the chlorinating capacity of MPO-I–Cl is stored.

Returning to the question of mass balance in the steady-state chlorination of taurine, if the chlorinating capacity is stored on an imidazole or amine side chain of the enzyme (shown below as an NH group attached to MPO), then mechanism II can be represented as:

$$\text{MPO}\!\!>\!\!\text{N-H} + H_2O_2 \xrightarrow{k_1} \text{MPO-I}\!\!>\!\!\text{N-H} + H_2O \tag{30}$$

$$\text{MPO-I}\!\!>\!\!\text{N-H} + Cl^- + H^+ \xrightarrow[H^+]{k_2} \text{MPO}\!\!>\!\!\text{N-Cl} + H_2O \tag{31}$$

$$\text{MPO}\!\!>\!\!\text{N-Cl} + \text{TauNH}_2 \xrightarrow{k_3} \text{MPO}\!\!>\!\!\text{N-H} + \text{TauNHCl} \tag{32}$$

$$\text{TauNH}_2 + H_2O_2 + Cl^- + H^+ = \text{TauNHCl} + 2H_2O \tag{33}$$

Thus the chlorinating enzyme species is represented as:

$$\text{MPO}\!\!>\!\!\text{N-Cl}$$

The second reaction, formation of the myeloperoxidase chlorinating intermediate, must consist of two steps: first, rate-controlling formation of either HOCl or Cl^+ within the enzyme active site, followed by rapid chlorination of an amino acid side chain of myeloperoxidase. In the absence of a rapidly reacting substrate such as taurine, dissociation of the MPO-I–Cl complex to MPO-I and Cl^-

$$\text{MPO}\!\!>\!\!\text{N-Cl} + H_2O \xrightarrow{k_{-2}} \text{MPO-I} + Cl^- + H^+ \tag{34}$$

---

[18] T. J. Fiedler, C. A. Davey, and R. E. Fenna, *J. Biol. Chem.* **275,** 11964 (2000).

provides a mechanism to conserve the chlorinating potential of the enzyme. Formation of MPO-I and Cl$^-$ occurs more rapidly than liberation of free HOCl

$$\underset{\text{MPO}}{\diagdown\text{N–Cl}} + H_2O \xrightarrow{k_4} \underset{\text{MPO}}{\diagdown\text{N–H}} + HOCl \qquad (35)$$

which would indiscriminately attack the enzyme.

Conclusions

The steady-state kinetics for the chlorination of taurine by myeloperoxidase provides strong evidence for the existence of a myeloperoxidase-bound chlorinating intermediate which is responsible for the chlorination of taurine, and presumably other primary amines. The overall picture which is emerging is that the chlorination capacity is stored on either an amine or imidazole side chain of the enzyme. It is the enzyme-bound chlorinating intermediate which chlorinates taurine, not free hypochlorous acid or Cl_2. Because of the electronic charge on monochlorotaurine it cannot diffuse through membranes. This restriction does not apply to other monochloroamines, which may be major reagents in bacterial killing in the phagosome.

Acknowledgments

We are indebted to Paul Furtmüller, Yuchiong Hsuanyu, and Christian Obinger, whose work on the transient-state kinetics of the reaction of myeloperoxidase with hydrogen peroxide and chloride ion provided additional data which (1) supported mechanism II for the chlorination of taurine, (2) indicated that the myeloperoxidase chlorinating intermediate has an optical spectrum similar to that of native myeloperoxidase, and (3) elucidated the mechanism of reaction of myeloperoxidase in the absence of a substrate to be chlorinated.[16] Figures 1–5 were reproduced with permission from *J. Biol. Chem.* **269**, 7950 (1994).

[26] Time-Resolved Resonance Raman Spectroscopy of Intermediates in Cytochrome Oxidase

By DENIS L. ROUSSEAU and SANGHWA HAN

Introduction

Cytochrome oxidase is the membrane-bound terminal enzyme in the electron transfer chain. Forms of the enzyme are found in essentially all organisms that utilize oxygen. In eukaryotic species the enzyme is located in the inner mitochondrial membrane where it accepts electrons from cytochrome c and brings about the four-electron reduction of oxygen to water. A redox-linked translocation of protons across the membrane contributing to the proton gradient, used in the generation of ATP, is associated with this process. One proton is vectorially translocated across the membrane for each electron that is transferred to the oxygen and another proton is used in a scalar fashion for the production of water. Thus, cytochrome oxidase serves the dual role of generating a proton gradient by coupling redox events to proton translocation and maintaining continued electron flow for oxidative phosphorylation by catalyzing the four-electron reduction of O_2 to H_2O.

The molecular mechanisms by which cytochrome oxidase carries out these functions has been under study for many years since understanding these mechanisms has wide applications in many different aspects of biochemistry, biophysics, and bioenergetics.[1,2] First, the important processes by which proteins are able to pump protons across a membrane against the gradient are poorly understood and yet form the essential step in the generation of ATP.[3] In cytochrome oxidase, nature has found a way to harness the redox energy released by the reduction of oxygen and utilize it to translocate protons against a gradient. Second, since the enzyme has four redox centers, this one protein serves as a valuable system in which many of the basic foundations of biophysics such as the influence of protein structure, redox potential, and donor–acceptor distances on electron transfer may be studied. Third, the enzyme is able to reduce dioxygen to water very efficiently without the generation of partially reduced deleterious oxygen radicals and ions such as superoxides, peroxides, and hydroxyl radicals. The understanding of the oxygen reduction mechanisms can help to clarify functional properties of other heme-containing proteins and enzymes, such as oxygenases and peroxidases, which also involve oxygen intermediates. Fourth, cytochrome oxidase has interest from a genetics standpoint since the mitochondrial enzyme contains 13 subunits, three of

---

[1] G. T. Babcock and M. Wikstrom, *Nature* **356**, 301 (1992).
[2] S. Ferguson-Miller and G. T. Babcock, *Chem. Rev.* **96**, 2669 (1996).
[3] S. M. Musser, M. H. Stowell, and S. I. Chan, *Adv. Enzymol. Relat. Areas Mol. Biol.* **71**, 79 (1995).

which are encoded in the mitochondrial genome and 10 of which are encoded in the cellular genome. This offers an interesting challenge of how these subunits become properly assembled in the mitochondrial membrane.

As an essential step in addressing these issues, it is necessary to determine the molecular basis for the four-electron reduction of molecular oxygen to water. This catalytic reaction occurs at a binuclear center formed by an a-type heme (heme a_3) and a copper atom (Cu_B) as shown in Fig. 1B. The electrons are transferred to the binuclear center from another heme group (heme a) which in turn receives its electrons from a copper center composed of two copper atoms (Cu_A) that hold one redox equivalent (Fig. 1A). The electrons from the bc_1 complex are shuttled to cytochrome oxidase by cytochrome c, which has a binding site near the Cu_A center of the oxidase. When fully reduced, cytochrome oxidase can hold four redox equivalents: one on Cu_A, one on heme a, one on heme a_3, and one on Cu_B. Thus, the full four-electron reduction of oxygen to water can occur in the enzyme for study in the laboratory without the complication of adding electrons during the catalytic process.

Methods

To investigate general enzymatic reaction mechanisms, flow techniques have been developed in which the enzyme and the substrate are mixed, thereby initiating the reaction, which is then followed in time by spectroscopic techniques or quenched by freezing for subsequent studies. The first flow methodologies were developed by Hartridge and Roughton in 1923.[4] In their continuous flow device the two solutions to be mixed flow from separate reservoirs to the mixer and then into an observation tube. The time evolution of the sample depends on the flow velocity and the distance between the mixer and the observation point. A major advance in the instrumentation occurred in 1951 when Chance developed stopped flow instrumentation in which the flow is terminated after the samples are mixed and a series of observations are made in order to follow the progression of the reaction.[5] Stopped flow instrumentation is now universally used in laboratories as the major technique to obtain kinetic information. With conventional commercial instrumentation the dead time for the solution mixing is in the few-milliseconds time range.

One would like to use the continuous or stopped flow methods to study the catalytic mechanism of cytochrome oxidase by simply mixing oxygen with the enzyme and then following the reaction to identify and characterize all of the intermediates. However, the catalytic reaction is complete in roughly 1 ms and therefore with conventional mixing techniques one is unable to detect the various intermediates. Two clever methods were developed to overcome this limitation, one for studies at cryogenic temperatures and one for studies at room temperature. For the cryogenic

---

[4] H. Hartridge and F. J. W. Roughton, *Proc. R. Soc. (London)* **A104**, 376 (1923).
[5] B. Chance, *Rev. Sci. Instr.* **22**, 619 (1951).

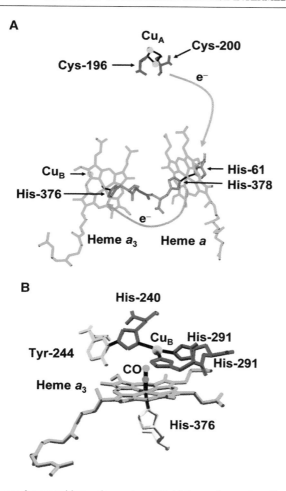

FIG. 1. Bovine cytochrome oxidase redox centers. (A) All four redox centers: Cu_A, heme a, heme a_3, and Cu_B. The pathway for the electron transfer is shown in which the electrons go from Cu_A to heme a to the binuclear center formed by heme a_3 and Cu_B. (B) The binuclear center which is the catalytic site of cytochrome oxidase. The heme a_3 iron atom is shown coordinated by CO. This is where oxygen binds during the enzymatic reaction. The Cu_B atom, which is 5 Å from the heme a_3 iron atom, is coordinated by three histidines, one of which (His-240) is coordinated to Tyr-244 through a posttranslational modification. The structure is from the Protein Data Bank (1OCO).

studies, Chance pioneered the triple trapping technique.[6] In that technique, CO-bound cytochrome oxidase is mixed with oxygen in a low-temperature glycerol slurry and then the temperature is lowered to about $-200°$ so as to prevent any spontaneous reactions from taking place. At the low temperature, the carbon monoxide

[6] B. Chance, C. Saronio, and J. Leigh, *Proc. Natl. Acad. Sci. U.S.A.* **72**, 1635 (1975).

FLOW - FLASH - PROBE

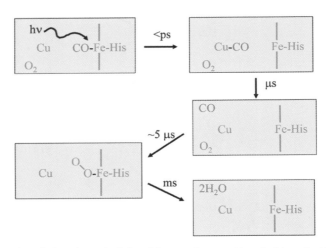

FIG. 2. The flow-flash-probe method of studying cytochrome oxidase. In this method CO is coordinated to the enzyme and it is mixed with oxygen saturated buffer. The CO is then photodissociated from the heme. This allows oxygen to bind to the heme within a few microseconds and the reaction is then probed from the microsecond to the millisecond time domain. Additional details are given in the text.

is photodissociated allowing the oxygen to bind. However, because the protein dynamics are completely restricted at the low temperature, the reaction will not progress to completion. Then, by a combination of varying the temperature and controlling the time at a given temperature, the various intermediates in the reaction sequence can be formed and they are trapped by subsequently lowering the temperature back down to a point at which the reaction will no longer proceed. In this way each of the intermediates may be stabilized and studied by a variety of different spectroscopies. A great deal of useful information concerning the catalytic intermediates was determined by Chance and collaborators by this technique.

To study the intermediates in the cytochrome oxidase reaction at room temperature, Greenwood and Gibson developed the flow–flash–probe technique.[7] In this method, illustrated in Fig. 2, CO-bound cytochrome oxidase is mixed with oxygen-saturated buffer in a stopped-flow or continuous-flow apparatus. The spontaneous replacement of CO by the oxygen is slow (a few seconds) so no reaction will occur in the inhibited enzyme. Woodruff and co-workers have shown that a pulse of light breaks the CO–iron bond in under 1 ps and the CO binds to the copper atom in the binuclear center.[8] After about 1 μs the CO spontaneously dissociates from

---

[7] Q. H. Gibson and C. Greenwood, *Biochem. J.* **86,** 541 (1965).

[8] W. H. Woodruff, Ó. Einarsdóttir, R. B. Dyer, K. A. Bagley, G. Palmer, S. J. Atherton, R. A. Goldbeck, T. D. Dawes, and D. S. Kliger, *Proc. Natl. Acad. Sci. U.S.A.* **88,** 2588 (1991).

the copper and diffuses away from the binuclear center allowing the oxygen to bind. The oxygen binding, and thereby initiation of the catalytic reaction, occurs within a few microseconds but the reaction does not go to completion until about 1 ms, so one can follow the various intermediates by monitoring the reaction in the microsecond to millisecond time regime. With this technique the reactions of all molecules of the enzyme are initiated in phase and the various intermediates can be followed as a function of time by spectroscopic methods.

The triple trapping and the flow-flash-probe methods in conjunction with optical absorption and electron paramagnetic resonance (EPR) spectroscopies have been very valuable for studying the mechanism of the reaction of oxygen and cytochrome oxidase.[9,10] However, the very broad transitions in the optical spectrum make it difficult to definitively identify and characterize the intermediates. On the other hand, EPR is limited by strong spin coupling among the metal centers resulting in spectral silence for most of the intermediates.[11] Resonance Raman scattering has developed into a very powerful technique for the study of heme proteins. With Raman scattering the vibrational spectra of the hemes are probed. The vibrational spectrum is rich in information about the hemes and thus it can be used to obtain quantitative structural and kinetic properties of heme proteins. Resonance Raman scattering is a particularly powerful technique for studying cytochrome oxidase because it allows the vibrational modes of the heme a and heme a_3 macrocycles to be studied independently.[12] Some modes are sensitive to the oxidation states of each heme and thereby are valuable for determining when an electron transfer event has taken place. Other modes are sensitive to the spin state of the central iron atom and allow the determination of properties conferred on the heme by various axial ligands. Still other modes are sensitive to the groups in the peripheral environment of the heme yielding information about their interactions with the protein surroundings. In addition, in ligand-bound forms additional modes are present that allow for the determination of the identity and structure of the ligand. By monitoring the changes in amplitude of these different vibrational modes, the kinetics of the various processes can be followed, making resonance Raman scattering a very rich and reliable technique to probe the enzymatic reaction.

The use of resonance Raman scattering for the study of the cytochrome oxidase intermediates was pioneered by Babcock in collaboration with Palmer and Woodruff in a seminal paper that was published in 1984.[13] They were able to detect changes in the high frequency region of the spectrum as a function of time

---

[9] G. M. Clore, L.-E. Andreasson, B. Karlsson, R. Aasa, and B. G. Malmstrom, *Biochem. J.* **185**, 139 (1980).
[10] B. C. Hill, C. Greenwood, and P. Nicholls, *Biochim. Biophys. Acta* **853**, 91 (1986).
[11] Z. K. Barnes and G. T. Babcock, *Biochemistry* **30**, 7597 (1991).
[12] Y. C. Ching, P. V. Argade, and D. L. Rousseau, *Biochemistry* **24**, 4938 (1985).
[13] G. T. Babcock, J. M. Jean, L. N. Johnston, G. Palmer, and W. H. Woodruff, *J. Am. Chem. Soc.* **106**, 8305 (1984).

during the reaction of the enzyme with oxygen. This paper set the stage for the later work that was done in parallel by his group, by Kitagawa's group, and by that of the present author. These three groups progressed very evenly in determining the various structures and kinetics of the intermediates in the reaction of the enzyme with oxygen by using resonance Raman scattering. However, the resonance Raman experiments were performed under different conditions in each laboratory, which in some cases has caused it to be difficult to make direct comparisons between the results because each has its own strengths and weaknesses. For example, the continuous flow method in conjunction with cw lasers used by Kitagawa's group[14-19] and by the present authors[20-24] gives a better signal-to-noise ratio than that of the pulsed laser technique used by Babcock's group.[25-28] However, with pulsed lasers more accurate time delays can be measured. With the recirculating artificial cardiovascular system used by Kitagawa and co-workers, excellent signal-to-noise is obtained by use of long integration times but it is difficult to generate a high concentration of molecular O_2, so the time course of the intermediates is not as well defined. In our work, it was felt that the continuous flow method without recirculation is a good compromise that gives sufficient signal-to-noise and time resolution to address the central issues.

The continuous flow apparatus used in our laboratory[24] is illustrated in Fig. 3. In this system the CO-bound cytochrome oxidase is mixed with oxygen-saturated buffer. The mixed sample flows into a narrow channel that is 250 μm by 250 μm in cross-section. The CO bound enzyme is then photodissociated allowing the oxygen to react. The second beam probes the spectrum. By changing the distance

---

[14] T. Ogura, S. Takahashi, K. Shinzawa-Itoh, S. Yoshikawa, and T. Kitagawa, *J. Biol. Chem.* **265,** 14721 (1990).

[15] T. Ogura, S. Yoshikawa, and T. Kitagawa, *Biochemistry* **28,** 8022 (1989).

[16] T. Ogura, S. Takahashi, K. Shinzawa-Itoh, S. Yoshikawa, and T. Kitagawa, *J. Am. Chem. Soc.* **112,** 5630 (1990).

[17] T. Ogura, S. Takahashi, S. Hirota, K. Shinzawa-Itoh, S. Yoshikawa, E. H. Appelman, and T. Kitagawa, *J. Am. Chem. Soc.* **115,** 8527 (1993).

[18] T. Ogura, S. Hirota, D. A. Proshlyakov, K. Shinzawa-Itoh, S. Yoshikawa, and T. Kitagawa, *J. Am. Chem. Soc.* **118,** 5443 (1996).

[19] D. A. Proshlyakov, T. Ogura, K. Shinzawa-Itoh, S. Yoshikawa, and T. Kitagawa, *Biochemistry* **35,** 76 (1996).

[20] S. H. Han, Y. C. Ching, and D. L. Rousseau, *Proc. Natl. Acad. Sci. U.S.A.* **87,** 8408 (1990).

[21] S. Han, Y. C. Ching, and D. L. Rousseau, *Nature* **348,** 89 (1990).

[22] S. W. Han, Y. C. Ching, and D. L. Rousseau, *Proc. Natl. Acad. Sci. U.S.A.* **87,** 2491 (1990).

[23] S. Han, Y. C. Ching, and D. L. Rousseau, *Biochemistry* **29,** 1380 (1990).

[24] S. Han, Y.-C. Ching, and D. L. Rousseau, *J. Am. Chem. Soc.* **112,** 9445 (1990).

[25] C. Varotsis and G. T. Babcock, *Biochemistry* **29,** 7357 (1990).

[26] C. Varotsis, W. H. Woodruff, and G. T. Babcock, *J. Biol. Chem.* **265,** 11131 (1990).

[27] C. Varotsis, Y. Zhang, E. H. Appelman, and G. T. Babcock, *Proc. Natl. Acad. Sci. U.S.A.* **90,** 237 (1993).

[28] C. Varotsis, G. T. Babcock, M. Lauraeus, and M. Wikstrom, *Biochim. Biophys. Acta* **1231,** 111 (1995).

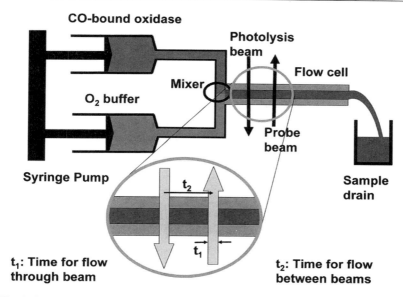

FIG. 3. Continuous flow apparatus for the study of the reaction of cytochrome oxidase with oxygen. The CO-bound enzyme is mixed with oxygen saturated buffer and passed into a 250 μm by 250 μm channel. The continuously flowing sample is photolyzed by one laser beam and subsequently probed by a second beam displaced from the first. Based on the flow rate and the separation between the laser beams the reaction can be followed from $\sim$10 μs to a few milliseconds.

between the two beams the intermediates in the reaction may be probed. When the two beams overlap, based on the typical flow rates and beam diameters, the reaction time is roughly 15 μs and by changing the relative positions between the laser beams, the kinetics of the formation and decay of the intermediates out to the millisecond time domain can be followed.

An understanding of the flow dynamics of the continuous flow device is necessary to be able to obtain reliable kinetic information from it. Ideally, one would like the solution to progress by "plug" flow, i.e., the solution flows down the channel as a plug so that the incremental velocity at the center is the same as that at the edge and is equal to the average velocity. However, solution flow in pipes generally becomes laminar in which the flow velocity distribution is parabolic. For laminar flow the velocity varies as the square of the distance from the side wall such that at the center it is twice the average velocity and it goes to zero at the vessel walls. To determine the flow characteristics in our system we have taken advantage of photodissociation of CO-bound cytochrome oxidase. If the sample stays in the laser beam for a long time the CO will be dissociated resulting in the ligand-free ferrous enzyme. On the other hand, if the flow velocity is very high less photodissociation will take place. Thus, by adjusting the laser power to obtain

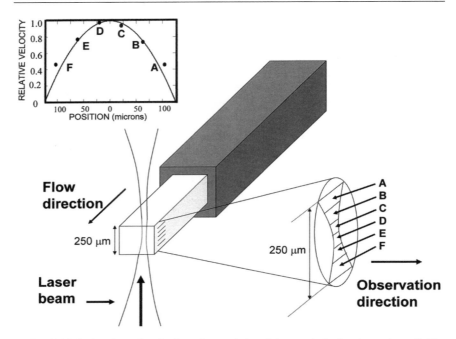

FIG. 4. Method to determine the flow characteristics of the sample in the observation cell. The image of the laser passing through the 250 μm cell is divided into six sections by binning the pixels of the CCD detector. CO-bound cytochrome oxidase is passed through the cell and the laser power is adjusted to result in partial photolysis. The degree of photolysis depends on the duration of the sample in the beam. Thus, the relative velocity of the sample at each section of the cell may be calculated. *Inset:* Relative velocity compared to the parabolic velocity distribution expected for laminar flow.

partial photodissociation, the degree of photodissociation can be used as a measure of the time in the beam, from which the relative velocity may be calculated.

To measure the flow characteristics, the pixels of the CCD detector were binned into six sections perpendicular to the flow such that the laser passage through the 250 μm cell was binned into segments, each about 40 μm long as illustrated in Fig. 4. The laser power was adjusted (lowered) to get partial photodissociation of the CO-bound enzyme and the high frequency resonance Raman spectrum was used to monitor the degree of photodissociation in the sample. The strong line at 1350–1380 cm^{-1} is an excellent marker for the state of the enzyme, occurring at 1374 cm^{-1} for the CO-bound heme a_3 and at 1359 cm^{-1} when it is ligand-free. Thus, from these data, shown in Fig. 5, we are able to obtain an estimate of the velocity distribution across the cell as shown in the inset in Fig. 4. Here, the measured velocity data are compared to the parabolic distribution of the velocity predicted for laminar flow. As can be seen from the data the flow at the center is close to laminar but it tails off at the edges. Therefore, in order to obtain the

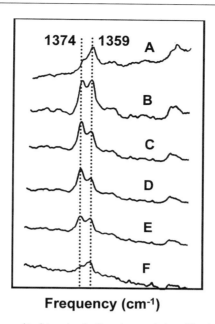

FIG. 5. The Raman line used to determine the flow characteristics of the observation cell as described in Fig. 4. The line at 1374 cm^{-1} results from the CO-bound enzyme and that at 1359 cm^{-1} results from the photodissociated species. When the flow is fast at the center of the cell, the amount of photodissociation is small whereas at the sides of the cell where the flow is slower the photodissociation is greater.

narrowest distribution of the data, the CCD detector can be binned to observe only the central region of the flow cell. Owing to the laminar flow characteristics, this gives several advantages. For example, in the center of the cell, the velocity (V_{max}) is twice the average velocity (V_{av}), which is the bulk velocity. Thus, by looking at the central region the shortest lifetimes may be examined. In practice, by using a height of ~180 μm of the 250 mm total path length, over 70% of the total possible intensity is available and the average velocity in the probed region is ~85% of V_{max}.

Results

For studying the reaction of cytochrome oxidase with oxygen by resonance Raman scattering, both the high and low frequency regions of the spectrum yield different and valuable information. The high frequency region (1000–1800 cm^{-1}) of the resonance Raman spectrum of heme proteins contains vibrational modes that are sensitive to the coordination, spin state, and oxidation state of the heme moiety.[29] For cytochrome c oxidase, the contributions to the spectrum from hemes

[29] S. Han, S. Takahashi, and D. L. Rousseau, *J. Biol. Chem.* **275,** 1910 (2000).

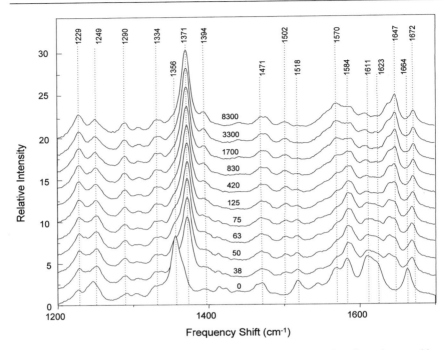

FIG. 6. High frequency region of the resonance Raman spectra of the reaction of cytochrome oxidase with oxygen. The reaction was followed from initiation to 8300 μs. The significance of the changes in the various lines are discussed in the text. All of the Raman data were obtained with 413.1 nm excitation. The enzyme concentration after mixing was 100 μM and the oxygen concentration was 700 μM. Typically each spectrum was accumulated for 30 sec.

a and a_3 overlap. However, several modes have been identified which have been demonstrated to be characteristic of each of the hemes and thereby can be used to monitor the properties of each without interference from the other.[12,20,30–33] In particular, the lines in the spectrum at 1518, 1611, 1623, and 1647 cm^{-1} can be used as markers of the oxidation state of heme a. The lines at 1518, 1611, and 1623 cm^{-1} originate from reduced heme a whereas the line at 1647 cm^{-1} originates from oxidized heme a.

The changes in the resonance Raman spectra of cytochrome oxidase in the high frequency region during the reaction of the enzyme with oxygen are shown in Fig. 6 for several different time points.[29] A rapid decrease in the intensity of the

[30] G. T. Babcock and I. Salmeen, *Biochemistry* **18**, 2493 (1979).
[31] G. T. Babcock, P. M. Callahan, M. R. Ondrias, and I. Salmeen, *Biochemistry* **20**, 959 (1981).
[32] I. Salmeen, L. Rimai, and G. Babcock, *Biochemistry* **17**, 800 (1978).
[33] P. V. Argade, Y. C. Ching, and D. L. Rousseau, *Biophys. J.* **50**, 613 (1986).

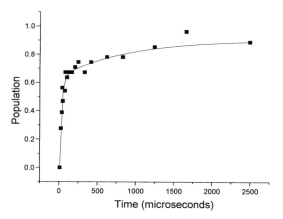

FIG. 7. Time dependence of the oxidation of heme a during the reaction of cytochrome oxidase with oxygen obtained from the high frequency resonance Raman spectra. The data may be fitted with two exponentials with rate constants of $\sim$40,000 s^{-1} and 1000 s^{-1}.

lines at 1518, 1611, and 1623 cm^{-1} and a corresponding increase in the intensity of the line at 1647 cm^{-1} occurs during the first $\sim$100 μs. Subsequently, the changes are more gradual, such that the formation of the fully oxidized spectrum of heme a nears completion at $\sim$3 ms. The time dependence of the oxidation of heme a, plotted in Fig. 7, obtained by monitoring the change in intensity of the line at 1518 cm^{-1}, clearly demonstrates that heme a becomes oxidized by at least two separate processes. A rapid process occurs in which the heme is partially oxidized with an apparent rate constant of $\sim$40,000 s^{-1} and it is followed by a slower process with an apparent rate constant of $\sim$1000 s^{-1}. Changes in the spin state, judged by the intensity of the low spin marker line at $\sim$1584 cm^{-1} and the high spin marker line at $\sim$1570 cm^{-1}, occur on the 0.1 to 10 ms time scale, with a rate constant of $\sim$1000 s^{-1}. These data demonstrate a significant growth of the high spin population in this time range. Because heme a is always low spin the change is attributed to heme a_3. Thus, the final product of the reaction has a high spin heme.[29]

In order to gain an understanding of the catalytic mechanism, it is necessary to relate the electron transfer events and electronic structure properties, determined from the high frequency resonance Raman spectrum, to the oxygen intermediates. In principle, from a comparison to well-studied heme protein chemistry, several intermediates are expected. The primary intermediate, referred to as compound A, is one in which the neutral oxygen is coordinated to the ferrous iron atom of heme a_3. However, rapid internal electron transfer from the iron atom to the oxygen is expected to generate a ferric–superoxide complex (Fe^{3+}–O_2^-) as reported in other heme proteins.[24] The addition of an electron to the iron–oxygen complex leads to an oxidation state at the peroxide level (Fe^{3+}–O_2^{2-}). When a third electron is added to the oxygen the O—O bond can be cleaved generating a free hydroxide (assuming a proton is present) and a ferric–oxene structure ($Fe^{3+}-O^-$), which through an internal electron transfer can convert to a ferryl–oxo species ($Fe^{4+}-O^{2-}$). Finally, when that oxygen receives a fourth electron another hydroxide is generated and the full reduction of the molecular oxygen is complete. Scheme 1 summarizes this progression.

$$[Fe^{3+}-O_2^-]_A + e^- \rightarrow [Fe^{3+}-O_2^{2-}] + e^- \rightarrow OH^- + [Fe^{4+}-O^{2-}]_F + e^- \rightarrow [Fe^{3+}-OH^-]_H$$

SCHEME 1

This scheme may be evaluated by measuring the low frequency region (100–1000 cm^{-1}) of the resonance Raman spectrum, where, in addition to the modes associated with the heme macrocycles, oxygen isotope-sensitive modes associated with the reduction intermediates are present during the reaction of the enzyme with oxygen. However, the iron–oxygen modes are very weak, so except for the primary Fe–O$_2$ intermediate they cannot be readily identified in the spectra. By using oxygen isotope difference spectra ($^{16}O_2-^{18}O_2$) the modes involving the oxygen can be identified as shown in Fig. 8 in which the laser excitation wavelength was 413.1 nm. Here all of the porphyrin lines, which are independent of the oxygen isotope, are absent from the oxygen isotope difference spectrum and only the modes involving motion of the oxygen atoms are present. At the earliest time only one line, at 586 cm^{-1}, is evident. As time evolves a new line appears at 786 cm^{-1} and as it decays another line grows in at 450 cm^{-1}. Thus, the technique allows for the following of the progression of the intermediates.

Based on the observed isotope dependence and a comparison with other heme proteins these lines may be assigned. The line at 568 cm^{-1}, which appears first, is the Fe–O$_2$ stretching mode of the primary intermediate.[22] The line at 786 cm^{-1} originates from the $Fe^{4+}-O^{2-}$ stretching mode of a ferryl–oxo moiety.[21] This species has a maximum in the optical absorption difference spectrum at 580 nm, termed compound "F." The line at 450 cm^{-1} in the resonance Raman spectrum originates from the Fe–OH stretching mode.[21] Thus, of the four intermediates presented in Scheme 1, only the peroxy species is not present in our data. The detection of a peroxy species has been elusive. For a long time, it was thought that the species, seen in the optical difference spectrum with a band at 607 nm, originated from an intermediate at the peroxy redox level. A species with a 607-nm band in the optical absorption difference spectrum has been reported to be formed in three different ways[34]: by the catalytic reaction of the fully reduced enzyme with oxygen,[35] by the

---

[34] M. I. Verkhovsky, J. E. Morgan, and M. Wikstrom, *Proc. Natl. Acad. Sci. U.S.A.* **93,** 12235 (1996).
[35] J. E. Morgan, M. I. Verkhovsky, and M. Wikstrom, *Biochemistry* **35,** 12235 (1996).

[26] RAMAN SPECTROSCOPY OF CYTOCHROME OXIDASE INTERMEDIATES 363

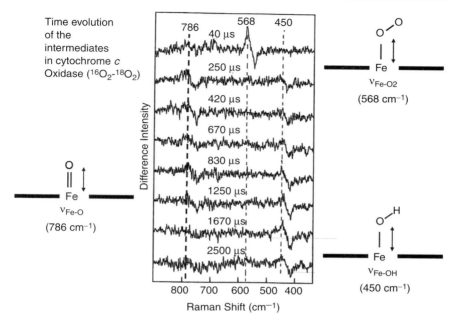

FIG. 8. Low frequency region of the resonance Raman difference spectra of the reaction of cytochrome oxidase with oxygen showing the progression of the oxygen intermediates. The data were obtained by determining the difference spectrum of the reaction with $^{18}O_2$ and with $^{16}O_2$. The assignment of the vibrational modes from each of the intermediates is illustrated.

catalytic reaction of the mixed valence enzyme with oxygen,[36] and by the addition of hydrogen peroxide to the oxidized enzyme.[37] The resonance Raman spectra of this intermediate species has also been detected and it displays a line at ∼801 cm^{-1}.

Weng and Baker originally proposed that the 607 nm species, also referred to as compound P, was actually a ferryl species in which the O—O bond had already been cleaved.[37] Kitagawa and co-workers confirmed the fact that the 607-nm species has only a single oxygen atom by an elegant series of mixed oxygen isotope experiments.[38] In those experiments, the 607 nm species was formed by adding isotopically labeled hydrogen peroxide ($H_2^{16}O_2$, $H_2^{18}O_2$, $H_2^{16}O^{18}O$) to the oxidized enzyme and the Raman spectra were measured. If the Raman mode detected at ∼801 cm^{-1} originated from a species with one oxygen atom, such as a ferryl–oxo intermediate, two lines would be expected, one from the ^{16}O adduct

---

[36] D. A. Proshlyakov, M. A. Pressler, and G. T. Babcock, *Proc. Natl. Acad. Sci. U.S.A.* **95**, 8020 (1998).
[37] L. C. Weng and G. M. Baker, *Biochemistry* **30**, 5727 (1991).
[38] D. A. Proshlyakov, T. Ogura, K. Shinzawa-Itoh, S. Yoshikawa, E. H. Appelman, and T. Kitagawa, *J. Biol. Chem.* **269**, 29385 (1994).

and one from the ^{18}O adduct. However, if the intermediate contained two oxygen atoms, such as a peroxy derivative, a third line from the $^{16}O-^{18}O$ moiety, with a frequency intermediate between the other two lines, is anticipated. The third line was not present in the spectra reported by Kitagawa and co-workers, proving that the intermediate contained only a single oxygen atom. These findings that the 607 nm species was not a peroxy intermediate, but instead was an intermediate in which the oxygen–oxygen bond was already cleaved, presented a dilemma because this species can be formed by adding oxygen to the mixed valence enzyme (Cu_A and heme a, oxidized; Cu_B and heme a_3, reduced) or by adding hydrogen peroxide to the fully oxidized enzyme. In both of these cases there are an insufficient number of electrons on the metal redox centers to cleave the O—O bond and therefore an additional redox center must be active.

To resolve this conundrum of the missing electron, it has been postulated that in these cases the extra redox equivalent is supplied by the tyrosine (Tyr-244) that is present in the Cu_B–heme a_3 binuclear site and forms the posttranslational modification with His-240 (see Fig. 1B). This tyrosine can donate both an electron and a proton to the iron–oxy complex. To test this hypothesis, Babcock and co-workers[39] formed the 607 nm species and added radioactive iodide to it. Iodide is known to label tyrosine radicals but not neutral tyrosine. By proteolytic treatment of the enzyme followed by HPLC separation they were able to show that Tyr-244 was labeled. The total amount of labeling was small presumably because it is difficult for the large iodide ion to reach the binuclear center. However, these experiments are the most definitive evidence for the formation of a tyrosine radical in this cytochrome oxidase intermediate.

From the resonance Raman measurements it is now clear that a true peroxy intermediate has not been detected under any conditions in cytochrome oxidase. Apparently, its lifetime is too short to allow it to be detected under the conditions used to measure the intermediates. In addition, the 607 nm species (compound P), which has a 801 cm^{-1} Raman line was not detected in the reaction of the fully reduced enzyme with oxygen in the experiments of Babcock and co-workers or in those of the present authors. However, it was detected in similar experiments reported by the Kitagawa group,[17] but the conditions were very different because of the recirculating system that he uses. Also, Wikstrom and co-workers detected the 607 nm species in the reaction of the fully reduced enzyme with oxygen at $-25°$, conditions under which the reaction is slowed down significantly to allow for its detection.[35] The hypothesis that compound P, in its formation, utilizes a redox equivalent from Tyr-244, which is adjacent to the oxygen binding site, suggests a major difference in the electrostatics of the binuclear center between compound P and compound F, the ferryl–oxo intermediate. In compound P all of the electrons that are transferred to the oxygen originate at the binuclear center, i.e., from the

[39] D. A. Proshlyakov, M. A. Pressler, C. DeMaso, J. F. Leykam, D. L. Dewitt, and G. T. Babcock, *Science* **290**, 1588 (2000).

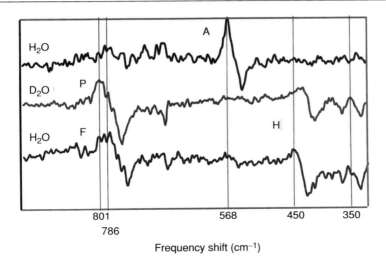

FIG. 9. Low frequency region of the resonance Raman difference spectra of the reaction of cytochrome oxidase with oxygen showing the difference between doing the reaction in deuterated versus protonated buffers. The top spectrum in protonated solvent was obtained at a reaction time of ~50 μs The bottom spectrum, also obtained in protonated solvent, is a sum of several spectra in the 300 to 1000 μs range. The middle spectrum obtained in deuterated solvent is a sum of spectra in the 1 to 2 ms range. The assignment of each of the vibrational modes as originating from compounds A, P, F, and H is indicated and these species are described in the text.

heme iron, from Cu_B, and from the tyrosine. However, in compound F, one electron is supplied from the heme a center. A principle of electroneutrality has been proposed in which whenever an electron enters the binuclear center a proton must accompany it so that the overall charge remains unchanged.[40] Thus, in compound P the proton and the electron both originate from the tyrosine that is in the binuclear center whereas for compound F a proton must enter the site from a proton channel to compensate for the electron charge that come from outside the binuclear center.

If the need for a compensating proton is indeed the difference between P and F then there should be a very large kinetic isotope effect. Indeed, Brzezinski and coworkers have found that the kinetic isotope effect for the P to F transition is ~2–5 and the transition involves two proton transfer steps: an internal proton transfer, postulated to be from Glu-286, to the binuclear center and a second step in which the proton-donating group is reprotonated.[41] In an effort to identify compound P during turnover, we measured the resonance Raman spectrum for the reaction of the fully reduced enzyme with oxygen in deuterated buffers as shown in Fig. 9. In the deuterated buffer the Fe—O stretching mode appears at 801 cm^{-1} confirming

[40] R. Mitchell and P. R. Rich, *Biochim. Biophys. Acta* **1186**, 19 (1994).
[41] M. Karpefors, P. Adelroth, A. Aagaard, I. A. Smirnova, and P. Brzezinski, *Israel J. Chem.* **39**, 427 (1999).

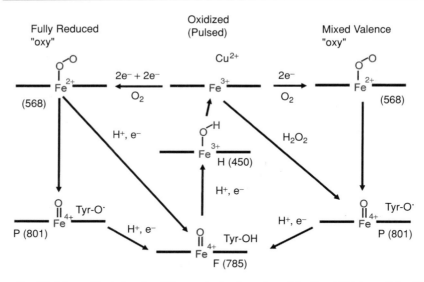

FIG. 10. Scheme illustrating the difference in the formation of the P and F intermediates. The reaction of the oxidized enzyme with H_2O_2 or the mixed valence enzyme with O_2 generates compound P in which Tyr-244 serves as a redox site by transferring an electron and a proton to the oxygen moiety. This same species can be generated from the fully reduced enzyme when proton transfer to the binuclear center is slowed, for example by placing the enzyme in deuterated solvent.

the presence of the P intermediate (i.e., the 607 nm species), whereas under the same conditions in protonated solvent the F intermediate is formed. The reaction sequence is summarized in Fig. 10. In this model compound P is formed whenever there is no electrostatic change at the binuclear center. Thus, it is formed in the reaction of oxygen with the mixed valence enzyme, in the reaction of hydrogen peroxide with the oxidized enzyme, and under conditions in which protons are not readily available to compensate for electron input into the binuclear center. Additional Raman experiments are in progress to test this model.

Conclusion

The reaction of cytochrome oxidase is very complex. However, great progress in its understanding has been made by resonance Raman studies. Most of the intermediates in the reaction scheme have been identified and their kinetics determined, and a simple catalytic cycle for the reaction as shown in Fig. 11 may be written. This reaction scheme is divided into two phases. In the oxidative phase the four-electron reduced enzyme reacts with oxygen and becomes fully oxidized. In the reductive phase the enzyme is rereduced. The first intermediate in the reaction is labeled as compound A and is considered to be a ferric-superoxide species. In the

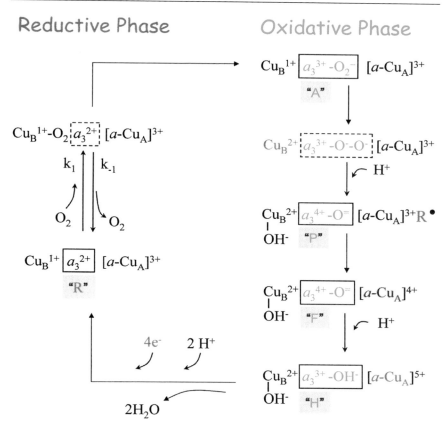

FIG. 11. A scheme showing the intermediates that are detected in the reaction of cytochrome oxidase with oxygen. The reaction is divided into two phases. In the oxidative phase the four-electron reduced enzyme reacts with oxygen and becomes fully oxidized. In the reductive phase the enzyme is rereduced. The description of the intermediates in this scheme is included in the text.

Raman spectrum this species has an iron–oxygen stretching mode at 568 cm^{-1}. The addition of an electron to compound A generates a peroxy species that decays too rapidly to be detected spectroscopically. Under certain conditions as described in the text compound P can be generated in which the O—O bond has been cleaved and a radical is formed on Tyr-244. Here this is illustrated by the R$^{\bullet}$. Compound P is characterized by an optical transition at $\sim$607 nm and an iron–oxygen stretching frequency at $\sim$801 cm^{-1}. Compound P decays to compound F which is also a ferryl–oxo species but in which there is no radical present. Compound F is characterized by an optical transition at $\sim$580 nm and an iron–oxygen stretching mode at $\sim$785 cm^{-1}. In the reaction of the fully reduced enzyme with oxygen in protonated solvent at neutral pH, the electron transfer from heme a measured by its oxidation (see Figs. 6 and 7), the decay of compound A and the formation of compound F all have the same rate constants ($\sim$40,000 s^{-1}). Associated with the remaining electron transfer from heme a, with a rate constant of $\sim$1000 s^{-1}, compound F decays into compound H, the hydroxyl intermediate, with an Fe—OH stretching frequency of 450 cm^{-1}. From the analysis of the high frequency resonance Raman spectra shown in Fig. 6, it was determined that the iron atom in this hydroxyl species has a high spin configuration.

The reaction cycle in Fig. 11 does not include the complex protonation events that are occurring in the enzyme during the catalytic cycle. In addition, clarification of the properties of compound P remains unresolved, especially that of its role under physiological conditions and if and how the tyrosine radical comes into play. In the model proposed here the formation of compound P in all cases is associated with the generation of a neutral tyrosine radical. However, analysis of optical data suggests that when compound P is formed during turnover of the fully reduced enzyme an electron is supplied from heme a and the tyrosine radical is not formed.[35] It is anticipated that future time-resolved resonance Raman scattering will play an essential role in resolving these issues.

Acknowledgment

This work is supported by NIH Grant #GM54806 (to D.L.R).

[27] Porphobilinogen Deaminase: Accumulation and Detection of Tetrapyrrole Intermediates Using Enzyme Immobilization

By ALCIRA BATLLE

Introduction

The functioning of living cells greatly depends on the action of thousands of different enzymes. Over the past 70 years, most of the studies on enzymes have been conducted using purified extracts, often in aqueous solutions, that is, under conditions far removed from those existing within the living cell. So, a practical approach to investigate intracellular enzymes under more natural circumstances has been to attach the soluble enzyme to some kind of solid supporting matrix.

Reports on chemical immobilization of proteins and enzymes first appeared in the late 1960s. Apart from biological considerations, interest in immobilized enzymes also stems from their possible use in many areas: (*i*) research, (*ii*) synthesis, (*iii*) analysis, (*iv*) biomedicine, e.g., in therapy and biosensors, (*v*) waste treatment, and (*vi*) industry.[1–4]

Insoluble enzyme derivatives can be used as model systems for the elucidation of reaction mechanisms, either by modifying the equilibrium conditions and removing one or more products of a partial reaction or otherwise by allowing the detection or isolation of intermediates which are very often difficult to obtain.

Binding enzymes or other ligands to supports can be achieved by covalent coupling, entrapment within a matrix, adsorption, or intermolecular cross-linking.[1] Of these methods the one of choice is the covalent binding of the enzyme to the solid matrix. Any water-insoluble carrier having a high surface density of functional groups would be good as an artificial matrix for enzymes; however, agarose has been the most popular because it meets nearly all the requirements of an ideal carrier for an enzyme. A change in chemical and physical properties might be imposed on the insoluble enzyme after its binding to the artificial support that should be determined.

We have been interested in the study of the kinetics and reaction mechanisms of the heme enzymes and in the use of immobilized enzymes and ligands for multiple applications.[3]

One of the most fascinating enzymes in the porphyrin pathway is porphobilinogenase (PBGase), a combination of two separate enzymes, one of which is more stable in heat than the other. The relatively heat-stable protein is PBG-deaminase (PBG-D, EC 4.3.1.8), also known as uroporphyrinogen I synthase (Uro-S) or hydroxymethylbilane synthase (HMB-S). It catalyzes the conversion of PBG into HMB, which ring-closes chemically to form uroporphyrinogen I (urogen I). The heat-labile enzyme is uroporphyrinogen III synthase (urogen III-S), cosynthase, or isomerase (EC 4.2.1.75), which after an intramolecular rearrangement of ring D converts HMB into urogen III, the physiological intermediate of hemes, chlorophylls, cytochromes, and corrins.

The mechanism of action of PBGase has been the subject of much speculation[5] and several years ago it was shown that *Escherichia coli* PBG-D contains a dipyrromethane cofactor covalently linked to the enzyme, which plays a key role in the PBG polymerization reaction.[6]

The possibility of accumulating different polypyrrylmethane intermediates in this reaction under normal conditions and also of investigating whether some kind

---

[1] K. Mosbach, *Sci. Am.* **224,** 26 (1971).
[2] R. Goldman, L. Goldstein, and E. Katchalski, in "Biochemical Aspects of Reactions on Solid Supports" (G. R. Stark, ed.), p. 1. Academic Press, New York, 1971.
[3] A. M. C. Batlle, E. A. Wider, and A. M. Stella, *Int. J. Biochem.* **9,** 407 (1978).
[4] L. F. Liang, Y. T. Li, and V. C. Yang, *J. Pharmaceut. Sci.* **89,** 979 (2000).
[5] A. M. C. Batlle and M. V. Rossetti, *Int. J. Biochem.* **8,** 251 (1977).
[6] P. M. Jordan and M. J. Warren, *FEBS Lett.* **225,** 87 (1987).

of association and/or interaction exist between PBG-D and cosynthase led us to attach purified PBGase and its isolated components from different sources to Sepharose.[7,8]

The detection of pyrrolic intermediates in the synthesis of urogen from the soybean callus enzymes under normal conditions was reported by Llambías and Batlle.[9,10] Additional evidence came from the *Euglena gracilis* enzymes,[11,12] indicating that the enzymatic polymerization of PBG occurs on the protein surface and that the di-, tri-, and tetrapyrrole intermediates might be bound to the enzyme and behave as intermediates thus enhancing final urogen formation. Interestingly, when free intermediates were added to the incubation mixture, inhibition of urogen synthesis was also found.

We will report here on the studies employing immobilized bovine liver PBG-D carried out to obtain further evidence regarding the formation and behavior of free and deaminase-bound tetrapyrrylmethane intermediates in the normal enzymatic condensation of PBG into urogens, as well as their involvement in urogen III formation.[13]

Materials and Methods

Materials

PBG is prepared according to Sancovich *et al.*[14] Cyanogen bromide is from Fluka, A. G. (Switzerland). Sephadex and Sepharose gels are from Pharmacia Fine Chemicals (Sweden). All other reagents are analytical reagent grade. All solutions are made in three times glass distilled, deionized water.

Bovine Liver PBG-D Purification

Unless otherwise indicated all operations were performed at 4°. (1) A homogenate (10%, w/v) from bovine liver is obtained in 0.25 M sucrose and (2) centrifuged for 30 min at 11,000g. (3) The pH of the 11,000g supernatant is taken to pH 5.0 by slowly adding glacial acetic acid. The mixture is stirred for 20 min and the precipitate formed is removed by centrifuging for 30 min at 11,000g. (4) The resulting supernatant is fractionated with solid ammonium sulfate; after

---

[7] M. V. Rossetti, V. E. Parera, and A. M. C. Batlle, *Acta Physiol. Lat.* **26,** 371 (1977).
[8] M. L. Kotler, A. A. Juknat, and A. M. C. Batlle, *Biotech. Appl. Biochem.* **13,** 173 (1991).
[9] E. B. C. Llambías and A. M. C. Batlle, *FEBS Lett.* **6,** 285 (1970).
[10] E. B. C. Llambías and A. M. C. Batlle, *Biochem. J.* **121,** 327 (1971).
[11] M. V. Rossetti and A. M. C. Batlle, *Int. J. Biochem.* **8,** 277 (1977).
[12] M. V. Rossetti, A. A. Juknat, and A. M. C. Batlle, *Int. J. Biochem.* **8,** 781 (1977).
[13] M. L. Kotler, A. A. Juknat, S. A. Fumagalli, and A. M. C. Batlle, *Biotech. Appl. Biochem.* **13,** 173 (1991).
[14] H. A. Sancovich, A. M. Ferramola, A. M. C. Batlle, and M. Grinstein, *Methods Enzymol.* **17,** 220 (1970).

stirring for 40 min the 45–70% saturation precipitate is collected by centrifugation for 10 min at 11,000g and then desalted through a Sephadex G-25 column (2 × 50 cm). The protein eluted is heated for 20 min at 65°; the precipitate formed is removed by centrifugation for 20 min at 11,000g and the supernatant concentrated by ammonium sulfate precipitation up to 70% saturation. The new precipitate is separated again by centrifugation, is dissolved in 50 mM Tris-HCl buffer pH 7.4, applied to a Sephadex G-100 column (2.5 × 57 cm), equilibrated and eluted with the same buffer Tris, at a flow rate of 25–30 ml/hr. Under these conditions a 300-fold purified PBG-D, with a specific activity of 8.4 nmol uroporphyrin/30 min/mg protein, yielding 95–100% Uro I is obtained.

Bovine Liver Cosynthase

The first four steps (homogenate, 11,000g supernatant, pH 5.0 supernatant, and 45–70% ammonium sulfate fraction) are the same as those described above for PBG-D purification. The protein supernatant from step 4 is precipitated further with solid ammonium sulfate to 70–90% saturation, the resulting precipitate is dissolved in 50 mM Tris-HCl buffer pH 7.4 and the solution applied to a Sephadex G-100 column (2.5 × 57 cm) equilibrated and eluted with the same buffer at the same flow rate as before. The protein eluates, when incubated with purified PBG-D, produce a 180-fold purified cosynthase with a specific activity of 2.2 nmol uroporphyrin III/30 min/mg protein, yielding between 90 and 100% uroporphyrin III.

Coupling of PBG-D to Sepharose

Besides the nature of the functional groups in both protein and support, the coupling yield and properties of the insolubilized enzyme are influenced by other variables such as density of the gel, its degree of activation, the amount of protein added, temperature, pH, and time of binding reaction. Therefore, optimal conditions for enzyme coupling are as follows: Sepharose is activated essentially by the method reported by Cuatrecasas.[15] Optimal binding and activation are obtained with 200 mg of cyanide bromide per ml of packed Sepharose. The activated Sepharose is suspended in cold (6°) 50 mM Tris-HCl buffer pH 8.0: maximal yield (70%) is attained by adding purified PBG-D solution at a ratio of 1 mg of enzyme/ml of Sepharose containing 20–30 mg protein. The mixture is left standing for 12–14 hr at 4° with slow but constant stirring. After this period immobilized PBG-D retains the highest activity (near 50% of the original). The Sepharose immobilized PBG-D is packed in a column and washed with Tris-HCl buffer pH 8.0 until no protein is found in the eluates. The amount of protein immobilized is calculated from the difference in the protein content of the enzyme in solution before and after coupling.

[15] P. Cuatrecasas, *J. Biol. Chem.* **245,** 3059 (1970).

PBG-D Assay

Soluble PBG-D. Unless otherwise stated, the standard incubation system contains enzyme preparation 0.7 mg, 50 mM Tris-HCl buffer pH 7.4, 110 μM PBG (220 nmol), and 0.4 ml of a mixture (1 : 1, v/v) of 0.6 M NaCl : 0.12 M MgCl$_2$, in a final volume of 2 ml. Incubations are aerobic, in the dark, with constant mechanical shaking, for 2 hr at 37°. At the end of incubation, concentrated HCl is added to reach a final concentration of 5% (w/v) to precipitate the protein. Removal of precipitate and measurement of porphyrins formed as well as remaining PBG in the supernatants are as described by Kotler *et al.*[16]

Insolubilized PBG-D. Unless indicated, the standard incubation system contains Sepharose-enzyme 2 ml (14.5 mg protein/ml gel), 66.4 μM PBG (332 nmol), and 1 ml of a mixture (1 : 1, v/v) of 0.6 M NaCl : 0.12 M MgCl$_2$ in a final volume of 5 ml. Incubations are aerobic, in the dark, with constant mechanical shaking, at 37° for the following times: 10, 20, 30, 60, and 90 min. After incubation, the mixtures are immediately cooled to 0° and quickly centrifuged for 5 min at 6000g at 4°; the pellets are washed once with 2 ml of cold 50 mM Tris-HCl buffer pH 7.4 and centrifuged again. The corresponding combined supernatants and the pellets thus obtained have been identified as S1, S2, S3, S4, and S5 and P1, P2, P3, P4, and P5, respectively. Measurements of porphyrins formed and remaining PBG in the supernatants are carried out as indicated before.[16] Results are shown in Table I (systems 1 to 5) and Fig. 1A.

Uroporphyrinogen III Synthesis Assay

To determine whether cosynthase can use as substrate the intermediate produced by the action of PBG-D on PBG, either free or still bound to the enzyme, the following experiments are performed. Immobilized PBG-D (124.4 mg protein— 14.64 mg protein/ml of gel) is resuspended in 50 mM Tris-HCl buffer pH 7.4 and incubated with 126 μM PBG (510 μg in 0.85 ml buffer) and 1 ml of the mixture 0.6 M NaCl : 0.12 M MgCl$_2$ (1 : 1, v/v), for 10 min at 37°. Incubation conditions and separation of the supernatant (S) and pellet (P) fractions are as described above. S and P in a final volume of 22.5 ml are incubated with soluble cosynthase (125 mg protein) for another 2 hr. A relation of PBG-D to cosynthase of 1 : 4 (mol : mol) is used to produce over 80% uroporphyrin III. Removal of protein and determination of porphyrins formed in the supernatant are carried out as indicated above.

Protein Determination

Protein concentration is measured by the method of Bradford[17] using bovine serum albumin (BSA) as standard.

---

[16] M. L. Kotler, S. A. Fumagalli, A. A. Juknat, and A. M. C. Batlle, *Int. J. Biochem.* **19,** 981 (1987).

[17] M. M. Bradford, *Anal. Biochem.* **72,** 248 (1976).

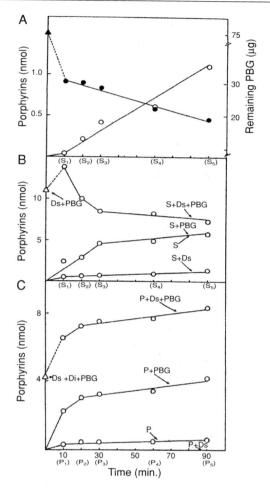

FIG. 1. Urogen I synthesis from free and enzyme-bound intermediates. Porphyrins (○), PBG (●). Incubations were carried out as described in Materials and Methods. Other conditions and abbreviations were as indicated in Table I.

Results and Discussion

Rate of Uroporphyrinogen Synthesis from Soluble and Sepharose-Bound PBG-Deaminase

When both PBGase and PBG-D from soybean callus[9,10] and *E. gracilis*[11] are incubated with PBG, a lag phase in urogen formation occurs, indicating the formation of polypyrrole intermediates. It has also been demonstrated that the

TABLE I
INCORPORATION OF FREE AND PBG-D BOUND INTERMEDIATES INTO UROPORPHYRINOGEN I[a]

	System	Porphyrins		Relative activity	PBG consumption (μg)
Number	Component	Total (nmol)	nmol porphyrin/ mg protein		
1	S1	0.043	0.0015	—	44.88
2	S2	0.209	0.0072	—	44.50
3	S3	0.403	0.0142	—	46.50
4	S4	0.611	0.0210	—	52.42
5	S5	1.110	0.0380	—	55.42
6	Di	1.58	0.054	—	58
7	Ds	10.498	1.74	100	49.55
8	S1	2.4	—	—	—
9	S2	2.9	—	—	—
10	S3	4.5	—	—	—
11	S4	4.9	—	—	—
12	S5	6.0	—	—	—
13	S1 + Ds	0.5	—	—	—
14	S2 + Ds	0.6	—	—	—
15	S3 + Ds	0.7	—	—	—
16	S4 + Ds	0.9	—	—	—
17	S5 + Ds	1.2	—	—	—
18	S1 + PBG	2.45	—	—	—
19	S2 + PBG	2.90	—	—	—
20	S3 + PBG	4.60	—	—	—
21	S4 + PBG	5.10	—	—	—
22	S5 + PBG	6.15	—	—	—
23	S1 + Ds + PBG	12.92	2.15	123	51.37
24	S2 + Ds + PBG	10.07	1.68	97	50.86
25	S3 + Ds + PBG	8.52	1.42	82	50.59
26	S4 + Ds + PBG	8.27	1.37	79	52.64
27	S5 + Ds + PBG	7.24	1.20	69	48.98
28	Ds + Di	4.22	0.12	100	63.58
29	P1	0.30	—	—	—
30	P2	0.45	—	—	—
31	P3	0.45	—	—	—
32	P4	0.50	—	—	—
33	P5	0.60	—	—	—
34	P1 + Ds	ND	—	—	—
35	P2 + Ds	ND	—	—	—
36	P3 + Ds	ND	—	—	—
37	P4 + Ds	ND	—	—	—
38	P5 + Ds	ND	—	—	—
39	P1 + PBG	2.20	—	—	—

TABLE I (continued)

System		Porphyrins			
Number	Component	Total (nmol)	nmol porphyrin/ mg protein	Relative activity	PBG consumption (μg)
40	P2 + PBG	3.00	—	—	—
41	P3 + PBG	3.20	—	—	—
42	P4 + PBG	3.40	—	—	—
43	P5 + PBG	4.15	—	—	—
44	P1 + Ds + PBG	6.469	0.184	153	59.52
45	P2 + Ds + PBG	7.235	0.206	172	59.68
46	P3 + Ds + PBG	7.498	0.214	179	61.66
47	P4 + Ds + PBG	7.677	0.219	183	61.97
48	P5 + Ds + PBG	8.403	0.240	240	61.32

[a] Incubation mixtures contained supernatant (0.5 ml) in 50 mM Tris-HCl buffer pH 7.4 and, when present, PBG (54 μg) and soluble PBG-D (6 mg) in a final volume of 3 ml; pellet (2 ml) in 50 mM Tris-HCl buffer pH 7.4 and, when present, PBG (75 μg) and soluble PBG-D (6 mg) in a final volume of 5 ml. The corresponding controls were also run. Incubations were carried out for 2 hr in the dark, at 37°, with vigorous shaking. All other experimental conditions are described in Materials and Methods. Each assay was performed in duplicate and the experiments were repeated three times. S: 6000g supernatant; P: 6000g precipitate; Ds: soluble PBG deaminase; Di: immobilized PBG deaminase; ND: not detectable. The isomer of urogen I formed was 95–100% type I.

product of the action of PBG-D on PBG, identified as HMB, is the substrate of cosynthase.[6]

The synthesis of porphyrins from soluble bovine liver PBGase and PBG-D increases linearly with time and no intermediates have been detected or isolated.[18] When we compared the properties of both soluble and Sepharose-bound PBG-D from bovine liver, we found that insoluble enzyme bound to Sepharose caused a change in the reaction rate. Porphyrins formed from the soluble enzyme increased linearly with time up to 4 hr incubation (data not shown), but a lag phase was observed for the immobilized PBG-D [Table I (systems 1 to 5) and Fig. 1A]. Up to the first 10 min, the rate of uroporphyrin I synthesis was only 4.3 pmol/min, then it increased linearly with time at a rate of 16.6 pmol porphyrins/min, nevertheless at a rate one-third lower than that of the soluble PBG-D. When we examined PBG consumption, no evidence of a lag phase was found and within the first 10 min as much as 60% of PBG was consumed. Thereafter, consumption occurred at a lower rate; however, after 90 min only 25% of the substrate remained in the system. As already reported for other sources,[19] these findings show a significant

[18] A. M. Stella, V. E. Parera, E. B. C. Llambías, and A. M. C. Batlle, *Biochim. Biophys. Acta* **252**, 481 (1971).

[19] M. L. Kotler, S. A. Fumagalli, A. A. Juknat, and A. M. C. Batlle, *Int. J. Biochem.* **19**, 981 (1987).

deviation from stoichiometry, with a resulting lower porphyrin synthesis and very likely the accumulation of polypyrrole intermediates in the range between PBG and urogens. It appeared therefore that the preparation of immobilized PBG-D could be useful for the study of the reaction mechanism for urogen synthesis by PBG polymerization.

Formation of Polypyrrolic Intermediates

To investigate the formation of polypyrrole intermediates between PBG and urogens, and their presence, either free in solution and/or bound to immobilized PBG-D, the following experiments have been performed. The supernatants and pellets obtained after incubating Sepharose-bound PBG-D with PBG and centrifuged at 6,000g as described in Materials and Methods, identified as S1, S2, S3, S4, S5 and P1, P2, P3, P4, P5, respectively, are treated as shown and detailed in Table I and Figs. 1B and 1C. From the total volume resulting from each supernatant, in a fraction of 0.5 ml, the porphyrins formed and PBG consumed are measured (systems 1 to 5, Table I). Another fraction of 0.5 ml is incubated without any addition (systems 8 to 12, Table I) under the conditions indicated for systems 6 to 48. In systems 13 to 17 (Table I) 0.5 ml of each supernatant is incubated with 1 ml of bovine soluble PBG-D. In systems 18 to 22 (Table I), 0.5 ml of each supernatant is incubated with 54 μg of PBG and in systems 23 to 27 (Table I), supernatants are incubated with soluble PBG-D and PBG. On the other hand, 2 ml of each pellet is incubated without additions (systems 29 to 33, Table I), with 1 ml of soluble PBG-D (systems 34 to 38, Table I), with 75 μg of PBG (systems 39 to 43, Table I), and with PBG-D and PBG (systems 44 to 48, Table I). Controls with Sepharose-enzyme (system 6), soluble PBG-D (system 7), and the mixture of both immobilized and soluble PBG-D (system 28) are run.

As already discussed, the results show evidence of a short lag, within the first 10 min (systems 1 to 5, Table I), in the synthesis of urogens and no lag in PBG consumption. In systems 8 to 12, without any additions (Table I and Fig. 1B) there is an increase in the formation of product, indicating that these porphyrins have been synthesized during the 2 hr incubation from the tetrapyrrylmethane intermediates. These intermediates should have already been present in the supernatants, since total porphyrins present in the corresponding controls (systems 1 to 5 and Fig. 1A) are only between 0.04 and 1.11 nmol. When supernatants are incubated with soluble PBG-D (systems 13 to 17 and Fig. 1B) an 80–85% decrease in urogen synthesis, is found when compared with systems 8 to 12. These results have been attributed to the binding of free tetrapyrrolic intermediates to the soluble enzyme, forming a relatively stable PBG-D–intermediate complex. When supernatants are incubated with PBG (systems 18 to 22 and Fig. 1B), the nonenzymatic cyclization of the intermediates into the final product urogen I increases with time. This is the same as that observed with systems 8 to 12, so, addition of PBG produces no changes. These results support further the presence of free intermediates in the supernatants. When the intermediates are incubated with soluble PBG-D and PBG (systems 23 to 27 and Fig. 1B) some interesting results are obtained. First, S1, the 10 min supernatant, containing the lowest concentration of intermediates, enhances the amount of urogen I (system 23) about 25%, when compared with control (system 7). Second, higher concentrations of free intermediates present in S2 to S5, increasingly inhibit urogen I formation (from 10 to 30%) (systems 24 to 27). Similar to systems 13 to 17, binding of free intermediates to PBG-D inhibits the enzyme activity. We have found that under normal incubation conditions, the *E. gracilis* PBG-D is able to form free polypyrrole intermediates, which do not behave as substrates but as actual intermediates in urogen I synthesis and that these intermediates strongly inhibit product formation.[11,12] Further support for this proposal has been found when a mixture of soluble and insoluble PBG-D is incubated (system 28). Here, activity is only 34% of the expected activity (system 6 plus 7), indicating that the free intermediates formed by both PBG-Ds can bind and inhibit the soluble enzyme.

When the 6000g pellets (P1 to P5) (systems 29 to 33 and Fig. 1C) are incubated without additions, uroporphyrin I formation increases with time, from only 0.3 to 0.6 nmol total, indicating the existence of a stable PBG-D–intermediate complex. When the same set of incubations are performed in the presence of soluble PBG-D (systems 34 to 38 and Fig. 1C) porphyrins are undetectable. These results can be explained on the same basis as those observed for the behavior of intermediates from systems 8 to 12 and 13 to 17 and results found for systems 39 to 43 and Fig. 1C. The polypyrrolic intermediates bound to the insoluble PBG-D are liberated to the solution in the presence of the substrate PBG, which is the only reason why the amount of total uroporphyrin I formed is above 7 times higher than that in controls (systems 29 to 33) and between 1.4- and 2.6-fold with respect to the control of immobilized PBG-D (system 6). These findings also agree with those of Battersby et al.,[20] who have demonstrated that binding of one PBG molecule to PBG-D after HMB synthesis leads to its release from the protein. Finally the presence of a PBG–polypyrrole intermediate complex in the pellet fractions P1 to P5 is further confirmed when these pellets are incubated with soluble enzyme and PBG (systems 44 to 48 and Fig. 1C); here the amount of urogen I synthesized is twice that in the control (system 28).

When we reanalyzed PBG consumption, in all systems where either the supernatants (systems 23 to 27) or the pellets (systems 44 to 48) were incubated with soluble PBG-D and PBG or in the controls (systems 6, 7, and 28), PBG between 10 and 20% remained showing that stoichiometric synthesis of urogens was not dependent on the amount of substrate consumed.[19]

---

[20] A. R. Battersby, C. J. R. Fookes, G. Hart, G. W. Matcham, and P. S. Pandey, *J. Chem. Soc. Perkin Trans.* **1**, 3041 (1983).

TABLE II
SYNTHESIS OF UROGEN III FROM FREE AND PBG-D BOUND
POLYPYRROLE INTERMEDIATES AND COSYNTHASE[a]

System		Uroporphyrins		
		Total (nmol)	Isomer type (%)	
Number	Components		I	III
1	Cosynthetase	—	—	—
2	S	3.00	100	0
3	P	2.10	100	0
4	S + Cosynthetase	3.22	25	75
5	P + Cosynthetase	29.51	20	80

[a] The preparation of the supernatant (S) and pellet (P), purified cosynthase, and the composition of the incubation systems are described in Materials and Methods. The corresponding controls were also run. Assays were carried out in duplicate and experiments repeated at least three times.

Immobilized PBG-D Forming Polypyrrolic Intermediates That Are Substrates for Cosynthase

Experiments employing fractions S and P, in the absence or presence of cosynthase, as shown in Table II, were performed to show that the polypyrrolic intermediates formed by immobilized PBG-D are substrates of cosynthase.

Cosynthase alone can use free as well as PBG-D-bound intermediates as substrates, forming 75–80% urogen III (systems 4 and 5). As expected, 100% urogen I is produced by the controls S and P without cosynthase (systems 2 and 3).

These findings are also indicate that the free intermediate released from insolubilized PBG-D (system 2) could be identified with the hydroxymethylbilane, which appears to be completely isomerized by cosynthase without change of activity (system 4). When the incorporation of the enzyme-bound intermediates into urogen III was examined, it was found that release of tetrapyrrolic intermediates was more efficient when cosynthase was present (system 5) yielding 15 times more porphyrins than the corresponding control (system 3). The intermediate bound to insoluble PBG-D may have a more reactive structure, such as that of an enzyme–methylenepyrrolenine complex (Fig. 2). There might also be an association and/or interaction between PBG-deaminase and cosynthetase, which would favor the transfer of the PBG-D bound intermediate to cosynthase, thus enhancing urogen III formation. However, using both soluble and immobilized PBG-D and cosynthase, no evidence has been obtained on the formation of a stable deaminase–cosynthase complex.[21]

---

[21] M. L. Kotler, A. A. Juknat, and A. M. C. Batlle, unpublished results, 1992.

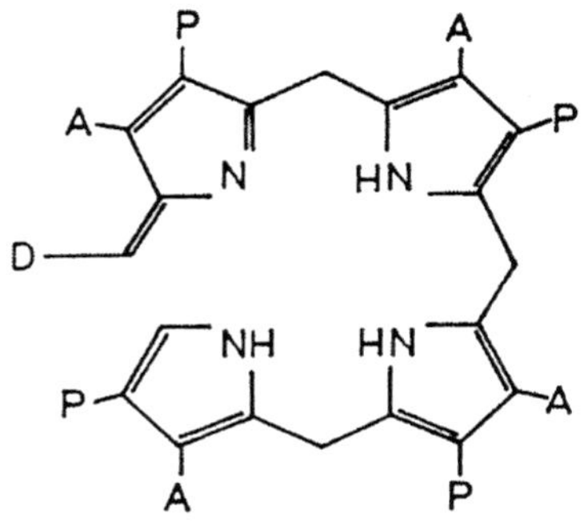

FIG. 2. Polypyrrole intermediate bound to PBG-D.

Conclusions

From the kinetics of the soybean callus PBGase and PBG-D[9,10] it was postulated that synthesis of urogens from PBG would occur in two steps: first, the formation of an intermediate tetrapyrrylmethane, identified with HMB,[22] and second, its final cyclization and production of urogen.

We have shown that immobilized bovine liver PBG-D leads to the detection and isolation of free and enzyme-bound linear and unrearranged tetrapyrrylmethane intermediates, and that their distribution and concentration are time dependent.

It has also been confirmed that free intermediates can bind the enzyme and that in doing so, they could act as inhibitors depending on their concentration. Incubations of the PBG-D bound intermediates with PBG-D and PBG increase porphyrin synthesis. These results clearly indicate that the intermediates bound to the enzyme actually behave as normal intermediates enhancing urogen formation. This is in keeping with previous findings obtained with the *E. gracilis* enzymes,[11,12] in which it was also reported that the enzyme-bound polypyrrole intermediates were only released when they had reached the length of the open linear tetrapyrrole and they were ready for cyclization to urogen.

It has been shown here that once the linear tetrapyrrylmethane intermediate has been synthesized, the binding of PBG to PBG-D leads to the rapid release of this HMB from the immobilized PBG-D, in agreement with the report of Battersby *et al.* using *E. gracilis* soluble PBG-D.[20]

It has been also demonstrated that both free and PBG-D-bound tetrapyrrylmethane intermediates were good substrates for soluble cosynthase, adding further support to our hypothesis.

The results presented here are of relevance for the potential usefulness of Sepharose-immobilized enzymes in the study of reaction mechanisms of enzymes.

---

[22] A. R. Battersby, C. J. R. Fookes, G. W. J. Matcham, E. McDonald, and K. E. Gustafson-Po Her, *J. Chem. Soc., Chem. Commun.*, 316 (1979).

Acknowledgments

The author wishes to thank Dr. M. L. Kotler, Dr. S. A. Fumagalli, Dr. A. A. Juknat, and her former Ph.D. students, for their most important collaboration in the accomplishment of this work. Part of the work is taken from the doctoral thesis submitted by M.L.K. for a Ph.D. degree from the University of Buenos Aires. This work was supported by grants from the CONICET and the University of Buenos Aires.

[28] Adenosylcobalamin-Dependent Glutamate Mutase: Pre-Steady-State Kinetic Methods for Investigating Reaction Mechanism

By HUNG-WEI CHIH, IPSITA ROYMOULIK, MARJA S. HUHTA, PRASHANTI MADHAVAPEDDI, and E. NEIL G. MARSH

Introduction

Glutamate mutase (EC 5.4.99.1) catalyzes the reversible interconversion of L-glutamate to L-*threo*-3-methylaspartate as the first step in the fermentation of L-glutamate by various species of clostridia [Eq. (1)].[1–6] It is one of a group of adenosylcobalamin (AdoCbl, coenzyme B_{12})-dependent isomerases that catalyze unusual isomerizations in which a hydrogen atom on one carbon atom is interchanged with an electron-withdrawing group, X, on an adjacent carbon [Eq. (2)].[7–9]

$$\text{L-glutamate} \rightleftharpoons \text{L-threo-3-methylaspartate} \quad (1)$$

$$\underset{C_1—C_2}{\overset{H\ \ X}{|\ \ |}} \rightleftharpoons \underset{C_1—C_2}{\overset{X\ \ H}{|\ \ |}} \quad (2)$$

---

[1] H. A. Barker, V. Rooze, F. Suzuki, and A. A. Iodice, *J. Biol. Chem.* **239**, 3260 (1964).
[2] W. Buckel and H. A. Barker, *J. Bacteriol.* **117**, 1248 (1974).
[3] H. A. Barker, *Methods Enzymol.* **113**, 121 (1985).
[4] D. E. Holloway and E. N. G. Marsh, *J. Biol. Chem.* **269**, 20425 (1994).
[5] H. P. Chen and E. N. G. Marsh, *Biochemistry* **36**, 14939 (1997).
[6] E. N. G. Marsh, *Bioorg. Chem.* **28**, 176 (2000).
[7] E. N. G. Marsh, *Bioessays* **17**, 431 (1995).
[8] R. Banerjee, *Chem. Biol.* **4**, 175 (1997).
[9] E. N. G. Marsh, *Essays Biochem.* **34**, 139 (1999).

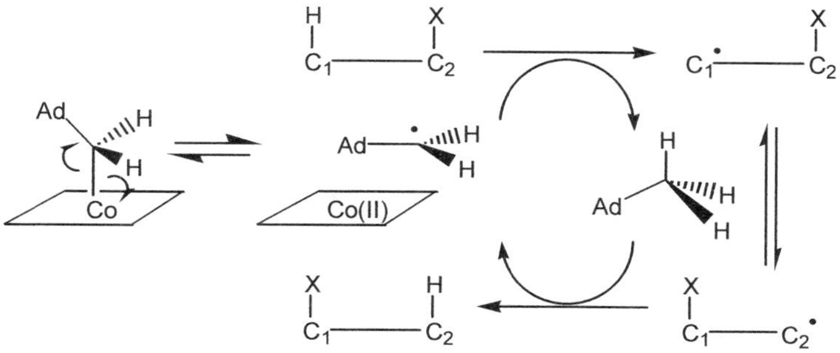

FIG. 1. Minimal mechanistic scheme for the rearrangement reactions catalyzed by AdoCbl-dependent isomerases.

X may be a hydroxyl or an amino group, as in the case of the reactions catalyzed by diol dehydrase[10] and ethanolamine deaminase,[11] or, most intriguingly, a carbon-containing fragment as in the case of the reactions catalyzed by glutamate mutase, 2-methyleneglutarate mutase, isobutyryl-CoA mutase, and methylmalonyl-CoA mutase.[12] Although the enzymes are mainly confined to various fermentation pathways in bacteria, methylmalonyl-CoA mutase is also an essential enzyme in animals where it plays a role in the metabolism of odd-chain fatty acids by converting methylmalonyl-CoA to succinyl-CoA.[13]

AdoCbl-dependent rearrangements proceed through a mechanism involving free radical intermediates as illustrated in Fig. 1. The catalytic cycle is initiated by homolysis of the unique cobalt–carbon bond of the coenzyme to generate a 5′-deoxyadenosyl radical. The adenosyl radical then abstracts the migrating hydrogen from the substrate to give a substrate radical and 5′-deoxyadenosine. Next, in a poorly understood step, the substrate radical undergoes rearrangement to form the product radical. The product radical then reabstracts a hydrogen from 5′-deoxyadenosine to give the product and regenerate the adenosyl radical. Finally, the adenosyl radical recombines with the cobalt to regenerate the coenzyme, thereby completing the catalytic cycle.

The B_{12}-dependent isomerases are one group of a larger, newly recognized class of enzymes that use carbon-based free radicals to catalyze a variety of chemical transformations on otherwise unreactive substrates.[7,14] In this context, glutamate mutase provides one of the simplest systems with which to study the phenomenon

---

[10] T. Toraya, *Cell. Mol. Life Sci.* **57**, 106 (2000).
[11] V. Bandarian and G. H. Reed, *Biochemistry* **39**, 12069 (2000).
[12] W. Buckel and B. T. Golding, *Chem. Soc. Rev.,* 329 (1996).
[13] C. L. Drennan, R. G. Matthews, D. S. Rosenblatt, F. D. Ledley, W. A. Fenton, and M. L. Ludwig, *Proc. Natl. Acad. Sci. U.S.A.* **93**, 5550 (1996).
[14] J. A. Stubbe, *Ann. Rev. Biochem.* **58**, 257 (1989).

of radical-mediated enzymatic catalysis: the substrates are small, stable molecules, the reaction is freely reversible, and the enzyme requires no cofactors other than AdoCbl.[4]

One important and poorly understood aspect of AdoCbl-dependent enzymes is their remarkable ability to stabilize what are normally considered highly reactive free radical species. Under conditions of steady-state turnover most AdoCbl enzymes accumulate significant concentrations of free radicals, leading to the conclusion that the protein must destabilize the cobalt–carbon bond of the coenzyme by about 30 kcal/mol.[15] Over the past few years we have undertaken a series of pre-steady-state kinetic experiments to clarify the mechanism of this unusual rearrangement and to examine the free energy profile for glutamate mutase.[16–19] These experiments, together with the recently determined crystal structure of the enzyme,[20] have begun to shed light on the question of how the enzyme stabilizes free radical species and directs them toward productive catalysis.

Kinetic Approach to Analyzing Reaction Mechanism

A detailed mechanistic scheme for the isomerization catalyzed by glutamate mutase is shown in Fig. 2. The reaction is freely reversible, and thus L-glutamate and L-*threo*-3-methylaspartate may be thought of as either substrates or products. Using a combination of UV–visible stopped-flow spectroscopy together with rapid chemical quench experiments, and by examining the effects of substituting the migrating hydrogen atoms with deuterium, we have been able to deduce information about each of the putative steps in this mechanism. Below we outline our approach.

The first step in the mechanism involves the substrate binding to the enzyme (species **I** and **VII** in Fig. 2). As this is a second-order process, it may be investigated by examining how varying the substrate concentration affects the pre-steady-state kinetics of the subsequent chemical step, e.g., cobalt–carbon bond homolysis. Plots of k_{obs} for the chemical step against substrate concentration can yield apparent dissociation constants for the enzyme:substrate complex and, in favorable circumstances, the on and off rates for the substrate. The first chemical step is homolytic fission of the Co–C bond of AdoCbl to produce adenosyl radical (species **II** and **VI**). This key step involves the formal reduction of octahedral Co(III) to a five-coordinate Co(II) species and results in extensive changes

---

[15] B. P. Hay and R. G. Finke, *J. Am. Chem. Soc.* **109**, 8012 (1987).
[16] E. N. G. Marsh and D. P. Ballou, *Biochemistry* **37**, 11864 (1998).
[17] H. W. Chih and E. N. G. Marsh, *Biochemistry* **38**, 13684 (1999).
[18] I. Roymoulik, N. Moon, W. R. Dunham, D. P. Ballou, and E. N. G. Marsh, *Biochemistry* **39**, 10340 (2000).
[19] H.-W. Chih and E. N. G. Marsh, *J. Am. Chem. Soc.* **122**, 10732 (2000).
[20] R. Reitzer, K. Gruber, G. Jogl, U. G. Wagner, H. Bothe, W. Buckel, and C. Kratky, *Structure* **7**, 891 (1999).

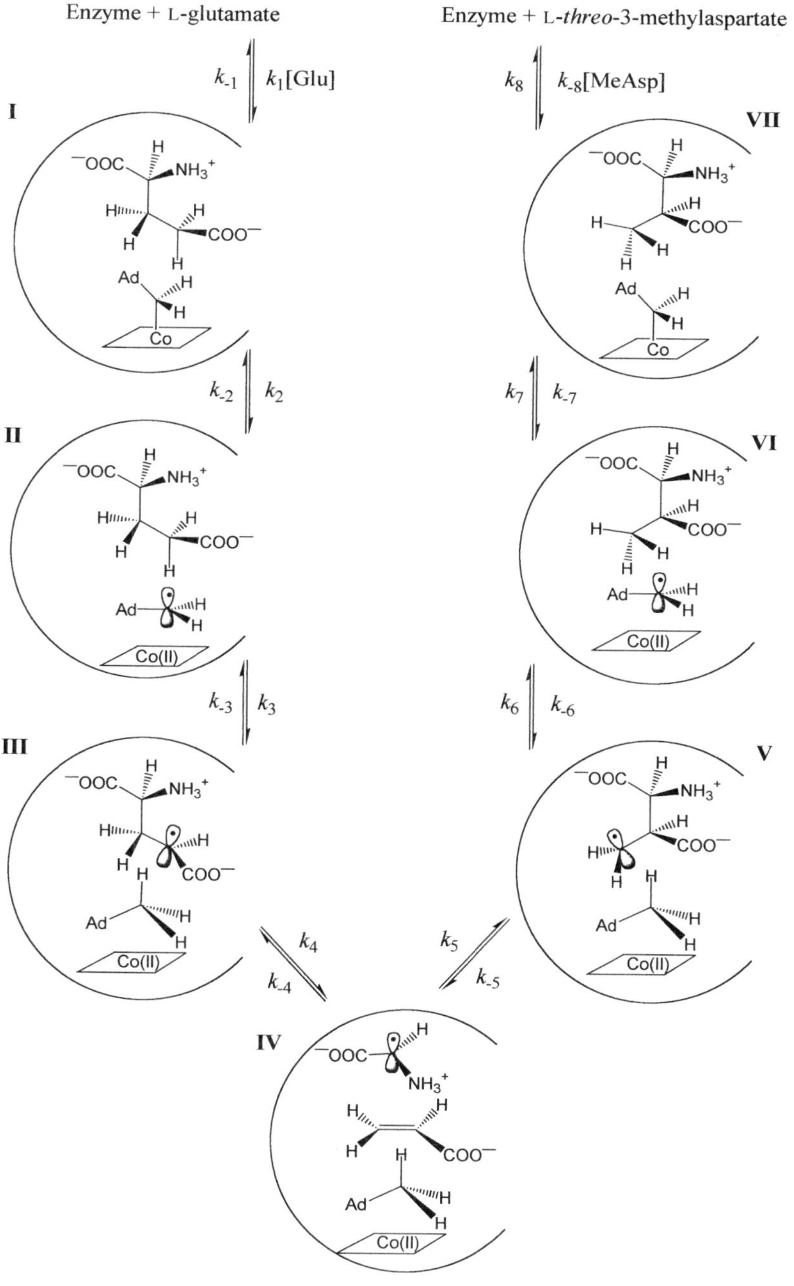

FIG. 2. The mechanistic and kinetic scheme for the rearrangement of L-glutamate to L-*threo*-3-methylaspartate catalyzed by glutamate mutase. Reprinted with permission from E. N. G. Marsh and D. P. Ballou, *Biochemistry* **37**, 11864 (1998). Copyright © 1998 American Chemical Society.

to the UV–visible spectrum of the coenzyme that allow Co–C bond homolysis to be followed by stopped-flow spectroscopy. The next step involves the transfer of hydrogen from glutamate (or methylaspartate) to form 5′-deoxyadenosine (5′-dA) and the substrate radical (species **III** and **V**). 5′-dA is a stable intermediate and we have detected its formation by rapid quench methods. This step is also subject to an isotope effect when appropriately deuterated substrates are used, and this has provided valuable insights into the reaction mechanism. The rearrangement of glutamyl radical to methylaspartyl radical is proposed to occur through the intermediacy of a glycyl radical and acrylate (species **IV**); here again, we have used rapid quench methodology to examine the formation of these intermediates. The final step is product release, which can often be rate determining in enzymatic reactions. We have investigated this step using rapid quench experiments to determine whether burst phase kinetics, indicative of slow product release, are observed.

The glutamate mutase-catalyzed reaction is readily reversible, with the equilibrium constant favoring glutamate by $\sim$12 to 1.[1] This has the advantage that the reaction can be examined in either direction, but introduces some complications in the interpretation of pre-steady-state kinetic data. The mechanism shown in Fig. 2 may be viewed as a series of eight coupled equilibria that requires 16 elementary rate constants to fully describe it. The addition of one or other substrate to the enzyme in essence perturbs the overall equilibrium, which then relaxes toward a new steady state and eventually, if the reaction is followed for long enough, toward true chemical equilibrium. Pre-steady-state experiments measure the relaxation time, τ, for an intermediate to reach a new equilibrium or steady-state concentration where $\tau = 1/k_{\text{obs}}$, the observed first-order rate constant.[21] In general, k_{obs} is determined by both the rates of reactions leading to the formation of a given intermediate and the rates of reactions leading to its breakdown, such that $k_{\text{obs}} = \Sigma(k_{\text{forward}} + k_{\text{reverse}})$. Therefore, k_{obs} may not necessarily correspond to a simple combination of the elementary rate constants used to describe the reaction in Fig. 2. It is important to bear this in mind when interpreting pre-steady-state kinetic data for enzyme-catalyzed reactions in which the equilibrium constant is close to one, as typified by glutamate mutase.

Although the wild-type enzyme is an E_2S_2 heterotetramer, the S subunits bind rather weakly and reversibly to the E_2 dimer. As a result, both the stoichiometry and the apparent K_d for AdoCbl, as well as the specific activity of the enzyme, are dependent on the relative concentrations of the two subunits.[4] To avoid this complication we engineered a glutamate mutase fusion protein in which the S subunit is joined to the C terminus of the E subunit through a flexible linker.[5] This $(E-S)_2$ dimer has similar steady-state kinetic properties to the wild-type enzyme, but importantly, neither the affinity for AdoCbl nor the turnover number for the enzyme depend on protein concentration. This enzyme is therefore much better

---

[21] C. A. Fierke and G. G. Hammes, *Methods Enzymol.* **249**, 3 (1995).

suited to characterization by pre-steady-state methods and was used in all the experiments described here.

Rapid Kinetic Methods

UV–Visible Stopped-Flow Spectroscopy

Experiments are performed at 10° with a Hi-Tech Scientific (U.K.) SF-61 stopped-flow apparatus controlled by KISS, a Kinetic Instruments Macintosh-based software suite. The temperature of the mixing chamber is controlled by a circulating water bath. The enzyme solution contains 125 μM glutamate mutase in 50 mM potassium phosphate buffer containing 1 mM EDTA and 10% (v/v) glycerol. Immediately before the experiment AdoCbl is added to a final concentration of 100 μM so that the effective concentration of holoenzyme is 100 μM (the K_d for AdoCbl is 2 μM; therefore these concentrations are sufficiently high that essentially all the coenzyme is bound by the protein). Solutions containing AdoCbl are handled so as to avoid exposure to bright light. The solution is placed in a glass tonometer and made anaerobic by repeated cycles of evacuation and flushing with purified argon. Substrates are dissolved in the same buffer as the enzyme, placed in glass syringes, and made anaerobic by bubbling purified argon through them for 10 min before use. Mixing in the stopped-flow apparatus dilutes both substrate and enzyme twofold, so that the concentration of holoenzyme in the measured reaction mixture is 50 μM.

In general, the reaction is monitored by following the decrease in absorbance at 530 nm that accompanies cobalt–carbon bond homolysis [although the reaction can equally well be followed at monitoring the increase in absorbance at 470 nm due to the formation of Cbl(II)]. For each concentration of substrate used, the data from at least three shots are averaged and fitted to either single or multiple parallel exponential functions to obtain rate constants using the program KISS. Secondary plots of data and curve fitting are performed using the Kaleidagraph program (Abelbeck Software).

Rapid Chemical Quench Experiments

Rapid quench experiments are performed at 10° with a Hi-Tech Scientific (UK) RQF-63 rapid mixing apparatus. The temperature of the mixing chamber is maintained at 10° using a circulating water bath. The enzyme solution contains 200 μM glutamate mutase in 50 mM potassium phosphate buffer, pH 7.0. Immediately before the experiment AdoCbl is added to a final concentration of 240 μM. The effective concentration of holoenzyme after mixing is 95 μM. The substrate solution contains either L-glutamate or L-*threo*-3-methylaspartate at twice their desired final concentration dissolved in the same buffer as the enzyme and also contains 200 μM L-leucine as the internal standard. Reactions are started by mixing 80 μl of enzyme solution with 80 µl of substrate solution. The solution is allowed to age for various times (5–1600 ms) before being quenched with a further 80 µl of 5% trifluoroacetic acid. Samples are stored at −20° prior to derivatization and HPLC analysis. Solutions containing AdoCbl are handled so as to avoid exposure to bright light.

Derivatization and HPLC of Amino Acids

The pH of the quenched reaction mixture (240 µl) is raised above 8.5 with 200 µl of 1 M NaHCO$_3$ and the solution treated with ≈5 mg of activated charcoal to remove AdoCbl. The charcoal is removed by centrifugation and 100 µl of a 100 mg/ml solution of dansyl chloride in acetone added to the supernatant. The sample is incubated at 45° for 1 hr to derivatize the substrate, product, and internal standard. Dansylated glutamate, methylaspartate, glycine, and leucine are separated by HPLC using a 25 cm C$_{18}$ reversed-phase column (Spherisorb S5 ODS2). The column is preequilibrated in 92.5% solvent A (25 mM potassium phosphate buffer containing 7% v/v acetonitrile and 3% v/v methanol) and 7.5% solvent B (70% v/v acetonitrile, 30% v/v methanol). Thirty to 50 µl of sample is injected onto the column and the derivatized amino acids eluted with an ascending gradient of solvent B as follows: 0–10 min, 7.5% B; 10–22 min, 7.5–50% B; 22–26 min, 50–80% B. The flow rate is 1.2 ml/min and compounds are detected by monitoring absorbance at 360 nm. The peaks areas are determined by computer integration and standardized relative to the leucine internal standard. The amount of product formed is calculated from standard curves constructed by derivatization and chromatography of known amounts of either glutamate or methylaspartate. In general, three portions of the quenched, derivatized sample are chromatographed separately, and the average of the standardized peak areas used to determine the amount of product formed.

Analysis of 5'-Deoxyadenosine (5'-dA)

Rapid quenched flow experiments are conducted and analyzed as described above but with the following modifications to the reaction and chromatography conditions. The enzyme solution contains 100 µM glutamate mutase and 120 µM AdoCbl so that the effective concentration of holoenzyme after mixing is 45 µM. The substrate solutions contain 20 µM L-tryptophan as an internal standard. 5'-dA, AdoCbl, and L-tryptophan are separated by HPLC using a 25 cm C$_{18}$ reversed-phase column (Spherisorb S5 ODS2). The column is preequilibrated in 0.1% trifluoroacetic acid (TFA) and eluted with a 20 ml linear gradient of 0–45% acetonitrile containing 0.1% trifluoroacetic acid. 5'-dA elutes at ≈10%, L-tryptophan at 40%, and AdoCbl at 45% acetonitrile. The peaks areas are determined by computer integration and standardized relative to the tryptophan internal standard. The amount of 5'-dA in the sample is calculated from standard curves obtained by chromatography of known amounts of 5'-dA.

Analysis of Acrylate and Glycine

Rapid quench flow experiments are performed by rapidly mixing 80 μl of a solution containing 200 μM glutamate mutase, 240 μM AdoCbl in 50 mM potassium phosphate buffer, pH 7.0, with an equal volume of 2 mM uniformly ^{14}C-labeled L-glutamate (specific activity 5800 dpm/nmol) containing 20 μM L-tryptophan and 200 μM L-leucine as internal standards. The solution is allowed to age for various times (5–400 ms) before being quenched with a further 80 μl of 5% TFA containing either glycine (0.4 mM final concentration) or acrylate (0.08 mM final concentration) as carrier. For the analysis of glycine samples are first treated with charcoal to remove AdoCbl before dansylation as described above. Acrylate is recovered directly after quenching by HPLC on a reversed-phase C_{18} column equilibrated in 0.1% TFA and eluted with an ascending gradient of acetonitrile.

Note on Experimental Design

In the design of rapid quench experiments we have found it very important to include an appropriate internal standard in the sample. A typical experiment to analyze the formation of methylaspartate, for example, involves initial mixing of enzyme and substrate, and subsequent quenching in the quench flow apparatus; the sample is then derivatized with dansyl chloride and finally chromatographed by reversed-phase HPLC. Errors can arise from small differences in the mixing volumes in the quench flow apparatus, pipetting errors during derivatization, and losses of sample on the HPLC column. When combined, these systematic errors can severely compromise the integrity of the experimental data. A good internal standard is one that does not interact with the enzyme, undergoes the same derivatization procedures as the sample, and chromatographs under the same conditions. In our experiments that analyze amino acids we typically use L-leucine as an internal standard. The amount of L-leucine in the sample is determined by comparison with a standard curve constructed using known concentrations of commercially available dansyl-leucine. This allows the losses during sample handling to be calculated and hence corrected.

Pre-Steady-State Formation of Cob(II)alamin

Experiments to follow the time course of Cbl(II) formation using stopped-flow spectroscopy have proved very informative. A set of stopped-flow traces obtained when the enzyme was reacted with various concentrations of L-*threo*-methylaspartate at 10° are shown in Fig. 3. Although not readily discernable from Fig. 3, there is an initial very fast decrease in absorbance that is nearly complete within the dead time of the spectrophotometer. Next, there is a further rapid decrease in absorbance at 530 nm, with a rate and amplitude that approach limiting values as substrate concentration increases. These first two phases are of

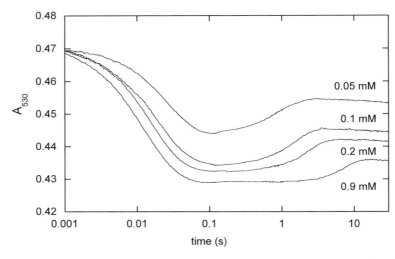

FIG. 3. The reaction of hologlutamate mutase (50 μM) with various concentrations of methylaspartate. The formation of Cbl(II) was followed monitoring the decrease in absorbance at 530 nm. *Note:* The time axis is plotted on a logarithmic scale. Reprinted with permission from E. N. G. Marsh and D. P. Ballou, *Biochemistry* **37**, 11864 (1998). Copyright © 1998 American Chemical Society.

approximately equal amplitude and are attributable to homolysis of the Co–C bond to form Cbl(II). The biphasic nature of the homolysis reaction is most likely due to negative cooperativity arising from the dimeric structure of the enzyme. Because the cobalt centers of the two active sites are separated by about 45 Å,[20] this suggests that the reaction is very sensitive to what are most likely subtle changes in protein structure involved in substrate binding.

Homolysis is followed by a period of time in which very little absorbance change occurs and that increases in length when more substrate is present. This phase can be attributed to steady-state turnover during which the enzyme is using the excess substrate. Note that in the top trace, where enzyme and substrate are equimolar, essentially no steady-state phase is observed. Finally, there is a slower partial recovery of absorbance that is also concentration dependent; at higher concentrations of substrate the recovery is slower and the amplitude of the recovery phase is diminished.

The recovery phase probably reflects the establishment of the final chemical equilibrium between methylaspartate, glutamate, and the various enzyme : substrate complexes. Because glutamate is bound about 10-fold less tightly than methylaspartate, and because the equilibrium favors glutamate by 12 : 1, once formed, some glutamate diffuses from the enzyme and decreases the total concentration of the enzyme : substrate complex, and hence the amount of enzyme in the Cbl(II) form. The ratios of methylaspartate to enzyme used were 1 : 1, 2 : 1, 4 : 1, and 18 : 1; thus at higher concentrations of methylaspartate multiple turnovers are needed to establish chemical equilibrium. Also, as the initial concentration of methylaspartate, and hence the final concentration of glutamate, increases, the equilibrium concentrations of the enzyme:methylaspartate and enzyme:glutamate complexes, both of which have some Cbl(II) character, also increase. This is evident from the kinetic traces that show a longer steady-state region with a greater final proportion of enzyme remaining in the Cbl(II) form at equilibrium with higher concentrations of substrates.

Stopped-flow experiments with various concentrations of L-glutamate gave very similar results (data not shown).[16] Again, there was evidence for negative cooperativity in the reactions of the two enzyme active sites. A very fast phase that accounted for about half of the total amplitude change occurred within the dead time of the spectrophotometer, followed by a slower, concentration-dependent phase, the rate constant for which tended toward an upper limit at as the glutamate concentration was increased. In contrast to the reaction with methylaspartate, no recovery of absorbance was observed at the end of the reaction. This may be understood by realizing that the reaction is starting close to equilibrium, since equilibrium favors glutamate, so that the concentration of glutamate changes very little during the course of the experiment.

Most interestingly, when the stopped-flow experiments were repeated with glutamate and methylaspartate in which deuterium replaced the migrating hydrogen, the rate of AdoCbl homolysis was significantly decreased. This is illustrated in Fig. 4 for the reaction of the enzyme with protiated and deuterated glutamate; with the deuterated substrate the reaction is slowed sufficiently that now both the fast and slow sites can be observed.

The fact that deuterated substrates react with the enzyme to induce AdoCbl homolysis at much slower rates than protiated substrates is at first sight surprising, since homolysis does not formally involve hydrogen abstraction from the substrate. However, as noted earlier, because the glutamate mutase-catalyzed reaction is readily reversible, the various steps in the isomerization should be considered as a series of coupled equilibria. Therefore, addition of the substrate perturbs the entire series of equilibria shown in Fig. 2. The establishment of the new equilibrium involves all of the rate constants in the system, including k_3, k_{-3}, k_6, and k_{-6}, which are sensitive to isotopic substitution. Thus, the isotope effect arises because the adenosyl radical reacts more slowly with the deuterated substrate so that the equilibrium is established more slowly. However, for a large isotope effect to be observed on the formation of Cbl(II), the formation of species **II** or **VI** in Fig. 2 must be both rapid *and* energetically unfavorable compared with the subsequent formation of species **III** or **V**. In other words, the adenosyl radical can exist only in very low concentration as a high energy intermediate.

An alternative explanation for the isotope effect is that Co–C bond cleavage and hydrogen abstraction occur in a single, concerted step, so that the 5'-deoxyadenosyl

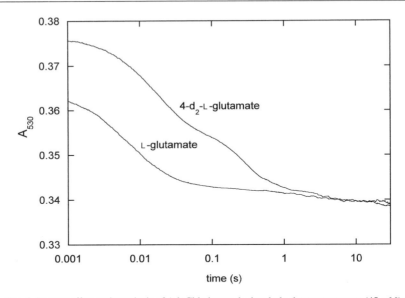

FIG. 4. Isotope effect on homolysis of AdoCbl observed when hologlutamate mutase (45 μM) was reacted with either protiated glutamate (4.5 mM) or deuterated glutamate (4.5 mM). With protiated substrate part of the reaction occurs within the dead time of the spectrophotometer. Reprinted with permission from E. N. G. Marsh and D. P. Ballou, *Biochemistry* **37**, 11864 (1998). Copyright © 1998 American Chemical Society.

radical, species **II** and **VI** in Fig. 2, are never actually formed. However, in the recently determined crystal structure of the enzyme the substrate binding site appears to be too far from the cobalt atom for a truly concerted reaction.[20] Therefore, whereas these experiments do not permit one to distinguish between kinetically coupled and concerted mechanisms for homolysis and hydrogen transfer, we consider the former mechanism more likely.

The isotope effects on homolysis (calculated for the slow site) are 28 and 35 with [4-^{2}H$_2$]glutamate and [^{2}H$_3$-*methyl*]methylaspartate, respectively.[16] These isotope effects are much larger than the steady-state effects[5] ($^DV = 4.0$ and 6.3 for glutamate and methylaspartate, respectively) and are much larger than would be expected simply from differences in zero-point vibrational energies of the reactive bonds. Most likely, they arise from quantum tunneling of the hydrogen from the substrate to the 5′-carbon of adenosine. Hydrogen tunneling is seen to be important in the mechanisms of an increasing number of enzymes that catalyze proton or hydride transfers,[22] and particularly in enzymes that catalyze hydrogen atom transfers. Tunneling has been demonstrated for AdoCbl-dependent

[22] A. Kohen and J. P. Klinman, *Acc. Chem. Res.* **31**, 397 (1998).

TABLE I
KINETIC DATA FOR GLUTAMATE MUTASE

Substrate	Kinetic parameter	Value	Deuterium isotope effect
L-Glutamate	k_{cat}	8.5 ± 0.3 s$^{-1\,a,d}$	3.9 ± 0.3^b
L-Glutamate	K_m	0.58 ± 0.08 mM^b	n.a.^{f}
L-Glutamate	k_{cat}/K_m	$(1.5 \pm 0.2) \times 10^4$ M^{-1}s^{-1}	4.0 ± 1.0^b
Methylaspartate	k_{cat}	7.6 ± 0.2 s$^{-1\,a,d}$	6.3 ± 0.5^b
Methylaspartate	K_m	0.14 ± 0.02 mM^b	n.a.^{f}
Methylaspartate	k_{cat}/K_m	$(5.4 \pm 0.8) \times 10^4$ M^{-1}s^{-1}	3.4 ± 0.7^b
L-Glutamate	K_d	0.61 ± 0.07 mM^c	n.a.^{f}
Methylaspartate	K_d	0.037 ± 0.005 mM^c	n.a.^{f}
L-Glutamate	$k_{obs}(\mathbf{I}\rightarrow\mathbf{II})^e$	97 ± 5 s$^{-1\,c}$	28^c
L-Glutamate	$k_{obs}(\mathbf{I}\rightarrow\mathbf{III})$	73 ± 8 s$^{-1\,a}$	n.d.^{f}
L-Glutamate	$k_{obs}(\mathbf{I}\rightarrow\mathbf{IV})$	35 ± 8 s$^{-1\,c}$	n.a.^{f}
Methylaspartate	$k_{obs}(\mathbf{VII}\rightarrow\mathbf{VI})$	80 ± 4 s$^{-1\,c}$	35^c
Methylaspartate	$k_{obs}(\mathbf{VII}\rightarrow\mathbf{V})$	64 ± 11 s$^{-1\,a}$	n.d.^{f}

^{a} Data taken from Ref. 17.
^{b} Data taken from Ref. 5.
^{c} Data taken from Ref. 16.
^{d} Slightly lower values for k_{cat} have also been reported using a coupled assay under slightly different conditions.^{5}
^{e} The Roman numerals associated with k_{obs} refer to the various enzyme intermediates illustrated in Fig. 2.
^{f} n.a.: Not applicable; n.d.: not determined.

methylmalonyl-CoA mutase, an enzyme that also exhibits very large deuterium isotope effects.[23] This suggests that coupling between Co–C bond cleavage and hydrogen abstraction involving quantum tunneling may be a general phenomenon in AdoCbl-dependent enzymes.

The kinetic data obtained in these stopped-flow experiments may be fitted and analyzed by standard methods to obtain apparent dissociation constants for glutamate and methylaspartate and the observed rate constants, k_{obs}, for Cbl(II) formation for each substrate. These kinetic data are summarized in Table I. From the magnitude of the overall absorbance change the equilibrium constant, K_{eq}, for Cbl(II) formation on the enzyme can be calculated. Knowing both k_{obs} and K_{eq} allows k_{obs} to be deconvoluted into the apparent forward and reverse rate constants for Cbl(II) formation. However, the isotope effect data show that $k_{forward}$ and $k_{reverse}$ do not correspond to k_2 and k_{-2} or k_7 and k_{-7} for the reaction scheme in Fig. 2, but better approximate the rate constants associated with the transfer of hydrogen between substrate and coenzyme (k_3 and k_{-3} or k_6 and k_{-6}).

[23] S. Chowdhury and R. Banerjee, *J. Am. Chem. Soc.* **122**, 5417 (2000).

Rapid Quench Flow Experiments

Rapid quench flow experiments might be considered the "pre-steady-state kinetic technique of last resort!" Typically, this technique uses much more enzyme than spectroscopic methods and has the added disadvantages that data are obtained discontinuously and that the analysis of each data point can be time consuming and labor-intensive. Nevertheless, with careful experimental design and by including appropriate internal standards, accurate kinetic data can be obtained using this technique for enzyme intermediates that could not otherwise be distinguished from each other spectroscopically. We have used rapid chemical quench methods on glutamate mutase to investigate the kinetics of 5'-deoxyadenosine (5'-dA) formation, the mechanism by which the substrate and product radicals interconvert, and to determine whether slow product release contributes to the overall rate of reaction.[17,19]

Kinetics of 5'-Deoxyadenosine Formation

We followed the formation of 5'-dA between 5 ms and 300 ms by which time the 5'-dA concentration had reached steady-state levels. 5'-dA was formed rapidly when the enzyme was reacted with either substrate and the rate of its appearance was adequately fitted by a single exponential curve as shown in Fig. 5. The coupled mechanism predicts that 5'-dA formation should exhibit the same kinetic behavior

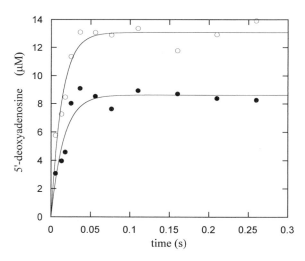

FIG. 5. Pre-steady-state formation of 5'-deoxyadenosine in the glutamate mutase reaction observed for the enzyme reacting with saturating concentrations of either glutamate (○) or methylaspartate (●). Reprinted with permission from H. W. Chih and E. N. G. Marsh, *Biochemistry* **38**, 13684 (1999). Copyright © 1999 American Chemical Society.

as Cbl(II) formation, and therefore one would expect that 5′-dA would also be formed with biphasic kinetics. The shortest time accessible by quench flow is ≈5 ms, which is much too slow to observe the "fast" site on the enzyme, and indeed, about half the reaction occurs within 5 ms. However, it is also possible that at very early times in the reaction the kinetic behavior of 5′-dA and Cbl(II) may be different. This case illustrates the limitations of the rapid quench flow technique for investigating complex kinetic phenomena that may involve more than a single rate constant. Nevertheless, the apparent rate constants calculated from these data for the formation of 5′-dA from AdoCbl and either glutamate or methylaspartate are in reasonable agreement with the rate constants (for the slow site) for AdoCbl homolysis obtained by spectroscopy (Table I). (The rate constants for 5′-dA formation are slightly slower than those for AdoCbl homolysis, but differences are probably not experimentally significant.) Thus, these experiments demonstrate that 5′-dA is a kinetically competent intermediate and the kinetics of formation are consistent with a kinetically coupled mechanism for Co–C bond homolysis and hydrogen abstraction.

The steady-state concentration of 5′-dA formed when the enzyme was saturated with L-glutamate corresponds to ∼30% of the enzyme active sites being in the free radical form, whereas in the presence of saturating methylaspartate the steady-state concentration of 5′-dA indicates that ∼20% of the enzyme is in the free radical form. The two experiments were performed with the same solution of enzyme, and therefore the differences in the steady-state concentration of 5′-dA most likely reflect real differences in the steady-state concentrations of enzyme-bound intermediates as opposed to differences in enzyme activity or concentration.

Identification of Acrylate and Glycyl Radical as Intermediates

The mechanism by which the glutamyl and methylaspartyl radicals interconvert is the least well understood aspect of the reaction. In the B_{12}-dependent rearrangements catalyzed by methylmalonyl-CoA mutase[24] and isobutyryl-CoA mutase,[25] the migrating carbon is the thioacyl carbon of the thioester with coenzyme A, whereas the rearrangement catalyzed by 2-methyleneglutarate mutase[26] involves the migration of a vinylic carbon. In these cases the migrating carbon is sp^2 hybridized, and this provides a low energy pathway for the radical rearrangement to occur through an associative mechanism involving a cyclopropyl intermediate.[27–29]

[24] N. H. Thoma, T. W. Meier, P. R. Evans, and P. F. Leadlay, *Biochemistry* **37**, 14386 (1998).
[25] B. S. Moore, R. Eisenberg, C. Weber, A. Bridges, D. Nanz, and J. A. Robinson, *J. Am. Chem. Soc.* **117**, 11285 (1995).
[26] B. Beatrix, O. Zelder, D. Linder, and W. Buckel, *Eur. J. Biochem.* **221**, 101 (1994).
[27] S. Ashwell, A. G. Davies, B. T. Golding, R. Haymotherwell, and S. Mwesigyekibende, *J. Chem. Soc. Chem. Commun.*, 1483 (1989).
[28] S. Wollowitz and J. Halpern, *J. Am. Chem. Soc.* **106**, 8319 (1984).
[29] D. M. Smith, B. T. Golding, and L. Radom, *J. Am. Chem. Soc.* **121**, 9388 (1999).

However, the carbon skeleton rearrangement catalyzed by glutamate mutase is unique in that the migrating carbon is sp^3 hybridized. A cyclopropyl intermediate cannot form in this case and, significantly, model studies in free solution have failed to demonstrate 1,2-migrations of sp^3 carbon atoms under radical conditions.[30,31]

We undertook rapid quench flow experiments to examine whether acrylate and glycyl radical (reductively trapped as glycine), the putative intermediates in the rearrangement, could be detected during turnover. In preliminary experiments radioactivity from ^{14}C-labeled L-glutamate could be detected in both carrier glycine and acrylate when the enzyme reaction was quenched with TFA after 400 ms, by which time the reaction had reached steady state. As expected, glycine and acrylate were formed in a 1:1 ratio; however, the total amounts of glycine and acrylate formed were surprisingly large, ~6% of enzyme active sites. This observation would suggest that during steady-state turnover glycyl radical comprised about one-quarter all of the free radical species present on the enzyme. This result appeared to contradict earlier EPR studies which demonstrated that the C-4 radical of glutamate is the major organic radical that accumulates on the enzyme,[32] and which failed to find substantive evidence for the C-2 radical of the putative glycyl intermediate.

We therefore considered the possibility that after denaturation of the protein by acid, the glutamyl radical, liberated from the active site, might *spontaneously* fragment to form acrylate and glycyl radical before being quenched by solvent or abstraction of hydrogen from the protein. To test this hypothesis we examined the effect of including increasing concentrations of dithiothreitol (DTT) in the quench solution as this thiol is an efficient reducing agent for organic radicals.[33] As the DTT concentration increased, the amount of acrylate *decreased*, which is consistent with DTT reacting with an intermediate that is formed *before* acrylate and glycyl radical. The limiting concentration of acrylate was about 1.5% that of enzyme active sites, and this may be taken as an upper estimate of the steady-state concentration of acrylate and glycyl radical during turnover.

Subsequently, the time course of acrylate formation was followed in experiments in which the holoenzyme was rapidly mixed with ^{14}C-labeled L-glutamate and the reaction quenched with 5% TFA containing high concentrations of DTT (Fig. 6). The formation of acrylate was described by first-order kinetics with $\tau = 29 \pm 6$ ms, $k_{obs} = 35 \pm 8$ s^{-1}, which is significantly faster than k_{cat}, but slower than the rate of 5'-dA formation (Table I). Thus, the kinetic data are consistent with acrylate being a kinetically competent intermediate that is formed subsequent to

---

[30] P. Dowd, S.-C. Choi, F. Duah, and C. Kaufman, *Tetrahedron* **44**, 2137 (1988).
[31] Y. Murakami, Y. Hisaeda, X.-M. Song, and T. Ohno, *J. Chem. Soc. Perkin Trans.* **2**, 1527 (1992).
[32] H. Bothe, D. J. Darley, S. P. Albracht, G. J. Gerfen, B. T. Golding, and W. Buckel, *Biochemistry* **37**, 4105 (1998).
[33] M. B. Goshe and V. E. Anderson, *Radiation Res.* **151**, 50 (1999).

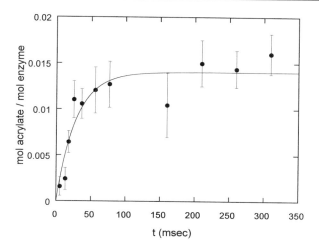

FIG. 6. Pre-steady-state formation of acrylate in the glutamate mutase reaction observed for the enzyme reacting with L-glutamate. Reprinted with permission from H.-W. Chih and E. N. G. Marsh, *J. Am. Chem. Soc.* **122,** 10732 (2000). Copyright © 2000 American Chemical Society.

5′-deoxyadenosine and glutamyl radical, which is an important requirement of the fragmentation–recombination mechanism.

The observation that glutamyl radical, once liberated from the enzyme, appeared to undergo spontaneous fragmentation to glycyl radical and acrylate was unexpected but is chemically quite reasonable. Fragmentation is entropically favorable and the C-2 radical of glycine is expected to be stabilized by delocalisation of the radical onto the carboxyl group and the amino nitrogen; indeed, several examples of enzymes that form protein-based glycyl radicals are known[34,35] (although in these cases glycine is incorporated into the pepide backbone and therefore the chemical stability of these radicals may be somewhat different). An important function of the enzyme must therefore be to prevent diffusion of the glycyl radical and acrylate out of the active site and maintain these intermediates in the correct relative orientations so that recombination of these fragments can occur and the reaction can proceed to form the rearranged product.

Pre-Steady-State Kinetics of Product Formation

So far we have discussed pre-steady-state kinetic experiments that have examined the various chemical steps in the mechanism laid out in Fig. 2. We noted earlier that although the intrinsic deuterium isotope effects for hydrogen transfer between

[34] A. Becker, K. Fritz-Wolf, W. Kabsch, J. Knappe, S. Schultz, and A. F. V. Wagner, *Nat. Struct. Biol.* **6,** 969 (1999).

[35] D. T. Logan, J. Andersson, B. M. Sjoberg, and P. Nordlund, *Science* **283,** 1499 (1999).

substrate and coenzyme are remarkably large, ~30, the isotope effects measured in the steady-state reaction of glutamate mutase with glutamate and methylaspartate are much smaller, 4 and 6.3, respectively.[5] This suggests that some other step (or steps) in the mechanism is partially suppressing the isotope effects and is therefore kinetically significant in the overall rate of the reaction.

Since coenzyme homolysis and hydrogen transfer between substrate and coenzyme are tightly coupled together, there remain three isotopically insensitive steps in the enzyme mechanism that could potentially suppress the isotope effects. One is the release of product from the enzyme (k_{-1} and k_8 in Fig. 2); the other two are associated with the rearrangement of the substrate radical to the product radical (k_4, k_{-4}, k_5, and k_{-5} in Fig. 2). We have used the rapid quench flow technique to examine the kinetics of the formation of both glutamate and methylaspartate at time intervals much shorter than $1/k_{cat}$. If product release is slow relative to the chemical steps in the mechanism then product formation should exhibit burst phase kinetics. In fact, as shown in Fig. 7, neither substrate elicits burst phase kinetics, consistent with product release being rapid compared with catalysis. This is unsurprising given that k_{cat} is relatively slow and that neither substrate is bound very tightly by the enzyme (the K_d values for glutamate and methylaspartate are 0.6 mM and 40 μM, respectively[16]).

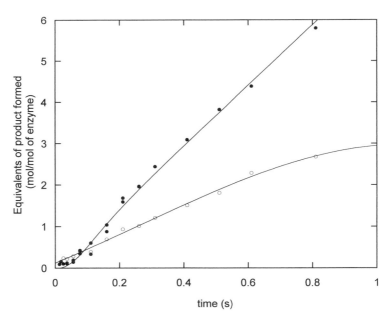

FIG. 7. Pre-steady-state formation of methylaspartate (○) and glutamate (●) observed when glutamate mutase was reacted with either 5 mM L-glutamate or 2 mM L-*threo*-3-methylaspartate, respectively.

With glutamate as the substrate, the plot of methylaspartate formation against time extrapolates linearly toward the origin at short times but shows curvature at longer times. This is associated with the approach of chemical equilibrium, which requires only 1/12 of the substrate molecules to be consumed; indeed, the curve extrapolates quite well to the theoretical number of turnovers needed to obtain chemical equilibrium. Interestingly, the apparent K_m for glutamate was much higher than that determined by steady-state methods using the usual coupled spectroscopic assay in which aspartate is converted to mesaconate by the action of aspartase.[3] This unexpected kinetic phenomenon probably arises as a result of the relatively high concentrations of enzyme (100 μM) relative to substrate (5–20 mM) that were needed to perform the experiment. Whereas we had previously measured the K_m for glutamate as 0.6 mM, in these experiments even 20 mM glutamate did not appear to saturate the enzyme. This may be explained by the rapid onset of product inhibition (as distinct from slow product release) caused by the high concentration of enzyme. Thus, after just one turnover the concentration of methylaspartate is ~100 μM, which is sufficient to cause a significant degree of product inhibition when one considers that the K_d for methylaspartate is about 40 μM.

Most interestingly, when methylaspartate is the substrate we observed a lag phase that is associated with the formation of glutamate (Fig. 7), indicating the formation and decay of an intermediate prior to the formation of glutamate. To test whether the lag phase is due to a slow substrate-binding step we repeated the experiment at 1 mM and 2 mM methylaspartate concentrations. The apparent rate constants associated with the lag phase were, within error, independent of substrate concentration. If substrate binding were contributing to the lag phase, one rate constant would be second order, and therefore doubling the concentration of substrate should increase this rate constant by a factor of 2, which was not seen.

More likely, the lag phase resulted from the accumulation and decay of a chemical intermediate (or intermediates) that is formed with an apparent rate constant $k_1^{app} \approx 22$ s^{-1} and that in a second slow step, $k_2^{app} \approx 22$ s^{-1}, is either converted to glutamate directly, or to another intermediate that is subsequently converted rapidly to glutamate. The identity of this intermediate is not known, but it may be that we are indirectly observing the formation and decay of the glycyl radical and acrylate intermediates. An important point, though, is that, regardless of its identity, the transient accumulation of the intermediate indicates that no single chemical step is cleanly rate determining in the rearrangement of methylaspartate to glutamate.

Free Energy Profile for Glutamate Mutase

The kinetic experiments described above have allowed us to construct a qualitative free energy profile for the glutamate mutase-catalyzed reaction, as shown

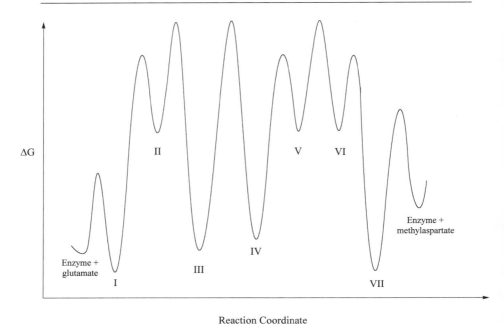

FIG. 8. Qualitative free energy profile for the glutamate mutase-catalyzed reaction.

in Fig. 8. Substrate binding and product dissociation appear to be rapid in either direction and do not limit the overall rate of catalysis. The homolysis of AdoCbl is also likely to be relatively rapid but energetically unfavorable and therefore kinetically coupled to the formation of the more stable substrate radicals. Hydrogen transfer probably involves an unusually high degree of tunneling. The rearrangement of substrate radical to product radical probably proceeds at a similar rate to the hydrogen transfer steps. This would account for the suppression of the unusually large intrinsic deuterium isotope effects, of ~30, associated with hydrogen transfer to the values of 4–6 measured in steady-state experiments.

Knowing k_{obs} for the formation of glycyl radical and acrylate from glutamate and AdoCbl (35 s^{-1}) and k_{obs} for the formation of glutamyl radical and 5′-dA from glutamate and AdoCbl (~80 s^{-1}), one can estimate k_{obs} for the fragmentation of glutamyl radical to glycyl radical and acrylate to be ~80 s^{-1}. This is consistent with the idea that the rearrangement of the substrate radical occurs at a similar rate to that at which it is generated. One step that remains unexplored is the fragmentation of methylaspartyl radical to glycyl radical and acrylate. Chemical reasoning suggests that this radical will be of similarly high energy to the 5′-deoxyadenosyl radical and its fragmentation to acylate and glycyl radical is expected to occur rapidly. Therefore in Fig. 8 we have set the energy barrier for the fragmentation of methylaspartyl radical relatively low.

Conclusions

Pre-steady-state kinetic studies have proved useful in investigating the chemical mechanism of the unusual rearrangement reaction catalyzed by glutamate mutase and have provided insights into the energetics of free radical formation. It is evident that one strategy employed by the protein is to couple an energetically unfavorable step such as Co–C bond homolysis to the formation of more stable substrate-based radicals. Nevertheless, the details of how the protein catalyzes homolysis of AdoCbl and stabilizes the substrate radicals to such a remarkable degree remain unclear. Now that the free energy profile and the crystal structure have been determined for the wild-type enzyme, the stage is set to address this question. The contribution of individual amino acid residues to catalysis can now be dissected with some precision, as the effect of mutations on both the structure of the protein and the free energy profile of the reaction may be determined.

Acknowledgments

The research described here has been supported by NIH Research Grant GM 59227 to E.N.G.M. We thank Dr. James Anderson for careful reading and comments on the manuscript.

[29] Ribonucleotide Reductase: Kinetic Methods for Demonstrating Radical Transfer Pathway in Protein R2 of Mouse Enzyme in Generation of Tyrosyl Free Radical

By ASTRID GRÄSLUND

Introduction

Ribonucleotide reduction is necessary for the biosynthesis of the deoxyribonucleotides needed for DNA synthesis and repair. The conversion to deoxyribonucleotides from the corresponding ribonucleotides is strictly regulated throughout the cell cycle, and this reaction may be regarded as a bottleneck for cell proliferation. Ribonucleotide reductase (RNR) [EC 1.17.4.1, ribonucleotide-diphosphate reductase] catalyzes the reduction of ribonucleotides. At least three different classes of ribonucleotide reductases have been described.[1–4] They all use metals and free radical chemistry for the seemingly simple reaction to reduce the 2′-hydroxyl group in the ribose ring. The iron-containing class I ribonucleotide reductases (RDR) contain a tyrosyl free radical in their active state.[5,6] The substrates for RDR are the nucleoside diphosphates. RDR is found, e.g., in mammalian cells as well as in aerobically growing *Escherichia coli* and is coded for by certain viruses.

Structure and Function of Class I Ribonucleotide Reductase

The class I RDR enzyme consists of two nonidentical homodimeric components, proteins R1 and R2. The substrate reaction takes place in the larger R1 component, and R1 is also where allosteric regulators for activity and specificity are bound. The smaller R2 component has the diiron sites and the nearby tyrosyl residues, where the stable free radicals necessary for enzyme activity are formed. The crystal structures of both proteins R1 and R2 have been solved separately[7–9] and a model of the R1/R2 complex has also been constructed. Figure 1 shows the iron/radical site in *E. coli* protein R2.[7] The two high spin ferric ions are connected by one μ-oxo and one μ-carboxylate bridge and are surrounded by carboxylate and histidine ligands. The iron pair is antiferromagnetically coupled and has a diamagnetic ground state. The side chain of Y122 (*E. coli* numbering) is the site of the stable free radical in the active enzyme.[10]

When the active site in RDR protein R1 binds substrate and the enzymatic reaction is about to begin, a long-range electron transfer is postulated to take place. This transfer results in the formation of an intermediate active site radical on a cysteinyl residue (*E. coli* C439), a prerequisite for catalytic activity. The catalytic reaction has been suggested to involve a series of coupled electron/proton or hydrogen atom (H·) transfer steps along an array of conserved hydrogen bonded

---

[1] P. Reichard, *Science* **260**, 1773 (1993).
[2] B.-M. Sjöberg, *Struct. Bonding* **88**, 139 (1997).
[3] J. Stubbe and W. A. van der Donk, *Chem. Rev.* **98**, 705 (1998).
[4] M. Sahlin and B.-M. Sjöberg, *in* "Subcellular Biochemistry," Vol. 35: "Enzyme-Catalyzed Electron and Radical Transfer" (A. Holzenburg and N. S. Scrutton, eds.), p. 405. Kluwer Academic/Plenum, New York, 2000.
[5] B.-M. Sjöberg, P. Reichard, A. Gräslund, and A. Ehrenberg, *J. Biol. Chem.* **253**, 6863 (1978).
[6] A. Gräslund and M. Sahlin, *Ann. Rev. Biophys. Biomol. Struct.* **25**, 259 (1996).
[7] P. Nordlund, B.-M. Sjöberg, and H. Eklund, *Nature* **345**, 593 (1990).
[8] P. Nordlund and H. Eklund, *J. Mol. Biol.* **232**, 123 (1993).
[9] U. Uhlin and H. Eklund, *Nature* **370**, 533 (1994).
[10] Å. Larsson and B.-M. Sjöberg, *EMBO J.* **5**, 2037 (1986).

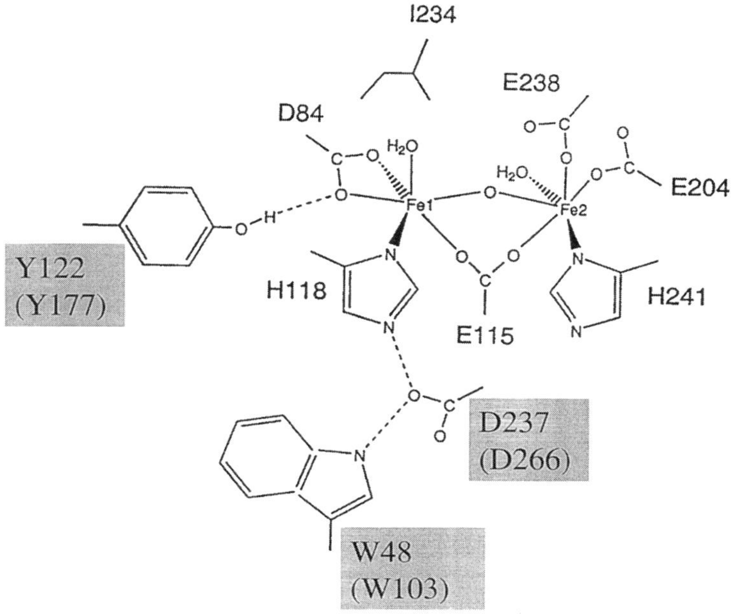

FIG. 1. The structure of the iron/tyrosyl radical site in met protein R2 of *E. coli* ribonucleotide reductase, with the iron ligands and some neighboring residues indicated [P. Nordlund, B.-M. Sjöberg, and H. Eklund, *Nature* **345**, 593 (1990); P. Nordlund and H. Eklund, *J. Mol. Biol.* **232**, 123 (1993)]. Residues D237 and W48 (mouse numbering D266 and W103 in parentheses) in the radical transfer pathway are highlighted, together with residue Y122 (mouse Y177) harboring the stable free radical.

residues in proteins R1 and R2[7–9,11,12] (Fig. 2). The radical transfer pathway (RTP) consists of (Fe^1)-His^{173}-Asp^{266}-Trp^{103} in protein R2 and Tyr^{738}-Tyr^{737}-Cys^{429} in R1 (mouse residue numbering). Mutations D266A and W103(Y/F) in mouse R2 lead to a complete loss of enzyme activity.[13] The corresponding residues in *E. coli* R2 have also been found to give essentially inactive enzyme when mutated.[2,14]

Mutagenesis studies strongly indicate that Tyr-370 in the mouse R2 protein (*E. coli* Tyr-356) links the radical transfer pathway between the R1 and R2 proteins.[15] However, Tyr-370 appears only to be involved in the catalytic reaction

[11] P. E. M. Siegbahn, M. R. A. Blomberg, and R. H. Crabtree, *Theor. Chem. Acc.* **97**, 289 (1997).
[12] A. Ehrenberg, in "Biological Physics: Third International Symposium" (H. Frauenfelder, G. Hummer, and R. Garcia, eds.), p. 163. American Inst. of Physics, Melville, NY, 1999.
[13] U. Rova, K. Goodtzova, R. Ingemarson, G. Behravan, A. Gräslund, and L. Thelander, *Biochemistry* **43**, 4267 (1995).
[14] M. Ekberg, M. Sahlin, M. Eriksson, and B.-M. Sjöberg, *J. Biol. Chem.* **271**, 20655 (1996).
[15] U. Rova, A. Adrait, S. Pötsch, A. Gräslund, and L. Thelander, *J. Biol. Chem.* **274**, 23746 (1999).

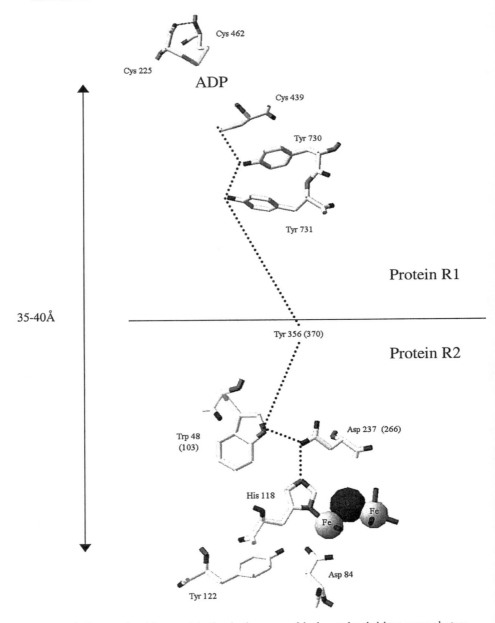

FIG. 2. Conserved residues participating in the proposed hydrogen bonded long-range electron transfer chain between substrate (adenine diphosphate, ADP) site in protein R1 and the iron/tyrosyl radical site in protein R2 of *Escherichia coli* ribonucleotide reductase [P. Nordlund, B.-M. Sjöberg,

and not in the reconstitution of the iron/tyrosyl free radical center (see below) in the isolated R2 protein. The explanation might be that Tyr-370 is close to the R2 protein flexible C-terminal tail which becomes rigid only after binding to the R1 protein.[16]

Redox States of Iron/Radical Site in Protein R2

The tyrosyl free radical in protein R2 is of the neutral phenoxy type,[17] formed by loss of one electron and one proton from the OH group of the tyrosyl residue (*E. coli* Y122, mouse Y177) neighboring the iron site. The free radical has a characteristic electron paramagnetic resonance (EPR) spectrum which varies to some extent between proteins from different species (Fig. 3). Different hyperfine couplings arise because of variations in the tyrosyl ring geometry in the different proteins. The tyrosyl radical is hydrogen bonded through its phenoxyl group in some isolated R2 proteins, but not in all.[17-21] It has been speculated that the hydrogen bonding may vary according to the functional state of the enzyme.[12,21] A variety of spectroscopic techniques have been used to characterize the iron/radical site, its formation, and its different redox states.

The iron/radical center of the mammalian R2 protein is much more labile than the corresponding structure of the *E. coli* protein. At physiological temperatures, iron is lost with a half-life of about 30 min in mouse R2. Therefore there is a continuous requirement of ferrous iron and oxygen to regenerate the iron-radical center of the mammalian R2 protein. This explains the observed sensitivity of

---

[16] P. Lycksell, R. Ingemarson, R. Davies, A. Gräslund, and L. Thelander, *Biochemistry* **33**, 2838 (1994).

[17] C. J. Bender, M. Sahlin, G. T. Babcock, B. A. Barry, T. K. Chandrasekhar, S. P. Salowe, J. Stubbe, B. Lindström, L. Peterson, A. Ehrenberg, and B.-M. Sjöberg, *J. Am. Chem. Soc.* **111**, 8076 (1989).

[18] P. Schmidt, K. Andersson, A.-L. Barra, L. Thelander, and A. Gräslund, *J. Biol. Chem.* **271**, 23615 (1996).

[19] P. Allard, A.-L. Barra, K. K. Andersson, P. P. Schmidt, M. Atta, and A. Gräslund, *J. Am. Chem. Soc.* **118**, 895 (1996).

[20] P. J. van Dam, J.-P. Willems, P. P. Schmidt, S. Pötsch, A.-L. Barra, W. R. Hagen, B. M. Hoffman, K. K. Andersson, and A. Gräslund, *J. Am. Chem. Soc.* **120**, 5080 (1998).

[21] A. Liu, A.-L. Barra, H. Rubin, G. Lu, and A. Gräslund, *J. Am. Chem. Soc.* **122**, 1974 (2000).

---

and H. Eklund, *Nature* **345**, 593 (1990); P. Nordlund and H. Eklund, *J. Mol. Biol.* **232**, 123 (1993); U. Uhlin and H. Eklund, *Nature* **370**, 533 (1994); P. E. M. Siegbahn, M. R. A. Blomberg, and R. H. Crabtree, *Theor. Chem. Acc.* **97**, 289 (1997); A. Ehrenberg, in "Biological Physics: Third International Symposium" (H. Frauenfelder, G. Hummer, and R. Garcia, eds.), p. 163. American Institute of Physics, Melville, NY, 1999; U. Rova, K. Goodtzova, R. Ingemarson, G. Behravan, A. Gräslund, and L. Thelander, *Biochemistry* **43**, 4267 (1995); M. Ekberg, M. Sahlin, M. Eriksson, and B.-M. Sjöberg, *J. Biol. Chem.* **271**, 20655 (1996); U. Rova, A. Adrait, S. Pötsch, A. Gräslund, and L. Thelander, *J. Biol. Chem.* **274**, 23746 (1999)]. Mouse protein R2 numbering is given in parentheses. The dotted lines indicate the proposed radical transfer pathway. Tyr-356 is not visible in the crystal structure of protein R2 because of dynamic disorder, but becomes ordered by interaction with R1 [P. Lycksell, R. Ingemarson, R. Davies, A. Gräslund, and L. Thelander, *Biochemistry* **33**, 2838 (1994)].

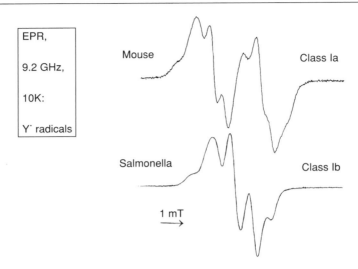

FIG. 3. X-band EPR spectra of the tyrosyl radical in protein R2 from (top) mouse, class Ia RNR, recorded at 20K; (bottom) *Salmonella typhimurium,* class Ib RNR, recorded at 20K.

mammalian cell DNA synthesis to iron deprivation, e.g., by strong iron chelators, or to low oxygen tension.[22-24]

Figure 4 summarizes observed redox states of protein R2 for *E. coli* or mouse RDR.[6] Active R2 contains the antiferromagnetically coupled EPR silent diferric site, and the tyrosyl radical which gives rise to the observable EPR spectrum. One-electron reduction of the tyrosyl radical results in the so-called met state of protein R2 (metR2).[25] Further one-electron reduction gives a mixed valent state, easily stabilized at room temperature in mouse RDR.[26] Even further reduction leads to the fully reduced diferrous state, which is stable under anaerobic conditions.[27] Reduced R2 spontaneously reacts with molecular oxygen to reconstitute active R2.[28,29] The apo state of protein R2 (apoR2) is devoid of iron and radical.

[22] L. Thelander, A. Gräslund, and M. Thelander, *Biochem. Biophys. Res. Commun.* **110,** 859 (1983).
[23] S. Nyholm, G. J. Mann, A. G. Johansson, R. J. Bergeron, A. Gräslund, and L. Thelander, *J. Biol. Chem.* **268,** 26200 (1993).
[24] C. E. Cooper, G. R. Lynagh, K. P. Hoyes, R. C. Hider, R. Cammack, and J. B. Porter, *J. Biol. Chem.* **271,** 20291 (1996).
[25] M. Sahlin, A. Gräslund, L. Petersson, A. Ehrenberg, and B.-M. Sjöberg, *Biochemistry* **28,** 2618 (1989).
[26] M. Atta, K. K. Andersson, R. Ingemarson, L. Thelander, and A. Gräslund, *J. Am. Chem. Soc.* **116,** 6429 (1994).
[27] M. Atta, N. Debaecker, K. K. Andersson, J.-M. Latour, L. Thelander, and A. Gräslund, *J. Bioinorg. Chem.* **1,** 210 (1996).
[28] C. L. Atkin, L. Thelander, P. Reichard, and G. Lang, *J. Biol. Chem.* **248,** 7464 (1973).
[29] L. Peterson, A. Gräslund, A. Ehrenberg, B.-M. Sjöberg, and P. Reichard, *J. Biol. Chem.* **255,** 6706 (1980).

Intermediate X (proposed characteristics)	Fe(III)-O^{2-}-Fe(IV)-TyrOH
Active R2	Fe(III)-O^{2-}-Fe(III)-TyrO•
Met R2	Fe(III)-O^{2-}-Fe(III)-TyrOH
Mixed valent R2	Fe(II)-OH^{-}-Fe(III)-TyrOH
Reduced R2	Fe(II)-------Fe(II)-TyrOH

FIG. 4. Overview of observed redox states of protein R2.

Formation of Iron/Radical Site in Protein R2

The formation of the iron/radical site in apoR2 is a complex reaction between ferrous iron and molecular oxygen in the protein environment.[28-30] Several intermediates have been observed and characterized in the *E. coli* wild-type and point mutant proteins in the so-called reconstitution reaction.[30-34] *In vitro* the reaction of one polypeptide chain in native mouse protein R2 may be summarized as[35,36]:

$$3Fe(II) + P\text{-Tyr-OH} + O_2 + H^+ \rightarrow$$
$$Fe(III)\text{-}O^{2-}\text{-}Fe(III)\text{-}P\text{-Tyr-O}\cdot + H_2O + Fe(III)$$

---

[30] J. M. Bollinger, D. E. Edmondson, B. H. Huynh, J. Filley, J. R. Norton, and J. Stubbe, *Science* **253**, 292 (1991).

[31] B. E. Sturgeon, D. Burdi, S. Chen, B. H. Huynh, D. E. Edmondson, J. Stubbe, and B. M. Hoffman, *J. Am. Chem. Soc.* **118**, 7551 (1996).

[32] J. P. Willems, H. I. Lee, D. Burdi, P. E. Doan, J. Stubbe, and B. M. Hoffman, *J. Am. Chem. Soc.* **119**, 9816 (1997).

[33] L. Que and J. Stubbe, *J. Am. Chem. Soc.* **120**, 849 (1998).

[34] D. Burdi, B. E. Sturgeon, W. H. Tong, J. Stubbe, and B. M. Hoffman, *J. Am. Chem. Soc.* **118**, 281 (1996).

[35] E. I. Ochiai, G. J. Mann, A. Gräslund, and L. Thelander, *J. Biol. Chem.* **265**, 15758 (1990).

[36] G. J. Mann, A. Gräslund, E.-I. Ochiai, R. Ingemarson, and L. Thelander, *Biochemistry* **30**, 1939 (1991).

Here P-Tyr-OH designates the protein with the normal Y122. The third ferrous iron ion, which may be replaced by another reductant, supplies an extra electron needed to form water.[35] It is not clear whether this iron has a specific binding site in the protein.

The time course of the reconstitution reaction and the intermediate states of the iron/radical site in the wild-type *E. coli* protein R2 have been studied by static and kinetic optical, EPR, and Mössbauer spectroscopies.[37,38] The time scale is subsecond or second at 5° and depends on the stoichiometry of available ferrous iron or other reducing agents (whether substoichiometric or in excess).

A proposal[39,40] regarding the different stages of the reconstitution reaction is presented in Fig. 5 (left-hand side). The scheme, based mainly on crystallographic and spectroscopic results on mutant R2 proteins, starts from the diferrous form of the protein, where the iron ions are bridged by the carboxylate ligands E115 and E238. Both irons are monodentate. In this proposed scheme O_2 binds initially to Fe2. Then a peroxide complex is formed and the ferrous ions are oxidized to ferric, at the same time as E238 undergoes a conformational change that makes Fe2 6-coordinate. Heterolytic cleavage of the peroxide gives a hypothetic high-valent iron–oxo species, not (yet) observed in the R2 protein, but equivalent with intermediate Q in methane monooxygenase.[41]

Electron/proton transfer from an external source via the radical transfer pathway[14,42] gives rise to the paramagnetic intermediate X, formally a Fe(III)–Fe(IV) center.[30–33] Intermediate X has an OH (or aqua) ligand on Fe1, originating from the molecular oxygen, and presumably an oxo bridge.[32] The μ-oxo bridge in the active radical containing form of the protein originates from the molecular oxygen.[43] The detailed electronic nature of the intermediate X is the subject of ongoing debate in the literature. Although formally a mixed-valent iron site, an alternative description of intermediate X as a diferric/free radical state, possibly with spin density shared between the iron ions and the ligands, may be considered (cf. also Ref. 30), as indicated in Fig. 5 (left-hand side).[40]

From experiments on the reconstitution of *E. coli* protein R2 it has been shown that the W48 residue, corresponding to mouse W103, harbors an intermediate

---

[37] J. M. Bollinger, Jr., W. Hang Tong, N. Ravi, B. Hahn Huynh, D. E. Edmondson, and J. Stubbe, *J. Am. Chem. Soc.* **116**, 8015 (1994).
[38] J. M. Bollinger, Jr., W. Hang Tong, N. Ravi, B. Hahn Huynh, D. E. Edmondson, and J. Stubbe, *J. Am. Chem. Soc.* **116**, 8042 (1994).
[39] M. E. Andersson, M. Högbom, A. Rinaldo-Matthis, K. K. Andersson, B.-M. Sjöberg, and P. Nordlund, *J. Am. Chem. Soc.* **121**, 2346 (1999).
[40] M. Assarsson, M. E. Andersson, M. Högbom, B. O. Persson, M. Sahlin, A.-L. Barra, B.-M. Sjöberg, P. Nordlund, and A. Gräslund, *J. Biol. Chem.* **276**, 26852 (2001).
[41] K. K. Andersson and A. Gräslund, *Adv. Inorg. Chem.* **43**, 359 (1995).
[42] P. P. Schmidt, U. Rova, B. Katterle, L. Thelander, and A. Gräslund, *J. Biol. Chem.* **273**, 21463 (1998).
[43] J. Ling, M. Sahlin, B.-M. Sjöberg, T. M. Loehr, and J. Sanders-Loehr, *J. Biol. Chem.* **269**, 5595 (1994).

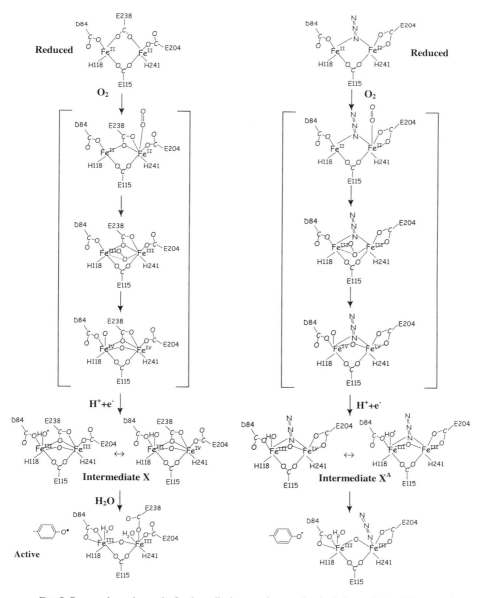

FIG. 5. Proposed reaction paths for the radical generation reaction in (left panel) the wild-type and (right panel) the mutant E238A + azide R2 proteins [M. Assarsson, M. E. Andersson, M. Högbom, B. O. Persson, M. Sahlin, A.-L. Barra, B.-M. Sjöberg, P. Nordlund, and A. Gräslund, *J. Biol. Chem.* **276**, 26852 (2001)]. Hypothetical intermediates for the initial oxygen bound forms, peroxide intermediates and high-valent Fe^{IV}–Fe^{IV} forms are indicated within brackets. Both wild-type and mutant proteins are proposed to employ an asymmetric mechanism of dioxygen cleavage, involving similar coordination environments, throughout the reaction.

cation radical during the reconstitution reaction.[44,45] An electron transfer process was proposed to give rise to intermediate X. This observation supports the role of residue 48 and the RTP in the electron transfer part of the radical reconstitution reaction, but does not answer the question whether, at what stage, and from where does a proton become involved in the formation and stabilization of intermediate X.

In the final step of the iron reconstitution reaction, the side chain of Y122, which is near but not a ligand to the iron, is oxidized to a free radical by X. This gives rise to the active state of protein R2 with a free radical on Y122. E238 becomes a monodentate ligand to Fe2 allowing an additional water ligand to bind to it. Both iron ions are now 6-coordinate, after D84 has become a bidentate ligand to Fe1.

The key feature of the proposed scheme[39,40] is that after oxygen has bound to the open coordination site on Fe2, filling up all of its coordination positions, only two free coordination positions remain on Fe1 to accommodate the products of oxygen cleavage. The outcome of the oxygen activation reaction, i.e., formation of a free radical on a nearby tyrosyl side chain, may be controlled by this geometry of the diiron site ("asymmetric" iron site control).

Available information supports the idea that the radical reconstitution reaction follows essentially the same route in the R2 proteins from *E. coli* and mouse. However, the time scale of the different stages of the reaction and their dependence on external conditions appears to vary significantly. This means that the experimental possibilities to observe the different stages may vary in enzymes from different origins. Therefore it is valuable to compare observations in different enzymes for a better understanding of the complex process.

Early on, the kinetics of the reconstitution of *E. coli* R2 with an excess of ferrous iron were studied in great detail.[37] Rapid freeze-quench EPR and Mössbauer studies have shown that intermediate X is a kinetically competent precursor of the tyrosyl radical. In a reaction at 5° the rate of formation of X is about 8 s^{-1}. This rate is mostly determined by binding of iron to the protein, since the rate was found to be as high as 60 s^{-1} after preloading the protein with ferrous iron.[33] The slowest step in both procedures was the formation of the tyrosyl radical from intermediate X, with a rate of about 1 s^{-1}.[30] This low rate may be due to protein control in a similar way as suggested for the radical transfer in the catalytic cycle.

New studies on the radical reconstitution reaction in the *E. coli* protein R2 iron ligand mutant E238A[40] have revealed that the presence of azide during the reaction restores the formation of both intermediate X and a transient tyrosyl radical. In the absence of azide, no significant amounts of paramagnetic species were observed

---

[44] J. Baldwin, C. Krebs, B. A. Ley, D. E. Edmondson, B. H. Huynh, and J. M. Bollinger, Jr., *J. Am. Chem. Soc.* **122,** 12195 (2000).

[45] C. Krebs, S. Chen, J. Baldwin, B. A. Ley, U. Patel, D. E. Edmondson, B. H. Huynh, and J. M. Bollinger, Jr., *J. Am. Chem. Soc.* **122,** 12207 (2000).

during the course of the iron reconstitution reaction. The kinetics of the reaction are, however, very different from the situation in the wild-type protein. The results show that while intermediate X is relatively rapidly formed in R2 E238A ($>30\,\text{s}^{-1}$ including the iron binding step), it gives rise only to slow formation of the tyrosyl radical ($0.04\,\text{s}^{-1}$). Intermediate X is observable in an unusually wide time window. The kinetics of the iron reconstitution reaction and tyrosyl radical formation in the wild-type and mutant E238A protein R2 of *E. coli* are also indicated in Fig. 5 (right-hand side). These new results[40] on the structure and reactivity of an azide complex of the R2 mutant E238A support the "asymmetric" iron site control of the oxygen activation mechanism as described above. No enzyme activity has, however, been found to be associated with this system, despite the appearance of the tyrosyl radical.

Rapid Freeze-Quench Apparatus

The iron reconstitution and free radical formation in protein R2 has been studied in native R2 proteins as well as in various mutants. The outcome as well as the kinetics of the reaction has been monitored by different spectroscopies. EPR spectroscopy has been the method of preference to observe the paramagnetic intermediates and end products of the reaction. The present article will focus on studies of the kinetics of the radical reconstitution reaction in mouse protein R2, probed by EPR and site-directed mutagenesis experiments.[42] The EPR studies of this system are preferentially performed at low temperatures. This is a common property of EPR signals from metal sites, and also of the free radicals in protein R2, since they interact more or less strongly with the metal site. The only exception known so far is the tyrosyl radical of the class Ib enzyme of *Mycobacterium tuberculosis*.[46] Its EPR spectrum is virtually identical at 10K and at room temperature, indicating a very weak interaction between the radical and the iron site. The normally "metallic" nature of the EPR signal of the tyrosyl radical in enzymes from other species will cause rapid relaxation and broad signals at room temperature, and better resolved signals with higher intensity are observed at low temperatures (77K or below). For the kinetic measurements it is convenient to stop the reaction by rapid freeze-quench and then store the samples at low temperatures until they are investigated, still at low temperature.

The system described here for the rapid freeze-quench (RFQ) work[47,48] is based on a System 1000 from Update Instruments with a Wiskind 4-grid mixer with high mixing efficiency. The experimental setup is shown schematically in

---

[46] A. Liu, S. Pötsch, A. Davydov, A.-L. Barra, H. Rubin, and A. Gräslund, *Biochemistry* **37**, 16369 (1998).
[47] R. C. Bray, *Biochem. J.* **81**, 189 (1961).
[48] R. C. Bray and R. Peterson, *Biochem. J.* **81**, 194 (1961).

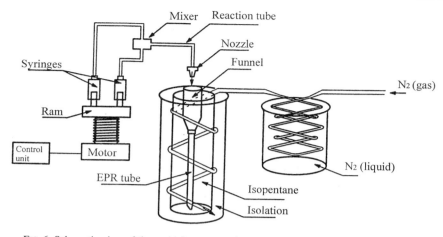

FIG. 6. Schematic view of the rapid freeze-quench apparatus used in making EPR samples. The system is based on System 1000 from Update Instruments.

Fig. 6. The prototype of the equipment was described in (Ref. 48). The cooling bath used to quench the reaction contains isopentane cooled with nitrogen gas and is kept below approximately $-110°$. Immersed in the isopentane is an EPR tube fitted with a funnel via a piece of rubber hose. For experiments with a reaction temperature of $5°$, a thermostated water cooling bath surrounds the syringes, the mixer, and the tubing. The two reactants are placed in the two 2-ml syringes. The nozzle is placed ~ 1 cm above the isopentane surface in the funnel. Every shot should ideally involve a ram displacement producing 200 μl sample. The packing of the sample in the EPR tube is done in the isopentane bath using stainless steel rods with Teflon tips, of smaller and larger diameter, used consecutively for efficient packing. The funnel is easily removed after packing the sample. After the packing, most of the isopentane is removed from the EPR tube by suction. The sample is then pumped by a water pump for approximately 0.5 hr, while stored in an n-pentane bath at approximately $-100°$ to remove the remaining isopentane more efficiently. Finally, samples are stored in liquid nitrogen.

Evaluation of Rapid Freeze-Quench Apparatus

Determination of Quenching Time. The dead time of the freeze-quench procedure is determined by studying the reaction of equine metmyoglobin (0.88 mM) with sodium azide (25 mM) at pH 7.8 in 20 mM Tris buffer and 0.1 M NaNO$_3$ at room temperature. The heme of myoglobin is in an $S = 5/2$ state which changes to $S = 1/2$ in the reaction with azide. The ratio between the EPR signals from the two spin states is determined at varying quenching times. The total dead time is

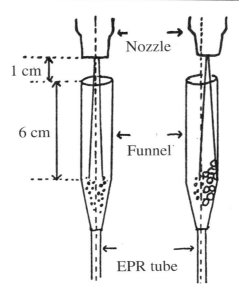

FIG. 7. Close-up view of EPR tube with attached funnel, together with the nozzle of the rapid freeze-quench apparatus.

estimated as 13 ms, composed of two about equal parts: the time for the sample jet to reach the surface of the isopentane, and the time for quenching of the reaction by freezing. Earlier work using a similar setup has reported a quenching time of about 5–6 ms at $-140°$, and 9–13 ms at $-100°$.[49] The rate constant for the metmyoglobin–azide reaction is 34 s^{-1} at room temperature,[48] which can be used to check the overall performance of the apparatus.

Reproducibility of Method. The main quantitative uncertainty in the methodology lies in the packing reproducibility. The packing efficiency may be defined as the amount of a stable sample that goes through the RFQ procedure compared to a conventionally frozen sample of the same concentration. In our experiments we estimate a packing efficiency of $50 \pm 10\%$ when comparing different experiments.

Special Considerations. The formation of small and reproducible crystals in the freezing process requires an isopentane depth of at least 6 cm. Otherwise the sample jet hits the wall of the funnel and large deformed crystals are formed. Therefore it is important to center the nozzle relative to the funnel, which has to be wide and deep enough (Fig. 7).

[49] D. P. Ballou and G. A. Palmer, *Anal. Chem.* **46,** 1248 (1974).

Reconstitution of Tyrosyl Radical in Mouse R2 Protein

The reconstitution reaction has been investigated by EPR in native mouse R2 protein and three point mutants, D266A, W103Y, and W103F on the R2 part of the radical transfer pathway[42] (for numbering of the residues in mouse R2, see Fig. 1). The time window is from about 20 ms and up. In contrast to the situation in *E. coli* R2, very little is seen by EPR of intermediate X in the reconstitution reactions in native mouse R2, despite varying conditions such as reaction times and temperature. The EPR spectra recorded during the native mouse R2 reconstitution reaction show directly the tyrosyl radical, indicating a very short-lived intermediate X in these reactions. Also in contrast to the case for *E. coli* R2,[33] the kinetics of tyrosyl radical formation in mouse R2 is not dependent on preloading of iron, indicating that iron binding is not a rate-limiting step in the mouse R2 reconstitution reaction. The kinetics has been studied as a function of reaction temperature, and an Arrhenius plot can be constructed (Fig. 8).

The kinetics of the radical reconstitution reaction were also studied in the radical transfer chain mutants.

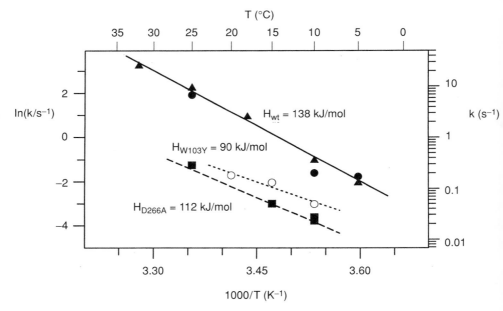

FIG. 8. Arrhenius plots of the rate of tyrosyl radical formation in native and radical transfer pathway mutant mouse R2 proteins after addition of anaerobic Fe^{2+} to air-exposed solutions of protein. The ratio of iron to protein was about 4. The rates of the mutant proteins are about 20-fold lower than that of the native protein at all temperatures. The slopes of the plots give the following similar activation enthalpies for the three proteins: native R2 protein, 138 kJ/mol; W103Y, 90 kJ/mol; D266A, 112 kJ/mol. From P. P. Schmidt, U. Rova, B. Katterle, L. Thelander, and A. Gräslund, *J. Biol. Chem.* **273**, 21463 (1998), with permission.

D266A: The residue Asp-266 has an intermediate position in the RTP. Replacing this residue with Ala should lead to a significant change of the hydrogen bonding pattern in the protein. It has been found that the kinetics of tyrosyl radical formation became about a factor of 20 slower.[42]

W103F: As already mentioned, residue Trp-103 (or *E. coli* Trp-48) has been found to be very important in the reconstitution reaction. In the mouse protein R2 W103F mutant we have found that there is no observable radical formation although a seemingly normal iron site is formed, in agreement with previous studies.[13,42] In contrast, work on the W48F mutant of the *E. coli* R2 protein has shown that intermediate X and the tyrosyl radical coexist on the 0.01–0.1 sec time scale in the reconstitution reaction.[45] This observation has been explained by the lack of "proper" electron transfer via Trp-48 for the formation of X. An explanation as to why, in the corresponding mouse R2 mutant W103F, no paramagnetic intermediates at all have been seen in the reconstitution reaction could be that the corresponding intermediates are too short-lived to be observed, or that magnetic interaction makes them EPR invisible.

W103Y: The W103Y mouse R2 mutant gives rise to a seemingly normal and stable tyrosyl radical, again with about a factor of 20 slower rate than in the native protein.[42] Clearly a Tyr residue replacing a Trp at this site can perform almost as well as the Trp in the reaction. Figure 8 also includes Arrhenius plots of rates of radical formation in mouse R2 D266A and W103Y.

From the observations on the reconstitution reaction in these point mutants we conclude that the radical formation requires transfer of a reducing equivalent from an external source. The reducing equivalent should be provided via the hydrogen bonded pathway that connects the iron site with the protein surface and is operative during the enzymatic reaction. Because the hydrogen bonds are the key property of the pathway, we propose that the electron transfer should be coupled to proton transfer in this reaction as in the catalytic process. Our interpretation is that in the mutants this native pathway is broken, and the electron must find alternative pathways which are not as efficient, hence the slower kinetics. The more recent observation of a Trp cation radical as an intermediate in the corresponding wild-type *E. coli* radical reconstitution reaction[44] does not contradict this interpretation, but simply leaves the question of the proton open. The direct involvement of the the pathway is strongly supported by the observation of the Trp radical in the wild-type *E. coli* protein.

It should also be noted that the Arrhenius plots show that although the rates of radical formation are lower in the mouse protein R2 mutants, the activation enthalpies are very similar to that of the native protein, indicating that the overall reaction mechanism is not significantly affected.

In the mouse R2 reconstitution experiments, minute amounts of an EPR spectrum corresponding to an intermediate X are observed toward higher reaction temperatures.[42] The appearance is not in good agreement with it being a kinetically competent precursor to the tyrosyl radical. However, a recent report[50] using different experimental conditions has shown considerable accumulation of X during reconstitution also of mouse R2, confirming the similar nature of the reconstitution reaction in *E. coli* and mouse R2 proteins.

Y177W,F,C: The reconstitution reaction has also been studied in another class of mouse protein R2 mutants, where the site of the stable tyrosyl radical has been changed to another residue. The protein R2 Y177W mutant gives rise to a transient tryptophan radical, presumably residing on Trp-177, identified and characterized by EPR and ENDOR spectroscopy.[51] It can be observed in a time window up to a few minutes at room temperature. The tryptophan free radical is an oxidized species, neutral and hydrogen bonded. In the two mutants Y177F and Y177C, no successor radicals are observed, but EPR signals corresponding to unusually long-living (several seconds) intermediates X are found.[51] They decay without giving rise to observable successor radicals.

Concluding Remarks

In conclusion, kinetic EPR experiments in proteins where a controlled redox process takes place can yield information about likely mechanisms operating in the process, illustrated here by observations on the radical reconstitution in protein R2 of mouse ribonucleotide reductase. The concept of coupled electron–proton transfer along a hydrogen–bonded chain is an attractive one to explain how the transfer is guided across long distances in the protein, perhaps minimizing damage to the protein and protecting the productive transfer by avoiding accidental losses of the unpaired electron. Kinetic EPR experiments using selected point mutants can help to evaluate the participation of particular residues in this process. The protein R2 part of the hydrogen-bonded radical transfer pathway in mouse ribonucleotide reductase, which participates in the enzymatic reaction, is strongly implicated to be involved also in the radical reconstitution reaction.

Acknowledgment

I thank Lars Thelander, Britt-Marie Sjöberg, and Anders Ehrenberg for valuable comments on the manuscript, and Susan Fridd and Maria Assarson for help with some of the figures. The work by Torbjörn Astlind, Susanna Sjöholm, and Peter P. Schmidt on the freeze-quench apparatus is gratefully acknowledged. Work on this topic in the author's laboratory was supported by the Swedish Natural Science Research Council, the Bank of Sweden Tercentenary Foundation, and the Carl Trygger Foundation.

[50] D. Yun, C. Krebs, G. Gupta, D. Iwig, B. Huynh, and J. M. Bollinger, Jr., *Biochemistry* **41**, 981 (2002).
[51] S. Pötsch, F. Lendzian, R. Ingemarson, A. Hörnberg, L. Thelander, W. Lubitz, G. Lassmann, and A. Gräslund, *J. Biol. Chem.* **274**, 17696 (1999).

[30] Galactose Oxidase: Probing Radical Mechanism with Ultrafast Radical Probe

By BRUCE P. BRANCHAUD and B. ELIZABETH TURNER

Introduction

The study of chemical reaction mechanisms can be broken down into a series of questions. First, one wants to know what chemical bonds were broken and made during the reaction or reactions. With that knowledge in hand one wants to know how the bonds were broken and made. Were concerted or stepwise processes involved? If there were stepwise processes, what kind of reactive intermediates were involved: cation, anion, and/or radical? Knowing the details of bond making and bond breaking outlined above, one finally wants to know the energetics of the mechanism. How high are the energy barriers for each step and how stable or unstable, and thus how transient, are intermediates along the reaction pathway?

The detection of radical intermediates in reaction mechanisms can be difficult. Many radicals are highly reactive species and have very short lifetimes. Many well-established methods exist to detect radical intermediates, including direct electron spin resonance (ESR) detection and spin-trapping methods. Still, there are many problems in enzyme reaction mechanisms where such methods are not effective in addressing mechanistic questions.

A radical at an enzyme active site will be short-lived if it is generated in the active site in close proximity to other radicals. This is not an uncommon situation. One selected example that illustrates this is in the reaction mechanism of flavin-containing monoamine oxidase enzymes. Electron transfer from the amine substrate to the flavin coenzyme generates a transient amine radical cation and flavin semiquinone radical. Further reactions can then lead to nonradical products and a normal catalytic turnover. This mechanism was determined through the use of radical clock radical-probing substrates to detect radical intermediates in the radical reaction mechanism.[1]

A radical clock is a radical rearrangement that occurs at a specific rate depending on the structure of the radical. Radical clocks can be applied to the study of enzyme reaction mechanisms as shown in Fig. 1. A substrate containing a radical-probing radical clock is applied to an enzyme. If a radical intermediate is formed either it can continue along the normal reaction pathway to form a normal product or it can undergo radical rearrangement to form a new radical. The rearranged radical can then undergo further reactions to provide radical-diagnostic products. Thus, although the radical is only produced transiently, it leaves its permanent

---

[1] R. B. Silverman, *Acc. Chem. Res.* **28,** 335 (1995).

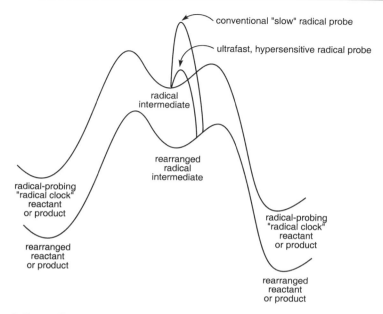

FIG. 1. Energy diagram illustrating the use of conventional "slow" radical probing radical clock substrates and ultrafast, hypersensitive radical probing substrates to detect radical intermediates.

signature in the structures of the rearranged products. A more detailed discussion of radical clocks can be found in a 1980 review of the first examples of radical clocks.[2] Many additional examples are described in a 1993 review.[3]

Rationale for Use of Ultrafast Radical Probes in Studies of Enzyme Reaction Mechanisms

The critical issue to consider in the use of radical-probing radical clocks as probes for radical intermediates in enzyme-catalyzed reactions is how fast the radical-probing radical rearrangement occurs relative to the competing rate of reaction of the radical intermediate in the normal enzyme-catalyzed reaction. Radical clocks rearrange at different rates. The prototype cyclopropylcarbinyl rearrangement (**1** to **2**) shown in Fig. 2 is a relatively slow one, with $k = 1.2 \times 10^8 \, \text{s}^{-1}$ at 25°.[3,4] In some cases such "slow" rearrangements are fast enough to detect radical intermediates if the normal competing enzyme-catalyzed reactions are slower than the radical-probing rearrangement. That is the case in the monoamine oxidase example mentioned above, where conventional "slow" probes could detect radical

---

[2] D. Griller and K. U. Ingold, *Acc. Chem. Res.* **13**, 317 (1980).
[3] M. Newcomb, *Tetrahedron* **49**, 1151 (1993).
[4] M. Newcomb and A. G. Glenn, *J. Am. Chem. Soc.* **111**, 275 (1989).

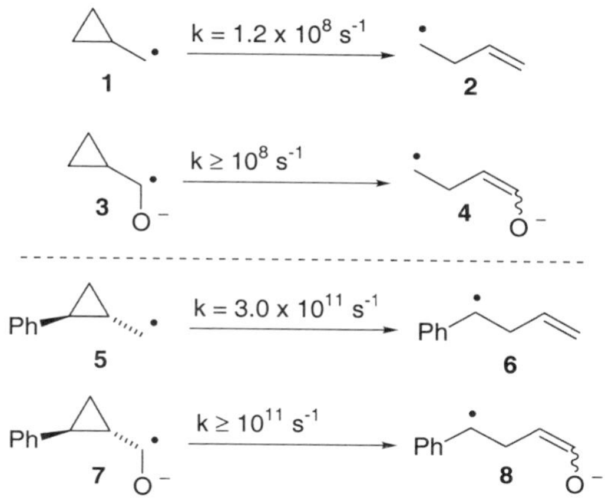

FIG. 2. Chemical structures and rates of rearrangement of representative conventional and ultrafast, hypersensitive radical probes.

intermediates.[1] In contrast to "slow" conventional radical-probing rearrangements, the ultrafast, hypersensitive rearrangement of **5** to **6**, with $k = 3.0 \times 10^{11}$ s^{-1} at 25°, is 2500 times faster than **1** to **2**.[3,5,6] From the energy diagram shown in Fig. 1, it is not unreasonable that the rearrangement of a "slow" conventional radical probe (**1** to **2**) might be too slow to compete with the reactions of the normal reaction pathway. Because many radical–radical reactions are exceptionally fast, with almost no barrier to reaction,[7] the normal reaction, which is often a radical–radical reaction, can be very, very fast. Thus, a reaction that proceeds through radical intermediates can still lead to only to normal product if the conventional radical probe is too slow to compete with the normal catalytic reaction. In contrast, the faster rearrangement of an ultrafast, hypersensitive radical probe (**5** to **6**) might be able to effectively compete with the normal reaction to provide some radical-diagnostic product. Such a situation in the use of radical probes has been experimentally determined in the case of cytochrome P450-catalyzed reactions, where conventional, "slow" radical probes fail to detect radical intermediates whereas ultrafast, hypersensitive radical probes do detect radical intermediates.[6]

Galactose Oxidase: A Radical Enzyme

Galactose oxidase (GOase; EC 1.1.3.9) from the filamentous wheat-rot fungus *Fusarium* spp.[8] catalyzes the oxidation of primary alcohols using molecular

---

[5] M. Newcomb, C. C. Johnson, M. B. Manek, and T. R. Varick, *J. Am. Chem. Soc.* **114**, 10915 (1992).
[6] M. Newcomb and P. H. Toy, *Acc. Chem. Res.* **33**, 449 (2000).
[7] M. J. Gibian and R. C. Corley, *Chem. Rev.* **73**, 441 (1973).
[8] Z. B. Ögel, D. Brayford, and M. McPherson, *Mycol. Res.* **73**, 474 (1994).

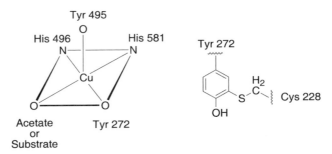

FIG. 3. Ligands to copper in galactose oxidase and the unusual thioether bond between Cys-228 and Tyr-272.

oxygen, producing aldehydes and hydrogen peroxide ($RCH_2OH + O_2 = RCHO + H_2O_2$).[9–11] The active site structure of GOase is unusual (Fig. 3) and was unique until a similar active site was discovered in glyoxal oxidase.[12] GOase contains two one-electron redox centers in the active site; a Cu(I)/Cu(II) center and a reduced tyrosine/oxidized tyrosine radical center (Tyr-272).[13,14] Tyr-272 is cross-linked to Cys-228 with a covalent C–S bond as shown in Figs. 3, 4, and 6.[14,15] GOase can exist in three distinct oxidation states.[14] Catalytically active enzyme is fully oxidized with Cu(II) and Tyr-272 radical. One-electron reduced enzyme is catalytically inactive, with Cu(II)/tyrosine and/or Cu(I)/tyrosine radical. Two-electron reduced enzyme, with Cu(I) and tyrosine, can be formed by interaction of an alcohol substrate with the enzyme under anaerobic conditions.[14] Two-electron reduced enzyme spontaneously oxidizes to the fully oxidized enzyme in the presence of O_2.[14] Fully oxidized and one-electron reduced forms are readily interconvertible under aerobic conditions using one-electron redox reagents.[16]

---

[9] D. J. Kosman, in "Copper Proteins and Copper Enzymes" (R. Lontie, ed.), Vol. 2, p. 1. CRC Press, Boca Raton, FL, 1984.

[10] G. A. Hamilton, in "Copper Proteins. Metals Ions in Biology" (T. G. Spiro, ed.), Vol. 3, p. 193. John Wiley & Sons, New York, 1981.

[11] M. J. Ettinger and D. J. Kosman, in "Copper Proteins. Metals Ions in Biology" (T. G. Spiro, ed.), Vol. 3, p. 219. John Wiley & Sons, New York, 1981.

[12] M. M. Whittaker, P. J. Kersten, N. Nakamura, J. Sanders-Loehr, E. S. Schweizer, and J. W. Whittaker, *J. Biol. Chem.* **271**, 681 (1996).

[13] G. T. Babcock, M. K. El-Deeb, P. O. Sandusky, M. M. Whittaker, and J. W. Whittaker, *J. Am. Chem. Soc.* **114**, 3727 (1992).

[14] M. W. Whittaker and J. W. Whittaker, *J. Biol. Chem.* **263**, 6074 (1988).

[15] N. Ito, S. E. Phillips, C. Stevens, Z. B. Ogel, M. J. McPherson, J. N. Keen, K. D. Yadav, and P. F. Knowles, *Nature* **350**, 87 (1991).

[16] M. P. Montague-Smith, R. M. Wachter, and B. P. Branchaud, *Anal. Biochem.* **207**, 353 (1992).

Strategy to Use Ultrafast, Hypersensitive Radical Probe with Galactose Oxidase

We have studied the reaction mechanism of galactose oxidase using radical-probing substrates. These studies have included the use of a round-trip radical probe,[17] the quadricyclylcarbinyl radical clock,[18] several β-haloethanol radical clocks,[19] and the ultrafast trans-2-phenylcyclopropylcarbinyl radical clock to be described in detail here.[20] Our rationale for using the ultrafast probe was as described above—to employ the most efficient method possible for the detection of radical intermediates in an enzyme-catalyzed reaction.

Galactose oxidase accepts only primary alcohols as substrates. Thus, alcohol **9** (Figs. 5 and 6) would be an ideal ultrafast, hypersensitive radical probe for galactose oxidase. The cyclopropylcarbinyl radical **1** rearranges by cyclopropane ring cleavage with a rate constant of $1.2 \times 10^8 \ s^{-1}$ at 25° (Fig. 2).[3,4] Ketyl radical anion **3** rearranges by cyclopropane ring cleavage with a rate constant that is at least that fast and probably faster (Fig. 2).[21] The *trans*-2-phenylcyclopropylcarbinyl radical **5** rearranges with a rate constant of $3.0 \times 10^{11} \ s^{-1}$ at 25°,[3,5,6] 2500 times faster than **1** (Fig. 2). Since **1** and **3** have similar rates of rearrangement, it is reasonable to assume that **7** will rearranged approximately as fast as **5**, at about $10^{11} \ s^{-1}$ (Fig. 2).

Thus, if a radical intermediate is formed in the oxidation of alcohol **9** (or **10**) by galactose oxidase, the ultrafast, hypersensitive radical rearrangement should detect a radical intermediate. Normally, one would analyze for rearranged products of the reaction. In the case of galactose oxidase, our work revealed that radical-probing substrates lead to turnover-mediated (mechanism-based) inactivation of the enzyme, as described below.

Reversible Turnover-Mediated (Mechanism-Based) Inactivation of Galactose Oxidase: Enzyme Inactivation Useful in Mechanistic Studies of Galactose Oxidase

As mentioned above, galactose oxidase exists in three redox states, catalytically active fully oxidized [Cu(II)/tyrosine radical], catalytically inactive one-electron reduced [Cu(II)/tyrosine and/or Cu(I)/tyrosine radical], and two-electron reduced [Cu(I)/tyrosine]. In all of our studies of galactose oxidase using radical-probing substrates it was found that the substrates were efficient turnover-mediated

---

[17] B. P. Branchaud, A. G. Glenn, and H. C. Stiasny, *J. Org. Chem.* **56**, 6656 (1991).
[18] B. P. Branchaud, M. P. Montague-Smith, D. J. Kosman, and F. R. McLaren, *J. Am. Chem. Soc.* **115**, 798 (1993).
[19] R. M. Wachter, M. P. Montague-Smith, and B. P. Branchaud, *J. Am. Chem. Soc.* **119**, 7743 (1997).
[20] B. E. Turner and B. P. Branchaud, *Bioorg. Med. Chem. Lett.* **9**, 3341 (1999).
[21] J. M. Tanko and J. P. Phillips, *J. Am. Chem. Soc.* **121**, 6078 (1999).

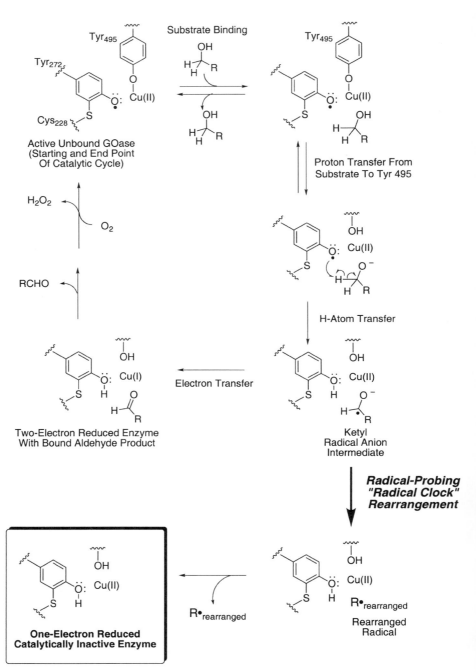

FIG. 4. Turnover-mediated (mechanism-based) inactivation of galactose oxidase: processing of radical-probing substrates by galactose oxidase produces one-electron reduced catalytically inactive enzyme.

Ph—△—CH₂OH   Ph—△—CD₂OH
    9    **10**

k_{inact} = 0.0026 s⁻¹ (± 25%) k_{inact} = 0.00082 s⁻¹ (± 13%)

K_i = 5.6 mM (± 63%) K_i = 5.1 mM (± 42%)

$k_{H(inact)} / k_{D(inact)}$ = 3.2

FIG. 5. Structures of **9** and **10**, substrates for galactose oxidase containing an ultrafast, hypersensitive radical-probing group.

(mechanism-based) inactivators of the enzyme.[18–20] In all cases the inactivation led to the one-electron-reduced, catalytically inactive form. Enzyme activity could be restored using one-electron oxidants.[18–20] These phenomena can be explained using the mechanism shown in Fig. 4. Cleavage of the C–H bond of the alcohol substrate by H-atom abstraction using the tyrosine radical would generate a ketyl radical anion intermediate. The ketyl radical anion can rearrange to generate a new radical, which would no longer be competent to complete a catalytic cycle. The enzyme would thus be trapped in the one-electron-reduced, catalytically inactive form.

Synthesis of Substrates

Synthesis of trans-(2)-Phenylcyclopropylcarbinol (9)

Commercially available *trans*-2-phenylcyclopropanecarboxylic acid is reduced to *trans*-2-phenylcyclopropylcarbinol using LiAlH₄ in anhydrous tetrahydrofuran (THF) following a standard procedure to prepare 9 (Fig. 5).[22,23] The product is obtained as a colorless oil in 84.5% yield. ¹H NMR (300 Hz, CDCl₃): δ 0.97 (m, 2 H, ring-CH₂), 1.45 (s, m, 2 H, OH, ring-CH in position 1), 1.82 (m, 1 H, ring-CH in position 2), 3.63 (m, 2 H, CH₂OH), 7.06–7.29 (m, 5 H, Ph) in agreement with previously reported ¹H NMR data.[24,25]

Synthesis of [α,α-²H₂]-trans-(2)-Phenylcyclopropylcarbinol (10)

(α,α-²H₂)-*trans*-2-Phenylcyclopropylcarbinol is obtained as for **9** above, substituting LiAlD₄ for LiAlH₄. The product is recovered in 87.4% yield with an isotopic enrichment of 98% ²H as determined by low-resolution mass spectroscopy.

---

[22] R. A. Sneen, K. M. Lewandowski, I. A. I. Taha, and B. R. Smith, *J. Am. Chem. Soc.* **83**, 4843 (1961).
[23] A. Burger and W. L. Yost, *J. Am. Chem. Soc.* **70**, 2198 (1948).
[24] R. Silverman and Y. Zelechonok, *J. Org. Chem.* **57**, 5785 (1992).
[25] O. Subbotin, *Org. Mag. Res.* **4**, 53 (1972).

The ^{1}H NMR is in agreement with previously reported data[24,25] except that the CH_2 of the CH_2OH is absent.

Isolation and Purification of Galactose Oxidase

Galactose oxidase (EC 1.1.3.9) is purified as reported[26] from the *Aspergillus nidulans* transformant, pGOF101, (a generous gift from Prof. M. McPherson, School of Biochemistry and Molecular Biology, University of Leeds, UK). A pool of enzyme from a single preparation is used for all kinetic experiments. Immediately prior to use the enzyme is activated using ferricyanide-bound QAE Sephadex resin as previously reported.[16]

Assay for Inactivation of Galactose Oxidase

Enzyme Activity Assay

Activity is measured spectrally by monitoring the conversion of 3-methoxybenzyl alcohol to the corresponding aldehyde at 314 nm as described previously.[27] Routinely, 0.043 to 0.094 nmol of activated GOase is injected into a cuvette with 80 mM 3-methoxybenzyl alcohol in 69 mM 1,4-piperazine bis(2-ethanosulfonic acid) (PIPES) pH 6.8 buffer/30% (v/v) tetrahydrofuran (THF), to make a total volume of 1 ml. The k_{obs} is determined from the slope of the absorbance time scan during the first 2.5 sec.

Inactivation Kinetics

Inactivator concentrations between 1 and 20 mM for **9** and **10** are used. The inactivator solutions contain 10% THF to solubilize **9** and **10**. The inactivator solutions are incubated at 25° with active enzyme; 50-μl aliquots are removed at specified time intervals (usually 1 min intervals) and assayed for residual GOase activity using the spectrophotometric assay described above. The concentration of the inactivator is diluted 20-fold when injecting the aliquot into a cuvette containing the assay mixture described above, thus stopping any further substrate-mediated enzyme inactivation. All experiments are replicated at least three times. Inactivation occurs by a saturable, pseudo first-order process, and k_{obs} values at various concentrations of inactivator are obtained from plots of log percent residual activity versus time. The constants k_{inact} and K_i are determined by computer-fitting the observed rate constants at various inactivator concentrations to a Michaelis–Menten type of equation.[28]

---

[26] A. J. Baron, C. Stevens, C. Wilmot, K. D. Seneviratne, V. Blakeley, D. M. Dooley, S. E. V. Phillips, P. F. Knowles, and M. J. McPherson, *J. Biol. Chem.* **269**, 25095 (1994).
[27] P. Tressel and D. Kosman, *Biochem. Biophys. Res. Commun.* **92**, 781 (1980).
[28] R. M. Wachter and B. P. Branchaud, *Biochemistry* **35**, 14425 (1996).

Reactivation of One-Electron Reduced GOase

Active GOase is incubated with 40 mM **9** in 10% THF/buffer. Aliquots are removed at specified times and the residual activity is measured using a standard oxygen electrode assay described previously.[28] When the residual activity levels off (at about 20%), 20 μl of 100 mM K$_3$Fe(CN)$_6$ is added to the incubation and the activity is found to be fully recovered. A standard for 100% activity is determined from a control incubation containing no inactivator.

Results

*Kinetics of Turnover-Dependent (Mechanism-Based) Inactivation Using **9** and **10***

Ultrafast radical-probing substrates **9** and **10** were concentration-and time-dependent inactivators of GOase. The data analysis provides a rate constant, k_{inact}, which is equivalent to k_{cat} in standard Michaelis–Menten kinetics. The data analysis also provides an apparent binding constant, K_i, which is equivalent to K_m in standard Michaelis–Menten kinetics. Compound **9** exhibited $k_{\text{inact}} = 0.0026$ s^{-1} ($\pm 25\%$) and $K_i = 5.6$ mM ($\pm 63\%$). Compound **10** exhibited $k_{\text{inact}} = 0.00082$ s^{-1} ($\pm 13\%$) and $K_i = 5.1$ mM ($\pm 42\%$).

*Primary Deuterium Isotope Effect on Turnover-Dependent (Mechanism-Based) Inactivation Using **9** and **10***

Inactivation of GOase by [α, α-H$_2$]-**9** compared to [α, α-^{2}H$_2$]-**10** displayed a primary deuterium isotope effect, $k_{\text{inact(H)}}/k_{\text{inact(D)}}$, of 3.2. This indicates that C–H bond cleavage is important in the rate-determining step for inactivation of the enzyme.

*Catalytic Turnover of **9** and **10***

Catalytic turnover of **9** and **10** could not be observed using an oxygen electrode assay described previously.[28]

*Reversible Turnover-Dependent (Mechanism-Based) Inactivation Using **9** and **10***

The inactivations were fully reversible. Addition of K$_3$Fe(CN)$_6$ to the incubation mixture led to complete recovery of initial activity.

*Substrate Protection from Turnover-Dependent (Mechanism-Based) Inactivation Using **9** and **10***

Addition of galactose to **9**/GOase incubations reduced the rate of inactivation, indicating competition for the active site.

Discussion

It is widely accepted that an enzyme inhibitor must meet several requirements to be regarded as a mechanism-based enzyme inactivator.[29–31] The main criteria are: (i) saturable, pseudo first-order kinetics, (ii) time-dependent loss of enzyme activity, (iii) inactivation slowed by addition of substrate (protection of the active site), and (iv) involvement of catalytic step.

The results for **9** and **10** display the key requirements for mechanism-based inactivation: criteria (i) and (ii) are demonstrated in the kinetics of the inactivation, with well-determined k_{inact} and K_i values for both **9** and **10**. Criterion (iii) is met by slower inactivation when galactose was added to inactivations using **9**. Criterion (iv) is met by the primary deuterium kinetic isotope effect on inactivation, since it is known that there is a significant primary deuterium isotope effect on turnover of normal substrates by GOase.[32]

The satisfaction of the above criteria implies that enzyme inactivation is a result of an enzyme-catalyzed reaction and that **9** and **10** are mechanism-based inactivators of GOase. The formation of one-electron reduced enzyme is consistent with the inactivation mechanism shown in Fig. 6. The ketyl radical anion intermediate **7**, formed by H-atom transfer from substrate to Tyr-272, can partition between two pathways, one leading to turnover and the other to inactivation. The extraordinarily rapid rate of rearrangement of ketyl radical anion **7** ensures that most of the reaction flux partitions into the inactivation pathway and little or none partitions into the turnover pathway. Efficient inactivation by **9** and **10** with no observable turnover indicates that the partition ratio, $(k_{\text{cat}} + k_{\text{inact}})/k_{\text{inact}}$, is approximately 1 for both **9** and **10**.

Previous work using β-haloethanol radical-probing radical clock substrates showed both turnover and inactivation.[19] The β-haloethanol probes used in that work had rearrangement rate constants approximately 10^9 s^{-1} or slightly greater, but less than the 10^{11} s^{-1} for the ultrafast, hypersensitive radical-probing radical rearrangement used in the work described here. Thus, it is entirely consistent that the slower β-haloethanol probes show both turnover and inactivation whereas the ultrafast probe shows no turnover and only very efficient mechanism-based inactivation. The ultrafast rearrangement is simply so fast that every radical intermediate that is formed partitions into the inactivation pathway and not into the turnover pathway. The correlation of the efficiency of radical rearrangement with the efficiency of enzyme inactivation using two different types of radical-probing rearrangements, β-haloethanols and the ultrafast probe described here, provides

---

[29] R. B. Silverman, *Methods Enzymol.* **249**, 240 (1995).
[30] C. Walsh, *Horizons Biochem. Biophys.* **3**, 36 (1977).
[31] M. A. Ator and P. R. Ortiz de Montellano, in "The Enzymes" (P. D. Boyer, ed.), 3rd Ed., Vol. 19, p. 213. Academic Press, New York, 1990.
[32] A. Marradufu, G. M. Cree, and A. S. Perlin, *Can. J. Chem.* **49**, 3429 (1971).

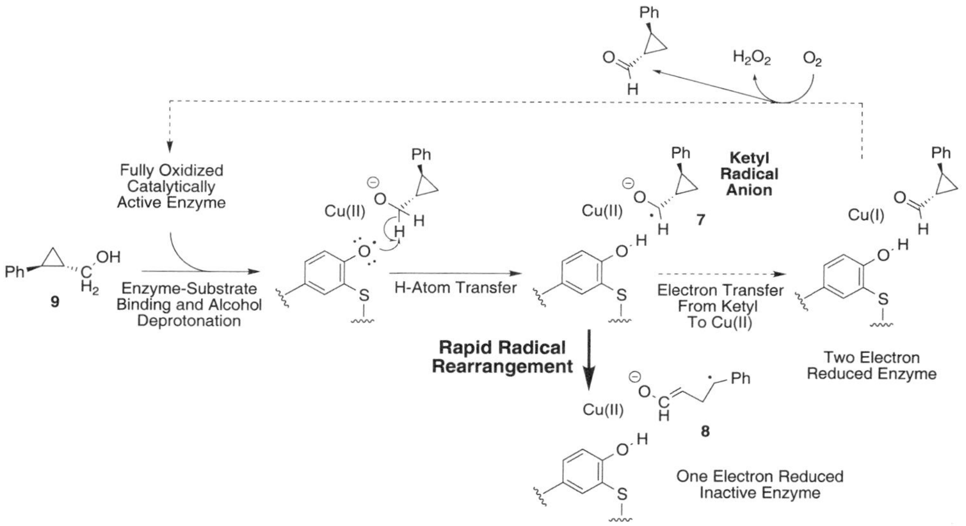

FIG. 6. A mechanism for the processing of ultrafast, hypersensitive radical probe **9** (and **10**) by galactose oxidase which is consistent with the experimental results.

compelling evidence for ketyl radical anion intermediates in the galactose oxidase catalytic mechanism.

Conclusion

Ultrafast, hypersensitive radical-probing radical rearrangements represent the ultimate in the detection of radical intermediates in enzyme-catalyzed reactions. Such probes can detect radical intermediates that cannot be detected by any other experimental technique. The ultrafast, hypersensitive *trans*-2-phenylcyclopropylcarbinyl radical rearrangement, **5** to **6**, has been applied to the study of several enzymes including galactose oxidase (described here), the chloroperoxidase from *Caldariomyces fumago*,[33,34] an alkane monooxygenase from *Pseudomonas oleovorans*,[35] and cytochromes P450.[6]

Acknowledgment

This work was supported by the National Science Foundation.

[33] A. Zaks and D. R. Dodds, *J. Am. Chem. Soc.* **117,** 10419 (1995).
[34] P. H. Toy, M. Newcomb, and L. P. Hager, *Chem. Res. Toxicol.* **11,** 816 (1998).
[35] H. Fu, M. Newcomb, and C.-H. Wong, *J. Am. Chem. Soc.* **113,** 5878 (1991).

[31] Kinetic Characterization of Transient Free Radical Intermediates in Reaction of Lysine 2,3-Aminomutase by EPR Lineshape Analysis

By Perry A. Frey, Christopher H. Chang, Marcus D. Ballinger, and George H. Reed

Introduction

Radical Reactions in Enzymology

Organic radicals are increasingly recognized as intermediates in complex enzymatic reactions.[1–10] Radicals have been detected in the steady states of most of the adenosylcobalamin-dependent reactions.[2–7,9,10] Although in most cases the structures of these radicals have not been fully established, the thiyl-protein radical in B_{12}-dependent ribonucleotide reductase has been characterized.[9,10] Radical intermediates in monooxygenase reactions such as those of cytochrome P450s and methane monooxygenase have been postulated. However, the assignment of radical mechanisms is controversial in these cases because of the indirect nature of the evidence for the participation of organic radicals.[11–13] Radicals have been observed and characterized in the reactions of a few oxygenases, notably lipoxygenase and prostaglandin synthase.[14,15]

---

[1] P. A. Frey, *Ann. Rev. Biochem.* **70**, 121 (2001).
[2] G. J. Gerfen, *in* "Chemistry and Biochemistry of B_{12}" (R. Banerjee, ed.), p. 165. Wiley, New York, 1999.
[3] J. Rétey, *in* "Chemistry and Biochemistry of B_{12}" (R. Banerjee, ed.), p. 271. Wiley, New York, 1999.
[4] R. Banerjee and S. Chowdhury, *in* "Chemistry and Biochemistry of B_{12}" (R. Banerjee, ed.), p. 707. Wiley, New York, 1999.
[5] T. Toraya, *in* "Chemistry and Biochemistry of B_{12}" (R. Banerjee, ed.), p. 783. Wiley, New York, 1999.
[6] W. Buckel, G. Bröker, H. Bothe, A. J. Pierek, and B. Golding, *in* "Chemistry and Biochemistry of B_{12}" (R. Banerjee, ed.), p. 757. Wiley, New York, 1999.
[7] G. H. Reed and V. Bandarian, *in* "Chemistry and Biochemistry of B_{12}" (R. Banerjee, ed.), p. 811. Wiley, New York, 1999.
[8] P. A. Frey and C. H. Chang, *in* "Chemistry and Biochemistry of B_{12}" (R. Banerjee, ed.), p. 835. Wiley, New York, 1999.
[9] S. Licht, G. J. Gerfen, and J. Stubbe, *Science* **271**, 477 (1996).
[10] G. J. Gerfen, S. Licht, J.-P. Willems, B. M. Hoffman, and J. Stubbe, *J. Am. Chem. Soc.* **118**, 8192 (1996).
[11] P. A. Frey, *Chem. Rev.* **90**, 1343 (1990).
[12] P. A. Frey, *in* "Comprehensive Natural Products Chemistry" (C. D. Poulter, ed.), Vol. 5, p. 205. Elsevier Science, Oxford, 1999.
[13] P. A. Frey, *Curr. Opin. Chem. Biol.* **1**, 347 (1997).
[14] M. J. Nelson, R. A. Cowling, and S. P. Seitz, *Biochemistry* **33**, 4966 (1994).

Because radicals are high-energy species, they often do not accumulate to spectroscopically detectable concentrations in the steady states of enzymatic reactions. Evidence of their presence is limited to observations of product profiles in the reactions of substrates that generate labile intermediates that can undergo fast, radical-mediated isomerizations to produce rearrangement products. The kinetic competence of such putative radicals cannot be unambiguously established.

The most direct evidence for a radical mechanism is the observation of transient radicals by electron paramagnetic resonance spectroscopy (EPR). In these cases, the question of kinetic competence can be addressed experimentally by rapid-mix freeze-quench EPR spectroscopy.[16] This technique allowed the kinetic competence of the thiyl-protein radical in the B_{12}-dependent ribonucleotide reductase to be proven.[9,10,16]

Free Radicals in Reaction of Lysine 2,3-Aminomutase

The reaction of lysine 2,3-aminomutase (LAM, EC 5.4.3.2) proceeds by the most completely characterized enzymatic radical mechanism.[1,17–26] LAM catalyzes the interconversion of L-lysine and L-β-lysine according to Eq. (1), where the migration of the amino group from C-2 to C-3 is accompanied by the countermigration of hydrogen from the 3-*pro-R* position of L-lysine to the 2-*pro-R*

$$^+H_3NCH_2CH_2CH_2\overset{H\ \ NH_3^+}{\underset{H\ \ H^*}{C-C}}COO^- \quad \rightleftharpoons \quad ^+H_3NCH_2CH_2CH_2\overset{H^*\ H}{\underset{^+H_3N\ \ H}{C-C}}COO^-$$

(1)

position of L-β-lysine.[27] This isomerization is typical of adenosylcobalamin-dependent reactions; however, the action of LAM does not require a

---

[15] G. Xiao, A. L. Tsai, G. Palmer, W. C. Boyar, P. J. Marshall, and R. J. Kulmacz, *Biochemistry* **36**, 1836 (1997).
[16] W. H. Orme-Johnson, H. Beinert, and R. J. Blakely, *J. Biol. Chem.* **249**, 2338 (1974).
[17] M. L. Moss and P. A. Frey, *J. Biol. Chem.* **262**, 14859 (1987).
[18] J. Baraniak, M. L. Moss, and P. A. Frey, *J. Biol. Chem.* **264**, 1357 (1989).
[19] M. L. Moss and P. A. Frey, *J. Biol. Chem.* **265**, 18112 (1990).
[20] M. D. Ballinger, G. H. Reed, and P. A. Frey, *Biochemistry* **31**, 949 (1992).
[21] M. D. Ballinger, P. A. Frey, and G. H. Reed, *Biochemistry* **31**, 10782 (1992).
[22] M. D. Ballinger, P. A. Frey, G. H. Reed, and R. LoBrutto, *Biochemistry* **34**, 10086 (1995).
[23] G. H. Reed and M. D. Ballinger, *Methods Enzymol.* **258**, 362 (1995).
[24] C. H. Chang, M. D. Ballinger, G. H. Reed, and P. A. Frey, *Biochemistry* **35**, 11081 (1996).
[25] W. Wu, K. W. Lieder, G. H. Reed, and P. A. Frey, *Biochemistry* **34**, 10532 (1995).
[26] J. Miller, V. Bandarian, G. H. Reed, and P. A. Frey, *Arch. Biochem. Biophys.* **387**, 281 (2001).
[27] D. J. Aberhart, S. J. Gould, H.-J. Lin, T. K. Thiruvengadam, and B. H. Weiller, *J. Am. Chem. Soc.* **105**, 5461 (1983).

B$_{12}$-coenzyme. Instead, LAM contains a [4Fe–4S] center and is activated by S-adenosyl-L-methionine (SAM). The 5′-deoxyadenosyl moiety of SAM mediates the transfer of hydrogen between C-3 of L-lysine and C-2 of L-β-lysine, just as it does in adenosylcobalamin-dependent reactions.[17,18]

LAM is a member of the radical SAM superfamily of proteins, which include within their amino acid sequences the unique iron-sulfide motif CxxxCxxC and a SAM-binding motif.[28] Other prominent members of this superfamily are SAM-dependent enzymes in which the 5′-deoxyadenosyl moiety functions as a hydrogen transfer agent, and they include pyruvate formate-lyase activases, bacterial anaerobic ribonucleotide reductases, biotin synthases, lipoyl synthases, and benzylsuccinate synthase activase.[28] The superfamily also includes many proteins of imprecisely known or unknown function.

In the reaction of LAM, SAM must be reversibly cleaved to the 5′-deoxyadenosyl radical in order to mediate hydrogen transfer and initiate free radical chemistry. This cleavage is brought about through a reversible chemical reaction between the [4Fe–4S] center and SAM at the active site.[29] The process can be described by Eq. (2), in which the highly reduced [4Fe–4S]$^+$ center reacts with SAM to transfer an electron, bind methionine, and generate the 5′-deoxyadenosyl radical.

(2)

Se-Adenosyl-L-selenomethionine is a good cofactor in place of SAM, and when it activates LAM, Se-XAFS reveals the direct coordination of Se in selenomethionine to the [4Fe–4S] center.[30]

On recognizing that SAM mediates hydrogen transfer, the radical mechanism in Fig. 1 could be written for the reaction of LAM.[17] Lysine is bound to LAM as its external aldimine with pyridoxal-5′-phosphate (PLP), and the 5′-deoxyadenosyl radical abstracts a hydrogen atom from the 3-pro-R position of the lysyl aldimine to form 5′-deoxyadenosine and lysyl radical **1**. Isomerization proceeds through internal radical addition to the aldimine to form the azacyclopropylcarbinyl radical **2**, a quasi-symmetric species, which opens in the forward direction to the β-lysyl radical **3**. Abstraction of a hydrogen atom from the methyl group of 5′-deoxyadenosine

---

[28] H. J. Sofia, G. Chen, B. G. Hetzler, J. F. Reyes-Spinola, and N. E. Miller, *Nucleic Acids Res.* **29**, 1097 (2001).

[29] K. W. Lieder, S. Booker, F. J. Ruzicka, H. Beinert, G. H. Reed, and P. A. Frey, *Biochemistry* **37**, 2578 (1998).

[30] N. J. Cosper, S. J. Booker, P. A. Frey, and R. A. Scott, *Biochemistry* **39**, 15668 (2000).

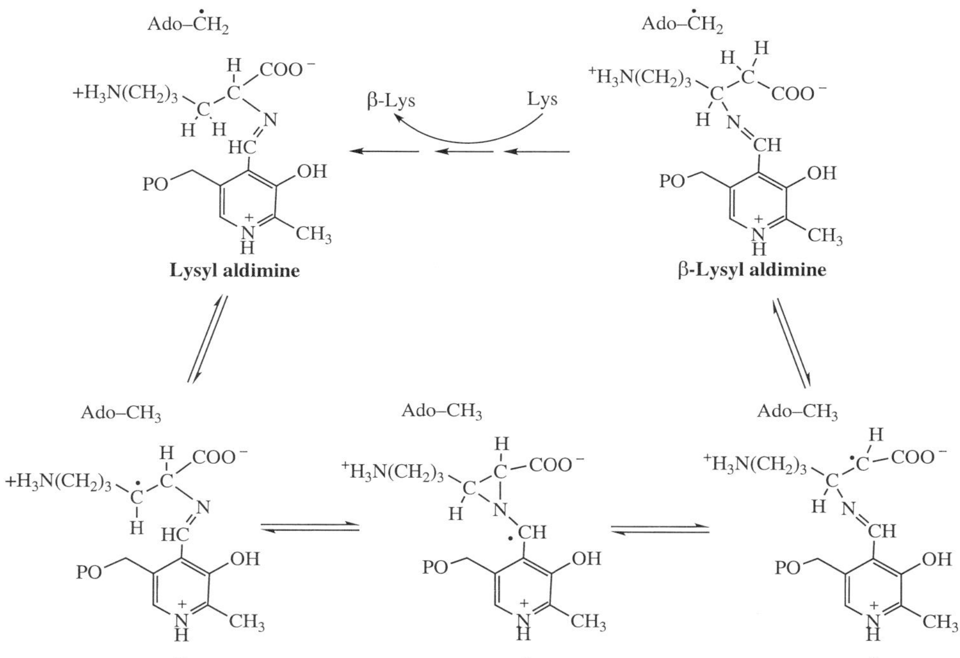

FIG. 1. Radicals in the mechanism of action of LAM. The 5′-deoxyadenosyl radical (Ado–CH$_2$˙) is generated in the reaction of SAM with the reduced [4Fe–4S] center in accordance with Eq. (2), and lysine is bound to the active site as its external aldimine with PLP. The 5′-deoxyadenosyl radical abstracts the 3-*pro-R* hydrogen atom from the lysyl side chain to produce 5′-deoxyadenosine and lysyl radical **1**. Internal radical addition to the imine linkage produces the azacyclopropylcarbinyl radical **2**, which opens in the forward direction to the β-lysyl radical **3**. Abstraction of a methyl hydrogen from 5′-deoxyadenosine produces the external aldimine of β-lysine with PLP and regenerates the 5′-deoxyadenosyl radical. Exchange of lysine with β-lysine in the external aldimine in several steps recharges the active site for a new round of catalysis.

produces the β-lysyl aldimine and regenerates the 5′-deoxyadenosyl radical. Reaction of L-lysine with β-lysyl aldimine in several enzymatic steps releases L-β-lysine and binds L-lysine for another cycle of catalysis.

Of the four organic free radicals in the mechanism, three have been observed by EPR spectroscopy. β-Lysyl radical **3** is the most stable radical in the mechanism when L-lysine is the substrate and SAM is the coenzyme, and this is the only radical observed in the steady state of the reaction.[20–22] When 4-thia-L-lysine is the substrate and SAM is the coenzyme, the most stable radical in the steady state is the 4-thia analog of the lysyl radical **1**, and this is the only radical observed by EPR.[25,26] When the enzyme is activated by 3′,4′-anhydroadenosyl-L-methionine (anSAM) in place of SAM, the most stable radical with L-lysine as the substrate is the

3′,4′-anhydro-5′-deoxyadenosyl radical, an allylic analog of the 5′-deoxyadenosyl radical, and it is the only radical observed in the steady state.[31,32] These three radicals have been characterized spectroscopically by analyzing the EPR spectra with the radicals labeled with deuterium or carbon-13 at various locations, as aids in focusing on the radical centers and the conformations of the radicals.

Kinetics of Radical Reactions

The spectroscopic observation and characterization of an organic radical in an enzyme proves the presence of the radical but not that it participates in the mechanism. The question of whether the radical is part of the mechanism or a peripheral species must be addressed. An essential property of any reaction intermediate is that it must be formed and react at a rate that is compatible with the overall rate of the reaction. That is, it must be kinetically competent to be an intermediate.

In enzymatic reactions that proceed by radical mechanisms, a method for examining the kinetic competence of a radical is rapid-mix freeze-quench EPR.[16] In the simplest application of this method, the enzyme is mixed in a fast mixing apparatus with the substrates and cofactors that are required to generate the free radical of interest. The reaction is allowed to proceed for a timed interval and then quenched as the sample is sprayed into a very cold organic solvent, such as liquid isopentane. The time from mixing to freezing may be as short as a few milliseconds. The resulting sample contains the reacting species in a chemically arrested state. At the low temperatures of the cryogenic solvent the chemical reactions are very slow in any case, and in the particular case of enzymatic solutions frozen as aqueous droplets translational motion of the molecules is arrested as well, so that the chemical reactions are halted. The snow can then be packed into EPR tubes for analysis by EPR spectroscopy. The snow samples can be collected quantitatively, so that integration of the EPR spectra from packed samples allows the total amount of radical in the sample to be measured. The progress curve of radical formation can be obtained by spray-freezing the enzymatic samples as a function of time and plotting the radical contents of the timed samples against time. Analysis of the progress curve gives the rate constant for radical formation. If this rate constant equals or exceeds the value of k_{cat} for the reaction, the radical is regarded as kinetically competent.

The rapid mixing machine generally consists of two or three syringes, the barrels of which are connected in variable configurations by tubes, with the plungers being operated by mechanical pressure exerted through one or more push bars. Depending on the complexity of the experiments, several configurations and mixing programs can be arranged. In the simplest experiment like that in the preceding paragraph, only two syringes and one push bar are required. The connection between the syringes and mixer is fixed in volume, but the flexible tubing containing

---

[31] O. Th. Magnusson, G. H. Reed, and P. A. Frey, *J. Am. Chem. Soc.* **121,** 9764 (1999).
[32] O. Th. Magnusson, G. H. Reed, and P. A. Frey, *Biochemistry* **40,** 7773 (2001).

the mixed sample can be varied in length. The longer the tubing containing the mixed sample, the longer the residence time between the mixer and spray tip, and the reaction time is proportional to the residence time. The reacting time varies with the length of the tube, and the reaction is quenched on being sprayed into the cryogenic solvent. The mechanical operations of commercial machines must be computer controlled to achieve the required accuracy in the timing of mixing and quenching.

Transient Kinetics of β-Lysyl Radical

Defining Kinetic Competence for LAM-Radical 3

The β-lysyl radical **3** in Fig. 1 is observed as an intense EPR signal in reaction mixtures consisting of LAM, SAM, and L-lysine frozen at 77K.[20–22] Although this radical could be observed and spectroscopically characterized, its kinetic characterization presented problems. In conventional rapid-mix freeze-quench experiments, no radical could be detected in less than 5 sec after mixing. Radical formation appeared to be much too slow to be catalytically relevant, yet the radical fulfilled other requirements for a catalytic intermediate. For example, its EPR signal intensity was maximal in the steady state 20–30 sec after mixing, and it decreased to a lower intensity at exactly the same rate as the approach to chemical equilibrium of free L-lysine and L-β-lysine.[20] This behavior is expected when both the substrate and product are chemically connected to the same intermediate species. Experiments traced the "anomaly" to the rate at which SAM activates LAM. The activation exhibits an induction phase that requires about 5 sec after mixing 20–100 μM LAM with 1 mM SAM and >50 mM L-lysine. Therefore, a simple rapid mixing of LAM with SAM cannot activate LAM within the time scale of enzymatic turnover; the value of k_{cat} is 24 s^{-1} for this enzyme at 21°.[24] Furthermore, preliminary activation of LAM by SAM in advance of the rapid-mix quench experiment proved to be impractical because of the instability of the activated enzyme to destructive side reactions in the absence of lysine.[24] Another approach to the issue of kinetic competence is required.

The structural characterization of lysyl radical **3** by EPR spectroscopy provided the tools that were needed to address the question of kinetic competence.[24] The clearest evidence of the structure of radical **3** are its EPR spectra generated by the use of lysine, [2-^{13}C]lysine, or [2-^{2}H]lysine as the substrate. The 2-^{13}C-label broadens the EPR signal, and the 2-^{2}H-label dramatically narrows the signal.[23] These phenomena are expected for and required by the structure of radical **3**. The nuclear hyperfine splitting constant for ^{13}C at the radical center is very large and will broaden the signal.[33] The nuclear hyperfine splitting constant for deuterium bonded

---

[33] J. A. Weil, J. R. Bolton, and J. E. Wertz, "Electron Paramagnetic Resonance." Wiley-Interscience, New York, 1994.

to a radical center is much smaller than for hydrogen bonded to the same center.[33] There is one hydrogen bonded to C2 of radical **3**, and its hyperfine coupling to the unpaired electron make a major contribution to the features and overall width of the EPR spectrum. Because the splitting constant with deuterium at this position is much smaller, the EPR signal is simplified and narrowed. These properties of radical **3**, which arose originally in its structural characterization, eventually faciltitated its kinetic characterization as an intermediate.

When the conventional rapid-mix quench technique is unsuitable, as it is for LAM, another kinetic criterion for kinetic competence can be adopted.[24] One can determine whether the intermediate turns over at the same rate as the enzymatic turnover number, k_{cat}. The turnover number of an isotopically labeled intermediate can be evaluated by measuring the rate of its exchange with the unlabeled substrate. This method is based on the implicit assumption that the only mechanism for isotopic exchange is transformation of the intermediate into the product (or substrate) by the usual catalytic route. This is a good and appropriate assumption. Then, measurement of the exchange rate in a pulse-chase experiment will test kinetic competence.

Advantage is gained in the case of LAM by the spectral properties of the isotopically labeled radical coupled with the operational flexibility of the modern rapid mixing apparatus. In a pulse-chase experiment, a suitably configured mixing machine can mix two solutions, hold the mixture for a specified time, and then mix with a second solution. The second mixture is the test reaction and can be freeze-quenched at varying times. Pulse-chase experiments with LAM and isotopically labeled L-lysine can address the question of kinetic competence. In the pulse, LAM, SAM, and isotopically labeled lysine are mixed and held for 5 sec to allow for the induction period for activation of LAM and maximize the radical concentration at the steady state. At this point in time the radical signal will be that of the isotopically labeled radical **3**. In the chase the solution is mixed with a very high concentration of unlabeled L-lysine, so that the isotopic label will be displaced from the enzyme and replaced with the spectroscopically distinct, unlabeled radical. Lineshape analysis of the EPR spectra of the freeze-quenched samples can give the fraction of unlabeled radical as a function of quenching time. The progress curve for radical transformation in the variably timed freeze quenched samples will allow the rate constant for the displacement of the isotopically labeled radical to be calculated. If the radical is kinetically competent, the rate constant for its displacement will be the same as the value of k_{cat} for the overall reaction in the direction of L-β-lysine formation.

In the reaction of LAM, either [2-^{13}C]lysine or [2-^{2}H]lysine can be used in the pulse, with unlabeled lysine in the chase. In initial experiments, the ^{13}C-label turned over very fast in trial experiments that indicated kinetic competency.[34]

---

[34] M. D. Ballinger, Ph.D. Dissertation, University of Wisconsin—Madison, 1993.

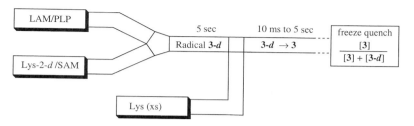

FIG. 2. Block diagram for the rapid-mix freeze analysis of the β-lysyl radical in the reaction of LAM.

Rapid-Mix Freeze-Quench Protocol for β-Lysyl Radical 3

The rate at which radical 3 in Fig. 1 turns over can be measured by the pulse-chase experiment illustrated by the block diagram in Fig. 2. The rapid-mix freeze-quench apparatus used in this laboratory is one from the Update Instruments Co. (Madison, WI) equipped with an Update Model 745 Syringe Ram Controller. Because LAM is sensitive to dioxygen and so is purified in the anaerobic chamber, the syringes are conditioned in the anaerobic chamber and then filled and sealed within the chamber (<1 ppm O_2). They are placed in the rapid-mix apparatus after cooling the isopentane bath, sample collection funnels, and anaerobic pot for freeze-quenching to $-140°$.

In the pulse, one syringe holds LAM (50–100 μM at pH 8.0) and 10 mM dithionite, which is required with SAM to activate LAM by reducing the [4Fe–4S] center. The second syringe holds 4 mM SAM, 160 mM DL-[2-^{2}H]lysine, and 40 mM Tris–SO_4 at pH 8.0. These are mixed in the first push and held for 5 sec to generate the steady-state concentration of the deuterated radical 3. The third syringe holds 450 mM L-lysine and 10 mM dithionite. It is mixed with the 5-sec-old steady-state reaction mixture in the second push, which mixes the 80 mM DL-[^{2}H]lysine (40 mM L-[^{2}H]lysine) with 225 mM L-lysine. (LAM does not interact with D-lysine in the racemate.) The reactions are freeze quenched at times varying upward from 10 ms to 5 sec. After 5 sec, the longest time before quenching, the exchange of 2-deuterolysyl radical (3-d) for the unlabeled radical (3) is at equilibrium, but the overall reaction is still in the steady state and far from equilibration between lysine and β-lysine.

After each shot for a timed point, the collection funnel and EPR tube are detached from the apparatus and set aside at $-140°$ to allow the crystals of quenched solution to settle toward the bottom of the EPR tube while another timed reaction is shot. When all of the timed shots are completed, the crystals in each EPR tube are packed into the bottom with the aid of a thin metal rod, the collection funnel is removed along with excess isopentane, and the tube is stored in liquid nitrogen for EPR analysis at a later time.

Turnover Rate Constant for β-Lysyl Radical 3

In conventional rapid-mix freeze-quench experiments the crystals of quenched samples should be collected as quantitatively as possible inside the EPR tubes. For this reason the freeze-quenching apparatus includes funnels filled with isopentane at $-140°$ connected to the EPR tubes. This allows for the frozen samples to be captured with a high degree of quantitation. In the method described here, the fraction of unlabeled radical is measured as a function of time in each sample, so that absolute quantitation in sample collection is not necessary. Sufficient sample must be collected to allow a good EPR spectrum to be obtained.

X-band EPR spectra of the quenched samples were obtained on a Varian E3 spectrometer equipped with liquid nitrogen immersion dewar. The spectrometer was interfaced with an IBM AT computer to allow for digital analysis of the spectra, each of which was acquired as an average from four 200-gauss scans. Sample spectra for the transformation of the 2-deutero-β-lysyl radical **3** (**3-d** in Fig. 2) into the unlabeled radical are shown in Fig. 3.[24] The initial narrow spectrum of the 2-deutero radical is clearly broadened to that of the equilibrium mixture of labeled and unlabeled radical during the course of the experiment.

To analyze the spectra quantitatively, baseline corrections were applied to each one to eliminate drift, and the spectra were integrated twice to obtain the radical concentrations. The EPR spectrum of each sample at time t, $[S'_{obs(t)}]$, was the summation of spectra for the unlabeled and 2-deutero versions of the β-lysyl radical **3**. To obtain the fraction of unlabeled β-lysyl radical **3** in each sample, a reference spectrum for the unlabeled β-lysyl radical **3**, $S'3_{ref}$, was required. A fraction (f) of $S'3_{ref}$ was subtracted from each observed spectrum $S'_{obs(t)}$ such that the difference $S'_{diff(t)}$ corresponded in lineshape exactly to that of the 2-deutero-β-lysyl radical **3** (**3-d** in Fig. 2). The fraction of unlabeled radical in each sample was then given by Eq. (3).

$$\frac{[\mathbf{3}]}{[\mathbf{3}]+[\mathbf{3\text{-}d}]} = \frac{1}{1 + \dfrac{\iint S'_{diff(t)}}{f \iint S'3_{ref}}} \tag{3}$$

A plot of the fraction of unlabeled radical **3** against time gave the progress curve for the exchange of the 2-deutero radical with unlabeled lysine to give the unlabeled radical, that is, the turnover rate constant for the radical. The progress curve described a first-order process, and the rate constant was $24 \pm 8\,\text{s}^{-1}$ at $21°$. An independent estimate of the value of k_{cat} at $21°$ gave $14\,\text{s}^{-1}$, a value that is lower than the turnover of the radical. The value of k_{cat} is likely to be low by about 50%, owing to the fact that only about one-half of the active sites are fully functional, based on analysis for iron and sulfide and assuming that the enzyme contains one [4Fe–4S] center per active site. Therefore, the turnover of radical **3** is as fast as that of the enzymatic reaction, and this radical is kinetically competent to be an intermediate.

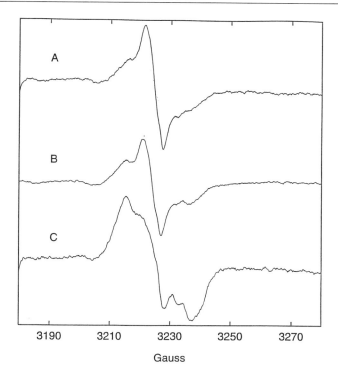

FIG. 3. Sample EPR spectra in the exchange of 2-deutero-β-lysyl radical 3 for unlabeled radical 3 at the active site of LAM. Spectrum A was obtained from the sample quenched at 13.6 ms and represents largely the spectrum of 2-deutero-radical 3. Spectrum B was obtained after 53 ms and shows the appearance of the broader spectrum of undeuterated radical 3. Spectrum C was obtained on the sample quenched at 5 sec and represents the spectrum of a mixture of 3 and 3-d arising from the mixture of 40 mM L-[2-^{2}H]lysine and 225 mM L-lysine. Adapted from C. H. Chang, M. D. Ballinger, G. H. Reed, and P. A. Frey, *Biochemistry* **34**, 10532 (1995). Reproduced with permission from the American Chemical Society.

The method described here should be generally applicable to the evaluation of kinetic competence for any enzymatic radical that can be labeled with deuterium or carbon-13 to perturb the EPR spectrum. Increasing numbers of such radicals are being observed in the rapidly expanding world of enzymatic radicals.

Acknowledgments

Research on enzymatic radical reactions in these laboratories is supported by Grants DK28607 from the National Institute of Diabetes and Digestive and Kidney Diseases (P.A.F.) and GM35752 from the National Institute of General Medical Sciences (G.H.R.).

[32] Demonstration of Peroxodiferric Intermediate in M-Ferritin Ferroxidase Reaction Using Rapid Freeze-Quench Mössbauer, Resonance Raman, and XAS Spectroscopies

By CARSTEN KREBS, DALE E. EDMONDSON, and BOI HANH HUYNH

Introduction

Ferritin is a ubiquitous protein found in various living organisms, such as bacteria, plants, insects, invertebrates, and vertebrates. It serves two important functions: First, it oxidizes Fe^{2+} ions to Fe^{3+} using molecular oxygen as the oxidant. This process, also known as the ferroxidase reaction, is important as free Fe^{2+} is very toxic because of its reactivity toward oxygen to form reactive oxygen species such as hydroxyl radicals and superoxide. The second function of ferritin is storage of Fe^{3+} in the form of a hydrated ferric oxide mineral core with various amounts of phosphate at the central cavity of the ferritin molecule. The ferritin molecule (molecular mass ca. 480 kDa) is comprised of 24 subunits, which are arranged as a spherical shell that spans a cavity with an inner diameter of 80 Å, where up to 4500 Fe^{3+} ions can be stored in the form of the mineral core.[1,2] Vertebrate ferritin molecules are composed of two different types of subunits, which differ by their rate of iron uptake and mineralization; they are termed H (fast) and L (slow). The distribution of the 24 subunits within a ferritin molecule is tissue-specific. For example, ferritins from brain and heart are rich in the H form, whereas ferritins from spleen and liver are rich in the L form. In amphibians, a third type of subunit has been observed, termed M, for which the kinetics of ferroxidation and mineralization are comparable to that exhibited by the H subunits. M subunits are homologous to H subunits: H-type and M-type ferritins from frog differ only by 19 residues.

Apoferritin can be reconstituted with Fe^{2+} in the presence of oxygen. Based on spectroscopic, kinetic, and site-directed mutagenesis studies on H-type ferritin a site for ferroxidation has been proposed, and the conserved residues E23, E58, H61, E103, and Q137 have been proposed to serve as Fe ligands.[1–3] This view has been supported by the crystal structures of H and M subunits in which the proposed ferroxidase site is occupied by two Tb^{3+} or by two Mg^{2+} ions, respectively.[4,5]

---

[1] G. S. Waldo and E. C. Theil, *in* "Comprehensive Supramolecular Chemistry" (K. S. Suslick, ed.), p. 65. Elsevier, Oxford, UK, 1996.
[2] P. M. Harrison and P. Arosio, *Biochim. Biophys. Acta* **1275,** 161 (1996).
[3] C. Krebs and B. H. Huynh, *in* "Iron Metabolism" (G. C. Ferreira, J. J. G. Moura, and R. Franco, eds.), p. 253. Wiley-VCH, Weinheim, 1999, and references therein.
[4] Y. Ha, D. Shi, G. W. Small, E. C. Theil, and N. M. Allewell, *J. Biol. Inorg. Chem.* **4,** 243 (1999).

These metal sites are structurally similar to the carboxylate-bridged nonheme diiron proteins, such as the R2 subunit of aerobic ribonucleotide reductase (R2), the hydroxylase component of soluble methane monooxygenase (MMOH), and the stearoyl-acyl carrier protein Δ^9-desaturase (Δ^9D). The commonalities include the folding of the protein as a four-helix bundle around the diiron site and the occurrence of the EXXH sequence among the Fe ligands. This sequence motif is observed twice in R2, MMO, and Δ^9D. In ferritin it is observed only once, however.

The first indication for an intermediate in the reaction of frog type-M ferritin with Fe^{2+} and O_2 (48 Fe/protein 24-mer) was observed by Fetter et al.[6] They studied this reaction by stopped-flow absorption spectroscopy and observed a transient absorbing species in the early phase of the reaction. Experiments, in which the time dependence of the absorption at 650 nm was followed, revealed a new intermediate which forms rapidly, accumulates to a maximum content after ca. 80 ms, and decays almost completely after ca. 2 sec. Fetter et al. also reported the kinetically resolved UV–Vis spectrum of this species, obtained by using the spectrometer in the diode array mode. They found that the new intermediate has a broad absorption with $\lambda_{max} = 650$ nm.

We have used rapid freeze-quench (RFQ) Mössbauer and stopped-flow absorption spectroscopies to monitor the kinetics of the ferroxidase reaction of recombinant apo-M ferritin from frog with Fe^{2+} and O_2 (36Fe/protein 24-mer).[7] An intermediate was observed that accumulates to high levels ($\geq 70\%$) at 25 ms reaction time. In order to further study the electronic and geometric structure of this intermediate we employed RFQ-XAS[8] and RFQ-resonance Raman spectroscopies.[9] Combining the results of all four methods the intermediate has been identified as a μ-1,2-peroxodiferric species, termed MFt_{peroxo}, with an unusually short Fe–Fe distance of 2.5 Å, and a small Fe–O–O angle of 107°.

Methods

Preparation of Solutions

All experiments described in this chapter are performed by a rapid and efficient mixing of two solutions. The first solution is an O_2-saturated solution of

---

[5] D. M. Lawson, P. J. Artymiuk, S. J. Yewdall, J. M. A. Smith, J. C. Livingstone, A. Treffry, A. Luzzago, S. Levi, P. Arosio, G. Cesareni, C. D. Thomas, W. V. Shaw, and P. M. Harrison, *Nature* **349,** 541 (1991).

[6] J. Fetter, J. Cohen, D. Danger, J. Sanders-Loehr, and E. C. Theil, *J. Biol. Inorg. Chem.* **2,** 652 (1997).

[7] A. S. Pereira, W. Small, C. Krebs, P. Tavares, D. E. Edmondson, E. C. Theil, and B. H. Huynh, *Biochemistry* **37,** 9871 (1998).

[8] J. Hwang, C. Krebs, B. H. Huynh, D. E. Edmondson, E. C. Theil, and J. E. Penner-Hahn, *Science* **287,** 122 (2000).

[9] P. Moënne-Loccoz, C. Krebs, K. Herlihy, D. E. Edmondson, E. C. Theil, B. H. Huynh, and T. M. Loehr, *Biochemistry* **38,** 5290 (1999).

recombinant apo-M ferritin from frog containing 0.2 M MOPS buffer and 0.2 M NaCl (pH 7.0). The protein is purified according to a published procedure.[6,10] The second solution is an O_2-saturated acidic (ca. 3.5 mM H_2SO_4) solution of $FeSO_4$ in 0.2 M NaCl. *Note:* In acidic solution Fe^{2+} is stable and is not oxidized to Fe^{3+} by O_2. For the RFQ-resonance Raman and stopped-flow absorption experiments natural abundance $FeSO_4$ is used; the iron content is assayed spectrophotometrically using ferrozine ($\varepsilon_{562} = 27,900\ M^{-1}cm^{-1}$). For the Mössbauer and XAS experiments ^{57}Fe-enriched $FeSO_4$ (>95% enrichment) is used. This solution is prepared as described.[6] In all experiments the Fe to protein (24-mer) ratio is kept constant at 36. The protein concentration (24-mer) after mixing is 25 μM for the stopped-flow experiments and 49 μM for the RFQ experiments.

Oxygenation of Solutions

For all RFQ-Mössbauer, RFQ-XAS, and stopped-flow experiments both solutions are oxygenated with natural abundance O_2 ($^{16}O_2$ is 99.5%) as described[11] with the exception that the solutions are maintained at room temperature. The resonance Raman samples are prepared not only with $^{16}O_2$ but also with $^{18}O_2$ (95 atom % ^{18}O) and with $^{16}O^{18}O$ (50 atom % ^{18}O). The latter contains a mixture of $^{16}O_2$, $^{18}O_2$, and $^{16}O^{18}O$ in a 1:1:2 ratio. ^{18}O-Enriched O_2 is analyzed for its isotopic distribution by Raman spectroscopy prior to use. It should be emphasized that in one experiment both solutions are oxygenated using O_2 with the same isotopic distribution.

Because of the expense of ^{18}O-enriched O_2 we use a different procedure for the oxygenation to minimize contamination with $^{16}O_2$. A three-way joint is connected as shown in Fig. 1 (a) to a tonometer containing the solution to be oxygenated, (b) to a cold finger and the storage bulb containing the oxygen gas, and (c) to a Schlenk line, which is connected to a vacuum pump and argon gas. In the first step, valves A and B are opened and the tonometer with the solution to be oxygenated is gently evacuated and charged with Ar gas in alternating cycles (total of 5 cycles) and finally gently evacuated. In the next step, valve B is closed, valve C is opened, and the cold finger is evacuated. Then valve C is closed, valve D is opened, the cold finger is cooled from the outside with liquid nitrogen, and the oxygen from the storage bulb is condensed inside the cold finger. After all oxygen is condensed in the cold finger, valve D is closed again. *Caution:* It is important to keep the level of the liquid nitrogen above the level of the liquefied oxygen inside the cold finger in order to prevent evaporation of the liquefied oxygen. To oxygenate the solution in the tonometer, valve A is closed, valves C and B are opened, the liquid nitrogen

---

[10] G. S. Waldo and E. C. Theil, *Biochemistry* **32**, 13262 (1993).

[11] J. M. Bollinger, W. H. Tong, N. Ravi, B. H. Huynh, D. E. Edmondson, and J. Stubbe, *Methods Enzymol.* **258**, 278 (1995).

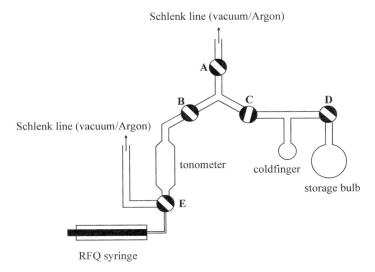

FIG. 1. Experimental setup for the oxygenation with ^{18}O-enriched O_2.

is moved away from the cold finger, and the liquefied oxygen is allowed to evaporate into the evacuated tonometer. The solution is incubated for 30 min, and the drive syringe for the freeze-quench experiment is connected to the tonometer. The three-way valve E is opened to connect the RFQ syringe to the Schlenk line. The hose connecting the RFQ syringe with the tonometer is evacuated and charged with argon gas (3 times). Then the valve E is opened to connect the tonometer with the RFQ syringe. The solution is transferred to the syringe by carefully pulling out the plunger. *Note:* As the plunger of the syringe is air-tight there is no exchange between the gas in the tonometer and the surrounding atmosphere. After the RFQ syringe has been loaded valve E is closed again. After the syringe has been loaded the O_2 gas from the tonometer is condensed back from the tonometer into the cold finger, and the procedure is repeated for the second solution of the RFQ experiment. Finally, the gas space of the tonometer is condensed back into the storage bulb.

Stopped-Flow Experiments

The stopped-flow experiments are performed using a Kinetic Instruments apparatus (Ann Arbor, MI). Time-dependent absorbance kinetic traces are collected at a fixed wavelength (650 nm) using a Nicolet 4094 digital oscilloscope. From an earlier stopped-flow absorption study it is known that the intermediate has a broad electronic transition with $\lambda_{max} = 650$ nm.[6] The kinetic data (4K for each trace) are transferred to a PC for analysis of the kinetic traces.

RFQ Experiments

The RFQ experiments are performed using an Update Instruments (Madison, WI) Model 100. Operation and calibration of the instrument used in this study have been described in detail.[11] Therefore we only present a short summary of the experimental setup. In an RFQ experiment two solutions are rapidly and efficiently mixed in a mixing chamber and passed through a calibrated hose (aging hose) before the reaction is quenched by spraying the aqueous reaction mixture into cold ($-140°$) isopentane. The ice crystals of the quenched solution are then packed into suitable holders for the various spectroscopic techniques. The reaction time is very well defined (the uncertainty is 5–10 ms, which is the estimated time for the reaction to be quenched in the cryosolvent) and is a function of the length of the aging hose and the flow speed for reaction transfer through the aging hose.

Preparation of RFQ Mössbauer and XAS Samples

The setup to prepare Mössbauer samples has been described in detail, and we have used exactly the same procedure.[11] Samples suitable for XAS spectroscopy are prepared by using the same protocol as for Mössbauer samples. However, a modified version of the Mössbauer sample holder is used. A drawing of this holder is shown in Fig. 2. The bottom of the cylindrical cup has an open slit (length:

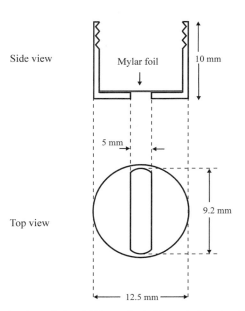

FIG. 2. Schematic drawing of the RFQ-sample holder suitable for Mössbauer and XAS spectroscopies.

0.92 cm, width: 0.5 cm) that is covered by glueing a round piece of Mylar foil to the bottom and at the inside of the cup (see Fig. 2). The Mylar foil is checked by XAS and Mössbauer spectroscopy for Fe impurities prior to use. This holder has the advantage of being suitable for both Mössbauer and XAS spectroscopy.

Preparation of RFQ Resonance Raman Samples

As both XAS and Mössbauer spectroscopy are sensitive to Fe under the conditions employed, the recorded spectra are not affected by the cryosolvent isopentane. The situation is different for resonance Raman spectroscopy. It has been found that isopentane is a strong Raman scatterer and interferes with signals originating from the sample. In order to minimize this interference isopentane has been replaced with liquid ethane as cryosolvent. Ethane has a melting point of $-183.2°$ and a boiling point of $-88.6°$, i.e., it is possible to maintain liquid ethane in a cold isopentane bath ($-155°$). Furthermore, liquid ethane and isopentane have similar heat capacities, which ensures rapid quenching within 5–10 ms.

Thick-wall quartz tubes (4 mm OD and 2 mm ID) of the same type used for preparing RFQ EPR samples are suitable sample holders for resonance Raman spectroscopy. Therefore, we use the sampling apparatus for preparing freeze-quench EPR samples.[11] The sampling apparatus (consisting of the quartz tube connected by latex tubing to a glass funnel) is filled with liquid ethane by the following procedure. *Note:* Ethane and isopentane vapors mixed with air are potentially explosive. Therefore, this procedure should be carried out carefully in a well-ventilated hood. The sampling apparatus is dipped empty into liquid nitrogen until the temperature is equilibrated. Then it is briefly removed from the liquid nitrogen to allow the liquid air that may have condensed at the inside of the quartz tube to evaporate. Wiping the outside of the tube can accelerate evaporation of the liquid air. The cold, empty sampling apparatus is again dipped into the liquid nitrogen and a gentle stream of ethane gas is immediately blown into the interior of the cold glass funnel, allowing the ethane to condense. The ethane gas flow is regulated so that the liquid nitrogen does not boil too vigorously. In order to prevent spilling of liquid nitrogen into the sampling setup the top rim of the glass funnel should be kept at least 1–2 cm above the level of the liquid nitrogen. After the sample holder has been filled, the ethane gas supply is turned off. During the filling time (about 5–10 min) part of the ethane solidifies. The sample holder is kept in liquid nitrogen until the ethane has completely solidified. Then it is removed from the liquid nitrogen bath and immediately placed into the cold isopentane bath, which is maintained at $-155°$. At this temperature, the solid ethane readily melts. *Note:* From our experience the initial melting process may be vigorous. Therefore the solidified ethane should be held in place (e.g., by pressing on the surface with a thin metal rod or a tweezer) in order to avoid spilling. Once the ethane is completely liquefied, the freeze-quench experiment is carried out by activating the ram drive of the RFQ unit and spraying the sample into the liquid ethane. The fine particles of the frozen sample are then packed as described.[11] After the sample is packed the liquid ethane is carefully transferred into a beaker, which has been prechilled in liquid nitrogen, and the beaker with liquid ethane is placed immediately into a well-ventilated hood and the discarded ethane is allowed to evaporate. To ensure complete removal of any liquid ethane from the sample the packed sample is maintained for several days at $-80°$ before storage in liquid nitrogen. Because this temperature is significantly higher than the temperature of the quenching bath it is important to make sure that the reaction intermediate is stable under these conditions. We found the ferritin peroxodiferric intermediate to be stable for days at $-80°$ as judged by Mössbauer spectroscopy.

Results

Kinetics of Ferroxidase Reaction Probed by RFQ Mössbauer Spectroscopy

Mössbauer spectroscopy, together with the RFQ method, can be used to obtain kinetic and mechanistic information about reactions involving Fe-containing proteins. Such a study involves preparation of a series of samples freeze-quenched at several well-defined reaction times by RFQ method, followed by Mössbauer investigation of the samples. Because the samples are prepared by quenching the reaction in midstream, multiple reaction species are expected to be present in the samples. The spectra of these samples are therefore rather complex. The data analysis involved is based on the following assumptions:

1. Each Mössbauer spectrum is the superposition of different subspectra, of which each represents a different Fe species contained in the sample. The ratio of the absorption area of a subspectrum to the total absorption area corresponds to the relative amount of that particular Fe species.

2. Each Fe species can be characterized by a set of parameters (isomer shift, quadrupole splitting, asymmetry parameter, magnetic hyperfine tensor, relaxation behavior, zero-field splitting parameters, and electronic spin state), and these parameters allow the calculation of Mössbauer subspectra for each species measured under different experimental conditions (temperature and externally applied magnetic field).

3. For a particular sample, all the Mössbauer spectra measured under different experimental conditions are superpositions of subspectra of absorption ratios corresponding to the same relative ratios of Fe species contained in the sample.

The goal of the analysis of such a Mössbauer time course is the deconvolution of the spectra into subspectra. Because the number of parameters involved in the analysis is large, all the spectra of a time course measured under different conditions have to be considered as a whole in order to obtain a unique solution.

The final solution is obtained by repeated iterations of trial solutions until a self-consistent solution is found that can explain all the spectra of all the samples recorded in various experimental conditions. From such a global analysis, the following information about each and every distinct Fe reaction species can be obtained:

1. The chemical nature of each Fe species (oxidation state, spin state, and possibly coordination number and type of ligand atoms) can be determined from the Mössbauer parameters.
2. The rate constants for formation and decay of each Fe species and sequence of events that happened during the reaction can be determined by fitting the temporal behaviors of the reaction species to a kinetic model.

We applied this technique to study the ferroxidase reaction catalyzed by M ferritin from frog. Samples were prepared by quenching the reaction of recombinant apo-M ferritin with Fe^{2+} and O_2 after 25 ms, 60 ms, 130 ms, 220 ms, 440 ms, and 1 sec (see Method). The Mössbauer spectrum of the 25-ms sample, shown in Fig. 3A, has three distinguishable features: two prominent sharp peaks at 0 mm/sec and 1.15 mm/sec accounting for most of the intensity, and a less intense, broad peak at 2.8 mm/sec. The latter is at a position consistent with that of the high-energy line of a quadrupole doublet arising from octahedrally coordinated Fe^{2+} in solution, which is presumably the starting material for the reaction. The low energy line of this component is overlapping with the line at ~ 0 mm/sec. Spectral deconvolution yields the following parameters for this species, which confirm its assignment as high-spin ferrous: $\delta = 1.31$ mm/sec, $\Delta E_Q = 3.0$ mm/sec. The spectral contribution of the Fe^{2+} is shown in Fig. 3A as a dashed line. As the high energy line of the Fe^{2+} does not overlap with any other spectral feature in any of the other spectra, its intensity can be estimated accurately for all time points.

The amount of Fe^{2+} as a function of the reaction time is shown in Fig. 4A. For $t = 0$ it is assumed that all the iron is present as Fe^{2+}. As can be seen, most of the Fe^{2+} is oxidized rapidly (within the first 25 ms). The simplest reaction scheme to describe this behavior is an irreversible, exponential decay. At least two exponential decay processes are required to simulate the time dependence of Fe^{2+} decay properly. The dashed line shown in Fig. 4A is a theoretical simulation for the biphasic decay of Fe^{2+} according to Eq. (1):

$$[Fe^{2+}](t) = A_{fast} \exp(-k_{fast}t) + A_{slow} \exp(-k_{slow}t) \qquad (1)$$

Approximately 85% of the Fe^{2+} is oxidized in a fast phase ($A_{fast} = 0.85$, $k_{fast} = 80$ s^{-1}), and the remaining 15% of the Fe^{2+} is oxidized more slowly ($A_{slow} = 0.15$, $k_{slow} = 1.7$ s^{-1}).

After removal of the contribution of the Fe^{2+} from the raw data, the remaining spectra were deconvoluted into several components. One of these components is

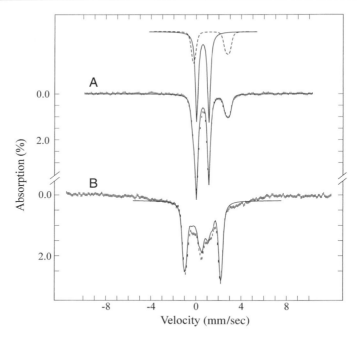

FIG. 3. Mössbauer spectra of the 25-ms sample recorded at 4.2K with a magnetic field of 50 mT (A) or 8 T (B) applied parallel to the γ beam. Spectrum A can be analyzed as a superposition of two components. One component (dashed line above the data) with parameters ($\Delta E_Q = 3.00$ mm/sec and $\delta = 1.31$ mm/sec) typical of high-spin ferrous ions accounts for ~31% of the total iron absorption and represents the not-yet-reacted Fe^{2+} ions. The other component (solid line above the data) accounts for ~69% of the total Fe absorption and is assigned to MFt_{peroxo}. The solid line plotted over the data in A is the superposition of these two components. The solid line in B is a theoretical simulation of MFt_{peroxo} using the ΔE_Q and δ values obtained from spectrum A with an asymmetry parameter $\eta = 1.3$ and assuming a diamagnetic ($S = 0$) ground state. [Adapted from A. S. Pereira, W. Small, C. Krebs, P. Tavares, D. E. Edmondson, E. C. Theil, and B. H. Huynh, *Biochemistry* **37,** 9871 (1998). Copyright © 1998 American Chemical Society.]

a reaction intermediate assigned to a peroxodiferric species (described below). The time-dependent accumulation of this intermediate is shown in Fig. 4B and has been analyzed with a simple kinetic model of two consecutive irreversible reactions described by Eq. (2):

$$[\text{intermediate}](t) = \frac{A_0 k_1}{k_2 - k_1}[\exp(-k_1 t) - \exp(-k_2 t)] \qquad (2)$$

In this formula A_0 corresponds to the initial fraction of Fe^{2+} that forms this intermediate, and k_1 and k_2 are the rates of its formation and decay, respectively. The following parameters were found from the analysis: $A_0 = 0.85$, $k_1 = 80$ s^{-1},

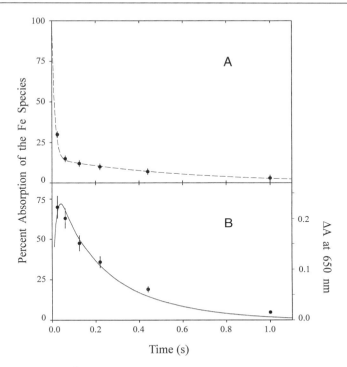

FIG. 4. Decay of the Fe^{2+} ions (A) and formation and decay of MFt_{peroxo} (B) in the early reaction of recombinant apo-M ferritin from frog with Fe^{2+} and O_2 (36 Fe/ferritin 24-mer). The left ordinate indicates the Mössbauer resonance absorption of the iron species with respect to the absorption of the total iron in the sample (100% absorption represents 36 Fe/ferritin 24-mer). The right ordinate in B represents the change in absorbance normalized for a path length of 1 cm. In A, the Mössbauer data are represented by filled circles and the dashed line is a theoretical calculation using Eq. (1) and parameters quoted in the text. In B, the filled circles are the Mössbauer data and the vertical bars are the estimated uncertainties. The solid line is the stopped-flow data representing the changes in absorbance at 650 nm as a function of the reaction time. [Adapted from A. S. Pereira, W. Small, C. Krebs, P. Tavares, D. E. Edmondson, E. C. Theil, and B. H. Huynh, *Biochemistry* **37**, 9871 (1998). Copyright © 1998 American Chemical Society.]

and $k_2 = 4.2$ s^{-1}. Because the fast phase of the Fe^{2+} decay and the formation of this intermediate exhibit the same kinetic parameters, this transient species is proposed to be the first intermediate resulting from the ferroxidation reaction of M ferritin. This species has an isomer shift of $\delta = 0.62$ mm/sec, which is indicative of a high-spin ferric species, and a quadrupole splitting of $\Delta E_Q = 1.08$ mm/sec. A theoretical simulation of this species is shown in Fig. 3A as a dotted line. The isomer shift is rather large for a high-spin ferric species, but similar to the values 0.63 mm/sec and 0.66 mm/sec observed for the peroxodiferric intermediates in the D84E variant of the R2 protein and MMOH, respectively. A spectrum of the 25-ms sample recorded at 4.2K in an external magnetic field of 80 kG applied parallel to the γ-beam (Fig. 3B) provides further information about the electronic structure of this intermediate. The solid line plotted over the experimental data is a theoretical simulation using the δ and ΔE_Q values obtained from the low-field spectrum (Fig. 3A), an asymmetry parameter $\eta = 1.3$, and assuming a diamagnetic ($S = 0$) ground state. The diamagnetic ground state can be explained by an antiferromagnetic coupling between two high-spin ferric sites ($S_1 = S_2 = 5/2$) and thus provides evidence supporting the intermediate as a dinuclear species. On the basis of the Mössbauer parameters, the diamagnetism, and the kinetic data indicating that it is the first intermediate formed in the reaction of M ferritin with Fe^{2+} and O_2, this intermediate is assigned to a peroxodiferric species, termed MFt_{peroxo}.

The global analysis of the Mössbauer data reveals further that several Fe species with similar isomer shifts $\delta = (0.50 \pm 0.03)$ mm/sec, but varying ΔE_Q values (ranging from 0.6 mm/sec to 2.4 mm/sec) form in concomitant with the decay of the peroxodiferric intermediate. These species are assigned to oxo/hydroxo-bridged diferric species based on their spectral parameters. Mössbauer spectra in an external field of 80 kG confirm this assignment. Diamagnetic ground states ($S = 0$) are observed for these species, which are again a consequence of antiferromagnetic coupling between the two high-spin ferric ions. In summary, RFQ-Mössbauer spectroscopy demonstrated that the majority (85%) of the added Fe^{2+} reacts rapidly with O_2 according to the following scheme.

$$\text{Ferrous ions} \xrightarrow{80\,s^{-1}} MFt_{peroxo} \xrightarrow{4.2\,s^{-1}} \text{oxo/hydroxo diferric species}$$

Kinetics of Ferroxidase Reaction Probed by Stopped-Flow Absorption Spectroscopy

RFQ-Mössbauer experiments has demonstrated the formation of an intermediate in the early stage of the reaction of M ferritin with Fe^{2+} and O_2. The kinetics of formation and decay of the putative peroxodiferric intermediate determined by RFQ-Mössbauer spectroscopy correlate well with the kinetic behavior of the "blue intermediate" ($\lambda_{max} = 650$ nm) observed by stopped-flow absorption spectroscopy by Fetter et al.[6] This implies that both spectroscopic features are associated with the same intermediate. The experimental conditions in these two studies were different, however. Fetter et al. studied the reaction of 48 Fe/M ferritin (24-mer) with a final protein concentration of 2 μM, whereas for the Mössbauer experiments a ratio of 36 Fe/M ferritin (24-mer) with a final protein concentration of 50 μM was used. Consequently, in order to correlate the results of both techniques, we studied this reaction by stopped-flow absorption spectroscopy under conditions similar to the ones employed for preparation of the RFQ-Mössbauer samples. Because of the increased viscosity of the protein solution, however, we could only

TABLE I
SPECTROSCOPIC PROPERTIES OF PEROXODIFERRIC INTERMEDIATES IN PROTEINS AND PEROXODIFERRIC COMPOUND

Protein or compound	δ (mm/sec)	ΔE_Q (mm/sec)	λ_{max} (nm)	ε (M^{-1}cm^{-1})	ν(O–O) (cm^{-1})	ν_{as}(Fe–O) (cm^{-1})	ν_s(Fe–O) (cm^{-1})	d(Fe–Fe) (Å)
M ferritin	0.62	1.08	650	1000	851	499	485	2.53
R2 variants[a]	0.63	1.58	700	1500	870	499	458	d
Δ^9D	0.68/0.64	1.90/1.06	700	1200	898	490	442	d
MMOH[b]	0.66	1.51	725	1800	d	d	d	d
[Fe$_2$(O$_2$)(carbox)$_2$L$_2$][c]	0.66	1.40	694	2650	888	e	415	4.0

[a] Spectroscopically similar peroxodiferric species have been observed in D84E-R2 and W48F/D84E-R2 variants.
[b] From *Methylococcus capsulatus* (Bath).
[c] Carbox, phenylacetate; L, hydrotris(3,5-isopropyl-1-pyrazolyl)borate.
[d] Not known.
[e] Not observed.

perform stopped-flow experiments with a final protein concentration of 25 μM. The absorbance change at 650 nm as a function of reaction time is shown as a solid line in Fig. 4B and coincides with the temporal behavior of the MFt$_{peroxo}$ determined by RFQ-Mössbauer spectroscopy (circles with error bars in Fig. 4B). Therefore it was concluded that both spectroscopic features arise from the same intermediate. As the concentration of MFt$_{peroxo}$ accumulated at a particular reaction time point can be determined from the Mössbauer data and the total Fe concentration (e.g., in the 25-ms sample, from the Mössbauer data that 70% of total iron is in the peroxodiferric state and from the final total Fe concentration of 1.8 mM, it follows that the concentration of peroxodiferric units in the 25-ms sample is 0.63 mM), it is possible to estimate the extinction coefficient of MFt$_{peroxo}$ by correlating its accumulation concentration with the absorbance change measured at the same reaction time point. Such an estimate yields $\varepsilon_{560} = 1000\ M^{-1}$cm^{-1} for MFt$_{peroxo}$. Similar electronic transitions are observed for a number of peroxodiferric protein intermediates and synthetic model systems (see Table I). On the basis of a detailed spectroscopic study on a model compound,[12] the 650-nm band observed in the reaction of M ferritin is assigned to a peroxo-to-iron charge transfer band.

RFQ-Mössbauer and stopped-flow absorption spectroscopies demonstrated that a peroxodiferric intermediate, termed MFt$_{peroxo}$, forms in the early stage of the ferroxidase reaction. The fact that this intermediate accumulates to high levels

[12] T. C. Brunold, N. Tamura, N. Kitajima, Y. Moro-Oka, and E. I. Solomon, *J. Am. Chem. Soc.* **120**, 5674 (1998).

(70% after 25 ms) allowed the use of other spectroscopic techniques to further characterize its geometric and electronic structure. In the next two sections we describe results obtained by RFQ-XAS and RFQ-resonance Raman spectroscopies.

Characterization of MFt$_{peroxo}$ by RFQ Resonance Raman Spectroscopy

Resonance Raman spectroscopy can provide valuable information for the characterization of transition metal–peroxide complexes, because it allows the determination of the binding mode of the peroxide ligand. The disadvantage of studying RFQ samples by resonance Raman spectroscopy is that these samples contain a substantial amount of cryosolvent, which may have Raman features. In particular, the commonly used RFQ cryosolvent, isopentane, is a strong Raman scatterer and interferes with the detection of weak signals of the sample. We found that samples prepared by using liquefied ethane as cryosolvent were suitable for resonance Raman spectroscopy. In order to identify features associated with MFt$_{peroxo}$ the spectra of a sample quenched at 25 ms (the time at which the MFt$_{peroxo}$ accumulates to a maximum) were studied. A sample quenched at 4 sec (the time at which the MFt$_{peroxo}$ has completely decayed) was used for baseline corrections. It is important to note that in order to obtain a proper baseline, the 4 sec sample should also be prepared by the RFQ method. For the resonance Raman experiments the 647.1-nm line of a Kr$^+$ laser was used for excitation since it corresponds to the maximum absorption of the peroxo-to-iron charge transfer band (650 nm) and therefore ν(O–O) and ν(Fe–O$_{peroxo}$) vibrations are expected to be resonance enhanced. The results are presented in Fig. 5.

The MFt$_{peroxo}$ exhibits a band at 851 cm^{-1} (Fig. 5A, right-hand spectrum), which is in the range of ν(O–O) vibrations of metal-coordinated peroxides. When the reaction is carried out using ^{18}O$_2$ the ν(O–O) vibration is shifted to 800 cm^{-1} (Fig. 5C, right-hand spectrum). The shift of -51 cm^{-1} is in good agreement with the calculated shift of -49 cm^{-1} for a diatomic O–O oscillator. The spectrum of a sample prepared with ^{16}O^{18}O shows a broad, unstructured band in the ν(O–O) region, which extends from 775 cm^{-1} to 875 cm^{-1} (Fig. 5B, right-hand spectrum). The maximum of this peak is between 800 cm^{-1} and 851 cm^{-1} as expected for a sample containing 50% ^{16}O^{18}O-peroxodiferric species. The broad signal, which results in part from the overlap of the spectral contributions of the ^{18}O$_2$- and ^{16}O$_2$-peroxodiferric species (25% each), does not allow spectral deconvolution.

At lower energies two overlapping bands at 485 cm^{-1} and 499 cm^{-1} (Fig. 5A, left-hand spectrum) are also observed, which are downshifted to 468 cm^{-1} and 487 cm^{-1} when ^{18}O$_2$ is used (Fig. 5C, left-hand spectrum). The spectrum of the peroxodiferric species prepared with ^{16}O^{18}O (Fig. 5B, left-hand spectrum) displays a broad cluster of bands in this region, as it is the superposition of multiple bands. On the basis of their energies and intensity ratio, these two isotope-sensitive vibrations are assigned to ν_s(Fe–O) and ν_{as}(Fe–O) vibrations, respectively. The

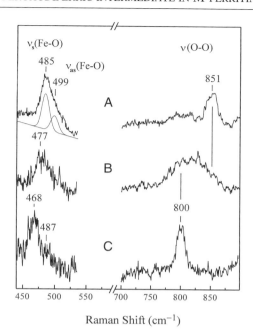

FIG. 5. Resonance Raman spectra of the 25-ms samples containing the intermediate MFt$_{peroxo}$. Samples were prepared with natural abundance O$_2$ (99.5% ^{16}O$_2$) (A), a 1 : 1 : 2 mixture of ^{16}O$_2$, ^{18}O$_2$, and ^{16}O^{18}O, respectively (B), and ^{18}O$_2$ (C). [Adapted from P. Moënne-Loccoz, C. Krebs, K. Herlihy, D. E. Edmondson, E. C. Theil, B. H. Huynh, and T. M. Loehr, *Biochemistry* **38**, 5290 (1999). Copyright © 1999 American Chemical Society.]

fact that two distinct Fe–O vibrations are observed indicates that the peroxide ligand bridges the two ferric ions. Qualitative arguments presented in the following suggest that the peroxide is bound in a μ-1,2-bridging mode.

1. The four most common O$_2$ binding modes observed in dimetal clusters are the μ-1,2-, μ-1,1-, μ-η_2:η_2-bridging modes and the end-on terminal mode (see Fig. 6). A number of peroxodiferric species have been characterized by resonance Raman spectroscopy (see Table I). The Raman features (number, position, and relative intensity of bands) observed for MFt$_{peroxo}$ are similar to those observed for the peroxodiferric species in W48F/D84E-R2 variant[13] and Δ^9D.[14] [The peroxodiferric species in W48F/D84E-R2 and Δ^9D are rather stable ($t_{1/2} > 2.5$ sec).

[13] P. Moënne-Loccoz, J. Baldwin, B. A. Ley, T. M. Loehr, and J. M. Bollinger, Jr., *Biochemistry* **37**, 14659 (1998).

[14] J. A. Broadwater, J. Ai, T. M. Loehr, J. Sanders-Loehr, and B. G. Fox, *Biochemistry* **37**, 14664 (1998).

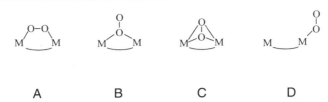

FIG. 6. Different binding modes for the peroxo ligand to coordinate to a dimetal unit: μ-1,2 (A), μ-1,1 (B), μ-η_2:η_2 (C), and end-on terminal (D).

The RFQ method was not required for trapping these reaction species for Raman investigations.] Both of these species have better resolved resonance Raman features and are μ-1,2-bridging. Furthermore, a number of μ-1,2-peroxodiferric model compounds, for which the binding mode was determin by X-ray crystallography, have ν(Fe–O) and ν(O–O) vibrations of similar energies.

2. On the basis of the significantly lower ν(O–O) frequency (749 cm^{-1}) observed for μ-η_2:η_2-peroxodicopper systems compared to those (830–890 cm^{-1}) of μ-1,1- or μ-1,2-peroxodicopper systems,[15] the μ-η_2:η_2-binding mode for MFt$_{peroxo}$ is considered unlikely as the observed ν(O–O) frequency for MFt$_{peroxo}$ is similar to those of known μ-1,2-peroxodiferric species (see Table I). Currently, no μ-η_2:η_2-peroxodiferric species is known.

3. In principle, the two remaining possibilities, μ-1,1 and μ-1,2, can be distinguished by the number and positions of bands in the spectrum of the ^{16}O,^{18}O-peroxodiferric species. The μ-1,1-binding mode is asymmetric with respect to the O atoms and therefore the ν(O–O) vibration is split into two lines. The splitting may be small, however, and the broadness of the ν(O–O) stretch observed in ^{16}O^{18}O-MFt$_{peroxo}$ does not allow the discrimination between these two possibilities. For the ν(Fe–O) vibrations the use of ^{16}O^{18}O has the following consequence: as μ-1,1-binding involves the same bridging O-atom, the resonance Raman features are expected to be similar [depending on the admixture of the ν(O–O) mode into the ν(Fe–O) modes] to the superposition of the ν(Fe–O) modes observed in the pure isotope (^{16}O$_2$ and ^{18}O$_2$) spectra in a 1 : 1 ratio. In contrast, the μ-1,2-binding mode involves both O atoms, and it is expected that the spectral features for this form are more similar to the average of the pure isotope data. The latter possibility is favored by spectral deconvolution of the data. Furthermore, the only well-characterized μ-1,1-peroxodimetal complex (a dicupric complex) has a very low ν_s(Cu–O) vibration. It should be kept in mind, however, that computational results indicate that the electronic structures of peroxodiferric and peroxodicupric systems are different, and insights gained from dicupric systems may not apply to diferric systems.

[15] D. E. Root, M. Mahroof-Tahir, K. D. Karlin, and E. I. Solomon, *Inorg. Chem.* **37**, 4838 (1998).

Brunold et al. reported a detailed experimental and computational study of a μ-1,2-peroxodiferric model compound.[12] They found that increasing the Fe–O–O angle results in an increase of the ν(O–O) frequency and a decrease of the ν_s(Fe–O) frequency. This increase of the frequency does not reflect a stronger O–O bond, but rather is due to mechanical coupling of these two vibrational modes. When this correlation is applied to the observed frequencies in MFt$_{peroxo}$, a small Fe–O–O angle of ca. 110° results.

Characterization of MFt$_{peroxo}$ by RFQ–XAS Spectroscopy

Further information about the structure of MFt$_{peroxo}$ was obtained by X-ray absorption spectroscopy (XAS), which includes X-ray absorption near edge structure (XANES) and extended X-ray absorption fine structure (EXAFS). In the XANES region the energy of the edge provides information about the average oxidation state of the iron species in the sample. The EXAFS region provides information about the number, distances, and types of atoms surrounding the iron. Two sets of RFQ samples were prepared for the XAS measurements and each set contains the following samples: (1) a sample in which M ferritin was reacted with Fe^{2+} for 25 ms in the absence of O_2; (2) two duplicate samples in which M ferritin was reacted with Fe^{2+} and O_2 for 25 ms; and (3) a sample in which M ferritin was reacted with Fe^{2+} and O_2 for 1 sec.

The first sample served as an anaerobic Fe^{2+} control sample, the two duplicate samples contained a maximum amount of MFt$_{peroxo}$, and in the last sample MFt$_{peroxo}$ has completely decayed. All samples were examined by Mössbauer spectroscopy before and after the XAS experiments to monitor sample integrity and to determine the amounts of the various species (Fe^{2+}, MFt$_{peroxo}$, oxo/hydroxo diferric clusters) in the samples.

For the Fe^{2+} control sample the edge energy is at ca. 7122 eV, which is typical for Fe^{2+}. In the 1-sec sample the edge is shifted by ca. 3 eV to higher energy, indicating the formation of ferric species, which is confirmed by Mössbauer quantitation (95% oxo/hydroxo diferric clusters, 5% Fe^{2+}). The edge energy of the 25-ms sample is slightly smaller than that of the 1-sec sample, indicating incomplete oxidation at 25 ms reaction time, and can be simulated as a 1 : 3 mixture of ferrous and ferric species. The 1 : 3 ratio correlates with the Mössbauer quantitation (70% MFt$_{peroxo}$ and 30% Fe^{2+}) and confirms that MFt$_{peroxo}$ is a ferric species.

The Fourier transformation of the EXAFS data is shown in Fig. 7. For all three samples the dominant peak at $R + \alpha \approx 1.55$ Å can be modeled as a Fe–O shell at ≈ 2.0 Å. In addition, the spectrum of the 25-ms sample shows a prominent peak at $R + \alpha \approx 2.25$ Å that can be well modeled as a Fe–Fe shell at 2.53 Å. The best fit parameters yield coordination numbers of 3 for O and 0.5 for Fe. The number 0.5 for Fe is in agreement with the diiron assignment of MFt$_{peroxo}$ based on the Mössbauer data.

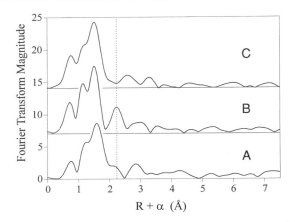

FIG. 7. Fourier transforms of the EXAFS data of an anaerobically prepared Fe^{2+} control sample (A), a 25-ms sample (B), and a 1-sec sample (C). The dashed line marks the center of the peak corresponding to the 2.53 Å Fe–Fe interaction that is specific to the 25-ms sample. [Adapted from J. Hwang, C. Krebs, B. H. Huynh, D. E. Edmondson, E. C. Theil, and J. E. Penner-Hahn, *Science* **287**, 122 (2000). Copyright © 2000 American Association for the Advancement of Science.]

An Fe–Fe distance of 2.53 Å is unexpectly short for μ-1,2-peroxodiferric species and has only been observed in a few Fe species. On the basis of model compound studies, the number of single atom bridges appears to be important for the Fe–Fe separation. Structurally characterized μ-1,2-peroxodiferric species have Fe–Fe separations ranging from 4 Å (no single atom bridge) to 3.1–3.5 Å (one single-atom bridge). Therefore it is proposed that MFt$_{peroxo}$ has two single-atom bridges in addition to the μ-1,2-peroxide ligand in order to accommodate the observed 2.53 Å Fe–Fe separation. Such a structural type is unprecedented in Fe chemistry, but has been observed in Mn chemistry. A compound with a Mn^{IV} $(\mu$-O$)_2(\mu$-1,2-O$_2)$Mn^{IV} core has a Mn–Mn separation of 2.53 Å[16] and is believed to be a good structural model for MFt$_{peroxo}$ (Fig. 8). As a consequence of the small Fe–Fe distance and the μ-1,2-binding mode of the peroxo ligand, the Fe–O–O angle is expected to be small. Assuming typical values for the Fe–O$_{peroxo}$ and O–O bond lengths, $\sim$1.9–2.0 Å and $\sim$1.4–1.45 Å, respectively, the angle is confined to be close to 107°. This value is in agreement with the Raman data, which suggested a value of $\sim$110° based on the low ν(O–O) frequency.

Conclusion

We have used rapid freeze-quench Mössbauer and stopped-flow absorption spectroscopies to study the reaction of recombinant apo-M ferritin from frog with

---

[16] U. Bossek, T. Weyhermüller, K. Wieghardt, B. Nuber, and J. Weiss, *J. Am. Chem. Soc.* **112**, 6387 (1990).

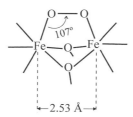

FIG. 8. Proposed core structure for MFt$_{peroxo}$.

Fe^{2+} (36 Fe/protein 24-mer) and O$_2$. The oxidation of Fe^{2+} was measured directly by Mössbauer spectroscopy and found to be biphasic: the majority of the Fe^{2+} (85%) is oxidized rapidly ($k = 80$ s^{-1}), and the remaining 15% is oxidized more slowly ($k = 1.7$ s^{-1}). Concomitantly with the fast decay of the ferrous an intermediate ($k_{formation} = 80$ s^{-1}) was observed, which decays ($k_{decay} = 4.2$ s^{-1}) to multiple oxo/hydroxo-bridged diferric species.

The rates of formation and decay of the intermediate determined by Mössbauer spectroscopy closely match those of the "blue intermediate" ($\lambda_{max} = 650$ nm) observed in stopped-flow experiments. Consequently, it was concluded that both techniques probe the same species. Based on the Mössbauer parameters ($\delta = 0.62$ mm/sec, $\Delta E_Q = 1.08$ nm/sec), the diamagnetic ground state (determined from Mössbauer), and the λ_{max} at 650 nm, this intermediate is proposed to be a peroxodiferric species, termed MFt$_{peroxo}$. As this intermediate accumulates to high levels ($\geq 70\%$) at 25 ms reaction time, it was possible to use the RFQ method to trap this intermediate for resonance Raman and XAS studies to further characterize the geometric structure of the intermediate. These studies confirmed the peroxodiferric assignment and revealed an unusual geometry for MFt$_{peroxo}$: EXAFS spectroscopy demonstrated a short Fe–Fe separation of 2.53 Å, and resonance Raman data are in accordance with a μ-1,2-bridging peroxo ligand. These structural features render the Fe–O–O angle to be small ($\sim 107°$). This unusually small angle has been proposed to strengthen the O–O bond and thus favor the release of H$_2$O$_2$, which is believed to be the product of ferritin ferroxidase reaction, over the cleavage of the O–O bond and formation of high-valent diiron intermediates as observed in R2 and MMOH.

In the following a short summary of the advantages and disadvantages of the four methods employed in our study is presented. Stopped-flow absorption and RFQ-Mössbauer are suitable to obtain time resolution of species participating in the reaction under investigation. The obvious limitations of RFQ-Mössbauer spectroscopy are that (1) only ^{57}Fe-enriched proteins can be studied, and (2) very concentrated solutions ([^{57}Fe] >1 mM after mixing) are required. If these two criteria are met (as in this study) RFQ-Mössbauer spectroscopy is extremely useful because it provides an accurate quantification and detailed characterization of all Fe species involved in the reaction. The Mössbauer data can be used to determine the nuclearity, electronic state, and kinetics of the Fe species, as well as the spin and oxidation state of the individual Fe atoms that form the Fe species. Detailed geometric information (bond length and bond angles) cannot be obtained from the Mössbauer data, however. Stopped-flow absorption spectroscopy has the advantage that it requires less concentrated solutions and that the data collection is fast. It is most efficient for investigating the effects of changing variables. The apparent disadvantage of this technique is that optical absorption bands are generally broad and difficult to interpret, and that multiple species, including initial reactants, reaction intermediates, and products, may all contribute to the same wavelength at which the reaction is being monitored. Also, because the molar absorptivity may not be known for newly discovered transient intermediates, they are difficult to quantify by the stopped-flow absorption technique.

The other two techniques described in this chapter, RFQ-XAS and RFQ-resonance Raman, can be used to obtain detailed information about the geometric and electronic structure of reaction intermediates. For RFQ-resonance Raman spectroscopy a major problem lies in the fact that the cryosolvent may be a good Raman scatterer, thereby possibly overshadowing true signals from the sample. We found liquefied ethane to be a suitable cryosolvent because it can be removed from the sample at $-80°$. As this temperature is significantly higher than the temperature of the quenching bath, it is important to make sure that the reaction intermediate is stable under these conditions. For the ferritin investigation presented here, Mössbauer spectroscopy was used to verify the stability of MFt_{peroxo} at $-80°$. For RFQ-XAS, a major drawback is the lack of selectivity. In the case of Fe-XAS, all Fe species present in a sample, regardless of their states, contribute to the XAS spectrum. As the spectral features of the different species are strongly overlapping, it is difficult to identify features associated with the species of interest. To achieve that, investigations of multiple samples of different compositions and applications of an independent method to quantify the sample compositions are required. In conclusion we would like to note that although each spectroscopic method has its own inherent shortcomings, a combination of different spectroscopic methods, as demonstrated here, can be an extremely valuable tool for the study of biological reactions involving Fe proteins. We would, however, also like to note that the RFQ method has an intrinsic time resolution on the order of milliseconds, and therefore cannot be used for the study of reactions occurring at a faster time scale.

Acknowledgment

We thank all our collaborators who are involved in the work reported here: Kara Herlihy, Jungwon Hwang, Pierre Moënne-Loccoz, Thomas M. Loehr, James E. Penner-Hahn, Alice S. Pereira, William Small, Pedro Tavares, and Elizabeth C. Theil.

[33] A Survey of Covalent, Ionic, and Radical Intermediates in Enzyme-Catalyzed Reactions

By R. Donald Allison and Daniel L. Purich

While the centenary of the first kinetic treatments[1,2] of the enzyme–substrate complex is quickly approaching, a fascinating truth is that the occurrence of covalent intermediates was not as readily accepted. Racker[3] first demonstrated that glyceraldehyde 3-phosphate reacted with an active-site thiol to form a thio-hemiacetal intermediate that subsequently underwent NAD^+-dependent oxidation to yield a enzyme-bound thiol ester intermediate; the acyl group is then transferred to inorganic phosphate, forming 1,3-bisphosphoglycerate. Although Warburg publicly ridiculed Racker's mechanism, and even referred to it as Racker's *Umweg* (detour), he later conceded that nature apparently preferred the *Umweg* over his proposal that P_i directly added to the aldehyde group, allowing subsequent oxidation to 1,3-bisphosphoglycerate. Today, we can amuse ourselves by this historical curiosity, because we have universally embraced the necessity of covalent catalysis in so many enzyme processes. These covalent intermediates facilitate catalysis by providing (*a*) a mechanism for preserving the group transfer potential in multistep reaction schemes, (*b*) a means for converting an atom, or group of atoms, into a better leaving group in nucleophilic reactions, (*c*) a physical way for transferring a reactant between topologically distant active sites on multifunctional enzymes, and (*d*) a capacity for creating timers, latches, and clamps—those specialized mechanoenzymes whose noncovalent binding interactions are maintained for as long as some particular covalent state of a mechanoprotein persists.[4] In addition to covalently attached intermediates, many enzymes form tight complexes with covalent reaction intermediates, while others mediate free radical reactions.

Thirty years ago, Bell and Koshland[5] summarized the evidence for the participation of covalent enzyme–substrate intermediates in some 60 enzyme reactions. In view of the focus of the present volume on the detection and/or characterization of reaction intermediates, we decided to compile a survey of enzymes that form covalent adducts/compounds, carbenium ions, and carbanions, as well as radicals. For brevity, we do not include examples of changes in metal ion oxidation state, even though some metal ions frequently form reaction intermediates by means

---

[1] A. J. Brown, *J. Chem. Soc.* **81**, 373 (1902).
[2] V. Henri, "Lois générales de l'action des diastases." Herman, Paris, 1903.
[3] E. Racker, "Mechanisms in Bioenergetics." Academic Press, New York, 1965.
[4] R. M. Bell and D. E. Koshland, Jr., *Science* **172**, 1253 (1971).
[5] D. L. Purich and R. D. Allison, "The Handbook of Biochemical Kinetics." Academic Press, New York, 2000.

of remarkably stable coordinately covalent and/or shared electron-pair covalent bonds. Table I presents alphabetically nearly 300 reactions and, where possible, Enzyme Commission classification numbers are included to avoid confusion. Our compilation is by no means exhaustive, especially when one considers the nearly 6000 known enzyme-catalyzed reactions.[6] Nor were we interested in evaluating the quality of experimental evidence supporting the catalytic competence of the reaction intermediates presented. Space restrictions prevent inclusion of more than a single literature citation for each enzyme intermediate, and we apologize in advance to those investigators whose names are unavoidably omitted.

In a sense, this chapter celebrates the intuition and inventiveness of those enzyme chemists who have enriched the general awareness of the diversity of covalent species that partake in enzyme processes.

[6] D. L. Purich and R. D. Allison, "The Enzyme Reference: A Guidebook to Enzyme Reactions, Nomenclature, and Methodology." Academic Press, New York, 2002.

TABLE I
EXAMPLES OF INTERMEDIATES IN ENZYME-CATALYZED REACTIONS

Enzyme	EC number	Intermediate/comment	Ref.[a]
Acetoacetate decarboxylase	[EC 4.1.1.4]	Imine adduct (lysine residue)	1
Acetylcholinesterase	[EC 3.1.1.7]	O-Acetylated enzyme (serine residue)	2
Acetyl-CoA acetyltransferase	[EC 2.3.1.9]	S-Acetylated enzyme (cysteine residue)	3
Acetyl-CoA acyltransferase	[EC 2.3.1.16]	S-Acetylated enzyme (cysteine residue)	4
Acetyl-CoA carboxylase	[EC 6.4.1.2]	N-Carboxybiotin and carboxyphosphate	5
Acetylenedicarboxylate decarboxylase	[EC 4.1.1.78]	Enzyme-bound oxaloacetate	6
Acid phosphatase	[EC 3.1.3.2]	Phospho-enzyme intermediate	7
Acyl-[acyl-carrier-protein] desaturase	[EC 1.14.19.2]	Peroxo-diferric intermediate	8
Acylaminoacyl-peptidase	[EC 3.4.19.1]	Acyl-enzyme intermediate	9
[Acyl-carrier protein] acetyltransferase	[EC 2.3.1.38]	Acetyl-enzyme intermediate	10
Adenosine deaminase	[EC 3.5.4.4]	Unstable hydrated purine adduct	11
S-Adenosylhomocysteine nucleosidase	[EC 3.2.2.9]	1,2-Epoxide intermediate	12
S-Adenosylmethionine decarboxylase	[EC 4.1.1.50]	Schiff base intermediate	13
Adenylosuccinate synthetase	[EC 6.3.4.4]	6-Phosphorylinosine 5′-monophosphate intermediate	14
Adenylylsulfate kinase	[EC 2.7.1.25]	Phosphoenzyme intermediate	15

TABLE I (*continued*)

Enzyme	EC number	Intermediate/comment	Ref.[a]
Adenylylsulfate reductase	[EC 1.8.99.2]	SO_3^{2-}-bound intermediate (cysteine residue)	16
D-Alanine aminotransferase	[EC 2.6.1.21]	Aldimine and ketimine intermediates	17
L-Alanine aminotransferase	[EC 2.6.1.2]	Aldimine and ketimine intermediates	18
L-Alanine dehydrogenase	[EC 1.4.1.1]	Carbinolamine intermediate	19
D-Alanine: poly (phosphoribitol) ligase	[EC 6.1.1.13]	D-Alanyl-AMP intermediate	20
Alanine racemase	[EC 5.1.1.1]	D-/L-Alanine aldimine intermediate	21
D-Alanyl-D-alanine synthetase	[EC 6.3.2.4]	Acyl phosphate intermediate	22
Alanyl-tRNA synthase	[EC 6.1.1.7]	Alanyl-adenylate intermediate	23
Alkaline phosphatase	[EC 3.1.3.1]	Phospho-enzyme intermediate	24
Amidase	[EC 3.5.1.4]	Acyl-enzyme intermediate	25
ω-Amidase	[EC 3.5.1.3]	Acyl-enzyme intermediate	26
Amidophosphoribosyl-transferase	[EC 2.4.2.14]	γ-Glutamyl thioester intermediate	27
Amine dehydrogenase	[EC 1.4.99.3]	Iminoquinone Schiff base between amine-substrate and cofactor (a semiquinone intermediate may also form)	28
Amine oxidase (copper-containing)	[EC 1.4.3.6]	Carbinolamine intermediate formed by attack of amine substrate at $5'$-position of enzyme-bound TOPA quinone; rearranges to Schiff base	28
D-Amino acid oxidase	[EC 1.4.3.3]	Carbanion intermediate	29
L-2-Aminoadipate-6-semialdehyde dehydrogenase	[EC 1.2.1.31]	Aminoadipoyl-AMP intermediate	30
p-Aminobenzoate synthase		γ-Glutamyl-enzyme intermediate	31
4-Aminobutyrate aminotransferase	[EC 2.6.1.19]	Aldimine and ketimine intermediates	32
1-Aminocyclopropane-1-carboxylate deaminase	[EC 4.1.99.4]	Aldimine (vinylglycyl-pyridoxal 5-phosphate)	33
(2-Aminoethyl) phosphonate: pyruvate aminotransferase	[EC 2.6.1.37]	Aldimine intermediate	34
2-Aminomuconate deaminase	[EC 3.5.99.5]	Imine intermediate	35
8-Amino-7-oxononanoate synthase	[EC 2.3.1.47]	External aldimine of L-alanine with bound cofactor; quinonoid intermediate forms second aldimine	36
2-Aminophenol oxidase	[EC 1.10.3.4]	O-Quinoneimine intermediate	37
Ammonia kinase	[EC 2.7.3.8]	Phosphoryl-enzyme intermediate	38
Anthranilate 1,2-dioxygenase	[EC 1.14.12.1]	Cyclic peroxide intermediate	39

continued

TABLE I (continued)

Enzyme	EC number	Intermediate/comment	Ref.[a]
Anthranilate 3-monooxygenase	[EC 1.14.13.35]	Imine and C4a-hydroperoxyflavin intermediate	40
Anthraniloyl-CoA monooxygenase	[EC 1.14.13.40]	Flavin hydroperoxide and 2-amino-5-hydroxybenzoyl-CoA	41
Arabinose-5-phosphate isomerase	[EC 5.3.1.13]	Enediol intermediate	42
Argininosuccinate lyase	[EC 4.3.2.1]	Transient carbanion intermediate	43
Argininosuccinate synthase	[EC 6.3.4.5]	Citrullinyl-adenylate intermediate	44
Arginyl-tRNA synthase	[EC 6.1.1.19]	Arginyl-adenylate intermediate	23
Aryl-aldehyde dehydrogenase (NADP$^+$)	[EC 1.2.1.30]	Acyl-adenylate intermediate	45
Arylamine N-acetyltransferase	[EC 2.3.1.5]	Acetyl-enzyme intermediate	46
Arylsulfate sulfotransferase	[EC 2.8.2.22]	Sulfo-enzyme intermediate (tyrosyl residue)	47
Ascorbate peroxidase	[EC 1.11.1.11]	Ferryl porphyrin π-cation radical intermediate	48
Asparagine synthase	[EC 6.3.5.4]	β-aspartyl-AMP intermediate	49
Asparaginyl-tRNA synthase	[EC 6.1.1.22]	Asparaginyl-adenylate intermediate	23
Aspartate aminotransferase	[EC 2.6.1.1]	Internal aldimine and ketimine intermediate	18
Aspartate 1-decarboxylase	[EC 4.1.1.11]	L-Aspartate Schiff base (enzyme pyruvoyl group)	50
Aspartate-semialdehyde dehydrogenase	[EC 1.2.1.11]	Thiohemiacetal and thiol ester intermediates	51
Aspartyl-tRNA synthase	[EC 6.1.1.12]	Aspartyl-adenylate intermediate	23
ATPase, K$^+$-transporting	[EC 3.6.3.12]	Acyl-phosphate intermediate	52
ATP:citrate (pro-3S)-lyase	[EC 4.1.3.8]	Phospho-enzyme intermediate	53
Benzoylformate decarboxylase	[EC 4.1.1.7]	Carbanion intermediate	54
Biotin synthase	[EC 2.8.1.6]	Dethiobiotin and adenosyl radical intermediates	55
Bisphosphoglycerate mutase	[EC 5.4.2.4]	Phosphoryl-enzyme intermediate (histidine residue)	56
Botryococcene synthase		Presqualene bisphosphate	57
Bromoperoxidase		Peroxyvanadium(V) species	58
Carbamoylphosphate synthase (ammonia)	[EC 6.3.4.16]	Carboxyphosphate intermediate	59
Carbamoylphosphate synthase (glutamine-hydrolyzing)	[EC 6.3.5.5]	Carboxyphosphate and γ-glutamyl thiol ester intermediate	59
3-Carboxyethylcatechol 2,3-dioxygenase	[EC 1.13.11.16]	Semiquinone-like intermediate	60
Carboxypeptidase C	[EC 3.4.16.5]	Acyl-enzyme intermediate	61
Carboxypeptidase Y	[EC 3.4.16.5]	Acyl-enzyme intermediate	61
Carnosine synthase	[EC 6.3.2.11]	Acyl-adenylate intermediate	62
Catechol 1,2-dioxygenase	[EC 1.13.11.1]	Fe(III)-superoxide intermediate and semiquinone-like intermediate	63

TABLE I (continued)

Enzyme	EC number	Intermediate/comment	Ref.[a]
CDP-Diacylglycerol:serine O-phosphatidyltransferase	[EC 2.7.8.8]	Phosphatidyl-enzyme intermediate	64
Chitinase	[EC 3.2.1.14]	Oxazoline ion intermediate	65
4-Chlorobenzoyl-CoA dehalogenase	[EC 3.8.1.7]	Meisenheimer intermediate formed by attack of active-site aspartyl residue on 4-position of benzoyl ring	66
Cholesterol oxidase	[EC 1.1.3.6]	Cholest-5-ene-3-one intermediate	67
Cholinesterase	[EC 3.1.1.8]	Acyl-enzyme intermediate (serine residue)	68
Chymotrypsin	[EC 3.4.21.1]	O-Acyl-enzyme intermediate (serine residue)	69
Coproporphyrinogen oxidase	[EC 1.3.3.3]	β-Hydroxypropionate-porphyrinogen intermediate	70
4-Cresol dehydrogenase (hydroxylating)	[EC 1.17.99.1]	Quinone methide intermediate	71
3-Cyanoalanine nitrilase	[EC 3.5.5.4]	L-Asparagine intermediate	72
Cyclohexanone monooxygenase	[EC 1.14.13.22]	C4a-Flavin-hydroxide intermediate	29
Cyclomaltodextrin glucanotransferase	[EC 2.4.1.19]	Covalent enzyme–substrate intermediate	73
Cyclosporin synthase		Amino acyl-AMP intermediates	74
Cystathionine β-lyase	[EC 4.4.1.8]	Aldimine and ketimine intermediates	75
Cysteine synthase	[EC 4.2.99.8]	α-Aminoacrylate intermediate	76
Cysteinyl-tRNA synthetase	[EC 6.1.1.16]	Cysteinyl-adenylate intermediate	23
2-Dehydro-3-deoxy-L-arabonate dehydratase	[EC 4.2.1.43]	Schiff base intermediate	77
5-Dehydro-4-deoxyglucarate dehydratase	[EC 4.2.1.41]	Schiff base intermediate	78
3-Dehydroquinate dehydratase	[EC 4.2.1.10]	Schiff base intermediate	77
2-Deoxyribose-5-phosphate aldolase	[EC 4.1.2.4]	Schiff base intermediate	77
2'-Deoxyuridylate hydroxymethyltransferase		Cysteinyl-C6-nucleotide adduct and enzyme-linked dCMP bonded to N-5 position of tetrahydrofolate via CH_2	79
Dethiobiotin synthase	[EC 6.3.3.3]	Carbamic-phosphoric anhydride intermediate	80
Dextransucrase	[EC 2.4.1.5]	Glycosyl-enzyme intermediate	65
Dichloromethane dehalogenase	[EC 4.5.1.3]	Halomethylthioether intermediate	81
Dihydrodipicolinate synthase	[EC 4.2.1.52]	Schiff base intermediates	82
Dihydrofolate synthase	[EC 6.3.2.12]	Acyl-phosphate intermediate	83
N-(2,3-Dihydroxybenzoyl) serine synthase	[EC 6.3.2.14]	2,3-Dihydroxybenzoyl-AMP intermediate	84
DNA ligase (NAD^+)	[EC 6.5.1.2]	AMP-enzyme intermediate (lysine residue)	85

continued

TABLE I (continued)

Enzyme	EC number	Intermediate/comment	Ref.[a]
Dopamine β-monooxygenase	[EC 1.14.17.1]	Substrate-derived benzylic radical	86
Elastase, pancreatic	[EC 3.4.21.36]	Acyl-enzyme intermediate (serine residue)	87
Endopeptidase E, pancreatic	[EC 3.4.21.70]	Acyl-enzyme intermediate (serine residue)	88
Enterobacter ribonuclease	[EC 3.1.27.6]	$2',3'$-Cyclic phosphate intermediate	89
Ethanolamine ammonia-lyase	[EC 4.3.1.7]	Cobalt(II) radical, anion radical, and acetaldehyde radical	90
Fatty acid synthase	[EC 2.3.1.85]	Acyl thiol ester intermediates	91
(−)-*endo*-Fenchol synthase	[EC 4.2.3.10]	($3R$)-Linalyl bisphosphate intermediate	92
Ficain	[EC 3.4.22.3]	Acyl-enzyme intermediate (cysteine residue)	93
Formamidopyrimidine-DNA glycosylase	[EC 3.2.2.23]	Schiff base intermediate	94
Formate C-acetyltransferase	[EC 2.3.1.54]	Glycyl radical and thiyl radical intermediates	95
[Formate acetyltransferase] activating enzyme	[EC 1.97.1.4]	$5'$-Deoxyadenosyl radical intermediate	96
Formyltetrahydrofolate synthase	[EC 6.3.4.3]	Formyl phosphate intermediate	97
β-Fructofuranosidase	[EC 3.2.1.26]	α-Fructofuranosylated-enzyme intermediate (active-site carboxyl group)	65
Fructose-2,6-bisphosphatase	[EC 3.1.3.46]	$N^{3'}$-Phosphohistidine intermediate	98
Fructose-1,6-bisphosphate aldolase	[EC 4.1.2.13]	Schiff base and enamine intermediates	77
Galactose oxidase	[EC 1.1.3.9]	Protein radical intermediate	99
β-Galactosidase	[EC 3.2.1.23]	Glycosyl-enzyme intermediate	100
GDP-Mannose 4,6-dehydratase	[EC 4.2.1.47]	GDP-4-keto-D-mannose intermediate	101
GDP-Mannose 3,5-epimerase	[EC 5.1.3.18]	Ene-diol(ate) intermediate	102
Geranoyl-CoA carboxylase	[EC 6.4.1.5]	Carboxybiotin intermediate	103
Geranyl-bisphosphate cyclase	[EC 5.5.1.8]	Allylic (−)-($3R$)-linalyl bisphosphate intermediate	104
Glucosamine acetyltransferase	[EC 2.3.1.3]	Acetyl-enzyme intermediate	105
Glucose oxidase	[EC 1.1.3.4]	Flavin C4a-hydroperoxide intermediate	29
Glucose-6-phosphatase	[EC 3.1.3.9]	Phosphoryl-enzyme intermediate (N^3-histidine residue)	106
Glucose-6-phosphate isomerase	[EC 5.3.1.9]	*cis*-Enediol intermediate	107
α-Glucosidase	[EC 3.2.1.20]	Oxocarbenium intermediate	65
β-Glucosidase	[EC 3.2.1.21]	α-Glucosyl enzyme intermediate	108
Glutaconate CoA-transferase	[EC 2.8.3.12]	CoA-Thiol ester intermediate (glutamate residue)	109

TABLE I (*continued*)

Enzyme	EC number	Intermediate/comment	Ref.[a]
Glutaconyl-CoA decarboxylase	[EC 4.1.1.70]	Carboxy-biotin intermediate	110
Glutamate decarboxylase	[EC 4.1.1.15]	Aldimine (with pyridoxal phosphate)	111
Glutamate racemase	[EC 5.1.1.3]	Carbanion intermediate	112
Glutamate synthase (NADPH)	[EC 1.4.1.13]	γ-Glutamyl thiolester intermediate (cysteine residue)	27
Glutamine synthase	[EC 6.3.1.2]	γ-Glutamyl-phosphate intermediate	113
Glutaminyl-tRNA synthase	[EC 6.1.1.18]	Glutaminyl-adenylate intermediate	114
γ-Glutamylcysteine synthase	[EC 6.3.2.2]	γ-Glutamyl-phosphate intermediate	115
γ-Glutamyl transpeptidase	[EC 2.3.2.2]	γ-Glutamyl-enzyme intermediate	116
Glutamyl-tRNA synthase	[EC 6.1.1.17]	Glutamyl-adenylate intermediate	23
Glutaryl-CoA dehydrogenase	[EC 1.3.99.7]	Glutaconyl-CoA intermediate	117
Glutathione synthase	[EC 6.3.2.3]	Acyl-phosphate intermediate	118
Glyceraldehyde-3-phosphate dehydrogenase	[EC 1.2.1.12]	Thiohemiacetal and thiolester intermediates	119
Glycerol dehydratase	[EC 4.2.1.30]	Radical anion intermediate	120
Glycine amidinotransferase	[EC 2.1.4.1]	Amidino-enzyme intermediate (cysteinyl residue)	121
Glycine N-choloyltransferase	[EC 2.3.1.65]	Cholyl-enzyme intermediate	122
Glycine reductase		*Se*-(Carboxymethyl)selenocysteinyl intermediate	123
Glycyl-tRNA synthase	[EC 6.1.1.14]	Glycyl-adenylate intermediate	23
GMP synthase	[EC 6.3.4.1]	2-*O*-Adenylyl-XMP intermediate	124
GMP synthase (glutamine-hydrolyzing)	[EC 6.3.5.2]	γ-Glutamyl thiolester intermediate	125
GTP guanlyltransferase	[EC 2.7.7.45]	Guanylyl-enzyme intermediate (histidine residue)	126
Haloacetate dehalogenase	[EC 3.8.1.3]	Ester intermediate	127
2-Haloacid dehalogenase	[EC 3.8.1.2]	Ester intermediate (aspartate residue)	128
Haloalkane dehalogenase	[EC 3.8.1.5]	Ester intermediate (aspartate residue)	129
Heparan-A-glucosaminide acetyltransferase	[EC 2.3.1.78]	Acetyl-enzyme intermediate	130
Heparosan-N-sulfate-glucuronate 5-epimerase	[EC 5.1.3.17]	Carbanion intermediate	131
Hexadecanal Dehydrogenase	[EC 1.2.1.42]	Thiolacetal adduct and thiol ester intermediate	132
Histamine N-methyltransferase	[EC 2.1.1.8]	Methylated enzyme intermediate	133
Histidine decarboxylase	[EC 4.1.1.22]	Schiff base (pyridoxal phosphate or a pyruvoyl residue)	134
Histidinol dehydrogenase	[EC 1.1.1.23]	Histidinal intermediate	135
Histidyl-tRNA synthetase	[EC 6.1.1.21]	Histidyl-adenylate intermediate	23
Homoserine *O*-acetyltransferase	[EC 2.3.1.31]	Acetyl-enzyme intermediate	136

continued

TABLE I (continued)

Enzyme	EC number	Intermediate/comment	Ref.[a]
Hydroxyacylglutathione hydrolase	[EC 3.1.2.6]	Acyl-imidazole intermediate	137
2'-Hydroxybenzalpyruvate aldolase		Schiff base and enamine intermediate	138
4-Hydroxybenzoate 3-monooxygenase	[EC 1.14.13.2]	C4a-Hydroperoxyflavin intermediate	29
4-Hydroxybutyryl-CoA dehydratase		4-Hydroxy-but-2-enoyl-CoA intermediate	139
3-Hydroxybutyryl-CoA epimerase	[EC 5.1.2.3]	2-*trans*-Enoyl-CoA intermediate	140
(*R*)-2-Hydroxyglutaryl-CoA dehydratase		Ketyl anion radical intermediate	141
4-Hydroxy-2-ketoglutarate aldolase	[EC 4.1.3.16]	Schiff-base intermediate	142
3-Hydroxy-3-methyl-glutaryl-CoA synthase	[EC 4.1.3.5]	Thiol ester intermediate (cysteine residue)	143
3-Hydroxy-2-methyl-pyridinecarboxylate dioxygenase	[EC 1.14.12.4]	Flavin hydroperoxide intermediate	144
4-Hydroxyphenylacetate 3-monooxygenase	[EC 1.14.13.3]	C4a-Flavin hydroperoxide intermediate	145
trans-L-3-Hydroxyproline dehydratase	[EC 4.2.1.77]	2,3-Dehydroproline intermediate	146
Hypoxanthine:guanine phosphoribosyl-transferase	[EC 2.4.2.8]	Ribooxocarbenium ion intermediate	147
Imidazoleglycerol-phosphate dehydratase	[EC 4.2.1.19]	Diazafulvene intermediate	148
Inorganic pyrophosphatase	[EC 3.6.1.1]	Pyrophosphoryl- and phosphoryl-enzyme intermediates	149
myo-Inositol oxygenase	[EC 1.13.99.1]	L-*myo*-Inosose-1 intermediate	150
myo-Inositol-1-phosphate synthase	[EC 5.5.1.4]	5-Dehydro-D-glucose 6-phosphate intermediate	151
Inulosucrase	[EC 2.4.1.9]	Fructosyl-enzyme intermediate	100
Isobutyryl-CoA mutase	[EC 5.4.99.13]	5'-Deoxyadenosyl- and butanoyl-CoA radical intermediates	152
Isocitrate dehydrogenase ($NADP^+$)	[EC 1.1.1.42]	Enzyme-bound oxalosuccinate intermediate	153
Isoleucyl-tRNA synthase	[EC 6.1.1.5]	Isoleucyl-adenylate intermediate	23
3-Ketoacid CoA-transferase	[EC 2.8.3.5]	Enzyme-B-bound thiol ester intermediate	154
3-Ketoacyl-[acyl-carrier protein] synthase	[EC 2.3.1.41]	Acyl-enzyme intermediate	155
α-Ketoglutarate synthase	[EC 1.2.7.3]	Radical intermediate	156
Kynureninase	[EC 3.7.1.3]	Pyruvate:pyridoxamine 5'-phosphate ketimine intermediate	157

TABLE I (*continued*)

Enzyme	EC number	Intermediate/comment	Ref.[a]
Lactoylglutathione lyase	[EC 4.4.1.5]	Thiohemiacetal and *cis*-enediolate intermediates	120
Leucine 2,3-aminomutase	[EC 5.4.3.7]	Cobalamin-mediated radical intermediates	120
Leucyl-tRNA synthase	[EC 6.1.1.4]	Leucyl-adenylate intermediate	23
Levansucrase	[EC 2.4.1.10]	Fructosyl-enzyme intermediate	158
Licheninase	[EC 3.2.1.73]	Glucosyl-enzyme intermediate	159
Long-chain-fatty-acid: luciferin-component ligase	[EC 6.2.1.19]	Acyl-adenylate intermediate	160
Lysine-2,3-aminomutase	[EC 5.4.3.2]	5′-Deoxyadenosyl radical & L-methionine intermediates	120
β-Lysine 5,6-aminomutase	[EC 5.4.3.3]	Schiff base (pyridoxal 5′-phosphate)	120
D-Lysine 5,6-aminomutase	[EC 5.4.3.4]	Schiff base and radical intermediates	120
Lysophospholipase	[EC 3.1.1.5]	Acyl-enzyme intermediate	161
Lysyl-tRNA synthase	[EC 6.1.1.6]	Lysyl-adenylate intermediate	23
Mandelate racemase	[EC 5.1.2.2]	Enolic intermediate	162
Melilotate 3-monooxygenase	[EC 1.14.13.4]	Flavin-C4a-hydroperoxide intermediate	29
Mercury(II) reductase	[EC 1.16.1.1]	Flavin-C4a-thiol intermediate	29
5,10-Methenyltetrahydrofolate synthase	[EC 6.3.3.2]	Phosphorylated-substrate intermediate	163
Methionyl-tRNA synthase	[EC 6.1.1.10]	Methionyl-adenylate intermediate	23
β-Methylaspartate mutase	[EC 5.4.99.1]	Adenosyl, β-methylaspartate, and glycyl radical intermediates	164
β-Methylcrotonyl-CoA carboxylase	[EC 6.4.1.4]	Carboxybiotin intermediate	165
2-Methyleneglutarate mutase	[EC 5.4.99.4]	5′-Deoxyadenosine, acrylate, and cyclopropylcarbinyl radical intermediates	120
Methylglyoxal synthase	[EC 4.2.3.3]	Ene-diol(ate) intermediate (collapses to a 2-hydroxy 2-propenal enol intermediate)	166
Methylmalonyl-CoA decarboxylase	[EC 4.1.1.41]	Carboxy-biotin intermediate	167
Methylmalonyl-CoA mutase	[EC 5.4.99.2]	5′-Deoxyadenosyl and succinyl-CoA radical intermediates	158
*M.Hha*I methyltransferase	[EC 2.1.1.73]	C4 adduct of cytosine (cysteine residue)	168
NADase	[EC 3.2.2.5]	ADP-ribosyl-enzyme intermediate	169
Nicotinate phosphoribosyltransferase	[EC 2.4.2.11]	Phospho-enzyme intermediate (histidine residue)	170
Nitric oxide synthase	[EC 1.14.13.39]	N^ω-Hydroxy-L-arginine intermediate	171
Nucleoside deoxyribosyltransferase	[EC 2.4.2.6]	Deoxyribosyl-enzyme intermediate (glutamate residue)	172
Nucleoside-bisphosphate kinase	[EC 2.7.4.6]	Phospho-enzyme intermediate (histidine residue)	173
Nucleoside phosphotransferase	[EC 2.7.1.77]	Phosphoryl-enzyme intermediate	174

continued

TABLE I (continued)

Enzyme	EC number	Intermediate/comment	Ref.[a]
D-Ornithine 4,5-aminomutase	[EC 5.4.3.5]	Schiff base and cobalamin-mediated radical intermediates	120
Pantothenoylcysteine decarboxylase	[EC 4.1.1.30]	Schiff base intermediate (pyruvoyl moiety)	175
Papain	[EC 3.4.22.2]	Thiol ester intermediate	176
Phenol 2-monooxygenase	[EC 1.14.13.7]	Flavin-C4a-hydroxide intermediate	177
Phenylalanine 4-monooxygenase	[EC 1.14.16.1]	C4a-Peroxytetrahydropterin intermediate	178
Phenylalanine racemase	[EC 5.1.1.11]	Aminoacyl-AMP and acyl-pantotheine intermediates	21
Phenylalanyl-tRNA synthase	[EC 6.1.1.20]	Phenylalanyl-AMP intermediate	179
Phosphatidylcholine:retinol O-acyltransferase	[EC 2.3.1.135]	Acyl-enzyme intermediate	180
Phosphatidylserine decarboxylase	[EC 4.1.1.65]	Schiff base (pyruvoyl moiety)	50
6-Phospho-2-dehydro-3-deoxygluconate aldolase	[EC 4.1.2.14]	Schiff base intermediate	77
Phosphodiesterase	[EC 3.1.4.1]	Phosphorylated intermediate (threonine residue)	181
6-Phosphofructo-2-kinase/fructose-2,6-bisphosphatase	[EC 2.7.1.105] [EC 3.1.3.46]	Phospho-enzyme intermediate (histidine residue)	182
6-Phospho-β-galactosidase	[EC 3.2.1.85]	Glycosyl-enzyme intermediate	183
Phosphogluconate dehydrogenase	[EC 1.1.1.44]	1,2-Enediol intermediate	184
Phosphoglucosamine mutase	[EC 5.4.2.10]	D-Glucosamine 1,6-bisphosphate intermediate	185
Phosphoglycolate phosphatase	[EC 3.1.3.18]	Phospho-enzyme intermediate	186
Phosphomannomutase	[EC 5.4.2.8]	Phospho-enzyme intermediate	187
Phosphonoacetaldehyde hydrolase	[EC 3.11.1.1]	Schiff base and β-aspartyl phosphate intermediates	188
Phosphopantothenoyl-cysteine decarboxylase	[EC 4.1.1.36]	Schiff base (pyruvoyl group)	50
Phosphoramidate:hexose phosphotransferase	[EC 2.7.1.62]	Phospho-enzyme intermediate	189
Phosphoribosylformyl-glycinamidine synthase	[EC 6.3.5.3]	γ-Glutamyl thiol ester intermediate	190
Phosphorylase	[EC 2.4.1.1]	Glycosyl-enzyme (or stabilized oxocarbenium ion) intermediate	65
Plasmin	[EC 3.4.21.7]	Acyl-enzyme intermediate	191
Porphobilinogen synthase	[EC 4.2.1.24]	Schiff base intermediate	77
D-Proline reductase (dithiol)	[EC 1.4.4.1]	Schiff base (pyruvoyl group)	192
Prolyl-tRNA synthase	[EC 6.1.1.15]	Prolyl-adenylate intermediate	23

TABLE I (continued)

Enzyme	EC number	Intermediate/comment	Ref.[a]
Propanediol dehydratase	[EC 4.2.1.28]	Substrate-derived radical intermediate	120
Propionyl-CoA carboxylase	[EC 6.4.1.3]	Carboxy-biotin intermediate	193, 194
Protamine kinase	[EC 2.7.1.70]	Phospho-enzyme intermediate	195
Protein-glutamine γ-glutamyltransferase	[EC 2.3.2.13]	Acyl-enzyme intermediate	196
Protein-N^π-phosphohistidine:sugar phosphotransferase	[EC 2.7.1.69]	Phospho-enzyme intermediate	197
Protein-tyrosine-phosphatase	[EC 3.1.3.48]	Phosphocysteinyl intermediate	198
Protochlorophyllide reductase	[EC 1.3.1.33]	Free-radical intermediate	199
Δ^1-Pyrroline-5-carboxylate synthase		γ-Glutamate semialdehyde intermediate	200
Pyruvate carboxylase	[EC 6.4.1.1]	Biotin and carboxy-phosphate intermediates	201
Pyruvate, orthophosphate dikinase	[EC 2.7.9.1]	Pyrophosphoryl- and phosphoryl-enzyme intermediates	202
Pyruvate oxidase	[EC 1.2.3.3]	Hydroxyethylthiamin pyrophosphate intermediate	203
Pyruvate synthase	[EC 1.2.7.1]	Stable radical intermediate	204
Pyruvate, water dikinase	[EC 2.7.9.2]	Phosphohistidine intermediate	205
Rhamnulose-1-phosphate aldolase	[EC 4.1.2.19]	Ene-diolate intermediate	152
Rhodanese	[EC 2.8.1.1]	Sulfur-enzyme intermediate	206
Ribonuclease, pancreatic	[EC 3.1.27.5]	Nucleoside-2′,3′-cyclic phosphodiester intermediates	207
Ribonuclease U2	[EC 3.1.27.4]	Nucleoside-2′,3′-cyclic phosphodiester intermediates	208
Ribonucleoside-bisphosphate reductase	[EC 1.17.4.1]	5′-Deoxyadenosyl radical intermediate	120
Ribonucleoside-triphosphate reductase	[EC 1.17.4.2]	Thiyl radical intermediate	209
Ribulose-1,5-bisphosphate carboxylase/oxygenase	[EC 4.1.1.39]	Carbanion and metal-stabilized enediol intermediates	152
D-Ribulose-5-phosphate 3-epimerase	[EC 5.1.3.1]	Enediolate intermediate	21
RNA-3′-phosphate cyclase	[EC 6.5.1.4]	Adenylyl-enzyme intermediate	210
Sabinene-hydrate synthase	[EC 4.2.3.11]	(3R)-Linalyl bisphosphate intermediate	211
Selenide, water dikinase	[EC 2.7.9.3]	Phosphoryl-enzyme intermediate	212
L-Seryl-tRNA selenium transferase	[EC 2.9.1.1]	2-Aminoacryloyl-tRNA aldimine intermediate	213
Seryl-tRNA synthase	[EC 6.1.1.11]	Serine-adenylate intermediate	23
Sialidases	[EC 3.2.1.18]	Glycosyl-enzyme intermediate	65
Steroid Δ-isomerase	[EC 5.3.3.1]	Dienolate intermediate	152
Succinyl-CoA synthetase	[EC 6.2.1.5]	Phospho-histidine and succinyl-phosphate intermediates	214

continued

TABLE I (*continued*)

Enzyme	EC number	Intermediate/comment	Ref.[a]
O-Succinylhomoserine (thiol)-lyase	[EC 4.2.99.9]	Pyridoxamine-vinylglyoxylate intermediate	215
Sucrose:1,6-α-glucan 3-α-glucosyltransferase		Glycosyl-enzyme intermediate	216
Sucrose phosphorylase	[EC 2.4.1.7]	Glucosyl-enzyme intermediate	217
Tagatose 3-epimerase		2-Ene-2,3-diol intermediate	218
Tartronate-semialdehyde synthase	[EC 4.1.1.47]	2-Hydroxymethylthiamin pyrophosphate intermediate	29
Thioglucosidase	[EC 3.2.3.1]	Glycosyl-enzyme intermediate (glutamate residue)	219
Threonyl-tRNA synthase	[EC 6.1.1.3]	Threonyl-adenylate intermediate	23
Thrombin	[EC 3.4.21.5]	Acyl-enzyme intermediate	220
Thymidylate synthase	[EC 2.1.1.45]	5,6-Dihydropyrimidine adduct (cysteine residue)	221
Transaldolase	[EC 2.2.1.2]	Schiff-base intermediate	222
Transketolase	[EC 2.2.1.1]	1,2-Dihydroxyalkyl-thiamin pyrophosphate intermediate	152
Trichodiene synthase	[EC 4.2.3.6]	Bisabolyl carbocation intermediate	223
Triose-phosphate isomerase	[EC 5.3.1.1]	*cis*-Enediolate intermediate	224
Trypsin	[EC 3.4.21.4]	Acyl-enzyme intermediate (serine residue)	225
Tryptophanase	[EC 4.1.99.1]	Aldimine and quinonoid intermediates	226
Tryptophan synthase	[EC 4.2.1.20]	Indole intermediate and aminoacrylate aldimine	227
Tryptophanyl-tRNA synthase	[EC 6.1.1.2]	Tryptophanyl adenylate intermediate	228
Tyrosyl-tRNA synthase	[EC 6.1.1.1]	Tyrosyl-adenylate intermediate	114
Ubiquitin carboxyl-terminal hydrolase	[EC 3.4.19.12]	Ubiquitinyl-enzyme intermediate	229
Ubiquitin:protein ligase	[EC 6.3.2.19]	Ubiquitin thiol ester (cysteine residue) and ubiquitin-AMP	230
UDP-N-acetylglucosamine 1-carboxy-vinyl-transferase	[EC 2.5.1.7]	Phosphothiolactoyl-enzyme intermediate	23
UDP-N-acetylglucosamine 2-epimerase	[EC 5.1.3.14]	Enzyme-bound 2-acetamidoglucal intermediate	21
UDP-N-acetylmuramate dehydrogenase	[EC 1.1.1.158]	Enediol intermediate	231
UDP-N-acetylmuramoyl-L-alanine synthase	[EC 6.3.2.8]	Acyl-phosphate intermediate	232
UDP-N-acetylmuramoyl-L-alanyl-D-glutamate synthase	[EC 6.3.2.9]	Acyl-phosphate intermediate	233
UDP-glucose:hexose-1-phosphate uridylyltransferase	[EC 2.7.7.12]	Uridylyl-enzyme intermediate	234
UDP-glucuronate 4-epimerase	[EC 5.1.3.6]	UDP-4-keto-D-glucuronate intermediate	235

TABLE I (continued)

Enzyme	EC number	Intermediate/comment	Ref.[a]
Urea carboxylase	[EC 6.3.4.6]	N-Carboxy-biotin intermediate	167
Valine dehydrogenase (NADP$^+$)	[EC 1.4.1.8]	Imine and carbinolamine intermediates	236
Valyl-tRNA synthase	[EC 6.1.1.9]	Valyl-adenylate intermediate	23
Vanillyl-alcohol oxidase	[EC 1.1.3.38]	5-(4'-Hydroxybenzyl)-FAD intermediate	237

[a] *Key to References:* (1) L. A. Highbarger, J. A. Gerlt, and G. L. Kenyon, *Biochemistry* **35,** 41 (1996); (2) D. Grisaru, M. Sternfeld, A. Eldor, D. Glick, and H. Soreq, *Eur. J. Biochem.* **264,** 672 (1999); (3) W. Huth, R. Jonas, I. Wunderlich, and W. Seubert, *Eur. J. Biochem.* **59,** 475 (1975); (4) S. Miyazawa, S. Furuta, T. Osumi, T. Hashimoto, and N. Ui, *J. Biochem.* **90,** 511 (1981); (5) R. W. Brownsey, R. Zhande, and A. N. Boone, *Biochem. Soc. Trans.* **25,** 1232 (1997); (6) E. W. Yamada and W. B. Jakoby, *J. Biol. Chem.* **233,** 706 (1958); (7) M. Igarashi, H. Takahashi, and N. Tsuyama, *Biochim. Biophys. Acta.* **220,** 85 (1970); (8) J. A. Broadwater, C. Achim, E. Munck, and B. G. Fox, *Biochemistry* **38,** 12197 (1999); (9) A. Scaloni, D. Barra, W. M. Jones, and J. M. Manning, *J. Biol. Chem.* **269,** 15076 (1994); (10) P. N. Lowe and S. Rhodes, *Biochem. J.* **250,** 789 (1988); (11) R. Wolfenden, *Ann. Rev. Biophys. Bioengin.* **5,** 271 (1976); (12) B. Allart, M. Gatel, D. Guillerm, and G. Guillerm, *Eur. J. Biochem.* **256,** 155 (1998); (13) H. Xiong, B. A. Stanley, and A. E. Pegg, *Biochemistry* **38,** 2462 (1999); (14) R. B. Honzatko, M. M. Stayton, and H. J. Fromm, *Adv. Enzymol.* **73,** 57 (1999); (15) C. Satishchandran, Y. N. Hickman, and G. D. Markham, *Biochemistry* **31,** 11684 (1992); (16) M. Weber, M. Suter, C. Brunold, and S. Kopriva, *Eur. J. Biochem.* **267,** 3647 (2000); (17) D. Peisach, D. M. Chipman, P. W. Van Ophem, J. M. Manning, and D. Ringe, *Biochemistry* **37,** 4958 (1998); (18) A. E. Braunstein, in "The Enzymes" (P. D. Boyer, ed.), 3rd Ed., Vol. 9, p. 379. Academic Press, New York, 1973; (19) P. M. Weiss, C.-Y. Chen, W. W. Cleland, and P. F. Cook, *Biochemistry* **27,** 4814 (1988); (20) M. Perego, P. Glaser, A. Minutello, M. A. Strauch, K. Leopold, and W. Fischer, *J. Biol. Chem.* **270,** 15598 (1995); (21) M. E. Tanner and G. L. Kenyon, in "Comprehensive Biological Catalysis: A Mechanistic Reference" (M. Sinnott, ed.), Vol. 2, p. 7. Academic Press, 1998; (22) C. T. Walsh, *Science* **261,** 308 and *Science* **262,** 164 (1993); (23) P. Schimmel, *Adv. Enzymol.* **63,** 233 (1990); (24) S. R. Jones, L. A. Kindman, and J. R. Knowles, *Nature* **275,** 564 (1978); (25) M. J. Woods, J. D. Findlater, and B. A. Orsi, *Biochim. Biophys. Acta* **567,** 225 (1979); (26) L. B. Hersh, *Biochemistry* **11,** 2251 (1972); (27) H. Zalkin, *Adv. Enzymol.* **66,** 203 (1993); (28) C. Anthony, in "Comprehensive Biological Catalysis: A Mechanistic Reference" (M. Sinnott, ed.), Vol. 3, p. 155. Academic Press, 1998; (29) B. A. Palfey and V. Massey, in "Comprehensive Biological Catalysis: A Mechanistic Reference" (M. Sinnott, ed.), Vol. 3, p. 83. Academic Press, 1998; (30) D. E. Ehmann, A. M. Gehring, and C. T. Walsh, *Biochemistry* **38,** 6171 (1999); (31) B. Roux and C. T. Walsh, *Biochemistry* **31,** 6904 (1992); (32) M. Maitre, L. Ciesielski, C. Cash, and P. Mandel, *Eur. J. Biochem.* **52,** 157 (1975); (33) K. Li, W. Du, N. L. S. Que, and H.-W. Liu, *J. Am. Chem. Soc.* **118,** 8763 (1996); (34) A. M. Lacoste, C. Dumora, L. Balas, F. Hammerschmidt, and J. Vercauteren, *Eur. J. Biochem.* **215,** 841 (1993); (35) Z. He and J. C. Spain, *J. Bacteriol.* **180,** 2502 (1998); (36) O. Ploux, O. Breyne, S. Carillon, and A. Marquet, *Eur. J. Biochem.* **259,** 63 (1999); (37) P. V. S. Rao and C. S. Vaidyanathan, *Arch. Biochem. Biophys.* **118,** 388 (1967); (38) M. J. Dowler and H. I. Nakada, *J. Biol. Chem.* **243,** 1434 (1968); (39) O. Hayaishi, M. Nozaki, and M. T. Abbott, in "The Enzymes" (P. D. Boyer, ed.), 3rd Ed., Vol. 12, p. 119. Academic Press, New York, 1975; (40) J. Powlowski, D. P. Ballou, and V. Massey, *J. Biol. Chem.* **265,** 4969 (1990); (41) B. Langkau and S. Ghisla, *Eur. J. Biochem.* **230,** 686 (1995); (42) E. C. Bigham, C. E. Gragg, W. R. Hall, J. E. Kelsey, W. R. Mallory, C. Richardson, C. Benedict, and P. H. Ray, *J. Med. Chem.* **27,** 717 (1984); (43) S. C. Kim and F. M. Raushel, *Biochemistry* **25,** 4744 (1986); (44) C. Ghose and F. M. Raushel, *Biochemistry* **24,** 5894 (1985); (45) G. G. Gross, *Eur. J. Biochem.* **31,** 585 (1972); (46) H. H. Andres, A. J. Klem, L. M. Schopfer, J. K. Harrison, and W. W. Weber, *J. Biol. Chem.* **263,** 7521 (1988); (47) K. Kobashi, D. H. Kim & T. Morikawa, *J. Protein Chem.* **6,** 237 (1987); (48) H. B. Dunford, in "Comprehensive Biological Catalysis: A Mechanistic Reference" (M. Sinnott, ed.), Vol. 3, p. 195. Academic Press, 1998; (49) P. M. Mehlhaff, C. A. Luehr, and S. M. Schuster, *Biochemistry* **24,** 1104 (1985); (50) P. D. Van Poelje and E. E. Snell, *Ann. Rev. Biochem.* **59,** 29 (1990); (51) W. E. Karsten and R. E. Viola, *Biochem. Biophys. Acta*

1077, 209 (1991); (52) A. Siebers and K. Altendorf, *J. Biol. Chem.* **264,** 5831 (1989); (53) B. Houston and H. G. Nimmo, *Biochem. J.* **224,** 437 (1984); (54) P. M. Weiss, G. A. Garcia, G. L. Kenyon, W. W. Cleland, and P. F. Cook, *Biochemistry* **27,** 2197 (1988); (55) T. P. Begley, J. Xi, C. Kinsland, S. Taylor, and F. Mclafferty, *Curr. Opin. Chem. Biol.* **3,** 623 (1999); (56) L. A. Fothergill-Gilmore and H. C. Watson, *Adv. Enzymol.* **62,** 227 (1989); (57) D. Zhang and C. D. Poulter, *J. Am. Chem. Soc.* **117,** 1641 (1995); (58) A. Butler, *in* "Comprehensive Biological Catalysis: A Mechanistic Reference" (M. Sinnott, ed.), Vol. 3, p. 427. Academic Press, 1998; (59) D. S. Kaseman and A. Meister, *Methods Enzymol.* **113,** 305 (1985); (60) F. Spence, G. J. Langley, and T. D. H. Bugg, *J. Amer. Chem. Soc.* **118,** 8336 (1996); (61) S. J. Remington and K. Breddam, *Methods Enzymol.* **244,** 462 (1994); (62) O. W. Griffith, *Ann. Rev. Biochem.* **55,** 855 (1986); (63) B. G. Fox, *in* "Comprehensive Biological Catalysis: A Mechanistic Reference" (M. Sinnott, ed.), Vol. 3, p. 261. Academic Press, 1998; (64) T. J. Larson and W. Dowhan, *Biochemistry* **15,** 5212 (1976); (65) G. Davies, M. L. Sinnott, and S. G. Withers, *in* "Comprehensive Biological Catalysis: A Mechanistic Reference" (M. Sinnott, ed.), Vol. 1, p. 119. Academic Press, 1998; (66) K. L. Taylor, R. Q. Liu, P. H. Liang, J. Price, D. Dunaway-Mariano, P. J. Tonge, J. Clarkson, and P. R. Carey, *Biochemistry* **34,** 13881 (1995). (67) J. MacLachlan, A. T. Wotherspoon, R. O. Ansell, and C. J. Brooks, *J. Steroid Biochem. Mol. Biol.* **72,** 169 (2000); (68) P. Taylor, *J. Biol. Chem.* **266,** 4025 (1991); (69) J. S. Fruton, *Adv. Enzymol.* **53,** 239 (1982); (70) M. Akhtar, *Ciba Found. Symp.* **180,** 131 (1994); (71) L. M. Cunane, Z. W. Chen, N. Shamala, F. S. Mathews, C. N. Cronin, and W. S. Mcintire *J. Mol. Biol.* **295,** 357 (2000); (72) H. Yanase, T. Sakai, and K. Tonomura, *Agric. Biol. Chem.* **47,** 473 (1983); (73) J. F. Robyt, *Essentials of Carbohydrate Chemistry,* Springer, New York, 1998; (74) J. Dittmann, R. M. Wenger, H. Kleinkauf, and A. Lawen, *J. Biol. Chem.* **269,** 2841 (1994); (75) T. Clausen, B. Laber, and A. Messerschmidt, *Biol. Chem.* **378,** 321 (1997); (76) C. H. Tai and P. F. Cook, *Adv. Enzymol.* **74,** 185 (2000); (77) K. N. Allen, *in* "Comprehensive Biological Catalysis: A Mechanistic Reference" (M. Sinnott, ed.), Vol. 2, p. 135. Academic Press, 1998; (78) R. Jeffcoat, H. Hassall, and S. Dagley, *Biochem. J.* **115,** 977 (1969); (79) K. L. Graves and L. W. Hardy, *Biochemistry* **33,** 13049 (1994); (80) H. Kack, K. J. Gibson, Y. Lindqvist, and G. Schneider, *Proc. Natl. Acad. Sci. U.S.A.* **95,** 5495 (1998); (81) F. A. Blocki, M. S. Logan, C. Baoli, and L. P. Wackett, *J. Biol. Chem.* **269,** 8826 (1994); (82) W. E. Karsten, *Biochemistry* **36,** 1730 (1997); (83) R. V. Banerjee, B. Shane, J. J. Mcguire, and J. K. Coward, *Biochemistry* **27,** 9062 (1988); (84) D. E. Ehmann, C. A. Shaw-Reid, H. C. Losey, and C. T. Walsh, *Proc. Natl. Acad. Sci. U.S.A.* **97,** 2509 (2000); (85) I. R. Lehman, *Science* **186,** 790 (1974); (87) G. Tian, J. A. Berry, and J. P. Klinman, *Biochemistry* **33,** 226 (1994); (88) X. Ding, B. F. Rasmussen, G. A. Petsko, and D. Ringe, *Biochemistry* **33,** 9285 (1994); (89) P. A. Mallory and J. Travis, *Biochemistry* **14,** 722 (1975); (90) T. P. Karpetsky, K. K. Shriver, and C. C. Levy, *J. Biol. Chem.* **255,** 2713 (1980); (89) P. A. Frey and G. H. Reed, *Adv. Enzymol.* **66,** 1 (1993); (91) S. J. Wakil and J. K. Stoops *in* "The Enzymes" (P. D. Boyer, ed.), 3rd Ed., Vol. 16, p. 3. Academic Press, New York, 1983; (92) H. Croteau, J. H. Miyazaki, and C. J. Wheeler, *Arch. Biochem. Biophys.* **269,** 507 (1989); (93) D. J. Buttle, *Methods Enzymol.* **244,** 639 (1994); (94) J. Tchou and A. P. Grollman, *J. Biol. Chem.* **270,** 11671 (1995); (95) H. Eklund and M. Fontecave, *Structure Fold. Des.* **7,** R257 (1999); (96) R. Külzer, T. Pils, R. Kappl, J. Hüttermann, and J. Knappe, *J. Biol. Chem.* **273,** 4897 (1998); (97) S. Song, H. Jahansouz, and R. H. Himes, *FEBS Lett.* **332,** 150 (1993); (98) D. A. Okar, D. H. Live, M. H. Devany, and A. J. Lange, *Biochemistry* **39,** 9754 (2000); (99) A. Messerschmidt, *in* "Comprehensive Biological Catalysis: A Mechanistic Reference" (M. Sinnott, ed.), Vol. 3, p. 401. Academic Press, 1998. (100) G. Mooser, *in* "The Enzymes" (P. D. Boyer, ed.), 3rd Ed., Vol. 20, p. 187. Academic Press, New York, 1992; (101) P. J. Oths, R. M. Mayer, and H. G. Floss, *Carbohydr. Res.* **198,** 91 (1990); (102) G. A. Barber, *J. Biol. Chem.* **254,** 7600 (1979); (103) X. Guan, T. Diez, T. K. Prasad, B. J. Nikolau, and E. S. Wurtele, *Arch. Biochem. Biophys.* **362,** 12 (1999); (104) R. Croteau, D. M. Satterwhite, D. E. Cane, and C. C. Chang, *J. Biol. Chem.* **261,** 13438 (1986); (105) K. J. Bame and L. H. Rome, *J. Biol. Chem.* **261,** 10127 (1986); (106) R. C. Nordlie, *Methods Enzymol.* **87,** 319 (1982); (107) S. H. Seeholzer, *Proc. Nat. Acad. Sci. U.S.A.* **90,** 1237 (1993); (108) S. G. Withers and L. F. Street, *J. Am. Chem. Soc.* **110,** 8551 (1988); (109) W. Buckel, U. Dorn, and R. Semmler, *Eur. J. Biochem.* **118,** 315 (1981); (110) B. Beatrix, K. Bendrat, S. Rospert, and W. Buckel, *Arch. Microbiol.* **154,** 362 (1990); (111) E. A. Boeker and E. E. Snell, *in* "The Enzymes" (P. D. Boyer, ed.), 3rd Ed., Vol. 6, p. 217. Academic Press, New York, 1972; (112) S. Glavas and M. E. Tanner, *Biochemistry* **38,** 4106 (1999); (113) J. A. Todhunter and D. L. Purich, *J. Biol. Chem.* **250,** 3505 (1975); (114) E. A. First, *in* "Comprehensive Biological Catalysis: A Mechanistic Reference" (M. Sinnott, ed.), Vol. 1, p. 573. Academic Press, 1998; (115) O. W. Griffith and R. J. Mulcahy, *Adv. Enzymol.* **73,** 209 (1999); (116) R. D. Allison, *Methods Enzymol.* **113,** 419 (1985); (117) T. M. Dwyer, K. S. Rao, S. I. Goodman and F. E. Frerman, *Biochemistry* **39,** 11488 (2000);

(118) O. W. Griffith, *Free Radic. Biol. Med.* **27,** 922 (1999); (119) E. Racker, *Mechanisms in Bioenergetics,* Academic Press, New York, 1965; (120) B. T. Golding and W. Buckel, in *Comprehensive Biological Catalysis: A Mechanistic Reference,* Vol. 3, p. 239, 1998; (121) E. Fritsche, A. Humm and R. Huber, *Eur. J. Biochem.* **247,** 483 (1997); (122) B. Czuba and D. A. Vessey, *J. Biol. Chem.* **255,** 5296 (1980); (123) T. C. Stadtman, *Ann. Rev. Biochem.* **65,** 83 (1996); (124) W. Von Der Saal, C. S. Crysler, and J. J. Villafranca, *Biochemistry* **24,** 5343 (1985); (125) J. Nakamura, K. Straub, J. Wu, and L. Lou, *J. Biol. Chem.* **270,** 23450 (1995); (126) J. L. Cartwright and A. G. Mclennan, *Arch. Biochem. Biophys.* **361,** 101 (1999); (127) J. Q. Liu, T. Kurihara, S. Ichiyama, M. Miyagi, S. Tsunasawa, H. Kawasaki, K. Soda, and N. Esaki, *J. Biol. Chem.* **273,** 30897 (1998); (128) I. S. Ridder, H. J. Rozeboom, K. H. Kalk, and B. W. Dijkstra, *J. Biol. Chem.* **274,** 30672 (1999); (129) J. Newman, T. S. Peat, R. Richard, L. Kan, P. E. Swanson, J. A. Affholter, I. H. Holmes, J. F. Schindler, C. J. Unkefer, and T. C. Terwilliger, *Biochemistry* **38,** 16105 (1999); (130) K. J. Bame and L. H. Rome, *J. Biol. Chem.* **261,** 10127 (1986); (131) A. Hagner-McWhirter, U. Lindahl, and J. P. Li, *Biochem. J.* **347,** 69 (2000); (132) R. C. Johnson and J. R. Gilbertson, *J. Biol. Chem.* **247,** 6991 (1972); (133) A. Thithapandha and V. H. Cohn, *Biochem. Pharmacol.* **27,** 263 (1978); (134) P. D. Van Poelje and E. E. Snell, *Ann. Rev. Biochem.* **59,** 29 (1990); (135) C. Grubmeyer and H. Teng, *Biochemistry* **38,** 7355 (1999); (136) T. L. Born, M. Franklin, and J. S. Blanchard, *Biochemistry* **39,** 8556 (2000); (137) D. L. Van Der Jagt, *Biochem. Soc. Trans.* **21,** 522 (1993); (138) A. E. Kuhm, H. Knackmuss, and A. Stolz, *J. Biol. Chem.* **268,** 9484 (1993); (139) U. Scherf and W. Buckel, *Eur. J. Biochem.* **215,** 421 (1993); (140) T. E. Smeland, J. Li, D. Cuebas, and H. Schulz, *Prog. Clin. Biol. Res.* **375,** 85 (1992); (141) U. Muller and W. Buckel, *Eur. J. Biochem.* **230,** 698 (1995); (142) W. A. Wood, in "The Enzymes" (P. D. Boyer, ed.), 3rd Ed., Vol. 7, p. 281. Academic Press, New York, 1972; (143) D. J. Creighton and N. S. R. K. Murthy, in "The Enzymes" (P. D. Boyer, ed.), 3rd Ed., Vol. 19, p. 323. Academic Press, New York, 1990; (144) P. Chaiyen, P. Brissette, D. P. Ballou, and V. Massey, *Biochemistry* **36,** 8060 (1997); (145) U. Arunachalam, V. Massey, and S. M. Miller, *J. Biol. Chem.* **269,** 150 (1994); (146) S. G. Ramaswamy, *Fed. Proc.* **42,** 2232 (1983); (147) S. P. Craig III and A. E. Eakin, *J. Biol. Chem.* **275,** 20231 (2000); (148) K. Gohda, Y. Kimura, I. Mori, D. Ohta, and T. Kikuchi, *Biochim. Biophys. Acta* **1385,** 107 (1998); (149) P. Heikinheimo, J. Lehtonen, A. Baytov, R. Lahti, B. S. Cooperman, and A. Goldman, *Structure* **4,** 1491 (1996); (150) N. I. Naber, J. S. Swan, and G. A. Hamilton, *Biochemistry* **25,** 7201 (1986); (151) A. L. Majumder, M. D. Johnson, and S. A. Henry, *Biochim. Biophys. Acta* **1348,** 245 (1997); (152) J. V. Schloss and M. S. Hixon, in "Comprehensive Biological Catalysis: A Mechanistic Reference" (M. Sinnott, ed.), Vol. 2, p. 43. Academic Press, 1998; (153) A. Whitty, C. A. Fierke, and W. P. Jencks, *Biochemistry* **34,** 11678 (1995); (154) G. D'Agnolo, I. S. Rosenfeld, and P. R. Vagelos, *J. Biol. Chem.* **250,** 5283 (1975); (155) L. Kerscher and D. Oesterhelt, *Eur. J. Biochem.* **116,** 587 (1981); (156) R. S. Phillips, B. Sundararaju, and S. V. Koushik, *Biochemistry* **37,** 8783 (1998); (157) A. D. Cameron, M. Ridderström, B. Olin, M. J. Kavarana, D. J. Creighton, and B. Mannervik, *Biochemistry* **38,** 13480 (1999); (158) L. Hernandez, J. Arrieta, C. Menendez, R. Vazquez, A. Coego, V. Suarez, G. Selman, M. F. Petit-Glatron, and R. Chambert, *Biochem. J.* **309,** 113 (1995); (159) C. Malet and A. Planas, *Biochemistry* **36,** 13838 (1997); (160) A. Rodriguez, L. Wall, S. Raptis, C. G. Zarkadas, and E. Meighen, *Biochim. Biophys. Acta* **964,** 266 (1988); (161) E. A. Dennis in "The Enzymes" (P. D. Boyer, ed.), 3rd Ed., Vol. 16, p. 307. Academic Press; New York, 1983; (162) S. L. Bearne and R. Wolfenden, *Biochemistry* **36,** 1646 (1997); (163) V. Schirch, in "Comprehensive Biological Catalysis: A Mechanistic Reference" (M. Sinnott, ed.), Vol. 1, p. 211. Academic Press, 1998; (164) H.-W. Chih and E. N. G. Marsh, *Biochemistry* **38,** 13684 (1999); (165) T. A. Diez, E. S. Wurtele, and B. J. Nikolau, *Arch. Biochem. Biophys.* **310,** 64 (1994); (166) D. Saadat and D. H. Harrison, *Biochemistry* **39,** 2950 (2000); (167) J. N. Earnhardt and D. N. Silverman, in "Comprehensive Biological Catalysis: A Mechanistic Reference" (M. Sinnott, ed.), Vol. 1, p. 495. Academic Press, 1998; (168) R. J. Roberts and X. Cheng, *Ann. Rev. Biochem.* **67,** 181 (1998); (169) L. J. Zatman, N. O. Kaplan, and S. P. Colowick, *J. Biol. Chem.* **200,** 197 (1953); (170) J. Gross, M. Rajavel, E. Segura, and C. Grubmeyer, *Biochemistry* **35,** 3917 (1996); (171) D. J. Stuehr and O. W. Griffith, *Adv. Enzymol.* **65,** 287 (1992); (172) S. A. Short, S. R. Armstrong, S. E. Ealick, and D. J. T. Porter, *J. Biol. Chem.* **271,** 4978 (1996); (173) E. Garces and W. W. Cleland, *Biochemistry* **8,** 633 (1969); (174) D. C. Prasher, M. C. Carr, E. H. Ives, T. C. Tsai, and P. A. Frey, *J. Biol. Chem.* **257,** 4931 (1982); (175) G. M. Brown, *J. Biol. Chem.* **226,** 651 (1957); (176) K. Brocklehurst, A. B. Watts, M. Patel, C. Verma, and E. W. Thomas, in "Comprehensive Biological Catalysis: A Mechanistic Reference" (M. Sinnott, ed.), Vol. 1, p. 381. Academic Press, 1998; (177) C. Enroth, H. Neujahr, G. Schneider, and Y. Lindqvist, *Structure* **6,** 605 (1998); (178) P. F. Fitzpatrick, *Adv. Enzymol.* **74,** 235 (2000); (179) F. Fasiolo and A. R. Fersht, *Eur. J. Biochem.*

85, 85 (1978); (180) Y. Q. Shi, I. Hubacek, and R. R. Rando, *Biochemistry* **32,** 1257 (1993); (181) J. S. Culp, H. J. Blytt, M. Hermodson, and L. G. Butler, *J. Biol. Chem.* **260,** 8320 (1985); (182) S. J. Pilkis, T. H. Claus, I. J. Kurland, and A. J. Lange, *Ann. Rev. Biochem.* **64,** 799 (1995); (183) P. Staedtler, S. Hoenig, R. Frank, S. G. Withers, and W. Hengstenberg, *Eur. J. Biochem.* **232,** 658 (1995); (184) C. C. Hwang, A. J. Berdis, W. E. Karsten, W. W. Cleland, and P. F. Cook, *Biochemistry* **37,** 12596 (1998); (185) L. Jolly, P. Ferrari, D. Blanot, J. van Heijenoort, F. Fassy, and D. Mengin-Lecreulx, *Eur. J. Biochem.* **262,** 202 (1999); (186) S. N. Seal and Z. B. Rose, *J. Biol. Chem.* **262,** 13496 (1987); (187) M. Pirard, Y. Achouri, J. F. Collet, E. Schollen, G. Matthijs, and E. Van Schaftingen, *Biochem. J.* **339,** 201 (1999); (188) A. S. Baker, M. J. Ciocci, W. W. Metcalf, J. Kim, P. C. Babbitt, B. L. Wanner, B. M. Martin, and D. Dunaway-Mariano, *Biochemistry* **37,** 9305 (1998); (189) J. R. Stevens-Clark, K. A. Conklin, A. Fujimoto, and R. A. Smith, *J. Biol. Chem.* **243,** 4474 (1968); (190) J. M. Buchanan, *Adv. Enzymol.* **39,** 91 (1973); (191) F. J. Castellino, *Chem. Rev.* **81,** 431 (1981); (192) P. A. Recsei and E. E. Snell, *Ann. Rev. Biochem.* **53,** 357 (1984); (193) A. W. Alberts and P. R. Vagelos, in "The Enzymes" (P. D. Boyer, ed.), 3rd Ed., Vol. 6, p. 37. Academic Press, New York, 1972; (194) J. Stubbe, S. Fish, and R. H. Abeles, *J. Biol. Chem.* **255,** 236 (1980); (195) A. G. Gabibov, S. N. Kochetkov, L. P. Sashchenko, I. V. Smirnov, and E. S. Severin, *Eur. J. Biochem.* **135,** 491 (1983); (196) J. E. Folk, *Methods Enzymol.* **87,** 36 (1982); (197) H. Hüdig and W. Hengstenberg, *FEBS Lett.* **114,** 103 (1980); (198) Z. Y. Zhang and J. E. Dixon, *Adv. Enzymol.* **68,** 1 (1994); (199) N. Lebedev and M. P. Timko, *Proc. Natl. Acad. Sci. U.S.A.* **96,** 9954 (1999); (200) J. J. Kramer, J. G. Henslee, Y. Wakabayashi, and M. E. Jones, *Methods Enzymol.* **113,** 113 (1985); (201) S. Jitrapakdee and J. C. Wallace, *Biochem. J.* **340,** 1 (1999); (202) S. H. Thrall and D. Dunaway-Mariano, *Biochemistry* **33,** 1103 (1994); (203) K. Tittmann, D. Proske, M. Spinka, S. Ghisla, R. Rudolph, G. Hubner, and G. Kern, *J. Biol. Chem.* **273,** 12929 (1998); (204) M. H. Charon, A. Volbeda, E. Chabriere, L. Pieulle, and J. C. Fontecilla-Camps, *Curr. Opin. Struct. Biol.* **9,** 663 (1999); (205) S. Narindrasorasak and W. A. Bridger, *J. Biol. Chem.* **252,** 3121 (1977); (206) D. A. Aird, R. L. Heinrikson, and J. Westley, *J. Biol. Chem.* **262,** 17327 (1987); (207) N. H. Williams, in "Comprehensive Biological Catalysis: A Mechanistic Reference" (M. Sinnott, ed.), Vol. 1, p. 543. Academic Press, 1998; (208) T. Uchida and F. Egami, in "The Enzymes" (P. D. Boyer, ed.), 3rd Ed., Vol. 4, p. 205. Academic Press, New York, 1971; (209) S. S. Licht, S. Booker, and J. Stubbe, *Biochemistry* **38,** 1221 (1999); (210) O. Vicente and W. Filipowicz, *Eur. J. Biochem.* **176,** 431 (1988); (211) T. W. Hallahan and R. Croteau, *Arch. Biochem. Biophys.* **269,** 313 (1989); (212) H. Walker, J. A. Ferretti, and T. C. Stadtman, *Proc. Natl. Acad. Sci. U.S.A.* **95,** 2180 (1998); (213) K. Forchhammer and A. Bock, *J. Biol. Chem.* **266,** 6324 (1991); (214) P. A. Frey, in "The Enzymes" (D. S. Sigman, ed.), 3rd Ed., Vol. 20, p. 141. Academic Press, New York, 1992; (215) P. Brzovic, E. L. Holbrook, R. C. Greene, and M. F. Dunn, *Biochemistry* **29,** 442 (1990); (216) G. Mooser, S. A. Hefta, R. J. Paxton, J. E. Snively, and T. D. Lee, *J. Biol. Chem.* **266,** 8916 (1991); (217) J. J. Mieyal and R. H. Abeles, in "The Enzymes" (P. D. Boyer, ed.), 3rd Ed., Vol. 7, p. 515. Academic Press, New York, 1972; (218) H. Itoh, H. Okaya, A. R. Khan, S. Tajima, S. Hayakawa, and K. Izumori, *Biosci. Biotech. Biochem.* **58,** 2168 (1994); (219) W. P. Burmeister, S. Cottaz, H. Driguez, R. Iori, S. Palmieri, and B. Henrissat, *Structure* **5,** 663 (1997); (220) L. J. Berliner, *Thrombin, Structure & Function,* Plenum Press, New York, 1992 (221) Y. Wataya and D. V. Santi, *J. Amer. Chem. Soc.* **99,** 4534 (1977); (222) B. L. Horecker, S. Pontremoli, C. Ricci, and T. Cheng, *Proc. Natl. Acad. Sci. U.S.A.* **47,** 1949 (1961); (223) R. A. Gibbs, in "Comprehensive Biological Catalysis: A Mechanistic Reference" (M. Sinnott, ed.), Vol. 1, p. 31. Academic Press, 1998; (224) I. A. Rose, *Methods Enzymol.* **87,** 84 (1982); (225) B. Keil, in "The Enzymes" (P. D. Boyer, ed.), 3rd Ed., Vol. 3, p. 250. Academic Press, New York, 1971; (226) M. Lee and R. S. Phillips, *Bioorg. Med. Chem.* **3,** 195 (1995); (227) E. W. Miles, *Adv. Enzymol.* **64,** 93 (1991); (228) M. Merle, V. Trezeguet, P. Graves, D. Andrews, K. H. Muench, and B. Labouesse, *Biochemistry* **25,** 1115 (1986); (229) C. M. Pickart and I. A. Rose, *J. Biol. Chem.* **261,** 10210 (1986); (230) A. Hershko and A. Ciechanover, *Ann. Rev. Biochem.* **67,** 425 (1998); (231) T. E. Benson, C. T. Walsh, and V. Massey, *Biochemistry* **36,** 796 (1997); (232) J. J. Emanuele, Jr., H. Jin, J. Yanchunas, Jr., and J. J. Villafrarca, *Biochemistry* **36,** 7264 (1997); (233) J. A. Bertrand, G. Auger, L. Martin, E. Fanchon, D. Blanot, D. Le Beller, J. Van Heijenoort, and O. Dideberg, *J. Mol. Biol.* **289,** 579 (1999); (234) P. A. Frey, L.-J. Wong, K.-F. Sheu, and S.-L. Yang, *Methods Enzymol.* **87,** 20 (1982); (235) R. Munoz, R. Lopez, M. De Frutos, and E. Garcia, *Mol. Microbiol.* **31,** 703 (1999); (236) A. R. Clarke and T. R. Dafforn, in "Comprehensive Biological Catalysis: A Mechanistic Reference" (M. Sinnott, ed.), Vol. 3, p. 1. Academic Press, 1998; (237) M. W. Fraaije, R. H. Van Den Heuvel, J. C. Roelofs, and W. J. Van Berkel, *Eur. J. Biochem.* **253,** 712 (1998).

Author Index

Numbers in parentheses are footnote reference numbers and indicate that an author's work is referred to although the name is not cited in the text.

A

Aagaard, A., 365
Aasa, R., 355
Abbott, M. T., 467
Abbott, S. J., 5, 6, 6(15)
Abeles, R. H., 11, 67, 114, 286, 470
Aberhart, D. J., 427
Abu-Soud, H. M., 322, 326, 326(11), 328, 328(23), 330, 331(16; 23), 335, 335(11)
Acher, F., 191
Achim, C., 467
Achouri, Y., 179, 184(13), 470
Adachi, H., 25
Adak, S., 335, 337
Adams, J., 285
Adams, W. S., 151
Adelroth, P., 365
Admiraal, S. J., 120, 120(29), 129
Adrait, A., 401, 403
Aebersold, R., 67, 68, 77, 81(13), 87, 87(28; 29; 35; 38; 42), 88, 89(42), 92, 93, 94(21; 28; 29; 38), 96, 96(35; 36), 98, 99, 100, 100(36), 101(35)
Affholter, J. A., 469
Agarwal, R. P., 118
Agner, K., 346
Agou, F., 129
Ahmed, A. I., 22
Ahn, D., 271
Ai, J., 449
Aird, B. A., 470
Akhtar, M., 468
Akita, S., 27
Akiyama, K., 191
Alber, T., 22
Alberts, A. W., 470
Albracht, S. P., 394
Alcock, N. W., 314
Alden, R. A., 23

Aldwin, L., 117, 118(18)
Alfonso, P., 36
Allart, B., 467
Allen, K. N., 468
Allen, M. H., 30, 36
Allewell, N. M., 436
Allison, R. D., 1, 2(3), 5(3), 8, 10, 13(33), 149, 168, 261, 455, 456, 468
Allman, S. L., 30, 35(29)
Almaula, N., 119
Alpin, R. T., 28
Als-Nielsen, J., 295
Altendorf, K., 468
Altschul, S. F., 179
Ames, B. N., 152
Amici, A., 149, 152, 152(5), 153(10), 154(5), 155, 156, 157(5)
Amory, A., 181
Amspacher, D. R., 295
Amyes, T. L., 167
Anderluzzi, D., 190
Andersen, J. N., 241, 244(15)
Anderson, D., 232
Anderson, G. A., 52
Anderson, K. S., 29, 30(19), 31(19), 50, 51(3), 60, 240
Anderson, M. E., 279, 284(15)
Anderson, M. S., 196
Anderson, P. M., 270
Anderson, V. E., 394
Andersson, I., 245
Andersson, J., 395
Andersson, K. K., 326, 329, 335(26), 404, 406, 408(39)
Andersson, M. E., 406, 407, 408(39; 40), 409(40)
Andreasson, L.-E., 355
Andres, H. H., 467
Andrews, A. T., 268
Andrews, D., 470

471

AUTHOR INDEX

Andrews, P. C., 345
Andrews, T., 69
Ansell, R. O., 468
Anthony, C., 467
Anthony, R. S., 6, 15
Antonov, V. K., 106, 117(3)
Aplin, R. T., 52
Appelman, E. H., 356, 363, 364(17)
Araiso, T., 329
Aravind, L., 179
Archambault, J., 179
Argade, P. V., 355, 360, 360(12)
Aricò, C. N., 36
Armstrong, S. R., 469
Arnold, E., 121, 122(10), 123(10)
Arosio, P., 436, 436(5), 437
Arrieta, J., 469
Artymiuk, P. J., 436(5), 437
Arunachalam, U., 469
Arvai, A. S., 204, 320, 321(4)
Ashford, J. S., 18
Ashton, D. S., 52
Ashwell, S., 393
Assarsson, M., 406, 407, 408(40), 409(40)
Atherton, S. J., 354
Atkin, C. L., 404, 405(28)
Ator, M. A., 424
Atta, M., 404
Auger, G., 191, 196, 470
Austin, R. H., 59
Averill, B. A., 177
Azerad, R., 191

B

Babbitt, P. C., 180, 470
Babcock, G. T., 351, 355, 360, 363, 364, 365, 418
Bachet, B., 310, 311(2)
Backer, J. M., 120, 120(50), 133
Badet, B., 262, 270
Badet-Denisot, M. A., 262
Bagley, K. A., 354
Bailly, V., 202
Bairoch, A., 85
Baker, A. S., 180, 470
Baker, G. M., 363
Bakkenist, A. R. J., 345
Balas, L., 467

Baldwin, J. E., 52, 408, 413(44; 45), 449
Bale, J. R., 18
Ballinger, M. D., 24, 426, 427, 429(20–22), 431(20–24), 432, 432(24), 435
Ballou, D. P., 33, 382, 383, 388, 389(16), 390, 390(16), 391(16), 396(16), 411, 467, 469
Balls, A. K., 106, 117(1)
Bally, M., 179
Bame, K. J., 468, 469
Banaszak, L. J., 132
Bandarian, V., 381, 426, 427, 429(26)
Banerjee, R., 380, 391, 426, 468
Banki, K., 197, 201
Bannworth, W., 240
Baoli, C., 468
Barak, I., 133
Baraniak, J., 427, 428(18)
Barber, G. A., 468
Barden, R. E., 6
Barford, D., 237, 238, 239, 239(1), 241, 241(6), 242(1), 244(6; 7; 15; 16; 21), 247, 247(21), 248(21; 25), 251(7)
Barinaga, C. J., 27
Barker, H. A., 380, 384(1), 397(3)
Barker, S. C., 52
Barman, T., 21
Barnes, L. D., 260
Barnes, Z. K., 355
Baron, A. J., 422
Barondeau, D. P., 296, 305(1), 306
Barra, A.-L., 406, 407, 408(40), 409, 409(40)
Barra, D., 467
Barshevskaya, T. N., 106, 117(3)
Barshop, B. A., 38
Bartlett, P. A., 279, 284, 284(18)
Bash, P. A., 25
Bates, C. J., 268
Batlle, A. M. C., 368, 369, 370, 372, 373, 377(11; 12; 19), 378, 379(9–12)
Batllori, X., 24
Battersby, A. R., 377, 379, 379(20)
Bauer, M. D., 28, 52
Bayly, R., 108
Baytov, A., 469
Bearne, S. L., 469
Beatrix, B., 393, 468
Bec, N., 322, 325(12), 326(12), 328, 328(12), 329, 329(12), 331(25), 335(12; 25; 26)
Becalski, J., 69
Becker, A., 395

AUTHOR INDEX   473

Becker, M. A., 225
Bedarker, S., 101
Beddell, C. R., 52
Beechem, J. M., 313
Beers, P. C., 252
Beevers, L., 295
Begley, T. P., 468
Behnke, M., 285
Behravan, G., 401, 403, 413(13)
Behrman, E. J., 279
Beijer, B., 295
Beinert, H., 427, 428, 430(16)
Bell, R. M., 6, 14(17), 168, 455
Bell, R. P., 117, 118(18)
Belrhali, H., 295
Benci, S., 235
Bender, A. T., 333
Bender, M. L., 42
Bendrat, K., 468
Benedict, C., 467
Benkovic, P. A., 316
Benkovic, S. J., 24, 28, 316
Bennett, A. J., 167
Benning, M. M., 26
Benson, D. E., 329
Benson, T. E., 191, 470
Benton, D. J., 314
Berdis, A. J., 470
Bergamini, M. V., 158
Berger, S. L., 176
Bergeron, R. J., 404
Berghuis, A. M., 131
Bergmeyer, H. U., 265, 267(47)
Berka, V., 337
Berkowitz, D. B., 286, 287(34)
Berliner, L. J., 470
Bernardino, D. D., 108
Bernatowicz, M. S., 118
Bernhard, S. A., 232
Bernt, E., 265, 267(47)
Berry, J. A., 468
Berthet-Colominas, C., 295
Bertrand, J. A., 191, 196, 470
Bettati, S., 235
Betzel, C., 25
Beynon, R. J., 52
Bhagwat, M., 205
Bhat, T. N., 25
Biallas, M. J., 279
Biemann, K., 186

Bienvenue, A., 329
Bigay, J., 130
Bigeleisen, J., 162
Biggs, J., 119
Bigham, E. C., 467
Billings, E. M., 285
Biondi, R. M., 123, 126, 129
Birck, C., 133
Bird, P., 87(39), 88
Birktoft, J. J., 23
Birsan, C., 87, 88(17), 95, 96(17)
Blackburn, G. M., 286
Blakeley, V., 422
Blakely, R. J., 427, 430(16)
Blanchard, C. Z., 295
Blanchard, J. S., 469
Blanche, F., 271
Blankenhorn, G., 202, 204(6), 206(6)
Blanot, D., 176, 189, 190, 191, 195, 195(22),
    196, 196(16), 470
Blätter, W. A., 6, 15(16)
Blexzunski, C. F., 36
Blocki, F. A., 468
Blomberg, M. R. A., 401, 403
Blonski, C., 197
Blow, D. M., 25
Blum, M. S., 29
Blytt, H. J., 470
Bock, A., 470
Bode, W., 262
Boehlein, S. K., 260, 261, 263, 268(12; 15; 44),
    271
Boeker, E. A., 468
Boelens, R., 132
Boerner, R. J., 52
Boernsen, K. O., 35
Boggs, S., 325, 328, 328(15), 329(15), 330,
    331(15), 335(15)
Bogyo, M., 285
Bohacek, R. S., 261
Bolin, J. T., 108
Bollinger, J. M., 405, 406(30), 408(30), 438,
    440(11), 441(11), 442(11)
Bollinger, J. M., Jr., 406, 408, 408(37),
    413(44; 45), 414, 449
Bolognesi, M., 262
Bolscher, B. G. J. M., 345
Bolton, J. R., 431, 432(33)
Bominaar, A., 127
Bono, V. H., 261

Booker, S., 428, 470
Booker, S. J., 428
Boone, A. N., 467
Booth, J. W., 260
Boretto, J., 120, 120(25), 127, 129(25)
Borkenhagen, L. F., 184
Born, T. L., 469
Borne, F., 179
Bossek, U., 452
Bothe, H., 382, 390(20), 394, 426
Bothner, B., 28, 35(12), 53
Bouhss, A., 176, 189, 191, 195, 195(22), 196
Bourdais, J., 126
Bovier-Lapierre, C., 108
Boyar, W. C., 426(15), 427
Boyd, A. S. F., 106
Boyer, P. D., 7, 9, 13, 132, 168, 169(5), 170(5), 171(5), 174(5), 181, 193, 256, 258(14)
Bradford, M. M., 372
Brajter-Toth, A., 316
Branchaud, B. P., 414, 418, 419, 421(18–20), 422, 422(16), 423(28), 424(19)
Brand, J., 29
Brannigan, J. A., 133, 263, 284, 285(24)
Brashear, W. T., 6, 7
Brass, H. J., 279
Braun, C., 68, 69(24), 71, 72(24), 73, 87(30), 88, 89
Braun, P. E., 175
Braunstein, A. E., 467
Bravdo, B.-A., 87, 94(20)
Bray, R. C., 409, 410(48), 411(48)
Brayer, G. D., 71, 88(53), 89, 95(53)
Brayford, D., 417
Breddam, K., 468
Brevet, A., 5
Brewer, H. B., 29
Breyne, O., 467
Brick, D., 227
Brick, P., 25
Bridger, W. A., 7, 132, 470
Bridges, A., 393
Briozzo, P., 120, 120(14), 122
Brissette, P., 469
Broadwater, J. A., 449, 467
Brocklehurst, K., 262, 263(29), 469
Brockway, R. A., 151
Brody, J. P., 59
Brody, R. S., 14
Bröker, G., 426

Brolin, S. E., 289
Brooke, G. S., 87(23), 88, 94(23)
Brooks, C. J., 468
Broome, J. D., 261
Brossmer, R., 163
Brown, A. J., 455
Brown, G. M., 469
Brown, N. R., 241, 244(16)
Brown, R. P., 28
Brown, R. P. A., 52
Brownsey, R. W., 467
Bruce, J. E., 52
Bruice, T. C., 316
Brumer, H., 72
Bruner, E., 337
Bruner, M., 159, 160, 165
Brünger, A. T., 102
Brunold, C., 467
Brunold, T. C., 447
Bruns, C., 26
Bruns, K., 53
Bruton, C. J., 18
Bryan, H. L., 106, 107(4), 117(4)
Bryk, R., 333
Brzezinski, P., 365
Brzovic, P. S., 321, 470
Brzozowski, A. M., 88(52), 89, 95(52)
Buchanan, J. M., 470
Buchwald, S. L., 14
Buckel, W., 380, 381, 382, 390(20), 393, 394, 426, 468, 469
Budarf, M. L., 186
Bugg, T. D. H., 106, 107, 108, 109, 109(8; 9), 111, 113(9), 117(9; 13), 189, 190, 190(2), 468
Burdi, D., 405, 406(31)
Burger, A., 421
Burgoyne, D. L., 87(26), 88
Burke, T. R., Jr., 286
Burkhard, P., 227
Burlingame, A. L., 176, 177
Burmeister, W. P., 87, 88(19), 470
Burner, U., 345, 346, 347(14)
Burns, M. R., 262, 268(35)
Burris, R. H., 298
Busch, R. D., 295
Butch, E., 39
Butler, A., 468
Butler, L. G., 470
Butterworth, S., 262

Buttle, D. J., 468
Byrne, C. R., 226
Byrne, W. L., 158

C

Cagen, L. M., 182
Calaycay, J., 52
Call, C. J., 64
Callahan, J. W., 87(26), 88
Callahan, P. M., 360
Callebaut, I., 65
Camafeita, E., 36
Cameron, A. D., 469
Cammack, R., 404
Campbell, A. S., 286
Canals, F., 24
Canard, B., 120, 120(25), 127, 129(25)
Cane, D. E., 468
Cantley, L. C., Jr., 7
Caplow, M., 265
Carey, P. R., 468
Carillon, S., 467
Caron, P. R., 202
Carr, M. C., 155, 469
Carr, S. A., 106, 107(4), 117(4)
Cartwright, J. L., 251, 252, 253(10), 254(10), 469
Case, R., 286
Cash, C., 467
Cassiman, J. J., 186
Castellino, F. J., 470
Castro, B., 310, 311(2)
Cerbelaud, E., 271
Cesareni, G., 436(5), 437
Chabin, R. M., 52, 190
Chabre, M., 130
Chabriere, E., 470
Chaffotte, A., 122
Chaiyen, P., 469
Chakrabarti, R., 261
Chakravarty, P. K., 295
Chambert, R., 469
Chan, C. C., 238
Chan, S.-I., 351
Chance, B., 32, 352, 353
Chang, C. C., 468
Chang, C. H., 24, 426, 427, 431(24), 432(24), 435

Chang, C. Y., 190
Chang, D. S., 274
Chang, P. K., 268
Chany, C. J., 72
Chaparian, M. G., 263, 265(40), 270(40)
Chapus, C., 108
Charles, C. H., 241
Charon, M. H., 305, 470
Chavez, R., 28, 35(12), 53
Cheah, E., 108
Chelsky, D., 14, 167
Chen, C. H., 30, 35(29), 197
Chen, C.-Y., 467
Chen, G., 428
Chen, H. P., 380, 384(5), 391(5), 396(5)
Chen, M. W., 271
Chen, R., 52
Chen, S., 405, 406(31), 408, 413(45)
Chen, Y.-L., 54
Chen, Y. R., 24
Chen, Z. W., 468
Cheng, A., 238
Cheng, T., 470
Cheng, X., 52, 469
Cherfils, J., 119, 120, 120(16; 32), 122, 129, 130
Chiadmi, M., 119, 120, 120(27; 49), 121, 127, 128, 133
Chih, H.-W., 380, 382, 391(17), 392, 392(17; 19), 395, 469
Ching, Y. C., 355, 356, 360, 360(12; 20), 361(23), 362(21; 22)
Chipman, D. M., 467
Chiusoli, G. P., 109, 113(14)
Cho, H., 179, 240
Chock, P. B., 18
Choi, S.-C., 394
Chow, J., 87(27), 88
Chowdhury, S., 391, 426
Christen, P., 227
Chun, K. Y., 24, 209, 216(11), 218, 219, 221, 222(11)
Ciechanover, A., 470
Ciesielski, L., 467
Ciocci, M. J., 180, 470
Cioni, P., 235
Clarke, A. R., 470
Clarkson, J., 468
Claus, T. H., 470
Clausen, T., 468

Cleland, W. W., 1, 2, 4(6), 5(2), 15, 153, 222, 270, 467, 468, 469, 470
Clepet, C., 179
Clifford, A. J., 31
Clore, G. M., 355
Coda, A., 262
Coego, A., 469
Cohen, J., 437, 438(6), 439(6), 446(6)
Cohen-Solal, M., 177
Cohn, M., 117, 118(18), 208
Cohn, V. H., 469
Colanduoni, J. A., 290
Coleman, D. E., 131
Collet, J.-F., 171, 174(10), 175, 177, 179, 179(5), 181(12), 182(5), 184(5; 11–14), 470
Collings, B. A., 54, 55(29)
Collins, K. D., 295
Collins, S., 238
Colloc'h, N., 310, 311(2)
Colowick, S. P., 7, 469
Colwell, R. E., 274
Combs, P., 295
Cong, P., 251, 260(2)
Conklin, K. A., 470
Connelly, G. P., 88(51), 89
Connolly, B. A., 14
Cook, P. F., 223, 227, 229, 230, 231(16), 233, 234, 234(22), 235, 235(22; 24), 467, 468, 470
Cooney, D. A., 261, 268(9)
Cooper, C. E., 404
Cooper, D. J., 52
Cooperman, B. S., 21, 469
Copeland, A., 232
Corbin, J. D., 180
Corley, R. C., 417
Cosper, N. J., 428
Cosson, M. P., 7
Cottaz, S., 87, 88(19), 470
Couture, M., 326, 328, 328(18)
Covey, T. R., 52, 53
Coward, J. K., 468
Cowling, R. A., 426
Crabtree, R. H., 401, 403
Craig, S. P. III, 469
Crane, B. R., 320, 321, 321(4)
Cree, G. M., 424
Creighton, D. J., 469
Creuzet, F., 24
Croce, C. M., 260

Cromlish, W., 238
Cronin, C. N., 468
Crooks, G. P., 29, 30(19), 31(19), 60
Croteau, R., 468, 470
Crowder, M. W., 177
Cruickshank, A. A., 285
Crysler, C. S., 469
Cuatrecasas, P., 371
Cuebas, D., 469
Culp, J. S., 106, 107(4), 117(4)
Cunane, L. M., 468
Curnow, A. W., 271
Curran, C. M. L., 337
Curti, B., 262, 270
Cusack, S., 295
Cutfield, J. F., 87(23), 88, 94(23)
Cygler, M., 108
Cyr, D. M., 25
Czuba, B., 469

D

Dabernat, S., 125, 126(23)
Daff, S., 326, 328, 330, 331(17)
Dafforn, T. R., 470
Dagley, S., 468
D'Agnolo, G., 469
Dahlquist, F. W., 161
Dalbey, R. E., 285
D'Alpaos, M., 36
Dalton, N. N., 163
Dalziel, K., 2
Dan, S., 87, 94(20)
Danger, D., 437, 438(6), 439(6), 446(6)
Darley, D. J., 394
Das, A. K., 238, 239(1), 242(1)
Dauter, M., 88(52), 89, 95, 95(52)
Dauter, Z., 95
Davenport, R. C., 25
Davey, C. A., 349
Davidson, A. R., 179
Davies, A. G., 393
Davies, D. R., 262
Davies, G. J., 64, 65, 66, 66(5), 82, 85, 87(24; 25; 34), 88, 88(52), 89, 94(24; 25; 34), 95, 95(52), 468
Davis, J. P., 177
Davis, R. D., 261
Davisson, V. J., 269, 270(53)

Davydov, A., 409
Dawes, T. D., 354
Dawson, J. H., 328
Dearborn, D. G., 203, 206(8), 207(11)
Debaecker, N., 404
De Boer, J. E. G., 345
De Frutos, M., 470
Degani, C., 168, 169(5), 170(5), 171(5), 174(5), 181, 193
Degn, H., 60
Dekel, M., 87, 94(20)
Dekker, H. L., 345
DeLacy, P., 52
Delepierre, M., 127
Delpierre, G., 175, 177, 179(5), 182(5), 184(5)
Demady, D. R., 333
DeMaso, C., 364
Dementin, S., 176, 189, 191, 195, 195(22), 196
Denisot, M.-A., 270
Dennis, E. A., 469
Denu, J. M., 39, 239, 240, 243
Deras, M. L., 269, 270(53)
DesJarlais, R. L., 106, 107(4), 117(4)
Deslongchamps, P., 82
Desmadril, M., 124
Desnuelle, P., 108
Desvages, G., 5
Deterre, P., 130
Devany, M. H., 468
Deville-Bonne, D., 118, 120, 120(25; 29), 124, 126, 126(21), 127, 129, 129(25)
DeWitt, D. L., 364
DeWolfe, R. H., 273
Dick, L. R., 285
Dickinson, T. A., 52
Dideberg, O., 191, 196, 470
Dietrich, H., 232
Diez, T. A., 468, 469
Differding, E., 215
Dijkhuizen, L., 75, 81, 82(47), 83, 87, 91(16)
Dijkstra, B. W., 74, 75, 78, 81, 82(47), 83, 84(39), 96, 108, 180, 469
Ding, J. L., 278
Ding, X., 468
Dion, R. L., 261
Discotto, L. F., 191
Distefano, M. D., 28, 31(11)
Dittmann, J., 468
Divne, C., 87(25), 88, 94(25)

Dixon, H. B. F., 202, 204(6), 206(6)
Dixon, J. E., 39, 181, 238, 239, 240, 241, 242, 242(20), 247(20), 470
Doan, P. E., 405, 406(32)
Dodds, D. R., 425
Dodson, G., 263, 284, 285(24)
Dodson, M. L., 202, 203, 204, 204(12), 206
Doi, M., 295
Dolphin, D., 87(27; 39), 88
Dooley, D. M., 422
Dorn, U., 468
Doublet, P., 192
Dougherty, T. J., 191
Douglas, D. J., 28, 30, 30(16), 50, 54, 55, 55(29), 56, 56(33), 57(33), 58, 59(33)
Douzou, P., 20
Dowd, P., 394
Dowhan, W., 468
Dowler, M. J., 467
Dowling, D. N., 108
Drennan, C. L., 381
Drenth, J., 262
Driguez, H., 470
Driscoll, J. S., 261
Dryhurst, G., 316
Du, W., 467
Dua, R. K., 107
Duah, F., 394
Dueker, S. R., 31
Duggleby, C. J., 108
Duggleby, H. J., 263, 284, 285(24)
Dugglesby, R. G., 161
Dumas, C., 119, 120, 120(11; 14; 49), 121, 122, 133
Dumora, C., 467
Dunaway-Mariano, D., 26, 180, 468, 470
Duncan, C. W., 52
Duncan, K., 290
Dunford, H. B., 329, 338, 339, 341, 342, 343, 344, 345, 345(2), 346, 347, 347(15), 350(16), 467
Dunham, W. R., 382
Dunn, M. F., 232, 233, 234(22), 235(22), 321, 470
Dupont, C., 87(34; 44), 88, 94(44), 95, 95(34)
Dwyer, T. M., 468
Dyer, R. B., 354
Dyson, H. J., 28

E

Eakin, A. E., 469
Ealick, S. E., 469
Earnhardt, J. N., 469
Eckstein, F., 14
Edmonds, C. G., 27
Edmondson, D. E., 270, 326, 329(19), 330, 334(19), 335(19), 405, 406, 406(30), 408, 408(30; 37), 413(44; 45), 436, 437, 438, 440(11), 441(11), 442(11), 444, 445, 449, 452
Edwards, J. O., 279
Egami, F., 470
Eggen, M.-J., 286, 287(34)
Egloff, M.-P., 238, 239(1), 242(1)
Eguchi, Y., 226
Ehmann, D. E., 467, 468
Ehrenberg, A., 401, 403, 404, 405(29)
Einarsdóttir, Ó., 354
Einspahr, H. M., 190
Eisenberg, R., 393
Eisenhardt, R. H., 32
Ekberg, M., 401, 403, 406(14)
Eklund, H., 400, 401, 401(7–9), 402, 403, 468
Elchebly, M., 238
El-Deeb, M. K., 418
Eldor, A., 467
El Hajji, M., 310, 311(2)
Elliott, W. H., 43
Ellis, B., 52
Elsas, L. J., 135
Eltis, L. D., 108
Emanuele, J. J., 190
Emanuele, J. J., Jr., 470
Emanuelli, M., 149, 152, 152(5), 153(10), 154(5), 155, 156, 157(5)
Engh, R. A., 262
Enkemann, S. A., 176
Enroth, C., 469
Epand, R. M., 265
Erickson, A. K., 281
Eriksson, M., 401, 403, 406(14)
Ervin, K. M., 190, 196(13)
Esaki, N., 52, 180, 469
Escalante-Semerena, J. C., 260
Escobar, W. A., 18
Ettinger, M. J., 418
Etzkorn, F. A., 191

Evans, D. R., 263, 265(40), 270(40)
Evans, P. R., 393
Eveland, S. S., 196
Evjen, G., 70, 88(50), 89

F

Faber, O. G., 75, 83
Fabrega, S., 65
Fabris, D., 28, 53
Fabry, B., 133
Fairbanks, L. D., 151
Fakhoury, S. A., 106, 107(4), 117(4)
Fales, H. M., 29
Falk, P. J., 190, 196(13)
Fan, C., 190
Fanchon, E., 191, 196, 470
Fasiolo, F., 469
Fassy, F., 470
Fauman, E. B., 238, 242
Fedele, D., 36
Feeney, R. E., 163, 202, 204(6), 206(6)
Feldman, P. L., 324
Feng, R., 52
Fenn, J. B., 27, 52
Fenna, R. E., 349
Fenton, W. A., 381
Ferguson-Miller, S., 351
Fernley, H. N., 158
Ferramola, A. M., 370
Ferrari, P., 191, 470
Ferretti, E., 152, 153(10)
Ferretti, J. A., 470
Fersht, A. R., 18, 19, 50, 57(1), 59(1), 295, 469
Fetter, J., 437, 438(6), 439(6), 446(6)
Fiedler, T. J., 349
Field, T. L., 134
Fierke, C. A., 384, 469
Filipowicz, W., 470
Filley, J., 405, 406(30), 408(30)
Finazzi-Agro, A., 24
Findlater, J. D., 467
Fink, A. L., 18, 20, 22, 22(71)
Fink, K., 151
Finke, R. G., 382
First, E. A., 468
Fischer, G. A., 268
Fischer, W., 467

Fish, S., 470
Fisher, A. J., 131
Fitzpatrick, P. F., 469
Fjeld, C. C., 243
Fleming, S. M., 106, 108, 117(13)
Fligge, T. A., 53
Flint, A. J., 238, 239, 241, 241(6), 244(6; 7; 21), 247, 247(21), 248(21; 25), 251(7)
Flores, O., 179
Floris, G., 24
Floris, R., 345
Floss, H. G., 468
Foglino, M., 179
Folk, J. E., 149, 470
Fontecave, M., 468
Fontecilla-Camps, J. C., 305, 470
Fookes, C. J. R., 377, 379, 379(20)
Forchhammer, K., 470
Forsyth, C. J., 295
Fothergill-Gilmore, L. A., 468
Fourme, P., 119
Foury, F., 181
Fox, B. G., 449, 467, 468
Fraaije, M. W., 470
François, J., 179
Frank, R., 87(31), 88, 470
Franken, E., 201
Franken, S. M., 108
Franklin, M., 469
Fraser, D. M., 296, 308
Fraser, M. E., 132
Freer, S. T., 23
Frerman, F. E., 468
Frey, M., 305
Frey, P. A., 14, 24, 27, 29, 36, 39(20), 44, 44(20), 46, 47, 48, 51, 64(5a), 134, 136, 136(6; 7), 137(14; 16), 140, 140(2; 13; 14; 16), 141(17), 145(18; 19), 146, 146(17), 147, 155, 251, 257, 259(15), 260(3), 426, 427, 427(1), 428, 428(17; 18), 429(20–22; 25; 26), 430, 431(20–22), 432(24), 435, 468, 469, 470
Fridovich-Keil, J. L., 135
Frieden, C., 18, 38
Friedmann, H. C., 182
Fritsche, E., 469
Fritz-Wolf, K., 395
Frolow, F., 108
Fromm, H. J., 1, 2(1), 3, 5(1), 7, 13, 15, 26, 467
Fronczek, F. R., 295

Fruton, J. S., 468
Fu, H., 425
Fujii, H., 151
Fujii, J., 275, 279, 279(8), 284(15)
Fujii, T., 180
Fujimoto, A., 470
Fujitaki, J. M., 256
Fukuchi, T., 119
Fumagalli, S. A., 370, 372, 373, 377(19)
Funderburk, L. H., 117, 118(18)
Furst, P., 176
Furtmüller, P. G., 345, 346, 347, 347(14), 350(16)
Furuta, S., 467

G

Gabibov, A. G., 470
Gachhui, R., 321, 326, 328, 328(23), 330, 331(23)
Gale, D. C., 52
Galperin, M. Y., 179
Gamesik, M. P., 106
Ganem, B., 52, 59(12)
Gao, J., 52
Garbeglio, M., 36
Garces, E., 2, 469
Garcia, E., 470
Garcia, G. A., 468
Garman, E., 250
Garrison, P. N., 260
Gaskell, S. J., 52
Gatel, M., 467
Gazynska, M., 285
Gebler, J. C., 87, 94(21), 96
Gee, K. R., 243
Geeganage, S., 134, 136, 140, 141(17), 145(18; 19), 146, 146(17)
Geeves, M. A., 122, 125(17)
Geeves, M. H., 20, 22(71)
Gefflaut, T., 197
Gegnas, L. D., 190
Gehring, A. M., 467
Gelb, M. H., 28, 35(7), 114, 286
Gerber, N. C., 333
Gerber, S. A., 28, 35(7)
Gerfen, G. J., 394, 426
Gergely, P., 197
Gerin, I., 179, 184(11)

Gerken, T. A., 203, 207(11)
Gerlt, J. A., 205, 467
Getzoff, E. D., 320, 321(4)
Ghisla, S., 467, 470
Ghose, C., 467
Ghosh, D. K., 321
Ghosh, S., 320, 321(1; 4)
Gibbons, B. H., 7
Gibbons, I. R., 7
Gibbs, R. A., 470
Gibian, M. J., 417
Gibson, K. J., 468
Gibson, Q. H., 32, 354
Gilbert, S. P., 18
Gilbertson, J. R., 469
Gileadi, O., 179
Gilkes, N. R., 87(35; 38), 88, 94(38), 96(35), 98, 99, 101(35), 102
Gillece-Castro, B. L., 176, 177
Gilles, A. M., 119
Gilman, A. G., 131
Ginodman, L. M., 106, 117(3)
Giriengl, H., 117
Glaser, L., 43, 44
Glaser, P., 119, 467
Glavas, S., 468
Glenn, A. G., 416, 419, 419(4)
Glick, D., 467
Glimcher, M. J., 177
Goessler, W., 321
Goffeau, A., 181
Gohda, K., 469
Goiten, R. K., 167
Goldbeck, R. A., 354
Goldberg, A. L., 285
Goldberg, J. D., 227
Golding, B., 381, 393, 394, 426, 469
Goldman, A., 108, 469
Goldman, R., 369
Goldstein, L., 369
Gonin, P., 124, 125, 126(23)
Goodman, S. I., 468
Goodtzova, K., 401, 403, 413(13)
Gorren, A. C., 322, 325(12), 326, 326(12), 328, 328(12), 329, 329(12), 331(25), 335(12; 25; 26)
Goshe, M. B., 394
Gouet, P., 133
Gould, S. J., 427
Grabowski, G. A., 87(28), 88, 94(28)

Grace, M. E., 87(28), 88, 94(28)
Graf, T., 60
Gragg, C. E., 467
Graham, R. W., 67, 76, 96
Grandas, A., 293
Gräslund, A., 399, 400, 401, 403, 404, 404(6), 405, 405(29), 406, 406(35), 407, 408(40), 409, 409(40; 42), 412, 412(42), 413(13; 42), 414, 414(42)
Graves, K. L., 468
Graves, P., 470
Green, B. G., 52
Green, B. N., 52
Greenblatt, J., 179
Greene, R. C., 470
Greenlee, W. J., 295
Greenwood, C., 354, 355
Grenier, L., 285
Gresser, M. J., 238
Griffin, P. R., 52
Griffin, R. G., 24
Griffith, O. W., 272, 273(3), 285(3), 290(3), 324, 468, 469
Griller, D., 416
Grindstaff, D. J., 30, 36
Grinstein, M., 370
Grisaru, D., 467
Grisham, M. B., 339
Grollman, A. P., 204, 468
Gross, G. G., 467
Gross, H. J., 163
Gross, J. W., 27, 29, 36, 39(20), 44, 44(20), 46, 47, 48, 51, 64(5a), 469
Gross, M., 149
Grossman, L., 202
Grubb, J. H., 87(43), 88, 94(43)
Grübel, G., 295
Gruber, K., 382, 390(20)
Grubmeyer, C., 469
Grunberg-Manago, J., 15
Guan, K.-L., 39, 181, 239, 240
Guan, X., 468
Guan, Y., 204
Guerreiro, C., 120, 120(25), 126, 127, 129, 129(25)
Guida, W. C., 261
Guidotti, G., 260
Guillerm, D., 467
Guillerm, G., 467
Gunther, M. R., 24

Gupta, G., 414
Guranowski, A., 260
Gurova, A. G., 106, 117(3)
Gusev, A. I., 30, 35, 36
Gustafson-Po Her, K. E., 379
Gutfreund, H., 32, 54

H

Ha, Y., 436
Hager, L. P., 329, 425
Hagishita, T., 227
Hagner-McWhirter, A., 469
Hahn Huynh, B., 406, 408(37)
Hajdu, J., 245
Hale, T. I., 227
Hall, W. R., 467
Hallahan, T. W., 470
Halpern, J., 393
Hama, H., 119
Hamilton, G. A., 418, 469
Hamm, H. E., 131
Hammerschmidt, F., 467
Hammes, G. G., 384
Hammond, S. M., 190
Han, S., 351, 356, 360, 361(23; 29), 362(21)
Han, S. H., 356, 360(20)
Han, S. W., 356, 362(22)
Handschumacher, R. E., 268
Hang Tong, W., 406, 408(37)
Hanlon, N., 241, 244(16)
Hansen, D. E., 14
Hanson, T. L., 3
Hara, S., 227, 229
Harding, C. O., 274
Hardman, K. D., 24
Hardy, L. W., 468
Harel, M., 108
Harley, E. H., 151
Harms, A., 43
Harris, S. R., 151
Harrison, D. H., 469
Harrison, D. J., 59
Harrison, J. K., 467
Harrison, P. M., 436, 436(5), 437
Hart, D. O., 72
Hart, G., 377, 379(20)
Hartley, B. S., 17, 18
Hartridge, H., 54, 352

Harvey, D. J., 30
Hashimoto, T., 467
Hassall, H., 468
Hassett, A., 14
Hasslacher, M., 117
Hata, Y., 180
Hatchikian, E. C., 305
Havenstein, J. D., 203
Hay, B. P., 382
Hayaishi, O., 467
Hayakawa, S., 470
Hayashi, N., 295
Haymotherwell, R., 393
He, S., 70, 72, 74, 78, 84(39), 87, 87(22; 32; 37; 43; 44), 88, 88(50), 89, 91(47; 48), 94(20; 22; 32; 37; 43; 44; 47; 48), 96
He, Z., 467
Heacock, D., 295
Hefta, S. A., 67, 81(18), 470
Hegeman, A. D., 29, 36, 39(20), 44, 44(20), 46, 47, 48
Heightman, T. D., 86(12), 87
Heikinheimo, P., 469
Heinrikson, R. L., 470
Heinzle, E., 28
Helander, E. R., 256
Hemann, C., 326, 328, 329(20), 330, 331(20), 334(20), 335(20)
Hemmens, B., 320, 321, 321(3)
Henderson, I. M. J., 107, 108, 109(9), 113(9), 117(9)
Hendrickson, C. L., 28
Hengge, A. C., 43, 222
Hengstenberg, W., 87(31), 88, 470
Henion, J. D., 52, 53, 59(12)
Henkin, T. M., 271
Henri, V., 455
Henrissat, B., 64, 65, 66, 85, 87, 88(19), 470
Henry, E. R., 312, 313, 314, 315
Henry, S. A., 469
Hensel, R. R., 37
Henslee, J. G., 470
Heo, C., 167
Her, J. H., 281
Hercules, D. M., 30, 35, 36
Herlihy, K., 437, 449
Hermodson, M., 470
Hernandez, L., 469
Herrmann, C., 21
Hers, H. G., 179

Herschlag, D. I., 120, 120(29), 129
Hersh, L. B., 210, 467
Hershko, A., 470
Hersperger, E., 119
Hess, G. P., 21
Hess, R. A., 222
Hetzler, B. G., 428
Hickman, Y. N., 467
Hider, R. C., 404
Higa, H. H., 163
Highbarger, L. A., 467
Hill, B. C., 355
Hille, R., 326, 328, 329(20), 330, 331(20), 334(20), 335(20)
Hillenkamp, F., 27
Himes, R. H., 468
Himms-Hagen, J., 238
Hiratake, J., 190, 272, 275, 276, 286(11), 289, 289(11), 290, 292
Hiromi, K., 50, 57(2)
Hirono, A., 151
Hirota, S., 355, 356, 364(17)
Hisaeda, Y., 394
Hisano, T., 180
Hixon, M. S., 469
Ho, H.-T., 14, 190, 196(13)
Ho, Y., 52
Hoenig, S., 87(31), 88, 470
Hofer, B., 108
Hoffman, B. M., 405, 406(31; 32), 426
Hofrichter, J., 312, 313, 314, 315
Hofstadler, S. A., 52, 64
Högbom, M., 406, 407, 408(40), 409(40)
Hogg, N., 24
Hohenester, E., 227
Holan, G., 69
Holbrook, E. L., 470
Holden, H. M., 26, 131, 136, 137(16), 140(16), 261, 262, 263, 269, 269(25), 270
Holladay, M. W., 209
Hollander, V. P., 158
Holloway, D. E., 380, 382(4), 384(4)
Holmes, I. H., 469
Holstege, F. C., 179
Homan, E. R., 261
Hong, K. W., 271
Honzatko, R. B., 24, 26, 467
Hoogland, H., 345
Horecker, B. L., 197, 198(1), 470
Horenstein, B. A., 159, 160, 165

Horiuchi, S., 279, 284(16)
Horn, B. R., 286, 292, 292(31)
Hörnberg, A., 414
Houck, B., 7
Houston, B., 468
Houston, C. T., 29, 36, 36(18), 39(18), 40, 41, 42
Howard, A. J., 263
Howard, S., 70, 71, 88, 88(50), 89, 91(48), 94(48)
Hoyes, K. P., 404
Hsiang, H. H., 107
Hsieh, F. Y. L., 52
Hsuanyu, Y., 347, 350(16)
Hu, A., 208
Huang, C. Y., 18
Huang, F. L., 252
Huang, J. T., 345
Huang, L. Q., 29, 30(19), 31(19), 60, 325, 328, 328(15), 329(15), 330, 331(15), 335(15)
Huang, W., 197, 200(7)
Huang, X., 270
Huang, Y., 260
Hubacek, I., 469
Hubbard, S. J., 52
Huber, R., 167, 262, 469
Hubner, G., 470
Hüdig, H., 470
Huebner, K., 260
Huhta, M. S., 380
Hultquist, D. E., 132, 256, 258(14)
Humm, A., 469
Hunt, D. F., 281
Hurshman, A. R., 320, 321(2), 326, 329(19), 330, 334(19), 335(19)
Hurst, J. K., 345
Huth, W., 467
Hüttermann, J., 468
Huynh, B. H., 326, 329(19), 330, 334(19), 335(19), 405, 406(30; 31), 408, 408(30), 413(44; 45), 414, 436, 437, 438, 440(11), 441(11), 442(11), 444, 445, 449, 452
Hwang, C. C., 470
Hwang, J., 437, 452
Hyland, L. J., 106, 107(4), 117(4)

I

Ichikawa, Y., 69
Ichimori, K., 326, 328, 330, 331(16)

Ichiyama, S., 52, 469
Iden, C. R., 204
Ido, Y., 27
Igarashi, M., 467
Ikeda, Y., 272, 273(4), 274(4), 275, 279, 279(8), 284(4; 15)
Ikeda-Saito, M., 345
In, Y., 295
Ingemarson, R., 401, 403, 404, 405, 413(13), 414
Ingold, K. U., 416
Ingram, S. W., 260
Inoue, M., 272, 275, 276, 279, 284(16), 286(11), 289, 289(11), 292, 295
Inouye, M., 119, 121, 122(10), 123(10)
Inouye, S., 119, 121, 122(10), 123(10)
Iodice, A. A., 380, 384(1)
Iori, R., 470
Irie, T., 290
Isaks, M., 279
Ishida, T., 295
Ishigooka, H., 108
Ishino, Y., 271
Isono, K., 191
Isupov, M. N., 262
Ito, E., 190, 286
Ito, N., 418
Itoh, H., 470
Ives, D. H., 155, 469
Iwahashi, H., 226
Iwanaga, T., 323, 324(13), 326, 326(13), 335(13)
Iwig, D., 414
Izumi, Y., 227, 231(16)
Izumori, K., 470

J

Jacobson, B. L., 190
Jacobson, T. M., 227, 229, 230
Jahansouz, H., 468
Jakes, R., 18
Jakoby, W. B., 467
James, M., 25, 132
Janin, J., 118, 119, 120, 120(11; 14–16; 25; 27; 29; 32; 49), 121, 122, 123, 124, 125, 126, 126(21; 23), 127, 128, 129, 129(25), 130, 133
Jansma, D. B., 179
Janson, C. A., 15

Jansonius, J. N., 227
Janssen, D. B., 180
Jantschko, W., 345
Jayaram, H. N., 261, 268(9)
Jean, J. M., 355
Jeanloz, R. W., 191
Jeffcoat, R., 468
Jefferson, M. M., 339
Jencks, W. P., 117, 118(18), 168, 210, 265, 469
Jennings, E. G., 179
Jensen, S. E., 25
Jentoft, J. E., 203, 207(11)
Jentoft, N., 203, 206(8), 207(11)
Jespersen, H. M., 66
Jezek, J., 194
Jia, J., 197, 198(6), 200, 200(7; 9), 201
Jia, Z., 239, 241(6), 244(6)
Jianmongkol, S., 333
Jin, H., 190, 470
Jitrapakdee, S., 470
Jogl, G., 382, 390(20)
Johansson, A. G., 404
Johns, K., 25, 87, 88(18), 89(18), 95(18), 96(18), 102, 102(18), 105(18)
Johnson, C. C., 417, 419(5)
Johnson, K. A., 18, 20, 21, 50, 51, 51(3), 321
Johnson, L. N., 241, 244(16)
Johnson, M. D., 469
Johnson, P. E., 5, 6, 6(15)
Johnson, R. C., 469
Johnston, L. N., 355
Jolly, L., 470
Jonas, R., 467
Jones, A. D., 31
Jones, B. E., 132
Jones, M. E., 470
Jones, M. T., 261, 268(9)
Jones, R. B., 30, 35(29)
Jones, S. R., 467
Jones, T. A., 87(25), 88, 94(25)
Jones, W. M., 467
Jordan, P. M., 369, 373(6)
Jorgensen, P. L., 95
Jörnvall, H., 43
Josephson, L., 7
Juknat, A. A., 370, 372, 373, 377(12; 19), 378, 379(12)
Jurlina, J. L., 208

K

Kabsch, W., 395
Kack, H., 468
Kahn, D., 133
Kahn, E. S., 30, 36, 43(26)
Kahn, K., 310, 312, 313, 314, 315, 317, 317(1), 318, 319(12)
Kajimoto, T., 69
Kalk, H. M., 262
Kalk, K. H., 75, 81, 83, 87, 91(16), 180, 469
Kalyanaraman, B., 24
Kamphuis, I. G., 262
Kan, L., 469
Kang, M. J., 28
Kapitannikov, Y. V., 106, 117(3)
Kaplan, N. O., 469
Kappl, R., 468
Karas, M., 27
Karkowsky, A. M., 158
Karlin, K. D., 450
Karlsson, B., 355
Karoui, H., 24
Karpefors, M., 365
Karpetsky, T. P., 468
Karplus, M., 25, 102
Karsten, W. E., 223, 227, 231(16), 468, 470
Kaseman, D. S., 468
Kassel, D. B., 52
Kast, J., 53
Katchalski, E., 369
Kato, H., 190, 290
Katterle, B., 406, 409(42), 412, 412(42), 413(42), 414(42)
Katunuma, N., 262
Kaufman, C., 394
Kavarana, M. J., 469
Kawasaki, H., 469
Kaziro, Y., 251, 260(1)
Kebarle, P., 52
Keen, J. N., 418
Keil, B., 470
Kelleher, N. L., 28
Keller, J. C., 247, 248(25)
Kelsey, J. E., 467
Kemal, C., 316
Kempton, J. B., 68, 69, 86(13), 87, 87(33; 42), 88, 89(33; 42)
Kennedy, B. P., 238
Kennedy, E. P., 184

Kent, D. E., 286
Kenyon, G. L., 467, 468
Kern, D., 133
Kern, G., 470
Kerscher, L., 469
Kersten, P. J., 418
Kezdy, F. J., 42
Khan, A. R., 470
Kharitonov, V. G., 337
Khedouri, E., 9
Khorana, H. B., 7
Khrisnaswamy, P. R., 9
Kijkstra, B. W., 87, 91(16)
Kikuchi, T., 469
Kilburn, D. G., 87(35), 88, 96(35), 98, 99, 101(35)
Kilby, B. A., 17
Kim, D. H., 467
Kim, J., 134, 136, 136(6), 140(13), 180, 202, 262, 268, 268(35), 470
Kim, S. C., 467
Kim, S. I., 271
Kim, T. K., 179
Kim, Y., 243
Kim, Y. C., 271
Kimura, N., 119
Kimura, Y., 469
Kindman, L. A., 467
King, R. C., 37
Kinsland, C., 468
Kirshenbaum, I., 163
Kitagawa, T., 356, 363, 364(17)
Kitajima, N., 447
Kitas, E., 240
Klebe, G., 261
Kleineidam, R. G., 163
Kleinkauf, H., 468
Klem, A. J., 467
Klem, T. J., 269, 270(53)
Klevit, R. E., 132
Kliger, D. S., 354
Klinman, J. P., 390, 468
Klosch, B., 321
Klunder, J. M., 285
Knackmuss, H., 469
Knappe, J., 395, 468
Knegtel, R. M., 75, 83
Knight, J. B., 59
Knight, W. B., 52
Knowles, J. R., 5, 6, 6(15), 14, 15(16), 418, 467

Knowles, P. F., 422
Knox, J. R., 190, 285
Kobashi, K., 467
Kobor, M. S., 179
Koch, G. L., 18
Kochetkov, S. N., 470
Koenigs, P. M., 28, 52
Koeppe, O. J., 9
Koerber, S. C., 232
Kogure, Y., 52
Kohara, Y., 191
Kohen, A., 390
Kolakowski, B. M., 30, 61, 62(41), 63(41)
Kolkmann, F., 286
Kollman, P. A., 263
Kolter, R., 191
Kominami, S., 323, 324(13), 326, 326(13), 335(13)
Komori, N., 30, 36, 43(26)
Kon, H., 226
Kondo, K., 286
Konermann, L., 28, 30, 30(16), 50, 54, 55, 55(29), 56, 56(33), 57(33), 58, 59, 59(33), 60(37), 61, 62(41), 63(41)
Konrad, M., 122, 125(17)
Koonin, E. V., 179
Kopriva, S., 467
Korey, S. R., 15
Kornberg, A., 7
Koshland, D. E., 65, 85
Koshland, D. E., Jr., 6, 9, 14(17), 168, 176, 177, 455
Kosman, D. J., 418, 419, 421(18), 422
Kotler, M. L., 370, 372, 373, 377(19), 378
Kotoh, M., 290
Koushik, S. V., 469
Kouyoumdjian, F., 179
Kowalski, A., 9
Kraft, W. G., 28, 52
Krahn, J. M., 262, 268, 268(35)
Kramer, J. J., 470
Krantz, A., 52
Kratky, C., 117, 382, 390(20)
Krause, J. M., 163
Kraut, J., 23
Krchnák, V., 194
Krebs, C., 326, 329(19), 330, 334(19), 335(19), 408, 413(44; 45), 414, 436, 437, 444, 445, 449, 452
Krebs, J. F., 28

Kredich, N. M., 225, 226, 227
Krinsky, N. I., 345
Krishnaraj, R., 240
Krishnaswamy, P. R., 171, 173(9)
Krivánek, J., 257
Krook, M., 43
Kucharczyk, N., 270
Kuhm, A. E., 469
Kulmacz, R. J., 426(15), 427
Külzer, R., 468
Kumagai, H., 275, 276
Kumar, R., 125, 126(23)
Kume, S., 177
Kunkel, T. A., 143, 248
Kurahashi, K., 135
Kurihara, T., 52, 180, 469
Kuriyan, J., 102
Kurland, I. J., 470
Kurokawa, I., 151
Kurtz, A. J., 202
Kustu, S., 133
Kuzmic, P., 313
Kwon, N. S., 324

L

Laber, B., 468
Labouesse, B., 470
Lacombe, M.-L., 119, 120, 120(11; 15), 121, 122, 123, 125, 126(23)
Lacoste, A. M., 467
Laemmli, U. K., 185
Lahti, R., 469
Lai, C. Y., 197
Lam, W. W. Y., 107, 108, 109(8)
Lambright, D. G., 131
Lambry, J.-C., 337
Lamden, L. A., 279, 284(18)
Lane, M. D., 210, 222(15)
Lane, W. S., 179
Lang, G., 404, 405(28)
Lange, A. J., 132, 468, 470
Lange, R., 322, 325(12), 326, 326(12), 328, 328(12), 329, 329(12), 331(25), 335(12; 25; 26)
Langer, L., 9
Langkau, B., 467
Langley, G. J., 106, 108, 111, 117(13), 468
Lapolla, A. A., 36

Large, R., 328
Larroque, C., 329
Larsen, B. S., 29
Larsen, K., 295
Larsen, T. M., 261
Larson, T. J., 468
Larsson, Å., 400, 401(9)
Lascu, I., 119, 120, 120(11; 14; 16; 27; 49), 121, 122, 124, 125, 126, 126(23), 127, 128, 133
Lascu, J., 124
Laslo, K., 28
Lassmann, G., 414
Latour, J.-M., 404
Lau, L. F., 241
Lauraeus, M., 356
Law, N., 29
Lawen, A., 468
Lawson, C. L., 81, 82(47)
Lawson, D. M., 436(5), 437
Lawson, S. L., 76, 96
Leadlay, P. F., 393
Learn, D. B., 348
Leary, J. A., 31
Leatherbarrow, R. J., 75
Lebedev, N., 470
Le Beller, D., 196, 470
Leber, A., 321
Leberman, R., 295
LeBras, G., 120, 120(14; 15; 49), 121, 122, 123, 133
Lecroisey, A., 127
Ledbetter, A. P., 328
Ledley, F. D., 381
Lee, E., 131
Lee, E. D., 53
Lee, H. C., 18
Lee, H. I., 405, 406(32)
Lee, M., 470
Lee, S.-L., 134
Lee, T. D., 52, 67, 81(18), 470
Lee, V. W. S., 54
Lee, Y. H., 132
Le Goffic, F., 270
Legrand, J.-F., 295
Lehman, I. R., 468
Lehmann, M., 295
Lehn, P., 65
Lehtonen, J., 469
Leigh, J., 353
Lemaire, H. G., 135

Lendzian, F., 414
Leopold, K., 467
Lerner, C. G., 119
Leslie, N. D., 135
Lester, W., 179
Leuhr, C. A., 467
Levi, S., 436(5), 437
Levinthow, L., 9
Levitt, D. G., 132
Levy, C. C., 468
Lewandowksi, K. M., 421
Lewis, E. R., 18
Lewis, M., 24
Lewis, R. J., 133
Lewis, W. S., 261
Ley, B. A., 408, 413(44; 45), 449
Leykam, J. F., 364
L'Hermite, G., 310, 311(2)
Li, J. P., 469
Li, K., 467
Li, Y.-F., 52, 180
Li, Y.-T., 52, 59(12), 369
Liang, L. F., 369
Liang, P. H., 468
Licht, S., 426, 470
Liebl, U., 337
Lieder, K. W., 427, 428, 429(25)
Lienhard, G. E., 13
Liger, D., 190, 191
Limbourg-Bouchon, B., 122
Lin, H.-J., 427
Lin, S., 167
Lin, Y., 64, 256
Lindahl, P. A., 296, 299, 305(1), 306, 307(2), 308
Lindahl, U., 469
Lindell, S. D., 295
Linder, D., 393
Linder, M. E., 131
Lindhorst, T., 73, 96
Lindqvist, Y., 197, 200, 200(7; 9), 468, 469
Ling, J., 406
Ling, V. W. K., 136, 145(18)
Linn, S., 202
Liotta, L. A., 119
Lipmann, D. J., 179
Lipscomb, W. N., 24
Littlechild, J. A., 262
Liu, A., 409
Liu, H.-W., 43, 467

Liu, J. J., 252, 254(9)
Liu, J. Q., 180, 469
Liu, K. K., 69
Liu, R.-Q., 26, 468
Live, D. H., 468
Livingstone, J. C., 436(5), 437
Llambías, E. B. C., 370, 373, 379(9; 10)
Lloyd, R. S., 202, 203, 204, 204(12), 206
Lobkovsky, E., 285
Loehr, T. M., 406, 437, 449
Logan, D. T., 395
Logan, M. S., 468
Lohse, D. L., 240
Lonberg-Holm, K. K., 32
Long, F. A., 161
Loo, J. A., 27, 52
Lopez, R., 470
Losel, R. M., 24
Losey, H. C., 468
Lou, L., 469
Low, N. H., 36
Lowe, P. N., 467
Loy, A. L., 238
Lu, W.-P., 28, 52
Lubitz, W., 414
Luchsinger, W. W., 9
Luck, S. D., 54
Ludwig, M. L., 381
Lueck, J. D., 7
Luginbühl, P., 133
Lusty, C. J., 263, 265(42), 270, 270(42)
Luthy, C. L., 316
Luzzago, A., 436(5), 437
Ly, H., 64, 66(4), 70, 88
Lynagh, G. R., 404

M

Ma, Y.-T., 285
Ma, Y. Z., 21
MacConnell, J. G., 29
Macfarlane, R. D., 37
MacGibbon, A. K. H., 232
MacGregor, E. A., 66
Mackenzie, L. F., 87(23–25; 34; 37), 88, 88(52), 89, 94(23; 25; 37), 95(34; 52)
Mackenzie, N. E., 106
MacLachlan, J., 468

MacLeod, A. M., 76, 96, 104(65)
MacPherson, H. T., 258
Madden, T. L., 179
Madhavapeddi, P., 380
Madsen, N. B., 73
Magaard, V. W., 106, 107(4), 117(4)
Magde, D., 337
Magni, G., 149, 152, 152(5), 153(10), 154(5), 155, 156, 157(5)
Magnusson, O. Th., 430
Mahroof-Tahir, M., 450
Mahuran, D. J., 107
Maitre, M., 467
Majumder, A. L., 469
Malet, C., 469
Mallory, P. A., 468
Mallory, W. R., 467
Malmstrom, B. G., 355
Malthouse, J. P. G., 106
Mammen, M., 52
Mancebo, H., 179
Mandel, P., 467
Manek, M. B., 417, 419(5)
Mangel, W. F., 25
Mann, G. J., 404, 405, 406(35)
Mann, M., 27, 37, 52
Mannervik, B., 469
Manning, J. M., 275, 467
Manuel, R. C., 204
Maples, K. R., 319
Marapaka, P., 206
Markham, G. D., 467
Marletta, M. A., 320, 321(2), 326, 329(19), 330, 333, 334(19), 335(19)
Marolewski, A. E., 24
Marquardt, A. C., 191
Marquardt, J. L., 191
Marquet, A., 467
Marquez, L. A., 339, 345, 345(2), 346, 347(15)
Marquez-Curtis, L. A., 338, 341, 342, 343, 344
Marradufu, A., 424
Marsh, E. N. G., 380, 382, 382(4; 5), 383, 384(4; 5), 388, 389(16), 390, 390(16), 391(5; 16; 17), 392, 392(17; 19), 395, 396(5; 16), 469
Marshall, P. J., 426(15), 427
Martasek, P., 24, 337
Martens, O., 271
Martin, B. M., 180, 470

Martin, J.-L., 337
Martin, L., 191, 196, 470
Martin, R., 28
Martin-Aragon, G., 261
Martinez, J. A., 261
Martino, P., 281
Martinson, K. H., 7
Marton, I., 87, 94(20)
Marty, M. L., 261
Masamune, S., 215
Mashhoon, N. J., 21
Mason, R. P., 24, 319
Massey, V., 467, 469, 470
Massière, F., 262
Masson, A., 190, 191
Masters, B. S. S., 24, 321, 328, 337
Matcham, G. W., 377, 379, 379(20)
Mathews, F. S., 468
Matson, D., 64
Matsui, K., 286
Matsumoto, H., 30, 36, 43(26)
Matthews, D. A., 23
Matthews, H. R., 127, 255, 256(12), 258(12), 259(12)
Matthews, R. G., 381
Matthijs, G., 179, 181(12), 184(12; 13), 186, 470
Mattia, K. M., 24
Mauk, A. G., 54
Maurin, L., 329
Mayaux, J. F., 271
Mayer, B., 320, 321, 321(3), 322, 325(12), 326, 326(11; 12), 328, 328(12), 329, 329(12), 331(25), 335(11; 12; 25; 26)
Mayer, C., 88, 91(47), 94(47)
Mayer, R. M., 468
Mayr, I., 262
Mazur, S. J., 202
McCarter, J. D., 67, 69, 70(29), 71, 87(26–28; 30), 88, 90(46), 91, 91(46), 94(28; 46; 56)
McClure, G. D., 234, 235(24)
McCullough, A. K., 206
McDermott, A. E., 24
McDonald, E., 379
McEwen, C. N., 29
Mcguire, J. J., 468
Mcintire, W. S., 468
McIntosh, L. P., 88(51; 53), 89, 95(53)
McLafferty, F., 468
McLaren, F. R., 419, 421(18)

McLennan, A. G., 251, 252, 253(10), 254(9; 10), 469
McMartin, C., 261
McMaster, J. S., 285
McMillan, K., 321, 328
McNeal, C. J., 37
McPherson, M. J., 417, 418, 422
Meade, A. L., 326, 328, 329(20), 330, 331(20), 334(20), 335(20)
Medda, R., 24
Meek, T. D., 106, 107(4), 117(4), 293
Mehlhaff, P. M., 467
Mehta, P. K., 227
Mei, B., 270
Meier, T. W., 393
Meighen, E., 469
Meinke, A., 87(38), 88, 94(38)
Meister, A., 9, 10, 13, 13(34), 171, 173(9), 190, 193, 195, 270, 275, 279, 279(8), 284(15), 290, 468
Mejean, V., 179
Melander, L., 167
Mellin, T. N., 295
Melo, A., 43
Ménard, R., 52, 262, 263(31), 270(31)
Méndez, E., 36
Mendola, C. E., 120, 120(50), 133
Menendez, C., 469
Meng, C. K., 27, 52
Mengin-Lecreulx, D., 189, 192, 196, 470
Mercer, R. S., 31
Merle, M., 470
Merola, F., 124
Messerschmidt, A., 468
Metcalf, W. W., 180, 470
Meyer, P., 119, 120, 120(25; 29), 127, 129, 129(25)
Miao, S., 77, 87(26; 28; 29; 36), 88, 93, 94(28; 29), 96(36), 100, 100(36)
Michaels, M. L., 204
Michaliszyn, E., 238
Michelson, A. M., 43
Midelfrot, C. F., 16
Mieyal, J. J., 11, 67, 470
Miles, E. W., 262, 470
Millen, W. A., 7
Miller, D. J., 190
Miller, J. H., 204, 427, 429(26)
Miller, N. E., 428
Miller, S. M., 469

Miller, W., 179
Milman, H. A., 261, 268(9)
Milne, G. W., 29
Milon, L., 125, 126(23)
Miner, C. S., 163
Minisci, F., 109, 113(14)
Minnich, M. D., 106, 107(4), 117(4)
Minoda, Y., 108
Minter, D. E., 227
Minutello, A., 467
Miosga, T., 201
Miran, S. G., 263
Misra, I., 211
Mitchell, R., 365
Mitsunaga, J., 295
Mitsunaga, T., 227
Miwa, S., 151
Miyagi, M., 180, 469
Miyate, H., 227
Miyazaki, J. H., 468
Miziorko, H. M., 24, 208, 209, 210, 211, 212, 214, 215, 215(9), 216(11), 217, 218, 219, 220, 221, 222(11; 15)
Mizumoto, K., 251, 260(1)
Moad, G., 316
Moënne-Loccoz, P., 437, 449
Moews, P. C., 190, 285
Mol, C. D., 204
Molineaux, S. M., 262
Mondesert, O., 39
Monroe, R. S., 226
Montague-Smith, M. P., 418, 419, 421(18; 19), 422(16), 424(19)
Moody, P. C. E., 263, 284, 285(24)
Moon, N., 382
Moore, B. S., 393
Moore, M. L., 106, 107(4), 117(4)
Moore, P., 314
Mooser, G., 67, 81(18), 468, 470
Moraga-Amador, D. A., 261
Morehouse, K. M., 24
Moreno, S., 39
Moréra, S., 119, 120, 120(11; 14–16; 27; 32; 49); 121, 122, 123, 127, 128, 129, 130, 133
Morgan, J. E., 362, 364(35)
Mori, I., 469
Morikawa, T., 467
Morino, Y., 279, 284(16)
Mornon, J. P., 65, 310, 311(2)
Moro-Oka, Y., 447

Morozov, Y. V., 223
Morr, M., 125, 126(23)
Morris, G. S., 151
Morris, R. G., 232
Morrison, J. F., 278
Mortenson, L. E., 298
Mosbach, K., 369
Mosi, R. M., 64, 74, 78, 81, 83, 84(39), 87, 91(16), 94, 96
Moss, M. L., 427, 428(17; 18)
Mothes, T., 36
Mottonen, J. M., 132
Mourey, L., 133
Moyer, R. W., 132, 256, 258(14)
Mozzarelli, A., 235
Muchova, K., 133
Mück, W., 53
Muddiman, D. C., 30, 35, 36
Mueller-Hill, B., 135
Muench, K. H., 470
Muijsers, A. O., 345
Mulcahy, R. T., 272, 273(3), 285(3), 290(3), 468
Muller, U., 469
Mullins, L. S., 17, 190
Mumford, R., 52
Munck, E., 467
Munoz, R., 470
Munoz-Dorado, J., 119, 121, 122(10), 123(10)
Munshi, C., 18
Murakami, Y., 394
Murthy, N. S. R. K., 469
Murzin, A. G., 263, 284, 285(24)
Musier-Forsyth, K., 295
Musil, D., 262
Musser, S. M., 351
Mwesigyekibende, S., 393
Myers, M. P., 241, 244(15)

N

Naber, N. I., 469
Nair, S. A., 21
Nakada, H. I., 467
Nakamura, J., 469
Nakamura, N., 418
Nakamura, T., 226
Nakasako, M., 180
Nalabolu, S. R., 227, 229, 230
Namchuk, M., 69, 87(30), 88, 95, 96

Nanz, D., 393
Narasimhan, C., 209, 211, 215(9)
Narindrasorasak, S., 470
Nathan, C. F., 324
Natori, H., 151
Neel, B. G., 238
Negrerie, M., 337
Neil, G., 261
Nelson, M. J., 426
Nelson, R. W., 30, 36
Neujahr, H., 469
Newcomb, M., 416, 417, 417(3), 419(3–6), 425, 425(6)
Newcomb, R. C., 273
Newman, J., 469
Nguyen, N. T., 88(53), 89, 95(53)
Nicholls, P., 355
Nichols, S. E., 256
Nikawa, J., 119
Nikolau, B. J., 468, 469
Nimmo, H. G., 468
Ning, J., 3
Nishihara, M., 180
Nitz, M., 87, 88(17), 96(17)
Noble, N. E. M., 241, 244(16)
Noel, J. P., 131
Noh, S. J., 239
Nohalle, M. J., 133
Nomizu, M., 286
Nomura, H., 180
Nordlie, R. C., 149, 150(2; 3), 468
Nordlund, P., 395, 400, 401, 401(7; 8), 402, 403, 406, 407, 408(39; 40), 409(40)
Normandin, D., 238
Northrop, D. B., 28, 53, 59, 161, 166
Norton, J. R., 405, 406(30), 408(30)
Notenboom, V., 87, 88(17), 95, 96(17), 103, 105(72)
Nozaki, M., 467
Nuber, B., 452
Numao, S., 70, 88(50), 89
Nuzzi, L., 262
Nyholm, S., 404

O

Obayashi, M., 286
Obinger, C., 345, 346, 347, 347(14), 350(16)
Obmolova, G., 262

Ochiai, E. I., 405, 406(35)
Ochoa, S., 15
Oda, J., 190, 290
Oesterhelt, D., 469
Ogata, G. M., 132
Ogawa, H., 180
Ögel, Z. B., 417, 418
Ogura, T., 356, 363, 364(17)
Oh, E. Y., 202
Ohno, T., 394
Ohshiro, T., 227, 231(16)
Ohta, D., 469
Ohta, M., 260
Oikawa, T. G., 252
Okar, D. A., 468
Okaya, H., 470
Okazaki, R., 43
Okazaki, T., 43
O'Leary, M. H., 262, 271
Oleschuk, R. D., 59
Olin, B., 469
Oliver, R. W. A., 52
Ollis, D. L., 108
Olson, T. W., 132
Omori, T., 108
Ondrias, M. R., 360
Oren, D. A., 121, 122(10), 123(10)
Orlowski, M., 158
Orme-Johnson, W. H., 298, 427, 430(16)
Orr, G. A., 5, 6, 6(15)
Orsi, B. A., 467
Ørsnes, H., 60
Ortiz de Montellano, P. R., 321, 333, 424
Osawa, Y., 333
Oshima, T., 197
Osumi, T., 467
Otaka, A., 286
Oths, P. J., 468
Owens, K. G., 37
Ozutsumi, M., 135

P

Packman, F., 135
Padiglia, A., 24
Paetzel, M., 285
Paglia, D. E., 151
Paiva, A. A., 29, 30(19), 31(19), 60
Palcic, M. M., 329

Palfey, B. A., 467
Palkovic, A., 316
Palm, D., 164
Palmer, G., 298, 354, 355, 411, 426(15), 427
Palmer, M. A., 215
Palmieri, S., 470
Pamiljans, V., 9, 171, 173(9)
Pandey, P. S., 377, 379(20)
Pannifer, A. D. B., 239, 244(7), 251(7)
Parera, V. E., 370, 373
Parikh, S. S., 204
Park, J. T., 189
Parkin, D. L., 161
Parkinson, J., 321
Parks, R. E. J., 118
Parquet, C., 176, 189, 190, 191, 195, 195(22), 196
Parrish, F. W., 208
Parry, R. J., 262, 268(35)
Parsons, S. M., 6, 7, 14, 167
Parsons, W. H., 295
Pascal, M., 158
Passeron, E., 123
Passeron, S., 123
Patchett, A. A., 295
Patel, M., 469
Patel, U., 408, 413(45)
Patte, J. C., 179
Paule, S., 28, 52
Paulson, J. C., 162, 163
Paxton, R. J., 67, 81(18), 470
Payette, P., 238
Payne, D. M., 281
Payne, M. A., 227
Pazur, J. H., 43
Peat, T. S., 469
Pedersen, J. Z., 24
Pederson, R. L., 69
Pedroso, E., 293
Pegg, A. E., 467
Peisach, D., 467
Penefsky, H. S., 266
Penner, M. H., 18
Penner-Hahn, J. E., 437, 452
Penninga, D., 75, 83
Peräkylä, M., 263
Perego, M., 467
Pereira, A. S., 437, 444, 445
Perie, J., 197
Perisic, O., 120, 120(50), 133

Perl, A., 197, 201
Perlin, A. S., 424
Persson, B. O., 43, 406, 407, 408(40), 409(40)
Peters, T. J., 278
Peterson, L., 404, 405(29)
Peterson, R., 409, 410(48), 411(48)
Petersson, L., 404
Pétillot, Y., 191
Petit-Glatron, M. F., 469
Petre, D., 271
Petsko, G. A., 22, 25, 468
Pfeffer, P. E., 208
Pfister, C., 130
Pflugrath, J. W., 25
Philips, J. P., 419
Philips, R. S., 107
Phillips, J. C., 263
Phillips, R. S., 469, 470
Phillips, S. E., 418, 422
Phung, Q., 28, 35(12), 53
Pickart, C. M., 169, 174(7), 470
Pierek, A. J., 426
Pieulle, L., 470
Pike, D. C., 295
Pilkis, S. J., 470
Pils, T., 468
Pirard, M., 175, 177, 179, 179(5), 181(12), 182(5), 184(5; 12; 13), 186, 470
Piras, C., 305
Pisano, J. J., 29
Plamondon, L., 285
Planas, A., 24, 469
Plat, H., 345
Plebani, M., 36
Pleogh, H., 285
Ploux, O., 467
Poje, M., 316
Poland, B. W., 26
Polikarpov, I., 262
Pompliano, D. L., 190, 196
Pontremoli, S., 470
Pop, R. D., 124
Popovic, T., 262
Porco, J. A., 69
Porter, D. J. T., 469
Porter, J. B., 404
Porumb, H., 124
Post, R. L., 177
Pötsch, S., 401, 403, 409, 414
Poulter, C. D., 468

Powers, S. G., 195
Powlowski, J., 467
Pradel, L.-A., 5
Prange, T., 310, 311(2)
Prasad, T. K., 468
Prasher, D. C., 155, 469
Pratt, R. F., 284, 285
Pratviel-Sosa, F., 191
Presecan, E., 119, 124
Pressler, M. A., 363, 364
Presta, A., 322, 326, 326(11), 328, 330, 331(16), 335(11)
Price, J., 26, 468
Pritchard, K. A., Jr., 24
Proctor, A., 35, 36
Proinov, I., 124
Proshlyakov, D. A., 356, 363, 364
Proske, D., 470
Przybylski, M., 53
Pucci, M. J., 191
Purich, D. L., 1, 2(3), 3, 4, 5(3), 8, 10, 13, 13(33), 15, 149, 168, 169, 171(8), 173, 173(8), 193, 195(34), 272, 273(1), 292(1), 455, 456, 468

Q

Que, L., 405, 406(33), 408(33), 412(33)
Que, N. L. S., 467
Quilico, A., 109, 113(14)

R

Raber, G., 321
Racker, E., 202, 455, 469
Radom, L., 393
Rafecas, F. J., 261
Raffaelli, N., 149, 152, 152(5), 153(10), 154(5), 155, 156, 157(5)
Raftery, M. A., 161
Rahil, J., 284, 285
Rajagopal, P., 132
Rajavel, M., 469
Ramachandran, C., 238
Ramaswamy, S. G., 469
Rand-Meir, T., 161
Rando, R. R., 469
Rao, G. S. J., 227, 229

Rao, K. S., 468
Rao, P. V. S., 467
Raptis, S., 469
Rasmussen, B. F., 468
Raushel, F. M., 17, 190, 262, 263, 269, 269(25), 270, 326, 328(23), 330, 331(23), 467
Ravi, N., 406, 408(37), 438, 440(11), 441(11), 442(11)
Ray, P. H., 467
Rayment, I., 131, 136, 137(14; 16), 140(13; 14; 16), 251, 260, 260(3), 261, 269
Recsei, P. A., 470
Reddy, S., 190
Reed, G. H., 24, 381, 426, 427, 428, 429(20–22; 25; 26), 430, 431(20–22), 432(24), 435, 468
Rees, D. C., 24
Rege, V., 227
Regelsberger, G., 345
Reich, N. O., 21
Reichard, P., 400, 404, 405(28; 29)
Reichardt, J. K. V., 135
Reichel, K. B., 4
Reid, T. W., 158
Reilly, J. P., 29, 36, 36(18), 39(18), 40, 41, 42
Reilly, P. J., 68
Reimann, E. M., 180
Reinberg, D., 179
Reinstein, J., 131
Reitzer, R., 382, 390(20)
Remington, S. J., 108, 468
René, L., 262
Rescigno, M., 262, 270
Rétey, J., 426
Reyes-Spinola, J. F., 428
Reznikoff, W. S., 134
Rhee, S., 262
Rhodes, S., 467
Ricci, C., 470
Rice, A. E., 243
Rich, D. H., 118, 209
Rich, P. R., 365
Richard, J. P., 14, 167
Richard, R., 469
Richards, N. G. J., 260, 261, 261(1), 263, 268(12; 44), 271, 271(15)
Richardson, C., 467
Richert, C., 36
Ridder, I. S., 180, 469
Ridderström, M., 469
Rider, M. H., 179, 184(11)

Riebel, P., 108
Rieger, R. A., 204
Rimai, L., 360
Rinaldo-Matthis, A., 406, 408(39)
Ringe, D., 25, 467, 468
Rishavy, M. A., 270
Risley, J. M., 208, 209, 212
Rizzi, M., 262
Roberts, G. D., 106, 107(4), 117(4)
Roberts, L. M., 296, 305(1), 306
Roberts, R. J., 469
Robertson, H. L., 52
Robertson, T. A., 106, 108, 117(13)
Robillard, G. T., 132
Robinson, A. K., 260
Robinson, J. A., 393
Robinson, V. J., 52
Robles, J., 293
Roby, W. G., 286, 292(31)
Robyt, J. F., 68, 468
Roder, H., 54
Rodriguez, A., 469
Rodriguez-Crespo, I., 321
Roelofs, J. C., 470
Roepstorff, P., 37
Roller, P. P., 286
Rollin, P., 87, 88(19)
Roman, L. J., 328, 337
Romaniuk, P. J., 14
Rome, L. H., 468, 469
Ronzio, R. A., 13, 290
Root, D. E., 450
Rooze, V., 380, 384(1)
Roque, J. P., 232
Rosa-Rodriguez, J. G., 263
Rose, D. R., 68, 84, 87, 88(17; 18), 89(18), 95, 95(18), 96(17; 18), 102, 102(18), 105(18)
Rose, I. A., 15, 16, 17, 169, 174(7), 470
Rose, U., 163
Rose, W. B., 290
Rose, Z. B., 179, 470
Rosell, F. I., 54
Rosenblatt, D. S., 381
Rosenbluth, R. J., 261
Rosenfeld, I. S., 469
Roskoski, R., 180
Rospert, S., 468
Ross, N. W., 175
Rossetti, M. V., 369, 370, 377(11; 12), 379(11; 12)

Rossolini, G. M., 177
Rossomando, A. J., 281
Roth, A., 295
Rotilio, G., 24
Roughton, F. J. W., 54, 352
Roushel, F. M., 328
Rousseau, D. L., 326, 328, 328(18), 335, 351, 355, 356, 360, 360(12; 20), 361(23; 29), 362(21; 22)
Rousseau, P., 133
Roustan, C., 5
Roux, B., 263, 267(41), 270(41), 467
Rova, U., 401, 403, 406, 409(42), 412, 412(42), 413(13; 42), 414(42)
Rowe, W. B., 13
Roymoulik, I., 380, 382
Rozeboom, H. J., 75, 83, 180, 469
Rubin, H., 409
Rudisill, D. E., 287
Rudolph, F. B., 293
Rudolph, R., 470
Ruggieri, S., 149, 152, 152(5), 153(10), 154(5), 155, 156, 157(5)
Rumsh, L. D., 106, 117(3)
Rupitz, K., 68, 76, 87(40; 42), 88, 89(42), 96, 104(65)
Rusche, K. M., 320, 321(2)
Russell, P., 39
Russell, W. K., 296
Rutter, R., 329
Ruzicka, F. J., 134, 136, 136(6), 137(16), 140(13; 16), 146, 147, 428
Ryan, J. A., 261
Rye, C., 64, 66(6), 67(6)

S

Saadat, D., 469
Sabini, E., 95
Sagami, I., 326, 328, 330, 331(17)
Sahlin, M., 400, 401, 403, 404, 404(6), 406, 406(14), 407, 408(40), 409(40)
Sahm, H., 197, 198(5; 6), 200, 200(7)
Saito, S., 135
Sakai, T., 468
Sakata, K., 272, 275, 276, 286(11), 289, 289(11), 292
Sakuma, T., 52
Sale, W. S., 7

Salih, E., 262, 263(29)
Salituro, F. G., 209
Salmeen, A., 241, 244(15)
Salmeen, I., 360
Salomaa, P., 161
Samama, J.-P., 133
Sammons, R. D., 14
Sampson, N. S., 284
Samuel, D., 281
Sanchez, A., 206
Sancovich, H. A., 370
Sanders, D. A., 176, 177
Sanders-Loehr, J., 406, 418, 437, 438(6), 439(6), 446(6), 449
Sandusky, P. O., 418
Santi, D. V., 470
Santolini, J., 337
Sanvoisin, J., 111
Sanz, G. F., 261
Sanz, M. A., 261
Saper, M. A., 238, 240, 242
Saraiva, M. C., 295
Sarfati, R., 124, 126, 126(21), 129
Sarfati, S., 120, 120(25), 126, 127, 129(25)
Sarma, A. D., 317, 318
Saronio, C., 353
Sashchenko, L. P., 470
Satishchandran, C., 467
Sato, H., 326, 328, 330, 331(17)
Satterwhite, D. M., 468
Saunders, W. H., 167
Sauve, A. A., 18
Scaloni, A., 467
Schaaff-Gerstenschläger, I., 201
Schaertl, S., 122, 125(17)
Schaffer, A. A., 179
Schaleger, L. L., 161
Schandle, V. B., 293
Schauer, R., 163
Scheek, R. M., 132
Scheele, J. S., 337
Scherf, U., 469
Schianchi, G., 235
Schiltz, M., 310, 311(2)
Schimmel, P., 467
Schindler, J. F., 469
Schippa, S., 177
Schirch, V., 469
Schlauderer, G., 122
Schlichting, I., 131

Schloss, H. G., 14
Schloss, J. V., 469
Schmacher, J., 133
Schmelter, T., 163
Schmid, K., 163
Schmidt, D. E., Jr., 107
Schmidt, K., 321
Schmidt, P. G., 118, 209
Schmidt, P. P., 326, 329, 335(26), 406, 409(42), 412, 412(42), 413(42), 414(42)
Schnackerz, K. D., 227, 229
Schneemann, A., 28, 35(12), 53
Schneider, B., 120, 120(25; 29), 123, 124, 126, 126(21), 127, 129, 129(25)
Schneider, G., 197, 198(6), 200, 200(7; 9), 201, 468, 469
Schnizer, H. G., 260, 263, 268(44)
Schofield, C. J., 28, 52
Schollen, E., 179, 184(13), 186, 470
Schopfer, L. M., 467
Schörken, U., 197, 198(5; 6), 200, 200(7; 9), 201
Schrag, J. D., 108
Schramm, V. L., 12, 18, 167
Schrammel, A., 328, 331(25), 335(25)
Schrock, R. D., 203, 204(12)
Schuber, F., 158
Schubert, H. L., 238, 242
Schülein, M., 87(24; 25), 88, 88(52), 89, 94(24; 25), 95, 95(52)
Schultz, S., 395
Schulz, G., 122
Schulz, H., 469
Schuster, S. M., 260, 261, 261(1), 263, 268(12; 44), 270(19), 271, 271(15), 467
Schwab, H., 117
Schwartz, B. L., 52
Schwartz, R. T., 163
Schweizer, E. S., 418
Scofield, M. A., 261
Scott, A. I., 106
Scott, C. R., 28, 35(7)
Scott, R. A., 428
Seah, S. Y. K., 108
Seal, S. N., 179, 470
Seaton, B. A., 25
Seaver, S. S., 295
Secemski, I. I., 13
Seddon, A. P., 193, 275
Seeholzer, S. H., 468
Seelig, G. F., 190

Segal, S., 134
Segura, E., 469
Seitz, S. P., 426
Sellam, O., 124, 126, 126(21)
Selman, G., 469
Sémériva, M., 5, 6, 6(15), 108
Semmler, R., 468
Seneviratne, K. D., 422
Seraglia, R., 36
Serfozo, P., 318
Setyawati, I., 87(45), 88
Seubert, W., 467
Severin, E. S., 470
Shabanowitz, J., 281
Shabarova, Z. A., 255
Shah, S., 52
Shamala, N., 468
Shane, B., 468
Sharkey, A. G., 35, 36
Sharma, V. S., 337
Shastry, M. C. R., 54
Shaw, E., 279
Shaw, W. V., 436(5), 437
Shaw-Reid, C. A., 468
Shearn, A., 119
Shemin, D., 210
Shen, Q., 286, 287(34)
Sheng, S., 261
Sheu, K.-F., 134, 136(7), 470
Sheu, K.-W., 14
Shevely, J. E., 67, 81(18)
Shi, D., 436
Shi, Y. Q., 469
Shiba, K., 295
Shimizu, E. T., 21
Shimizu, T., 326, 328, 330, 331(17)
Shin, W., 296, 299, 307(2)
Shinzawa-Itoh, K., 356, 363, 364(17)
Shivley, J. E., 52
Shoelson, S. E., 286
Shoemaker, R. K., 286, 287(34)
Shoku, Y., 295
Short, S. A., 469
Shoseyov, O., 87, 94(20)
Shriver, K. K., 468
Shuey, E. W., 43
Shum, K., 70, 88
Shuman, S., 251, 260(2)
Shushan, B., 52
Sidhu, G., 88(53), 89, 95(53)

Siebers, A., 468
Siegbahn, P. E. M., 401, 403
Siegele, D. A., 191
Sielecki, A., 25
Sierks, M. R., 66
Sigal, G. B., 52
Sigler, P. B., 131
Sikorski, J. A., 50, 51(3)
Silman, I., 108
Silver, B. L., 281
Silverman, D. N., 469
Silverman, R. B., 415, 417(1), 421, 422(24), 424
Silverstein, E., 6, 8, 15(26)
Sim, S. S., 107
Simcox, P. D., 6
Simmonds, H. A., 151
Simmons, D. A., 30, 54, 61
Simmons, J. W. III, 229, 230
Simon, E. J., 210
Simon, H., 164
Simon, M., 67
Simpson, F. B., 28, 53, 59
Sims, P. F., 72
Singer, P. T., 25
Sinnott, M. L., 64, 66(5), 72, 85, 87, 167, 468
Sinskey, A. J., 215
Sipe, H. J., Jr., 24
Siprashvili, Z., 260
Siuzdak, G., 28, 35(12), 53
Sjöberg, B.-M., 395, 400, 401, 401(2; 7), 402,
    403, 404, 405(29), 406, 406(14), 407,
    408(39; 40), 409(40)
Skarstedt, M. T., 6, 8, 15(26)
Sleep, J., 21
Sligar, S. G., 329
Sly, W. S., 87(43), 88, 94(43)
Small, G. W., 436
Small, W., 437, 444, 445
Smeland, T. E., 469
Smirnov, I. V., 470
Smirnova, I. A., 365
Smith, B. R., 421
Smith, C. A., 131
Smith, D. M., 393
Smith, G., 31
Smith, G. D., 278
Smith, J. L., 260, 262, 263, 268, 268(35), 269,
    269(20), 270(53), 279, 284, 285(24)
Smith, J. M., 262, 436(5), 437
Smith, M., 252

Smith, R., 131
Smith, R. A., 52, 256, 470
Smith, R. D., 27, 52, 64
Smith, T. K., 275, 279(8)
Smyth, M. S., 286
Sneen, R. A., 421
Snell, E. E., 468, 469, 470
Sniekus, V., 108
Snively, J. E., 470
Soda, K., 180, 469
Sofia, H. J., 428
Sogbein, O. O., 54
Solioz, M., 176
Soll, D., 271
Solomon, E. I., 447, 450
Sondek, J., 131
Song, H., 241, 244(16)
Song, S., 468
Song, X.-M., 394
Sono, M., 328
Soreq, H., 467
Soubrier, F., 271
Souchard, I. J., 87
Souza-e-Silva, U., 162
Spain, J. C., 467
Spector, L. B., 6, 15
Spence, F., 468
Spinka, M., 470
Sporns, P., 36
Sprang, S. R., 131
Sprenger, G. A., 197, 198(5; 6), 200, 200(7; 9), 201
Sproat, B., 295
Stadtman, T. C., 469, 470
Staedtler, P., 87(31), 88, 470
Stafford, P. R., 296, 307(2)
Stanley, B. A., 467
Stark, G. R., 203, 295
Stayton, M. M., 467
Steeg, P. S., 119
Stein, R. L., 285
Steiner, A. W., 256
Stella, A. M., 369, 373
Stelte, B., 175
Sternfeld, M., 467
Stevens, C., 418, 422
Stevens-Clark, J. R., 470
Stevis, P., 177
Stewart, J. D., 28, 260, 261, 263, 268(44)
Stiasny, H. C., 419

Stock, A. M., 132, 176, 177
Stock, J. B., 132
Stoker, P. W., 271
Stole, E., 275
Stoll, D., 87(32), 88, 94(32)
Stolz, A., 469
Stombaugh, N. A., 298
Stoops, J. K., 468
Storer, A. C., 52, 262, 263(31), 270(31)
Stothers, J. B., 208
Stowell, M. H., 351
Straka, R., 194
Strambini, G. B., 235
Straub, K., 469
Strauch, M. A., 467
Street, I. P., 68, 87(33; 39–42), 88, 89(33; 41; 42), 468
Strokopytov, B., 75, 83
Strominger, J. L., 43, 190
Strongin, R. M., 295
Stroobant, V., 171, 174(10), 175, 177, 179, 179(5), 182(5), 184(5; 14), 187(14)
Stroud, R. M., 25
Struehr, D. J., 320, 321(4)
Strupp, C., 28, 35(12), 53
Strynadka, N. C., 25
Stubbe, J. A., 381, 400, 405, 406, 406(30–33), 408(30; 33; 37), 412(33), 426, 438, 440(11), 441(11), 442(11), 470
Stuckey, J. A., 238, 242
Stuehr, D. J., 320, 321, 321(1), 322, 324, 325, 326, 326(11), 328, 328(15; 18; 23), 329(15; 20), 330, 331(15; 16; 20; 23), 334(20), 335, 335(11; 15; 20), 337, 469
Sturgeon, B. E., 24, 405, 406(31)
Sturgill, T. W., 281
Suarez, V., 469
Subbotin, O., 421, 422(25)
Sugimura, K., 108
Suizdak, G., 28
Sukalski, K. A., 149, 150(3)
Sullivan, P. A., 87(23), 88, 94(23)
Sulzenbacher, G., 87(25; 34), 88, 94(25; 34), 95
Sun, H., 241
Sun, Y., 28, 52
Sundararaju, B., 469
Sundquist, J. E., 298
Suslick, K. S., 329
Sussman, J. L., 108
Sussman, M. R., 43

Suter, M., 467
Sutoh, K., 25, 131
Suzuki, F., 380, 384(1)
Suzuki, H., 275, 276
Svaren, J. P., 114, 286
Svensson, B., 66, 81(10)
Swan, J. S., 469
Swanson, P. E., 469
Swarte, M. B. A., 262
Sweet, R. M., 25
Swiderek, K. M., 52
Switzer, R. L., 6
Swyryd, E. A., 295

T

Taha, I. A. I., 421
Tai, C.-H., 227, 229, 230, 233, 234, 234(22), 235(22), 468
Tainer, J. A., 204
Tajima, S., 470
Takahashi, H., 467
Takahashi, S., 356, 360, 361(29), 364(17)
Takayama, S., 28
Tamura, N., 447
Tan, A. K., 18
Tanabe, T., 227
Tanaka, K., 27
Tanaka, Y., 227
Tanford, C., 203
Tang, K., 30, 35(29)
Tang, W.-J. Y., 7
Tang, X., 39
Taniguchi, N., 272, 273(4), 274(4), 275, 279, 279(8), 284(4; 15)
Tanko, J. M., 419
Tanner, M. E., 190, 196(16), 467, 468
Tao, B. Y., 68
Tata, P., 35, 36
Tatusov, R. L., 179
Tavares, P., 437, 444, 445
Taylor, B. L., 6
Taylor, E. W., 21
Taylor, J.-S., 206
Taylor, K. L., 26, 468
Taylor, P., 468
Taylor, S., 468
Taylor, W. P., 29, 36, 36(18), 39(18), 40, 41, 42
Tchou, J., 468

Teng, H., 469
Teplyakov, A., 262
Terracina, G., 108
Terwilliger, T. C., 469
Tesmer, J. J. G., 269, 270(53)
Thaller, M. C., 177
Thatcher, G. R. J., 286
Theil, E. C., 436, 437, 438, 438(6), 439(6), 444, 445, 446(6), 449, 452
Thelander, L., 401, 403, 404, 405, 405(28), 406, 406(35), 409(42), 412, 412(42), 413(13; 42), 414, 414(42)
Thelander, M., 404
Thirumoorthy, R., 261
Thiruvengadam, T. K., 427
Thithapandha, A., 469
Thoden, J. B., 131, 136, 137(16), 140(16), 261, 262, 263, 269, 269(25), 270
Tholey, A., 28
Thoma, N. H., 393
Thomas, C. D., 436(5), 437
Thomas, E. L., 339, 348
Thomas, E. W., 469
Thomas, M. G., 260
Thompson, T. B., 260
Thorell, S., 197, 201
Thornberry, N. A., 262
Thorson, J. S., 43
Thrall, S. H., 470
Thurston, E. L., 37
Tian, G., 468
Tiganis, T., 241, 244(21), 247(21), 248(21)
Tilton, R. F., Jr., 29, 30(19), 31(19), 60
Timko, M. P., 470
Timmis, K. N., 108
Tipton, P. A., 310, 312, 313, 314, 315, 317, 317(1), 318
Tittmann, K., 470
Todhunter, J. A., 4, 10, 13, 13(33), 15, 169, 171(8), 173, 173(8), 193, 195(34), 468
Tokutake, N., 290
Toledo, D. L., 25
Tollersrud, O. K., 70, 88(50), 89
Tomaszek, T. A., 106, 107(4), 117(4)
Tomchick, D. R., 263, 268, 284, 285(24)
Tomkins, G. M., 225, 227
Tong, W. H., 405, 438, 440(11), 441(11), 442(11)
Tong, X., 52
Tonge, P. J., 468

Tonks, N. K., 238, 239, 241, 241(6), 244(6; 7; 15; 21), 247, 247(21), 248(21; 25), 251(7)
Tonomura, K., 468
Toraya, T., 381, 426
Tortorella, D., 285
Towatari, T., 262
Toy, P. H., 417, 419(6), 425, 425(6)
Toyoshima, C., 180
Trainer, J. A., 320, 321(4)
Traldi, P., 36
Travers, F., 21
Travis, J., 468
Travo, P., 158
Treffry, A., 436(5), 437
Tremblay, M. L., 238
Tressel, P., 422
Trezequet, V., 470
Trigalo, F., 191
Trimbur, D., 67, 76, 96
Troll, H., 119
Trumbore, M. W., 176
Tsai, A. L., 426(15), 427
Tsai, T. C., 155, 469
Tsernoglou, D., 22
Tso, J. Y., 262
Tsolas, O., 197, 198(1)
Tsou, C. L., 278
Tsunasawa, S., 52, 469
Tsuyama, N., 467
Tuhácková, Z., 257
Tukalo, M., 295
Tull, D., 76, 86(14), 87, 87(35; 36; 38), 88, 88(18), 89(18), 93, 94(38), 95(18), 96, 96(18; 35; 36), 98, 99, 100, 100(36), 101(35), 102, 102(18), 104(65), 105(18)
Turecek, F., 28, 35(7)
Turk, D., 262
Turk, V., 262
Turner, B. E., 414, 419, 421(20)
Turner, R. M., 295

U

Uchida, T., 470
Udgaonkar, J. B., 21
Udseth, H. R., 27
Ueda, H., 295
Uhlin, U., 400, 401(9), 403
Ui, N., 467
Uitdehaag, J., 74, 78, 81, 83, 84(39), 87, 91(16), 96
Umland, T. C., 25
Unkefer, C. J., 469
Urberg, M., 262
Urey, H. C., 117, 118(18)
Urey, M., 117, 118(18)
Utter, M. F., 6

V

Vaccari, S., 235
Vaganay, S., 190, 191, 196(16)
Vagelos, P. R., 469, 470
Vaidyanathan, C. S., 467
Valentine, K. M., 208
Valentine, W. N., 151
Van Berkel, W. J., 470
van Cleemput, M., 262
Van Den Heuvel, R. H., 470
Vandercammen, A., 179
van der Donk, W. A., 400
Van Der Jagt, D. L., 469
van der Veen, B. A., 81, 83, 87, 91(16)
Van Etten, R. L., 39, 177, 208, 209, 212
Van Heeke, G., 261, 270(19)
Van Heijenoort, J., 176, 189, 190, 191, 192, 195, 195(22), 196, 196(16), 470
van Kuilenburg, A., 345
van Lier, J. E., 329
Van Nuland, N. A. J., 132
Vanoni, M. A., 262, 270
Van Ophem, P. W., 467
Van Poelje, P. D., 469
van Riel, C., 345
Van Schaftingen, E., 171, 174(10), 175, 177, 179, 179(5), 181(12), 182(5), 184(5; 11–14), 186, 470
van Thoai, N., 5
Varick, T. R., 417, 419(5)
Varotsis, C., 356
Varrot, A., 88(52), 89, 95(52)
Vasella, A., 86(12), 87, 88(19)
Vasquez-Vivar, J., 24
Vazquez, R., 469
Veiga-da-Cunha, M., 179, 184(11)
Vener, A. V., 43
Venkataramanan, R., 35, 36, 202

Vercauteren, J., 467
Verkhovsky, M. I., 362, 364(35)
Verly, W. G., 202
Verma, C., 469
Véron, M., 119, 120, 120(11; 14; 16; 25; 29; 49), 121, 122, 124, 126, 126(21), 127, 129, 129(25), 133
Verschueren, K. H. G., 108
Vessey, D. A., 469
Vestal, M. L., 30, 36
Vestling, M. M., 29, 36, 39(20), 44, 44(20), 46, 47, 48
Vicente, O., 470
Vierstra, R. D., 43
Vijayalakshmi, J., 240
Viladot, J. L., 24
Villafranca, J. J., 17, 190, 208, 290, 293, 469, 470
Vinarov, D. A., 24, 208, 209, 212, 214, 215, 215(9), 216(11), 217, 218, 219, 220, 221, 222(11)
Vincent, J. B., 177
Viola, R. E., 468
Vishwanath, A., 59
Vocadlo, D. J., 87(37), 88, 91(47), 92, 94(37; 47)
Voelker, C., 322, 325(12), 326(12), 328, 328(12), 329(12), 335(12)
Voet, J. G., 67
Vogel, P. D., 24
Volbeda, A., 305, 470
Volk, K. S., 190, 196(13)
Volkman, B. F., 133
Volkova, L. I., 106, 117(3)
Von Der Saal, W., 469
Vonica, A., 119
Vonrheim, C., 122
Vorm, O., 37
Vos, M. H., 337
Vuletich, J. L., 333

W

Wachs, T., 52
Wachter, R. M., 418, 419, 421(19), 422, 422(16), 423(28), 424(19)
Wackett, L. P., 468
Waddell, S. T., 190
Wagner, A. F. V., 395
Wagner, E., 274
Wagner, U. G., 117, 382, 390(20)
Wakabayashi, U., 470
Wakarchuk, W. W., 76, 96
Waki, H., 27
Wakil, S. J., 468
Waldo, G. S., 436, 438
Waldrop, G. L., 295
Waley, S. G., 52
Walker, E., 191
Walker, H., 470
Wall, L., 469
Wallace, J. C., 470
Wallet, V., 119
Walsh, C. T., 24, 28, 189, 190, 190(2), 191, 215, 240, 263, 267(41), 270(41), 278, 290, 424, 467, 468, 470
Walsh, C. T., Jr., 6
Walworth, E. S., 261
Wang, J., 36, 335
Wang, Q., 67, 76, 87(38), 88, 94(38), 96, 326, 328, 329(20), 330, 331(20), 334(20), 335, 335(20)
Wang, R., 176
Wang, Y., 242, 242(20), 247(20)
Wang, Z.-Q., 320, 326, 328, 329(20), 330, 331(20), 334(20), 335(20)
Wanner, B. L., 180, 470
Ward, K. A., 226
Warner, A., 252
Warren, M. S., 24, 369, 373(6)
Warren, R. A. J., 67, 68, 76, 87, 87(32; 35; 38; 42), 88, 88(17), 89(42), 94(32; 38), 95, 96, 96(17; 35), 98, 99, 101(35), 104(65)
Watanabe, K., 119
Wataya, Y., 470
Watson, H. C., 468
Watts, A. B., 469
Webb, M. R., 14, 17
Webb, P. A., 120, 120(50), 133
Weber, C., 393
Weber, M. J., 281, 467
Weber, W. W., 467
Webster, D. R., 151
Wedding, R. T., 227
Wedekind, J. E., 136, 137(14), 140(13; 14), 251, 260(3)
Wedler, F. C., 13, 42, 286, 292, 292(31)
Wei, C.-C., 320, 326, 328, 329(20), 330, 331(20), 334(20), 335(20)
Wei, J., 28, 35(12), 53

Wei, Y.-F., 127, 255, 256(12), 258(12), 259(12)
Weil, J. A., 431, 432(33)
Weiller, B. H., 427
Weinberger, S. R., 35
Weinstein, J., 162, 163
Weiss, J., 452
Weiss, P. M., 467, 468
Welham, K. J., 52
Weller, M., 181
Weller, V. A., 28, 31(11)
Wellner, D., 275
Wellner, V., 9, 270
Wemmer, D. E., 133
Weng, L. C., 363
Wenger, R. M., 468
Werner, E. R., 326, 328, 331(25), 335(25)
Wertz, J. E., 431, 432(33)
Wesenberg, G., 26, 269
Westley, J., 470
Wever, R., 345
Weyhermüller, T., 452
Wheeler, C. J., 468
White, A., 87, 88(18), 89(18), 95(18), 96(18), 102, 102(18), 105(18)
Whitehouse, C. M., 27, 52
Whitesides, G. M., 52
Whittaker, J. W., 418
Whittaker, M. M., 418
Whittaker, M. W., 418
Whitten, J. P., 287
Whitty, A., 469
Wicki, J., 68, 84
Wider, E. A., 369
Widlanski, T. S., 29, 36, 36(18), 39(18), 40, 41, 42
Wieghardt, K., 452
Wikstrom, M., 351, 356, 362, 364(35)
Wild, K., 274
Wilkinson, A. J., 133
Willems, J. P., 405, 406(32), 426
Willenbrock, F., 262, 263(29)
Williams, N. H., 470
Williams, P. A., 108, 274
Williams, R. L., 120, 120(50), 121, 122(10), 123(10), 133
Williams, S. F., 215
Williams, S. J., 88(54), 89
Willson, M., 197
Wilmot, C., 422

Wilson, I. B., 158, 265
Wilson, K. S., 87(34), 88, 95, 95(34)
Winfield, C. J., 108
Winkler, D. A., 69
Winkler, W., 271
Wise, J. G., 24
Wiseman, J., 324
Withers, S. G., 28, 30(16), 55, 56, 56(33), 57(33), 58, 59(33), 64, 66, 66(3–6), 67, 67(12), 68, 69, 70, 70(29), 71, 72, 73, 74, 76, 77, 78, 81, 81(13), 83, 84, 84(39), 85, 86(13; 14; 16), 87, 87(22–45), 88, 88(17; 18; 50–54), 89, 89(11; 18; 33; 41), 90(46), 91, 91(8; 11; 46–48), 92, 93, 94(11; 20–25; 28; 29; 32; 37; 38; 43; 44; 46; 48; 56), 95, 95(18; 34; 52; 53), 96, 96(17; 18; 35; 36), 98, 99, 100, 100(36), 101(35), 102, 102(18), 104(65), 105(18), 468, 470
Witzel, H., 175
Wo, Y. Y., 177
Woehl, E. U., 233, 234(22), 235(22)
Wöldike, H. F., 87(25), 88, 94(25)
Wolf, G., 286
Wolfenden, R., 12, 467, 469
Wolff, D. J., 333
Wolff, J. A., 274
Wolfsberg, M., 162
Wollowitz, S., 393
Wolodko, W. T., 132
Womack, F. C., 7
Wong, A. W., 70, 87(43), 88, 94(43)
Wong, C.-H., 28, 69, 425
Wong, K. K., 190
Wong, L.-J., 134, 136(7), 140(2), 470
Wong, S. F., 27, 52
Woo, S., 135
Wood, H. N., 106, 117(1)
Wood, W. A., 469
Woodruff, W. H., 354, 355, 356
Woods, M. J., 467
Wotherspoon, A. T., 468
Wrenn, R. F., 38
Wu, C., 320, 321, 321(4)
Wu, J., 28, 469
Wu, L., 39, 239, 242
Wu, W., 427, 429(25)
Wunderlich, I., 467
Wurtele, E. S., 468, 469
Wyttenbach, C., 9

X

Xi, J., 468
Xiang, H., 26
Xiao, G., 426(15), 427
Xiong, H., 467
Xu, N., 64
Xu, Y., 119, 120, 120(15; 32), 122, 123, 125, 126(23), 129, 130
Xu, Y. W., 124, 126, 126(21)

Y

Yadav, K. D., 418
Yamada, E. W., 467
Yamada, H., 227
Yamazaki, T., 323, 324(13), 326, 326(13), 335(13)
Yanase, H., 468
Yanchunas, J., Jr., 190, 470
Yang, G., 26
Yang, S.-L., 134, 136(7), 257, 259(15), 470
Yang, V. C., 369
Yanofsky, C., 262
Yaremchuk, A., 295
Yavashev, L. P., 106, 117(3)
Ye, Q. Z., 24
Yeh, P., 271
Yeung, W., 87(27), 88
Yewdall, S. J., 436(5), 437
Yip, B., 293
Yokoyama, C., 227
Yoshida, T., 27, 227
Yoshida, Y., 27
Yoshikawa, S., 356, 363, 364(17)
Yost, W. L., 421
Young, A. P., 262
Young, D. M., 261, 268(9)
Young, R. A., 179
Yuan, R., 271
Yun, D., 414

Z

Zaks, A., 425
Zalkin, H., 260, 262, 268, 268(35), 269(50), 270, 279, 467
Zanetti, G., 262, 270
Zaoral, M., 194
Zarkadas, C. G., 469
Zarkowsky, H., 44
Zatman, L. J., 469
Zawadzke, L. E., 24, 190
Zechel, D. L., 28, 30(16), 55, 56, 56(33), 57(33), 58, 59(33), 64, 66, 66(3), 67(12), 85, 87(44), 88, 89(11), 91(8; 11), 94(11; 44)
Zeikus, G. J., 87(37), 88, 94(37)
Zelder, O., 393
Zelechonok, Y., 421, 422(24)
Zeng, B., 190
Zeppezauer, M., 232
Zewe, V., 3
Zhande, R., 467
Zhang, D., 468
Zhang, J., 179
Zhang, S., 87(26), 88
Zhang, Y., 356
Zhang, Z.-Y., 39, 43, 177, 179, 238, 239, 240, 241, 242, 242(20), 247(14; 20), 470
Zhao, K. Y., 193
Zhao, Y., 43, 239
Zharkov, D. O., 204
Zhong, Z., 69
Zhou, G., 39
Zhou, M. M., 177
Zhu, A., 88
Zimmerle, C. T., 38
Zimmermann, F., 201
Ziser, L., 77, 87(29; 45), 88, 88(53), 89, 94(29), 95(53)
Zucic, D., 262
Zumft, W. G., 298

Subject Index

A

Acetate kinase, stereochemical studies of intermediates, 15
Acetyl-CoA synthase, redox titration of *Clostridium thermoaceticum* enzyme, 307–308
O-Acetylserine sulfhydrylase
    catalytic reaction, 225
    isoforms, 225–227
    kinetic mechanism, 227
    purification, 227
    stopped-flow studies, 232–235
    ultraviolet–visible spectroscopy studies, 228–230
Acyl-phosphate intermediates, *see also* Phosphoaspartate intermediate
    ATP-dependent ligases, 293–295
    phosphate efficiency as leaving group, 168
    protein distribution, 179
    sodium borohydride reduction
        CheY, 176
        collagen α_2 chains, 177
        complications and controls, 174
        Mur synthetases, *see* Mur synthetases
        myelin proteolipid protein, 175
        noncovalently-bound intermediates, 170–173
        nucleoside phosphotransferase, 175–176
        phosphomannomutase, 175
        phosphorylated enzyme compounds, 170
        principles, 169
        prothymosin α, 176
        separation and detection of amino-alcohol reduction products, 173–174
        ubiquitin carboxyl-terminal hydrolase, 174–175
        vanadate-sensitive membrane ATPase, 176–177
Adenosylcobalamin, *see* Glutamate mutase
Adenylosuccinate synthetase, X-ray crystallography detection of intermediates, 26–27

Alkaline phosphatase, phosphoryl-enzyme intermediate, 158
Asparagine synthetase B
    catalytic reactions
        asparagine biosynthesis, 261, 263
        glutamine hydrolysis, 261, 263
    classification and mechanisms of glutamine-dependent amidotransferases, 269–271
    crystal structure and catalytic residues, 261–263
    glutaminase assay, 267
    γ-glutamyl thioester intermediate in glutaminase reaction
        C1A mutant studies, 268
        formation mechanism, 263
        formation rates, 269
        glutamine concentration response, 268
        [^{14}C]glutamine intermediate
            deacylation rate, 266
            isolation, 265–266
            quantification, 267–268
            stability assessment, 266
        hydroxylamine trapping, 263, 265
        pH-dependence of formation, 268
    inhibitor therapy in humans, 260–261
    rate-limiting step, 271
ATP phosphoribosyltransferase
    partial exchange reaction studies of intermediates, 6–7
    stereochemical studies of intermediates, 14

B

Burst kinetics, studies of covalent intermediates, 17–20

C

Carbon monoxide dehydrogenase, *see* Acetyl-CoA synthase
Carboxypeptidase A, X-ray crystallography detection of intermediates, 24–25

Cellulomonas fimi exoglycanase
  active site nucleophile
    labeling and kinetic analysis, 96–97
    mass spectrometry of labeled peptide, 98–101
    X-ray crystallography
      mechanism-based inhibitor complex, 101–103
      mutant enzyme bound to substrate, 103–105
  structure, 96
Cex, *see Cellulomonas fimi* exoglycanase
CGTase, *see* Cyclodextrin glucanotransferase
CheY, acyl-phosphate reduction with sodium borohydride, 176
4-Chlorobenzoyl-CoA dehalogenase, X-ray crystallography detection of intermediates, 26
Chymotrypsin, burst kinetics, 17, 19
Collagen, α_2 chain acyl-phosphate reduction with sodium borohydride, 177
Cryoenzymology
  α-glycosidase glycosyl-enzyme intermediate trapping, 67–68
  rapid freeze-quench techniques, *see* Rapid quench
  trapping of covalent intermediates, 20, 22–23
Cyclodextrin glucanotransferase
  covalent intermediate trapping, 73–74
  α-glycosyl fluoride substrate kinetics
    assay, 74–75
    Glu257Gln mutant studies, 75–77
    kinetic parameters, 75–76
  mass spectrometry of covalent intermediate
    nucleophile-containing peptide identification, 77–79, 81
    nucleophile identification, 81
    sample preparation, 77, 79
    spectra acquistion, 77–78
  X-ray crystallography of β-linked glycosyl-enzyme intermediate, 81–84
Cysteinyl-phosphate intermediate, *see* Protein tyrosine phosphatase
Cytochrome oxidase
  electron paramagnetic resonance, 355
  flow studies
    flow–flash–probe technique, 354–355
    historical perspective, 352
    triple trappng technique, 353–354
  function, 351
  redox centers, 352
  research questions, 351–352
  resonance Raman spectroscopy of intermediates
    compound P-to-F transition, 365–366
    continuous flow apparatus, 356–358
    continuous-wave versus pulsed lasers, 356
    data collection, 358–359
    high-frequency region changes upon oxygen reaction, 359–362
    low-frequency region changes and line assignment, 362–364
    mechanistic scheme and reaction cycle, 366–368
    rationale, 355–356

D

dTDP-glucose-4,6-dehydratase
  mechanism, 43–44
  rapid mix-quench mass spectrometry analysis, 43–49

E

Electron paramagnetic resonance
  covalent intermediate detection, 23–24
  lysine 2,3-aminomutase, rapid freeze-quench electron paramagnetic resonance studies of β-lysyl radical
    apparatus, 430–431, 433
    kinetic competence, 431–432
    mixing conditions and data collection, 433
    principles, 430
    structural characterization, 431–432
    turnover rate constant, 434–435
  metalloenzyme redox titration, 299, 305, 307
  ribonucleotide reductase iron/radical site in R2 protein, rapid freeze-quench electron paramagnetic resonance studies
    apparatus and performance, 409–411
    intermediate X formation, 408–409, 412–414
    quenching time determination, 410–411
    reproducibility, 411
    tyrosyl radical reconstitution in wild-type and mutant proteins, 412–414

SUBJECT INDEX

Endonuclease V, T4
　mechanism of reaction, 204–205
　Schiff base intermediate
　　methylation of enzyme amines, 202–204, 206–207
　　pK_a of lysine, 203–204
　　precedents, 202
　　sodium borohydride reduction, 205–206
　　sodium cyanoborohydride reduction, 205–206
EPR, see Electron paramagnetic resonance

F

Ferritin
　functions, 436
　iron ligands, 436–437
　peroxodiferric intermediate studies in M-ferritin
　　evidence for intermediate, 437
　　rapid freeze-quench Mössbauer spectroscopy
　　　advantages and limitations, 453–454
　　　ferroxidase reaction kinetics, 442–446, 453
　　　instrumentation, 440
　　　sample preparation, 440–441
　　rapid freeze-quench resonance Raman spectroscopy
　　　advantages and limitations, 454
　　　binding mode of peroxide ligand, 448–451
　　　sample preparation, 441–442
　　　structure elucidation, 437, 451
　　rapid freeze-quench X-ray absorption spectroscopy
　　　advantages and limitations, 454
　　　sample preparation, 440–441
　　　structure elucidation, 437, 451–452
　　solutions
　　　oxygenation, 438–439
　　　preparation, 437–438
　　stopped-flow spectroscopy
　　　apparatus, 439
　　　ferroxidase reaction kinetics, 446–448
　subunits and tissue distribution, 436
FixJ, phosphohistidine intermediate, 133
Fructose-2,6-bisphosphatase, phosphohistidine intermediate, 132

G

Galactose oxidase
　active site structure, 418
　assay, 422
　purification, 422
　reaction, 417–418
　redox states, 419
　substrates containing ultrafast, hypersensitive radical probes
　　catalytic turnover, 423
　　inactivation kinetics, 422
　　mechanism-based inactivation
　　　criteria, 424
　　　isotope effect, 423
　　　kinetics, 423
　　　overview, 419–421
　　　reversible turnover-dependent inactivation, 423
　　　substrate protection, 423
　　radical rearrangement rates, 419, 424–425
　　reactivation of one-electron reduced enzyme, 423
　　synthesis
　　　trans-(2)-phenylcyclocarbinol, 421
　　　$[\alpha,\alpha\text{-}^2H_2]$-trans-(2)-phenylcyclocarbinol, 421–422
　　types, 419, 421
Galactose-1-phosphate uridylyltransferase
　kinetic characterization of mutants
　　C160A, 143–145
　　Q168N, 143–145
　　Q168R, 145–146
　　S161A, 143–145
　kinetic mechanism, 134
　mutation and galactosemia, 134–135, 145–146
　rapid quench studies, 141–143
　structure, 135–136
　uridylyl-enzyme complex
　　hydrolysis kinetics, 146–148
　　X-ray crystal structure of enzyme and uridylyl complex, 136–140
GalT, see Galactose-1-phosphate uridylyltransferase
GDP:GTP guanylyltransferase
　function, 251–252
　guanylate covalent reaction intermediate
　　amino acid identification

high-performance liquid chromatography of phosphoamino acids, 258–260
histidyl residue identification, 260
phosphamino acid standard preparation, 256
sodium periodate treatment, 256–257
thin-layer chromatography of phosphoamino acids, 257–258
enzyme distribution, 251, 260
preparation, 252–254
stability analysis, 254–255
gem-Diol intermediate, *see* 2-Hydroxy-6-keto-nona-2,4-diene 1,9-dioic acid 5,6-hydrolase
GGT, *see* γ-Glutamyltranspeptidase
Glucose-6-phosphatase, phosphoryl-enzyme intermediate, 149
Glutamate mutase
  adenosylcobalamin function and mechanism, 380–382, 384
  free energy profile of reaction, 397–399
  kinetic scheme, 382, 384
  rapid quench
    acrylate intermediate analysis, 387, 393–395
    amino acid derivatization and high-performance liquid chromatography, 386
    controls, 387
    5′-deoxyadenosine formation analysis, 386, 392–393
    glutamate and methylaspartate formation analysis, 396–397
    glycyl radical intermediate analysis, 387, 393–395
    internal standards, 387
    mixing conditions, 385–386
  rate-limiting step, 396
  reaction, 380
  stopped-flow spectroscopy
    cob(II)alamin formation studies, 387–391
    data analysis, 391
    data collection, 385
    isotope effects in adenosylcobalamin homolysis, 389–391, 396
Glutaminase, *see* Asparagine synthetase B
Glutamine-dependent amidotransferases, classification and mechanisms, 269–271

Glutamine synthetase
  analogs of covalent intermediates, 12–14
  function, 272
  γ-glutamyl-phosphate intermediate
    2-amino-6,6-difluoro-5-oxo-6-phosphonohexanoic acid probing
      characterization, 289
      inhibition studies, 289–290, 293
      rationale, 286
      synthesis, 286–288
    2-amino-4-(fluorophosphono)butanoic acid probing, 290–292
    mechanism overview, 272–274
    4-(phosphonoacetyl)-L-α-aminobutyrate inhibition, 285–286, 292
    side reactions and mechanism elucidation, 8–10
    sodium borohydride reduction of glutamyl phosphate intermediate, 193
    stereochemical studies of intermediates, 17
γ-Glutamylcysteine synthase
  function, 272
  γ-glutamyl-phosphate intermediate
    2-amino-6,6-difluoro-5-oxo-6-phosphonohexanoic acid probing
      characterization, 289
      inhibition studies, 289–290, 293
      rationale, 286
      synthesis, 286–288
    2-amino-4-(fluorophosphono)butanoic acid probing, 290–292
    mechanism overview, 272–274
γ-Glutamyl peptide ligases, *see* Glutamine synthetase; γ-Glutamylcysteine synthase
γ-Glutamyltranspeptidase
  2-amino-4-(fluorophosphono)butanoic acid hydrolase inhibition assay and kinetic analysis, 278–280
  mass spectrometry analysis of modified threonine residue, 280–281, 283–284
  mechanism-based inhibition, 274–275
  stability, 277
  synthesis, 275–277
  function, 272
  γ-glutamyl-enzyme intermediate
    mechanism overview, 272–274
    trapping, 278
  inhibitors, 274–275
  structure of *Escherichia coli* enzyme, 280
  substrate specificity, 284

SUBJECT INDEX

Glyceraldehyde-3-phosphate dehydrogenase,
  Racker's mechanism, 455
α-Glycosidases
  classification, 64–65
  double-displacement reaction of retaining
    enzymes, 65–66
  glycosyl-enzyme intermediate trapping
    cryoenzymology, 67–68
    inhibitors and mechanism-based
      inactivators
        2-deoxy-2-fluoroglycosides, 68–69
        5-fluoroglycosides, 69–70
    rapid quench, 67
  single-displacement reaction of inverting
    enzymes, 65
  X-ray crystal structures, 67
β-Glycosidases, see also Cellulomonas fimi
    exoglycanase
  classification, 84–85
  double-displacement reaction of retaining
    enzymes, 84–86
  glycosyl-enzyme intermediate trapping
    2-deoxy-2-fluoroglycosides, 87–90
    5-fluoroglycosides, 90–91
    mass spectrometry of labeled peptides,
      91–94
    site-directed mutant crystal structures,
      94–96
  isotope effects, 86–87
  single-displacement reaction of inverting
    enzymes, 84–85
  transition state, 86

H

Haldane–Pauling transition-state stabilization
  model, 12
Histidine phosphocarrier protein,
  phosphohistidine intermediate, 132–133
HPr, see Histidine phosphocarrier protein
2-Hydroxy-6-keto-nona-2,4-diene 1,9-dioic acid
    5,6-hydrolase
  activity assay, 110
  acyl enzyme intermediate trapping, 108
  carbon–carbon cleavage mechanism, 108
  gem-diol intermediate studies
    $H_2^{18}O$ incorporation into reaction products,
      110–113
    4-keto-nona-1,9-dioic acid noncleavable
      analog

  inhibition mechanism, 114
  isotope exchange studies, 114–117
  ^{18}O labeling, 111
  time-dependent inhibition, 113–114
  purification from recombinant Escherichia
    coli, 109–110
  stereochemistry, 108
3-Hydroxy-3-methylglutaryl-CoA synthase,
    carbon-13 nuclear magnetic resonance
    isotope shift studies
  acetyl-enzyme reaction intermediate
    preparation, 211
  shift detection, 209, 213–216
  acylation-impaired mutant studies,
    216–220
  calibration standards, 212–213
  instrumentation, 209
  magnitude of shifts, 211–212
  mechanism of ^{18}O-induced shift, 208
  mechanistic interpretations, 220–223
  spectra acquisition, 210
  synthesis of carbon-13 compounds
    acetoacetyl-CoA, 210–211
    acetyl-CoA, 210
    3-hydroxy-3-methylglutaryl-CoA,
      211

I

Immobilized enzymes
  applications, 368–369
  porphobilinogen deaminase,
    see Porphobilinogen deaminase
Intermediates, see also specific enzymes and
    intermediates
  analogs of covalent intermediates, 12–14
  burst kinetics studies of covalent
    intermediates, 17–20
  catalysis facilitation mechanisms, 1, 455
  chemical trapping of covalent intermediates,
    10–12
  cryoenzymology, 20, 22–23
  initial rate kinetics for covalent intermediate
    detection, 1–4
  partial exchange reaction studies, 4–8
  rapid quench, covalent intermediate studies,
    20–21
  side reaction studies, 8–10
  spectroscopic detection, 23–24
  stereochemical clues, 14–17

table of covalent, ionic, and radical enzyme
      intermediates, 456–467
   X-ray crystallography detection, 23–27
Isotope effect
   galactose oxidase, 423
   glutamate mutase adenosylcobalamin
      homolysis, 389–391, 396
   β-glycosidases, 86–87
   $\alpha(2\rightarrow 6)$-sialyltransferase kinetic isotope
      effects
      calculations, 162
      CMP-NeuAc solvolysis versus transfer,
         160
      solvent deuterium isotope effect
         measurements, 162–163
      transfer measurements, 161–162
      UMP-NeuAc as substrate, 164–166
Isotope scrambling, stereochemical studies of
   intermediates, 16–17

K

Kinetic mechanism
   considerations for partial exchange reaction
      studies of intermediates, 4–8
   initial rate kinetics for covalent intermediate
      detection, 2–4

L

β-Lactamase
   phosphonic acid monoester inactivation,
      284–285
   X-ray crystallography detection of
      intermediates, 25
LAM, *see* Lysine 2,3-aminomutase
Lineweaver–Burke plot, initial rate kinetics for
   covalent intermediate detection, 2–4
Lysine 2,3-aminomutase
   mechanism and free radicals, 427–430
   rapid freeze-quench electron paramagnetic
      resonance studies of β-lysyl radical
      apparatus, 430–431, 433
      kinetic competence, 431–432
      mixing conditions and data collection,
         433
      principles, 430
      structural characterization, 431–432
      turnover rate constant, 434–435

M

Mass spectrometry
   advantages and limitations in enzyme reaction
      analysis, 30–31
   electrospray ionization mass spectrometry
      with on-line continuous-flow mixing
      apparatus, 54–55
      flow regime consideratons, 59–60
      rationale, 52–53
      temporal resolution, 59
      xylanase studies, 55–57, 59
   electrospray ionization
      interfering substances, 64
      overview, 27–30, 52–54
   γ-glutamyltranspeptidase, analysis of
      modified threonine residue, 280–281,
      283–284
   β-glycosidase labeled active site peptides
      Cellulomonas fimi exoglycanase, 98–101
      overview, 91–94
   matrix-assisted laser desorption ionization
      overview, 27–30
   membrane inlet mass spectrometer
      continuous flow studies, 59
      stopped-flow studies, 61
   rapid quench analysis
      apparatus, 32
      dTDP-glucose-4,6-dehydratase studies,
         43–49
      integration of sample peaks, 38, 40
      matrix-assisted laser desorption ionization,
         33, 34
      matrix selection, 35
      mixing, 32
      quantitative analysis, 35, 36, 46
      quenching, 33
      sample
         application, 36–37
         purification, 34
         segregation during crystallization, 37
      stabilization of intermediates, 33, 45
      Stp1 studies, 39–43
   steady-state analysis of enzymatic reactions,
      28, 31
   stopped-flow electrospray ionization mass
      spectrometry
      apparatus, 60–62
      chlorophyll α demetallation as test
         reaction, 62–63

SUBJECT INDEX

prospects, 63
temporal resolution, 63
Metalloenzyme redox titration
  acetyl-CoA synthase from *Clostridium thermoaceticum,* 307–308
  advantages and limitations, 308–309
  anaerobic conditions, 300–301
  candidate redox systems, 302–303
  customization for specific enzymes, 300
  multiple titrations, 301
  NiFe hydrogenase from *Desulfovibrio gigas,* 305, 307
  overview, 296
  protein purity and concentration, 301–302
  rationale, 296–297
  reduction potential ascertainment, 298
  region of interest for titration, 298
  requirements, 297–298
  samping frequency, 301
  selection of reductant or oxidant, 299
  simulations, 303–305
  spectroscopic monitoring, 299
MhpC, *see* 2-Hydroxy-6-keto-nona-2,4-diene 1,9-dioic acid 5,6-hydrolase
Mössbauer spectroscopy
  peroxodiferric intermediate in M-ferritin, rapid freeze-quench studies
    advantages and limitations, 453–454
    ferroxidase reaction kinetics, 442–446, 453
    instrumentation, 440
    sample preparation, 440–441
  ribonucleotide reductase iron/radical site in R2 protein, 408–409
MPO, *see* Myeloperoxidase
MS, *see* Mass spectrometry
Mur synthetases
  acyl-phosphate reduction with sodium borohydride
    high-voltage paper electrophoresis, 193
    materials, 191
    MurC, 176, 193–196
    MurD, 176, 193–196
    reduction conditions, 192–193
  catalytic reactions, 189
  gene cloning and overexpression
    MurA, 191–192
    MurB, 192
  mechanism, 189–191
  types, 189
  UDP-[^{14}C]MurNAc synthesis, 192

Myelin proteolipid protein, acyl-phosphate reduction with sodium borohydride, 175
Myeloperoxidase
  catalase activity, 346
  chloride binding sites, 349
  chlorinating intermediate
    dissociation, 349–350
    rate constant for formation, 347–348
    side chain binding, 350
    structure, 348–349
  reaction, 338–339
  taurine chlorination
    complications of proposed mechanism findings, 345–346
    ionization of taurine, 339
    nonenzymatic reaction, 340
    pH optimum, 340
    proposed mechanisms and equations, 340–343
    second-order rate constant, 347
    spectra of intermediates, 339
    steady-state studies, 340–343

N

NiFe hydrogenase, redox titration of *Desulfovibrio gigas* enzyme, 305, 307
Nitric oxide synthase
  rapid quench analysis
    arginine hydroxylation kinetics, 331, 333–334
    chromatography of ^{14}C-labeled reaction products, 324
    data analysis, 324
    instrumentation, 323
    mixing conditions, 323
    tetrahydrobiopterin radical formation during arginine hydroxylation, 334
  reaction, 320
  reduced protein preparation, 321–322
  reductive titration of inducible isoform, 325
  single-turnover studies
    advantages, 320
    prospects, 338
  stopped-flow spectroscopy
    arginine and oxygen reaction with reduced enzyme, 326–331
    data analysis, 324
    heme transitions during oxygen binding and hydroxyarginine oxidation, 334–338

rapid-scanning spectrometer, 322–323
tetrahydrobiopterin effects, 331
structure, 320
NMR, see Nuclear magnetic resonance
NOS, see Nitric oxide synthase
NtrC, phosphohistidine intermediate, 133
Nuclear magnetic resonance
  covalent intermediate detection, 23–24
  3-hydroxy-3-methylglutaryl-CoA synthase, carbon-13 nuclear magnetic resonance isotope shift studies
    acetyl-enzyme reaction intermediate preparation, 211
    shift detection, 209, 213–216
    acylation-impaired mutant studies, 216–220
    calibration standards, 212–213
    instrumentation, 209
    magnitude of shifts, 211–212
    mechanism of ^{18}O-induced shift, 208
    mechanistic interpretations, 220–223
    spectra acquisition, 210
    synthesis of carbon-13 compounds
      acetoacetyl-CoA, 210–211
      acetyl-CoA, 210
      3-hydroxy-3-methylglutaryl-CoA, 211
Nucleoside-diphosphate kinase
  dideoxy analos as substrates, 125–127
  phosphohistidine intermediate
    equilibrium constant for formation, 124
    hydrolysis, 127
    kinetics of phosphorylation and dephosphorylation, 125
    preparation, 124–125
    tryptophan fluorescence assay, 124
    X-ray crystallography
      aluminum fluoride binding, 129–131
      beryllium fluoride binding, 129–131
      dead-end complexes, 128–131
      dipotassium phosphoramidate-treated enzyme, 127–128
      mutant enzyme, 119
  reaction steps, 118
  substrate specificity, 118
  three-dimensional structure
    active site and ligand complexes, 122–124
    crystal structure types, 119–121
    ferredoxin fold, 122
    subunit fold, 121
  types, 119

Nucleoside phosphotransferase, acyl-phosphate reduction with sodium borohydride, 175–176
Nucleotide phosphoramidate intermediate, see GDP:GTP guanylyltransferase

O

OASS, see O-Acetylserine sulfhydrylase

P

Pepsin, general base mechanism, 106–107, 118
trans-(2)-Phenylcyclocarbinol
  deuterated compound synthesis, 421–422
  synthesis, 421
Phosphoaspartate intermediate
  enzymes, see Phosphomannomutase; Phosphoserine phosphatase
  identification
    candidate peptide approach, 187–188
    gel electrophoresis of radiolabeled protein, 181, 185
    mass spectrometry, 182, 186–187
    protein preparation, 183–184
    radiolabeled substrate
      preparation, 184
      protein incorporation, 184–185
    sodium borohydride reduction, 181–182, 185–186
    stability of bond, 180–181, 185
    stoichiometry of phosphorylation considerations, 182–183
Phosphoglycerate kinase, partial exchange reaction studies of intermediates, 5–6
Phosphohistidine intermediate, see FixJ; Fructose-2,6-bisphosphatase; Histidine phosphocarrier protein; NtrC; Nucleoside-diphosphate kinase; Succinyl-CoA synthetase
Phosphomannomutase
  acyl-phosphate reduction with sodium borohydride, 175
  phosphoaspartate identification
    candidate peptide approach, 187–188
    gel electrophoresis of radiolabeled protein, 181, 185
    mass spectrometry, 182, 186–187

protein preparation, 183–184
radiolabeled substrate
preparation, 184
protein incorporation, 184–185
sodium borohydride reduction, 181–182, 185–186
stability of bond, 180–181, 185
stoichiometry of phosphorylation considerations, 182–183
Phosphoserine phosphatase
DXDXT/V N-terminal motif proteins, 177–179
haloacid dehalogenase sequence motifs, 179–180
phosphoaspartate identification
candidate peptide approach, 187–188
gel electrophoresis of radiolabeled protein, 181, 185
mass spectrometry, 182, 186–187
protein preparation, 183–184
radiolabeled substrate
preparation, 184
protein incorporation, 184–185
sodium borohydride reduction, 181–182, 185–186
stability of bond, 180–181, 185
stoichiometry of phosphorylation considerations, 182–183
phosphoryl-enzyme intermediate, 158
PLP, *see* Pyridoxal phosphate
Porphobilinogenase
components, *see* Porphobilinogen deaminase; Uroporphyrinogen III synthase
mechanism of action, 369–370
Porphobilinogen deaminase
assays, 372
cofactor, 369
immobilized enzyme studies of intermediates
coupling to Sepharose, 371
methylenepyrrolenine complex structure, 378–379
polypyrrolic enzyme intermediate
formation incubation conditions, 373–375
urogen synthesis kinetics, 375–377
uroporphyrinogen III synthase substrate formation, 378–379
uroporphyrinogen synthesis rates versus soluble enzymes, 373
purification from bovine liver, 370–371

Protein tyrosine phosphatase
cysteinyl-phosphate intermediate
formation, 240, 242
hydrolysis rate, 242
site-directed mutagenesis studies, 241, 243–244, 251
X-ray crystallography
caged compound utilization, 245–246
crystal preparation and selection, 249
data collection, 250–251
freezing of crystals, 250
limitations, 245
substrate diffusion, 246, 249–250
trapping principles, 247–248
functions, 237–238
mechanism, 239–242
pH optimum, 242–243
purification of recombinant PTP1B, 248–249
structure of PTP1B, 238–239
Prothymosin α, acyl-phosphate reduction with sodium borohydride, 176
PTP, *see* Protein tyrosine phosphatase
Pyridoxal phosphate
enzymes, *see* O-Acetylserine sulfhydrylase; Serine–Glyoxylate aminotransferase
quinonoid intermediate, 224
Schiff base mechanism, 223–224
Pyrimidine nucleotidases
activity assays
nucleotidase activity, 152
phosphotransferase activity, 152–153
branched mechanism evidence, 150–151, 159
chemotherapy agents as substrates, 149, 151–152
inhibitor studies of phosphoransferase, 153–155
kinetic mechanism, 150, 155
substrate specificity
nucleotidase, 153
overview, 151
phosphoransferase, 155–157
thermodynamics, 158–159
types, 149, 151

Q

Quenched-flow, *see* Rapid quench

R

Radical clock
  enzyme reaction mechanism studies, 415–416
  rearrangement rates, 416–417
Rapid quench
  covalent intermediate studies, 20–21
  glutamate mutase
    acrylate intermediate analysis, 387, 393–395
    amino acid derivatization and high-performance liquid chromatography, 386
    controls, 387
    5′-deoxyadenosine formation analysis, 386, 392–393
    glutamate and methylaspartate formation analysis, 396–397
    glycyl radical intermediate analysis, 387, 393–395
    internal standards, 387
    mixing conditions, 385–386
  α-glycosidase glycosyl-enzyme intermediate trapping, 67
  limitations, 51
  lysine 2,3-aminomutase, rapid freeze-quench electron paramagnetic resonance studies of β-lysyl radical
    apparatus, 430–431, 433
    kinetic competence, 431–432
    mixing conditions and data collection, 433
    principles, 430
    structural characterization, 431–432
    turnover rate constant, 434–435
  mass spectrometry analysis
    apparatus, 32
    dTDP-glucose-4,6-dehydratase studies, 43–49
    integration of sample peaks, 38, 40
    matrix-assisted laser desorption ionization, 33, 34
    matrix selection, 35
    mixing, 32
    quantitative analysis, 35, 36, 46
    quenching, 33
    sample
      application, 36–37
      purification, 34
      segregation during crystallization, 37
    stabilization of intermediates, 33, 45
    Stp1 studies, 39–43
  nitric oxide synthase
    arginine hydroxylation kinetics, 331, 333–334
    chromatography of ^{14}C-labeled reaction products, 324
    data analysis, 324
    instrumentation, 323
    mixing conditions, 323
    tetrahydrobiopterin radical formation during arginine hydroxylation, 334
  peroxodiferric intermediate studies in M-ferritin
    evidence for intermediate, 437
    rapid freeze-quench Mössbauer spectroscopy
      advantages and limitations, 453–454
      ferroxidase reaction kinetics, 442–446, 453
      instrumentation, 440
      sample preparation, 440–441
    rapid freeze-quench resonance Raman spectroscopy
      advantages and limitations, 454
      binding mode of peroxide ligand, 448–451
      sample preparation, 441–442
      structure elucidation, 437, 451
    rapid freeze-quench X-ray absorption spectroscopy
      advantages and limitations, 454
      sample preparation, 440–441
      structure elucidation, 437, 451–452
    solutions
      oxygenation, 438–439
      preparation, 437–438
  principles, 51
  ribonucleotide reductase iron/radical site in R2 protein, rapid freeze-quench electron paramagnetic resonance studies
    apparatus and performance, 409–411
    intermediate X formation, 408–409, 412–414
    quenching time determination, 410–411
    reproducibility, 411
    tyrosyl radical reconstitution in wild-type and mutant proteins, 412–414
RDR, see Ribonucleotide reductase

SUBJECT INDEX

Redox titration, see Metalloenzyme redox titration
Resonance Raman spectroscopy
   cytochrome oxidase intermediates
      compound P-to-F transition, 365–366
      continuous flow apparatus, 356–358
      continuous-wave versus pulsed lasers, 356
      data collection, 358–359
      high-frequency region changes upon oxygen reaction, 359–362
      low-frequency region changes and line assignment, 362–364
      mechanistic scheme and reaction cycle, 366–368
      rationale, 355–356
   peroxodiferric intermediate in M-ferritin, rapid freeze-quench studies
      advantages and limitations, 454
      binding mode of peroxide ligand, 448–451
      sample preparation, 441–442
      structure elucidation, 437, 451
Ribonucleotide reductase
   function, 399–400
   iron/radical site in R2 protein
      formation, 405–406, 408–409
      Mössbauer spectroscopy studies, 408–409
      rapid freeze-quench electron paramagnetic resonance studies
         apparatus and performance, 409–411
         intermediate X formation, 408–409, 412–414
         quenching time determination, 410–411
         reproducibility, 411
         tyrosyl radical reconstitution in wild-type and mutant proteins, 412–414
      redox states, 403–404
   structure and mechanism of class I enzyme, 400–401, 403

S

Serine–glyoxylate aminotransferase
   catalytic reaction, 227
   kinetic mechanism, 227
   purification, 227
   stopped-flow studies, 232–233, 235–237
   ultraviolet–visible spectroscopy studies, 228, 230–232

$\alpha(2\rightarrow 6)$-Sialyltransferase
   catalytic reaction, 159–160
   kinetic isotope effects
      calculations, 162
      CMP-NeuAc solvolysis versus transfer, 160
      solvent deuterium isotope effect measurements, 162–163
      transfer measurements, 161–162
      transition state structure, 166–167
   UMP-NeuAc as substrate
      isotopic labeling, 164–165, 168
      kinetic isotope effects, 164–166
      kinetic parameters, 163
      mechanistic inferences, 166–167
      pH optimum, 163
Sodium borohydride
   acyl-phosphate reduction, see Acyl-phosphate intermediates
   dihydroxyacetone–lysine intermediate reduction, see Transaldolase B
   phosphoaspartate reduction, see Phosphoaspartate intermediate
Stopped-flow
   O-acetylserine sulfhydrylase studies, 232–235
   electrospray ionization mass spectrometry
      apparatus, 60–62
      chlorophyll α demetallation as test reaction, 62–63
      prospects, 63
      temporal resolution, 63
   glutamate mutase
      cob(II)alamin formation studies, 387–391
      data analysis, 391
      data collection, 385
      isotope effects in adenosylcobalamin homolysis, 389–391, 396
   historical perspective, 352
   limitations, 51
   membrane inlet mass spectrometer studies, 61
   nitric oxide synthase
      arginine and oxygen reaction with reduced enzyme, 326–331
      data analysis, 324
      heme transitions during oxygen binding and hydroxyarginine oxidation, 334–338
      rapid-scanning spectrometer, 322–323
      tetrahydrobiopterin effects, 331

peroxodiferric intermediate studies in
    M-ferritin
  apparatus, 439
  ferroxidase reaction kinetics, 446–448
  principles, 50–51
  serine–glyoxylate aminotransferase studies,
    232–233, 235–237
  steady-state study rationale, 340
  urate oxidase
    absorbance changes, 311
    global data fitting, 313–314
    5-hydroxyisorate fluorescence detection,
      315–316
    intermediate absorption spectra, 316
    kinetic phases of reaction, 311–312
    rate constant assignment, 314–315
    singular value decomposition analysis,
      312–313
Stp1
  function, 39
  mechanism, 39–40
  rapid mix-quench mass spectrometry
    analysis, 39–43
Subtilisin, X-ray crystallography detection of
  intermediates, 23
Succinyl-CoA synthetase
  partial exchange reaction studies of
    intermediates, 7
  phosphohistidine intermediate,
    131–132
Sucrose phosphorylase, chemical trapping of
  covalent intermediates, 11

T

Taurine, *see* Myeloperoxidase
Tetrahydrobiopterin, *see* Nitric oxide synthase
Transaldolase B
  activity assay, 198
  catalytic reaction, 197
  dihydroxyacetone Schiff base intermediate
    sodium borohydride reduction, 199–200
    X-ray crystallography, 200–201
  purification of recombinant protein from
    Escherichia coli, 198
α-Transglycosylases, *see also* Cyclodextrin
  glucanotransferase
  classification, 64–65
  double-displacement reaction of retaining
    enzymes, 65–66

  glycosyl-enzyme intermediate trapping
    cryoenzymology, 67–68
    4-deoxyglycoside substrates, 73
    rapid quench, 67
  single-displacement reaction of inverting
    enzymes, 65
  X-ray crystal structures, 67
Triose-phosphate isomerase, X-ray
  crystallography detection of intermediates,
  25–26
Trypsin, X-ray crystallography detection of
  intermediates, 25
Tyrosyl-tRNA synthetase, X-ray crystallography
  detection of intermediates, 25

U

Ubiquitin carboxyl-terminal hydrolase,
  acyl-phosphate reduction with sodium
  borohydride, 174–175
Urate oxidase
  dehydrourate intermediate evidence, 316–318
  energetics, 319
  function, 310
  mechanism, 310–311, 319
  pH-dependence of inhibition, 317
  stopped-flow spectroscopy
    absorbance changes, 311
    global data fitting, 313–314
    5-hydroxyisorate fluorescence detection,
      315–316
    intermediate absorption spectra, 316
    kinetic phases of reaction, 311–312
    rate constant assignment, 314–315
    singular value decomposition analysis,
      312–313
  urate hydroperoxide intermediate evidence,
    317–318
Uroporphyrinogen III synthase
  assay, 372
  function, 369
  porphobilinogen deaminase immobilization
    for substrate formation, 378–379
  purification from bovine liver, 371

V

Vanadate-sensitive membrane ATPase,
  acyl-phosphate reduction with sodium
  borohydride, 176–177

X

XAS, *see* X-ray absorption spectroscopy
X-ray absorption spectroscopy, peroxodiferric intermediate in M-ferritin, rapid freeze-quench studies
  advantages and limitations, 454
  sample preparation, 440–441
  structure elucidation, 437, 451–452
X-ray crystallography
  cyclodextrin glucanotransferase, β-linked glycosyl-enzyme intermediate, 81–84
  galactose-1-phosphate uridylyltransferase structure and uridylyl complex, 136–140
  β-glycosidase covalent intermediate studies
    Cellulomonas fimi exoglycanase mechanism-based inhibitor complex, 101–103
    mutant enzyme bound to substrate, 103–105
    overview, 94–96
  nucleoside-diphosphate kinase
    active site and ligand complexes, 122–124
    crystal structure types, 119–121
    ferredoxin fold, 122
    phosphohistidine intermediate
      aluminum fluoride binding, 129–131
      beryllium fluoride binding, 129–131
      dead-end complexes, 128–131
      dipotassum phosphoramidate-treated enzyme, 127–128
      mutant enzyme, 119
    subunit fold, 121
  protein tyrosine phosphatase
    cysteinyl-phosphate intermediate
      caged compound utilization, 245–246
      crystal preparation and selection, 249
      data collection, 250–251
      freezing of crystals, 250
      limitations, 245
      substrate diffusion, 246, 249–250
      trapping principles, 247–248
  transaldolase B dihydroxyacetone Schiff base intermediate, 200–201

ISBN 0-12-182257-5

90051